Food Engineering Series

Springer's *Food Engineering Series* is essential to the Food Engineering profession, providing exceptional texts in areas that are necessary for the understanding and development of this constantly evolving discipline. The titles are primarily reference-oriented, targeted to a wide audience including food, mechanical, chemical, and electrical engineers, as well as food scientists and technologists working in the food industry, academia, regulatory industry, or in the design of food manufacturing plants or specialized equipment.

More information about this series at http://www.springer.com/series/5996

Alan L. Kelly • Lotte Bach Larsen
Editors

Agents of Change

Enzymes in Milk and Dairy Products

Springer

Editors
Alan L. Kelly
School of Food and Nutritional Sciences
University College Cork
Cork, Cork, Ireland

Lotte Bach Larsen
Department of Food Science
Aarhus University
Aarhus, Denmark

ISSN 1571-0297
Food Engineering Series
ISBN 978-3-030-55481-1 ISBN 978-3-030-55482-8 (eBook)
https://doi.org/10.1007/978-3-030-55482-8

This Springer imprint is published by the registered company Springer Nature Switzerland AG
The registered company address is: Gewerbestrasse 11, 6330 Cham, Switzerland

Preface

Milk has been converted into dairy products for millennia, firstly perhaps by accident with the discoveries of transformation of milk to curds and whey by animal-derived coagulants and the benefits of fermentation by adventitious microorganisms. In recent decades, the application of scientific study to what, in many cases, were traditional processes has illustrated the roles of enzymes in many different aspects of dairy products, whether deliberate (in the use of rennet to make cheese), natural and desirable (the antimicrobial activities of many enzymes) or natural and undesirable (bacterial spoilage or development of lipolytic rancidity). There is no doubt that milk and dairy products comprise complex enzyme systems, and enzymes truly are agents of change for many dairy products; in the current parlance, one could be tempted to use the 'omics' terminology to refer to the dairy 'enzome'.

These enzymes are typically considered in most textbooks and reference works on dairy science and technology, and there have been conferences dedicated to the topic, such as the First International Conference on Indigenous Milk Enzymes, organised by the International Dairy Federation and held in Cork, Ireland, in April 2005 (which the co-editors of this book were both involved in organising).

However, there has been no reference book solely dedicated to this topic, despite all aspects of milk and dairy product enzymology being active current areas of research (e.g., in the area of applications of cross-linking enzyme or the characterisation of psychrotrophic bacterial enzymes). Thus, the objective of this book is to gather in one place a collection of diverse articles on all the elements of this complex element of dairy science, written by experts in their relevant areas, such that the result is a reference which reflects the very state of current knowledge of the field and should be of use for many years to come for those involved in, studying or applying in industry dairy science and processing.

As editors, we would like to firstly thank Daniel Falatko at Springer Nature for encouraging and accepting our proposal for this book and to Sofia Valsendur for support and assistance in the production process of the manuscript. We would next like to thank all the authors whose work makes up this book—we were delighted that so many true experts on their own field accepted our invitation to contribute to this project, and thanks to their excellent contributions that this will hopefully be a valuable and unique resource in a very important area of dairy science for many years to come; we also thank our authors for their cooperation and patience during the preparation of this book and their shared enthusiasm for the project, which made our job as editors a pleasure.

We also would like to thank Dylan Kelly and, in particular, Ciara Tobin for their local editorial help in manuscript preparation and formatting, Jørgen Gamborg Andersen for fruitful discussions and many colleagues for useful and helpful discussions and suggestions.

Cork, Ireland
Aarhus, Denmark
May 2020

Alan L. Kelly
Lotte Bach Larsen

Contents

Chapter 1
Enzymology of Milk and Dairy Products: Overview

Patrick Fox

1.1 Introduction

Milk and dairy products are inextricably linked with enzymes; the involvement of enzymes falls into six main areas:

1. Biosynthesis of milk constituents in the mammary gland;
2. Indigenous enzymes in milk, i.e., enzymes which enter milk from mammary tissue, including somatic cells, or the blood;
3. Endogenous enzymes, i.e., enzymes secreted in milk or dairy products by contaminating microorganisms or microorganisms added as cultures;
4. Exogenous enzymes, i.e., enzymes added to milk or dairy products to induce specific changes,
5. Use as analytical agents, which are very sensitive and which avoid the need to purify the target substrate;
6. Enzymes involved in the catabolism of the constituents of milk or dairy products.

Aspects 1 and 6 are considered to be outside the scope of this book. In addition, one of the principal proteins in milk, α-lactalbumin, is an enzyme modifier, involved in the biosynthesis of the milk sugar, lactose, but will not be considered. Aspects 2, 3, 4 and 5 will be considered briefly to provide an introduction to the chapters that follow.

P. Fox (✉)
School of Food and Nutritional Sciences, University College Cork, Cork, Ireland
e-mail: pff@ucc.ie

A. L. Kelly, L. B. Larsen (eds.), *Agents of Change*, Food Engineering Series,
https://doi.org/10.1007/978-3-030-55482-8_1

1.2 Indigenous Enzymes of Bovine Milk

At least 60 indigenous enzymes have been reported in normal bovine milk. They arise from three principal sources:

(a) The blood *via* defective mammary cell membranes;
(b) Secretory cell cytoplasm, some of which is occasionally entrapped within fat globules by the encircling fat globule membrane (MFGM);
(c) The MFGM itself, the outer layers of which are derived from the apical membrane of the secretory cells, which, in turn, originates from the Golgi membranes.

Thus, most enzymes enter milk due to peculiarities of the mechanism by which milk constituents, especially the fat globules, are excreted from the secretory cells. Milk does not contain substrates for many of the enzymes present, while others are inactive in milk owing to unsuitable environmental conditions, e.g., pH.

Several indigenous milk enzymes are technologically significant from many viewpoints:

1. Deterioration [lipase (potentially, the most significant enzyme in milk), proteinases such as plasmin, acid phosphatase and xanthine oxidoreductase] or preservation (sulphydryl oxidase, superoxide dismutase) of milk quality;
2. As indices of the thermal history of milk, e.g., alkaline phosphatase, γ-glutamyl transferase and lactoperoxidase;
3. As indices of mastitic infection; the concentration of several enzymes increases on mastitic infection, especially catalase, *N*-acetyl-β-D-glucosaminidase and acid phosphatase;
4. Antimicrobial activity: lysozyme and lactoperoxidase (which is exploited as a component of the lactoperoxidase-thiocyanate system for the cold pasteurization of milk);
5. As a potential commercial source of enzymes, e.g.., ribonuclease and lactoperoxidase.

With a few exceptions (e.g., lysozyme and lactoperoxidase), the indigenous milk enzymes do not have a beneficial effect on the nutritional or organoleptic attributes of milk, and hence their inactivation by heat is one of the objectives of many dairy processes.

Lactoperoxidase in milk was demonstrated by C. Arnold in 1881, i.e., 20 years before the name "enzyme" was introduced and 40 years before the proteinaceous nature of enzymes was demonstrated in 1926, when J. B. Sumner, at Cornell University, demonstrated that crystalline urease, isolated from jack beans, had enzymatic activity. By 1902, the following enzymes had been reported in milk: lactoperoxidase, diastase (amylase), proteinase, lipase, arylesterase (arylester hydrolase) and fibrinferment (thrombin). Progressively, many more enzymes were isolated and characterized; a very extensive literature on this topic is available and has been reviewed periodically (see Fox and Kelly 2006a, b, 2008 and chapters in this book). The principal indigenous enzymes in milk and their catalytic activity are listed in

Table 1.1 Principal indigenous enzymes of significance to milk

Enzyme	Significance
Lipase	Off-flavours in milk, flavour development in some cheese varieties
Proteinase (plasmin)	Reduced storage stability of UHT products, cheese ripening
Catalase	Index of mastitis, pro-oxidant
Lysozyme	Bacteriocidal agent
Xanthine oxidoreductase	Pro-oxidant, cheese ripening
Sulphydryl oxidase	Amelioration of cooked flavour
Superoxide dismutase	Antioxidant
Lactoperoxidase	Index of pasteurization, bactericidal agent, pro-oxidant
Alkaline phosphomonoesterase	Index of pasteurization
Acid phosphomonoesterase	Reduce heat stability of milk, cheese ripening
Ribonuclease	Milk is a very rich of source of RNase, which is similar to pancreatic RNase
α-Amylase	Starch hydrolysis

Table 1.1. The indigenous enzymes in milk were the subject of an International Dairy Federation Symposium in Cork, Ireland, in 2005 (IDF 2006).

Being a liquid, it is relatively easier to isolate enzymes from milk than from solids such as muscle or plant tissue; furthermore, many of the indigenous enzymes in milk are in true solution, whereas those in solids are localized on membranes and must be solubilized, which may cause denaturation. There were many reasons for the interest in the indigenous enzymes in milk, including the development of off-flavours and, especially in the 1920s and 1930s, the need for a simple reliable indicator of the heat treatment of milk, when pasteurization of milk became mandatory.

In this chapter, the occurrence, distribution, isolation and characterization of the principal indigenous enzymes will be discussed briefly, with an emphasis on their commercial significance in milk and dairy products. The review will focus on bovine milk, with reference to the principal activities in the milk of other important species. Studies on the enzymes in non-bovine milk have been concerned mainly with quantifying the activity and the effects of various animal-related factors, storage and heat treatments thereon. Few of the enzymes in non-bovine milk have been isolated and characterized, and it is assumed that these enzymes are similar to the corresponding enzymes in bovine milk. The milk of all species probably contains the same range of enzymes as bovine milk. Not surprisingly, many of the enzymes in human milk have been studied extensively and their significance in human nutrition highlighted. Human milk contains a very high level of lysozyme (~4% of total protein), a bile salts-activated lipase (which the milk of other species lacks) and α-amylase. Reviews on the indigenous enzymes in human, buffalo, ovine, caprine, equine, and porcine milk have been included in Chap. 10 of Fox et al. (2015) and in later chapters of this book. A list of the minor enzymes in milk and their functions was reported by Fox et al. (2015).

1.3 Endogenous Enzymes in Milk

Milk is a good medium for growth of many microbial species, which may have positive or negative consequences. In milk, the growth of lactic acid bacteria (LAB; *Lactobacillus, Streptococcus or Lactococcus)* produces lactic acid, resulting in coagulation of the casein and the formation of a gel which can be converted into one of a range of fermented milk products or acid-curd cheese (~20% of all cheese). Enzymes released from dead LAB cells are major contributors to the maturation of rennet-coagulated cheeses.

The acidification of milk by the growth of an added culture of LAB is an essential step in the production of rennet-coagulated cheeses. The LAB die in the cheese curd during ripening but their intracellular enzymes, especially peptidases, are released into the curd and make a major contribution to flavour development in the cheese during ripening. Non-starter lactic acid bacteria (NSLAB), principally mesophilic *Lactobacillus* spp., enter cheese curd from the milk and/or environment and grow from low numbers initially to 10^7–10^8 per g of cheese; these are major contributors to ripening, especially in cheese made from raw milk.

Psychrotrophic bacteria, especially *Pseudomonas* species, growing in milk secrete extracellular enzymes, especially proteinases, which are very heat-stable and cause flavour and textural defects in cheese and gelation of ultra-high temperature sterilized products. Coliform bacteria produce various compounds which cause off-flavours in milk and products produced therefrom and CO_2 which causes holes or slits in solid products, e.g., cheese.

Moulds are potent sources of proteinases and lipase and are major contributors to the ripening of some cheese varieties, especially the blue cheeses and the surface mould varieties, e.g., Camembert and Brie. However, mould spores in the environment may contaminate the surface of other cheese varieties and cause defects in the appearance and flavour of cheese.

Yeasts produce ethanol and CO_2 and cause flavour and textural defects for liquid and solid products.

1.4 Exogenous Enzymes

1.4.1 Rennets

An exogenous enzyme preparation, rennet, containing the enzyme, chymosin, produced from the stomachs of slaughtered young animals, has been used in dairy technology for the manufacture of cheese, since pre-historic times. The use of chymosin to coagulate milk for cheesemaking was obvious from milk coagulum in the vomit of infants but rennet preparations from plants, e.g., figs and thistle, have also been used since ancient times.

Rennets are probably the principal exogenous enzyme used in food processing. Increased cheese production caused an insufficient supply of calf rennet in the 1950s and alternative rennets, "rennet substitutes", were introduced. The first rennet substitutes were pepsins produced from porcine, bovine and, to a limited extent, from chicken or fish gastric tissue. Later, acid proteinases from *Rhizomucor miehei, R. pusillus* or *Endothia parasitica* became available and were widely used and are still used to a limited extent.

The gene for calf chymosin was cloned in yeast and *E. coli* about 1970 and these rennets are now widely used; this was probably the first enzyme from a genetically engineered microorganism used in the food industry. The gene for camel chymosin has also been cloned; this enzyme is less proteolytic than calf chymosin and is preferred for certain situations. The primary function of a rennet is coagulation of the casein in milk, through specific limited proteolysis of the micelle-stabilizing protein, κ-casein. Most of the rennet is removed in the whey but some in retained in the curd (the amount retained depends, *inter alia*, on the moisture content of the curd and the curd cooking temperature; it is about 10% for Cheddar cheese). The rennet retained in the cheese curd causes limited proteolysis in the cheese during ripening and contributes to flavour, and especially, to development of texture and functionality.

The rennet used for some cheese varieties, especially Italian varieties, is prepared from the stomachs of young calves, kids or lambs slaughtered after suckling; the milk-filled stomach is macerated. In addition to chymosin, this rennet contains a lipase, pre-gastric esterase, which is secreted by a gland at the base of the tongue; this is stimulated by suckling and the secretion is washed into the stomach with the milk. This lipase causes the release of short and medium chain-length fatty acids, which are esterified at the Sn3 position of milk triglycerides, resulting in the characteristic pungent/piquant flavour of the cheese. Fungal lipases have been assessed as substitutes for pre-gastric esterase but have been found unsuitable.

Cheese ripening is a slow, expensive and not fully controlled process; consequently, there is interest in accelerating ripening. Various approaches have been investigated to accelerate ripening, including a higher ripening temperature (especially for Cheddar-type cheese, which is usually ripened at 6–8 °C), the use of exogenous proteinases and peptidases, of modified starters (e.g., heat-shocked or lactose-negative strains), genetically engineered starters or starter adjuncts The possible use of exogenous proteinases and peptidases attracted considerable attention for a period but uniform distribution of the enzymes in the cheese curd is a problem; microencapsulation of enzymes offers a possible solution. Exogenous proteinases/peptidases are not used commercially in natural cheeses, but are being used to produce "Enzyme Modified Cheese (ECM, Chap. 16)" for use in processed cheese, cheese dips and sauces. Selected genetically modified and adjunct cultures appear to be more promising than free enzymes.

Protein hydrolysates are used as flavourings in soups and gravies and in dietetic foods. They are generally prepared from soy protein, gluten, milk, meat or fish proteins by acid hydrolysis. Neutralization results in a high salt content which is acceptable for certain applications but may be unsuitable for dietetic foods and food supplements. Furthermore, acid hydrolysis causes total or partial destruction of some amino acids. Partial enzymatic hydrolysis is a viable alternative for some applications but bitterness due to hydrophobic peptides is frequently encountered; such bitterness may be eliminated, or at least reduced to an acceptable level, by treatment with activated carbon, carboxypeptidase, aminopeptidase, ultrafiltration or hydrophobic chromatography. Caseins yield very bitter hydrolysates but this problem may be minimized by the judicious selection of the proteinase(s) and by using exopeptidases (especially aminopeptidases) together with the proteinase.

The functional properties of milk proteins may be improved by limited proteolysis. Acid-soluble casein, free of off-flavour and suitable for incorporation into beverages and other acid foods, can be produced by limited proteolysis. The antigenicity of casein is destroyed by proteolysis and the hydrolysate is suitable for use in milk protein-based foods for infants allergic to bovine milk formulations. Controlled proteolysis improves the meltability of directly-acidified cheese but excessive proteolysis causes bitterness. Partial proteolysis of heat-coagulated whey protein, which is insoluble and has very poor functional properties, yields a product that is almost completely soluble above pH 6, although the product is slightly bitter. Limited proteolysis of whey protein concentrate increases its heat stability and its specific foam volume but reduces its emulsifying capacity and foam stability.

A novel, potentially very significant application of proteinases in milk protein technology is the production of biologically active peptides. Carefully selected proteinases of known specificity are required for such applications but the resulting products have high added value.

1.4.2 β-Galactosidase (Lactase)

The second most widely used exogenous enzyme in the dairy industry is β-galactosidase (lactase) for the hydrolysis of lactose in milk, whey or permeate. Lactose is a reducing disaccharide, O-β-D–galactopyranosyl–α-D-glucopyranose, which is unique to milk. Lactose has a number of undesirable characteristics (e.g., it has low solubility but once in solution it is difficult to crystallize, has low sweetness, and is hydroscopic under certain conditions); these properties limit its applications.

Disaccharides, including lactose, are not absorbed in the small intestine and must be hydrolysed to monosaccharides; lactose is hydrolysed by β-galactosidase (lactase). Most human infants secrete an adequate amount of β-galactosidase to hydrolyze all ingested lactose but, as the infant ages, the secretion becomes inadequate (after the age of about 10 years) and most people, except north-western Europeans and some African tribes, become lactose-intolerant, suffering diarrhoea and flatulence after its consumption. The problem may be solved by hydrolysing the lactose in milk by β-galactosidase, which may be performed at the dairy or domestically. The use of β-galactosidase for this purpose is not as widespread as hoped for in regions of the world where lactose intolerance is most widespread, such as South East Asia and Africa, because the consumption of milk is very limited in those regions for other reasons.

Other uses of β-galactosidase are the production of glucose-galactose syrups and galactooligosaccharides. Glucose–galactose syrups are much sweeter than lactose and could be used as such in the food industry, but may not be cost-competitive with sucrose or glucose syrups produced from starch. β-Galactosidase functions as a transferase as well as a hydrolase and conditions may be chosen so that transferase activity dominates and the principal products are galactooligosaccharides which have desirable functional properties in foods.

1.4.3 Lipases

As indicated above, lipases are used to modify the flavour of certain cheese varieties. Lipases are also used to hydrolyze milk fat for a variety of uses in the confectionary, candy, chocolate, sauce and snack food industries and there is interest in using immobilized lipases to modify fat flavours for such applications. Enzymatic interesterification of milk lipids to modify rheological properties is also feasible.

1.4.4 Lysozyme

Lysozyme has been isolated from the milk of several species; human and equine milks are especially rich sources. In view of its anti-bacterial activity, the large difference in the lysozyme content between human and bovine milks may have significance in infant nutrition. It is claimed that supplementation of baby food formulae based on cows' milk with egg white lysozyme is beneficial, especially for premature babies, but views on this are not unanimous, and it is not used commercially.

Nitrate is added to many cheese varieties to prevent the growth of *Clostridium tyrobutyricum* which causes off-flavours and late gas blowing. However, the use of nitrate in foods is considered to be undesirable because of its involvement in nitrosamine formation, and many countries have reduced the permitted level or prohibited its use. Lysozyme, which inhibits the growth of vegetative cells of *Cl. tyrobutyricum* and hinders the germination of its spores, is an alternative to nitrate for the control of late gas blowing in cheese. Lysozyme also kills *Listeria* spp. Co-immobilized lysozyme may be used for self-sanitizing immobilized enzyme columns.

1.4.5 Catalase

Hydrogen peroxide is a very effective chemical sterilant and although it causes some damage to the physico-chemical properties and nutritional value of milk protein, principally by oxidizing methionine, it is used as a milk preservative, especially in warm regions lacking refrigeration. Excess H_2O_2 may be reduced following treatment by soluble exogenous catalase (from beef liver, *Aspergillus niger* or *Micrococcus lysodeiktieus*). Immobilized catalase has been investigated for this purpose but the immobilized enzyme is rather unstable.

Catalase is frequently used together with glucose oxidase in many of the food applications of the latter; however, the principal potential application of glucose oxidase in dairy technology is for the *in situ* production of H_2O_2, for which the presence of catalase is obviously undesirable.

1.4.6 Glucose Oxidase

Glucose oxidase (GO) catalyses the oxidation of glucose to gluconic acid *via* gluconic acid-δ-lactone, with H_2O_2 as a by-product. The H_2O_2 formed is normally reduced by catalase present as a contaminant in commercial GO preparations (from *Penicillum notatum, Penicillum glaucum* or *Aspergillus niger*) or added separately. Glucose oxidase, which has a pH optimum of ~5.5, is highly specific for D-glucose and may be used to assay specifically for D-glucose in the presence of other sugars.

In the food industry, glucose oxidase has four principal applications:

1. *Removal of residual trace levels of glucose*: This application, which is particularly useful for the treatment of egg white prior to dehydration (although alternative procedures using yeast fermentation are used more commonly), is of little, if any, significance in dairy technology;
2. *Removal of trace levels of oxygen*: Traces of oxygen in wines and fruit juices cause discolouration and/or oxidation of ascorbic acid. Chemical reducing agents may be used to scavenge oxygen but enzymatic treatment with GO may be preferred. Glucose oxidase has been proposed as an antioxidant for high-fat products such as mayonnaise, butter and whole milk powder but it does not appear to be widely used for this purpose, probably because of cost *vis-à-vis* chemical antioxidants (if permitted) and the relative effectiveness of inert gas flushing in preventing lipid oxidation in canned milk powder;
3. *Generation of H_2O_2 in situ*: The H_2O_2 generated by glucose oxidase has a direct bactericidal effect (which appears to be a useful side-effect of GO applied to egg products) but its bactericidal properties can be much more effectively exploited as a component of the lactoperoxidase/H_2O_2/SCN^- system. Glucose required for GO activity may be added or produced by the action of ß-galactosidase on lactose (both β-galactosidase and glucose oxidase have been immobilized on porous glass beads). H_2O_2 may also be generated *in situ* by the action of xanthine oxidoreductase on added hypoxanthine. It is likely that exogenous H_2O_2 will be used in such applications rather than H_2O_2 generated by glucose oxidase or xanthine oxidase.
4. *Production of acid in situ*: Direct acidification of dairy products, particularly Cottage and Mozzarella cheeses, is fairly common. Acidification is normally performed by addition of acid or acidogen (usually gluconic acid-δ-lactone) or by a combination of acid and acidogen. *In situ* production of gluconic acid from

added glucose or from glucose produced *in situ* from lactose by β-galactosidase or from added sucrose by invertase. Immobilized glucose oxidase has been investigated but is not used commercially for the direct acidification of milk. Production of lactobionic acid from lactose by lactose dehydrogenase has also been proposed for the direct acidification of dairy and other foods.

1.5 Exogenous Enzymes in Food Analysis

Exogenous enzymes have several applications in food analysis (see Fox et al. 2015). One of the principal attractions of enzymes as analytical reagents is their specificity which eliminates the need for extensive clean-up of the sample and makes it possible to quantify separately closely-related molecules, e.g., D- and L-glucose, D- and L-lactic acid. Enzymatic assays can be very sensitive; some can detect concentrations at the picomole level. Enzymes can be immobilized as enzyme electrodes and as such can be used continuously to monitor changes in the concentration of a substrate in a product stream. Disadvantages of the use of enzymes as analytical reagents are their relatively high cost, especially when few samples are to be analysed, relatively poor stability (due to denaturation or inhibition) and the need to use highly purified enzymes.

Enzymes are rarely used by industrial food laboratories but find regular application in more specialized analytical or research laboratories. There are alternative chemical and/or physical methods, especially some form of chromatography, for all these applications, but extensive clean-up and perhaps concentration may be required.

The use of luciferase to quantify ATP in milk is the principle of modern rapid methods for assessing the bacteriological quality of milk based on the production of ATP by bacteria. Such methods have been automated and mechanized.

An indirect application of enzymes in analysis is as a marker or label in enzyme-linked immunosorbent assays (ELISA). In an ELISA, the enzyme does not react with the analyte; instead, an antibody is raised against the analyte (antigen or hapten) and labelled with easily-assayed enzyme, usually a phosphatase or a peroxidase. The enzyme activity is proportional to the amount of antibody in the system, which in turn is proportional, directly or indirectly depending on the arrangement used, to the amount of antigen present. ELISA may be used in competitive and non-competitive, each of which may be used in either of two modes.

Examples of the use of ELISA in dairy analyses include:

- Quantifying denaturation of ß-lactoglobulin in milk products;
- Detection and quantitation of adulteration of milk from one species with that from other species, e.g., ovine milk by bovine milk;
- Authentication of cheese, e.g., ovine milk cheese;
- Detection and quantitation of bacterial enzymes in milk, e.g., from psychrotrophic bacteria;
- Quantification of antibiotics.

1.6 Conclusion

Milk and dairy products contain a wide range of enzymes, from different sources and with different significance and consideration, as well as control strategies. The chapters that follow will expand on this brief introduction and give a comprehensive overview of these complex systems.

References

Fox PF, Kelly AL (2006a) Indigenous enzymes in milk: overview and historical aspects – part 1. Int Dairy J 16:500–516

Fox PF, Kelly AL (2006b) Indigenous enzymes in milk: overview and historical aspects – part 2. Int Dairy J 16:517–532

Fox PF, Kelly AL (2008) The principal indigenous enzymes in milk. In: Koutinas A, Pandey A, Larroche C (eds) Current topics on bioprocesses in food industries. Asiatech Publishers, New Delhi, pp 4–29

Fox PF, Uniacke-Lowe T, McSweeney PLH, O'Mahony JA (2015) Enzymology of milk and milk products. In: Dairy chemistry and biochemistry, 2nd edn. Springer, Cham, pp 377–414

Further Reading

Birch GG, Blakeborough N, Barker K (1981) Enzymes and food processing. Elsevier, London

Chandrasekaran M (2018) Enzymes in food and beverage processing. CRC Press, Boca Raton, FL

Fox PF (1991) Food enzymology, vol 1 & 2. Elsevier Applied Science Publishers, London

IDF (2006) First IDF symposium on indigenous enzymes in milk. Int Dairy J 16:#6

McKellar RC (1989) Enzymes of psychrotropha in raw foods. CRC Press, Boca Raton, FL

O'Mahony JA, Fox PF, Kelly AL (2013) Indigenous enzymes in milk. In: McSweeney PLH, Fox PF (eds) Advanced dairy chemistry, volume 1A, proteins: basic aspects, 4th edn. Springer, New York, pp 337–385

Panesar, P.J., Marwahs, SS, and Chopra, H.K. (2010). Enzymes in food processing and potential applications. I.K. International Publishing House, New Delhi

Ray RC, Rosell CM (2018) Microbial enzyme technology in food applications. CRC Press, Boca Raton, FL

Schwimmer S (1981) Source book of food enzymology, 2nd edn. AVI Publishing, Westport, CT

Whitaker JR (1993) Principles of enzymology for the food sciences, 2nd edn. CRC Press, Boca Raton, FL

Whitaker JR, Voragen AGJ, Wong DWS (2003) Handbook of food enzymology. Marcel Dekker, New York

Whitehurst RJ, Van Oort M (eds) (2009) Enzymes in food technology, 2nd edn. Wiley-Blackwell, Hoboken, NJ

Chapter 2
The Plasmin System in Milk and Dairy Products

Thomas C. France, James A. O'Mahony, and Alan L. Kelly

2.1 Introduction

The presence of a proteolytic enzyme in fresh, raw milk was first reported by Babcock and Russell (1897) as they studied the decomposition of skim milk and proteolysis of cheese during ripening. The observed proteolysis was attributed to an indigenous enzyme, galactase, with other researchers reaching a similar conclusion in time (Thatcher and Dahlberg 1917; Somer 1938). Other researchers, however, concluded that the observed proteolysis was most likely from bacterial enzymes present in milk (see Humbert and Alais 1979). Later work verified that the proteolytic activity in milk was not a result of bacterial proteases, but due to a naturally present, indigenous milk proteinase, subsequently named plasmin (Harper et al. 1960; Fox 1982). Bovine plasmin is an alkaline serine proteinase identical to plasmin isolated from blood when compared kinetically (Bastian and Brown 1996), and is analogous to human plasmin in terms of pH stability, inhibition patterns and amino acid sequence (Kaminogawa et al. 1972; Reimerdes et al. 1981). In blood, the physiological role of plasmin is to break down blood clots; in milk, however, its role is not as clear.

2.2 Plasmin and Plasminogen

Bovine plasmin (EC 3.4.21.7), the principal indigenous milk proteinase, displays trypsin-like activity (Harper et al. 1960; Kaminogawa et al. 1969; Reimerdes et al. 1975) with similarities in its activity characteristics (e.g., sensitivity to inhibitors). Plasmin (48,000 Da) exhibits optimum activity at pH 7.4–7.5 and a temperature of

T. C. France · J. A. O'Mahony (✉) · A. L. Kelly
School of Food and Nutritional Sciences, University College Cork, Cork, Ireland
e-mail: 114345886@umail.ucc.ie; sa.omahony@ucc.ie; a.kelly@ucc.ie

A. L. Kelly, L. B. Larsen (eds.), *Agents of Change*, Food Engineering Series,
https://doi.org/10.1007/978-3-030-55482-8_2

37 °C (Warner and Polis 1945; Kaminogawa et al. 1969, 1972; Richter et al. 1979). Hence, the physiological conditions in the mammary glands of bovines are almost ideal for plasmin activity.

Bovine plasminogen, the inactive zymogen of plasmin, is a glycosylated protein containing 786 amino acid residues with a molecular weight of 88,092 Da, based on its amino acid sequence and carbohydrate content (Schaller et al. 1985). Bovine plasminogen consists of two carbohydrate moieties and therefore contains two specific glycosylation sites, like human plasminogen (Schaller et al. 1985). Plasminogen contains five characteristic, triple-loop structures, known as kringles, each stabilised by three disulphide bonds, as shown in Fig. 2.1. These kringles contain lysine-binding sites which possess different affinities for lysine but are all critical in the regulation of activation of plasminogen to plasmin, as shown in human plasminogen (Markus et al. 1978; Wiman and Collen 1978; Collen 1980). The affinity for lysine residues and ω-aminocarboxylic acids such as ε-amino-n-caproic acid (EACA) has

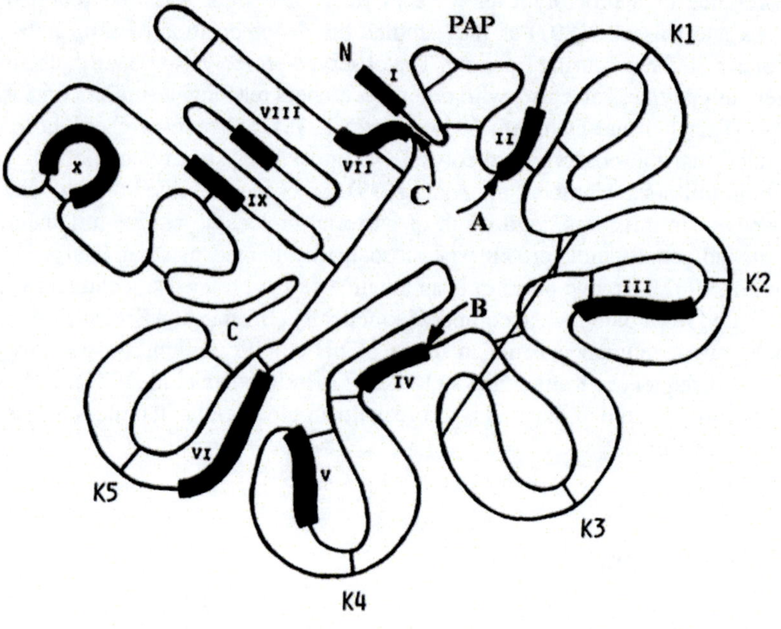

Fig. 2.1 Domain structure of plasminogen isolated from bovine milk. The sequences identified by Benfeldt et al. (1995), approximately 15% of the sequence, are represented by bold lines and Roman numerals. Disulphide bridges are predicted according to homology with human plasminogen. The three arrows, A, B and C, correspond to the sites cleaved by plasmin or by plasminogen activators. Arrow C: The activation cleavage site between kringle 5 and the catalytic domain, cleaved by plasminogen activators (t-PA and u-PA). Arrow A: The autocatalytic cleavage site by plasmin (intra or intermolecular) between Lys_{77}-Arg_{78}, releasing the preactivation peptide (PAP). Arrow B: Plasmin cleavage site (intra or intermolecular) in bovine system, creating midi-plasmin (kringle 4 + kringle 5 + catalytic domain) which is probably the most abundant form of active, bovine plasmin (modified from Benfeldt et al. 1995)

been attributed to the structure of these kringles (Wiman and Wallén 1977; Sottrup-Jensen et al. 1978).

The concentration of plasmin in bovine milk is influenced by a number of environmental factors, in addition to the various processing conditions to which milk is subjected (discussed in Sect. 2.5). Richardson and Pearce (1981) found pasteurised, bovine milk to contain 0.55–2.75 μg plasminogen/ml and 0.14–0.73 μg plasmin/ml, determined based on an enzymatic activity assay. These results were found to be in agreement with those of previous studies (Dulley 1972; Halpaap et al. 1977), with the concentration of plasminogen decreasing and plasmin increasing during storage as plasminogen is converted to plasmin by plasminogen activators. A later study by Benfeldt et al. (1995) quantified the amount of plasminogen present in skimmed milk using an enzyme-linked immunosorbent assay (ELISA), and reported values of 1.5 μg plasminogen/ml, comparable to those previously reported.

2.2.1 *Proteolytic Specificity of Plasmin on Milk Proteins*

Plasmin has an affinity for lysine and arginine residues, hydrolysing Lys-X and Arg-X bonds, with a preference for peptide bonds containing lysine. Plasmin acts on all caseins; however, the susceptibility of the caseins to plasmin action varies, with β-casein being the principal substrate for plasmin, followed by α_{s2}-casein.

2.2.1.1 β-casein

β-Casein is a single polypeptide chain consisting of 209 amino acid residues (Huppertz 2013) which is hydrolysed by plasmin, producing γ-caseins and proteose peptones (Snoeren and van Riel 1979; Andrews and Alichanidis 1983). Plasmin preferentially cleaves β-casein at three specific sites: Lys_{28}-Lys_{29}, Lys_{105}-His_{106} and Lys_{107}-Glu_{108}, with Lys_{29}-Ile_{30}, Lys_{113}-Tyr_{114} and Arg_{183}-Asp_{184} also being cleaved by plasmin, but to a lesser extent (Fox et al. 1994). The cleavage of the three primary sites produces three γ-caseins; γ_1 (β-CN f29-209), γ_2 (β-CN f106-209) and γ_3 (β-CN f108-209), in addition to the corresponding proteose peptones: proteose peptone (PP) 8_{fast} (β-CN f1-28), PP 5 (β-CN f1-105/107) and PP 8_{slow} (β-CN f29-105/107) (Ribadeau-Dumas et al. 1972; O'Mahony et al. 2013).

2.2.1.2 α_{s2}-casein

α_{s2}-casein contains a greater number of plasmin-sensitive peptide bonds compared to β-casein. Plasmin cleaves α_{s2}-casein at eight specific sites, seven of type Lys-X and one Arg-X: Lys_{21}-Gln_{22}, Lys_{24}-Asn_{25}, Arg_{114}-Asn_{115}, Lys_{149}-Lys_{150}, Lys_{150}-Thr_{151}, Lys_{181}-Thr_{182}, Lys_{188}-Ala_{189} and Lys_{197}-Thr_{198} (Le Bars and Gripon 1989; Visser et al. 1989). The hydrolysis of α_{s2}-casein by plasmin results in the formation of approximately 14 peptides, with three of the peptides arising from the C-terminus region displaying high hydrophobicity, and therefore potentially being bitter peptides (α_{s2}-CN f198-207, f182-207 and f189-207) (Le Bars and Gripon 1989).

2.2.1.3 α_{s1}-casein

Plasmin has also been shown to act on α_{s1}-casein, albeit to a lesser extent than β- and α_{s2}-casein. Aimutis and Eigel (1982) found the action of plasmin on α_{s1}-casein resulted in the formation of λ-casein, a minor casein component. McSweeney et al. (1993) found the principal cleavage sites in α_{s1}-casein to be Arg_{22}-Phe_{23}, Arg_{90}-Tyr_{91}, Lys_{102}-Lys_{103}, Lys_{103}-Tyr_{104}, Lys_{105}-Val_{106}, Lys_{124}-Glu_{125} and Arg_{151}-Gln_{152}, with up to 11 additional cleavage sites.

2.2.1.4 κ-casein

κ-casein is very resistant to plasmin, despite κ-casein containing several lysyl residues; Chen and Ledford (1971) reported that κ-casein was unaffected by plasmin. Eigel (1977) reported no hydrolysis of κ-casein when incubated in the presence of bovine plasmin at 37 °C, conditions which resulted in the complete hydrolysis of α_{s1}-casein. However, Andrews and Alichanidis (1983) reported little hydrolysis of κ-casein by indigenous plasmin in pasteurised milk incubated at 37 °C for 7 days, with 4% of the peptides produced originating from κ-casein. Pihlanto-Leppälä et al. (1993) reported the release of some peptides from the N-terminus of κ-casein by plasmin; however, the specificity of plasmin on κ-casein is yet to be determined. Differences in results are most likely due to differences in experimental conditions of the studies, but indicate that limited hydrolysis of κ-casein can occur when exposed to adequate amounts of plasmin for sufficient periods of time.

2.2.1.5 Whey Proteins

Due to the tight, globular structure of whey proteins, they are less susceptible to plasmin action than caseins (Kelly and McSweeney 2003). A number of studies have reported that β-lactoglobulin and α-lactalbumin are not hydrolysed by plasmin (Chen and Ledford 1971; Yamauchi and Kaminogawa 1972; Aslam and Hurley 1997). However, considerable hydrolysis of β-lactoglobulin by plasmin was reported by Caessens et al. (1999a, b), with significant disulphide-sulphydryl exchange occurring during hydrolysis. Aslam and Hurley (1997) identified peptides produced by plasmin-mediated hydrolysis of lactoferrin during involution of the mammary gland. The susceptibility of β-lactoglobulin and α-lactalbumin to plasmin-mediated hydrolysis can be affected by photooxidation and lactosylation (Dalsgaard et al. 2007, 2008), causing changes in the structure of the protein and the modification of lysine residues.

2.3 The Plasmin System in Bovine Milk

The plasmin system in milk is a complex protease-protease inhibitor system consisting of the active enzyme (plasmin), the inactive zymogen (plasminogen), plasminogen activators (which convert plasminogen to plasmin), plasminogen activator inhibitors and plasmin inhibitors (with the latter both able to inhibit the activity of plasmin) (Fig. 2.2).

Plasminogen is not expressed in the bovine mammary gland or mammary cells (Berglund et al. 1995); therefore, the origin of plasmin and plasminogen in milk is the cow's blood. Plasmin is transferred to bovine milk from blood as the inactive zymogen, plasminogen. As the mammary glands have the ideal physiological conditions (i.e., correct temperature and pH), there is often sufficient time for plasminogen to be activated and for plasmin to act on its substrate within the mammary gland (Schaar 1985; Kelly and Fox 2006) or during storage (Driessen and van der Waals 1978). The occurrence of γ-caseins, the products of casein hydrolysis as mentioned above, directly after milking is an indication of plasminogen activation and plasmin activity within the udder. The conversion of plasminogen to plasmin is controlled by two main types of plasminogen activators present in milk (Heegaard et al. 1993), urokinase-type plasminogen activators (u-PA) and tissue-type plasminogen activators (t-PA). Plasminogen to plasmin ratios can be used to express the extent of activation of plasminogen to plasmin at different times (e.g., different stages of processing or lactation). Plasminogen activator inhibitors and plasmin inhibitors both act to reduce plasmin activity; however, their effect on the plasmin system is often dependent on processing conditions such as thermal treatments (see Sect. 2.5).

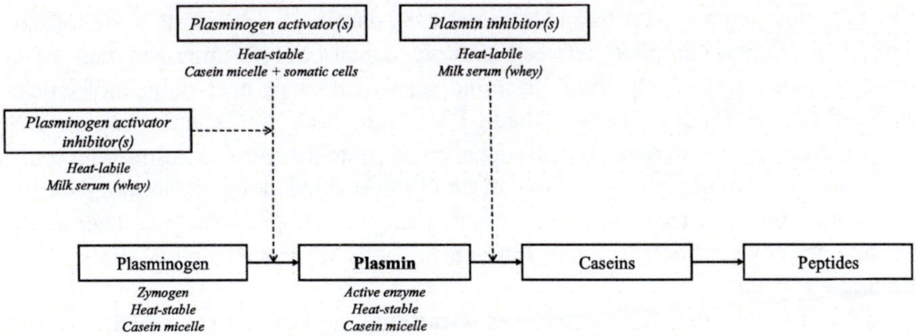

Fig. 2.2 Schematic representation of the plasmin system in bovine milk

2.3.1 Activators of the Plasmin System

Richardson (1983a) found the concentration of plasminogen in milk to decrease during storage at 37 °C for 3 days, with a subsequent increase in plasmin concentration. This phenomenon is caused by the presence of plasminogen activators, which hydrolyse the inactive zymogen (plasminogen) to plasmin. Since plasmin and plasminogen pass from the blood into bovine milk, plasminogen activators and inhibitors present in blood also leak into milk. As plasminogen activators convert plasminogen to plasmin and therefore control the plasmin activity in milk, there has been considerable interest in developing a better understanding of them; particularly t-PA and u-PA, the main plasminogen activator types present in both blood and bovine milk (Lu and Nielsen 1993a; Bastian and Brown 1996; Nielsen 2002).

u-PA is the best characterised plasminogen activator, consisting of two subunits linked by a disulphide bond. It has been shown by researchers, including Ismail et al. (2006), that u-PA causes greater plasminogen activation than t-PA. Ismali et al. (2006) observed that a small amount of u-PA can cause more plasminogen activation than that caused by a larger amount of t-PA. Thus, urokinase is often the preferred plasminogen activator for the determination of plasminogen in milk. The activities of u-PA and t-PA in milk have been shown to increase in the presence of casein (Markus et al. 1993; Politis et al. 1995), while the presence of fibrin increases the activity of t-PA (Karlan et al. 1987). Markus et al. (1993) and Politis et al. (1995) reported that all the casein fractions (α-, β- and κ-casein) enhanced plasminogen activation, depending on their relative concentration, with the α-casein fraction inducing the greatest rate of plasminogen activation by both u-PA and t-PA. A later study conducted by Heegaard et al. (1997) reported that α_{s2}-casein, specifically dimeric α_{s2}-casein, resulted in a concentration-dependent enhancement of plasminogen activation by t-PA. From ligand-blotting experiments, the authors showed that both plasminogen and t-PA bound to dimeric α_{s2}-casein, suggesting that the formation of a ternary complex between dimeric α_{s2}-casein, plasminogen and t-PA enhances plasminogen activation. Plasminogen activators are heat-stable molecules, with u-PA being more heat-stable than t-PA (Prado et al. 2007), and often survive heat treatments used in dairy processing such as pasteurisation and ultra-high temperature (UHT) treatments. Lu and Nielsen (1993a) showed that plasminogen activators may be even more heat-stable than plasmin and plasminogen. Therefore, plasmin activity can increase on heat treatment, as inhibitors of the plasmin system are often heat-labile.

t-PA and u-PA are serine proteinases (Grufferty and Fox 1988a), which activate bovine plasminogen by hydrolysing the Arg_{557}-Ile_{558} bond between kringle 5 and the catalytic domain (arrow c in Fig. 2.1) (Schaller et al. 1985; Benfeldt et al. 1995). Bovine plasminogen contains an N-terminal aspartic acid and therefore native bovine plasminogen is commonly referred to as Asp-plasminogen (Fig. 2.3). Plasmin autocatalytically cleaves off the first 77 residues of bovine plasminogen at the Lys_{77}-Arg_{78} bond, leaving an N-terminus arginine (Schaller et al. 1985), producing the proteolytically-modified Arg-plasminogen. However, plasmin cannot cleave

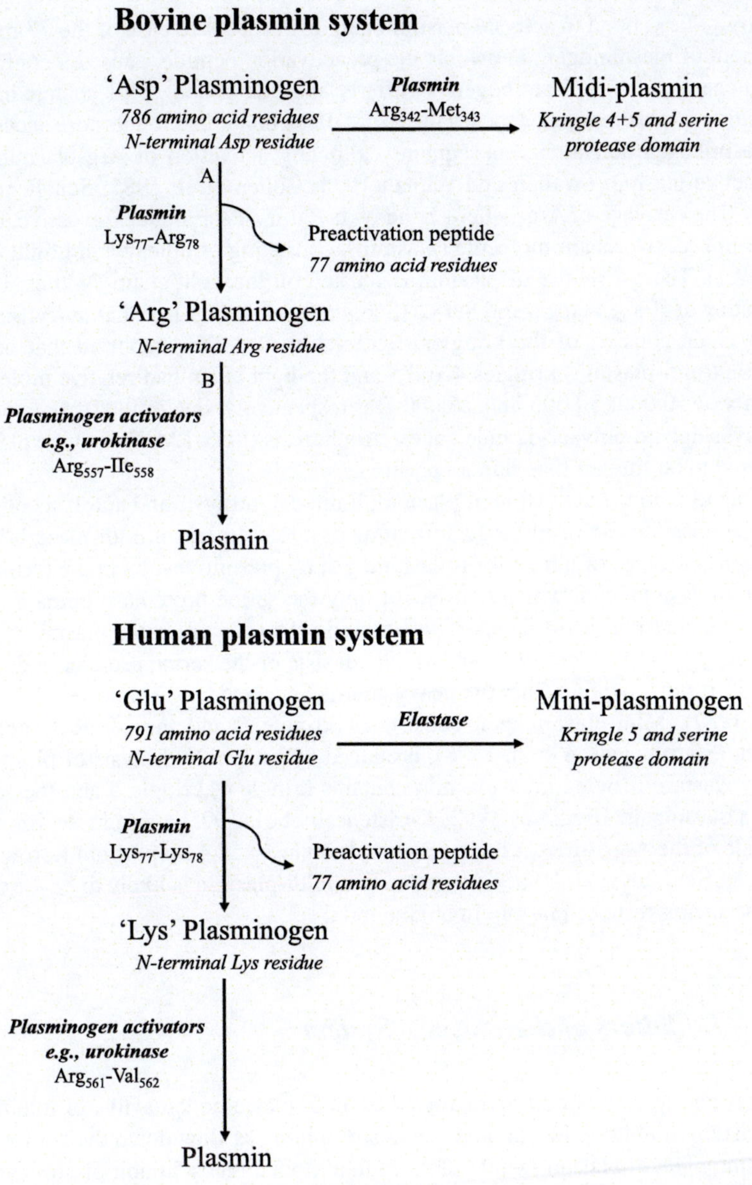

Fig. 2.3 Schematic representation of the activation of bovine and human plasminogen as a comparison. Cleavage of A can occur before (inter-molecular) or after (intra-molecular) cleavage at B. Details of cleavage sites in human plasminogen obtained from Castellino and Ploplis (2005) and Booth and Backmann (2006)

the Arg_{557}-Ile_{558} bond to activate plasminogen. Instead, the release of the 77-residue fragment of plasminogen, known as the preactivation peptide, causes a conformational change in Arg-plasminogen (Grufferty and Fox 1988a). This change in conformation of plasminogen exposes the Arg_{557}-Ile_{558} bond, making it more accessible to plasminogen activators, subsequently allowing activation of Arg-plasminogen into active plasmin (Wallén and Wiman 1975; Collen et al. 1985; Schaller et al. 1985). The cleavage of Arg_{557}-Ile_{558} bond by u-PA or t-PA produces an active, disulphide-linked, two-chain molecule (plasmin), consisting of a heavy and light chain (Fig. 2.1). The active site of plasmin is located on the light chain (Wallén 1978), consisting of His_{598}, Asp_{641}, and Ser_{736}, (Schaller et al. 1985), like that of trypsin; the heavy chain consists of the kringles. Benfeldt et al. (1995) reported that bovine plasmin (midi-plasmin) kringles 4 and 5 and the light chain had relative molecular weights of 50,000, 30,000 and 26,000 Da, respectively, determined from sodium dodecyl sulphate-polyacrylamide gel electrophoresis (SDS-PAGE); bovine plasmin appeared to be smaller than human plasmin.

In addition to the activation of plasmin, limited hydrolysis and autolysis of plasminogen/plasmin can result in the formation of micro-, mini- or midi-plasmin/plasminogen. Cleavage of human Lys-plasminogen by plasmin results in the formation of micro-plasminogen, which consists of only the serine proteinase domain (light chain) (Shi and Wu 1988). In addition, autolytic cleavage of human plasmin results in the formation of micro-plasmin, which consists of the serine proteinase domain and a 31 residue peptide from the heavy chain connected by disulphide bonds (Wu et al. 1987). Mini-plasminogen consists of kringle 5 and the serine proteinase domain (Sottrup-Jensen et al. 1978), produced by limited hydrolysis of plasminogen by elastase. Bovine midi-plasmin contains kringle 4, kringle 5 and the serine proteinase domain (Benfeldt 1992; Christensen et al. 1995b), produced from the cleavage of the Arg_{342}-Met_{343} bond between kringle 3 and 4 by plasmin (arrow b in Fig. 2.1); Benfeldt et al. (1995) suggested that midi-plasmin is likely to be the naturally occurring form of plasmin in bovine milk.

2.3.2 *Inhibitors of the Plasmin System*

The plasmin system in milk is regulated by both plasminogen activator inhibitors and plasmin inhibitors. Plasminogen activator inhibitors slow down the conversion of plasminogen to plasmin, while plasmin inhibitors directly inhibit plasmin activity. Plasmin system inhibitors are present in the serum phase of milk (whey) (Reimerdes et al. 1976; Korycha-Dahl et al. 1983), and their activity is affected by pH and heat treatments to a greater extent than the activators of the plasmin system. The mechanism of action of proteinase inhibitors, many of which belong to the serpin family, is not yet fully understood. Despite this, it is known that exposed, reactive-site loops in the C-terminus of the serpin interact with the active site of plasmin, potentially blocking or altering the shape of the active site (Christensen and Sottrup-Jensen 1994).

Unlike plasminogen activators, less information is available on the inhibitors of the plasmin system in milk. However, it can be assumed that the plasmin inhibitors controlling the plasmin system in milk are similar to those in the blood stream, as the source of the plasmin system is the blood stream. Many of the serine proteinase inhibitors that have been studied in detail are trypsin inhibitors (Reimerdes et al. 1976). Precetti et al. (1997) partially purified two serine proteinase inhibitors from bovine milk, plasminogen activator inhibitor (plasminogen activator inhibitor 1) and a plasmin inhibitor (α_2-antiplasmin), finding both inhibitors to be present in several forms, which they attributed to the possible formation of complexes with other milk proteins.

Several blood-derived proteinase inhibitors have been detected in bovine milk, many of which are serpins, including C1-inhibitor (C1 Inh), antithrombin-III, inter-α-trypsin inhibitor, α_2-macroglobulin (α_2-M), α_1-proteinase inhibitor and α_2-antiplasmin (α_2-AP) (Lindberg 1979; Weber and Nielsen 1991; Christensen and Sottrup-Jensen 1992; Christensen et al. 1995a). The principal inhibitor of plasmin is considered to be α_2-AP (Wiman and Collen 1978; Levi et al. 1993). α_2-AP is a single-chain glycoprotein consisting of two disulphide bridges with a molecular weight of 70,000 Da. Bovine α_2-AP shares close homology with human α_2-AP, especially at the reactive site and N-terminus (Christensen and Sottrup-Jensen 1992). The effective plasmin-inhibiting ability may be due to the strong complex formed between plasmin and α_2-AP through the lysine-binding sites of plasmin (Wiman and Collen 1978). Christensen and Sottrup-Jensen (1994) isolated a proteinase inhibitor of trypsin (TI) from bovine plasma and milk, which had the ability to inhibit plasmin in addition to trypsin, even though it showed little similarity to α_2-AP. Interactions between plasminogen activators and plasminogen activator inhibitors, as well as between plasmin inhibitors and plasmin, provide a complex mechanism for the control of plasmin activity in bovine milk.

2.4 Distribution of Plasmin System Components in Milk

The components of the plasmin system are distributed between the serum phase (whey) and the micellar phase (casein) of milk (Bastian and Brown 1996; Politis et al. 1992). The distribution of plasminogen activators and plasminogen activator inhibitors in bovine milk is known; however, there have been many conflicting results over the years as to which plasminogen activator is primarily associated with casein micelles.

Lu and Nielsen (1993a) indicated that plasminogen activators u-PA and t-PA are associated with the casein micelle, while other studies have suggested that both u-PA and t-PA are associated with somatic cells (Politis et al. 1991; Zachos et al. 1992). Many researchers have agreed that the plasminogen activators, u-PA and t-PA, are predominantly associated with somatic cells (*via* a urokinase receptor) and casein micelles, respectively (Heegaard et al. 1993, 1994a, b; White et al. 1995); Heegaard et al. (1994b) assessed the binding of t-PA to native casein micelles and purified preparations of the individual caseins; they concluded that t-PA bound to dimeric α_{s2}-casein and multimeric κ-casein, suggesting that t-PA binds to casein micelles *via* electrostatic interactions. White et al. (1995) isolated the plasminogen activators from both the casein micelles and somatic cells, confirming that t-PA is the main plasminogen activator associated with micelles and that u-PA is associated with the somatic cells. Ismail et al. (2006) showed u-PA activity to be highest in milk with high somatic cell count (SCC), confirming results from previous researchers that u-PA is associated with somatic cells. Results also confirmed that both u-PA and t-PA associate with caseins, and that u-PA has the ability to dissociate from somatic cells and re-associate with micelles.

Unlike plasminogen activators, plasmin inhibitors are found only within the serum phase (whey) (Reimerdes et al. 1976; Korycha-Dahl et al. 1983; Christensen and Sottrup-Jensen 1994; Christensen et al. 1995a; Precetti et al. 1997). Korycha-Dahl et al. (1983) found that plasminogen activators were only associated with casein micelles while inhibitors of both plasmin and plasminogen activators were found solely in the serum phase. Both plasmin and plasminogen are primarily associated with the casein micelles (Kaminogawa et al. 1972; Korycha-Dahl et al. 1983; Richardson 1983a). Politis et al. (1992) found that 82% of total plasminogen was associated with casein micelles, while the remainder was present in the serum phase. The conversion of plasminogen to plasmin does not cause dissociation of plasmin from the casein matrix (Baer et al. 1994); therefore, the presence of plasmin in whey is due to certain factors causing its dissociation from the micelle (see Sect. 2.4.1). Numerous researchers have conducted plasmin activity assays on different milk fractions, all finding plasmin-casein interactions (Reimerdes et al. 1975; Snoeren and van Riel 1979; Fox 1981; Politis et al. 1992). Hofmann et al. (1979) found plasmin associated with the milk fat globule membrane (MFGM): components of the MFGM displayed protease activity, with plasmin being the dominant protease present. Politis et al. (1992) also found plasminogen to be present within the MFGM; however, the membrane-associated plasminogen is believed to be a result of small amounts of casein present within the MFGM, as plasminogen is strongly associated with casein micelles; this was later confirmed by Benfeldt et al. (1995).

Early studies showed that plasmin remains associated with casein micelles after centrifugation (Reimerdes and Klostermeyer 1974), isoelectric precipitation (acid casein) (Warner and Polis 1945; Kiermeier and Semper 1960) and rennet coagulation (i.e., rennet casein) (Richardson and Elston 1984), indicating a strong plasmin-casein bond. Plasminogen, plasmin and plasminogen activators bind to casein micelles through lysine-binding sites and, to a lesser extent, through electrostatic interactions (Grufferty and Fox 1988a; Baer et al. 1994). In blood, lysine-binding is also important for the plasminogen-fibrin complex (Wiman and Collen 1978). The amount of plasminogen bound to the casein micelles depends on the type of casein, with Baer et al. (1994) observing that immobilised α-casein bound twice as much plasminogen as immobilised β- and κ-casein.

2.4.1 *Factors Affecting the Distribution of Plasmin System Components Within Milk*

The distribution of plasmin system components within the different milk fractions is affected by several environmental and non-environmental factors, some of which cause complete dissociation from the micelle. The addition of lysine or 6-aminohexanoic acid/ε-amino-n-caproic acid (EACA) (a lysine derivative) at concentrations up to 50 mM dissociates plasmin from casein (Korycha-Dahl et al. 1983; Richardson 1983a); however, at this concentration, it can increase the rate of plasminogen activation, as both ligands bind to plasminogen, altering its conformation, thereby enhancing cleavage of the preactivation peptide (Walther et al. 1975; Searls 1980).

Plasmin-casein association is strongly influenced by pH. The acidification of fresh bovine milk to pH 4.6–4.7 caused almost complete dissociation of plasmin from the casein micelles, with plasmin beginning to dissociate from the micelle at pH 5.7 (Richardson and Elston 1984). Grufferty and Fox (1988b) also found that, at pH 4.6, most of the plasmin was dissociated; however, they observed no reduction in plasmin-micelle activity between pH 6.6 and pH 4.8. The discrepancies observed lie in the preparation of the casein micelles, as Richardson and Elston (1984) isolated casein by isoelectric precipitation before assaying the casein for plasmin activity, while Grufferty and Fox (1988b) utilised ultracentrifugation to isolate the casein micelles. The dissociation of plasmin from the casein micelle to the serum phase can potentially cause negative effects on the quality of food products containing whey protein ingredients due to residual plasmin activity. Hayes and Nielsen (2000) and Crudden and Kelly (2003) studied plasmin activity in acid and sweet whey, finding significantly higher plasmin activity in acid whey compared to sweet whey. Hayes and Nielsen (2000) suggested the higher plasmin activity in acid whey may be due to the disruption of the casein micelles at the low pH. Crudden and Kelly (2003) analysed plasminogen-derived activity in whey, finding low levels of plasminogen in acid whey, suggesting that activation of plasminogen may have partly contributed to the high plasmin activities. Crudden and Kelly (2003) also reported an inverse relationship between plasmin activity in whey and the rate of acidification. The effect of pH on plasmin dissociation can help explain the higher levels of plasmin observed in acid whey (44 μg/g protein), compared to sweet whey (4 μg/g protein) (Ismail and Nielsen 2010).

The addition of NaCl to milk has also been reported to cause the dissociation of plasmin from casein micelles (Johnson 1974; Grufferty 1986). The effect of temperature on plasmin dissociation is unclear, with conflicting evidence regarding its effect. A report by Reimerdes and Herlitz (1979) found an increase in plasmin activity in milk stored at 4 °C. However, Donnelly and Barry (1983) concluded that temperature had no significant influence on plasmin-casein interactions, with the work of Grufferty and Fox (1988b) supporting such findings.

2.5 Factors Affecting Plasmin Activity and Consequent Proteolysis

The concentration of plasmin and plasminogen in bovine milk is affected by numerous factors including environmental factors (Politis et al. 1989a), processing conditions (Korycha-Dahl et al. 1983; Kelly and Foley 1997; Aaltonen and Ollikainen 2011; García-Risco et al. 2003), storage conditions (Crudden et al. 2005a) and the presence of microbial proteases (Nielsen 2002). The plasmin system is complex, involving a multitude of interactions that ultimately result in the activation of plasminogen to plasmin. The system is made more complex by the interference of milk components with plasmin and plasminogen under different conditions (pH, storage temperature, heating). The conditions mentioned ultimately affect the kinetics of the plasmin system, altering the rate/extent of plasminogen activation, or hydrolysis of caseins by plasmin. The effect of these conditions and the subsequent interactions with the different plasmin system components are discussed below.

2.5.1 Mastitis

Mastitis is an inflammation of the mammary gland caused by injury to the udder, trauma or bacterial infection (Oliver et al. 2011). Plasmin activity is higher in milk during mastitis (Barry and Donnelly 1981; de Rham and Andrews 1982; Andrews 1983a; Schaar and Funke 1986; Politis et al. 1989a). Barry and Donnelly (1981) observed that the increased proteolytic activity observed in mastitic milk was primarily due to plasmin, with proteinases from other sources (such as leukocytes) representing only a small fraction of total proteolytic activity. Studies conducted by de Rham and Andrews (1982) and Andrews (1983a) also reported that plasmin activity increased in mastitic milk, but did not have as great a contribution to total proteolysis as previously reported by Barry and Donnelly (1981). Politis et al. (1989b) reported an increase in the concentration of plasmin and plasminogen, by 105 and 74%, respectively, as the SCC increased from less than 250,000/ml to over 1,000,000/ml. Previous work also showed an increase in plasmin activity as SCC increased from 100,000 to 1,300,000/ml (Politis et al. 1989a).

The increased plasmin activity in mastitic milk can be attributed to a number of factors, often due to the leakage of blood components into milk as a result of a compromised blood-milk barrier. During mastitis, there is an increase in the SCC (i.e., white blood cells) of milk to help combat the infection/inflammation. Leucocytes contain plasminogen activators (Christman et al. 1977), which they can secrete; therefore, the secretion of these plasminogen activators may contribute to some of the increased plasmin activity observed in mastitic milk. Plasminogen activators are also produced by activated neutrophils and macrophages (Mullins and Rohlich 1983), which can increase the rate at which plasminogen is converted to plasmin in mastitic milk, thereby increasing plasmin activity. Heegaard et al. (1994a) reported a 10- to 20-fold increase in the content of plasminogen activators in mastitic milk. The authors also observed an increase in both u-PA and t-PA in mastitic milk; somatic cells appeared to have a larger number of u-PA receptors compared to that in uninfected milk. The large increase in SCC during mastitis, coinciding with an increase in plasminogen activators, helps explain increased plasmin activity in mastitic milk. Bacterial mastitis can also contribute to increased plasmin activity, as specific bacteria such as *Streptococcus uberis* can contain/secrete plasminogen activators (discussed in Sect. 2.5.9).

2.5.2 *Stage of Lactation*

Throughout lactation, the level of plasmin activity fluctuates. Late lactation is associated with a higher level of plasmin activity (Barry and Donnelly 1980; Andrews 1983a; Donnelly and Barry 1983; Richardson 1983b; Schaar 1985; Politis et al. 1989a, b), resulting in increased proteolysis of caseins. A study by Barry and Donnelly et al. (1980) showed that there is less β- and α_{s1}-casein and more γ-casein in late-lactation milk than in mid-lactation milk, indicative of higher plasmin activities. The reason for this increased plasmin activity in late-lactation milk has been disputed between researchers. Richardson (1983b) reported that plasminogen and plasmin concentration in late-lactation milk increased, and suggested that the increased plasmin activity observed was due to more plasmin entering the mammary gland, rather than increased activation of plasminogen. On the other hand, Korycha-Dahl et al. (1983) reported that the increased plasmin activity is likely due to increased plasminogen levels, observing a twofold increase in plasminogen concentrations in late-lactation milk compared to mid-lactation. Schaar (1985) reported similar findings to Richardson (1983b), observing a gradual increase in plasmin activity with advancing lactation. Politis et al. (1989b) reported an increase in plasmin activity in late-lactation milk; however, they concluded that this increase was predominately due to an increase in plasminogen activators. They observed a decrease in the ratio of plasminogen to plasmin from early lactation (6.55) to late lactation (3.29). Bastian et al. (1991a) found plasmin activity to increase steadily at the beginning of lactation, before increasing rapidly towards the end of lactation, with a corresponding increase in the percentage of plasminogen activated.

These studies indicate that there is an increase in the rate of plasminogen activation to plasmin in late-lactation milk, potentially due to increased plasminogen activators, and suggested that the leakage of plasmin from the blood into milk increases in early lactation but remains constant throughout lactation. It is known that there is an increase in the permeability of epithelial cells in mammary glands during late lactation, resulting in increased permeability of somatic cells (known to contain plasminogen activators) and other constituents originating from blood. Therefore, it is likely that increased plasminogen activation, caused by increased permeability of mammary gland tissue due to a compromised blood-milk barrier, have a significant impact on the higher plasmin activity seen in late-lactation milk.

2.5.3 Age of Cow

Plasmin activity is higher in milk from older cows (Schaar 1985; Politis et al. 1989a; Bastian et al. 1991a) but, although plasmin activity increases in milk from older cows, plasminogen levels remain relatively constant. Therefore, similar to late-lactation milk, the increase in plasmin activity is likely due to increased permeability of the secretory tissue, increasing the presence of plasminogen activators. In addition to this, older cows are likely to have experienced more mastitis in their lifetime; it has been reported that plasmin activity in milk does not return to levels prior to the onset of mastitis (Saeman et al. 1988), and therefore this may be responsible for the higher plasmin activity in milk from older cows.

2.5.4 Breed of Cow

The concentration of plasmin in milk can vary depending on the breed of the cow. Richardson (1983b) found milk of Holstein-Friesian cows (0.27–0.53 mg/L) to have a higher plasmin activity than milk of Jersey cows (0.15–0.37 mg/L). Schaar (1985) reported similar findings, with Jersey cows producing milk with the lowest plasmin activity, while cross-bred Swedish Friesian-Swedish Red and White cows possessed the highest activities. However, when the effect of casein content was considered, it was found that the differences in plasmin between breeds was mainly due to differences in their relative amount of casein. Schaar (1985) reported a negative correlation between casein content and plasmin activity, previously observed by Richardson and Pearce (1981), and suggested that this finding was due to the competition between casein and the synthetic substrate used for the analysis of the enzyme. A study by Bastian et al. (1991b) reported that casein interferes with plasmin assays when synthetic substrates are used, acting as a competitive inhibitor. However, they found that, by reducing the casein to substrate ratio, such interference could be avoided. Bastian et al. (1991a) found no significant differences in the plasmin activity of milk collected from Jersey and Holstein-Friesian cows once the casein to substrate ratio was adjusted to prevent any interference by casein.

2.5.5 pH of Milk

The activity of plasmin strongly depends on the pH of its environment (i.e., milk, cheese). There have been several studies conducted to determine the optimum pH of plasmin over the years (Kiermeier and Semper 1960; Carini and Bozzolati 1970; Chen and Ledford 1971; Kaminogawa et al. 1972), with many reporting different optimum pH ranges (see Humbert and Alais 1979). It is now generally accepted that the optimum pH for plasmin is between 7.5–8 (Fox 1981), but it is recognised that plasmin remains active and stable over a broad pH range. Decreasing the pH causes a decrease in the activity of plasmin as it moves away from its optimum, as well as causing the dissociation of plasmin from the micelle (as discussed in Sect. 2.4.1).

2.5.6 Heat Treatment

Plasmin is a thermostable enzyme capable of surviving many of the heat treatments applied to dairy products (Kiermeier and Semper 1960; Dulley 1972; Prado et al. 2006). The severity of the heat treatment of milk depends on various factors including the desired shelf life of the product and the end use of the milk. The wide range of heat treatments that can be applied to milk, in addition to plasmin system components varying in their heat stabilities, further highlights the complexity of this system. Pasteurisation is a mild heat treatment (72 °C for 15–30 s) employed in the dairy industry to kill vegetative pathogenic microorganisms, thereby extending the shelf-life of the milk. Noomen (1975) reported an increase in the rate of proteolysis of 30–40% and 8–24% after milk had been heated to 75 °C for 15 s and 63 °C for 30 min, respectively. Richardson (1983a) and Korycha-Dahl et al. (1983) reported that pasteurisation of fresh milk at 72 °C for 15 s decreased plasmin content by 17 and 10%, respectively. Although there may be a reduction in the plasmin content on heating, plasmin activity after such a heat treatment increases (de Rham and Andrews 1982; Korycha-Dahl et al. 1983; Richardson 1983a). Richardson (1983a) suggested that the increase in plasmin activity was due to the inactivation of plasminogen activator inhibitors (Prado et al. 2006), resulting in increased plasminogen activation, as the decrease in the concentration of plasminogen was greater than the increase in plasmin activity.

Although the increase in plasmin activity on mild heat treatments has been attributed to the inactivation of plasminogen activator inhibitors and plasmin inhibitors, other factors such as improved efficiency of activation of denatured plasminogen or the thermal stabilities of the plasminogen activators (u-PA and t-PA), could contribute to the increased plasmin activity (Ismail and Nielsen 2010). The denaturation temperature of plasminogen has been reported to range from 50–62 °C (Castellino et al. 1981; Burbrink and Hayes 2006). In this temperature range, plasminogen becomes more accessible to the action of plasmin, due to conformational changes and the partial unfolding of the kringles, exposing the Lys_{77}-Arg_{78} bond (Burbrink and Hayes 2006). Prado et al. (2006) found almost complete inactivation of plasminogen activator inhibitors on heating milk to 75 °C for 15 s, with plasmin inhibitors being less affected by the heat treatment. A later study by Prado et al. (2007) explored the difference in the thermal stabilities of the plasminogen activators u-PA and t-PA. u-PA displayed greater heat stability than t-PA when milk was heated to 85 °C for 30 s, with t-PA losing half of its activity, while u-PA lost only 30%. The sensitivity of plasmin inhibitors to mild heat treatments, in addition to the higher heat stability of plasminogen activators, may explain the increased plasmin activities in milk subjected to pasteurisation-like conditions.

As discussed earlier (Sect. 2.4.1), plasmin activity in acid whey is higher than in sweet whey, with Crudden et al. (2005b) having assessed the thermal behaviour of plasmin in acid and sweet whey. Plasmin inactivation, in both acid and sweet whey, followed first-order inactivation kinetics, which was consistent with the findings of plasmin inactivation in buffers, model milk systems and milk (Alichanidis et al. 1986; Rollema and Poll 1986; Kennedy and Kelly 1997; Saint Denis et al. 2001a). Crudden et al. (2005b) reported D-values (the time required for a 90% reduction in activity) of 108 ± 29 min and 0.021 ± 0.006 min for the inactivation of plasmin in acid and sweet whey at 90 °C, respectively. In comparison, D-values for plasmin inactivation in milk of 59 s at 90 °C (Kennedy and Kelly 1997) or 71 s at 90 °C (Saint Denis et al. 2001a) had previously been reported. Crudden et al. (2005b) attributed the higher heat stability of plasmin in acid whey to reduced sulphydryl/disulphide interchange between plasmin and β-lactoglobulin, due to higher heat stability of β-lactoglobulin in acid whey, and to structural changes in plasmin.

More severe heat treatments applied to milk include UHT processing and sterilisation. In these processes, the heat treatment applied, involves time-temperature combinations of 135–150 °C for 1–5 s and 115–120 °C for 12–20 min for UHT and sterilisation, respectively (Spreer 1998). Driessen and van der Waals (1978) reported D-values of 7 s for plasmin at 142.5 °C, indicating that plasmin is capable of surviving most UHT treatments. They found that, in order to fully inactivate plasmin, temperatures as high as 142 °C for 18 s were needed. de Koning et al. (1985) found no proteolysis in UHT-sterilised, evaporated milk produced from milk that was preheated for 3 min at 120 °C. A later study by Saint Denis et al. (2001a) reported D-values of 13 and 11 s for indigenous plasmin and plasminogen in bovine milk, respectively, at 140 °C. Studies have also shown that plasminogen activators not only survive mild heat treatments but also survive UHT treatments (Kelly and Foley 1997); D-values of 16 s at 140 °C have been reported for plasminogen activators in bovine milk. Kelly and Foley (1997) concluded that the increase in plasminogen activation in UHT milk with a high SCC was most likely due to u-PA. Later work by Prado et al. (2007) supported these conclusions, reporting that u-PA is more heat stable than t-PA.

It can be concluded that the plasmin system is not completely inhibited by high temperature treatments; the system in which plasmin occurs (i.e., milk or acid whey) also has an influence on its thermal stability. The presence of remaining plasminogen and plasminogen activators, in addition to the destruction of plasmin inhibitors, indicates that some residual plasmin activity remains after heat treatment. The studies mentioned above give an indication of the complexity with which plasmin activity is influenced by heat treatments, and how the shelf-life of milk-based products, from an enzymatic perspective, can ultimately be influenced by the specific heat treatment applied.

2.5.7 Presence of β-Lactoglobulin

As discussed previously (Sect. 2.5.6), heat treatment has a significant impact on the plasmin system, with the components of the system varying in their thermal stabilities. However, heating also results in the interaction of plasmin system components with milk constituents. Evidence in the literature states that whey proteins have the ability to reduce plasmin activity (Rollema and Poll 1986; Bastian et al. 1993; Grufferty and Fox 1986; Kelly and Foley 1997; Enright and Kelly 1999; van Asselt et al. 2008). A study by Alichanidis et al. (1986) showed that, in the absence of β-lactoglobulin, both plasmin and plasminogen are heat-stable, indicating that β-lactoglobulin interacts with plasmin. β-Lactoglobulin, the predominant whey protein in bovine milk, contains a free sulphydryl (-SH) group buried within its globular structure. On heating to >55 °C, the 3-dimensional structure begins to unfold, exposing the sulphydryl group on Cys_{121} (Sawyer 2003). The subsequent decrease in plasmin activity is believed to be due to the free sulphydryl group of β-lactoglobulin interacting with the disulphide bonds which link subunits within the plasmin molecule *via* thiol disulphide interchange (Grufferty and Fox 1988a; Kelly and Foley 1997). Chemical modification of β-lactoglobulin, such that there are no free -SH groups, results in no inhibition of plasmin, verifying the role of -SH groups in plasmin inhibition. It has been shown that, at temperatures lower than 65 °C, plasmin is reversibly inactivated, only becoming irreversibly inactivated once temperatures surpass 65 °C (Metwalli et al. 1998).

Both native β-lactoglobulin (Kaminogawa et al. 1972; Rollema and Poll 1986; Bastian et al. 1993) and denatured β-lactoglobulin (Snoeren et al. 1980; Bastian et al. 1993) inhibit the action of plasmin, with some studies reporting native β-lactoglobulin to have a greater inhibitory effect (Bastian et al. 1993). Rollema and Poll (1986), on the other hand, showed that heat denatured β-lactoglobulin inhibited plasmin activity to a greater extent than native β-lactoglobulin, in a solution consisting of casein micelles suspended in simulated skim milk ultrafiltrate (SMUF). β-Lactoglobulin has also been shown to affect plasminogen. The conformational change that occurs in plasminogen on heating may promote interactions with denatured β-lactoglobulin (Burbrink and Hayes 2006). Burbrink and Hayes (2006) observed a decrease in plasminogen-derived plasmin at high temperatures (>62 °C) and concluded that the formation of plasminogen-β-lactoglobulin complexes was responsible; β-lactoglobulin, when bound to plasminogen, can interfere with the activation of plasminogen to plasmin, in addition to affecting the accessibility of plasminogen activators to plasminogen.

In a milk system comprising of proteins with free sulphydryl groups (i.e., β-lactoglobulin), plasminogen activators have been found to be as sensitive as plasmin and plasminogen on heating (Saint Denis et al. 2001a). Rippel et al. (2004) studied the effect of native and denatured proteins on plasminogen activators, concluding that both α-lactalbumin and β-lactoglobulin increased plasminogen activator activity. Kinetics experiments indicated that both native and denatured α-lactalbumin, and denatured β-lactoglobulin, increased the catalytic efficiency of u-PA, suggesting that the whey proteins have the ability to facilitate the release of u-PA from active plasmin - a phenomenon which occurs much more slowly in their absence.

The effect of β-lactoglobulin on the plasmin system depends on the concentration of native or denatured β-lactoglobulin present, and its effect on plasmin inactivation is reduced during long heat treatments, as the presence of -SH groups available for disulphide bonding decreases.

2.5.8 *Storage Temperature of Milk*

Many dairy-based products are refrigerated (≤5 °C) to reduce microbial growth and extend shelf life, while some products in which microbial growth is not of concern are stored at room temperature (20–25 °C). The temperature at which these dairy based products are stored can have a significant impact on plasmin activity and can influence the extent of proteolysis.

2.5.8.1 Cold Storage

The temperature at which milk is stored prior to processing can greatly affect plasmin activity and the quality of products made from the resulting milk. In industry, milk is primarily stored at low temperatures (2–5 °C) prior to processing. Milk is also rapidly cooled from 37 to <5 °C on the farm immediately after milking, and stored at this temperature until collection. Proteolysis of caseins by plasmin action before processing can impact products made from that milk (i.e., proteolysis prior to cheesemaking can reduce cheese yield). After pasteurisation, milk is also cooled and stored at refrigeration temperatures for extended periods of time. The extended storage of milk at low temperatures can therefore affect milk quality by affecting the plasmin system, and also due to endogenous proteinases produced as a result of the growth of psychrotrophic microorganisms.

The optimum temperature for plasmin activity is 37 °C; however, plasmin can remain active at cold temperatures. A study by Donnelly and Barry (1983) characterised the rate of conversion of β-casein to γ-casein at 4 and 37 °C, finding the activity in milk stored at 37 °C to be approximately 20 times higher than that stored at 4 °C. However, a previous study indicated that plasmin activity in milk stored at 4 °C is higher than that stored at 26 °C (Reimerdes and Herlitz 1979). They came to the conclusion that the higher activity at 4 °C was due to the dissociation of β-casein and plasmin from the micelle into the serum phase, thus allowing increased proteolysis due to closer proximity of plasmin and its substrate. Subsequent studies have confirmed that, during cold storage, dissociation of β-casein from the micelle does occur (Davies and Law 1983; Dalgleish and Law 1989). Crudden et al. (2005a) studied plasmin activity in raw milk at 37, 20 and 5 °C, reporting plasmin activity to be highest at 37 °C initially, but decreasing more rapidly during storage than at the other temperatures. This was in agreement with the findings of Huppertz et al. (2004a), who studied plasmin activity of high pressure-treated bovine milk stored at 37 °C, attributing the decrease to the autolysis or self-hydrolysis of plasmin. The greatest degree of dissociation of β-casein was found at 5 °C; however, hydrolysis of β-casein was highest at 37 °C, with higher amounts of intact β-casein at the lower temperatures. Like plasmin, plasminogen activators can be active at refrigeration temperatures. The initial high plasmin activity at 37 °C, corresponding to high initial proteolysis, was attributed to conversion of plasminogen to plasmin by plasminogen activators.

Schroeder et al. (2008) evaluated the effects of both cold storage temperature and calcium on the plasmin system. Milk was stored at 2.2, 3.3 and 4.4 °C, with the latter resulting in higher plasmin activity and plasminogen activation. Under cold storage conditions, the solubilisation of colloidal calcium phosphate within casein micelles occurs, increasing the concentration of calcium (Ca) in the serum phase (Fox et al. 2015). The higher serum Ca level found at the coldest temperature (2.2 °C) corresponded to decreased plasminogen activation (Schroeder et al. 2008). Subsequent reaction kinetics experiments showed that calcium had no impact on plasmin activity kinetics; however, Ca was found to lower the affinity of plasminogen activators for plasminogen, potentially through a competitive inhibition mechanism, ultimately reducing plasminogen activation and plasmin activity at the lowest temperature (2.2 °C).

2.5.8.2 Warm (Room Temperature) Storage

The storage of milk and milk-based products at room temperature is not commonly practised due to both microbiological concerns and plasmin activity. UHT milk and sterilised milk-based products are normally stored at room temperature. Plasmin activity in these products can be high, as plasminogen activator inhibitors are already inactivated due to the high heat treatment applied to such products; in addition, room temperature (20–25 °C) conditions are close to the optimum temperature (37 °C) for plasminogen activation and plasmin activity. This can ultimately lead to more extensive casein hydrolysis, resulting in impaired functional properties due to reduced intact protein and quality issues, e.g., bitterness and gelation of UHT milk (see Sect. 2.7.2).

2.5.9 Microbial Proteases

Microbial proteases (often metalloproteases) produced by psychrotrophic microorganisms can cause proteolysis during refrigerated storage (Nielsen 2002) (see Chap. 11). Over the years there have been several studies indicating that the presence of microbial proteases can affect the milk enzyme system. Some bacteria have the ability to enhance plasminogen activation or, in some instances, actually produce plasminogen activators during growth. *Streptococcus uberis* has been shown to secrete proteinases capable of activating bovine plasminogen; Leigh (1994) purified a 57,000 Da plasminogen activator from culture filtrates of *S. uberis* strain 0140J. Johnsen et al. (1999) identified a 45,000 Da bovine plasminogen activator secreted from *Streptococcus uberis* strain SK880, which was isolated from a case of mastitis. Plasminogen activators produced by *Streptococcus uberis* such as PauA and PauB are thought to enhance proteolysis in the mammary gland during mastitis, thereby providing the bacterium with essential nutrients through the hydrolysis of caseins (Rosey et al. 1999; Ward and Leigh 2002). Larson et al. (2006) showed that a protease produced by *Bacillus polymyxa* has the ability to act as a plasminogen activator by cleaving bovine plasminogen to produce a product with plasmin-like activity.

Pseudomonas strains produce heat-stable proteases of concern to the food industry, as they are capable of surviving pasteurisation. Pseudomonas strains can grow at refrigeration temperatures and have been shown to disrupt the plasmin system (Fajardo-Lira et al. 2000). The growth of pseudomonas resulted in an initial increase in plasmin in the whey fraction and a subsequent decrease in the casein fraction. The authors concluded that the extracellular proteases secreted from *Pseudomonas fluorescens* caused the release of plasmin from the casein micelles. Further studies into proteases produced by *Pseudomonas fluorescens* M3/6 have shown that it can enhance plasminogen activator activity, therefore increasing plasmin activity (Frohbieter et al. 2005). The presence of M3/6 proteases increased bovine plasminogen activation 2.5-fold and increased the catalytic efficiency of bovine u-PA fourfold. Some microbial proteases can have the opposite effect, acting as inhibitors of the plasmin system. *Escherichia coli* produces ecotin, a periplasmic protein, which has the ability to inhibit u-PA (Eggers et al. 2004).

These studies suggest that extracellular bacterial proteases have the ability to interact and interfere with the plasmin system, altering the rate of proteolysis and the location of plasmin, potentially affecting the quality of milk, cheese and other foods that utilize these ingredients in functional applications.

2.5.10 Membrane Filtration

The filtration of milk can greatly influence plasmin activity (Benfeldt 2006). Ultrafiltration (UF) of milk prior to cheese making is commonly practised in the dairy industry (Soodam and Guinee 2018). Studies have shown that cheese produced from UF milk exhibits slower proteolytic action during storage, with reduced hydrolysis of β- and α_s-caseins (Lelievre and Lawrence 1988; Benfeldt 2006). This has been attributed to increased levels of native whey protein in UF-cheeses compared to traditional cheese, which inhibits plasmin, thereby affecting cheese maturation. Benfeldt (2006) found lower plasmin activity in cheese produced from UF milk, and several reasons were given as to the lower plasmin activity, including the increased incorporation of plasmin inhibitors into the cheese matrix, and the inactivation of plasminogen activators during UF concentration.

Microfiltration (MF) can also affect the plasmin system due to the separation of certain components within the plasmin system. The MF of milk often involves the separation of casein from whey proteins. Whey contains many inhibitors of the plasmin system and therefore alters the plasmin system equilibrium (see Sects. 2.3.2 and 2.4). Aaltonen and Ollikainen (2011) studied the effect of MF on the activities of plasmin and plasminogen in the retentate of MF skim milk. As expected, plasmin and plasminogen were retained in the retentate, as they are both associated with casein micelles (see Sect. 2.4). Subsequent diafiltration resulted in additional removal of serum proteins from the retentate, resulting in increased activation of plasmin. Plasminogen levels decreased due to increased conversion of plasminogen to plasmin, as inhibitors of plasmin and plasminogen activators were removed. Consequently, proteolysis of casein was enhanced, with an accumulation of proteolysis products in the retentate. During diafiltration, the concentration of β-lactoglobulin in retentate decreased, resulting in increased plasmin activity. Aaltonen and Ollikainen (2011) found that plasmin activity in MF retentate was strongly correlated with the concentration of β-lactoglobulin, with plasmin activity increasing with decreasing levels of β-lactoglobulin.

2.5.11 High Pressure

The treatment of milk with new and emerging technologies is becoming more popular, with high pressure treatment gaining increased consumer acceptance as a food processing technique. High pressure treatment of milk is a non-thermal treatment which can extend shelf-life by inactivating microorganisms and denaturing enzymes (Patterson et al. 1995; García-Risco et al. 1998) without negatively affecting the sensory and nutritional qualities (Cheftel 1995). García-Risco et al. (1998) subjected milk to high pressure (400 MPa at 25 °C for 30 min) and monitored microbiological and biochemical changes during subsequent cold storage. Plasmin was not fully inactivated by the pressure treatment, as indicated by the formation of γ-caseins on storage; however, they estimated that the treatment caused a 25% reduction in total plasmin activity. Rademacher et al. (1998) found pressures as high as 800 MPa for 8 min were required to completely inactivate plasmin in milk, with a treatment at 500 MPa for 90 min and 600 MPa for 10 min only causing 50% inactivation of plasmin. A later study by García-Risco et al. (2003) gave similar findings, with the pressure treatments not causing the inactivation of plasmin and only decreasing total plasmin activity by 20–30% after plasminogen activation. Garía-Risco et al. (2000) showed that pressurisation at higher temperatures (400 MPa at 60 °C for 15 min) caused greater inactivation of plasmin (86.5% reduction), thereby reducing proteolysis and improving the overall organoleptic properties of the milk. The decreased plasmin activity cannot be attributed to one single phenomenon, but rather to a series of complex changes upon pressurisation, besides the effects on plasmin and plasminogen directly. Casein micelles, when subjected to very high pressure, can dissociate (López-Fandiño et al. 1998; Needs et al. 2000a, b), due to the disruption of hydrophobic and electrostatic interactions (Mozhaev et al. 1996), which may result in the dissociation of plasmin and plasminogen from the micelle into the serum phase. This would result in increased contact with inhibitors of the plasmin system present in the serum, resulting in enhanced inactivation and lower activity (Korycha-Dahl et al. 1983; Politis et al. 1992).

In addition to the aforementioned effects, high pressure treatment causes the denaturation of whey proteins (Garía-Risco et al. 2000; Scollard et al. 2000a, b; García-Risco et al. 2003; Huppertz et al. 2004a, b), which have been shown to act as proteinase inhibitors. The denaturation of β-lactoglobulin during high pressure treatment exposes a free -SH group that can interact with plasmin, causing irreversible inactivation of plasmin. Scollard et al. (2000a) showed that, in a buffer system, plasmin is capable of resisting pressures of 600 MPa at 20 °C for 20 min in the absence of β-lactoglobulin, but plasmin activity is reduced at 400 MPa in its presence. Scollard et al. (2000b) found that applied pressures of ≥400 MPa reduced plasmin activity in milk. However, they observed increased proteolysis in milk subjected to pressures of 300–500 MPa, possibly due to the changes in the structure of casein micelles and dissociation of some casein from the micelle, increasing the surface area of protein for proteolysis; Huppertz et al. (2004a) reported similar findings, with increased proteolysis in milk treated at 300–400 MPa also suggesting the more extensive proteolysis was due to high pressure-induced disruption of casein micelles.

Although high pressure treatment can inactivate microorganisms while maintaining the nutritive and sensory profiles of milk, the effect on plasmin activity, micelle integrity and subsequent proteolysis in milk could be significant, from a quality and sensory perspective, for the manufacture of products from high-pressure-treated milk.

2.6 Characterisation of Plasmin and Plasminogen

Plasminogen-derived activity is defined as the plasmin activity resulting from the addition of a defined amount of urokinase. There are various methods used to determine both plasminogen-derived and plasmin activity, using both indirect and direct methods. These methods vary in their sensitivity, sample preparation and accuracy. Multiple analytical techniques are also used to observe plasmin-mediated proteolysis, as well as identifying and quantifying proteolysis products.

2.6.1 Assay Methods for Plasmin and Plasminogen

There are several methods used for determining plasmin and plasminogen-derived activity in milk, varying from monitoring degradation of milk proteins, to fluorometric and chromogenic assays. Methods originally employed for determining plasmin activity involved monitoring degradation of milk proteins, such as quantifying the rate of formation of γ-caseins. Driessen and van der Waals (1978) determined plasmin activity in sterilised milk by monitoring γ-casein formation using polyacrylamide gel electrophoresis (PAGE) and densitometry. Andrews (1983a) employed similar techniques, using quantitative PAGE on gel rods and qualitative PAGE on gel slabs to monitor plasmin activity in mastitic milk. The increase in the number of free amino acids has also been used to determine plasmin activity (Chism et al. 1979; McKellar 1981); milk was treated with trichloroacetic acid (TCA) and filtered, and the filtrate was analysed for free amino acids by monitoring absorbance at 420 nm, with an increase in free amino acids (µmoles/ml milk) indicating increased proteolysis and therefore higher plasmin activity (McKellar 1981).

Dulley (1972) determined plasmin activity by incubating plasmin and casein (substrate) at 37 °C for 16 h before precipitating out the protein with 10% TCA. The supernatant obtained after centrifugation was assayed for the amino acid tyrosine and tyrosine-containing peptides by monitoring absorbance at 750 nm using Folin-Ciocalteu phenol reagent and alkaline copper, as described by Lowry et al. (1951). The increase in absorbance at 750 nm was used as a measure of proteolysis and therefore plasmin activity.

Other researchers have used labelled proteins such as ^{14}C-methylated β-casein to determine plasmin activity (Donnelly et al. 1980; Donnelly and Barry 1983). Reductive methylation of β-casein provides an ideal substrate for monitoring plasmin-like activities with the proteolysis product, ^{14}C γ-casein, easily detected by diethylaminoethyl (DEAE)-cellulose chromatography. Igarashi (1989) used DEAE-cellulose chromatography to separate γ-caseins from milk before measuring absorbance at 280 nm to determine the rate of formation of γ-caseins.

Andrews (1982) developed a novel method for detecting and measuring both proteinase and proteinase inhibitor activities. Substrate (casein) was immobilised to Sepharose 4B which was linked *via* glutaraldehyde or *m*-maleimidobenzoyl N-hydroxysuccinimide ester to β-glucosidase, a reporter enzyme resistant to the action of plasmin. The cleavage of the substrate by plasmin releases the reporter enzyme into the supernatant, where it acts on its substrate molecules. The action of the reporter enzyme results in a significant amplification of the initial hydrolysis caused by plasmin, ultimately increasing the sensitivity of the analysis. Many of these assays described require extensive sample preparation and are not specific for plasmin (e.g., many other proteinases can cause the formation of peptides and free amino acids).

Pierzchala et al. (1979) developed and described a new fluorogenic peptide, 7-(N-succinoylalanylphenylalanyllysylamido)-4-methylcoumarin trifluoroacetate salt, for the detection of plasmin in cells. Richardson and Pearce (1981) modified the method described by Pierzchala et al. (1979) to develop a rapid, fluorometric, amidolytic assay for the detection of plasmin in dairy products using N-succinyl-L-alanyl-L-phenylalanyl-L-lysyl-7-amido-4-methyl coumarin, a non-fluorescent substrate. Plasmin hydrolyses the lysine-coumarin bond releasing 7-amido-4-methyl coumarin (AMC), a fluorescent product. Fluorescence intensity, using excitation and emission wavelengths of 380 and 460 nm, respectively, is proportional to plasmin activity. Recordings of intensity over time are taken at regular intervals with one AMC unit of plasmin activity releasing one nanomole AMC per min. Using this method, they reported plasmin activities of 0.037–0.195 AMC units/ml, and plasmin and plasminogen concentrations of 0.14–0.73 μg/ml and 0.55–2.75 μg/ml, respectively, in pasteurised bovine milk.

Chromogenic assays are also used to characterise proteolytic activity in milk. Reimerdes et al. (1979) experienced low sensitivity when using the chromogenic substrate, amino acid-4-nitroanilides. The chromogenic substrate H-D-Valyl-L-Leucyl-L-Lysine-*p*-nitroanilide dihydrochloride (S-2251) has been used for the detection of plasmin associated with casein fractions (Snoeren and van Riel 1979; Rollema and Poll 1986). Plasmin hydrolyses the lysine-nitroanilide bond, releasing *p*-nitroanilide, a chromogenic product; absorbance at 405 nm is used to detect *p*-nitroanilide, with a change in absorbance per unit of time taken as a measure of plasmin activity (Politis et al. 1993). Plasminogen concentrations of between 0.7–2.4 mg/L were reported for skim milk, which agrees with those reported by Richardson and Pearce (1981), who used a fluorogenic substrate.

For accurate measurement of plasmin, the use of synthetic substrates is often advised due to their purity, defined structures and their ability to bind to fluorescent and chromogenic groups. In addition, the proteolytic products of these substrates are easy to detect and the specificity of enzymes for the substrate is often well-defined. The advantages of these fluorometric and chromogenic assays over the previously presented methods include reduced sample preparation and increased sensitivity and limits of detection. However, it has been reported that whey proteins and caseins present in milk can interfere with these methods (Schaar 1985; Bastian et al. 1991b), leading to the underestimation of true plasmin levels.

Plasminogen can also be detected by fluorometric and chromogenic assays. Plasminogen is activated by the addition of urokinase, a plasminogen activator, which converts all plasminogen present to plasmin (Politis et al. 1993). Sufficient time, along with the correct temperature, allows plasminogen present to be converted to plasmin, which can be then measured with the two methods described previously. The amount of plasminogen present can be calculated by subtracting plasmin activity with no added urokinase (i.e., only plasmin) from plasmin activity with added urokinase (i.e., plasmin + plasminogen).

$$[\text{Plasmin} + \text{plasminogen}\left(\text{urokinase added}\right)] - [\text{plasmin}\left(\text{no urokinase added}\right)] = \text{plasminogen}$$

Plasmin and plasminogen can also be determined in skim milk or for different fractions of milk (i.e., whey (serum) and casein micelles). Determination of plasmin and plasminogen in milk serum and casein fractions involves defatting milk, ultra-centrifuging (100,000 × g for 1 h at 4 °C) and obtaining the supernatant (milk serum) and pellet (casein micelles) (Korycha-Dahl et al. 1983). As previously discussed, casein micelles can interfere with some assays that use synthetic peptides. Therefore, plasmin and plasminogen can be dissociated from the casein micelles into the milk serum using sodium citrate (Politis et al. 1989a) or ε-aminocaproic acid (EACA) (Korycha-Dahl et al. 1983; Politis et al. 1989b). However, as many of the inhibitors of the plasmin system reside in the serum phase, measurement of plasmin activity in the milk serum could be questionable. To avoid this issue, Politis et al. (1993) ultra-centrifuged skim milk prior to EACA addition to obtain a casein pellet with plasmin and plasminogen still associated; the pellet was then reconstituted and EACA was added, resulting in the dissociation of plasmin and plasminogen into a whey protein-free serum.

Collin et al. (1988) determined plasmin using ELISA. They reported higher plasmin activities than those determined by Richardson and Pearce (1981), attributing these differences to the role of caseins and other inhibitors in fluorometric and chromogenic assays, which utilize synthetic substrates. Later studies by Politis et al. (1992) and Benfeldt et al. (1995) determined the concentration of plasminogen in bovine milk using ELSIA methodology, reporting results comparable to those of Richardson and Pearce (1981) and one another.

Many of these assays require the clarification of samples, due to interference of some milk components (i.e., casein and fat globules) with plasmin and plasminogen-derived activities. However, these clarification/purification steps can alter the true plasmin activity in the sample, leading to over or under-estimation of plasmin or plasminogen-derived activities. Therefore, assays such as the ELISA described above have been developed to try overcome the interference of some milk components. Ozen et al. (2003) successfully utilised Fourier-transform infrared spectroscopy (FTIR), combined with multivariate statistical analysis, to measure and differentiate between plasminogen and plasmin concentrations in the presence/absence of whey proteins and casein. These studies indicate that the plasmin enzyme system can be studied in the presence of casein and whey proteins.

In addition to the more "traditional" fluorometric and chromogenic assays, researchers have optimised methods for specific products. Rauh et al. (2014a) optimised a spectrophotometric method for the measurement of both plasmin and plasminogen-derived activity in dairy products, specifically in turbid, fat-containing products, combining spectrophotometric and sample preparation methods from Rollema and Poll (1986) and Saint Denis et al. (2001b), respectively. A reproducible assay was optimised, allowing for the measurement of cheese extracts, whole milk and micellar casein solutions, with improved repeatability and levels of detection, compared to previous spectrophotometric assays.

As the turbidity of dairy products can affect spectrophotometric measurements, and clarification steps (i.e., filtration and centrifugation) are time-consuming and may impact the results, the use of a 'clarifying reagent' may provide a solution. The clarifying reagent can be used at the end of enzymatic colorimetric assays, retarding the reaction and producing a transparent solution due to the disruption of fat globules and solubilisation of casein micelles (Saint Denis et al. 2001a, b). This allows for direct spectrophotometric measurements without the need for extensive sample preparation (Saint Denis et al. 2001a; Humbert et al. 2006).

2.6.2 *Assay Methods for Activators and Inhibitors of the Plasmin System*

Several different plasminogen activator measurements in blood and cells have been developed (Rånby et al. 1982; Verheijen et al. 1982). However, as described by Lu and Nielsen (1993b), these assays often need to be adapted to measure plasminogen activators in bovine milk, for various reasons. Plasmin, plasminogen and plasminogen activators are all associated with casein micelles, and therefore plasmin and plasminogen are difficult to remove in the purification of plasminogen activators. The concentration of plasminogen activators in bovine milk is also very low, and therefore highly sensitive assays are required for their detection.

Lu and Nielsen (1993b) developed a colorimetric assay for the measurement of plasminogen activators using the synthetic substrate Spectrozyme™, which formed a yellow chromophore *p*-nitroanilide on cleavage by plasmin, which could be quantified by measuring absorbance at 405 nm by ELISA (Baldi et al. 1993; Fajardo-Lira et al. 2000). The measurement of plasminogen activators is thus based around plasmin activity resulting from the activation of plasminogen by plasminogen activators. In order to eliminate the effect of turbidity from casein micelles, plasminogen activator activity was calculated from the difference in absorbance measured at 405 nm (from *p*-nitroanilide and turbidity) and 490 nm (turbidity only). They also reported the use of casein-plasminogen SDS-PAGE assay to visualise and determine the molecular weight of plasminogen activators. These assays were developed by modifying techniques used to quantify plasminogen activators in blood and other sources.

Saint Denis et al. (2001a, b) assessed plasminogen activator activity by measuring the conversion of exogenous inactive plasminogen into active plasmin, adapting methods previously described by Lu and Nielsen (1993a, b) and Baldi et al. (1993). Untreated milk sample containing plasminogen activators was added to a solution consisting of human plasminogen, starting the reaction. The increase in plasmin activity due to plasminogen activators was quantified using the synthetic substrate N-succinyl-L-alanyl-L-phenylalanyl-L-lysyl-7-amido-4-methyl coumarin; the increase in the AMC fluorescent product was measured using a spectrofluorometer with excitation and emission wavelengths set to 370 and 440 nm, respectively.

The measurement of plasmin and plasminogen activator inhibitors has involved adaptations from plasmin activator activity assays (Lu and Nielsen 1993b; Precetti et al. 1997). The first step in determining plasmin and plasminogen inhibitors involves their partial isolation from bovine milk (Precetti et al. 1997). Determining their activities involves indirect assays which assess the differences in activity in their absence and presence, therefore calculating activities by difference. Prado et al. (2006) measured plasminogen activator inhibitor activity using a direct assay for u-PA, with the substrate Spectrozyme™ UK. Plasmin inhibitor activity was measured using a similar method, with plasmin assays being carried out according to those described by Fajardo-Lira et al. (2000) using Spectrozyme™ PL as a substrate (as outlined above). The plasminogen activator inhibitory activity and plasmin inhibitor activity was then calculated as follows:

$$\text{Inhibitory Activity}_{\mu PA/PL} = \text{Activity}_{\mu PA/PL} - \text{Activity}_{(\mu PA+PAI)or(PL+PI)}$$

2.6.3 Monitoring of Plasmin-Mediated Proteolysis

There are several ways in which plasmin-mediated proteolysis can be monitored using numerous analytical techniques.

2.6.3.1 Soluble Protein in Supernatant

An easy, but indirect, method to determine the action of plasmin is to measure the protein content of the supernatant following the addition of 12% TCA to the milk sample (Kelly and Foley 1997). An increase in the level of TCA-soluble nitrogen over storage gives an indication of the extent of plasmin action, due to an increase in proteolysis products (i.e., free amino acids, small peptides, proteose peptones). The protein content of the supernatant after centrifugation can be measured using techniques such as the Kjeldahl or Lowry method (Lowry et al. 1951). However, this technique is relatively crude and does not take into consideration the action of other indigenous proteinases present which produce such products (e.g., PP).

2.6.3.2 Analysis of Free Amino Groups

Free amino groups can be determined by mixing TCA-soluble extracts (Sect. 2.6.3.1) with 1 M potassium borate buffer (pH 9.2) and 5 mM trinitrobenzenesulfonic acid (TNBS). After the mixture is incubated at 25 °C for 30 min, 2 M monobasic sodium phosphate (containing 18 mM sodium sulphite) is added (McKellar 1981; Chove et al. 2011). Absorbance is then read at 420 nm and converted to micromoles of free amino groups/ml milk using a standard curve. The increase in the concentration of TCA-soluble free amino groups per millilitre of milk can then be used as an indication of proteolysis.

2.6.3.3 Urea-Polyacrylamide Gel Electrophoresis

As previously mentioned, the principal substrate for plasmin in dairy products is β-casein. The hydrolysis of β-casein yields γ-caseins and proteose peptones. The formation of γ-caseins, in addition to the degradation of α_s-caseins, can be monitored using urea-Polyacrylamide Gel Electrophoresis (urea-PAGE) (Andrews 1983b; Kelly and Foley 1997; Enright et al. 1999). Urea-PAGE allows qualitative analysis of plasmin-mediated proteolysis over time, giving an indication of the plasmin activity in that sample. It therefore provides additional information regarding plasmin activity measurements, providing information on the rate of proteolysis of each specific substrate, allowing comparisons between the proteolytic activity and specificity of different samples.

2.6.3.4 Reversed Phase-High Performance Liquid Chromatography

Reversed phase-high performance liquid chromatography (RP-HPLC) is a highly sensitive technique that provides both qualitative and quantitative analysis of plasmin mediated-proteolysis. pH 4.6-soluble extracts of the samples are generally run on HPLC (Kelly and Foley 1997; Enright et al. 1999), with samples acidified to pH 4.6, generally with acetic acid, and centrifuged to obtain a supernatant containing proteolytic products such as proteose peptones, peptides (i.e., α_s-casein peptide fragments) and γ-caseins. Plasmin action can also be monitored by measuring the levels of intact casein in milk (Visser et al. 1991; Enright et al. 1999), with lower levels of intact casein (particularly β-casein) indicating higher proteolytic activities. The proportion of intact casein and proteolysis products can be quantified by determining peak area, allowing for improved comparisons between the proteolytic profile of samples/products.

2.6.3.5 Mass Spectrometry

The specific peptide fragments produced by plasmin-mediated hydrolysis can be identified and quantified using mass spectrometry (Rauh et al. 2014a). Although generally an expensive technique compared to other methods used to monitor proteolysis, mass spectrometry can provide very specific information, such as the rate of formation of specific peptides at different storage times; therefore, information on the bonds most susceptible to cleavage within the protein can be obtained.

2.7 Applications of Plasmin and Its Significance in Various Dairy Products

Plasmin can have both positive and negative impacts on a wide range of dairy products. Some of these effects of plasmin on the quality of dairy products are outlined here.

2.7.1 Cheese

Plasmin can have numerous effects on the cheesemaking process and on the sensory properties of the resulting cheese. There have been several studies on the effects of plasmin-mediated proteolysis on curd formation and ripening of cheese during maturation (Fox 1989; Farkye and Fox 1991, 1992; Bastian and Brown 1996).

2.7.1.1 Cheese-making Properties

It is well known that late-lactation milk results in poor clotting properties, with weaker gel formation, resulting in longer clotting times and reduced curd firmness (Grufferty and Fox 1988c; Mara et al. 1998). It has been reported that plasmin can be attributed to the poor clotting properties of cheese milk due to increased proteolysis of caseins (Okigbo et al. 1985). Farkye (1986) found that the addition of plasmin to milk did not alter rennet coagulation times, but resulted in poor curd firming rates and reduced curd firmness (Grufferty and Fox 1988c). Bastian et al. (1991a), however, found no relationship between milk clotting properties and plasmin in milk, although 90% of milk samples contained less than 300,000 somatic cells/ml. As high SCCs are often associated with increased plasmin levels (see Sect. 2.5.1), the effect of high SCCs and plasmin activity on milk coagulation parameters was not explored. Mara et al. (1998) reported similar findings to Farkye (1986) and Grufferty and Fox (1988c), observing that the addition of plasmin to milk prior to rennet coagulation increased curd firming time and reduced the final curd firmness. In general, these results suggest that, although plasmin may not affect rennet coagulation times, it has a significant impact on the gel strength and resulting curd firmness.

The moisture content and yield of cheese have also been reported to be affected by plasmin-mediated hydrolysis of caseins, but to a lesser extent than the properties discussed above, despite their great importance from both an economic and quality perspective. A number of studies have found that cheese made from late-lactation milk and milk with high SCCs have higher moisture contents; many have partially attributed this to plasmin-mediated hydrolysis of caseins, resulting in higher levels of serum casein, reduced syneresis and therefore higher moisture levels (O'Keeffe 1984; Barbano et al. 1991). These studies involved the analysis of late-lactation milk or milk with high SCCs with no plasmin addition; therefore, the role of plasmin in contributing to the high moisture observed in these studies was only hypothesised. A study conducted by Mara et al. (1998), in which different concentrations of plasmin were added to milk prior to rennet coagulation, found curd moisture to be unaffected by the addition of plasmin to milk; the addition of plasmin also increased protein losses, with increased protein levels in whey, thus reducing cheese yield, which was attributed to increased hydrolysis of caseins.

These studies indicate that plasmin can have a significant effect on milk coagulation and cheese yield; however, the need to further investigate the influence of high plasmin levels on such parameters remains.

2.7.1.2 Cheese Ripening

Unlike the role of plasmin in milk coagulation and cheese yield, there have been numerous studies which examine the importance of plasmin during cheese ripening. Plasmin is of great importance to the organoleptic properties of several cheeses (Ollikainen and Kivelä 1989; Farkye and Fox 1991, 1992; Farkye and Landkrammer 1992), producing peptides which are precursors for the production of free amino acids by starter bacteria and enzymes (i.e., peptidases) (Fox 1989). Farkye and Fox (1991) demonstrated the importance of plasmin in cheese ripening, as the addition of a plasmin inhibitor to curds at salting resulted in significantly lower proteolysis of β-casein compared to cheese with no added plasmin inhibitor.

The contribution of plasmin to cheese ripening fluctuates between different cheese varieties and depends on factors such as the pH of the cheese during maturation, cooking temperature of the coagulum and ripening temperature (Richardson and Pearce 1981; Lawrence et al. 1987; Farkye and Fox 1990; Bastian and Brown 1996). Richardson and Pearce (1981) found plasmin activity to be significantly lower in Cheddar cheese compared to Swiss cheese, a finding confirmed by Farkye and Fox (1990), who found Cheddar cheese to have much lower plasmin activity when compared to other cheese varieties. Lawrence et al. (1987) attributed differences in plasmin activities in different cheese varieties to the different pH at drainage of the cheeses. The later study by Farkye and Fox (1990) showed that plasmin activity in rennet curds drained at different pH values were identical, suggesting that pH at drainage does not affect plasmin activity. These findings were in agreement with the results of Grufferty and Fox (1988b), who found the association of plasmin with casein micelles to be independent of pH between pH 4.8 and 6.6. If the results of Grufferty and Fox (1988b) are correct, the pH range relevant to rennet coagulated cheeses (i.e., >5.0) at drainage should not affect plasmin retention within the curd. The final pH of many cheese varieties also does not favour plasmin activity, with pH values much lower than the pH optimum of plasmin (pH 7.5); however, plasmin action can become more significant in cheese varieties in which pH increases during ripening (e.g., blue cheeses) (Fox 1989).

The pH at drainage during cheese manufacture may not account for the different plasmin activities observed in different varieties of cheese; Fox (1989) suggested that some differences between cheese varieties may be due to different salting procedures, as salt causes the dissociation of plasmin from the micelle, resulting in the loss of some plasmin into the whey. Farkye and Fox (1990), however, disproved this theory, finding that plasmin activity in Cheddar, Emmental and Cheshire cheese was not influenced by the method of salting. They did, however, find that increasing cooking temperatures during the manufacture of the cheeses caused higher plasmin activities.

As higher cooking temperatures are used for the manufacture of Swiss cheese (~52 °C) compared to other cheeses such as Cheddar (~38 °C), cooking temperature is likely to contribute to the different plasmin activities observed between cheese varieties. The contribution of plasmin to primary proteolysis is most significant in high-cooked cheese varieties, as the coagulant is extensively denatured (Sheehan et al. 2007). Increasing cooking temperature increases plasmin activity in the resulting cheese; Choi et al. (2006) observed increased plasmin, plasminogen and urokinase-type PA activities extracted from curds cooked at higher temperatures, with Somers and Kelly (2002) concluding that the higher plasmin activity is due to increased activation of plasminogen. Heat treatment of milk prior to cheese manufacture can also impact proteolysis during cheese ripening. Benfeldt et al. (1997) subjected milk to different heat treatments prior to cheesemaking, and reported that plasmin activity in the resulting cheese decreased as the temperature and holding time increased. Reduced casein hydrolysis during ripening of cheese made from heated milks was attributed to reduced plasmin activity, due to thermal inactivation of plasminogen activators, and interactions between β-lactoglobulin and components of the plasmin system on heating (see Sect. 2.5.7).

Plasmin activity in cheese is also strongly influenced by the ripening temperature. As mentioned in earlier sections, the optimum temperature for plasmin activity is 37 °C. As some cheese varieties are ripened at low temperatures, the rate of plasmin activity, and therefore proteolysis, is reduced. Ripening temperatures for Cheddar (6–8 °C) and Dutch varieties (12–14 °C) are much lower than the optimum temperature for the action of plasmin, with ripening temperatures reaching as low as 4 °C (Fox et al. 1996). Increasing ripening temperatures increases the rate of proteolysis (Fox et al. 1996; Feeney et al. 2001). This has gained a lot of interest, especially in Cheddar cheese, as ripening is an expensive and time-consuming process. However, as not all biochemical reactions increase at the same rate with increasing temperature, high temperatures can cause limitations such as unbalanced flavour profiles (Fox et al. 1996).

Plasmin-mediated hydrolysis decreases during cheese ripening, often due to variations in plasmin activity (Bastian and Brown 1996; Benfeldt et al. 1997). Therefore, some researchers have added plasmin (Farkye and Fox 1992) and plasminogen activators (i.e., urokinase) (Bastian et al. 1991a, 1997) to cheese milk to maintain high plasmin activity throughout cheese ripening; these studies resulted in increased proteolysis of the cheese with improved flavour and quality of some cheese. Studies have also shown that the addition of exogenous plasmin to accelerate cheese ripening is possible while avoiding defects such as bitterness and off-flavours (Farkye and Fox 1992; Scherze et al. 1994; Kelly 1995). Upadhyay et al.

(2004) added a preparation of streptokinase (a plasminogen activator produced by *Streptococcus uberis*) to cheese milk, reporting increased plasmin activity and proteolysis in the resulting cheese compared to cheese with no added streptokinase. A later study by Upadhyay et al. (2006) accelerated the ripening of Cheddar-type cheese using a streptokinase-producing strain of *Lactococcus lactis.* Using this strain, they found an approximately three-fold increase in plasmin activity compared to cheese manufactured with a control strain; this resulted in increased hydrolysis of β-casein, with higher levels of γ_1-, γ_2- and γ_3-caseins. The addition of exogenous plasminogen activators to cheese milk results in higher plasmin activity in the resulting cheese due to an increase in the rate of conversion of plasminogen to plasmin. Many of these results suggest that the addition of exogenous plasmin to cheese milk could be used to accelerate cheese ripening. Although the addition of exogenous plasmin is effective, from an economic perspective it is not practical, due to its high cost.

2.7.2 *Milk-Based UHT Products*

UHT treatment of milk involves heating milk at very high temperatures (~140 °C) for short periods of time (4–6 s) (Rauh et al. 2014b). Plasmin is a heat-stable enzyme capable of surviving many UHT treatments and therefore plasmin-mediated proteolysis can occur in many UHT-treated dairy beverages. UHT-treated milk provides an ideal environment for the action of remaining plasmin, as they are often stored at room temperature (close to the optimum temperature for plasmin activity). Storage temperature, processing conditions (i.e., time-temperature combinations), other ingredients used in the product and plasmin levels in the milk prior to processing, have all been shown to be major factors influencing age gelation (Manji et al. 1986; Ismail and Nielsen 2010; Rauh et al. 2014b, c; Anema 2017). Kohlmann et al. (1988) found that the addition of low levels of plasmin and plasminogen to UHT-treated milk resulted in greater proteolytic activity and faster milk gelation, respectively, compared to milk with no added enzyme or zymogen. Kelly and Foley (1997) found the addition of potassium iodate to milk prior to UHT treatment resulted in considerably higher plasmin activity during storage, compared to UHT milk with no added potassium iodate. As discussed in earlier sections, denatured β-lactoglobulin can form complexes with plasmin via disulphide bonds, inactivating plasmin. Therefore, these results suggest that denaturation of β-lactoglobulin has a significant role in reducing plasmin activity in UHT-treated milk. Studies by Rauh et al. (2014b, c) confirmed that plasmin action contributes to the age gelation of UHT milk. An increase in plasmin activity, with a simultaneous decrease in plasminogen-derived activity was observed during storage, with a corresponding increase in casein hydrolysis. Rauh et al. (2014c) suggested that the hydrolysis of casein-calcium phosphate and casein-casein interaction sites destabilises the casein micelles, causing age gelation in UHT milk. In both studies, age gelation occurred with extensive hydrolysis of β- and α_s-caseins (plasmin's principal substrates), indicating that plasmin contributes to the age gelation of UHT milk.

Plasmin-mediated proteolysis of caseins has been shown to produce a soft and fragile gel in UHT-treated milk on storage (Kelly and Foley 1997; Zhang et al. 2018) (see Chap. 13). Zhang et al. (2018) added plasmin to skim UHT milk-based model systems, finding that plasmin induced the formation of soft gels when approximately 60% of both β- and α_{s1}-casein were hydrolysed and the degree of hydrolysis reached 2.1%.

Milk processed by direct and indirect UHT heating systems differs in terms of their susceptibility to plasmin activity and age gelation. Studies have shown that plasmin activity is lower in indirect-treated UHT milk, compared to direct-treated milk (Corradini and Pecchini 1981; Manji et al. 1986). Kohlmann et al. (1988) observed UHT milk processed by direct heating, with added plasminogen (0.3 mg protein/L), to gel after 155 days, while milk processed by indirect heating took longer to gel (195 days). Manji et al. (1986) reported directly processed milk to contain 37 and 19% of the original plasminogen and plasmin activities, respectively, while indirect heat-treated milk contained 19% of the original plasminogen activity and no plasmin activity. The longer heating-up and cooling-down times in indirectly heat-treated milk is believed to contribute to the higher degree of plasmin loss (Stoeckel et al. 2016). Therefore, in many direct-heating treatments, a preheating step is implemented to adequately reduce plasmin activity in the final UHT-treated product and ensure a suitable shelf life is achieved (Newstead et al. 2006; van Asselt et al. 2008), in addition to stabilising β-lactoglobulin before high temperature heating to 135–150 °C (Datta and Deeth 2001). Anema (2017) found that proteolysis could be prevented, and gelation slowed, by increasing pre-heating severity (90 °C for 60 or 120 s), increasing UHT holding times (143 °C for 15 or 30 s), and combinations of the two. However, these conditions do not reflect those currently used, as such extensive heat treatment would cause quality issues such as 'cooked' flavours. These results suggest that mild heat treatments before UHT treatments may induce plasmin-β-lactoglobulin interactions, reducing plasmin's ability to hydrolyse caseins, and therefore reducing plasmin-mediated gelation.

The high heat treatment applied to UHT milk and dairy-based beverages can result in changes in protein conformation, protein aggregation and Maillard reactions (van Boekel 1998). These physicochemical changes can ultimately affect the action of plasmin and alter the susceptibility of caseins to plasmin-mediated hydrolysis. The susceptibility of caseins to plasmin action can be affected by micellar structure. Enright and Kelly (1999) found that increasing the severity of heat treatments applied to milk resulted in lower levels of plasmin-mediated proteolysis, and suggested that the binding of denatured proteins to the surface of casein micelles results in steric blockage of plasmin to susceptible peptide bonds (i.e., Lys-X bonds), thus reducing hydrolysis. In addition to the modification of micellar structure through the adsorption of whey proteins, the susceptibility of caseins to plasmin can also be affected by the modification of lysine residues on caseins (i.e., lactosylation) (Bhatt et al. 2014, 2017). Lactosylation is the first step in the Maillard reaction in which lactose and the ε-amino groups of lysine residues react to form lactulosyllysine (the Amadori rearrangement product) (van Boekel 1998). Dalsgaard et al. (2007) examined the effect of lactosylation on five different milk proteins (α-casein, κ-casein, β-casein, β-lactoglobulin and lactoferrin), finding that lactosylation reduced hydrolysis of these proteins by plasmin. A later study by Bhatt et al. (2014) studied the effect of lactosylation on plasmin-induced hydrolysis of β-casein, finding that lactosylation resulted in the modification of the ε-amino groups of lysine residues, making casein unrecognisable to the substrate-binding pocket of plasmin, thereby reducing hydrolysis of β-casein by plasmin.

The role of plasmin in the destabilization of UHT milk-based products is likely due to the loss of colloidal stability of the casein micelles due to extensive plasmin-mediated hydrolysis of the caseins. However, the effect of plasmin on UHT products is complicated and depends on various factors, with processing conditions, storage temperature and initial plasmin levels (depending on stage of lactation, mastitis, age of the cow, etc.) being some of the most important.

2.7.3 High Protein Milk Products

The use of milk protein ingredients in the formulation of products and beverages is widely practised. However, depending on the composition and production method of the protein ingredient, the presence of plasmin in these ingredients must also be considered, as it can potentially affect the functional and stability properties of such final products. These ingredients are often added to products to provide functionality such as foaming, gelation and emulsification and for nutritional purposes (e.g., to support high protein label claims). However, these properties are largely influenced by the structure and physicochemical properties of the proteins. Plasmin-mediated hydrolysis can subsequently affect the functional properties of these protein ingredients, in addition to potentially hydrolysing other proteins present in the formulated product.

2.7.3.1 Whey Protein-Containing Ingredients

The presence of plasmin in whey is generally lower than that of milk, as plasmin and plasminogen are associated with the casein micelles. However, the dissociation of plasmin into the serum phase (whey) could be of significance, as it could affect the quality of food products such as high protein beverages, which incorporate whey products such as whey protein concentrate. As previously discussed in Sect. 2.4.1, plasmin activity in acid whey is higher than that of sweet whey; therefore, food products that incorporate whey protein manufactured from acid whey may experience more issues than if whey proteins manufactured from sweet whey were used. Hayes and Nielsen (2000) reported lower plasmin activities in whey protein isolates (WPIs) than in sweet- and acid whey-derived protein concentrates, attributing the lower levels observed to the ion-exchange process used in the manufacture of WPIs. They also observed variability in plasmin activity associated with the commercial whey protein products, with the highest plasmin activity being from an acid whey derived protein concentrate ingredient (~80% protein). Therefore, the source of whey used to manufacture protein ingredients, in addition to processing conditions, are likely to cause variations in plasmin activity in these ingredients.

The presence of plasmin in whey protein ingredients and whey-derived products may have negative effects on products that utilise these ingredients for nutritional or functional purposes.

2.7.3.2 Casein-Containing Ingredients

The use of casein-containing milk protein ingredients (e.g., sodium caseinate, milk protein concentrate/isolate and micellar casein concentrate/isolate), like whey protein ingredients, are widely applied in formulation of food systems (e.g., beverages, coffee creamers, yoghurt, etc.). The presence of residual plasmin activity in these ingredients can lead to flavour and stability issues. However, there is limited knowledge on plasmin activity in the different milk protein ingredients mentioned above and how their manufacturing process affects plasmin activity. Gazi et al. (2014) measured plasmin and plasminogen-derived activity in a range of milk protein ingredients following reconstitution; the study included 14 milk protein concentrate, 2 micellar casein isolate, 19 sodium caseinate and 13 calcium caseinate samples. They found higher levels of plasmin and a higher occurrence of plasmin in micellar casein isolate and milk protein concentrate, compared to sodium and calcium caseinates. They observed the activation of plasminogen to plasmin during storage, which resulted in extensive plasmin-mediated casein hydrolysis. Micellar casein ingredients are often produced by the microfiltration of skim milk using a 0.1-μm pore size membrane, resulting in the permeation of whey proteins, soluble constituents and low molecular weight nitrogenous compounds (i.e., the serum phase) through the membrane (into the permeate). Intact casein micelles are retained by the membrane (retentate) where they are further concentrated by diafiltration, removing additional serum phase components. Therefore, the process of manufacturing micellar casein ingredients essentially removes the inhibitors of the plasmin system, present in the serum phase, while plasmin, plasminogen and plasminogen activators are concentrated with the casein micelles. Such ingredients can therefore possess high plasmin activities which can impact their functionality when added as an ingredient to other products.

The results from this study suggest that residual plasmin activity should be an important factor when selecting milk protein ingredients for food formulation.

2.8 Conclusions and Areas for Additional Research

Knowledge of the plasmin system and factors affecting the different plasmin system components continues to grow. However, there remain gaps in our understanding of the plasmin system in various dairy products. In addition to this, little research has been conducted on residual plasmin activity in milk protein ingredients, although these ingredients are widely used in dairy and non-dairy products for their functional properties, where residual plasmin activity could have negative consequences for the resulting quality and functional properties of that product.

Acknowledgements The authors would like to acknowledge the Dairy Processing Technology Centre (DPTC), an Enterprise Ireland initiative, for financial support and permission to publish this work.

References

Aaltonen T, Ollikainen P (2011) Effect of microfiltration of milk on plasmin activity. Int Dairy J 21:193–197

Aimutis WR, Eigel WN (1982) Identification of λ-casein as plasmin-derived fragments of bovine α_{s1}-casein. J Dairy Sci 65:175–181

Alichanidis E, Wrathall JHM, Andrews AT (1986) Heat stability of plasmin (milk proteinase) and plasminogen. J Dairy Res 53:259–269

Andrews AT (1982) A new approach to the general detection and measurement of proteinase and proteinase inhibitor activities. Biochim Biophys Acta 708:194–202

Andrews AT (1983a) Breakdown of caseins by proteinases in bovine milks with high somatic cell counts arising from mastitis or infusion with bacterial endotoxin. J Dairy Res 50:57–66

Andrews AT (1983b) Proteinases in normal bovine milk and their action on caseins. J Dairy Res 50:45–55

Andrews AT, Alichanidis E (1983) Proteolysis of caseins and the proteose-peptone fraction of bovine milk. J Dairy Res 50:275–290

Anema SG (2017) Storage stability and age gelation of reconstituted ultra-high temperature skim milk. Int Dairy J 75:56–67

Aslam M, Hurley WL (1997) Proteolysis of milk proteins during involution of the bovine mammary gland. J Dairy Sci 80:2004–2010

Babcock SM, Russell HL (1897) Unorganized ferments of milk: A new factor in the ripening of cheese. In: Fourteenth annual report of the Wisconsin agricultural experiment station, pp 161–193

Baer A, Ryba I, Collin JC (1994) Binding of bovine plasminogen to immobilized casein and its activation thereon. Int Dairy J 4:597–616

Baldi A, Politis I, Barbano DM, Gorewit RC (1993) Primary characterization and determination of plasminogen activator in bovine milk. J Dairy Sci 76(Supp 1):92

Barbano DM, Rasmussen RR, Lynch JM (1991) Influence of milk somatic cell count and milk age on cheese yield. J Dairy Sci 74:369–388

Barry JG, Donnelly WJ (1980) Casein compositional studies: I. The composition of casein from Friesian herd milks. J Dairy Res 47:71–81

Barry JG, Donnelly WJ (1981) Casein compositional studies: II. The effect of secretory disturbance on casein composition in freshly drawn and aged bovine milks. J Dairy Res 48:437–446

Bastian ED, Brown RJ (1996) Plasmin in milk and dairy products: an update. Int Dairy J 6:435–457

Bastian ED, Brown RJ, Ernstrom CA (1991a) Plasmin activity and milk coagulation. J Dairy Sci 74:3677–3685
Bastian ED, Brown RJ, Ernstrom CA (1991b) Casein interference in bovine plasmin assays using a synthetic substrate. J Dairy Sci 74:4119–4124
Bastian ED, Hansen KG, Brown RJ (1993) Inhibition of plasmin by β-lactoglobulin using casein and a synthetic substrate. J Dairy Sci 76:3354–3361
Bastian ED, Lo CG, David KMM (1997) Plasminogen activation in cheese milk: influence on Swiss cheese ripening. J Dairy Sci 80:245–251
Benfeldt C (1992) Isolation and characterization of plasminogen and fragments of plasminogen from bovine milk. Thesis, Department of Molecular Biology, University of Aarhus, Denmark
Benfeldt C (2006) Ultrafiltration of cheese milk: effect on plasmin activity and proteolysis during cheese ripening. Int Dairy J 16:600–608
Benfeldt C, Larsen LB, Rasmussen JT, Andreasen PA, Petersen TE (1995) Isolation and characterization of plasminogen and plasmin from bovine milk. Int Dairy J 5:577–592
Benfeldt C, Sørensen J, Ellegård KH, Petersen TE (1997) Heat treatment of cheese milk: effect on plasmin activity and proteolysis during cheese ripening. Int Dairy J 7:723–731
Berglund L, Andersen MD, Petersen TE (1995) Cloning and characterization of the bovine plasminogen cDNA. Int Dairy J 5:593–603
Bhatt H, Cucheval A, Coker C, Patel H, Carr A, Bennett R (2014) Effect of lactosylation on plasmin-induced hydrolysis of β-casein. Int Dairy J 38:213–218
Bhatt H, Cucheval A, Coker C, Patel H, Carr A, Bennett R (2017) Effect of micellar structure of casein and its modification on plasmin-induced hydrolysis. Int Dairy J 75:75–82
Booth NA, Backmann F (2006). Plasminogen-plasmin system. In: Colman RW, Marder VJ, Clowes AW, George, JN, Goldhaber SZ (eds) Hemostasis and thrombosis: basic principles and clinical practice, 5th edn. Lippincott Williams & Wilkins, Philadelphia, pp 335–364
Burbrink CN, Hayes KD (2006) Effect of thermal treatment on the activation of bovine plasminogen. Int Dairy J 16:580–585
Caessens PWJR, Visser S, Gruppen H, Voragen AGJ (1999a) β-Lactoglobulin hydrolysis. 1. Peptide composition and functional properties of hydrolysates obtained by the action of plasmin, trypsin, and *Staphylococcus aureus* V8 protease. J Agric Food Chem 47:2973–2979
Caessens PWJR, Daamen WF, Gruppen H, Visser S, Voragen AGJ (1999b) β-lactoglobulin hydrolysis. 2. Peptide identification, SH/SS exchange, and functional properties of hydrolysate fractions formed by the action of plasmin. J Agric Food Chem 47:2980–2990
Carini S, Bozzolati M (1970) Milk proteases. I. Proteolysis of casein. Sci Tecn Lattiero Casearia 21:277–290
Castellino FJ, Ploplis VA, Powell JR, Strickland DK (1981) The existence of independent domain structures in human Lys77-plasminogen. J Biol Chem 256:4778–4782
Castellino FJ, Ploplis VA (2005) Structure and function of the plasminogen/plasmin system. Thromb Haemost 93:647–654
Cheftel JC (1995) High-pressure, microbial inactivation and food preservation. Food Sci Technol Int 1:75–90
Chen JH, Ledford RA (1971) Purification and characterization of milk protease. J Dairy Sci 54:763
Chism GW, Huang AE, Marshall JA (1979) Sensitive assay for proteases in sterile milk. J Dairy Sci 62:1798–1800
Choi LH, Were LM, Nielsen SS (2006) Effects of incubation temperature and salt concentration on plasminogen activators in cheese curd. Int Dairy J 16:609–618
Chove LM, Grandison AS, Lewis MJ (2011) Comparison of methods for analysis of proteolysis by plasmin in milk. J Dairy Res 78:184–190
Christensen S, Sottrup-Jensen L (1992) Bovine α_2-antiplasmin N-terminal and reactive site sequence. FEBS Lett 312:100–104
Christensen S, Sottrup-Jensen L (1994) Characterization of two serpins from bovine plasma and milk. Biochem J 303:383–390

Christensen S, Wiegers T, Hermansen J, Sottrup-Jensen L (1995a) Plasma-derived protease inhibitors in bovine milk. Int Dairy J 5:439–449

Christensen S, Sottrup-Jensen L, Christensen U (1995b) Stopped-flow fluorescence kinetics of bovine α_2-antiplasmin inhibition of bovine midiplasmin. Biochem J 305:97–102

Christman JK, Silverstein SC, Acs G (1977) Proteinases in mammalian cells and tissues. North Holland Publishing Company, New York, pp 91–149

Collen D (1980) On the regulation and control of fibrinolysis. Thromb Haemost 43:77–89

Collen D, Lijnen HR, Verstraete M (1985) Thrombolysis. Biological and therapeutic properties of new thrombolytic agents. Churchill Livingstone, New York

Collin JC, Compagnone P, Ryba I, Baer A (1988) Determination of plasmin (alkaline milk proteinase) and chymosin in milk products by the ELISA assay. Lait 68:235–239

Corradini C, Pecchini G (1981) Effect on proteinases of different UHT treatments. Neth Milk Dairy J 35:393–395

Crudden A, Kelly AL (2003) Studies of plasmin activity in whey. Int Dairy J 13:987–993

Crudden A, Fox PF, Kelly AL (2005a) Factors affecting the hydrolytic action of plasmin in milk. Int Dairy J 15:305–313

Crudden A, Oliveira JC, Kelly AL (2005b) Kinetic studies of the thermal inactivation of plasmin in acid or sweet whey. Int Dairy J 15:1245–1253

Dalgleish DG, Law AJR (1989) pH-induced dissociation of bovine casein micelles. II. Mineral solubilization and its relation to casein release. J Dairy Res 56:727–735

Dalsgaard TK, Nielsen JH, Larsen LB (2007) Proteolysis of milk proteins lactosylated in model systems. Mol Nutr Food Res 51:404–414

Dalsgaard TK, Heegaard CW, Larsen LB (2008) Plasmin digestion of photooxidized milk proteins. J Dairy Sci 91:2175–2183

Datta N, Deeth HC (2001) Age gelation of UHT milk—a review. Food Bioprod Process 79:197–210

Davies DT, Law AJR (1983) Variation in the protein composition of bovine casein micelles and serum casein in relation to micellar size and milk temperature. J Dairy Res 50:67–75

de Koning PJ, Kaper J, Rollema HS, Driessen FM (1985) Age-thinning and gelation in unconcentrated and concentrated UHT-sterilized skim milk; effect of native milk proteinase. Neth Milk Dairy J 39:71–87

de Rham O, Andrews AT (1982) Qualitative and quantitative determination of proteolysis in mastitic milks. J Dairy Res 49:587–596

Donnelly WJ, Barry JG (1983) Casein compositional studies. III. Changes in Irish milk for manufacturing and role of milk proteinase. J Dairy Res 50:433–441

Donnelly WJ, Barry JG, Richardson T (1980) ^{14}C-methylated β-casein as a substrate for plasmin, and its application to the study of milk protein transformations. Biochim Biophys Acta 626:117–126

Driessen FM, van der Waals CB (1978) Inactivation of native milk proteinase by heat treatment. Neth Milk Dairy J 32:245–254

Dulley JR (1972) Bovine milk protease. J Dairy Res 39:1–9

Eggers CT, Murray IA, Valerie A, Craik CS (2004) The periplasmic serine protease inhibitor ecotin protects bacteria against neutrophil elastase. Biochem J 379:107–118

Eigel WN (1977) Effect of bovine plasmin on α_{s1}-B and κ-A caseins. J Dairy Sci 60:1399–1403

Enright E, Kelly AL (1999) The influence of heat treatment of milk on susceptibility of casein to proteolytic attack by plasmin. Milchwissenschaft 54:491–493

Enright E, Bland AP, Needs EC, Kelly AL (1999) Proteolysis and physicochemical changes in milk on storage as affected by UHT treatment, plasmin activity and KIO3 addition. Int Dairy J 9:581–591

Fajardo-Lira C, Oria M, Hayes KD, Nielsen SS (2000) Effect of psychrotrophic bacteria and of an isolated protease from *Pseudomonas fluorescens* M3/6 on the plasmin system of fresh milk. J Dairy Sci 83:2190–2199

Farkye NY (1986) Role of chymosin and porcine pepsin in Cheddar cheese ripening. PhD Dissertation, Utah State University, Logan, UT

Farkye NY, Fox PF (1990) Observations on plasmin activity in cheese. J Dairy Res 57:413–418
Farkye NY, Fox PF (1991) Preliminary study on the contribution of plasmin to proteolysis in Cheddar cheese: cheese containing plasmin inhibitor, 6-aminohexanoic acid. J Agric Food Chem 39:786–788
Farkye N, Fox PF (1992) Contribution of plasmin to Cheddar cheese ripening: effect of added plasmin. J Dairy Res 59:209–216
Farkye NY, Landkrammer CF (1992) Milk plasmin activity influence of cheddar cheese quality during ripening. J Food Sci 57:622–628
Feeney EP, Fox PF, Guinee TP (2001) Effect of ripening temperature on the quality of low moisture Mozzarella cheese: 1. Composition and proteolysis. Lait 81:463–474
Fox PF (1981) Proteinases in dairy technology. Neth Milk Dairy J 35:233–253
Fox PF (1982) Proteolysis in milk and dairy products. Biochem Soc Trans 10:282–284
Fox PF (1989) Proteolysis during cheese manufacture and ripening. J Dairy Sci 72:1379–1400
Fox PF, Singh TK, McSweeney PLH (1994) Proteolysis in cheese during ripening. In: Andrews AT, Varley J (eds) Biochemistry of milk products. Royal Society of Chemistry, Cambridge, pp 1–31
Fox PF, Wallace JM, Morgan S, Lynch CM, Niland EJ, Tobin J (1996) Acceleration of cheese ripening. Ant Leeuwenhoek 70:271–297
Fox PF, Uniacke-Lowe T, McSweeney PLH, O'Mahony JA (2015) Salts of milk. In: Dairy chemistry and biochemistry, 2nd edn. Springer, Cham, pp 241–270
Frohbieter KA, Ismail B, Nielsen SS, Hayes KD (2005) Effects of *Pseudomonas fluorescens* M3/6 bacterial protease on plasmin system and plasminogen activation. J Dairy Sci 88:3392–3401
García-Risco MR, Cortés E, Carrascosa AV, López-Fandiño R (1998) Microbiological and chemical changes in high-pressure-treated milk during refrigerated storage. J Food Prot 61:735–737
García-Risco MR, Recio I, Molina E, López-Fandiño R (2003) Plasmin activity in pressurized milk. J Dairy Sci 3:728–734
Garía-Risco MR, Olano A, Ramos M, López-Fandiño R (2000) Micellar changes induced by high pressure. Influence in the proteolytic activity and organoleptic properties of milk. J Dairy Sci 83:2184–2189
Gazi I, Vilalva IC, Huppertz T (2014) Plasmin activity and proteolysis in milk protein ingredients. Int Dairy J 38:208–212
Grufferty MB (1986) Alkaline milk proteinase: some of its characteristics and its influence, and that of milk salts, on some processing properties of milk. Doctoral dissertation, PhD thesis, National University of Ireland
Grufferty MB, Fox PF (1986) Potassium iodate-induced proteolysis in ultra heat treated milk during storage: the role of β-lactoglobulin and plasmin. J Dairy Res 53:601–613
Grufferty MB, Fox PF (1988a) Milk alkaline proteinase. J Dairy Res 55:609–630
Grufferty MB, Fox PF (1988b) Factors affecting the release of plasmin activity from casein micelles. N Z J Dairy Sci Technol 23:153–163
Grufferty MB, Fox PF (1988c) Functional properties of casein hydrolyzed by alkaline milk proteinase. N Z J Dairy Sci Technol 23:95–108
Halpaap I, Reimerdes EH, Klostermeyer H (1977) Milk proteinases. 6: comparative isolation of plasminogen from bovine blood and a proteinase from cow's milk. Milchwissenschaft 32:341–346
Harper WJ, Robertson JA, Gould IA (1960) Observations on milk protease. J Dairy Sci 43:1850–1851
Hayes KD, Nielsen SS (2000) Plasmin levels in fresh milk whey and commercial whey protein products. J Dairy Sci 83:387–394
Heegaard CW, Christensen T, Andreasen PA (1993) t-PA binds to casein micelles in bovine milk. In: Proc mol cell biol plasminogen activation, Cold Spring Harbor, New York, p 64
Heegaard CW, Christensen T, Rasmussen MD, Benfeldt C, Jensen NE, Sejrsen K, Petersen TE, Andreasen PA (1994a) Plasminogen activators in bovine milk during mastitis, an inflammatory disease. Fibrinolysis 8:22–30

Heegaard CW, Rasmussen LK, Andreasen PA (1994b) The plasminogen activation system in bovine milk: differential localization of tissue-type plasminogen activator and urokinase in milk fractions is caused by binding to casein and urokinase receptor. Biochim Biophys Acta 1222:45–55

Heegaard CW, Andreasen PA, Petersen TE, Rasmussen LK (1997) Binding of plasminogen and tissue-type plasminogen activator to dimeric α_{s2}-casein accelerates plasmin generation. Fibrinol Proteol 11:29–36

Hofmann CJ, Keenan TW, Eigel WN (1979) Association of plasminogen with bovine milk fat globule membrane. Int J Biochem 10:909–917

Humbert G, Alais C (1979) Review of the progress of dairy science: the milk proteinase system. J Dairy Res 46:559–571

Humbert G, Guingamp MF, Linden G, Gaillard JL (2006) The clarifying reagent, or how to make the analysis of milk and dairy products easier. J Dairy Res 73:464–471

Huppertz T (2013) Chemistry of the caseins. In: McSweeney PLH, Fox PF (eds) Advanced dairy chemistry: vol 1A: proteins: basic aspects, 4th edn. Springer, Boston, pp 135–160

Huppertz T, Fox PF, Kelly AL (2004a) Plasmin activity and proteolysis in high pressure-treated bovine milk. Lait 84:297–304

Huppertz T, Fox PF, Kelly AL (2004b) High pressure treatment of bovine milk: effects on casein micelles and whey proteins. J Dairy Res 71:97–106

Igarashi Y (1989) A method for determination of γ-casein and its use for investigating proteolysis in bovine milk. J Dairy Res 56:619–629

Ismail B, Nielsen SS (2010) Invited review: plasmin protease in milk: current knowledge and relevance to dairy industry. J Dairy Sci 93:4999–5009

Ismail B, Choi LH, Were LM, Nielsen SS (2006) Activity and nature of plasminogen activators associated with the casein micelle. J Dairy Sci 89:3285–3295

Johnsen LB, Poulsen K, Kilian M, Petersen TE (1999) Purification and cloning of a streptokinase from *Streptococcus uberis*. Infect Immun 67:1072–1078

Johnson HA (1974) The composition of milk. In: Webb BH, Johnson AH, Alford JA (eds) Fundamentals of dairy chemistry, 2nd edn. AVI Publishing Co. Inc, Westport, pp 1–57

Kaminogawa S, Yamauchi K, Tsugo T (1969) Properties of milk protease concentrated from acid-precipitated casein. Jpn J Zootech Sci 40:559–565

Kaminogawa S, Mizobuchi H, Yamauchi K (1972) Comparison of bovine milk protease with plasmin. Agric Biol Chem 36:2163–2167

Karlan BY, Clark AS, Littlefield BA (1987) A highly sensitive chromogenic microtiter plate assay for plasminogen activators which quantitatively discriminates between the urokinase and tissue-type activators. Biochem Biophys Res Commun 142:147–154

Kelly AL (1995) Variations in total and differential milk somatic cell counts and plasmin levels and their role in proteolysis and quality of milk and cheese. PhD Thesis, National University of Ireland, Cork

Kelly AL, Foley J (1997) Proteolysis and storage stability of UHT milk as influenced by milk plasmin activity, plasmin/β-lactoglobulin complexation, plasminogen activation and somatic cell count. Int Dairy J 7:411–420

Kelly AL, Fox PF (2006) Indigenous enzymes in milk: a synopsis of future research requirements. Int Dairy J 16:707–715

Kelly AL, McSweeney PLH (2003) Indigenous proteinases in milk. In: Fox PF, McSweeney PLH (eds) Advanced dairy chemistry: vol 1. Proteins, 3rd edn. Plenum Press, New York, pp 495–521

Kennedy A, Kelly AL (1997) The influence of somatic cell count on the heat stability of bovine milk plasmin activity. Int Dairy J 7:717–721

Kiermeier F, Semper G (1960) Incidence of a proteolytic enzyme and a trypsin-inhibitor in cows' milk. Part I. Proteolytic activity. Z Lebensm-Unters-Forsch 111:282–307

Kohlmann KL, Nielsen SS, Ladisch MR (1988) Effect of serine proteolytic enzymes (trypsin and plasmin), trypsin inhibitor, and plasminogen activator addition to ultra-high temperature processed milk. J Dairy Sci 71:1728–1739

Korycha-Dahl M, Dumas BR, Chene N, Martal J (1983) Plasmin activity in milk. J Dairy Sci 66:704–711

Larson NK, Ismail B, Nielsen SS, Hayes KD (2006) Activity of *Bacillus polymyxa* protease on components of the plasmin system in milk. Int Dairy J *16*:586–592

Lawrence RC, Creamer LK, Gilles J (1987) Texture development during cheese ripening. J Dairy Sci 70:1748–1760

Le Bars D, Gripon JC (1989) Specificity of plasmin towards bovine α_{s2}-casein. J Dairy Res 56:817–821

Leigh JA (1994) Purification of a plasminogen activator from *Streptococcus uberis*. FEMS Microbiol Lett 118:153–158

Lelievre J, Lawrence RC (1988) Manufacture of cheese from milk concentrated by ultrafiltration. J Dairy Res 55:465–478

Levi M, Roem D, Kamp AM, de Boer JP, Hack CE, ten Cate JW (1993) Assessment of the relative contribution of different protease inhibitors to the inhibition of plasmin in vivo. Thromb Haemost 70:141–146

Lindberg T (1979) Protease inhibitors in human milk. Pediatr Res 13:969–972

López-Fandiño R, De la Fuente MA, Ramos M, Olano A (1998) Distribution of minerals and proteins between the soluble and colloidal phases of pressurized milks from different species. J Dairy Res 65:69–78

Lowry OH, Rosebrough NJ, Farr AL, Randall RJ (1951) Protein measurement with the Folin phenol reagent. J Biol Chem 193:265–275

Lu DD, Nielsen SS (1993a) Isolation and characterization of native bovine milk plasminogen activators. J Dairy Sci 76:3369–3383

Lu DD, Nielsen SS (1993b) Assays for native plasminogen activators in bovine milk. J Dairy Sci 76:3362–3368

Manji B, Kakuda Y, Arnott DR (1986) Effect of storage temperature on age gelation of ultra-high temperature milk processed by direct and indirect heating systems. J Dairy Sci 69:2994–3001

Mara O, Roupie C, Duffy A, Kelly AL (1998) The curd-forming properties of milk as affected by the action of plasmin. Int Dairy J 8:807–812

Markus G, DePasquale JL, Wissler FC (1978) Quantitative determination of the binding of ε-aminocaproic acid to native plasminogen. J Biol Chem 253:727–732

Markus G, Hitt S, Harvey SR, Tritsch GL (1993) Casein, a powerful enhancer of the rate of plasminogen activation. Fibrinolysis 7:229–236

McKellar RC (1981) Development of off-flavors in ultra-high temperature and pasteurized milk as a function of proteolysis. J Dairy Sci 64:2138–2145

McSweeney PLH, Olson NF, Fox PF, Healy A, Højrup P (1993) Proteolytic specificity of plasmin on bovine α_{s1}-Casein. Food Biotechnol *7*:143–158

Metwalli AAM, de Jongh HHJ, van Boekel MAJS (1998) Heat inactivation of bovine plasmin. Int Dairy J 8:47–56

Mozhaev VV, Heremans K, Frank J, Masson P, Balny C (1996) High pressure effects on protein structure and function. Proteins 24:81–91

Mullins DE, Rohlich ST (1983) The role of proteinase in cellular invasiveness. Biochim Biophys Acta 695:177–214

Needs EC, Capellas M, Bland AP, Manoj P, MacDougal D, Paul G (2000a) Comparison of heat and pressure treatments of skim milk, fortified with whey protein concentrate, for set yogurt preparation: effects on milk proteins and gel structure. J Dairy Res 67:329–348

Needs EC, Stenning RA, Gill AL, Ferragut V, Rich GT (2000b) High-pressure treatment of milk: effects on casein micelle structure and on enzymic coagulation. J Dairy Res 67:31–42

Newstead DF, Paterson G, Anema SG, Coker CJ, Wewala AR (2006) Plasmin activity in direct-steam-injection UHT-processed reconstituted milk: effects of preheat treatment. Int Dairy J 16:573–579

Nielsen SS (2002) Plasmin system and microbial proteases in milk: characteristics, roles, and relationship. J Agric Food Chem 50:6628–6634

Noomen A (1975) Proteolytic activity of milk protease in raw and pasteurized cow's milk. Neth Milk Dairy J 29:153–161

O'Mahony JA, Fox PF, Kelly AL (2013) Indigenous enzymes of milk. In: McSweeney PLH, Fox PF (eds) Advanced dairy chemistry: vol 1A: proteins: basic aspects, 4th edn. Springer, Boston, MA, pp 337–385

O'Keeffe AM (1984) Seasonal and lactational influences on moisture content of Cheddar cheese. Ir J Food Sci Technol 8:27–37

Okigbo LM, Richardson GH, Brown RJ, Ernstrom CA (1985) Casein composition of cow's milk of different chymosin coagulation properties. J Dairy Sci 68:1887–1892

Oliver SP, Pighetti GM, Almeida RA (2011) Enviromental pathogens. In: Fuquay JW, Fox PF, McSweeney PLH (eds) Encyclopedia of dairy sciences, vol 3, 2nd edn. Academic Press, Oxford, p 415

Ollikainen P, Kivelä T (1989) The importance of plasmin in swiss-type cheese ripening. Milchwissenschaft 44:204–206

Ozen BF, Hayes KD, Mauer LJ (2003) Measurement of plasminogen concentration and differentiation of plasmin and plasminogen using Fourier-transform infrared spectroscopy. Int Dairy J 13:441–446

Patterson MF, Quinn M, Simpson R, Gilmour A (1995) Sensitivity of vegetative pathogens to high hydrostatic pressure treatment in phosphate-buffered saline and foods. J Food Prot 58:524–529

Pierzchala PA, Dorn CP, Zimmerman M (1979) A new fluorogenic substrate for plasmin. Biochem J 183:555–559

Pihlanto-Leppälä A, Pahkala E, Antila V (1993) Hydrolysis of κ-casein in solution by chymosin, plasmin, trypsin and *Lactobacillus* proteinases. Agric Sci Finl 2:489–496

Politis I, Hang KNK, Giroux RN (1989a) Environmental factors affecting plasmin activity in milk. J Dairy Sci 72:1713–1718

Politis I, Lachance E, Block E, Turner JD (1989b) Plasmin and plasminogen in bovine milk: a relationship with involution? J Dairy Sci 72:900–906

Politis I, Zhao X, McBride BW, Burton JH, Turner JD (1991) Plasminogen activator production by bovine milk macrophages and blood monocytes. Am J Vet Res 52:1208–1213

Politis I, Barbano DM, Gorewit RC (1992) Distribution of plasminogen and plasmin in fractions of bovine milk. J Dairy Sci 75:1402–1410

Politis I, Zavizion B, Barbano DM, Gorewit RC (1993) Enzymatic assay for the combined determination of plasmin plus plasminogen in milk: revisited. J Dairy Sci 76:1260–1267

Politis I, White JH, Zavizion B, Goldberg JJ, Guo MR, Kindstedt P (1995) Effect of individual caseins on plasminogen activation by bovine urokinase-type and tissue-type plasminogen activators. J Dairy Sci 78:484–490

Prado BM, Sombers SE, Ismail B, Hayes KD (2006) Effect of heat treatment on the activity of inhibitors of plasmin and plasminogen activators in milk. Int Dairy J 16:593–599

Prado BM, Ismail B, Ramos O, Hayes KD (2007) Thermal stability of plasminogen activators and plasminogen activation in heated milk. Int Dairy J 17:1028–1033

Precetti AS, Oria MP, Nielsen SS (1997) Presence in bovine milk of two protease inhibitors of the plasmin system. J Dairy Sci 80:1490–1496

Rademacher B, Pfeiffer B, Kessler HG (1998) Inactivation of microorganisms and enzymes in pressure-treated raw milk. In: Isaaacs NS (ed) High pressure food science, bioscience and chemistry. The Royal Society of Chemistry, Cambridge, pp 145–151

Rånby M, Norrman B, Wallén P (1982) A sensitive assay for tissue plasminogen activator. Thromb Res 27:743–749

Rauh VM, Bakman M, Ipsen R, Paulsson M, Kelly AL, Larsen LB, Hammershøj M (2014a) The determination of plasmin and plasminogen-derived activity in turbid samples from various dairy products using an optimised spectrophotometric method. Int Dairy J 38:74–80

Rauh VM, Sundgren A, Bakman M, Ipsen R, Paulsson M, Larsen LB, Hammershøj M (2014b) Plasmin activity as a possible cause for age gelation in UHT milk produced by direct steam infusion. Int Dairy J 38:199–207

Rauh VM, Johansen LB, Ipsen R, Paulsson M, Larsen LB, Hammershøj M (2014c) Plasmin activity in UHT milk: relationship between proteolysis, age gelation, and bitterness. J Agric Food Chem 62:6852–6860

Reimerdes EH, Herlitz E (1979) The formation of γ-caseins during cooling of raw milk. J Dairy Res 46:219–221

Reimerdes EH, Klostermeyer H (1974) Milchproteasen. I. Micellenassoziierte protease-anreicherung und hydrolyse von caseinen. Milchwissenschaft 29:517–523

Reimerdes EH, Mrowetz G, Klostermeyer H (1975) Milchproteasen. IV. Differenzierung der micellenassoziierten Aktivitat mit Aminosaure-4-nitroaniliden. Milchwissenschaft 30:271–274

Reimerdes EH, Klostermeyer H, Sayk E (1976) Milch-Proteinasen. VII. Fraktionierung von komponenten des proteinasen-inhibitor-systems der milch. Milchwissenschaft 31:329–334

Reimerdes EH, Petersen F, Kielwein G (1979) Milk proteinases. 9. Proteolytic activity profiles of casein micelles, milk serum, blood serum and *Pseudomonas fluorescens*. Milchwissenschaft 34:548–551

Reimerdes EH, Halpaap J, Klostermeyer H (1981) Milk proteases. 8. Comparative characterization of plasmin from cow's blood with a serine proteinase from cow's milk. Milchwissenschaft 36:19–22

Ribadeau-Dumas B, Brignon G, Grosclaude F, Mercier JC (1972) Structure primaire de la caséine β bovine: séquence complète. Eur J Biochem 25:505–514

Richardson BC (1983a) Proteinases of bovine milk and the effect of pasteurization on their activity. N Z J Dairy Sci Technol 18:233–245

Richardson BC (1983b) Variation of the concentration of plasmin and plasminogen in bovine milk with lactation. N Z J Dairy Sci Technol 18:247–252

Richardson BC, Elston PD (1984) Plasmin activity in commercial caseins and caseinates. N Z J Dairy Sci Technol 19:63–67

Richardson BC, Pearce KN (1981) Determination of plasmin in dairy products. N Z J Dairy Sci Technol 16:209–220

Richter RL, Schmidt RH, Smith KL, Mull LE, Henry SL (1979) Proteolytic activity in ultra-pasteurized, aseptically packaged whipping cream. J Food Prot 42:43–45

Rippel KM, Nielsen SS, Hayes KD (2004) Effects of native and denatured whey proteins on plasminogen activator activity. J Dairy Sci 87:2344–2350

Rollema HS, Poll JK (1986) The alkaline milk proteinase system: kinetics and mechanism of heat-inactivation. Milchwissenschaft 41:536–540

Rosey EL, Lincoln RA, Ward PN, Yancey RJ Jr, Leigh JA (1999) PauA: a novel plasminogen activator from *Streptococcus uberis*. FEMS Microbiol Lett 178:27–33

Saeman AI, Verdi RJ, Galton DM, Barbano DM (1988) Effect of mastitis on proteolytic activity in bovine milk. J Dairy Sci 71:505–512

Saint Denis T, Humbert G, Gaillard JL (2001a) Heat inactivation of native plasmin, plasminogen and plasminogen activators in bovine milk: a revisited study. Lait 81:715–729

Saint Denis T, Humbert G, Gaillard JL (2001b) Enzymatic assays for native plasmin, plasminogen and plasminogen activators in bovine milk. J Dairy Res 68:437–449

Sawyer L (2003) β-lactoglobulin. In: Fox PF, McSweeney PLH (eds) Advanced dairy chemistry, vol 1: proteins, 3rd edn. Springer, Boston, pp 319–386

Schaar J (1985) Plasmin activity and proteose-peptone content of individual milks. J Dairy Res 52:369–378

Schaar J, Funke H (1986) Effect of subclinical mastitis on milk plasminogen and plasmin compared with that on sodium, antitrypsin and N-acetyl-β-D-glucosaminidase. J Dairy Res 53:515–528

Schaller J, Moser PW, Dannegger-Müller GAK, Rösselet SJ, Kämpfer U, Rickli EE (1985) Complete amino acid sequence of bovine plasminogen: comparison with human plasminogen. Eur J Biochem 149:267–278

Scherze I, Sienkiewcz T, Krenkel K (1994) Studies on proteolytic degradation of caseins. 2. Influence of plasmin on the proteolysis in Gouda cheese. Milchwissenschaft 49:564–569

Schroeder DL, Nielsen SS, Hayes KD (2008) The effect of raw milk storage temperature on plasmin activity and plasminogen activation in pasteurized milk. Int Dairy J 18:114–119

Scollard PG, Beresford TP, Murphy PM, Kelly AL (2000a) Barostability of milk plasmin activity. Lait 80:609–619

Scollard PG, Beresford TP, Needs EC, Murphy PM, Kelly AL (2000b) Plasmin activity, β-lactoglobulin denaturation and proteolysis in high pressure treated milk. Int Dairy J 10:835–841

Searls DB (1980) An improved colorimetric assay for plasminogen activator. Anal Biochem 107:64–70

Sheehan JJ, Oliveira JC, Kelly AL, Mc Sweeney PLH (2007) Effect of cook temperature on primary proteolysis and predicted residual chymosin activity of a semi-hard cheese manufactured using thermophilic cultures. Int Dairy J 17:826–834

Shi GY, Wu HL (1988) Isolation and characterization of microplasminogen. A low molecular weight form of plasminogen. J Biol Chem 263:17071–17075

Snoeren THM, van Riel JAM (1979) Milk proteinase, its isolation and action of α_{s2}-and β-casein. Milchwissenschaft 34:528–531

Snoeren THM, van Riel JAM, Both P (1980) Some properties of a milk proteinase isolated from UHT milk. Zuivelzicht 72:42–43

Somer HH (1938) Market milk and related products. Olsen Publishing Co., Milwuakee

Somers JM, Kelly AL (2002) Contribution of plasmin to primary proteolysis during ripening of cheese: effect of milk heat treatment and cheese cooking temperature. Lait 82:181–191

Soodam K, Guinee TP (2018) The case for milk protein standardisation using membrane filtration for improving cheese consistency and quality. Int Dairy J 71:277–291

Sottrup-Jensen L, Claeys H, Zajdel M, Petersen TE, Magnusson S (1978) The primary structure of human plasminogen: Isolation of the two lysine-binding fragments and one 'mini'-plasminogen (MW. 38000) by elastase catalysed, specific limited proteolysis. In: Davidson JF, Rowan RM, Samama MM, Desnoyers PC (eds) Progress in chemical fibrinolysis and thrombolysis, vol 3. Raven Press, New York, pp 191–209

Spreer E (1998) Milk and dairy product technology. Marcel Dekker, New York

Stoeckel M, Lidolt M, Stressler T, Fischer L, Wenning M, Hinrichs J (2016) Heat stability of indigenous milk plasmin and proteases from *Pseudomonas*: a challenge in the production of ultra-high temperature milk products. Int Dairy J 61:250–261

Thatcher RW, Dahlberg AC (1917) Enzymes of milk and butter. J Agric Res 11:437

Upadhyay VK, Sousa MJ, Ravn P, Israelsen H, Kelly AL, McSweeney PLH (2004) Use of exogenous streptokinase to accelerate proteolysis in Cheddar cheese during ripening. Lait 84:527–538

Upadhyay VK, Ravn P, Israelsen H, Sousa MJ, Kelly AL, McSweeney PLH (2006) Acceleration of proteolysis during ripening of Cheddar-type cheese using of a streptokinase-producing strain of *Lactococcus*. J Dairy Res 73:70–73

van Asselt AJ, Sweere APJ, Rollema HS, De Jong P (2008) Extreme high-temperature treatment of milk with respect to plasmin inactivation. Int Dairy J 18:531–538

van Boekel MAJS (1998) Effect of heating on maillard reactions in milk. Food Chem 62:403–414

Verheijen JH, Mullaart E, Chang GTG, Kluft C, Wijngaards G (1982) A simple, sensitive spectrophotometric assay for extrinsic (tissue-type) plasminogen activator applicable to measurements in plasma. Thromb Haemost 47:266–269

Visser S, Slangen KJ, Alting AC, Vreeman HJ (1989) Specificity of plasmin in its action on bovine α_{s2}-casein. Milchwissenschaft 44:335–339

Visser S, Slangen CJ, Rollema HS (1991) Phenotyping of bovine milk proteins by reversed-phase high-performance liquid chromatography. J Chromatogr A 548:361–370

Wallén P (1978) Chemistry of plasminogen and plasminogen activation. In: Davidson JF, Desnoyers PC, Samama MM (eds) Progress in chemical fibrinolysis and thrombolysis, vol 3. Raven Press, New York, pp 167–181

Wallén P, Wiman B (1975) On the generation of intermediate plasminogen and its significance for activation. In: Reich E, Rifkin DB, Shaw E (eds) Proteases and biological control. Cold Spring Harbour Laboratory, New York, pp 291–303

Walther PJ, Hill RL, McKee PA (1975) The importance of the preactivation peptide in the two-stage mechanism of human plasminogen activation. J Biol Chem 250:5926–5933

Ward PN, Leigh JA (2002) Characterization of PauB, a novel broad-spectrum plasminogen activator from *Streptococcus uberis*. J Bacteriol 184:119–125

Warner RC, Polis E (1945) On the presence of a proteolytic enzyme in casein. J Am Chem Soc 67:529–532

Weber BA, Nielsen SS (1991) Isolation and partial characterization of a native serine-type protease inhibitor from bovine milk. J Dairy Sci 74:764–771

White JH, Zavizion B, O'Hare K, Gilmore J, Guo MR, Kindstedt P, Politis I (1995) Distribution of plasminogen activator in different fractions of bovine milk. J Dairy Res 62:115–122

Wiman B, Collen D (1978) Molecular mechanism of physiological fibrinolysis. Nature 272:549–550

Wiman B, Wallén P (1977) The specific interaction between plasminogen and fibrin. A physiological role of the lysine binding site in plasminogen. Thromb Res 1:213–222

Wu HL, Shi GY, Wohl RC, Bender ML (1987) Structure and formation of microplasmin. Proc Natl Acad Sci U S A 84:8793–8795

Yamauchi K, Kaminogawa S (1972) Decomposition of milk proteins by milk protease. Agric Biol Chem 36:249–254

Zachos T, Politis I, Gorewit RC, Barbano DM (1992) Effect of mastitis on plasminogen activator activity of milk somatic cells. J Dairy Res 59:461–467

Zhang C, Bijl E, Hettinga K (2018) Destabilization of UHT milk by protease AprX from *Pseudomonas fluorescens* and plasmin. Food Chem 263:127–134

Chapter 3
Lysosomal and Other Indigenous Non-plasmin Proteases in Bovine Milk

Lotte Bach Larsen, Søren Drud-Heydary Nielsen, Lizandra Paludetti, and Alan L. Kelly

3.1 Introduction

While the presence of a proteolytic enzyme in milk has been reported since 1897, and broadly accepted since around 1960, it was long thought that this was due solely to plasmin activity (Kelly et al. 2006). In good quality milk from healthy animals, this is likely to be the dominant source of indigenous (i.e., non-bacterial) proteolytic activity, but since the 1970s there have been reports of other proteases, with the acid protease cathepsin D being identified by Kaminogawa and Yamauchi (1972).

Proteases in milk can originate from a number of sources, one of which is blood (from where plasmin derives, together with its entire system of activators, inhibitors and its zymogen, plasminogen; see Chap. 2) and another one of which is the lysosomes of white blood cells in milk (somatic cells) or even mammary epithelial cells. Growing evidence of heightened level and diversity of proteolytic enzymes in milk as somatic cell count increases (e.g., during mastitis infection) caused more focus on lysosomes as enzyme sources. The nature, characteristics and significance of these lysosomal proteases is a main focus of this chapter. A list of the milk proteases described in this chapter and some of their characteristics are shown in Table 3.1. It should be noted that, even though this list is based on information from bovine and

L. B. Larsen (✉) · S. D.-H. Nielsen
Aarhus University, Department of Food Science, Faculty of Technical Sciences, Aarhus N, Denmark
e-mail: lbl@food.au.dk; sodn@food.au.dk

L. Paludetti
School of Food and Nutritional Sciences, University College Cork, Cork, Ireland

Teagasc Food Research Centre Moorepark, Fermoy, Co. Cork, Ireland

A. L. Kelly
School of Food and Nutritional Sciences, University College Cork, Cork, Ireland
e-mail: a.kelly@ucc.ie

A. L. Kelly, L. B. Larsen (eds.), *Agents of Change*, Food Engineering Series,
https://doi.org/10.1007/978-3-030-55482-8_3

Table 3.1 Selected list of lysosomal and other non-plasmin proteases or protease activities reported in milk

Protease	Protease class	Presence in	Comments
Cathepsin B[a, b, c, d]	Cysteine	Macrophages, PMNs, epithelial cells	Requires thiols or reductive agents for activity, lysosomal, mainly carboxy dipeptidase, some endoproteinase activity
Cathepsin C[a,d,e]	Cysteine	PMNs	Requires thiols or reductive agents for activity, lysosomal, alternative name dipeptidyl peptidase 1
Cathepsin D[a,d–f]	Aspartic	Macrophages, PMNs, epithelial cells	Lysosomal, endoprotease
Cathepsin E[e]	Aspartic	Neutrophils	Lysosomal, endoprotease
Cathepsin G[a,d,e]	Serine	Macrophages, PMNs	Lysosomal, endoprotease
Cathepsin H[a,d,e,g]	Cysteine	Macrophages	Requires thiols or reductive agents for activity, lysosomal, aminopeptidase and endoprotease
Cathepsin L[a,d,e]	Cysteine	PMNs, epithelial cells	Requires thiols or reductive agents for activity, lysosomal, endoprotease
Cathepsin K[d,e]	Cysteine	Unknown	Requires thiols or reductive agents for activity, lysosomal
Cathepsin S[a,d,e,g]	Cysteine	Macrophages, PMNs	Requires thiols or reductive agents for activity, lysosomal
Cathepsin Z[d,h]	Cysteine	Unknown	Requires thiols or reductive agents for activity, lysosomal
PMN elastase[a,d,e,i]	Serine	Macrophages, PMNs, lymphocytes	Lysosomal
Kallikrein[j,k]	Serine	Many types of tissues	Secreted/extracellular
Other amino- and carboxypeptidases[h,l–o]	Mixed	Not known	Not very characterized

[a]Talukder and Ahmed (2017); [b]Magboul et al. (2001); [c]Zeece et al. (1992); [d]Wickramasinghe et al. (2012); [e]Stepaniak (2004); [f]Larsen et al. (1993); [g]D'Alessandro et al. (2010); [h]D'Alessandro et al. (2011); [i]Dallas et al. (2015); [j]Peeters et al. (1976); [k]Zeitlin and Eshraghi (2002); [l]Reimerdes et al. (1979); [m]Wedholm et al. (2008); [n]Molinari et al. (2012); [o]Hinz et al. (2010)

other milks, the list is likely not exhaustive even for bovine milk, as the list of detectable proteases is growing and linked with the use of more and more sensitive detection methods, e.g., mass spectrometry (MS).

It is known that the lysosomes of bovine somatic cells, which can be found in milk, contain a wide complement of proteolytic enzymes which is then also expected to be present in milk (due to leakage, release or lysis of cells, before or during processing). However, direct evidence for the activity of many of these enzymes, with the exceptions of cathepsin D (assayed activity) and cathepsin B (immunological detection) is scarce, and in many cases conclusions about their likely presence and activity have been inferred from comparison of the results of studies of their specificities against the caseins with cleavage sites responsible for the generation of peptides found in milk, or the results of inhibitor studies.

3.2 Origin of Non-plasmin Proteolytic Enzymes in Bovine Milk

The origin of the non-plasmin proteolytic enzymes in bovine milk is in many cases not clear, and may be complex, i.e., their presence can originate from different sources, cell types or even cellular organelles. As several of the proteolytic enzymes observed in bovine milk are reported to be lysosomal (cathepsins B, C, D, E, G, H, L, S and Z and polymorphonuclear (PNM) leucocyte derived elastase) (Stepaniak 2004) it is natural to look at this compartment as main cellular source and its secretion mechanism to understand the protease profile of bovine milk. Lysosomes are known as powerful degradative cellular compartments participating in molecular turn-over, especially of proteins, for recycling of building blocks, also reflected in the cathepsin name (Greek, "kathepsin" meaning "to digest") of many of these enzymes.

Apart from the lysosomal proteases, a range of apparently non-lysosomal proteases, including kallikrein, amino-and carboxypeptidases, have also been reported in bovine milk, and may originate from other sources.

A strong association between somatic cells or somatic cell count (SCC) and proteolytic activity in milk has been shown (Somers et al. 2003). Therefore, it is evident that somatic cells, and especially their lysosomes, are potentially important sources of these non-plasmin proteases. Somatic cells constitute of different cell types, comprising polymorphonuclear (PMN) leucocytes, also called neutrophils, macrophages and lymphocytes, as well as a small amount of mammary epithelial cells (Larsen et al. 2004), with macrophages being dominant in milk from healthy cows, and with the level of PMN leucocytes increasing dramatically by infection.

Even though it has, in many cases, not been completely determined which enzymes precisely can be secreted by which of these cell types, a compilation of knowledge is reported in Table 3.1. In milk, the proteases can be present as a result of secretion from viable or damaged somatic cells, which may result in different profiles of active or pro-forms of the leaked or secreted enzyme profiles. Alternatively, they can be secreted from the mammary epithelial cells lining the udder or by viable epithelial cells present in milk, or even leaked from such dead milk epithelial cells. Cultured human mammary cells have been shown to secrete low amounts of procathepsin D (Capony et al. 1989), but the situation for bovine milk is still not really known. Viable cells like macrophages and PMN leucocytes can actively secrete proenzyme molecules in response to inflammation, and this is a likely mechanism for secretion of at least proenzymes into milk, while active forms of the milk proteases may result from leakage from dead cells, i.e., macrophages, PMN leucocytes or even milk epithelial cells (Larsen et al. 2006). Accordingly, the presence of and ratio between active and inactive forms of the bovine milk proteases may be influenced by both cellular secretion pathways, SCC level and composition, and milk cell viability.

3.3 The Lysosomal and Other Non-plasmin Proteases in Bovine Milk

Members of the milk cathepsin family include aspartic, serine and cysteine proteases types (Table 3.1). Cathepsins are typically active at a low pH, reflecting the pH around 4–5 within the lysosomes. The specificities of the cathepsins and the other non-plasmin milk proteases reflects the type of protease, and the resulting activity is very much dependent on presence of forms (active/proenzymes), activation mechanisms and requirements, as well as on pH optimum and other requirements for activity (e.g., requirements for reducing environment, presence or absence of natural inhibitors). This section will cover the cathepsins and other non-plasmin proteases identified in bovine milk, as well as some examples of these enzymes found in milk from other species, when evidence from the bovine system is lacking. Characteristics in relation to molecular organizations and structural features, presence of pro/mature forms, and presence in different milk fractions will be covered.

Lysosomal proteinases have been extensively studied, as they are actively hydrolysing milk caseins and consequently can affect milk processing properties and sensorial quality of dairy products (Li et al. 2014). The lysosomes from somatic cells contain a range of enzymes, such as cathepsins (e.g. D, B, H, L, S and Z) and elastase. These proteases are typically secreted in their inactive form and must be cleaved at a specific peptide bond by a protease activator to become the active enzyme (Dallas et al. 2015). According to Grieve and Kitchen (1985), caseins are broadly degraded by leukocyte proteinases in the following order: α_{S1}- > β->> κ-casein, but studies of the specificities of individual enzymes will be discussed below.

3.3.1 Cathepsin B

The molecular structure of bovine milk cathepsin B shows a molecular mass of 21–30 kDa (Magboul et al. 2001; D'Alessandro et al. 2011). The bovine milk enzyme has not been purified, and knowledge is therefore based only on immunoblotting mobility and identification. Therefore it is also not clear if proenzymes of cathepsin B are present in bovine milk, but measurable activity in the presence of reducing agents is evident. It is known from the mammalian system that lysosomal cysteine proteases are synthesized as pre-proenzymes with molecular masses of 37–55 kDa and are later converted to mature, catalytically active, forms with molecular masses of 21–30 kDa by means of other lysosomal proteases, involving both cathepsin D, aminopeptidases and cathepsin B itself (Isidoh and Kominami 2002). The catalytically active forms of bovine cathepsin B is normally considered to consist of a mixture of a 30 kDa single chain form and its processed form, consisting of a heavy chain (25 kDa) + a light chain (5 kDa). Both the single chained and the two-chained forms contain four disulphide bonds, of which one is responsible for holding together the two-chained form in its unreduced form (30 kDa) (Kirschke et al. 1998). The immunoreactive form detected in milk corresponds to the position of the heavy chain of mature cathepsin B from bovine spleen (25–30 kDa in the gel system used), and a further band with apparent mass of 42 kDa, which may correspond to procathepsin B (Magboul et al. 2001), though this still needs to be verified. It has been shown by expression studies of the isolated somatic cell pool consisting of lymphocytes, PMNs leucocytes, macrophages and exfoliated epithelial cells that cathepsin B was expressed by these bovine cells in milk (Wickramasinghe et al. 2012). Cleavage sites compatible with cleavage specificities of cathepsin B were identified in a peptidomic study of milk from cows infused with lipoteichoic acid from *Staphylococcus aureus* (Larsen et al. 2010).

Cathepsin B is primarily a carboxy-dipeptidase, exhibiting some endoproteinase activity. The first observation of cysteine protease activity in bovine milk was measured in whey (Suzuki and Katoh 1990), with it being able to degrade caseins in the presence of a reducing agent. Later, cathepsin B was identified in bovine milk (Magboul et al. 2001), using a combination of chromogenic substrate cleaved in the presence of reducing agents and immunoblotting using antibodies raised against bovine cathepsin B. Cysteine protease activity was partly purified by preparative column chromatography, and cathepsin B identified in one of the partly purified fractions. The partly purified fraction containing cathepsin B was demonstrated to be able to cleave both α_{S1}- and β-casein and retained approximately 20% of its activity after heating at 72 °C for 30 s (Magboul et al. 2001).

According to Considine et al. (2004), cathepsin B has a broad specificity against caseins. Those authors reported extensive hydrolysis of α_{S1}-casein and β-casein, as 35 and 32 sites were cleaved in those proteins. The enzyme appeared to have a preference for bonds incorporating the amino acids Leu, Val, Gln, Pro and Ser. This study also showed that some cleavage sites were similar to those of chymosin activity on α_{S1}-casein and β-casein. For example, the Phe^{23}-Phe^{24} bond is a cleavage site of both cathepsin D, cathepsin B and also chymosin on α_{S1}-casein. This suggests that cathepsin B, like cathepsin D, may contribute to chymosin-like activity in dairy products, possibly enabling release of the α_{S1}-I-casein fragment. Cathepsin B activity in bovine milk has been shown to correlate with SCC, have pH optimum at 6.0, and is inactivated at pH above 7 (Suzuki and Katoh 1990; Kirschke et al. 1998; O'Driscoll et al. 1999).

3.3.2 *Cathepsin C: Dipeptidyl Peptidase 1*

Cathepsin C, also called dipeptidyl peptidase 1, is a cysteine protease of a molecular mass around 200 kDa and consists of 8 subunits of each 24 kDa (Bohley and Seglen 1992). It has a pH optimum of 5–6. The activity of cathepsin C involves the cleavage of di-peptides from the N-terminus of proteins and peptides, and stops at N-terminal lysine, arginine or proline in P1' or P2' positions. Cathepsin C has been inferred to be present in milk (Talukder and Ahmed 2017), due to its presence in PMN leucocytes but has not to date been demonstrated in milk. Cathepsin C functions as a key enzyme in the activation of serine proteases in neutrophils, like cathepsin G and elastase (Dahl et al. 2001).

3.3.3 *Cathepsin D*

Acid protease activity in bovine milk with pH optimum 3.0–5.0 was first reported by Kaminogawa and Yamauchi (1972), and was later demonstrated to be identical to cathepsin D by N-terminal sequencing of the protease purified from bovine milk (Larsen et al. 1993). Cathepsin D is an aspartyl protease, like chymosin, and is also, in analogy with chymosin, able to cleave caseins, including weak milk clotting activity through cleavage of the Phe^{104}-Met^{105} peptide bond in κ-casein (McSweeney et al. 1995; Larsen et al. 1996). The properties of bovine milk cathepsin D were reviewed by Hurley et al. (2000). The protease is part of a cathepsin D system, and present in five different molecular forms of different molecular masses (Larsen and Petersen 1995), of which the 45 kDa procathepsin D form was predominant (Fig. 3.1).

The predominance of the pro-forms over the active forms indicate that bovine milk cathepsin D is mainly present as its preform, procathepsin D, and thereby probably is secreted as such from either PMN leucocytes, macrophages and/or

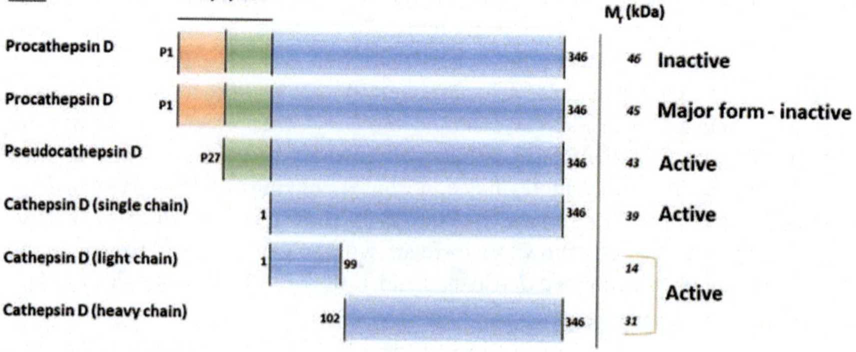

Fig. 3.1 Molecular forms of cathepsin D identified in bovine milk

epithelial cells, present in bovine milk, or from the cells lining the alveoli or ducts in the mammary gland itself; it is at present not clear which. Furthermore, it is not clear whether the mature enzyme itself is actually present in bovine milk, or whether it is induced as a consequence of the purification procedure, as this involved steps at low pH (pH 3.5), which may allow activation to occur.

The processing steps and activation of procathepsin D has been relatively extensively studied, still without clear evidence of precise sequence of steps and involved proteases. However, a synthesis of present stage of knowledge involves the following steps. The molecule is synthesized as pre-procathepsin D, and after initial cleavage of the signal peptide, the 45–46 kDa procathepsin D (Fig. 3.1) molecules have been found to be converted into intermediate forms, which in the bovine system has been shown to result in 43 kDa pseudocathepsin D (Fig. 3.1, Larsen et al. 1993). *In vivo* this presumably occurs in pre-lysosomal compartments, but this pseudo-form was also shown to result when incubating purified procathepsin D at pH 3.5.

The acid-induced autoactivation of isolated procathepsin D from bovine milk yields the intermediate form, pseudocathepsin D, as a result of a cleavage of the $LeuP^{26}$-$IleP^{27}$ peptide bond in the propart (P numbering refers to position in the 44-residues propart) (Larsen et al. 1993). A two-phase mechanism for the autoactivation of procathepsin D into proteolytically active forms has been suggested (Wittlin et al. 1999), involving an initial phase where local denaturation of the propeptide from its coverage within the active site cleft of a few molecules of procathepsin D produces a few pseudocathepsin D molecules. This is followed by a second phase, where the process is accelerated by further autoactivation of more procathepsin D molecules catalyzed by the peudocathepsin D enzyme formed.

In vivo it has been shown that ultimate activation of procathepsin D is dependent on lysosomal cysteine proteases (Wittlin et al. 1999). *In vitro* activation studies using procathepsin D isolated from bovine milk and added cysteine lysosomal proteases, alone or in combinations, of cathepsins C and L showed that the addition of this combination of lysosomal proteases was able to generate further processed forms of pseudo-cathepsin D, but not the mature form (Fig. 3.2), which therefore presumably is dependent on further proteases (LB Larsen, unpublished results). Cathepsins L and C were selected as they respectively represent both a lysosomal cysteine endoproteinase activity and a dipeptidyl peptidase which may be able to cleave dipeptides from the N-terminus of the propeptide. The second residue of the

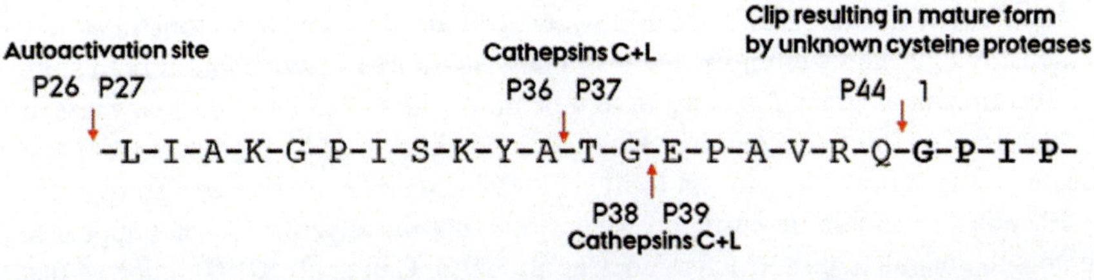

Fig. 3.2 Positions of identified *in vitro* cleavage sites in the propeptide part as generated by the added lysosomal proteases, cathepsins C + L, in combination with the position of the autoactivation site and the final clip site resulting in mature cathepsin D

mature cathepsin D is a proline (Fig. 3.2), which is notoriously difficult to cleave by many proteases, and therefore the ultimately cleaving protease could be a kind of aminopeptidase or amino-di-peptidase.

Varying among species, but apparently not cell type, the single chain molecule with the mature N-terminus generated by this activation mechanism may be cleaved further by the lysosomal cysteine proteinases (Gieselmann et al. 1985) into two chains, a 15 kDa light chain derived from the N-terminus, and a 30 kDa C-terminally derived heavy chain by excision of a few residues. Mature bovine cathepsin D exists in the lysosomes as a mixture of single and two-chain forms in approximately equal amounts, as also reflected in the presence of different forms in bovine milk (Fig. 3.1). The reason for this differential clipping of the mature molecules in the bovine system is not known.

Generally, according to predictive databases (adopted from http://bioware.ucd.ie/~enzpred/patterns.html) cathepsin D is characterized by preferential cleavage between two hydrophobic amino acids. It prefers to cleave bonds where Alanine, Valine, Leucine, Isoleucine, Proline, Methionine, Phenyalanine or Tryptophan is at the P1 site, and Alanine, Valine, Leucine, Isoleucine, Proline, Methionine, or Phenyalanine is at the P1' sites.

The degradation of milk proteins by cathepsin D has been investigated extensively (McSweeney et al. 1995; Hinz et al. 2012), and peptides potentially originating from cathepsin D activity have been identified in high cell count milk (Wedholm et al. 2008; Hinz et al. 2012). This proteinase preferentially hydrolyses α_{S1}-casein and its specificity on α_{S1}, β- and κ-caseins is rather similar to that of chymosin (Kaminogawa and Yamauchi 1972; Kaminogawa et al. 1980). Larsen et al. (1996) observed similar degradation patterns of milk caseins when comparing cathepsin D from bovine spleen and procathepsin D purified from milk; both enzymes could cleave the four caseins (α_{S1}-, α_{S2}-, β- and κ-caseins) and α-lactalbumin in solution at pH 5.0. The authors also observed that the enzyme's cleavage sites in α_{S1}- and β-caseins were similar with those of chymosin. Both enzymes cleaved Phe^{23}-Phe^{24} bond of α_{S1}-caseins, producing α_{S1}-I-casein, as well as the Leu^{149}-Phe^{150} bond. Five out of 13 peptide bonds hydrolysed by cathepsin D in β-casein are also cleaved by chymosin; those five common cleavage sites were Leu^{58}-Val^{59}, Leu^{127}-Thr^{128}, Trp^{143}-Met^{144}, Leu^{165}-Ser^{166} and Leu^{192}-Tyr^{193}. A combined list of identified peptides released from bovine caseins by the addition of cathepsin D from the two studies (McSweeney et al. 1995; Larsen et al. 1996) is shown in Table 3.2.

Kaminogawa et al. (1980) and McSweeney et al. (1995) also demonstrated that cathepsin D purified from milk had a similar specificity to chymosin. In both studies, the enzyme degraded α_{S1}-casein to a peptide with the same molecular mass as α_{S1}-casein f24–199 (α_{S1}-I-casein). This peptide is known to be produced from α_{S1}-casein by the action of chymosin (Hill et al. 1974). Hinz et al. (2012) also reported hydrolysis of κ-casein by purified Cathepsin D, but the enyzme does not appear to be able to coagulate milk (McSweeney et al. 1995). Gan et al. (2019) reported that cathepsin D from human milk hydrolysed human α-lactalbumin, particularly at lower pH values (around 3), such as might be found under gastric conditions in the infant gut.

Table 3.2 Peptides identified in model studies of bovine caseins cleaved by cathepsin D[a]

Protein	Peptide
α_{s1}-CN	RPKHPIKHQGLPQEVLNENLLRF
	VAPFPEVFGKEKVNELSKDIGSESTEDQAMEDIKQMEAESISSSEEIVPNSVEQKHIQKEDVPSERYLGYLEQL
	LRLKKYKVPQLEIVPNSAEERLHSMKEGIHAQQKEPMIGVNQELAYFYPEL
	FRQFYQLDAYPSGAWYYVPLGTQYTDAPSFSDIPNPIGSENSEKTTMPLW
	RPKHPIKHQGLPQEVLNENLLRFF
	VAPFPEVFGKEKVNELSKDIGSESTEDQAMEDIKQMEAESISSSEEIVPNSVEQKHIQKEDVPSERYLGYLEQLLRLKKYKVPQL
	-EIVPNSAEERLHSMKEGIHAQQKEPMIGVNQELAYFYPELFRQFYQLDAYPSGAWYYVPLGTQYTDAPSFSDIPNPIGSENSEKTTMPLW
	FVAPFPEVFGKEKVNEL
	FRQF
	FRQFYQL
	DAYPSGAW
	YQLDAY
	YQLDAYPSGAW
	YYVPLGTQYTDAPSFSDIPNPIGSENSEKTTMPLW
β-CN	RELEELNVPGEIVESLSSSEESITRINKKIEKFQSEEQQQTEDELQDKIHPF
	AQTQSL
	VYPFPGPIPNSLPQNIPPLTQTP
	VVVPPFLQPEVMGVS
	KVKEAMAPKHKEMPFPKYPVEPFTESQSL
	TL
	TDVENLHLPLPLLQSW
	MHQPHQPLPPTVMF
	PPQS
	VLSL
	SQSKVLPVPQKAVPYPQRDMPIQAFL
	YQEPVLGPVRGPF
	PIIV

(continued)

Table 3.2 (continued)

Protein	Peptide
α_{s2}-CN	KNTMEHVSSSEESIISQETYKQEKNMAINPSKENLCSTFCKEVVRNANEEEYSIGSSSEESAEVATEEVKITVDDKHYQKALNEINQFYQKFPQYLQYL
	YQGPIVLNPWDQVKRNAVPITPTL
	NREQLSTSEENSKKTVDMESTEVFTKKTKLTEEEKNRLNFLKKISQRYQKFALPQYL
	KT
	VYQHQKAMKPWIQPKTKVIPYVRYL
κ-CN	QEQNQEQPIRCEKDERFFSDKIAKYIPIQYVL
	SRYPSYGLNYYQQKPVALINNQFLPYPYYAKPAAVRSPAQILQWQVL
	SNTVPAKSCQAQPTTMARHPHPHLSF
	MAIPPKKNQDKTEIPTINTIASGEPTSTPTTEAVESTVATLEDSPEVIESPPEINTVQVTSTAV
α-Lactalbumin	EQLTKCEVFRELKDLKGYGGVSLPEWVCTTFHTSGYDTQAIVQNNDSTEYGL
	FQINNKIWCKDDQNPHSSNICNISCDKFLDDDLTDDIMCVKKILDKVGINYW
	LAHKALCSEKLDQWLCEKL

[a]According to McSweeney et al., 1995; Larsen et al., 1996. CN, casein

3.3.4 Cathepsin E

Cathepsin E is a mammalian, lysosomal aspartic protease, and mentioned in relation to milk in a review by Stepaniak (2004). Cathepsin E is present in neutrophils, and therefore potentially present in milk and secreted by somatic cells, but this remains to be demonstrated. It has been reported to have a low pH optimum, at 3.0–3.2, and a molecular mass of approximately 42 kDa, and may be present in some instances as composed of two identical subunits in the mature form (Yonezawa et al. 1987; Athauda et al. 1991). Cathepsin E is, similar cathepsin D, synthesized as a proenzyme (Athauda et al. 1991), but the exact presence of mature and/or pro-cathepsin E in bovine milk, and that of other species, awaits further studies.

3.3.5 Other Lysosomal Cysteine Proteases

A range of cathepsins of the cysteine protease type have been referred to be present or potentially present in milk (Table 3.1), as they are part of the lysosome systems of mammalian cells, including somatic cells, though only a limited number of these have actually been shown through molecular means to be present. These will be briefly presented here in alphabetical order.

Cathepsin H has been identified in human milk (D'Alessandro et al. 2011), and further reported as milk protease, probably derived from lysosomes in milk macrophages (Stepaniak 2004; Talukder and Ahmed 2017), though this awaits definite confirmation, while the enzyme's presence in bovine milk also awaits further studies. Expression of cathepsin H by bovine somatic cells has been demonstrated (Wickramasinghe et al. 2012). Cathepsin H is known to be an aminopeptidase, cleaving residues from the N-terminus of proteins and peptides, as well as having some endoproteinase activity as well. Peptides compatible with cathepsin H activity were identified in a peptidomic study of milk from cows infused with lipoteichoic acid from *Staphylococcus aureus* (Hinz et al. 2012). The molecular mass has been reported to be around 26 kDa, with subunits of 22 + 4 kDa, and a pH optimum around 6 (Barrett and Kirschke 1981).

Cathepsin L was likewise reported as a milk lysosomal protease (Stepaniak 2004; Talukder and Ahmed 2017), probably derived from milk PMNs and epithelial cells, but has not been demonstrated in milk. Cathepsin L is a powerful mammalian endoprotease, but with apparently more restricted specificity than, e.g., cathepsin B, and is able to cleave casein (Barrett and Kirschke 1981). Cathepsin L is synthesized as a proenzyme, with an approximately 30 -kDa mature form, with subunits of 25 + 5 kDa (Isidoh and Kominami 1994), and was interestingly shown to be secreted by cultured mammary gland epithelial cells in its proenzyme form (Recklies and Mort 1985). Cathepsin L is probably present in bovine milk, but its presence has not been confirmed (Kelly et al. 2006).

Cathepsin K was likewise mentioned as a milk lysosomal cysteine protease (Stepaniak 2004). Potential responsible cell type is not known, and its presence in milk also awaits demonstration. Cathepsin K was shown to be expressed by bovine somatic cells and is therefore also likely to be present in milk, at least at protein level if not in an active form (Wickramasinghe et al. 2012).

Cathepsin S was identified in human milk (D'Alessandro et al. 2011), and furthermore mentioned (Stepaniak 2004; Talukder and Ahmed 2017) as being potentially secreted into milk by macrophages and PMNs. This is further strengthened by the demonstration of expression of cathepsin S by isolated bovine somatic cells (Wickramasinghe et al. 2012). Cathepsin S is an approximately 24 kDa single-chain cysteine protease with a pH optimum of 5–6, able to cleave casein with endoprotease actitvity (Bohley and Seglen 1992). Like the other lysosomal proteases it is synthesized as a proenzyme (Wiederanders et al. 1992).

Cathepsin Z was identified in bovine milk (D'Alessandro et al. 2011), but more information awaits further studies. Cathepsin Z was identified as being expressed by bovine somatic cells (Wickramasinghe et al. 2012) and is therefore also likely to be present.

3.3.6 Cathepsin G

Cathepsin G may be present in bovine milk, but presence has not been confirmed (Kelly et al. 2006). Cathepsin G is a 24–26 kDa serine protease (Watorek et al. 1993) reported (Stepaniak 2004; Talukder and Ahmed 2017) as potentially secreted into milk by macrophages and PMNs, a likelihood supported by the observation that Cathepsin G is expressed by the bovine milk somatic cell pool (Wickramasinghe et al. 2012). Cathepsin G, together with PNM elastase, represent the two major enzymes present in and secreted by neutrophils. The proteolytic specificity of cathepsin G on bovine α_{S1}- and β-casein have been reported (Considine et al. 2002), and it was found to cleave at least 16 sites in α_{S1}-casein and at least 21 sites in β-casein, and thereby to have a quite broad specificity, including the ability to generate the α_{S1}-casein-I fragment by cleavage at the Phe^{23}-Phe^{24} peptide bond.

3.3.7 PMN Elastase

Elastase, a serine proteinase with a molecular weight of 24–30 kDa, is one of the main enzymes found in polymorphonuclear leucocytes (Verdi and Barbano 1991; Travis and Fritz 1991), and is expressed by bovine milk somatic cells (Wickramasinghe et al. 2012). Elastase has not been directly identified in bovine milk, although there have been reports of assayed activity (as discussed later in this chapter); however, it has been detected in polymorphonuclear neutrophils in bovine milk in experimentally-induced mastitis. Thus, this enzyme could contribute to hydrolysis of casein in milk at both neutral and acid pH, particularly milk with high SCC (Verdi and Barbano 1991). This enzyme digests a wide variety of protein substrates and has a preferred specificity for bonds involving uncharged, non-aromatic amino acids, such as Ala, Val, Leu, Ile, Gly and Ser (Lewis et al. 1956; Naughton and Sanger 1961).

Considine et al. (1999) and Considine et al. (2000) investigated the proteolytic specificity of elastase on bovine α_{S1}-casein and β-casein, respectively; showing that it has a broad cleavage specificity. Elastase cleaves at least 25 sites on α_{S1}-casein and 19 sites on β-casein, mainly towards the C- and N-terminus of the molecule. Some sites cleaved on β-casein are identical to or near those cleaved by plasmin, chymosin or cell envelope-associated proteinases of several strains of *Lactococcus*. For example, elastase cleaved the bond Ala^{189}-Phe^{190}, which is also cleaved by chymosin. Elastase also cleaved two sites (Ile^{26}-Asn^{27} and Glu^{108}-Met^{109}) close to the three principal plasmin cleavage sites (Lys^{28}-Lys^{29}, Lys^{105}-His^{106} and Lys^{107}-Glu^{108}) on β-casein. Therefore, it is possible that indigenous elastase in milk may contribute significantly to the breakdown of milk proteins.

3.3.8 Kallikrein

Kallikreins (KLK) are a group of serine proteases with either trypsin-like or chymotrypsin-like specificity, initially described in the 1930s. Kallikreins are found in blood plasma and organ tissue, mainly exocrine organs, as well as in the secretions from these organs. Human KLK is the most widely studied, and been identified in human kidneys, urine, genial and sweat glands. Bovine KLK has been studied to a much lesser extent, but bovine KLK has been isolated from the bovine mammary gland (Peeters et al. 1976) and tissue KLK activity measured in the bovine udder (Zeitlin and Eshraghi 2002). Plasma KLK (secreted into the circulatory system from the pancreas where it is produced) and tissue KLK differ in their physical properties and specificities. In humans, the tissue KLKs consist of 15 proteases (KLK1 – KLK15). Trypsin-like KLKs (plasma KLK and tissue KLK2, −4, −6, −8, −10, −12, −13 and −15) cleave at the C-terminal of arginine or lysine residues. Chymotrypsin-like KLKs (KLK3, −7 and −9) cleave at the C-terminal of phenylalanine, tyrosine and tryptophan. The remaining KLKs show both trypsin- and chymotrypsin-like specificity.

KLKs are synthesized and secreted in their zymogen form and can be activated by cleaving off their pro-peptide. In blood, plasma KLKs are important in the contact activation system triggering other plasma proteolytic system such as those in blood coagulation. Plasma KLKs are also able to activate plasmin by hydrolyzing plasminogen into its active form, plasmin. Plasmin is known to be present in bovine milk, as covered in Chap. 2. However, to potentially counteract this, milk has been shown to contain KLK inhibitors, such as plasma serine protease inhibitor, kallistatin and plasma protease C1 inhibitor (D'Alessandro et al. 2011).

The zymogen and active form of KLK have been quantified in human milk using ELISA assays and their activity was confirmed using proteolytic activity assays and peptidomic analysis (Demers-Mathieu et al. 2017; Nielsen et al. 2017a). In bovine milk, plasma KLK has been identified (Heegaard et al. 1994a, b) in studies on mastitic milk. However, further studies are needed to validate this.

3.3.9 Amino-and Carboxypeptidases

Amino- and carboxypeptidases are exo-proteases that cleave off amino acids from the N- or C-terminals, respectively, of proteins and peptides. In addition to exopeptidases cleaving off one residue at a time, di-peptidases or tri-peptidases are also known. Examples of this include cathepsin B (carboxy di-peptidase), cathepsin C (amino-dipeptidase) and cathepsin H (aminopeptidase) activities, as described earlier. However, in addition to these, several more amino- or carboxypeptidases are potentially present in milk, as many different types of such enzymes have been shown to be present in mammalian lysosomes (Bohley and Seglen 1992). The first evidence of aminopeptidases in bovine milk (whey) was reported by Reimerdes et al. (1979). The further presence of amino- and carboxypeptidases have been shown by ELISA or proteomics-based investigations (Demers-Mathieu et al. 2017). Further evidence for amino- and carboxypeptidase activities in bovine milk was provided by Wedholm et al. (2008) who, using peptidomics, observed trimming of both N- and C-terminals of identified peptides from milk with different SCC, as well as by peptidomic analysis of human milk (Nielsen et al. 2017a). Peptides demonstrating both amino- and carboxypeptidase trimming were shown in a peptidomic study of milk from cows infused with lipoteichoic acid from *Staphylococcus aureus* (Hinz et al. 2012). More specific proteases have included cytosol aminopeptidase and carboxypeptidase B2 in human milk (Molinari et al. 2012), which have also been demonstrated in bovine milk (D'Alessandro et al. 2011). Currently, more knowledge on the exact amino- and carboxypeptidases in bovine milk awaits further studies, both on additional identification of proteases present, as well as their forms (mature or pre-forms).

3.4 Measurement of Lysosomal and Other Non-plasmin Proteases in Milk

One approach to evaluation of the elative contribution of lysosomal enzymes to proteolysis in milk is to down-regulate the otherwise dominating activity of plasmin (certainly in low SCC samples) through addition of a suitable inhibitor; using this approach, Somers et al. (2003) reported much greater heterogeneity of protease activities (including acid and cysteine proteases, likely including cathepsins D and

B) in high somatic cell count (SCC) milk than in low SCC milk, and noted that such activities were greatly reduced by pasteurisation, suggesting that such enzymes were more heat-sensitive than plasmin.

Later, Santillo et al. (2009) assayed several lysosomal protease activities in caprine milk by use of specific substrates and also incubation in the presence of inhibitors such as aprotinin (to inhibit plasmin) and pepstatin (to inhibit cathepsin D); their results indicated that the majority of proteolytic activity was due to serine proteinases, likely plasmin but also perhaps elastase. Albenzio et al. (2009), measuring protease activities in ovine milk, showed that elastase activity increased through lactation, whereas activity of cathepsins peaked in mid-lactation.

A further factor through which elevated SCC in milk could influence proteolysis is through upregulation of plasmin activity through the action of cell-associated plasminogen activators (PA), specifically tissue-type PA (t-PA; Zachos et al. 1992). Verdi and Barbano (1991) showed that somatic cells recovered from high SCC milk were capable to converting plasminogen to plasmin, with this activity being inactivated by pasteurisation. Kelly and Foley (1997) reported higher rates of activation of plasminogen during storage of UHT milk made from raw milk with high SCC compared to that made from low SCC milk.

Recently, peptidomic studies have contributed to the knowledge on enzyme activities present in milk (Nielsen et al. 2017b). Wedhom et al. (2008) and Guerrero et al. (2015) used this to decipher enzyme activity of indigenous enzymes in bovine milk hydrolysing milk proteins in normal and mastitic milk. Online tools for prediction of possible responsible cleavage enzymes based on identified peptides are available (http://bioware.ucd.ie/~enzpred/Enzpred.php; http://www.proteasix.org). These predictions are based on the C and N-terminal cleavage sites from position P4 to P4' (Klein et al. 2013; Vijayakumar et al. 2012). Determination of enzyme concentration in milk has further been done by Enzyme-Linked Immunosorbent Assay (ELISA), but these can not necessarily differentiate between active and inactive forms (Benfeldt et al. 1995 Larsen et al. 1996). More advanced techniques using mass spectrometry are now able to quantify low abundant proteins in complex samples, using technologies such as multiple-reaction monitoring. However, so far, no published study has used this technique.

3.5 Thermal Inactivation

Heat inactivation of these enzymes in milk has been only studied in detail for cathepsin D (Kelly et al. 2006). In 1972, Kaminogawa & Yamauchi reported that cathepsin D was completely inactivated by heating at 70 °C for 10 min. Larsen et al. (2000) detected cathepsin D activity in pasteurized milk (72 °C for 15 or 60 s) and milk heated to temperature as high as 99 °C for up to 1 min. However, even though activity was still detectable, it decreased as temperatures increased. Hayes et al. (2001) investigated the thermal inactivation kinetics of bovine cathcpsin D and also observed decreases in the enzyme activity with increasing temperatures. The authors reported a survival rate of 45% of the enzyme when subjected to processes such as heating at 55 °C for 30 min during manufacture of high-cook cheese varieties, and a survival rate of 8% when subjected to HTST pasteurization. Therefore, cathepsin D may still be active in dairy products after manufacture, albeit at low levels, and may contribute to proteolysis.

3.6 Impact on Dairy Products

The quality of dairy products made from milk with elevated somatic cell count (SCC), particularly in cases of mastitis, but also during late lactation, has long been judged to be inferior to that made from low SCC milk, and so the potential contribution of lysosomal enzymes to such changes has been a topic of interest for many years.

A primary question in such considerations is the global impact of cell enzymes on the caseins, through which effects on dairy product characteristics could be mediated. A relatively simple way to study the total impact of somatic cell proteinases on proteolysis of casein is to recover the cells (typically by centrifugation) and add them to casein, followed by incubation. An early study of this type was by Verdi and Barbano (1988), who recovered somatic cells from bovine milk and incubated them with casein and showed, by inclusion of a plasmin inhibitor (6-aminohexanoic acid), that they possessed non-plasmin proteolytic activity, which partly survived a heat treatment similar to pasteurisation. Later, Verdi and Barbano (1991) recovered cells from bovine milk (high SCC) and blood, and tested their activity against β-casein at pH 6.6 and pH 5.2; both types of cells were proteolytically active, with such activity being higher at pH 6.6 than at pH 5.2, and the milk-derived cells being more active (perhaps because of differences in cell types present).

Hachana et al. (2010), also using this approach, reported a preference of cell-associated proteinases for β-, followed by α_s- and κ-casein, when proteolysis in UHT-treated milk was examined. Considine et al. (2002) reported that purified lysosomal elastase could itself coagulate milk, while cathepsin B influenced the rheology of rennet gels and whole somatic cells did not influence rennet coagulation when added to milk. Marti-De Olives et al. (2015) reported that proteolysis levels in ovine milk increased with SCC, and also noted that rennet coagulation properties of such milk progressively deteriorated with increasing SCC.

A number of studies of the relationship between SCC and cheese-making properties and cheese quality are summarised in Table 3.3. In general, higher SCC is associated with impaired cheese yield and milk component recovery in cheese more than cheese composition or ripening, which was also broadly the conclusion of a meta-analysis by Geary et al. (2013), who also reported a trend towards higher moisture levels in cheese at higher milk SCC.

In an attempt to isolate the specific contribution of lysosomal enzymes to cheese ripening, Kelly (1999) produced starter- and rennet-free miniature cheese from milk of low and high SCC and found elevated proteolysis, not consistent with a system

Table 3.3 Studies of cheese-making properties of milk with elevated somatic cell count (SCC)

Cheese type	SCC ranges compared (cells/mL)	Principal findings	Reference
Cheddar	< 500,000 and > 500,000	High SCC milk gave cheese with higher moisture content and proteolysis, and lower flavour and body scores	Rogers and Mitchell (1994)
Colonial	< 500,000 and > 500,000	Reduced yield and increased losses in whey with higher SCC, but similar composition	Mattiello et al. (2018)
Cottage	83,000 and 872,000 (experimentally-induced mastitis)	High SCC led to higher protein losses in whey and faster proteolysis during cheese storage	Klei et al. (1998)
Minas Frescal	125,000, 437,000, 1,053,000	Poorer firmness and sensory scores for cheese made from high SCC milk, and increased loss of protein in whey.	Andreatta et al. (2009a)
Mozzarella	<200,000, 400,000, >800,000	Higher meltability in cheese made from high SCC milk, but few other differences	Andreatta et al. (2009b)
Mozzarella	< 115,000, 422,000, >987,000	Similar composition across SCC groups, but higher pH with higher SCC and greater proteolysis during ripening	Hachana et al. (2018)
Prato	>200,000 and >700,000	Lower acceptance of cheese from high SCC milk due to flavour and texture defects	Vianna et al. (2008)
Semi-soft caprine milk cheese	<500,000, 500,000 – 1,000,000, >1,000,000	No differences in yield, but lower sensory scores and poorer texture for cheese made from milk with higher SCC	Chen et al. (2010)

dominated by plasmin and cathepsin D and characterised in part by an accumulation of hydrophobic peptides, in the latter cheese.

Few studies have examined specific lysosomal enzymes in cheese or other products; detected cathepsin D and procathepsin D in Quarg cheese by immunoblotting with specific antibodies, and confirmed activity using a specific enzyme assay and the identification of peptides in cheese extracts; interestingly, it appeared that such activity at least partially survived pasteurisation of the milk (Hurley et al. 2000b). The likely contribution of cathepsin D to proteolysis in cheese was reviewed by Hurley et al. (2000a).

Some studies have evaluated the correlation of individual cell types with protease activity in milk, on the basis that different cell types have different compliments of proteases in their lysosomes; for example, Albenzio et al. (2015) studied caprine milk with levels from <700,000 cells/ml (levels being generally higher than in bovine milk) to >1,500,000 cells/ml, and reported higher levels of polymorphonuclear leucocytes (PMN) in high SCC milk, with differences in peptide profiles generated when sodium caseinate was incubated with cells with different relative proportions of PMN and macrophages. Wickstrom et al. (2009) also associated milk with higher SCC with decreased casein level and more proteolysis but, when considering PMN level specifically, did not find this to be a better predictor of proteolysis than total SCC. Albenzio et al. (2009) incubated cells recovered from ovine milk with casein and reported that PMN-associated elastase led to very significant breakdown of casein, and proposed that increased proteolysis in late lactation milk was associated with higher elastase activity in such milk. The relationship between SCC, PMN level, their associated proteinases, and milk composition was reviewed by Le Roux et al. (2003). In that review, several mechanisms by which PMN can influence proteolysis in milk were identified:

1. direct hydrolysis of caseins, through lysosomal enzymes such as collagenase, cathepsins B, C, D, G and L, elastase and proteinase 3, which either occur due to enzyme release or ingestion and digestion of milk constituents;
2. indirect relationship with plasmin activity, through activation or degradation of the extracellular matrix in the mammary gland, allowing an influx of plasmin from blood;

It is noted that several PMN proteases have acidic optima (cathepsins B, C, D and L) and may not be active in the neutral pH environment of milk, except in the immediate periphery of PMN, where more acidic conditions may be found; other neutral proteases such as elastase and cathepsin G may be released into milk through degranulation of PMN (Le Roux et al. 2003).

In recent years, proteomic and peptidomic tools have allowed more sophisticated analysis of the protein and proteolysis patterns in milk of elevated SCC; using such approaches, Guerrero et al. (2015) inferred, through analysis of cleavage sites of peptides, increased activity of cathepsin D and elastase in milk from cows with subclinical mastitis. Similarly, Khaldi et al. (2014) determined likely activity of elastase and cathepsin D in human milk, leading to production of a peptide sequence with putative antimicrobial activity. By studying the peptidomic profile of Holstein cows at peak lactation, Dallas et al. (2014) identified peptides which suggested the activity of plasmin, cathepsins B and D and elastase. Wedholm et al. (2008) identified peptide cleavage sites consistent with plasmin, cathepsins B and D and elastase in milk of different SCC, with the relative significance of plasmin decreasing as SCC increased.

The pathways of proteolysis in high SCC milk may be further complicated by changes in enzyme-inhibitor balance and changes in substrate affecting hydrolysis patterns. For example, Zhang et al. (2015) reported higher levels of protease inhibitors in high SCC bovine milk, which may suggest a protective mechanism against proteolysis in such cases. In addition, Pinto et al. (2013) suggested that dephosphorylation by alkaline phosphatase may influence the proteolysis pathways in ovine milk during mastitis.

A further complicating factor in evaluating the specific contribution of lysosomal enzymes has been that milk in which their level is elevated (i.e., high SCC milk) also typically differs from better quality milk in other ways, such as lower lactose level, altered casein/whey protein ratio, and altered mineral balance, frequently leading to elevated pH. These then act as confounding factors, making isolating the effect of cell-derived enzymes alone challenging (Li et al. 2014). The most robust approach to evaluating the specific isolated effect of the cellular enzyme complement involves recovering the cells from high SCC milk, either by centrifugation or by filtration, and adding these to low SCC milk to study hydrolysis patterns and product quality. Following this approach, Marino et al. (2005) recovered cells by centrifugation from mastitic milk, added these to low SCC milk, and made miniature cheese; cells contributed directly to proteolysis in the cheese (although this effect was reduced by pasteurisation) and also to increasing cheese moisture content.

Li et al. (2017) recovered somatic cells by microfiltration, added these to low SCC milk, and made Swiss-type cheese; cheese made from milk with added cells, even at levels of 900,000 cells/ml, had only minor increases in proteolysis, but more notable increases in lipolysis, suggesting a merit in future studies to examination of the role of lipases from somatic cells in dairy product quality. A similar approach was used by Sanchez-Machias et al. (2013) for somatic cells recovered from caprine milk and added to milk for production of miniature cheeses; they also reported increased proteolysis, at a level dependent on whether the milk plus cells was pasteurised or not, but increased lipolysis with added cells independent of pasteurization, further supporting the presence of a heat-stable lysosomal lipase.

There have been fewer reports of the effects of SCC or lysosomal enzymes on products other than cheese. Hachana et al. (2012) reported increased viscosity and decreased casein content of yoghurt made from milk as SCC increased, and noted that yoghurt made from milk with elevated SCC (1,150,000 cells/mL) had increased levels of both proteolysis and lipolysis; broadly similar findings were reported for yoghurt by Vivar-Quintana et al. (2006) and Fernandes et al. (2007).

The effect of SCC on the flavour of pasteurised milk was studied by Ma et al. (2000) and Santos et al. (2003), who reported earlier detection of sensory defects in milk with elevated SCC. Fernandes et al. (2008a, b) studied proteolysis in UHT milk made from milk with different SCC, and reported higher proteolysis during storage of milk (particularly of β- and α_{s1}-casein) which initially had higher SCC. High SCC was specifically linked to faster gelation of UHT milk by Auldist et al. (1996), and Gaucher et al. (2008) reported cleavage sites consistent with the activity of cathepsins B, D and G and elastase in UHT milk during storage. Recent studies have employed peptidomics to study peptide profiles in bovine milk products, like cheese and infant formulae (Sforza et al. 2012; Lambers et al. 2015).

3.7 Studies of Indigenous Non-plasmin Milk Proteases in Cases of Induced Mastitis

As discussed in earlier sections, much has been learned about the activities of lysosomal enzymes from the study of proteolysis in milk of naturally varying SCC. A number of studies have controlled this scenario more systematically by artificially inducing mastitis followed by milk collection and enzymatic studies. In an early study of this kind, Saeman et al. (1988) induced mastitis by infusion of an inoculum of *Streptococcus agalactiae* into bovine udders, and found that this resulted in significant increases in global proteolytic activity, not all of which could be attributed to plasmin; non-plasmin activity levels returned to pre-infection levels after the infections were cured, and appeared to thus reflect changes in milk SCC.

As another example of this approach, Michelutti et al. (2007) induced mastitis using *Staphylococcus aureus* and found significant increases in proteolysis (measured through proteose peptone content) in the early stages of infection, but not during a chronic phase (up to 28 days post-infection); the authors also noted that the response differed from that elicited by an *Escherichia coli* infection. Haddadi et al. (2006) reported significant proteolysis associated with PMN, and likely due to cathepsins, elastase and collagenase, in experimentally-induced *E. coli* mastitis, while Moussaoui et al. (2004) also reported increased levels of proteolysis in such cases, and Moussaoui et al. (2003) attributed such increases to the activity of PMN elastase. Larsen et al. (2004) induced mastitis using *Streptococcus uberis* and measured increased activity of cathepsin D and level of cathepsin B (measured by immunoblotting), leading to significant increases in levels of proteolysis in mastitic milk.

An even more controlled approach is to use isolated bacterial components to elicit a mastitis-like response, with the advantage here being the lack of other bacterial-derived enzymes and components possibly found in the milk, while the extent of increase in SCC can be as high as for whole-cell infections. Le Roux et al. (2004) induced mastitis using lipopolysaccharide from *E. coli* and reported a pattern of proteolysis largely dominated by plasmin, but with at least one new product during the peak of the infection that appeared to be associated with lysosomal enzymes.

Hinz et al. (2012) used a peptidomic approach to determine enzyme activities in milk of cows in which a mastitic reaction had been induced by infusion of lipoteichoic acid from *Staphylococcus aureus*, and determined activities of cathepsins B and D and elastase, as well as potentially amino- and carboxypeptidase activities (possibly indicating the activity of cathepsin H, not previously reported). In a subsequent study, Hinz et al. (2012) followed a similar approach using lipopolysaccharide from *E. coli* and reported significant increases in proteolysis of all caseins, and electrophoretic patterns (on one- and two-dimensional gels) and identified fragments consistent with activity of plasmin and cathepsin D.

3.8 Biological Significance

In many cases, the functions of the lysosomal or other non-plasmin proteases in milk is not clear. Initial speculations involved potential spill-over from protein synthesis by the mammary gland, but the linkage with SCC and a growing body of evidence have indicated a wide variety of possible biological functions, though it still needs more verification to settle definitely. The potential physiological significance for cathepsin D has been briefly reviewed by Hurley et al. (2000a). Since then, the understanding of the significance of the milk enzymes, including lysosomal and non-plasmin proteases, centers around the obvious functional advantage is their possible role in pre-digestion of milk proteins, which could be of benefit for the newborn baby or calf (Nielsen et al. 2017a; Nielsen et al. 2018). Milk protease activation in the infant or calf stomach could indicate that milk proteases can directly contribute to infant and calf health, development and protection (Dallas and German 2017). Furthermore, the proteases could have functions in the mammary gland. The cathepsins in milk may be involved in degradation of extracellular matrix, tissue remodelling, generation of bioactive peptides, antigen processing and presentation etc. (Uchiyama et al. 1994; Hurley et al. 2000a).

Other potential physiological roles include the possible functions of bioactive peptides generated by milk proteases and released from the milk proteins, and especially the caseins (Bhattacharya et al. 2019). These peptides can have an array of biological functions, but their activity depends on whether they reach their site of action, as well as concentrations needed for action. The information on published bioactive peptides deriving from milk protein of all species have been collected by Nielsen et al. (2017c) and made available online through the Milk Bioactive Peptide Database (MBPDB, http://mbpdb.nws.oregonstate.edu/). In this database, 713 entries of unique peptides from bovine milk proteins have been described, which was far more than another other species. However, not all of these can be assigned to a certain known enzyme activity. The predominance of bioactive peptides from bovine milk proteins are likely not due to a biological reality, but derives from the fact that bioactive peptides from bovine milk have been more comprehensively studied.

The activity of the indigenous enzyme cathepsin D has already been confirmed in bovine milk, and its potential cleavage site in milk proteins has been identified. The peptides generated by the activity of this enzyme might have biological functionality. Searching the MBPDB (Nielsen et al. 2017c) revealed that seven peptides would be released by cathepsin D activity that are identical to known bioactive peptides, two of these had been identified with multiple functions (antihypertensive, antimicrobial and immunoregulatory activity). In addition to the identical sequences, two additional peptides were identified as being highly homologous with known bioactive peptides (more than 80% sequence homology), and one of these was also found to have more than one biological function (Table 3.4).

3.9 Future Aspects and Missing Links

As outlined here, there is still much more to do and understand regarding the presence, activity, function and significance of lysosomal and other non-plasmin proteases in bovine milk. The identification of lysosomal proteases awaits further studies, as many have not been molecularly demonstrated in the milk, but remain speculative. Expression studies using somatic cells isolated from milk are interesting as

Table 3.4 Peptides deriving from cathepsin D protease activity in bovine milk with sequence similarity to known bioactive peptides

Peptides released by cathepsin D activity	Protein	Bioactive peptide	Function	Alignment %
MAIPPKKNQDKTEIPTINTIASGEPTSTPTTEAVESTV ATLEDSPEVIESPPEINTVQVTSTAV	κ-CN	MAIPPKKNQDKTEIPTINTIASGEPTSTPTTEAVESTV ATLEDSPEVIESPPEINTVQVTSTAV	Immunomodulatory	100
			Antimicrobial	100
			Antihypertensive	100
RPKHPIKHQGLPQEVLNENLLRF	α_{S1}-CN	RPKHPIKHQGLPQEVLNENLLRF	Antimicrobial	100
			Immunomodulatory	100
RPKHPIKHQGLPQEVLNENLLRFF	α_{S1}-CN	RPKHPIKHQGLPQEVLNENLLRFF	Antimicrobial	100
DAYPSGAW	α_{S1}-CN	DAYPSGAW	Antihypertensive	100
YYVPLGTQYTDAPSFSDIPNPIGSENSEKTTMPLW	α_{S1}-CN	GTQYTDAPSFSDIPNPIGSENSEKTTMPLW	Antihypertensive	86
VYQHQKAMKPWIQPKTKVIPYVRYL	α_{S2}-CN	VYQHQKAMKPWIQPKTKVIPYVRYL	Antimicrobial	100
		TVYQHQKAMKPWIQPKTKVIPYVRYL	Antimicrobial	96
		KTVYQHQKAMKPWIQPKTKVIPYVRYL	Antimicrobial	93
		LKTVYQHQKAMKPWIQPKTKVIPYVRYL	Antimicrobial	89
		VDQHQKAMKPWTQPKTNAIPYVRYL	Antimicrobial	84
		KAMKPWIQPKTKVIPYVRYL	Antimicrobial	80
KVKEAMAPKHKEMPFPKYPVEPFTESQSL	β-CN	GVSKVKEAMAPKHKEMPFPKYPVEPFTESQ	Immunomodulatory	90
			Opioid	90
YQEPVLGPVRGPF	β -CN	YQEPVLGPVRGPFPI	Antimicrobial	87
		EPVLGPVRGPFP	Antihypertensive	85
		YQEPVLGPVRG	Antihypertensive	85

they unequivocally demonstrate the expression of the proteases from the milk somatic cell pool, which contains lymphocytes, neutrophils, macrophages and exfoliated epithelial cells. Further studies are needed to show the exact molecular forms of the proteases actually present, especially pro-forms vs. active forms. This aspect is of especial importance in relation to understanding the physiological effects of the milk proteases, e.g., in relation to calf health, effects in consumers (such as infants receiving infant formulae) and implications for dairy product quality. Not least, the potential impact and benefit for the cow herself in relation to udder health is also very interesting, as bioactive peptides have been shown to be generated by the milk proteases from the somatic cell pool which have potential antimicrobial and immunoregulative properties and therefore potentially form part of a feed-back loop. A function of milk proteases in the remodeling of the udder during lactation stages and involution has also been suggested and warrants more studies.

New methods, like multiple reaction monitoring-based proteomics could be candidates for such revisiting for demonstrations and permitting of absolute quantification of milk proteases and the distributions of their various forms present at different conditions. One such warranted category of further studies is the further identifications and understanding of significances of amino- and carboxypeptidases and their eventual relation to processes like formation of aroma components and contribution to generation of free amino acids in milk and dairy products.

References

Albenzio M, Santillo A, Caroprese M, D'Angelo F, Marino R, Sevi A (2009) Role of endogenous enzymes in proteolysis of sheep milk. J Dairy Sci 92(1):79–86

Albenzio M, Santillo A, Kelly AL, Caroprese M, Marino R, Sevi A (2015) Activities of indigenous proteolytic enzymes in caprine milk of different somatic cell counts. J Dairy Sci 98:7587–7594

Andreatta E, Fernendes AM, de Santos MV, Mussarelli C, Marques MC, de Oliveira CAF (2009a) Composition, functional properties and sensory characteristics of Mozzarella cheese manufactured from different somatic cell counts in milk. Brazlian Archives Biol Technol 52:1235–1242

Andreatta E, Fernendes AM, de Santos MV, Mussarelli C, Marques MC, Gigante ML, de Oliveira CAF (2009b) Quality of minas fresal cheese manufactured from milk with different somatic cell counts. Pesquicsa Agropecuraria Brasileira 44:320–326

Athauda SBP, Takahashi T, Kageyama T, Takahashy K (1991) Auticatalytic processing of procathepsin E to cathepsin E and their structural differences. Biochem Biophys Res Commun 175:152–158

Auldist MJ, Coats ST, Sutherland BJ, Clarke PH, McDowell, GH, Rogers GL (1996) Effect of somatic cell count and stage of lactation on the quality of full cream milk powder. Aust J Dairy Technol 51:94–98

Barrett AJ, Kirschke H (1981) Cathepsin B, cathepsin H and cathepsin L. Meth Enzymol 80:535–561

Benfeldt C, Larsen LB, Rasmussen JT, Andreasen PA, Petersen TE (1995) Isolation and characterization of plasminogen and plasmin from bovine milk. Int Dairy J 5:577–592

Bohley P, Seglen PO (1992) Proteases and proteolysis in the lysosomes. Experientia 48:151–157

Bhattacharya M, Salcedo J, Robinson RC, Henrick BM, Barile D (2019) Peptidomic and glycomic profiling of commercial dairy products: identification, quantification and potential bioactivities. NPJ Sci Food 3(4):1–13

Capony F, Rougeot C, Montcourrier P, Cavailles V, Salazar G, Rochefort H (1989) Increased secretion, altered processing and glycosylation of procathepsin D in human mammary cancer cells. Cancer Res 49:3904–3909

Chen SX, Wang JZ, Van Kessel JS, Ren FZ, Zeng SS (2010) Effect of somatic cell count in goat milk on yield, sensory quality and fatty acid profile of semisoft cheese. J Dairy Sci 93:1345–1354

Considine T, Healy A, Kelly AL, McSweeney PHL (2004) Hydrolysis of bovine caseins by cathepsin B, a cysteine proteinase indigenous to milk. Int Dairy J 14:117–124

Considine T, Healy A, Kelly A, McSweeney PLH (2000) Proteolytic specificity of elastase on bovine α_{S1}-casein. Food Chem 69:19–26

Considine T, Healy A, Kelly AL, McSweeney PLH (1999) Proteolytic specificity of elastase on bovine β-casein. Food Chem 66:463–470

Considine T, Geary S, Kelly AL, McSweeney PLH (2002) Proteolytic specificity of cathepsin G on bovine αs1-and β-caseins. Food Chem 76(1):59–67

D'Alessandro A, Scaloni A, Zolla L (2010) Human milk proteins:An interactomics and updated functional overview. J Proteome Res 9(7):3339–3373

D'Alessandro A, Zolla L, Scaloni A (2011) The bovine milk proteome: cherishing, nourishing and fostering molecular complexity. An interactomics and functional overview. Mol BioSyst 7:579–597

Dahl SW, Halkier T, Lauritzen C, Dolenc I, Pedersen J, Turk V, Turk B (2001) Human recombinant pro-dipeptidyl peptidase I (cathepsin C) can be activated by cathepsins L and S but not by autocatalytic processing. Biochemist 40:1671–1678

Dallas DC, German JB (2017) Enzymes in human milk. Nestle Nutr Inst Workshop Ser 88:129–136

Dallas DC, Guerrero A, Parker EA, Garay LA, Bhandari A, Lebrilla CB, Barile D, German JB (2014). Peptidomic Profile of Milk of Holstein Cows at Peak Lactation. J Agric Food Chem 62(1):58–65

Dallas DC, Murray NM, Gan J (2015) Proteolytic systems in milk: perspectives on the evolutionary function within the mammary gland and the infant. J Mammary Gland Biol Neoplasia 20:133–147.

Demers-Mathieu V, Nielsen SD, Underwood MA, Borghese R, Dallas DC (2017) Analysis of milk from mothers who delivered prematurely reveals few changes in proteases and protease inhibitors across gestational age at birth and infant postnatal age. J Nutr 147:1152–1159

Fernandes AM, Bovo F, Moretti TS, Oosim RE, de Lima CG, Oliveira CAF (2008b) Relationship between the somatic cell count in raw milk and the casein fractions of UHT milk. Aust J Dairy Technol 63:45–49.

Fernandes AM, Moretti TS, Bovo F, de Lima CG, Oliveira CAF (2008a) Effect of somatic cell counts on lipolysis, proteolysis and apparent viscosity of UHT milk during storage. Int J Dairy Technol 61:327–332.

Fernandes AM, Oliveira CAF, Lima CG (2007) Effects of somatic cell counts in milk on physical and chemical characteristics of yoghurt. Int Dairy J 17:111–115.

Gan J, Zheng JY, Krishnakumar N, Gounatilleke E, Lebrilla CB, Barile D, German JB (2019). Selective proteolysis of alpha-lactalbumin by endogenous enzymes of human milk at acidic pH. Mol Nutr Food Res 63(18):1900259

Gaucher I, Molle D, Gagnaire V, Gaucheron F (2008) Effects of storage temperature on physico-chemical characteristics of semi-skimmed UHT milk. Food Hydrocolloids 22:130–143

Geary U, Lopez-Villalobos N, O'Brien B, Garrick DJ, Shalloo L (2013) Meta-analysis to investigate relationships between somatic cell count and raw milk composition, Cheddar cheese processing characteristics and cheese composition. Irish J Ag Food Res 52:119–133

Gieselmann V, Hasilik A, von Figura K (1985). Processing of human cathepsin D in lysosomes in vitro. J Biol Chem. 260(5):3215–3220

Grieve A, Kitchen BJ (1985) Proteolysis in milk: the significance of proteinases from milk leucocytes and a comparison of the action of leucocytes, bacterial and natural milk proteinases on caseins. J Dairy Res 52:101–112

Guerrero A, Dallas DC, Contreras S, Bhandari A, Canovas A, Islas-Trejo A, Medrano JF, Parker EA, Wang M, Hettinga K, Chee S, German JB, Barile D, Lebrilla CB (2015) Peptidomic analysis of healthy and subclinically mastitic bovine milk. Int Dairy J 46:46–52

Hachana Y, Kraiem K, Paape M (2010). Effect of plasmin, milk somatic cells and psychrotrophic bacteria on casein fractions of ultra high temperature treated milk. Food Sci Technol Res 16(1):79–86

Haddadi K, Prin-Mathhieu C, Moissaoui F, Faure GC, Vangroenweghe H, Burvenich C, Le Roux Y (2006) Polymorphonuclear neutrophils and Escherichia coli proteases involved in proteolysis of casein during experimental E. coli mastitus. Int Dairy J 16:639–647

Hachana Y, Paape M (2012) Physical and chemical characteristics of yoghurt produced from whole milk with different levels of somatic cell counts. Int J Food Sci Nutr 63:303–309

Hachana Y, Znaiai A, M'Hadami N (2018) Effect of somatic cell count on milk composition and Mozzarella cheese quality. Acta Alimentaria 47:88–96.

Hayes MC, Hurley MJ, Larsen LL, Heegard CW, Magboul AAA, Oliveira JC, McSweeney PLH, Kelly AL (2001) Thermal inactivation kinetics of bovine cathepsin D. J Dairy Res 68:267–276

Heegaard CW, Christensen T, Rasmussen MD, Benfeldt C, Jensen NE, Sejrsen K, Petersen TE, Andreasen PA (1994a) Plasminogen activators in bovine milk during mastitis, an inflammatory disease. Fibrinolysis 8:22–30

Heegaard CW, Rasmussen LK, Andreasen PA (1994b) The plasminogen activation system in bovine milk: differential localization of tissue-type plasminogen activator and urokinase in milk fractions is caused by binding to casein and urokinase receptor. Biochim Biophys Acta 1222:45–55

Hill RD, Lahav E, Givol D (1974) A rennin-sensitive bond in α_{S1} B-casein. J Dairy Res 41:147–153

Hinz K, Larsen LB, Wellnitz O, Bruckmaier RM, Kelly AL (2012). Proteolytic and proteomic changes in milk at quarter level following infusion with *E. coli* lipopolysaccharide. J Dairy Sci 95: 1655–1666

Hurley MJ, Larsen LB, Kelly AL, McSweeney PLH (2000a) The milk acid proteinase: a review. Int Dairy J 10:673–681

Hurley MJ, Larsen LB, Kelly AL, McSweeney PLH (2000b) Cathepsin D activity in quarg. Int Dairy J 10:453–458

Isidoh K, Kominami E (1994) Multi-step processing of procathepsin L in vitro. FEBS Lett 352:281–284

Isidoh K, Kominami E (2002) Processing and activation of lysosomal proteinases. Biol Chem 383:1827–1831

Kaminogawa S, Yamauchi K (1972) Acid protease of bovine milk. Agric Biol Chem 36:2351–2356

Kaminogawa S, Yamauchi K, Miyazawa S, Koga Y (1980) Degradation of casein components by acid protease of bovine milk. J Dairy Sci 63:701–704

Kelly AL (1999). Effect of plasmin and somatic cell enzymes on proteolysis in aseptic starter and rennet free model cheeses. Milchwissenschaft 54:249–252

Kelly AL, Foley J (1997) Proteolysis and storage stability of UHT milk as influenced by milk plasmin activity, plasmin/β-lactoglobulin complexation, plasminogen activation and somatic cell count. Int Dairy J 7(6–7):411–420

Kelly AL, O'Flaherty F, Fox PF (2006) Indigenous proteolytic enzymes in milk: a brief overview of the present state of knowledge. Int Dairy J 16:563–572

Khaldi N, Vijayakumar V, Dallas DC, Guerrero A, Wickramasinghe S, Smilowitz JT, Medrano J F, Lebrilla CB, Shields DC, German JB (2014) Predicting the important enzymes in human breast milk digestion. J Agric Food Chem 62(29):7225–7232

Kirschke H, Barrett AJ, Rawlings ND (1998) Lysosomal cysteine proteinases, 2nd edn. Oxford University Press, Oxford

Klei L, Yun J, Sapru A, Lynch J, Barbano D, Sears P, Galton D (1998) Effects of milk somatic cell count on cottage cheese yield and quality. J Dairy Sci 81:1205–1213

Klein J, Eales J, Zurbig P, Vlahou A, Mischak H, Stevens R (2013) Proteasix: a tool for automated and large-scale prediction of proteases involved in naturally occurring peptide generation. Proteomics 13:1077–1082

Lambers TT, Gloerich J, van Hoffen E, Alkema W, Hondmann DH, van Tol EA (2015) Clustering analyses in peptidomics revealed that peptide profiles of infant formulae are descriptive. Food Sci Nutr 3(1):81–90

Larsen LB, Benfeldt C, Rasmussen LK, Petersen TE (1996) Bovine milk procathepsin D and cathepsin D: coagulation and milk protein degradation. J Dairy Res 63:119–130

Larsen LB, Boisen A, Peterson TE (1993) Procathepsin D cannot autoactivate to cathepsin D at acid pH. FEBS Lett 319:54–58

Larsen LB, Hinz K, Jørgensen ALW, Møller HS, Wellnitz O, Bruckmaier RM and Kelly AL (2010) Proteomic and Peptidomic Study of Proteolysis in Quarter Milk after Infusion with Lipoteichoic Acid from Staphylococcus aureus. J Dairy Sci 93:5613–5626

Larsen LB, McSweeney PLH, Hayes MG, Andersen JB, Ingvartsen KL, Kelly AL (2006) Variation in activity and heterogeneity of bovine milk proteases with stage of lactation and somatic cell count. Int Dairy J 16:1–8

Larsen LB, Rasmussen MD, Bjerring M, Nielsen JH (2004) Proteases and protein degradation in milk from cows infected with Stretococcus uberis. Int Dairy J 14:899–907

Larsen LB, Petersen TE (1995) Identification of five molecular forms of cathepsin D in bovine milk. In: Takahashi (ed) Aspartic proteinases: structure, function and biomedical implications. Plenum Press, New York, pp 279–283

Larsen LB, Wium H, Benfeldt C, Heegaard CW, Ardo Y, Qvist KB, Petersen TE (2000) Bovine milk procathepsin D: presence and activity in heated milk and in extracts of rennet-free UF-feta cheese. Int Dairy J 10:67–73

Le Roux Y, Laurent F, Moussaoui F, Le Roux Y (2003) Polymorphonuclear proteolytic activity and milk compositional change. Vet Res 34(5):629–645

Lewis UJ, Williams DE, Brink NG (1956) Pancreatic elastase: purification, properties, and function. J Biol Chem 222:705–720

Li N, Richoux R, Leconte N, Bevilacqua C, Maillard MB, Parayre S, Aubert-Frogerais L, Warlouzel J, Moya-Leclair E, Denis Cm Martin P, Gagnaire V (2017) Somatic cell recovery by microfiltration technologies: a novel strategy to study the actual impact of somatic cells on cheese matrix. Int Dairy J 65:5–13.

Li N, Richoux R, Boutinaud M, Martin P (2014) Role of somatic cells on dairy processes and products: a review. Dairy Sci Technol 34:517–538

Ma Y, Ryan C, Barbano DM, Galton DM, Rudan MA, Boor KJ (2000) Effects of somatic cell count on quality and shelf life of pasteurized fluid milk. J Dairy Sci 83:254–274.

McSweeney PLH, Fox PF, Olson NF (1995) Proteolysis of bovine caseins by cathepsin D: preliminary observations and comparison with chymosin. Int Dairy J 5:321–336

Magboul AAA, Larsen LB, McSweeney PLH, Kelly AL (2001) Cysteine protease activity in bovine milk. Int Dairy J 11:865–872

Marino R, Considine T, Sevi A, McSweeney PLH, Kelly AL (2005) Contribution of proteolytic activity associated with somatic cells in milk to cheese ripening. Int Dairy J 15:1026–1033.

Martí-De Olives A, Navarro-Ríos MJ, Rubert-Alemán J, Fernández N, Molina MP (2015) Composition, proteolysis indices and coagulating properties of ewe milk as affected by bulk tank somatic cell count. J Dairy Res. 82(3):344–349

Mattiello CA, Silveira SM, Carli F, Cunha A, Alessio DRM, Pelizza A, Cardozo LL, Neto AT (2018) Industrial yield, manufacturing efficiency and physical and chemical characteristics of colonial cheese produced from milk with two levels of somatic cells. A`rquivo Brasileiro De Dedicine Veterinaria E Zootecnia 70:1916–1924

Molinari CE, Casadio YS, Hartmann BT, Livk A, Bringans S, Arthur PG, Hartmann PE (2012) Proteome mapping of human skim milk proteins in term and preterm milk. J Proteome Res 11:1696–1714

Moussaoui F, Vangroenweghe H, Haddadi K, Le Roux Y, Laurent F, Duchateau L, Burvenich C (2004) Proteolysis in milk during experimental Escherichia coli mastitis. J Dairy Sci 87:2923–2931

Moussaoui F, Laurent F, Girardet JM, Humbert G, Gaillard JL, Le Roux Y (2003) Characterization and proteolytic origins of specific peptides appearing during lipopolysaccharide experimental mastitis. J Dairy Sci 86:1163–1170

Michelutti I, Haddadi K, Le Roux Y (2007) Staphylococcus aureus related mammary infection in cows: correlation between somatic cell count and proteolysis during early and chronic phase of lactation. J Animal Feed Sci 16:117–129

Naughton AS, Sanger F (1961) Purification and specificity of pancreatic elastase. Biochem J 78:156–163

Nielsen SD, Beverly RL, Dallas DC (2017a) Milk proteins are predigested within the human mammary gland. J Mammary Gland Biol Neoplasia 22:251–261

Nielsen SD, Beverly RL, Dallas DC (2017b) Peptides released from foremilk and hindmilk proteins by breast milk proteases are highly similar. Front Nutr 4.

Nielsen SD, Beverly RL, Qu YY, Dallas DC (2017c) Milk bioactive peptide database: a comprehensive database of milk protein-derived bioactive peptides and novel visualization. Food Chem 232:673–682

Nielsen SD, Beverly RL, Underwood MA, Dallas DC (2018) Release of functional peptides from mother's milk and fortifier proteins in the premature infant stomach. PLoS One 13:e0208204

O'Driscoll BM, Rattray FP, McSweeney PLH, Kelly AL (1999) Protease activities in raw milk using a synthetic heptapeptide substrate. J Food Sci 64:606–611

Pinto G, Caira S, Nicolai MA, Mauriello R, Cuollo M, Pirisi A, Piredda G, Chianese L, Addeo F (2013) Proteolysis and partial dephosphorylation of casein are affected by high somatic cell counts in sheep milk. Food Res Int 53:510–521.

Peeters G, Verbeke R, Houvenaghel A, Reynaert R (1976) Isolation of kallikrein from mammary gland of cows. Q J Exp Physiol Cogn Med Sci 61:1–14

Rogers SA, Mitchell, GE (1994) The relationship between somatic cell count, composition and manufacturing properties of bulk milk. 69. Cheddar cheese and skim milk yogurt. Aust J Dairy Technol 49:70–74

Recklies AD, Mort JS (1985) Rat mammary gland in culture secretes a stable high molecular weight for of cathepsin L. Biochem Biophys Res Commun 131(1):402–407

Reimerdes EH, Petersen F, Kielwein G (1979) Milchproteinasen 9. Proteinasespektren von Caseinmicellen, milchserum, Rinderblutserum und Pseudomonas flourescens. Milchwissenschaft 34:548–551

Saeman AI, Verdi RJ, Galton DM, Barbano DM (1988) Effect of mastitis on proteolytic activity in bovine milk. J Dairy Sci 71:505–512

Santos MV, Ma Y, Barbano DM (2003) Effect of somatic cell count on proteolysis and lipolysis in pasteurized fluid milk during shelf-life storage. J Dairy Sci 86:2491–2503.

Santillo A, Kelly AL, Palermo C, Sevia A, Albenzio M (2009) Role of indigenous enzymes in proteolysis of casein in caprine milk. Int Dairy J 19(11):655–660

Sanchez-Machias D, Morales-De la Nunez A,Torres A, Hernandez-Castellano LE, Jiminez-Flores R, Castro N, Argguello A (2013) Effects of addition of somatic cells to caprine milk on cheese quality. Int Dairy J 29:61–67.

Somers JM, O'Brien B, Meaney WJ, Kelly AL (2003) Heterogeneity of proteolytic enzyme activities in milk samples of different somatic cell count. J Dairy Res 70(45–5057):153–171

Stepaniak L (2004) Dairy enzymology. Int J Dairy Technol 57:153–171

Sforza S, Cavatorta V, Lambertini F, Galaverna G, Dossena A, Marchelli R (2012) Cheese peptidomics: a detailed study on the evolution of the oligopeptide fraction in Parmigiano-Reggiano cheese from curd to 24 months of aging. J Dairy Sci 95(7):3514–3526

Suzuki J, Katoh N (1990) Cysteine protease in bovine milk capable of hydrolysing casein as the substrate and elevation of the activity during the course of mastitis. Jap J Vet Sci 52:947–954

Talukder M, Ahmed HMM (2017) Effect of somatic cell count on dairy products: a review. Asian J Med Biol Res 3(1):1–9

Travis J, Fritz H (1991) Potential problems in designing elastase inhibitors for therapy. Am Rev Respir Dis 143:1412–1415

Uchiyama Y, Waguri S, Sato N, Watanabe T, Ishido K, Komonami E (1994) Review: cell and tissue distribution of lysosomal cysteine proteinases, cathepsins B, H and L, and their biological roles. Acta Histochem Cytochem 27(4):287–308

Verdi DJ, Barbano RM (1988) Preliminary Investigation of the Properties of Somatic Cell Proteases. J Dairy Sci 71(2):534–538

Verdi RJ, Barbano D (1991) Properties of proteases from milk somatic cells and blood leukocytes. J Dairy Sci 74:2077–2081

Vivar-Quintana AM, De La Mano EB, Revilla I (2006) Relationship between somatic cell counts and the properties of yoghurt made from ewes' milk. Int Dairy J 16:262–267.

Vijayakumar V, Guerrero AN, Davey N, Lebrilla CB, Shields DC, Khaldi N (2012) Enzymepredictor: a tool for predicting and visualizing enzymatic cleavages of digested proteins. J Proteome Res 11:6056–6065

Vianna PCB, Mazal G, Santos MV, Bolini HMA, Gigante ML (2008) Microbial and sensory changes throughout the ripening of Prato cheese made from milk with different levels of somatic cells. J Dairy Sci 91:1743–1750

Watorek W, van Halbeek H, Travis J (1993) The isoforms of human neutrophil elastase and cathepsin G differ in their carbohydrate side chain structures. Biol Chem 374:383–393

Wedholm A, Møller HS, Lindmark-Månsson H, Rasmussen MD, Andrén A, Larsen LB (2008). Identification of peptides in milk as a result of proteolysis at different levels of somatic cell counts using LC MALDI MS/MS detection. J Dairy Res. 75(1):76–83

Wickramasinghe S, Rincon G, Islas-Trejo A, Medrano JF (2012). Transcriptional profiling of bovine milk using RNA sequencing. BMC Genomics 13:45–59

Wickström E, Persson-Waller K, Lindmark-Månsson H, Ostensson K, Sternesjö A (2009) Relationship between somatic cell count, polymorphonuclear leucocyte count and quality parameters in bovine bulk tank milk. J Dairy Res. 76(2):195–201

Wiederanders B, Brömme D, Kirschke H, von Figura KV, Schmidt B, Peters C (1992) Phylogenitic conservation of cysteine proteinases. J Biol Chem 267(19):13708–13713

Wittlin S, Rösel J, Hofman F, Stover DR (1999) Mechanisms and kinetics of procathepsin D activation. Eur J Biochem 265:384–393

Yonezawa S, Tanaka T, Miyauche T (1987) Cathepsin E from rat neutrophils: its properties and possible relations to cathepsin D-like and cathepsin E-like acid proteinases. Arch Biochem Biophys 256:499–508

Zachos T, Politis I, Gorewit R, Barbano D (1992) Effect of mastitis on plasminogen activator activity of milk somatic cells. J Dairy Res 59(4):461–467

Zeece MG, Woods TL, Keen MA, Reville WJ (1992) Role of proteinases and inhibitors in postmortem muscle protein degradation. In: Proceedings of the 45th Reciprocal Meat Conference, Colorado. National Liverstock and Meat Board, Chicago, pp 51–61

Zeitlin IJ, Eshraghi HR (2002) The release and vascular action of bradykinin in the isolated perfused bovine udder. J Physiol 543:221–231

Zhang L, Boeren S, van Hooijdonk ACM; Vervoort, JM; Hettinga, KA (2015) A proteomic perspective on the changes in milk proteins due to high somatic cell count. J Dairy Sci 98(8):5339–5351

Chapter 4
Phosphatases in Milk

Nana Y. Farkye

4.1 Introduction

The presence of a phosphatase activity in milk – first described as a phosphorus compound (Kay 1925) – which was later identified as alkaline phosphatase – was first reported by Graham and Kay (1933). Bovine milk contains two principal indigenous phosphatases, namely alkaline and acid phosphatases, as well as other phosphatases such as ribonuclease. Only milk alkaline phosphatase (ALP) (EC 3.1.3.1) and milk acid phosphatase (ACP) (EC 3.1.3.2) are reviewed here, because other phosphatases present in milk have not been shown to have any technological significance in milk or milk products.

4.2 Alkaline Phosphatase

ALP occurs in all mammalian milks, at levels that vary considerably between species. The average ALP activities (measured by colorimetric assay and expressed as micrograms of phenol liberated per unit time from disodium phenlyphosphate by 1 mL of milk [ISO 3356|IDF 63; 2009]) in raw caprine, bovine and ovine milk were reported to be 165, 1562 and 3512 μg mL^{-1} phenol, respectively (Klotz et al. 2008) with corresponding detection limits of 0.485, 0.051 and 0.023%, respectively (Klotz et al. 2008). Using a fluorometric assay (ISO 1186-1; 2006), in which ALP activity is expressed in milliunits per liter (mU L^{-1}) of sample, Lorenzen et al. (2010) reported highest mean ALP activity in ovine milk (1414 U L^{-1} > bovine milk

N. Y. Farkye (✉)
Animal Science Department, California Polytechnic State University,
San Luis Obispo, CA, USA
e-mail: nfarkye@calpoly.edu

A. L. Kelly, L. B. Larsen (eds.), *Agents of Change*, Food Engineering Series,
https://doi.org/10.1007/978-3-030-55482-8_4

(774 U L^{-1}) > caprine milk (67 U L^{-1}). Also, ALP activity varies seasonally and among individual species. ALP activity is very high in colostrum, falls to a minimum within 1–2 weeks after parturition and reaches a constant level after about 25 weeks of lactation. Chávarri et al. (1998) reported that ALP activity in raw commercial ovine milk increased steadily to >twofold between January (1.7 U mL^{-1}) and June (3.75 U mL^{-1}).

ALP is a glycoprotein containing sialic acid. Electrophoresis of the isolated ALP from bovine milk showed three isozymes (α, β and γ). The α-form is predominant in skim milk; the β-form is found mostly in MFGM, while the γ-form is present in both skim milk and MFGM (Peereboom 1970).

Its activity is used as an index of the efficiency of adequate pasteurization of milk because ALP is slightly more resistant to heat, and conditions more severe than the temperature and time required to inactivate *Mycobacterium avium* subsp. *paratuberculosis*, which was the most heat-stable pathogen known in milk at the time of introduction of pasteurization in the dairy industry, although *Coxiella burnetti* is the current target organism (Cerf and Condron 2006). The ALP test is of great significance to public health as a means of checking the thoroughness of pasteurization of milk or the addition or mixing of raw milk to pasteurized milk or milk products.

4.3 Origin, Isolation and Characterization of ALP in Milk

ALP is found in varying amounts in many mammalian tissues and secretions. The best characterized mammal ALPs are of intestinal or placental origin, because these organs are rich sources of the enzyme. Indigenous milk ALP does not belong to the intestinal or placental phosphatase groups but is identical to the enzyme present in the mammary gland (Folley and Kay 1936).

In bovine milk, ALP is distributed between cream (~45%) and skim milk (~55%). Of the amount in skim milk phase, ~70% is associated with whey membrane particles (Silanikove and Shapiro 2007). In cream, ALP is concentrated in the milk fat globule membrane (MFGM) where it is associated with microsomes (which were identified as phospholipid particles (Morton 1953) and later as lipoprotein vesicles or extracellular vesicles (Patton and Keenan 1975) which are currently known as extracellular vesicles or exosomes), i.e., membrane vesicles with diameter 40–100 nm (Witwer and Théry 2019).

Because the MFGM is released into buttermilk on phase inversion during butter making, buttermilk is most frequently used as starting material for the purification of ALP. Upadhyay and Verma (2015) described a three-step approach to purification of ALP from raw milk. Similar approaches have been reported by others (see review by Fox and Kelly 2006). Buttermilk is treated with n-butanol (29% v/v) to break the lipoprotein complex of the milk microsomes to release ALP. A single-step extraction using 8% (v/v) n-butanol at room temperature extracts >90% of the ALP activity from bovine MFGM. The K_M values and substrate specificity of ALP do not change following extraction with 8% (v/v) n-butanol. The solubilized enzyme is purified by pH adjustment and repeated precipitation with acetone or ammonium sulphate, followed by the removal of low molecular weight impurities by treatment with activated charcoal. A combination of gel filtration on Sephadex G-200 and chromatography on DEAE cellulose is then used to further purify ALP, resulting in a yield of 28% and 7440-fold purification. The combination of gel filtration and ion-exchange chromatography is most frequently used to purify ALP after butanol extraction. Also, ALP may be purified by affinity chromatography on Concanavalin A-Sepharose 4B. Crude ALP can be obtained from skim milk by electroelution from native polyacrylamide gels or by sequential chromatography on Macro-Prep High Q, Sephacryl S-200 and Concanavalin A agarose. Electroelution is a relatively fast process that results in a higher yield (23.4%) than column chromatography (1.6%). Most recently, Farzi-Khajeh et al. (2019) reported about 14-fold purification efficiency of ALP using modified magnetic Fe_3O_4 nanoparticles (magnetic nanocarrier) technology in which sodium hexametaphosphate affinity ligand is attached to a nanocarrier with an18-atom linker for improved separation of ALP.

ALP may also be released from phospholipids by treating milk with phosphatidyl inositol-specific phospholipase C (Low and Finean 1977) because it is bound to microsomal membranes of the mammary gland and MFGM by linkage to phosphatidyl inositol.

ALP is a phosphomonoesterase with optimum activity in the pH range 9–10.5, depending on the type of buffer used in the assay mixture. Lorient and Linden (1976) found that the pH of optimum activity of bovine milk ALP depended on the complexity of substrate, being 10.1 for p-nitrophenylphosphate, 8.6 for phosphoserine, 8.0 for phosvitin and 6.8 for casein. ALP catalyzes the reaction:

$$\text{orthophosphoric monoester} + H_2O \rightarrow \text{an alcohol} + \text{orthophosphate}$$

ALP is thought to be like inorganic pyrophosphatase (EC 3.6.1.1), which dephosphorylates pyrophosphate according to the equation:

$$\text{Pyrophosphate} + H_2O \rightarrow 2\text{orthophosphates}$$

ALP is stable between pH 7.0 and 10.0 at 24 h × 5 °C. However, at 24 h × 33 °C, ALP was most stable at 8.0–9.0 (Kuzuya et al. 1980).

Thermal inactivation kinetics of ALP in raw bovine milk show D_{60}-values of 19.47, 23.0 and 27.2 min and z-values of 4.9 and 6.7 °C (Andrews 1992; Claeys et al. 2002; Levieux et al. 2007) whereas the calculated z-value for ALP in raw equine milk was 5.31 °C (Marchand et al. 2009).

The molecular mass of ALP human, bovine and ovine milks are very similar, i.e., 160 kDA, 190 kDa and 160 KDa, respectively. Bovine milk ALP is a dimer of two identical 85-kDa subunits (Linden and Alias 1978). There is no 3D model for milk ALP (Dumitrașcu et al. 2015). However, ALP isolated from *E.coli* consists of 449 amino acid residues (Bradshaw et al. 1981) and has two intramolecular disulfide bonds (Cys168-Cys178) and (Cys286-Cys336) per monomer (Michio et al. 1997).

Each catalytic site of enzyme contains three divalent ions, i.e., two Zn^{2+} and one Mg^{2+}, which are necessary for enzymatic activity. ALP is strongly but reversibly inhibited by divalent ion (e.g., Ca^{2+}, Be^{2+} and Cu^{2+}) and chelators (e.g., EDTA); it is activated by Zn^{2+}, Mg^{2+}, Mn^{2+}, Co^{2+}, Cd^{2+} and Ni^{2+} (Linden and Alias 1978; Kuzuya et al. 1980). Activation of the apoenzyme may be used as a sensitive assay for available zinc in foods. Inorganic orthophosphates are strong competitive inhibitors of ALP using *para*-nitrophenylphosphate as substrate (Kuzuya et al. 1980). Veterinary drugs have mixed effect on inhibition of ALP (Bilen et al. 2015). Known uncompetitive inhibitors of ALP are L-Homoarginine (Lin and Fishman 1972) and Levamisole (Khodaparast-Sharifi and Snow 1989).

ALP dephosphorylates phosphoproteins, including casein in buffer—maximally at pH 6.5–7.0 (Lorient and Linden 1976; Jasinska et al. 1985); however, it does not dephosphorylate caseins in milk, probably due to inhibition of the enzyme by the high level of orthophosphate in milk. β-Lactoglobulin has a minor inhibitory effect on ALP (Jasinska et al. 1985) which is probably of no real technological importance.

4.4 Assay Methods for ALP Activity

Biochemical assays used to detect ALP are older colorimetric or newer fluorometric or chemiluminescent substrates that are quenched by a covalently attached organophosphate molecule. The unit of measurement for the colorimetric assays is μg of phenol per mL of milk, reflecting the amount of phenol reactant organically extracted from the ortho-phenyl phosphate substrate. Colorimetric assays and the phenol measurement method for ALP to indicate pasteurization effectiveness were adopted as public health standards for pasteurization in the 1950s through 1980s. The colorimetric methods have a limit of detection of approximately 0.05–0.2% residual or contaminated raw bovine milk in pasteurized milk.

The inorganic phosphate content of raw milk increases during storage due to enzymatic hydrolysis of the phosphoric esters by phosphatase. Therefore, most of the tests for phosphatase in milk are based on the principle that ALP will hydrolyze monophosphate esters at an appropriate temperature and pH, thereby liberating compounds that can be detected by color development. The intensity of color developed is proportional to the activity of the enzyme. The usual substrates are phenyl phosphate, para-nitrophenyl phosphate or phenolphthalein phosphate, which are hydrolyzed to inorganic phosphate and phenol, para-nitrophenol or phenolphthalein, respectively. The release of inorganic phosphate may be assayed, but the secondary products are usually determined. Phenol is colorless but forms a colored complex on reaction with one of several reagents, e.g., 2,6-dibromo- or dichloroquinonechloroimide, with which it forms a blue complex. Para-Nitrophenol is yellow and phenolphthalein is pink at the alkaline assay pH of 10 and hence they are easily quantified.

In the fluorometric or chemiluminescent methods, ALP hydrolyzes the organophosphate from the substrate and produces a photoactivated product that is detected by instruments. The principle involved in both assays is use of an enzyme photoactivated substrate (EPAS). The unit of measurement in EPAS assays is milliunits (mU) of enzyme activity per L milk, where one mU is defined as the amount of enzyme that catalyzes 1 ng of specific substrate hydrolyzed per min per L solution.

A quantitative assay for measuring residual ALP activity in dairy products using the FLM 300 Fluorophos ALP Test System based on mixing a sample of the milk product with an aromatic monophosphoric ester and measuring the resulting fluorescence (excitation at 440 nm, emission at >505 nm). The method has been adopted by IDF and AOAC (see below). The assay claims to detect up to 0.003% (v/v) raw milk contamination (Rocco 1990; Advanced Instruments 2019). A modified chemiluminescent method for ALP assay with detection limit of 20 to 50 mU L^{-1} has been reported (Albillos et al. 2011). A potentiometric method for assaying both acid and alkaline phosphatases in milk has been described (Hassan et al. 2009).

There are three approved methods for ALP assay described by the Food and Drug Administration (FDA), as follows:

1. the fluorescent Fluorophos method;
2. the chemiluminescent method Paslite (Charm Sciences) using a NovaLUM with temperature compensation (Chemi-Lum), and;
3. the Paslite chemiluminescent method using a Charm II scintillation analyzer (Chemi-6600).

The International Dairy Federation (IDF) also has three standard methods for determining ALP activity as follows:

1. Standard 63 (2009), using sodium phenylphosphate as substrate and quantitative determination by the `indophenol-reaction';
2. Standard 209 (2007), using a chemiluminescent (EPAS) method;
3. IDF Standard 155–1 and 2 (2013, 2016), which specify a fluorometric method for determination of ALP activity in milk and milk-based drinks and cheese.

The Association of Official Analytical Chemists (AOAC) specifies two methods for determination of ALP activity as follows:

1. AOAC Method 972.17 for residual phosphatase in milk, where the milk is incubated with phenolphthalein monophosphate in carbonate buffer. Free phenolphthalein liberated by residual ALP activity is measured directly using a spectrophotometer.
2. AOAC Method 991.24, in which ALP activity in fluid dairy products (whole milk, skim milk and chocolate milk) is measured by a continuous fluorometric direct kinetic assay. A non-fluorescent aromatic monophosphoric ester substrate is hydrolyzed by ALP to give a highly fluorescent product.

Chemical principles involved in the detection and measurement of ALP activity are the same for all dairy products, but different dairy products require modifications of the method used because of their different physical properties, composition, and particularly buffering capacity.

Klotz et al. (2008) reported that detection limit of ALP in bovine milk, using the official ALP methods in Canada (colorimetric assay; MFO-3) and in the United States (Fluorophos), were 0.8 μg mL^{-1} phenol for the MFO-3 method and 43 mU L^{-1} for the Fluorophos method. They also reported that the limit of quantitation was 2.02 μg mL^{-1} phenol for the MFO-3 method and 85 mU L^{-1} for the Fluorophos method. Klotz et al. (2008) concluded that, although both methods can be used to determine ALP activity in bovine, caprine and ovine milk, the Fluorophos assay was superior to the colorimetric method because it is more sensitive and requires less time to complete analysis.

4.5 Significance of ALP in Dairy Products

ALP is an important indigenous milk enzyme because of its technological significance. The application of ALP test in milk and milk products has been reviewed by Rankin et al. (2010). ALP activity of raw bovine and caprine milks is variable and is generally higher (~600,000–1,000,000 mU L^{-1}) in raw bovine milk than raw caprine milk (~20,000 mU L^{-1}) (Rola and Sosnowski 2010). The assay of milk ALP activity is used as an indicator for adequate pasteurization of dairy products (Sanders 1949) although the enzyme may not be completely inactivated by HTST (72 °C × 15 s) pasteurization (Linden and Alias 1978). Fadiloglu et al. (2004) reported complete inactivation of ALP in pasteurized milk heated to 70 and 80 °C but 28 and 15% of ALP activity present when milk was heated to 50 and 60 °C, respectively for 120 min. Inactivation time for raw milk was reduced nearly 18-fold by increasing temperature from 50 to 70 °C.

The U.S. and EU public health authorities limit for ALP in pasteurized dairy beverages is 350 mU L^{-1}. Fluid milk and milk products that have ALP activity below the legal limit (i.e., <1 mg mL^{-1} phenol in pasteurized milk by the Schaar rapid method, <350 mU L^{-1} for fluid milk products, and < 500 mU L^{-1} for other products by the fluorometric or equivalent method) are deemed adequately pasteurized and safe for human consumption.

Fifteen laboratories from eight countries (Salter and Fitchen 2006) measured ALP in cow, goat, sheep, and buffalo milk, skim milk, 20% fat cream, and 2% fat chocolate milk using the chemiluminescence method (Charm PasLite). At ALP levels of 350 and 500 mU L^{-1}, the average relative standard deviation for repeatability (RSDr) was 7.5%, and the average relative standard deviation of reproducibility was (RSDr) 15%. For ALP at 100 and 50 mU L^{-1}, the average RSDr values were 10.5 and 12.6%, respectively, and the average RSDr values were 18 and 25%, respectively. The authors reported the limit of detection as 20 mU L^{-1}.

Egger et al. (2016) conducted a study of ALP activity in ~150 commercial cheese samples from France, Italy and Switzerland classified into three different groups (soft, semi-hard and hard cheese) and milk thermal treatment (unheated, thermized, pasteurized), and proposed a limit for ALP activity of 10 mU g^{-1} for pasteurized milk cheeses.

Pellegrino and Rosi (2008) reported that extra-hard cheeses had ALP activity of 4 mU g^{-1}, semi-hard cheeses at 0.8 mU g^{-1}, and pressed cheeses at 0.6–0.9 mU g^{-1}. In French soft cheese (Camembert), mean ALP activities were 2.83 mU g^{-1}, 729 mU g^{-1}, 930 mU g^{-1}, respectively, for cheese made from pasteurized, thermized milk and microfiltered milk, confirming the effects of heat treatment of milk on ALP activity. ALP activity in raw bovine milk cheeses ranged 1900–2200 mU g^{-1} (Desbourdes et al. 2008). Mean ALP activities in Brie from pasteurized and thermized milk were, respectively, 2.93 and 83 mU g^{-1} (Desbourdes et al. 2008). Other values reported for ALP activity in cheeses are: ~ 1.45 mU g^{-1} for some soft cheeses, 0.83 mU g^{-1} for cottage cheese, and 0.39 mU g^{-1} for mozzarella made from bovine milk. Mean ALP activity for cottage cheese from caprine milk was 1.75 mU g^{-1} and 2.55 mU g^{-1} for hard cheeses from caprine milk (Rola and Sosnowski 2010; Sosnowski et al. 2016).

Shakeel-ur-Rehman et al. (2006) added ALP to pasteurized milk for the manufacture of Cheddar-type cheese and reported differences in HPLC peptide patterns in the water-soluble nitrogen of ripened cheeses made with and without exogenous ALP, suggesting that the presence of ALP in cheese may play a role in ripening.

Mistry (1989) added equal quantities of ALP as raw milk to pasteurized skim milk and UF skim milk retentates (10.01 and 16.61% protein) and measured residual ALP activity in both milks after heating to 60 °C for 20, 30, and 40 min and to 63 °C for 10, 20, and 30 min. Heat inactivation of ALP was more rapid in the 16.61% protein retentate than in non-UF skim milk and also in 10.01% protein retentate, at both 60 and 63 °C and for all holding periods. Addition of lactose to 16.61% protein retentate increased the heat resistance of ALP, suggesting that lactose removal during UF alters the heat inactivation characteristics of ALP, and thereby raising questions about the effectiveness of using residual ALP activity to measure the efficiency of pasteurization of highly concentrated UF milk.

FDA guidelines for screening cheeses for residual phosphatase activity (Ziobro 2001) stipulate that cheeses made from bovine milk may be considered adulterated when ALP levels are:

(a) Higher than 5 μg/0.25 g (20 μg phenol equivalents per g cheese) in one or more subsamples for brick cheese, semi-soft (21 CFR 133.187) and semi-soft part-skim cheeses;
(b) Higher than 4 μg/0.25 g (16 μg phenol equivalents per g) in one or more subsamples for Limburger cheese;
(c) Higher than 3 μg/0.25 g (12 μg phenol equivalents per g) in one or more subsamples for all other cheeses or;
(d) For dairy products other than cheese, ALP levels higher than or equal to 2.0 μg phenol equivalents per g in one or more subsamples.

Although FDA guidelines for screening cheese for residual phosphatase activity stipulates a negative test to indicate adequate pasteurization of cheese milk, certain cheeses are manufactured with starter cultures that produce ALP, and they may thereby show a positive ALP test. Rosenthal et al. (1996) reported that *Penicillium roqueforti* mould exhibits ALP activity, suggesting that blue mold-ripened cheeses made from properly pasteurized milk may test positive for ALP.

Microfiltration of skim milk at 55 °C does not affect ALP activity; hence, cheeses made from microfiltered milk show high ALP activity. The ALP activity of Camembert made from microfiltered milk was 930 mU g^{-1}. ALP loses >99.9% of activity when skim milk is heated at 70 °C for 15 s. ALP activity in spray-dried milk powder decreases with increasing outlet temperature in the spray-drier, but significant reactivation of the enzyme occurs in the critical region of 70–80% total solids.

The presence of NaCl influences ALP activity in butter. Most unsalted butter made from pasteurized cream has little or no ALP activity, but salted butters made from pasteurized cream always show some activity and often give a positive ALP test.

Although ALP is used universally as an indicator for adequate pasteurization of milk, it may not be the most appropriate enzyme for this purpose because: (1) ALP may reactivate under certain conditions, which complicates interpretation of the test results; (2) the enzyme appears to be fully inactivated by sub-pasteurization conditions (70 °C × 16 s); and (3) the relationship between initial activity of ALP and the pasteurization equivalent (PE) is less linear than similar relationships for other milk enzymes (e.g., lactoperoxidase or γ-glutamyl transpeptidase activities) that have been proposed as better indices for adequate pasteurization of milk (McKellar et al. 1991; Blel et al. 2002; Lorenzen et al. 2010).

ALP is only partially inactivated by high-pressure treatment of milk (200–400 Pa for 15–120 min); therefore, the ALP test is not a suitable indicator of the effectiveness of the high-pressure treatment process (an alternative for the pasteurization of milk). Although ALP can dephosphorylate casein under suitable conditions, the technological significance of casein dephosphorylation by ALP is unknown. The use of ALP inactivation as a process indicator for the efficiency of microbial inactivation of HP homogenization (200–300 MPa) of milk has been suggested (Picarta et al. 2006). Measurement of ALP activity in milk that is subjected to ultrasound treatment to eliminate spoilage organisms has been suggested (Cameron et al. 2009).

Proteolysis is a major contributor to the development of flavor and texture in cheese during ripening. Most of the small water-soluble peptides in cheese are from the N-terminal half of α_{S1}- or β-casein; many of these peptides are phosphorylated and show evidence that they have been acted on by phosphatase, i.e., they are partially dephosphorylated. In cheese made from pasteurized milk, indigenous acid phosphatase or bacterial phosphatase are probably responsible for dephosphorylation, but in raw milk cheese, e.g., Parmigiano Reggiano or Grana Padano, ALP appears to be more important. Subjecting raw bovine milk to high pressure CO_2 treatment (20 MPa at 50 °C for 50 min) results in total inactivation of ALP (Liao et al. 2019). Ceni et al. (2016) reported ALP inactivation of 94.5% when supercritical CO_2 was added to milk at a mass ratio of 0.05 at 70 °C and pressure of 80 bar (8 MPa) for 30 min.

4.6 Reactivation of ALP

The reactivation of ALP has been extensively studied (Wright and Tramer 1953a, b, 1954, 1956; Lyster and Aschaffenburg 1962; Murthy et al. 1975) due to the observation that (UHT)-treated milk that is phosphatase-negative immediately after processing becomes phosphatase-positive during storage. Lorenzen et al. (2011) found a significant increase in ALP activity in UHT-treated milk stored at 20 °C over the duration of its shelf life.

The ALP test was originally designed to effectively determine the adequacy of low-temperature long time (LTLT, 63 °C × 30 min) batch pasteurization (Painter and Bradley Jr 1997). However, milk heated to temperatures ranging from 82 to 180 °C or cream heated at 74–180 °C for a short time (e.g., HTST and UHT) acquired ALP activity during storage at 4–40 °C, indicating partial reactivation of the enzyme. In general, reactivation of ALP increases with pasteurization temperature and decrease in holding time. In addition, the presence of Mg^{2+}, Zn^{2+} and NaCl increases reactivation of ALP (Richardson et al. 1964). It has been suggested that Mg^{2+} or Zn^{2+} causes a conformational change in the denatured enzyme that is necessary for renaturation (Peereboom 1970).

Sulphydryl (SH) groups from other proteins (e.g., β-lactoglobulin) are essential for the reactivation of ALP (Peereboom 1970; Stănciuc et al. 2011). This may be a reason for phosphatase reactivation in UHT milk but not in HTST milk. The role of SH groups, which are supplied by denatured whey proteins, is thought to be the chelation of heavy metals, which would otherwise bind to -SH groups (on Cys residues) of the enzyme, thus preventing renaturation. Reports suggest that mercury and cadmium inhibit reactivation. The maximum reactivation of ALP occurs when skim milk, cream or buttermilk is heated to 104 °C followed by incubation at 34 °C; reactivation is high at pH 6.5 (Murthy et al. 1975). The mean activation energy (E_a) for reactivation of ALP in milk that had been heated to 87.8 or 104.4 °C for 6 s and incubated in the presence of Mg^{2+} (27.4 mM) is 24.108 kJ mol^{-1}, while samples incubated without Mg^{2+} had a mean E_a of 22.646 kJ mol^{-1} (Murthy et al. 1975). The reactivation of ALP is spontaneous and is inversely proportional to homogenization pressure but is independent of fat content. Because the reactivation of ALP results in a false positive phosphatase test, it raises doubts about the reliability of the phosphatase test for determining the adequacy of pasteurization of milk. Methods for distinguishing between renatured and residual native ALP are based on the increase in phosphatase activity resulting from the addition of Mg^{2+} (magnesium acetate for 1 h at 34 °C) to the reaction mixture followed by a 6× dilution.

McKellar et al. (1988) found that ALP activity persisted in 32% fat cream heated up to 75 °C and was reactivated in cream pasteurized at 75 °C × 16 s. When butter made from cream pasteurized at 85 °C × 16 s was melted at 40 °C, a positive ALP reaction was found, suggesting reactivation due to temperature abuse. Addition of 0.1 mM EDTA to freshly pasteurized cream minimized reactivation of ALP due to subsequent temperature abuse. McKellar et al. (1988) also noted that reactivation of ALP when butter was incubated for 1 h at 30 °C, and thereby recommended using a solid sample method for routine analysis of ALP in butter. Peereboom (1970) found that a combination of proteins (0.75% casein +0.75% β-lactoglobulin) with 0.12 M $MgCl_2$ strongly reactivates denaturated ALP.

4.7 Acid Phosphatase

Milk also contains acid phosphatase (ACP), the activity of which is much lower, i.e., ~2% of the level of ALP, and is most active at pH 4.5–5.0 (Mullen 1950; Andrews 1974). The activity of ACP in normal bovine milk ranges from 2.6×10^{-4} to 2.6×10^{-3} IU mL^{-1} (note units different to those for ALP). The concentration of ACP reaches a maximum 5–6 days *post-partum*, then decreases and remains low until the end of lactation. The activity of ACP in ewes' milk increases fourfold (17 mU mL^{-1}) in early lactation and then remains constant to the end of lactation (Chávarri et al. 1998). Some research has shown that milk from healthy cows contains one type of ACP, while mastitic milk may contain two additional ACPs (Andrews and Alichanidis 1975b), which do not appear to have been further characterised. The activity of ACP is about 4–10 times higher in milk from cows with mastitis than in normal bovine milk.

ACP is heat stable; 70% of its activity remained in skim milk and cream after heating at 65 °C for 15 min (Bingham et al. 1961) suggesting that it survives batch pasteurization of milk. Complete inactivation of ACP in milk requires heating at 88 °C × 30 min (Larsen and Parada 1988). No activity of ACP is lost during heating of milk at pH 6.7 at 100 °C for 5 s, while 90% activity is lost when it is heated at 100 °C for 20 s. In-container sterilization or UHT treatment completely inactivates ACP. Thermal denaturation of ACP follows first order kinetics and is unaffected by casein. Activity of ACP is rapidly lost when exposed to visible light. The D-value of ACP ranges from 4.8 s at 100 °C to 612.4 min at 65 °C. The z values (temperature change required to produce a ten-fold reduction in D-value) of ACP were 6.6 °C and 27.6 °C, respectively at 75 °C and 85 °C (Griffiths 1986).

4.8 Distribution, Isolation and Characterization of ACP

ACP is found mostly in the skim milk phase (50–70%) and the rest is found in the cream phase, associated with the MFGM (Flynn et al. 1998). The first step in the isolation of ACP from skim milk or cream involves using an acidic ion-exchange resin, which gives about a 300-fold increase in the specific activity (IU mg^{-1}) of the enzyme. The second step is usually gel filtration (Sephadex), which gives a 30-fold increase in the specific activity of the crude ACP isolated in the first step (Andrews 1974; Andrews and Pallavicini 1973). A homogeneous ACP, with a specific activity of over 30 IU mg^{-1}, can be obtained by subjecting the ACP isolated in the first two steps to sequential separations on ion exchange, on cellulose phosphate and affinity chromatography.

ACP is a phosphomonoesterase that hydrolyses aromatic phosphomonoesterase, pyrophosphates (ADP/ATP), polyphosphates and phosphoserine residues of phosphoproteins. ACP has isoelectric pH of ~7.9 and optimally active at pH 4.9 (Andrews and Pallavicini 1973). It is activated by Mn^{2+} and reducing agents and is strongly inhibited by fluoride, iodoacetate, oxidizing agents and heavy metals (Andrews and Pallavicini 1973). Bovine caseins are competitive inhibitors of ACP, in the order α_S (α_{S1}+ α_{S2}) > β > κ, when assayed on p-nitrophenyl phosphate. ACP is a single polypeptide chain with a molecular weight of about 42 kDa, and purified ACP is a glycoprotein containing 2 mol galactose, 2 mol mannose and 4 mol N-acetyl glucosamine mol^{-1} enzyme (Andrew 1976). The enzyme contains low levels of proline and histidine and relatively high amounts of lysine and arginine and devoid of methionine (Andrews and Pallavicini 1973). An Arrhenius plot (based on 50% inactivation) for ACP in the presence of 1% casein at pH 6.7, 5.2 or 4.9 shows an activation energy (E_a) of 245, 270 or 297 kJ mol^{-1}, respectively, suggesting that the stability of ACP to thermal denaturation increases with decreasing pH (Andrews 1974).

4.9 Significance of ACP

The higher heat stability and lower pH optimum of ACP compared with ALP influence the significance of ACP in dairy processing. The significance of ACP in cheese has recently been reviewed (Akuzawa and Fox 2004). The dephosphorylation of caseins by ACP activity may reduce the heat stability of dairy products. The caseins are phosphoproteins and are good substrates for ACP. The micellar integrity of caseins is lost on cleavage of phosphate groups from the serine residues of casein by ACP. Active ACP has been found in Cheddar cheese, suggesting that ACP activity may influence cheese flavor via its effect on proteolysis (Dully and Kitchen 1973). Andrews and Alichanidis (1975a) found that starter bacteria did not significantly contribute to ACP activity in Cheddar cheese although Hynek et al. (1999) found differences in ACP activity in Gouda cheese made with different cultures, which also influenced sensory attributes of the cheese during 120 days ripening. Several partially dephosphorylated phosphopeptides, probably resulting from phosphatase activity, have been isolated from cheese (De Noni et al. 1997; Adt et al. 2011). ACP from ewes' milk has been shown to retain 16% of its activity when assayed at cheese-ripening temperature (8–12 °C) compared with its activity at 37 °C. The activity of ACP is reported to increase two-fold during 180 days of ripening in ovine milk cheeses made in summer months compared to cheese made in winter and spring (Chávarri et al. 1998).

4.10 Conclusions

The presence of both acid and alkaline phosphatases in milk and their role in dairy science and technology have been reviewed. While the technological significance of ALP is well known and it is used as an indicator for adequate pasteurization of milk to enhance the microbiological safety of dairy product consumption, the significance of ACP appears to be more nutritional.

References

Adt I, Dupas C, Boutrou R, Oulahal N, Noel C, Mollé C, Jouvet T, Degraeve P (2011) Identification of caseinophosphopeptides generated through in vitro gastro-intestinal digestion of Beaufort cheese. Int Dairy J 21:129–134

Advanced Instruments (2019). https://www.aicompanies.com/flm300-fluorophos-alp-test-system/accessed. Accessed 21 Apr 2019

Akuzawa R, Fox PF (2004) Acid phosphatase in cheese. Anim Sci J 75:385–391

Albillos SM, Reddy R, Salter R (2011) Evaluation of alkaline phosphatase detection in dairy products using a modified rapid chemiluminescent method and official methods. J Food Prot 74:1144–1154

Andrews AT (1974) Bovine milk acid phosphatase II. Bonding to casein substrates and heat inactivation studies. J Dairy Res 41:229–237

Andrew AT (1976) Bovine milk acid phosphatase: III. Purification and characterisation of the enzyme. Biochim Biophy Acta (BBA) - Protein Struct 434:345–353

Andrews AT (1992) Indigenous enzymes in milk, phosphatases. In: Fox PF (ed) Advanced dairy chemistry, Proteins, vol 1, 2nd edn. Elsevier Applied Science, London, pp 320–331

Andrews AT, Alichanidis E (1975a) Acid phosphatase activity in cheese and starters. J Dairy Res 42:327–339

Andrews AT, Alichanidis E (1975b) The acid phosphatases of bovine leucocytes, plasma and the milk of healthy and mastitic cows. J Dairy Res 42:391–400

Andrews AT, Pallavicini C (1973) Bovine milk acid phosphatase. I. Some kinetics studies and other properties using a partially purified preparation. Biochim Biophys Acta (BBA) – Enzymology 321:197–209

Bradshaw RA, Cancedda F, Ericsson LH, Neumann PA, Piccoli SP, Schlesinger MJ, Shriefer K, Walsh KA (1981) Amino acid sequence of Escherichia coli alkaline phosphatase. Proc Natl Acad Sci U S A 78:3473–3477

Bilen C, Gençer N, Arslan O (2015) Purification of alkaline phosphatase from bovine milk and investigation of inhibitory effects of some veterinary drugs on enzyme activity. Hacettepe J Biol Chem 43(3):195–203

Bingham EW, Jasewicz J, Zittle CA (1961) Acid phosphatase in cream and skimmilk. J Dairy Sci 44:1247–1256

Blel M, Guingamp M-F, Gaillard J-L, Humbert G (2002) Studies on the thermal sensitivity of γ-glutamyl transpeptidase measured with a modified test procedure and compared with that of alkaline phosphatase and lactoperoxidase in milk. Lait 82:555–566

Cameron M, McMaster LD, Britz TJ (2009) Impact of ultrasound on dairy spoilage microbes and milk components. Dairy Sci Technol 89:83–98

Ceni G, Silva MF, Valério C Jr, Cansian RL, Oliveira JV, Rosa CD, Mazutti MA (2016) Continuous inactivation of alkaline phosphatase and Escherichia coli in milk using compressed carbon dioxide as inactivating agent. J CO2 Utilization 13:24–28

Cerf O, Condron R (2006) *Coxiella burnetii* and milk pasteurization: an early application of the precautionary principle? Epidemiol Infect 134:946–995

Chávarri F, Santisteban A, Virto M, de Renobales M (1998) Alkaline phosphatase, acid phosphatase, lactoperoxidase, and lipoprotein lipase activities in industrial ewe's milk and cheese. J Agric Food Chem 46:2926–2932

Claeys WL, Van Loey AM, Hendrickx ME (2002) Kinetics of alkaline phosphatase and lactoperoxidase inactivation, and of beta-lactoglobulin denaturation in milk with different fat content. J Dairy Res 69:541–553

De Noni I, Pellegrino L, Resmini P (1997) Ferranti P. About presence of free phosphoserine in ripened cheese and in enzymatic hydrolysate of casein Nahrung 41:268–273

Desbourdes C, Nicolas M, Boitelle AC (2008) Phosphatase activity in cheese. AFSSA, CRL Milk and Milk products. In: 11th Workshop of the EU CRL for Milk and Milk Products, Vienna

Dully JR, Kitchen BJ (1973) The acid phosphatases of Cheddar cheese. Aust J Dairy Technol 28:114–116

Dumitrașcu L, Stănciuc N, Aprodu I, Ciuciu A-M, Alexe P, Bahrim GE (2015) Monitoring the heat-induced structural changes of alkaline phosphatase by molecular modeling, fluorescence spectroscopy and inactivation kinetics investigations. J Food Sci Technol 52(10):6290–6300

Egger L, Nicolas M, Pellegrino L (2016) Alkaline phosphatase activity in cheese as a tracer for cheese milk pasteurization. LWT - Food Sci Technol 65:963–968

Farzi-Khajeh H, Jafarib B, Safaa KD, Dastmalchi S (2019) Magnetic iron oxide nanoparticles modified with vanadate and phosphate salts for purification of alkaline phosphatase from the bovine skim milk. Coll Surfaces B: Biointerfaces 175:644–653

Flynn NKR, Akuzawa R, Fox PF (1998) Characterization of isozymes of acid phosphatase isolated from bovine milk. In: Abstracts 28th Food Science and Technology Research Conference. University College, Cork, p 165

Folley SJ, Kay HD (1936) Variations in the phosphomonoesterase content of the milk of the cow in relation to the progress of lactation. Enzymologia 1:48–54

Fox PF, Kelly AL (2006) Indigenous enzymes in milk: Overview and historical aspects - Part 2. Int Dairy J 16:517–532

Graham WR, Kay HD (1933) Phosphorus compounds in milk. V. The phosphorus partition in milk with preliminary observations on milk phosphatase. J Dairy Res 5:54–62

Griffiths M (1986) Use of milk enzymes as indices of heat treatment. J Food Prot 49:696–705

Hassan SSM, Sayour HEM, Kamel AH (2009) A simple potentiometric method for determination of acid and alkaline phosphatase enzymes in biological fluids and dairy products using a nitrophenyl phosphate plastic membrane sensor. Analytical Chimica Acta 640:75–81

Hynek R, Sajdok J, Drab V (1999) Activity of acid phosphatase in Gouda cheese made with different mesophilic starters. Czech J Food Sci 17(2):55–60

International Dairy Federation (2007) Milk and milk-based drinks — Determination of alkaline phosphatase activity — Enzymatic photo-activated system (EPAS) method. In: International Standard ISO 22160. IDF 209. International Dairy Federation, Brussels

International Dairy Federation (2009) Milk - Determination of alkaline phosphatase. In: International Standard ISO 3356. IDF 63, 2nd edn. International Dairy Federation, Brussels

International Dairy Federation (2013) Milk and milk products. Determination of alkaline phosphatase activity. Part 1. Fluorometric methods for milk and milk-based drinks. In: International Standard ISO 11816-1 IDF 155-1. International Dairy Federation, Brussels

International Dairy Federation (2016) Milk and milk products. Determination of alkaline phosphatase activity. Part 2. Fluorometric methods for cheese. International Standard ISO 11816-2 IDF 155-2, 2nd edn. International Dairy Federation, Brussels

Jasinska B, Kleczkowski K, Michalak W (1985) Influence of β-lactoglobulin on milk alkaline phosphatase activity toward the main milk caseins. J Dairy Sci 68:2172–2175

Kay HD (1925) Some phosphorus compounds of milk. The presence of milk of organic acid-soluble phosphorus compounds. Biochem J 19:433–446

Khodaparast-Sharifi SH, Snow LD (1989) Levamisole inhibition of alkaline phosphatase and 5′-nucleotidase of bovine milk fat globule membranes. Int J Biochem 21(4):401–405
Klotz V, Hill A, Warriner K, Griffiths M, Odumeru J (2008) Assessment of the colorimetric and fluorometric assays for alkaline phosphatase activity in cow's, goat's, and sheep's milk. J Food Prot 71(9):1884–1888
Kuzuya Y, Kanamaru Y, Tanahashi T (1980) Partial purification and some properties of bovine milk alkaline phosphatase. Jpn J Zootech Sci 51(7):501–510
Larsen RF, Parada JL (1988) Acid phosphatases in some cheeses and starters. Sci Des Aliments 8:285–294
Levieux D, Geneix N, Levieux A (2007) Inactivation-denaturation kinetics of bovine milk alkaline phosphatase during mild heating as determined by using a monoclonal antibody-based immunoassay. J Dairy Res 74:296–301
Liao H, Zhong K, Huc X, Liao X (2019) Effect of high-pressure carbon dioxide on alkaline phosphatase activity and quality characteristics of raw bovine milk. Innovative Food Sci Emerging Technol 52:457–462
Lin C-W, Fishman WH (1972) L-Homoarginine. An organ specific uncompetitive inhibitor of human liver and bone alkaline phosphohydralases. J Biol Chem 247:3082–3087
Linden G, Alias C (1978) Alkaline phosphatase in human, cow and sheep milks: molecular and catalytic properties and metal ion action. Ann Biol Anim Bioch Biophys 18(3):749–758
Lorenzen PC, Clawin-Rasecker I, Einhoff K, Hammer P, Hartmann R, Hoffmann W, Martin D, Molkentin J, Walte HG, Devrese M (2011) A survey of the quality of extended shelf life (ESL) milk in relation to HTST and UHT milk. Int J Dairy Technol 64:166–178
Lorenzen PC, Martin D, Clawin-Rädecker I, Barth K, Knappstein K (2010) Activities of alkaline phosphatase, γ-glutamyltransferase and lactoperoxidase in cow, sheep and goat's milk in relation to heat treatment. Small Ruminant Res 89:18–23
Lorient D, Linden G (1976) Dephosphorylation of bovine casein by milk alkaline phosphatase. J Dairy Res 43:19–26
Low MG, Finean JB (1977) Release of alkaline phosphatase from membranes by a phosphatidylinositol-specific phospholipase C. Biochem J 167:281–284
Lyster RLJ, Aschaffenburg R (1962) The reactivation of milk alkaline phosphatase after heat treatment. J Dairy Res 29:21–35
Marchand S, Merchiers M, Messens W, Coudijzer K, De Block J (2009) Thermal inactivation kinetics of alkaline phosphatase in equine milk. Int Dairy J 19:763–767
McKellar RC, Cholette H, Emmons DB (1988) Modification of the alkaline phosphatase assay for butter. Can Inst Food Sci Technol J 21:97–101
McKellar RC, Emmons DB, Farber J (1991) Gamma-glutamyl transpeptidase in milk and butter as an indicator of heat treatment. Int Dairy J 1:241–251
Michio S, Satoshi K, Tohru Y, Koreaki I (1997) Roles of disulfide bond in bacterial alkaline phosphatase. J Biol Chem 272:6174–6178
Mistry VV (1989) Thermal inactivation characteristics of alkaline phosphatase in ultrafiltered Milk. J Dairy Sci 72:1112–1117
Morton RK (1953) Alkaline phosphatase of milk. Association of the enzyme with a particulate lipoprotein complex. Biochem J 55:786–795
Mullen JEC (1950) The acid phosphatase of cows' milk: I Some properties of the enzyme. J Dairy Res 17:288–305
Murthy GK, Cox S, Kaylor L (1975) Reactivation of alkaline phosphatase in ultra high-temperature, short-time processed liquid milk products. J Dairy Sci 59:1699–1710
Painter CJ, Bradley RL Jr (1997) Residual alkaline phosphatase activity in milks subjected to various time-temperature treatments. J Food Prot 60:525–530
Patton S, Keenan TW (1975) The milk fat globule membrane. Biochem Biophys Acta 415:273–309
Pellegrino L, Rosi V (2008) Alkaline phosphatase: legal limits for cheese. Study on Italian cheeses. In: 11th Workshop of the EU CRL for Milk and Milk Products, Vienna

Peereboom JWC (1970) Studies on alkaline milk phosphatase III. A theory for the mechanism of reactivation of alkaline milk phosphatase in a model system. Fette Seifen Anstrichmittel 72:301–308

Picarta L, Thiebaud M, Rene M, Guiraud JP, Cheftel JC, Dumaya E (2006) Effects of high pressure homogenisation of raw bovine milk on alkaline phosphatase and microbial inactivation. A comparison with continuous short-time thermal treatments. J Dairy Res 73:454–463

Rankin SA, Christiansen A, Lee W, Banavara DS, Lopez-Hernandez A (2010) Invited review: the application of alkaline phosphatase assays for the validation of milk product pasteurization. J Dairy Sci 93:5538–5551

Richardson LA, McFarren EF, Campbell JE (1964) Phosphatase reactivation. J Dairy Sci 47:205–210

Rocco RM (1990) Fluorometric analysis of alkaline phosphatase in fluid dairy products. J Food Prot 53:588–591

Rola JG, Sosnowski M (2010) Determination of alkaline phosphatase activity in milk and milk products by fluorimetric method. Bull Vet Inst Pulawy 54:537–542

Rosenthal I, Bernstein S, Rosen B (1996) Alkaline phosphatase activity in *Penicillium roqueforti* and in blue-veined cheeses. J Dairy Sci 79:16–19

Sanders SP (1949) The phosphatase test for pasteurization of dairy products. C. R XIIe Congr Int Lait, Stockholm 2:757–765

Salter RS, Fitchen J (2006) Evaluation of a chemiluminescence method for measuring alkaline phosphatase activity in whole milk of multiple species and bovine dairy drinks: Interlaboratory study. J AOAC Int 89:1061–1070

Shakeel-Ur-Rehman, Farkye NY, Yim B (2006) A preliminary study on the role of alkaline phosphatase in cheese ripening. Int Dairy J 16:697–700

Silanikove N, Shapiro F (2007) Distribution of xanthine oxidase and xanthine dehydrogenase activity in bovine milk: physiological and technological implications. Int Dairy J 17:1188–1194

Sosnowski M, Jolanta G, Rola JG, Osek J (2016) Alkaline phosphatase activity and microbiological quality of heat-treated goat milk and cheeses. Small Rum Res 136:132–136

Stănciuc N, Ardelean A, Diaconu V, Râpeanu G, Stanciu S, Nicolau A (2011) Kinetic and thermodynamic parameters of alkaline phosphatase and γ–glutamyl transferase inactivation in bovine milk. Dairy Sci Techol 91:701–717

Upadhyay LSB, Verma N (2015) A three step approach for the purification of alkaline phosphatase from non-pasteurized milk. J Food Sci Technol 52(5):3140–3146

Witwer KW, Théry C (2019) Extracellular vesicles or exosomes? On primacy, precision, and popularity influencing a choice of nomenclature. J Extracell Vesicles 8(1):1648167.

Wright RC, Tramer J (1953a) Reactivation of milk phosphatase following heat treatment I. J Dairy Res 20:177–188

Wright RC, Tramer J (1953b) Reactivation of milk phosphatase following heat treatment II. J Dairy Res 20:258–273

Wright RC, Tramer J (1954) Reactivation of milk phosphatase following heat treatment III. J Dairy Res 21:37–49

Wright RC, Tramer J (1956) Reactivation of milk phosphatase following heat treatment: IV the influence of certain metallic ions. J Dairy Res 23:248–256

Ziobro GC (2001) Bacteriological Analytical Manual Chapter 27 Screening Method for Phosphatase (Residual) in Cheese. https://www.fda.gov/Food/FoodScienceResearch/LaboratoryMethods/ucm073603.htm. Accessed 21 Apr 2019

Chapter 5
Antimicrobial Enzymes in Milk, and Their Role in Human Milk

Nidhi Bansal, Jie Zhang, and Zhengzheng Zou

5.1 Introduction

Antimicrobial enzymes are widespread in nature, where they play a critical role in defending living organisms from bacterial attack. The unique significance of antimicrobial enzymes in milk to the health and growth of newborn mammals has been well documented. The immune system of newborns, especially in preterm babies, is immature and they are at a higher risk of infection. The antimicrobial enzymes are passively transferred from mother to infants *via* milk and constitute an important element of the infant's primal immunity. For many species, the milk-derived antimicrobial system is crucial for survival of the infants (Van Hooijdonk et al. 2000). In addition, the possibility of isolating these bioactive enzymes has been investigated as a potential strategy for the prevention and treatment of gastrointestinal diseases in humans (Korhonen et al. 2000).

Antimicrobial enzymes alone, with potentiating non-enzyme substances, or with physical processing, can prevent the growth of microorganisms effectively in several ways. They can do this by:

1. producing reaction products harmless to humans or animals that are bacteriostatic or bactericidal to the microorganisms;
2. destroying an outer membrane, cell wall, or cell membrane component, or causing a change in permeability or a physical disruption of the cell wall and/or cell membrane, resulting in the death of the cell;
3. completely utilising a necessary nutrient (such as oxygen) and making it unavailable for microbial growth;
4. inactivating an enzyme essential for microbial growth (Scott 1988).

N. Bansal (✉) · J. Zhang · Z. Zou
School of Agriculture and Food Sciences, The University of Queensland,
St Lucia, QLD, Australia
e-mail: n.bansal@uq.edu.au

A. L. Kelly, L. B. Larsen (eds.), *Agents of Change*, Food Engineering Series,
https://doi.org/10.1007/978-3-030-55482-8_5

According to the mechanism of antimicrobial effect, two classes of enzymes are relevant, namely, the hydrolases and the oxidoreductases. The substrates for the hydrolases are key structural components of the cell walls of microorganisms; degradation of these components inactivates the cells.

The bacterial cell wall, in particular the peptidoglycan component [a carbohydrate backbone consisting of β-1-4-linked residues of *N*-acetyl-D-glucosamine and *N*-acetyl-muramic acid interlinked by peptide crosslinks], is responsible for the rigidity of the bacterial cell. Degradation of the cell wall may result in cell lysis. Peptidoglycan is present in the cell wall of both Gram-positive and Gram-negative bacteria, but the molecular structure varies from microorganism to microorganism. The type of peptidoglycan and the cell-wall structure determine the resistance of the bacterial cell to the action of lytic enzymes. The cell envelope of Gram-negative bacteria is both structurally and functionally more complex than that of Gram-positive bacteria. In addition to the cytoplasmic membrane, an outer membrane (outside the cell wall), consisting of lipopolysaccharides, phospholipids, lipoproteins, and proteins, surrounds the cell of the Gram-negative bacteria. Because of the function of the outer membrane of Gram-negative bacteria as a barrier to permeation, Gram-negative bacteria appear to be less sensitive to bacteriolytic enzymes than Gram-positive bacteria. Gram-negative bacteria may be susceptible to cell-wall-degrading enzymes once the outer membrane has been destroyed by physical or chemical treatments.

Bacteriolytic enzymes are generally grouped into three different classes: *N*-acetylhexosaminidses that catalyse the cleavage of the β (1–4)-glucosidic linkages in the carbohydrate backbone of the peptidoglycan, *N*-acetylmuramy-L-alanine amidases that catalyse the cleavage between the carbohydrate moiety and the peptide moiety of the peptidoglycan, and endopeptidases that hydrolyze the peptide bonds in the peptide crosslinks of the peptidoglycan. Oxidoreductases exert their effect by the *in situ* generation of reactive molecules that can destroy vital proteins in the cell (Fuglsang et al. 1995).

These enzymes are now increasingly being exploited against microbial systems because many chemical preservatives are becoming unacceptable to consumers. A few enzymes are already used on a limited scale, as will become apparent, but the cost incurred in producing them by extraction from their natural sources is a severe limitation on their widespread use. However, with the aid of modern gene technology, such enzymes could be cloned, and the prospects of producing them cost effectively are now realistic, but yet not widely exploited (Thallinger et al. 2013).

The purpose of this chapter is to describe the antibacterial properties and mode of action of antimicrobial enzymes in human milk, and to compare with other species. This will improve our understanding of the role of these enzymes in antibacterial defence strategies in nature and provide a basis for the development and improvement of applications of these enzymes as antibacterial agents.

5.2 Antimicrobial Enzymes in Human Milk

There are several antimicrobial enzymes identified in human milk. Xanthine oxidase (XO) and lactoperoxidase (LPO) together constitute an antimicrobial 'oxidative enzyme system' essential for early immunity. Lysozyme is another major antimicrobial enzyme. The main role of lipases in milk is aiding digestion, particularly for preterm infants, when their digestive system is immature. However, there is evidence that free fatty acids and monoglycerides released from milk triglycerides during lipase lipolysis have antimicrobial activity. Milk-borne lipase also has anti-viral activity (Ballard and Morrow 2013). This chapter discusses the general characteristics, antimicrobial mechanism, factors affecting activity and activity assay methods for these enzymes.

5.2.1 Lysozyme

5.2.1.1 General Characteristics of Lysozyme

Lysozyme (E.C. 3.2.1.17), also known as N-acetylmuramidase or muranidase, is a hydrolytic enzyme that cleaves the β-1,4 bond between the C-1 of N-acetylmuramic acid and the C-4 of N-acetylglucosamine of peptidoglycan polymers in the bacterial cell wall, thereby lysing sensitive bacteria (Jolles and Jolles 1984). It was first identified by Fleming in 1921, when a droplet from Fleming's nose accidentally fell into a petri dish containing microorganisms. Fleming noticed the reaction between the drop and the bacteria and successfully isolated this protein. In 1922, he went on to characterize this enzyme isolated from hen egg white (Fleming 1922), which is the main commercial source of lysozyme nowadays. Some years later, this enzyme was detected in the milk of several species and it was found that human milk is a comparatively rich source of lysozyme. The content of lysozyme in bovine milk was controversial since some researchers, including Fleming (1922), found lysozyme in bovine milk, while others did not (Vakil et al. 1969). Lysozyme has been isolated from the milk of a wider range of species than any other milk enzyme; this may indicate the potential importance of lysozyme as a natural protective agent in milk.

Lysozyme was firstly purified from human milk by Jolles and Jolles (1961) and then the purification method was improved in further studies (Jolles and Jolles 1984; Jolles 1967). It was reported that the average content of lysozyme in human milk was 39 mg/100 mL using 117 samples, which was about 3000 times higher than that of bovine milk (Chandan et al. 1964). Lysozyme could also be found in the milk of donkey, horse, dog, sow, cat, rat, rabbit, llama and rhesus monkey, while no lysozyme or only traces were found in milk of other species, such as bovine, goat, sheep, camel and guinea pig (Chandan et al. 1965, 1968; Benkerroum 2008). It is notable that the high lysozyme activity in donkey and horse milk may lead to their high antibacterial activity.

While lysozymes in milk of various species are generally similar, there are some differences even between closely related species, such as cow and buffalo (McSweeney and Fox 2003). The optimal pH of activity of lysozyme from human and bovine milk is 6.35 and 7.9, respectively (Parry Jr et al. 1969). Bovine milk lysozyme has a molecular weight of ~18 kDa compared to ~15 kDa for human milk and egg white lysozyme and its amino acid composition and immunological properties are considerably different from those of the latter two lysozymes (Eitenmiller et al. 1971, 1974). According to the complete amino acid sequence study by Jollès et al. (1972), human milk lysozyme contains 130 amino acid residues, compared to 129 in egg white lysozyme, the extra residue in the former being Val_{100}. Horse milk lysozyme has the same number of amino acid residues as egg white lysozyme (McKenzie and Shaw 1985). The three-dimensional structure of egg white lysozyme was reported by Blake and others (Blake et al. 1965; Johnson 1998).

5.2.1.2 Mechanism of Antibacterial Activity of Lysozyme

The most important physiological role of lysozyme is to act as an antibacterial agent, evidence of which has been available since 1922 (Fleming 1922). The mechanism of antibacterial activity of milk lysozyme, as part of innate defence system, is well elucidated. It could either degrade sensitive bacterial cell wall components independently or participate in immunological reactions as a component (Varaldo et al. 1989; Masschalck et al. 2002). The inhibitory concentration of lysozyme varies depending on the target microorganism and environmental parameters in the reaction medium, and may be as low as 10 μg/mL, which is below the concentration of lysozyme in several types of mammalian milk (Masschalck et al. 2001).

Milk is a complex mixture of various components, which could influence the ultimate activity of lysozyme. In particular, lactoferrin was reported to have a synergistic effect on its antimicrobial activity (León-Sicairos et al. 2006). Lactoferrin first binds tightly to components of the outer cell membrane, lipopolysaccharides, of Gram-negative bacteria, and creates holes in the membrane, through which lysozyme then enters the glycomatrix of the bacteria, degrading it and effectively killing the pathogens. In addition to antimicrobial activity, lysozyme has some other biological functions too, such as immunomodulatory, antiviral and anti-inflammatory activity (Huopalahti et al. 2007). Due to these significant functions, yet lack of availability, recent research has focused on transgenic production of lysozyme to meet the requirements of the food industry (Yang et al. 2011).

5.2.1.3 Factors Affecting Antimicrobial Activity of Lysozyme

Lysozyme activity in milk can be affected by many factors during storage and processing in human milk banks and the dairy industry. Although early studies (Björkstén et al. 1980) suggest that lysozyme activity in human milk could be preserved during freezing, more recent studies have shown that prolonged frozen storage (at −20 °C for 1 month) reduces its activity considerably (Akinbi et al. 2010; Chang et al. 2013). Lysozyme is expected to be heat-labile around neutral pH, which is the pH range of the human milk (pH 7–7.4) (Morriss Jr et al. 1986). Several reports confirmed that 21–74% of lysozyme content was reduced after low temperature long time (LTLT pasteurization) (Koenig et al. 2005; Viazis et al. 2007; Czank et al. 2009). The retention of lysozyme activity in human milk was 75%, when irradiated with a UV-C dosage of 4683 J/L (Christen et al. 2013). The effect of high pressure processing on human milk lysozyme has been investigated by several researchers and they found that lysozyme activity was maintained after treatment at 400 MPa for 120 min (Viazis et al. 2007) and at 650 MPa for 30 min (Sousa et al. 2014; Mayayo et al. 2016). However, microwave heating at high energy caused a marked decrease in human milk lysozyme activity (Quan et al. 1992). The effect of other pasteurization methods, such as pulsed electric field and ultra-sonication, on lysozyme activity in milk should be further investigated.

5.2.1.4 Assay Methods for Lysozyme Activity

The presence and activity of lysozyme is normally assayed by the lysis of a culture of *M. lysodeikticus*, measured by a decrease in turbidity (Manas et al. 2006), but it can also be assayed by enzyme-linked immunosorbent assay techniques using monoclonal or polyclonal antibodies (Rauch et al. 1990; Yoshida et al. 1991; Kerkaert et al. 2010; Schneider et al. 2010), reversed-phase high-performance liquid chromatography with fluorescence detection (Pellegrino and Tirelli 2000), liquid chromatography-mass spectrometry (LC-MS), immune-capture mass spectrometry or surface enhanced mass spectrometry (Schneider et al. 2010). In recent years, the increasing incidence of cases describing allergic reactions to lysozyme present in food products has refocused attention on comparison and development of rapid, specific, sensitive and reliable methods for the detection and quantification of lysozyme in food products (Kerkaert et al. 2010; Jiménez-Saiz et al. 2011; Schneider et al. 2011).

5.2.2 *Lactoperoxidase*

5.2.2.1 General Characteristics of Lactoperoxidase

Lactoperoxidase [EC 1.11.1.7] (LPO) belongs to the family of peroxidases and is widely distributed in the salivary, mammary and lachrymal glands, as well as secretions such as saliva, milk and tears (Wolfson and Sumner 1993; Kussendrager and Van Hooijdonk 2000; Seifu et al. 2005). The peroxidases from various tissues are similar from a chemical and immunological perspective (Tenovuo 1985) and the one from milk is named lactoperoxidase. It is present in all mammalian milk types that have been analysed so far, including bovine and human milk. Bovine lactoperoxidase has been widely investigated while studies on human lactoperoxidase are limited due to its low content in human milk (Sousa et al. 2014). This enzyme plays an important role in preventing maternal mastitis and developing the infants' defence system against pathogenic microorganisms (Naidu 2000). Therefore, it has a great physiological importance in human milk.

LPO purified from human milk showed two bands with molecular masses of 80 and 100 kDa on SDS-polyacrylamide gel electrophoresis and immunoblotting, indicating heterogeneity of the LPO molecule (Shin et al. 2000; Shin et al. 2001).

Bovine LPO, on the other hand, is made up of a single polypeptide chain containing 612 amino acid residues with a molecular weight of 78 kDa (Cals et al. 1991). It has a high isoelectric point, of 9.6. This glycoprotein contains about 10% carbohydrate and at least ten variants of bovine LPO are known. The enzymatic activity of these variants is not significantly different (Rosen 1986). The iron content of bovine LPO is 0.07%, equivalent to one iron atom per LPO molecule (Rosen 1986). A calcium ion is also strongly bound to LPO to stabilise its molecular conformation (Booth et al. 1989). The heme moiety is located in the catalytic site of the enzyme and is comprised of a heterocyclic ring of porphyrin linked to itself and holding one iron atom and a calcium ion in the center. LPO has a maximum absorbance at 412 nm (Rosen 1986; Kussendrager and Van Hooijdonk 2000).

In bovine milk, LPO is the second most abundant enzyme after xanthine oxidase (DeWit and Van Hooydonk 1996). Its concentration is around 30 mg/L constituting about 1% of whey proteins. Human milk LPO activity is quite low and varies from 0.06 to 0.97 units/mL (Wolfson and Sumner 1993); its about 20 times lower than the LPO activity in bovine milk (1.2–19.4 units/mL) (Gothefors and Marklund 1975). Human milk LPO concentration has been reported as 0.77 ± 0.38 mg/L as determined by ELISA analysis of the whey fraction (Shin et al. 2001).

5.2.2.2 Mechanism of Antibacterial Activity of Lactoperoxidase

Lactoperoxidase (LPO), combined with thiocyanate and hydrogen peroxide, forms the lactoperoxidase (LPO) system, which has a bacteriostatic and bactericidal effect on a wide range of Gram-positive organisms, such as *Listeria monocytogenes* and *Staphylococcus aureus,* and Gram-negative microorganisms such as *Escherichia coli* (Wolfson and Sumner 1993).

LPO catalyses the conversion of halides and pseudohalides to hypohalides and hypothiocynous acids (Demba et al. 2018). LPO catalyses a complex series of reactions, as reflected by its structural complexity, that leads to production of antimicrobial compounds. In its ground state, the protoporphyrin ring of the active site contains iron in its ferric (Fe^{3+}) form. When H_2O_2 is present, it reacts with this Fe^{3+} which binds one oxygen atom, then releasing a water molecule. A transient oxidised form of the enzyme called Compound I, containing Fe=O in the active site, is formed, which then may enter either a 'halogenation' or a 'peroxidation' cycle (Bafort et al. 2015). In the halogenation cycle, the Fe=O undergoes disproportionation in the presence of a halide ($Cl^-/Br^-/I^-$) or a pseudohalide ion, e.g., thiocyanate

(SCN^-), the preferred physiological substrate for LPO (Bafort et al. 2015). The Fe-bound oxygen is transferred to form hypothiocyanite ($OSCN^-$) or another hypohalide ion and the LPO haem returns to its Fe^{3+} ground state. The overall reaction is shown in equation below (Bafort et al. 2015):

$$H_2O_2 + SCN^- \rightarrow OSCN^- + H_2O$$

It is these products (such as $OSCN^-$) that have broad-spectrum antibacterial activity through inhibition of growth, oxygen uptake and lactic acid production.

The LPO system can either kill or inhibit the growth and metabolism of a variety of microorganisms by altering their cellular components, such as the cell wall, cytoplasmic membrane, glycolytic enzymes and nucleic acids. The effectiveness of this system varies with different groups of bacteria, which can probably be explained by the different cell wall structure and barrier properties of these bacteria (DeWit and Van Hooydonk 1996). When the system is provided with sufficient lactoperoxidase, the antimicrobial activity is exhibited even at low temperature (0–5 °C) and low pH (5 or less). The antimicrobial effect of the LPO system may be interfered by the medium, such as the small molecules present in brain heart infusion broth (Klebanoff et al. 1966). Although controversy exists over whether bacteria in their stationary phase of growth are more sensitive to LPO system, the growth stage of bacteria does impact the results of inactivation (Reiter et al. 1976; Purdy et al. 1983). In addition, anaerobic bacteria seem to be more susceptible to the LPO system than aerobic ones (Carlsson et al. 1983). The antimicrobial effect of the LPO system in bovine milk has been widely studied, while the data are limited for human milk.

5.2.2.3 Factors Affecting Antimicrobial Activity of Lactoperoxidase

Holder pasterurisation has been shown to inactivate human milk LPO (Tully et al. 2001; Akinbi et al. 2010). Recently, we studied the effect of different heating temperatures on LPO activity in human milk. Human milk samples were thermally treated at 40 °C, 50 °C, 55 °C and 62.5 °C for 30 min (i.e., 'low-temperature-long-time', LTLT), or 72.5 °C for 15 s, or 127 °C for 15 s (i.e., 'high-temperature-short-time', HTST or 'ultra-high temperature', UHT), to replicate commonly-used pasteurisation methods. LPO activity retention in human milk treated at or below 55 °C for 30 min was at least 60%; however, <6% LPO activity was retained in samples heated at or above 62.5 °C for 30 min.

On the other hand, bovine LPO is one of the most heat-stable enzymes in milk. It has been reported that normal pasteurization (62.5 °C, 30 min or 72 °C, 15 s) of milk does not inactivate LPO in milk (Korhonen 1980; Marks et al. 2001). LPO is partially inactivated by heating at 74 °C for a short time, but the remaining activity is still sufficient to catalyse the reactions to generate antibacterial activity (Wolfson and Sumner 1993). The complete inactivation of bovine LPO has reported for heating at 80 °C for 2.5 s (Korhonen 1980) or at 78 °C for 15 s (DeWit and Van Hooydonk 1996). Inactivation of the LPO activity is very temperature-sensitive; Griffiths

(Griffiths 1986) reported that it had $D_{70°C}$ of 940 s and a z-value of 5.4 °C. The heat stability of LPO decreases under acidic conditions (pH 5.3) and is influenced by the concentration of calcium ions (Kussendrager and Van Hooijdonk 2000). However, LPO has maximum enzyme activity at pH 5–6. In addition, LPO is rather resistant to proteolytic enzymes and is not inactivated by trypsin, thermolysin and the gastric juice of infants (Rosen 1986). However, it is sensitive to light, due to the presence of riboflavin (DeWit and Van Hooydonk 1996).

Although information about effect of high pressure processing of human milk on LPO is limited, studies on bovine milk suggest that it is resistant to pressures up to 400 MPa (Huppertz et al. 2006).

5.2.2.4 Assay Methods for Lactoperoxidase Activity

LPO activity is usually assayed spectrophotometrically using 2,2′-azino-bis(3-ethylbenzthiazoline)-6-sulfonic acid (ABTS)/pyrogallol/guaiacol as substrate (Elfakharany et al. 2016; Tayefi-Nasrabadi et al. 2011; Atasever et al. 2013), although high sensitivity measurements cannot be achieved. This is particularly problematic for the assessment of LPO activities in milk samples other than bovine milk, as the activities in those milk samples may be too low to be detected reliably by spectrophotometric assays. For example, the LPO activity in human colostrum could not be quantified using the ABTS method (Sousa et al. 2014). Another colorimetric method using guiacol and para-phenylenediamine was developed, with the advantage of applying clarifying agent to solubilize casein micelles and fat globules that should be removed by precipitation and filtration in other methods (Blel et al. 2001). Thiocyanate (SCN-) is the substrate of this system and it activates the system at a concentration of 15 ppm.

We have recently developed a high-throughput microplate methodology for a highly sensitive micro-plate fluorescent assay to determine LPO activities in various milk samples. The method is based on oxidation of Amplex® Red (1-(3,7-dihydrox yphenoxazin-10-yl) ethanone, AR) by LPO to the red-fluorescent oxidation product resorufin which is measured in a temperature-controlled spectrofluorimeter, following the automatic addition of hydrogen peroxide (H_2O_2). The limit of detection (LOD) of the assay was 0.071 U/L and inter-run and intra-run precision values with CVs < 10% in high, medium and low activity samples were obtained, indicating high assay sensitivity and reproducibility (manuscript submitted to Food Chemistry).

5.2.3 Xanthine Oxidase

5.2.3.1 General Characteristics of Xanthine Oxidase

Xanthine Oxidase (XO) is a molybdoprotein that consists of two identical subunits, each of which contains one molybdenum (Mo), one flavin adenine dinucleotide (FAD) and two non-identical ironsulphur redox centres (Fe-S) (Harrison 2002). XO and xanthine dehydrogenase (XDH) are two forms of the xanthine oxidoreductase (XOR) enzyme. The enzyme is synthesized as XDH, but can be readily converted irreversibly to XO by oxidation of sulfhydryl residues, or by proteolysis. It seems that the enzyme in milk only exists in the XO form, while the other form, xanthine dehydrogenase (XDH), is mainly intracellular (Godber et al. 1997). Hence only XO is discussed in this chapter. XO purified from human milk showed a single major band on SDS-PAGE, corresponding to a MW of approx. 150,000 Da.

XO is widely distributed throughout various mammalian tissues, especially in liver and small intestine (Pritsos 2000). It is reported that XO activity exists in the milk of all mammals studied, with lower activity in human milk than in bovine milk (Godber et al. 1997). XO is a major protein component of the milk fat globule membrane surrounding the fat droplets, which is highly expressed in mammary epithelial cells.

As shown in Fig. 5.1, XO catalyses hypoxanthine to xanthine and further to uric acid accompanied with the reduction of oxygen to H_2O_2 (Enroth et al. 2000). The Mo centre plays a critical role in this reaction by donating electrons, changing from Mo (VI) to Mo (IV) (Okamoto et al. 2004; Brondino et al. 2006). These electrons are then transmitted to oxygen at the FAD site through Fe_2S_2 cluster (Harrison 2004).

5.2.3.2 Mechanism of Antibacterial Activity of Xanthine Oxidase

The house-keeping enzyme xanthine oxidoreductase (XOR) has been studied intensively over the past 100 years, yet the complexity of its *in vivo* function is still poorly understood. XO has been the focus of considerable research as the terminal enzyme of purine catabolism, from hypoxanthine to xanthine and xanthine to uric acid. However, there is increasing evidence that XO plays an important role in innate immunity. This enzyme interacts with a wide range of substrates and is capable of reducing oxygen to generate the reactive oxygen species (ROS), superoxide and hydrogen peroxide. It can also reduce nitrite, yielding reactive nitrogen species (RNS), such as nitric oxide and peroxynitrite, which has recently been show to be an important signaling molecule (Vorbach et al. 2003). By its involvement in producing ROS and RNS, XO plays an important role in giving antimicrobial protection to the neonatal gut. XO also participates in the LPO system in milk, since LPO

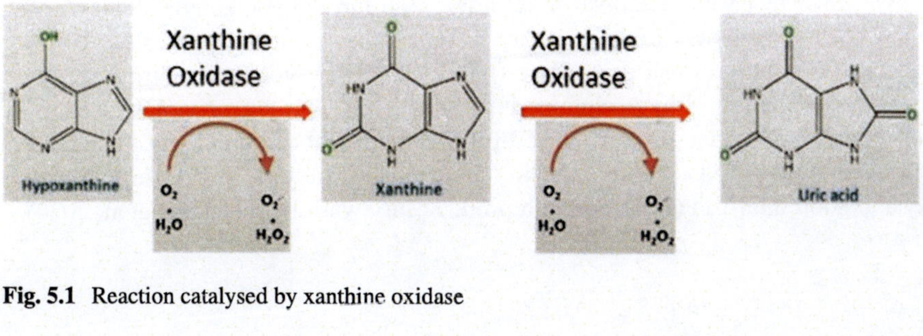

Fig. 5.1 Reaction catalysed by xanthine oxidase

catalyses the reaction of peroxide with thiocyanate (SCN^-) to produce the potent anti-bacterial hypothiocyanite ($OSCN^-$) ion, which inactivates crucial enzyme and protein systems in pathogens through oxidation of free SH-groups (Touch et al. 2004; Welk et al. 2009; Yener et al. 2009; Fox and Kelly 2006; Sermon et al. 2005).

On the basis of the above properties, a role for XO has been proposed in innate immunity, specifically in the inflammatory response and in anti-microbial defence of the gastrointestinal tract. This hypothesis has been supported by the effect of XO on bacteria inactivation in the rat gastrointenstinal tract (Van Den Munckhof et al. 1995). *In vivo*, XO purified from bovine milk is capable of inhibiting the growth of several bacteria, including *Staphylococcus aureus* (Harrison 2006). It has also been found that human milk inhibits the growth of *Salmonella enteritidis* and *E.coli* with the supplementation of hypoxanthine (Stevens et al. 2000).

5.2.3.3 Factors Affecting Antimicrobial Activity of Xanthine Oxidase

XO is sensitive to heat treatments normally used for pasteurising human milk (62.5 °C for 30 min, Holder pasteurisation). XO activity has been reported to be low/zero in thermally-pasteurised human milk (Tully et al. 2001, Akinbi et al. 2010). Another study (Al-Shehri et al. 2015) reported that XO activity was undetectable (<0.1 U/L) in infant formula and pasteurised bovine and human milk. We found that the percent retention of XO activity in human milk drastically reduces to <14% upon heating at and above 62.5 °C for 30 min. Bovine xanthine oxidase, however, has been shown to maintain its apparent affinity and activity for its substrate when bovine milk was pasteurised using batch or HTST pasteurisation, but it was inactivated by ultra-high temperature (UHT) treatment (Ozturk et al. 2019).

5.2.3.4 Assay Methods for Xanthine Oxidase Activity

XO activity is usually assayed spectrophotometrically at 290 nm by the production of uric acid, using xanthine or hypoxanthine as substrate, although very high sensitivity cannot be achieved (Kostic et al. 2015). Fluorometric methods are also available, e.g., the pterin assay of XO (Baghiani et al. 2003), but the extinction coefficient of the fluorescent product is relatively low.

We have recently developed a sensitive high-throughput kinetic fluorescence assay for measurement of XO activity in milk samples, using a coupled enzyme reaction with horseradish peroxidase (HRP). The H_2O_2 generated during the XO catalysed oxidation of hypoxanthine is utilised by HRP to turn Amplex® Red to a red-fluorescent oxidation product, resorufin; the amount of fluorescence generated corresponds to the XO activity. The limit of detection was 0.0003 U/L for XO activity and good inter-run and intra-run reproducibility was obtained (Zou et al. 2020).

5.2.4 *The XO-LPO "Oxidative Enzyme System' in Human Milk and Its Activation by Infant Saliva*

We recently showed that when human milk mixes with infant saliva, XO is activated by high levels of salivary hypoxanthine and xanthine to produce H_2O_2 (Al-Shehri et al. 2015). This in turn activates LPO to produce bacteriostatic free radicals (Fig. 5.2). We showed that, in human infant saliva, the median concentrations of the XO substrates hypoxanthine and xanthine were ten-fold higher (27 and 19 μM, respectively) compared to that of adults (2.1 and 1.7 μM, $p < 0.05$). We confirmed that fresh breastmilk contains ~25 μM H_2O_2 (median). We demonstrated that equal mixing of neonatal saliva and breastmilk in vitro reactivates XO to generate >40 μM H_2O_2, sustained >1 hour. These concentrations, present during breast-feeding, were sufficient to inhibit *in vitro* opportunistic pathogens such as *Staphylococcus aureus* and *Streptococcus* species, but was tolerated by beneficial commensal bacteria such as *Lactobacillus* spp. (Al-Shehri et al. 2015; Sweeney et al. 2018). Our work demonstrates that the complex metabolic interactions of infant saliva with human milk constitute a previously unrecognised element of the infant's primal immunity.

5.2.5 *Polyamine Oxidases*

5.2.5.1 General Characteristics of Polyamine Oxidase

Polyamine oxidase (PAO, EC 1.5.3.11) is a flavoprotein, which plays a key role in polyamine catabolism. It is abundantly expressed in most tissues and biological fluids of vertebrates (Takao and Shirahata 2011) and has been reported to be present in mature human milk (Bjelakovic et al. 2012). It catalyses the oxidation of the secondary amino groups of spermine and spermidine. The spermine is firstly converted to spermidine and then to putrescine by PAO. The biodegradation of putrescine is carried out by diamine oxidase (DAO, EC 1.4.3.22), with generation of γ-Aminobutyric acid or malondialdehyde. Using O_2, the substrates spermin/spermidine are oxidised by PAO to form equimolar amounts of spermidine/putrescine,

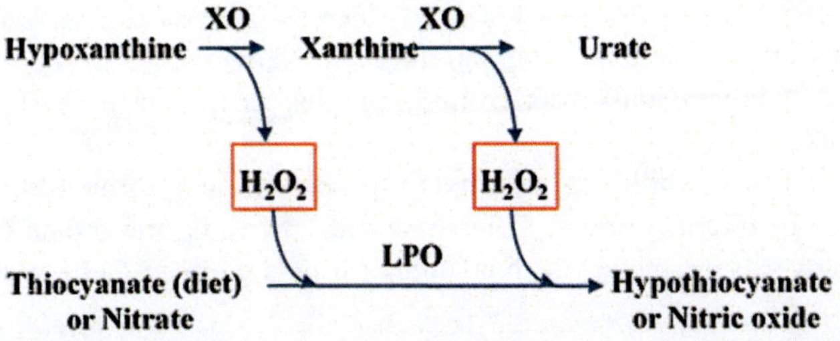

Fig. 5.2 XO-LPO 'oxidative enzyme system' in milk and infant saliva. XO: xanthine oxidase, LPO: lactoperoxidase

3-aminopropanal and hydrogen peroxide (H_2O_2). Human PAO was firstly cloned and characterised in 2001 (Wang et al. 2001). The human PAO gene encompasses seven exons and six introns; the protein contains 555 amino acids, with an apparent molecular weight of 61.9 kDa (Wang et al. 2001).

5.2.5.2 Mechanism of Antibacterial Activity of Polyamine Oxidase

Polyamines, mainly spermine, spermidine and to a less content putrescine have been shown to be critical for mammalian cell proliferation, differentiation and migration. They also play a regulatory role on structure of intestinal epithelial cell membrane and decrease mucosal permeability to antigenic proteins (Vujcic et al. 2003; Yamashiro et al. 1989). It was proposed that the increased PAO activity in human milk during the first month of lactation enhances the maternal milk polyamine supply and improves gut maturation of suckling babies (Bjelakovic et al. 2012).

As the substrates spermine/spermidine are regular constituents of milk (Sanguansermsri et al. 1974), the metabolites generated from PAO-catalysed oxidation reactions of spermine/spermidine, particularly H_2O_2, induce oxidative stress and contribute to the protective effects of milk. Moreover, the H_2O_2 produced by PAO can be further utilised by indigenous milk LPO for generation of bacteriostatic hypothiocyanite; thus, PAO may supplement the antimicrobial role of XO in the XO-LPO "oxidative enzyme system" in human milk described above. Although DAO is another H_2O_2-generating enzyme indigenously present in milk, the low concentration of this enzyme as well as its substrate, putrescine in human milk might limit its bacteriostatic activity (Bjelakovic et al. 2012). We propose an antimicrobial mechanism that involves XO, PAO and LPO as shown in Fig. 5.3.

5.2.5.3 Factors Affecting Activity of Polyamine Oxidase

PAO activity has been successfully identified in human and bovine milk (Morgan and Toothill 1985). In human milk, the activity decreased for the first 2 days after delivery and then stayed fairly constant from the second to the tenth day (Morgan and Toothill 1985). In another study, a steady increase in PAO activity was observed during first month of lactation in human milk (Bjelakovic et al. 2012). The level of PAO activity in bovine milk was around half that in human milk (Morgan and Toothill 1985).

The PAO in bovine milk was reported to be heat resistant through the manufacturing process of infant formula. Compared with raw milk, more than 60% of the enzyme activity was retained in the final infant formula (Gomez-Gallego et al. 2016).

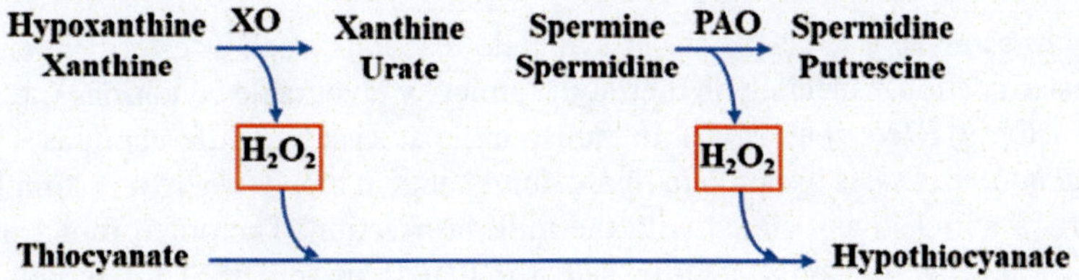

Fig. 5.3 Proposed antimicrobial mechanism of XO, PAO and LPO in milk. XO, xanthine oxidase; PAO, polyamine oxidase; LPO, lactoperoxidase

5.2.5.4 Assay Methods for Polyamine Oxidase

Natural polyamines or fluorinated polyamines can be used as substrates for PAO activity measurement. The oxidation products are usually measured by high-performance liquid chromatography (HPLC) and post-column labelling with *o*-phthalaldehyde or precolumn dansyl-derivatization needs to be conducted for the detection (Halline and Brasitus 1990; Vujcic et al. 2002). Determination of stable-isotope-labelled polyamines has also been successfully applied for PAO activity measurement using mass spectrometry (MS) (Niitsu 2002). Besides detection of consumption of polyamine substrates and generation of polyamine products, methods that quantify hydrogen peroxide (H_2O_2) (Goodwin et al. 2011) or aldehyde (Morgan 1998) are also used for PAO activity assay. However, aldehyde is not produced by certain plant PAOs.

The influence of aminoaldehyde dehydrogenase, which might be present in the sample and oxidises 3-aminopropanal needs to be taken into consideration when measuring formation of the aldehyde (Morgan 1998). The highly sensitive H_2O_2/fluorescence method, which couples H_2O_2 production by PAO to horseradish peroxidase (HRP) catalysed fluorescent reaction has been widely used. Luminol was used as the HRP substrate to generate excited 3-aminophthalate ion (extinction coefficient = 7650 $M^{-1}.cm^{-1}$) in the coupled reaction (Goodwin et al. 2011). The Amplex® Red assay we recently developed (Zou et al. 2020) for XO activity measurement has been successfully adapted for PAO activity determination in milk samples. The extinction coefficient of the oxidation product resorufin is 58,000 $M^{-1}.cm^{-1}$ and the limit of detection of the assay was determined to be 0.0007 U/L (1 U of PAO activity was defined as the production of 1 μmol of H_2O_2 per min, using spermine as the substrate).

5.2.6 *Lipases*

5.2.6.1 General Characteristics of Lipases

Lipases (EC 3.1.1.3) are defined as a class of hydrolases capable of hydrolyzing insoluble tri-, di-, and monoglycerides into glycerol and free fatty acids (Shuai et al. 2017). Besides hydrolytic properties, lipases have gained interest as a result of their ability to catalyze a large variety of synthetic reactions, such as esterification and transesterification, especially in thermodynamically favourable conditions (i.e., low water activity) (Reis et al. 2009). In human milk, at least two different lipases have been identified. One is lipoprotein lipase, found both in blood, where it is stimulated by serum, and also associated with the milk-fat fraction. The other, found in the aqueous portion of milk, is bile salt-stimulated lipase (BSSL) (Hernell and Olivecrona 1974b; Hernell and Olivecrona 1974a).

BSSL is generated by the mammary gland and comprises ~1% of the total human milk protein. The amount of BSSL secreted through lactation by mothers is relatively large (~100 mg/L milk), for both term (Hernell and Blackberg 1983) and preterm infants (Hamosh 1983). The larger part of BSSL can be found in milk and the lesser part is secreted by the infant's exocrine pancreas. Human BSSL has 722 amino acid residues and a total molecular mass of ~105 kDa, which includes 15–20% carbohydrate (Nilsson et al. 1990).

BSSL has the capacity to hydrolyze a number of ester substrates, while the form of the substrates, such as monomeric, micellar, or emulsified form, has almost no impact on enzyme activity. In addition to its non-specificity for substrates, BSSL shows no positional specificity, which means BSSL can hydrolyze all three ester bonds in the triglycerides. A dominant feature of BSSL is that its activity is activated by bile salts, especially for long-chain triglycerides (Hernell and Olivecrona 1974a).

BSSL, as well as gastric and colipase-dependent pancreatic lipases, are responsible for the efficient absorption and utilisation of human milk fat in the newborn infant (Piemontese et al. 2012). Gastric lipase is essential for initiating digestion of milk triglycerides in the neonate stomach. BSSL then further hydrolyses the products of triglyceride digestion from monoglyceride to glycerol and free fatty acid, which can be readily absorbed in the intestine. Moreover, BSSL may hydrolyze a variety of other substrates such as cholesteryl esters, fat-soluble vitamin esters and long-chain polyunsaturated fatty acids ($C > 18$) in micelles to promote their use in neonatal digestion (Hernell and Bläckberg 1994). As it is often the case in the neonatal period, if the intraluminal bile salt concentrations are low, then the end product pattern of free glycerol and free fatty acids may be favourable for absorption (Wang et al. 1983). It has been shown that the inactivation of BSSL by pasteurization of human milk reduces fat absorption by as much as 30% (Williamson et al. 1978).

In addition to its key role in fat digestion, it has been reported that BSSL is a major contributor that enables the absorption of long-chain polyunsaturated fatty acids (Hernell et al. 1993). These fatty acids are fundamental not only as structural components in the brain and the retina, but also as the precursors for eicosanoids. *In vivo*, colipase-dependent lipase has been reported to have limited activity on the hydrolysis of esters with long-chain polyunsaturated fatty acids. However, BSSL does not show discrimination between these and other fatty acids. In addition, some fatty acids released by BSSL may promote an infant's health by producing an antibacterial effect (Hamosh 1995).

Another lipase that has been identified in human milk is lipoprotein lipase (LPL), which has a much lower activity than that of BSSL. LPL exists as a homodimer of 450 amino acid residues, with a molecular mass of ~90 kDa. The optimal temperature and pH of LPL is 37 °C and ~ 9. LPL catalyzes hydrolysis of triglycerides in blood and milk, as well as very low density lipoprotein (VLDL) in blood. Almost all LPL activity in human milk is in the cream, while in bovine milk only 30% of LPL activity is in the cream and the rest is found in casein micelles in the skim milk portion (Hernell and Olivecrona 1974b). A detailed review of lipases in milk and dairy products can be found in Chap. 11 of this book.

5.2.6.2 Mechanism of Antibacterial Activity of Lipases

Lipases have been reported to have antimicrobial effects during storage at 4 °C or digestion in the gastrointestinal tract (Liao et al. 1984; Isaacs et al. 1986). This was also proven by another study where infant formula with added lipase produced antiviral and antibacterial activity, although the effect varied with the types of lipase and bacteria or virus (Isaacs et al. 1992). One working theory behind this phenomenon is that free fatty acids and monoglycerides released from milk triglycerides during lipase lipolysis have antimicrobial activity, which can be duplicated using purified fatty acids and monoglycerides (Thormar et al. 1987). The short-chain (butyric, caproic, and caprylic) and the long-chain (palmitic and stearic) fatty acids in milk have little or no antiviral effect even at extremely high concentration, while medium-chain saturated and long-chain unsaturated fatty acids are all antiviral (Thormar et al. 1987).

Some other theories related to the antimicrobial activity of BSSL itself have been proposed by other researchers. One component in BSSL, Lewis X, could bind to dentritic cell-specific ICAM3-grabbing nonintegrin (DC-SIGN) and inhibit the transfer of human immunodeficiency virus (HIV) -1 to cluster of differentiation (CD) 4+ T cells (Naarding et al. 2006). Furthermore, human milk BSSL, which contains α1–2 linked fucose, can play a role as a decoy receptor and inhibit norovirus attachment to gastroduodenal tissue (Ruvoen-Clouet et al. 2006). Also, BSSL prevents the binding of *Strepthococcus* mutans to saliva-coated hydroxyapatite *in vitro* (a model for tooth enamel) (Niemi et al. 2009).

LPL also plays a key role in antimicrobial activity. Although BSSL activity is several hundred-fold greater than that of LPL in human milk, samples with relatively high levels of LPL developed antimicrobial activity during storage at 4 °C (Isaacs et al. 1986). Human milk with naturally low LPL activity achieved the same antimicrobial activity with endogenous LPL activity when purified LPL was added. It was also reported that bovine milk, which contains LPL but not BSSL, exhibits antimicrobial activity during storage (Isaacs et al. 1986).

5.2.6.3 Factors Affecting Activity of Lipases

Bile salt stimulated lipase (BSSL) activity is highly heat-labile (Baro et al. 2011). Holder pasteurisation, commonly used for pasteurisation of human milk, destroys BSSL activity completely (Hamosh 2001). It was also reported that the lipase activity was completely lost even after 1 min when heating at 62.5 °C (Hernell and

Olivecrona 1974b). However, there is disagreement over the effect of high temperature short time (HTST) on BSSL activity. Baro et al. (Baro et al. 2011) found that the lipase activity and integrity was preserved after HTST treatment, while we have recently found that BSSL was relatively unaffected by HTST pasteurization (72 °C for 15 s), and our result was consistent with those of others (Farkye and Imafidon 1995). We showed that the residual activities of LPL and BSSL in human milk were reduced dramatically from 82% to 21%, and 87% to 2%, respectively when the heating temperature was increased from 50 to 55 °C. No BSSL activity was detected in bovine, camel and caprine milk or in infant formula (unpublished data).

5.2.6.4 Assay Methods for Lipases

A number of measurement methods, based on the release of either fatty acids or glycerol from acylglycerols or on the hydrolysis of synthetic fatty acid esters, have been proposed to assay lipase. Lipase activity is traditionally measured using sodium hydroxide to titrate the acidity generated from the release of fatty acids. However, the sensitivity of this method is so low that only activities higher than 1 μmol/min can be detected (Bier 1955; Stotz 1955; Stadler et al. 1995). Another spectrophotometric method for lipase activity measurement involves the release of *p*-nitrophenol from *p*-nitrophenyl esters. The limitation of this method is that these esters can also be hydrolysed by non-specific esterases and proteases often found in biological samples. Spectroscopic assays have been developed to determine the increase in turbidity when fatty acids released by lipase activity are precipitated with the addition of calcium (Tigerstrom and Stelmaschuk 1989). This method is reported as being 36 times more sensitive than the titrimetric method (Bier 1955; Stotz 1955) and four times more sensitive than the spectrophotometric method (Winkler and Stuckmann 1979). However, this method is less sensitive when lipase activity has to be measured in turbid solutions, such as milk.

Recently, we measured the activity of LPL and BSSL in human milk using a flourometric method that measures fluorescent methylresorufin. LPL was detected directly by reaction of milk with the substrate, while BSSL was detected after the addition of bile salts. For LPL, the method showed linear activity from 65 to 523 U/L and the detection limit was 65 U/L. For BSSL, the linear range was from 10 to 267 U/L and the detection limit was18 U/L. In human milk, the mean activity of LPL was 523 ± 4 U/L and the mean activity of BSSL was 2670 ± 53 U/L.

5.2.7 N-Acetyl-β-D-Glucosaminidase (NAGase)

5.2.7.1 General Characteristics of NAGase and Its Activity

N-Acetyl-ß-D-glucosamindase (NAGase: EC 3.2.1.30) is a lysosomal enzyme and its activity usually has been used as an indicator of mastitis (Kitchen et al. 1978). This enzyme is secreted in large quantities in milk by mammary gland during involution and inflammation. The NAGase enzyme has also been found in other secretions, such as uterine fluids. The specific function of NAGase in milk is still unknown; however, some research has suggested that NAGase may exhibit some antimicrobial activity (Hussain et al. 1992).

There is a relationship between the presence of pathogens in the udder and NAGase levels in milk. Marked increases in NAGase activity resulting from the presence of major mastitis pathogens have been observed. Since NAGase has been found in uterine fluids, it has been suggested that NAGase may have a role in the bactericidal function of the uterus as well. Researchers have studied the bactericidal effect of NAGase on several common potential bacterial pathogens that infect the cow. Of these pathogens, *Actionmyces pyogenes*, *Staphylococcus aureus*, *Streptococcus agalactiae*, and *Pseudomonas aeroginosa* were inhibited by NAGase, while *Escherichia coli* and *Enterobacter aerogenes* were not inhibited (Hussain et al. 1992). Although these results cannot be directly extrapolated to bacterial strains that cause mastitis, they do lend support for such an antimicrobial function of NAGase in the mammary gland. The mechanism of antimicrobial activity of NAGase is not clear. NAGase is not heat-stable and is inactivated by HTST pasteurization (Andrews et al. 1987).

5.2.7.2 Assay Method for Activity of NAGase

A spectrofluorometric method to detect NAGase activity was developed using 4-methylumbelliferyl-N-acetyl-β-d-glucosaminide as substrate, which is considered to be superior to another substrate, *p*-nitrophenyl-N-aceltyl-β-d-glucosaminide (Kitchen et al. 1978). The analysis is conducted at 37 °C and the release of 4-methylumbelliferone fluorometerically measured using maximal excitation at 365 nm and maximal emission at 450 nm. The authors also proposed measurement of NAGase activity using the spectrofluorometric method as a diagnostic tool for detecting mastitis in cows. A high level of NAGase activity was found in mammary gland secretory cells and appeared to be the major source of the enzyme in milk. NAGase activity from other sources (white blood cells, blood serum) was only a minor proportion of the total activity in milk. Thus, NAGase test was proposed as a means of mastitis diagnosis in animals. Later on, Chagunda et al. (2006) compared the activities of L-lactate dehydrogenase (LDH) and NAGase in bovine milk as indicators of non-specific mastitis and found LDH activity to be better than NAGase activity at classifying clinically mastitic cows as sick.

5.2.8 Platelet-Activating Factor (PAF) Acetylhydrolase

5.2.8.1 General Characteristics, Mechanism and Activity of PAF Acetylhydrolase

Platelet-activating factor (PAF), 1-*O*-alkyl-2-acetyl-*sn*-glycero-3-phosphocholine, is known as a potent pro-inflammatory agent that could lead to necrotizing enterocolitis (NEC) in babies (Caplan et al. 1994; Caplan and MacKendrick 1994). PAF transmits signals by binding to a G-protein-coupled receptor and then activating downstream events that contribute to epithelial apoptosis, leukocyte activation, and the initiation of NEC in infants (Caplan et al. 1992; Hsueh et al. 1994; Sun et al. 1994; Prescott et al. 2000). The enzyme that inactivates PAF, PAF acetylhydrolase (PAF-AH), shows a specificity for the hydrolysis of sn-2 acyl group of PAF and structurally related phospholipids, thereby abolishing the signal and regulating inflammatory response (Stafforini et al. 1987; Tjoelker et al. 1995a; Karasawa 2006).

PAF-AH (EC.3.1.1.47) is a Ca^{2+} independent enzyme belonging to group 7 of the family of phospholipase A_2 (Six and Dennis 2000). There are three isoforms of PAF-AH: plasma PAF-AH, also known as lipoprotein-associated phospholipase A_2 (Lp-PLA_2), and intracellular Types I and II PAF-AH (Arai 2002). Although plasma PAF-AH was first discovered in plasma (Blank et al. 1981), it has also been found in other extracellular fluids such as milk and the urine (Moya et al. 1994; Moon et al. 2003). This enzyme in human milk has been identified as the plasma type PAF-AH according to the inhibitor studies (Furukawa et al. 1993). Plasma PAF-AH has a molecular weight of 43 kDa with a monomer polypeptide. The amino acid sequence of human plasma PAF-AH is 441 amino acid residues and contains a Gly-X-Ser-X-Gly (GXSXG) motif, which is found in esterases, lipases and other hydrolase superfamily of enzymes (Tjoelker et al. 1995a; Tjoelker et al. 1995b).

Most PAF-AH activity (97%) in milk is associated with aqueous phase, while little activity is present in the cream or centrifugal pellet fraction. This enzyme is stable at pH values between 4 and 9, which indicates its importance in retaining its activity in small intestine after transport through the stomach of the neonate. PAF-AH activity in milk from various species was determined and it was found that milk from humans, pigs, goats and rats contain similar activity, which is around 10 nmol/min/ml in colostrum and 4 nmol/min/mL after 1 week postpartum. The activity in ovine milk is low, while it is undetectable in bovine milk (Furukawa et al. 1993). Due to its significance to the neonate, PAF-AH activity in human milk was measured at different gestational ages. The activity of PAF-AH was similar in colostrum of term and preterm (between 26 and 32 weeks of gestation) deliveries, at 2.7 nmol/min/mL and 3.0 nmol/min/mL, respectively. Mothers who delivered between 33 and 36 weeks of gestation had higher PAF-AH activity (5.6 nmol/min/mL) than the former two groups (Moya et al. 1994). Recombinant PAF-AH showed significant effect on decreasing the bacterial number of *Salmonella Typhimurium* and *Escherichia coli* after the cecal ligation and puncture procedure (Teixeira-da-Cunha et al. 2013).

5.2.8.2 Assay Method for Activity of PAF Acetylhydrolase

PAF-AH activity has been assayed in a mixture of Tris-HCl, bovine serum albumin, nonradiolabeled PAF and [^{3}H]PAF, as well as human milk. After incubation, the reaction was terminated by addition of trichloroacetic acid. The activity of PAF-AH was quantified by the release of [^{3}H]acetate using liquid scintillation spectroscopy (Miwa et al. 1988; Furukawa et al. 1993; Moya et al. 1994).

5.3 Summary

Milk contains several important immunoprotective and nutritional biomolecules. The antimicrobial enzymes present in milk, alone or with cofactors, can prevent the growth of microorganisms through a variety of mechanisms. The variety and cocentration of these antimicrobial enzymes indigenously present in milk show huge interspecies variation. Perhaps the availablity of such antimicrobial factors in milk is governed by the physioloical and immunological needs of the neonate of a particular species. These antimicrobial enzymes can inactivate (or inhibit the growth of) the unwanted microorganisms through several mechanisms. For example, lysozyme could attack the structure of bacterial cell wall, thus physically killing the bacteria, while lactoperoxidase, xanthine oxidase and polyamine oxidase cooperate with each other to generate antimicrobial products. Although the exact antibacterial mechanism of lipase is unclear, it has been shown that some free fatty acids as well as monoglycerides generated by lipolytic reaction of lipase could inhibit the growth of bacteria. Bile salt stimulated lipase (BSSL), a form of lipase in human milk, may play a greater role in antimicrobial activity than lipoprotein lipase (LPL). Together, these antimicrobial enzymes in maternal milk constitute an important element of the infant's primal immunity. Additional clinical evidence of a positive role of these natural biochemical mechanisms in improving infants' health outcomes is needed. Clinical research-based supplementation of infant formula preparations with immunoprotective biomolecules inherently present in maternal milk can improve feeding solutions for infants where sufficient maternal milk is unavailable.

A great deal is known about the structure and properties of most of these enzymes. Perhaps future research can focus on development of sensitive, high throughput methods for accurately analysing low concentrations of the minor enzymes in some types of milk, such as human milk. Development of presevration systems that harness these indegenous antimcirbial enzymes to extend the shelf-life of milk can also be of commercial significance, for bovine milk when refrigeration is not readily available (such as the protocol developed by The UN Food and Agriculture Organization in collaboration with the World Health Organization) and for human milk in donor milk banks.

References

Akinbi H, Meinzen-Derr J, Auer C, Ma Y, Pullum D, Kusano R, Reszka KJ, Zimmerly K (2010) Alterations in the host defense properties of human milk following prolonged storage or pasteurization. J Pediatr Gastroenterol Nutr 51(3):347–352

Al-Shehri SS, Knox CL, Liley HG, Cowley DM, Wright JR, Henman MG, Hewavitharana AK, Charles BG, Shaw PN, Sweeney EL, Duley JA (2015) Breastmilk-saliva interactions boost innate immunity by regulating the Oral microbiome in early infancy. PLoS One 10(9):e0135047

Andrews AT, Anderson M, Goodenough PW (1987) A study of the heat stabilities of a number of indigenous milk enzymes. J Dairy Res 54(2):237–246

Arai H (2002) Platelet-activating factor acetylhydrolase. Prostaglandins Other Lipid Mediat 68:83–94

Atasever A, Ozdemir H, Gulcin M, Kufrevioglu OI (2013) One-step purification of lactoperoxidase from bovine milk by affinity chromatography. Food Chem 136(2):864–870

Bafort F, Parisi O, Perraudin JP, Jijakli MH (2015) Mode of action of lactoperoxidase as related to its antimicrobial activity: a review. Enzyme Res 2014:517164

Baghiani A, Harrison R, Benboubetra M (2003) Purification and partial characterisation of camel Milk xanthine oxidoreductase. Arch Int Physiol 111(5):407–414

Ballard O, Morrow AL (2013) Human milk composition: nutrients and bioactive factors. Pediatr Clin 60(1):49–74

Baro C, Giribaldi M, Arslanoglu S, Giuffrida M, Dellavalle G, Conti A, Tonetto P, Biasini A, Coscia A, Fabris C (2011) Effect of two pasteurization methods on the protein content of human milk. Front Biosci (Elite Ed) 3:818–829

Benkerroum N (2008) Antimicrobial activity of lysozyme with special relevance to milk. Afr J Biotechnol 7(25):4856–4867

Bier M (1955) [106] lipases: RCOOR′+ H2O→ RCOOH+ R′ OH. Methods Enzymol.

Bjelakovic L, Kocic G, Bjelakovic B, Najman S, Stojanovic D, Jonovic M, Pop-Trajkovic Z (2012) Polyamine oxidase and diamine oxidase activities in human milk during the first month of lactation. Iran J Pediatr 22:218–222

Björkstén B, Burman LG, De Château P, Fredrikzon B, Gothefors L, Hernell O (1980) Collecting and banking human milk: to heat or not to heat? Br Med J 281(6243):765–769

Blake C, Koenig D, Mair G, North A, Phillips D, Sarma V (1965) Structure of hen egg-white lysozyme: a three-dimensional Fourier synthesis at 2 Å resolution. Nature 206(4986):757–761

Blank ML, Lee T, Fitzgerald V, Snyder F (1981) A specific acetylhydrolase for 1-alkyl-2-acetyl-sn-glycero-3-phosphocholine (a hypotensive and platelet-activating lipid). J Biol Chem 256(1):175–178

Blel M, Guingamp M-F, Gaillard J-L, Humbert G (2001) Improvement of a method for the measurement of lactoperoxidase activity in milk. Int Dairy J 11(10):795–799

Booth KS, Kimura S, Lee HC, Ikeda-Saito M, Caughey WS (1989) Bovine myeloperoxidase and lactoperoxidase each contain a high affinity site for calcium. Biochem Biophys Res Commun 160(2):897–902

Brondino CD, Romão MJ, Moura I, Moura JJ (2006) Molybdenum and tungsten enzymes: the xanthine oxidase family. Curr Opin Chem Biol 10(2):109–114

Cals MM, Mailliart P, Brignon G, Anglade P, Dumas BR (1991) Primary structure of bovine lactoperoxidase, a fourth member of a mammalian heme peroxidase family. FEBS J 198(3):733–739

Caplan MS, Hedlund E, Adler L, Hsueh W (1994) Role of asphyxia and feeding in a neonatal rat model of necrotizing enterocolitis. Pediatr Pathol 14(6):1017–1028

Caplan MS, Kelly A, Hsueh W (1992) Endotoxin and hypoxia-induced intestinal necrosis in rats: the role of platelet activating factor. Pediatr Res 31(5):428–433

Caplan MS, MacKendrick W (1994) Inflammatory mediators and intestinal injury. Clin Perinatol 21(2):235–246

Carlsson J, Iwami Y, Yamada T (1983) Hydrogen peroxide excretion by oral streptococci and effect of lactoperoxidase-thiocyanate-hydrogen peroxide. Infect Immun 40(1):70–80

Chagunda MG, Larsen T, Bjerring M, Ingvartsen KL (2006) L-lactate dehydrogenase and N-acetyl-beta-D-glucosaminidase activities in bovine milk as indicators of non-specific mastitis. J Dairy Res 73(4):431–440

Chandan RC, Parry RM, Shahani KM (1965) Purification and some properties of bovine milk lysozyme. Biochim Biophys Acta 110(2):389–398

Chandan RC, Parry RM, Shahani KM (1968) Lysozyme, lipase, and ribonuclease in Milk of various Species1, 2. J Dairy Sci 51(4):606–607

Chandan RC, Shahani KM, Holly RG (1964) Lysozyme content of human Milk. Nature 204:76–77

Chang J-C, Chen C-H, Fang L-J, Tsai C-R, Chang Y-C, Wang T-M (2013) Influence of prolonged storage process, pasteurization, and heat treatment on biologically-active human milk proteins. Pediatr Neonatol 54(6):360–366

Christen L, Lai CT, Hartmann B, Hartmann PE, Geddes DT (2013) The effect of UV-C pasteurization on bacteriostatic properties and immunological proteins of donor human Milk. PLoS One 8(12):e85867

Czank C, Prime DK, Hartmann B, Simmer K, Hartmann PE (2009) Retention of the immunological proteins of pasteurized human milk in relation to pasteurizer design and practice. Pediatr Res 66(4):374–379

Demba S, Eszter T, Aaron G, Balázs R (2018) Antimicrobial actions of dual oxidases and lactoperoxidase. J Microbiol 56(6):373–386

DeWit JN, Van Hooydonk ACM (1996) Structure, functions and applications of lactoperoxidase in natural antimicrobial systems. Neth Milk Dairy J 50:227–244

Eitenmiller R, Barnhart H, Shahani K (1971) Effect of γ-irradiation on bovine and human milk lysozymes. J Food Sci 36(7):1127–1130

Eitenmiller R, Friend BA, Shahani K, Ball EM (1974) Immunochemical studies on human and bovine milk lysozymes. J Food Sci 39(5):930–933

Elfakharany E, Uversky V, Redwan E (2016) Comparative analysis of the antiviral activity of camel, bovine, and human Lactoperoxidases against herpes simplex virus type 1. Appl Biochem Biotechnol 182(1):1–17

Enroth C, Eger BT, Okamoto K, Nishino T, Nishino T, Pai EF (2000) Crystal structures of bovine milk xanthine dehydrogenase and xanthine oxidase: structure-based mechanism of conversion. Proc Natl Acad Sci 97(20):10723–10728

Farkye N, Imafidon G (1995) Thermal denaturation of indigenous milk enzymes. In: Fox PF (ed) Heat-induced changes in Milk. International Dairy Federation, Brussels, pp 331–348

Fleming A (1922) On a remarkable bacteriolytic element found in tissues and secretions. Proc R Soc London Ser B 93(653):306–317

Fuglsang CC, Johansen C, Christgau S, Adler-Nissen J (1995) Antimicrobial enzymes: applications and future potential in the food industry. Trends Food Sci Technol 6(12):390–396

Furukawa M, Narahara H, Yasuda K, Johnston JM (1993) Presence of platelet-activating factor-acetylhydrolase in milk. J Lipid Res 34(9):1603–1609

Godber B, Sanders S, Harrison R, Eisenthal R, Bray RC (1997) > or = 95% of xanthine oxidase in human milk is present as the demolybdo form, lacking molybdopterin. Biochem Soc Trans 25(3):519s

Gomez-Gallego C, Recio I, Gomez-Gomez V, Ortuno I, Bernal MJ, Ros G, Periago MJ (2016) Effect of processing on polyamine content and bioactive peptides released after in vitro gastrointestinal digestion of infant formulas. J Dairy Sci 99(2):924–932

Goodwin AC, Murray-Stewart TR, Casero RA Jr (2011) A simple assay for mammalian spermine oxidase: a polyamine catabolic enzyme implicated in drug response and disease. Methods Mol Biol (Clifton, NJ) 720:173–181

Gothefors L, Marklund S (1975) Lactoperoxidase activity in human milk and in saliva of newborn infants. Infect Immun 11(6):1210–1215

Griffiths MW (1986) Use of Milk enzymes as indices of heat treatment. J Food Prot 49(9):696–705

Halline AG, Brasitus TA (1990) Reversed-phase high-performance liquid chromatographic method for the measurement of polyamine oxidase activity. J Chromatogr B Biomed Sci Appl 533(C):187–194

Hamosh M (2001) Bioactive factors in human milk. Pediatr Clin N Am 48(1):69–86

Hamosh M (1983) Bile-salt-stimulated lipase of human milk and fat digestion in the preterm infant. J Pediatr Gastroenterol Nutr 2(Suppl 1):S248–S251

Hamosh M (1995) Enzymes in human milk-C. In: Jensen RG (ed) Handbook of milk composition. Academic Press, Cambridge, pp 388–427

Harrison R (2002) Structure and function of xanthine oxidoreductase: where are we now? Free Radic Biol Med 33(6):774–797

Harrison R (2004) Physiological roles of xanthine oxidoreductase. Drug Metab Rev 36(2):363–375
Harrison R (2006) Milk xanthine oxidase: properties and physiological roles. Int Dairy J 16(6):546–554
Hernell O, Blackberg L (1983) Bile-salt-stimulated lipase of human milk and lipid digestion in the neonatal period. J Pediatr Gastroenterol Nutr 2(Suppl 1):S242–S247
Hernell O, Bläckberg L (1994) Human milk bile salt-stimulated lipase: functional and molecular aspects. J Pediatr 125(5):S56–S61
Hernell O, Blackberg L, Chen Q, Sternby B, Nilsson A (1993) Does the bile salt-stimulated lipase of human milk have a role in the use of the milk long-chain polyunsaturated fatty acids? J Pediatr Gastroenterol Nutr 16(4):426–431
Hernell O, Olivecrona T (1974a) Human milk lipases II. Bile salt-stimulated lipase. Biochim Biophys Acta (BBA) 369(2):234–244
Hernell O, Olivecrona T (1974b) Human milk lipases. I Serum stimulated lipase. J Lipid Res 15(4):367–374
Hsueh W, Caplan M, Sun X, Tan X, MacKendrick W, Gonzalez-Crussi F (1994) Platelet-activating factor, tumor necrosis factor, hypoxia and necrotizing enterocolitis. Acta Paediatr 83:11–17
Huopalahti R, Anton M, López-Fandiño R, Schade R (2007) Bioactive egg compounds. Springer, Berlin
Huppertz T, Fox PF, de Kruif KG, Kelly AL (2006) High pressure-induced changes in bovine milk proteins: a review. Biochim Biophys Acta 1764(3):593–598
Hussain A, Daniel R, Frost A (1992) The bactericidal effect of N-acetyl-β-D-glucosaminidase on bacteria. Vet Microbiol 32(1):75–80
Isaacs CE, Litov RE, Marie P, Thormar H (1992) Addition of lipases to infant formulas produces antiviral and antibacterial activity. J Nutr Biochem 3(6):304–308
Isaacs CE, Thormar H, Pessolano T (1986) Membrane-disruptive effect of human milk: inactivation of enveloped viruses. J Infect Dis 154(6):966–971
Jiménez-Saiz R, Martos G, Carrillo W, López-Fandiño R, Molina E (2011) Susceptibility of lysozyme to in-vitro digestion and immunoreactivity of its digests. Food Chem 127(4):1719–1726
Johnson LN (1998) The early history of lysozyme. Nat Struct Mol Biol 5(11):942–944
Jollès J, van Leemputten E, Mouton A, Jollès P (1972) Amino acid sequence of Guinea-hen egg-white lysozyme. Biochim Biophys Acta (BBA) 257(2):497–510
Jolles P (1967) Relationship between chemical structure and biological activity of hen egg-white lysozyme and lysozymes of different species. Proc R Soc London Ser B 167(1009):350–364
Jolles P, Jolles J (1961) Lysozyme from human Milk. Nature 192:1187–1188
Jolles P, Jolles J (1984) What's new in lysozyme research? Always a model system, today as yesterday. Mol Cell Biochem 63(2):165–189
Karasawa K (2006) Clinical aspects of plasma platelet-activating factor-acetylhydrolase. Biochim Biophys Acta (BBA) 1761(11):1359–1372
Kerkaert B, Mestdagh F, De Meulenaer B (2010) Detection of hen's egg white lysozyme in food: comparison between a sensitive HPLC and a commercial ELISA method. Food Chem 120(2):580–584
Kitchen B, Middleton G, Salmon M (1978) Bovine milk N-acetyl-β-D-glucosaminidase and its significance in the detection of abnormal udder secretions. J Dairy Res 45(1):15–20
Klebanoff S, Clem W, Luebke R (1966) The peroxidase-thiocyanate-hydrogen peroxide antimicrobial system. Biochim Biophys Acta (BBA) 117(1):63–72
Koenig Á, Diniz EMDA, Barbosa SFC, Vaz FAC (2005) Immunologic factors in human Milk: the effects of gestational age and pasteurization. J Hum Lact 21(4):439–443
Korhonen H (1980) A new method for preserving raw milk-the lactoperoxidase antibacterial system. World Anim Rev 35:23–29
Korhonen H, Marnila P, Gill H (2000) Milk immunoglobulins and complement factors. Br J Nutr 84(S1):75–80
Kostic DA, Dimitrijević DS, Stojanović GS, Palić IR, Đorđević AS, Ickovski JD (2015) Xanthine oxidase: isolation, assays of activity, and inhibition. J Chem 2015:1–8

Kussendrager KD, Van Hooijdonk A (2000) Lactoperoxidase: physico-chemical properties, occurrence, mechanism of action and applications. Br J Nutr 84(S1):19–25

León-Sicairos N, López-Soto F, Reyes-López M, Godínez-Vargas D, Ordaz-Pichardo C, De La Garza M (2006) Amoebicidal activity of milk, apo-lactoferrin, sIgA and lysozyme. Clin Med Res 4(2):106–113

Liao TH, Hamosh P, Hamosh M (1984) Fat digestion by lingual lipase: mechanism of lipolysis in the stomach and upper small intestine. Pediatr Res 18(5):402–409

Manas P, Munoz B, Sanz D, Condon S (2006) Inactivation of lysozyme by ultrasonic waves under pressure at different temperatures. Enzym Microb Technol 39(6):1177–1182

Marks N, Grandison A, Lewis M (2001) Challenge testing of the lactoperoxidase system in pasteurized milk. J Appl Microbiol 91(4):735–741

Masschalck B, Deckers D, Michiels CW (2002) Lytic and nonlytic mechanism of inactivation of gram-positive bacteria by lysozyme under atmospheric and high hydrostatic pressure. J Food Prot 65(12):1916–1923

Masschalck B, Van Houdt R, Van Haver EG, Michiels CW (2001) Inactivation of gram-negative bacteria by lysozyme, denatured lysozyme, and lysozyme-derived peptides under high hydrostatic pressure. Appl Environ Microbiol 67(1):339–344

Mayayo C, Montserrat M, Ramos SJ, Martínez-Lorenzo MJ, Calvo M, Sánchez L, Pérez MD (2016) Effect of high pressure and heat treatments on IgA immunoreactivity and lysozyme activity in human milk. Eur Food Res Technol 242(6):891–898

McKenzie H, Shaw D (1985) The amino acid sequence of equine milk lysozyme. Biochem Int 10(1):23–31

McSweeney PL, Fox PF (2003) Advanced dairy chemistry, vol 1. Springer, New York

Miwa M, Miyake T, Yamanaka T, Sugatani J, Suzuki Y, Sakata S, Araki Y, Matsumoto M (1988) Characterization of serum platelet-activating factor (PAF) acetylhydrolase. Correlation between deficiency of serum PAF acetylhydrolase and respiratory symptoms in asthmatic children. J Clin Invest 82(6):1983–1991

Moon TC, Kim MS, Lee SJ, Lee TY, Kwon SH, Baek SH, Chang HW (2003) Detection and characterization of 45 kDa platelet activating factor acetylhydrolase in cerebrospinal fluid of children with meningitis. Arch Pharm Res 26(7):554–558

Morgan DML (1998) Measurement of aldehyde formation as an assay for polyamine oxidase activity. In: Morgan DML (ed) Polyamine protocols. Humana Press, Totowa, NJ, pp 105–108

Morgan DML, Toothill VJ (1985) Polyamine oxidase activity in human and bovine milk. Biochem Soc Trans 13(2):352–353

Morriss FH Jr, Brewer ED, Spedale SB, Riddle L, Temple DM, Caprioli RM, West MS (1986) Relationship of human milk pH during course of lactation to concentrations of citrate and fatty acids. Pediatrics 78(3):458–464

Moya FR, Eguchi H, Zhao B, Furukawa M, Sfeir J, Osorio M, Ogawa Y, Johnston JM (1994) Platelet-activating factor acetylhydrolase in term and preterm human milk: a preliminary report. J Pediatr Gastroenterol Nutr 19(2):236–239

Naarding MA, Dirac AM, Ludwig IS, Speijer D, Lindquist S, Vestman E-L, Stax MJ, Geijtenbeek TBH, Pollakis G, Hernell O (2006) Bile salt-stimulated lipase from human milk binds DC-SIGN and inhibits human immunodeficiency virus type 1 transfer to CD4+ T cells. Antimicrob Agents Chemother 50(10):3367–3374

Naidu A (2000) Natural food antimicrobial systems. CRC press, Boca Raton

Niemi LD, Hernell O, Johansson I (2009) Human milk compounds inhibiting adhesion of mutans streptococci to host ligand-coated hydroxyapatite in vitro. Caries Res 43(3):171–178

Niitsu M (2002) Simultaneous determination of endogenous and orally administered 15N-labeled polyamines in rat organs. Anal Biochem 301(2):255–260

Nilsson J, Blackberg L, Carlsson P, Enerback S, Hernell O, Bjursell G (1990) cDNA cloning of human-milk bile-salt-stimulated lipase and evidence for its identity to pancreatic carboxylic ester hydrolase. Eur J Biochem 192(2):543–550

Okamoto K, Matsumoto K, Hille R, Eger BT, Pai EF, Nishino T (2004) The crystal structure of xanthine oxidoreductase during catalysis: implications for reaction mechanism and enzyme inhibition. Proc Natl Acad Sci U S A 101(21):7931–7936

Ozturk G, German JB, de Moura BJMLN (2019) Effects of industrial heat treatments on the kinetics of inactivation of antimicrobial bovine milk xanthine oxidase. NPJ Sci Food 3(1):13

Fox PF, Kelly AL (2006) Indigenous enzymes in milk: overview and historical aspects—part 1. Int Dairy J 16:500–516

Parry R Jr, Chandan R, Shahani K (1969) Isolation and characterization of human milk lysozyme. Arch Biochem Biophys 130:59–65

Pellegrino L, Tirelli A (2000) A sensitive HPLC method to detect hen's egg white lysozyme in milk and dairy products. Int Dairy J 10(7):435–442

Piemontese P, Roncada P, Soggiu A, Bonizzi L, Budelli A, Agostoni C, Giannì M, Roggero P, Mosca F (2012) 357 human Milk bile salt-stimulated lipase: the dynamic changes during lactation. Arch Dis Child 97(Suppl 2):A105–A105

Prescott SM, Zimmerman GA, Stafforini DM, McIntyre TM (2000) Platelet-activating factor and related lipid mediators. Annu Rev Biochem 69(1):419–445

Pritsos CA (2000) Cellular distribution, metabolism and regulation of the xanthine oxidoreductase enzyme system. Chem Biol Interact 129(1–2):195–208

Purdy M, Tenovuo J, Pruitt K, White W (1983) Effect of growth phase and cell envelope structure on susceptibility of Salmonella typhimurium to the lactoperoxidase-thiocyanate-hydrogen peroxide system. Infect Immun 39(3):1187–1195

Quan R, Yang C, Rubinstein S, Lewiston NJ, Sunshine P, Stevenson DK, Kerner JA (1992) Effects of microwave radiation on anti-infective factors in human Milk. Pediatrics 89(4):667

Rauch P, Hochel I, Kaš J (1990) Sandwich enzyme immunoassay of hen egg lysozyme in foods. J Food Sci 55(1):103–105

Reis P, Holmberg K, Watzke H, Leser M, Miller R (2009) Lipases at interfaces: a review. Adv Colloid Interf Sci 147:237–250

Reiter B, Marshall VM, BjörckL RCG (1976) Nonspecific bactericidal activity of the lactoperoxidases-thiocyanate-hydrogen peroxide system of milk against Escherichia coli and some gram-negative pathogens. Infect Immun 13(3):800–807

Rosen H (1986) The Lactoperoxidase system: chemistry and biological significance. (book review), vol 61. Dekker, New York

Ruvoen-Clouet N, Mas E, Marionneau S, Guillon P, Lombardo D, Le Pendu J (2006) Bile-salt-stimulated lipase and mucins from milk of 'secretor' mothers inhibit the binding of Norwalk virus capsids to their carbohydrate ligands. Biochem J 393(Pt 3):627–634

Sanguansermsri J, György P, Zilliken F (1974) Polyamines in human and cow's milk. Am J Clin Nutr 27(8):859–865

Schneider N, Becker C-M, Pischetsrieder M (2010) Analysis of lysozyme in cheese by immunocapture mass spectrometry. J Chromatogr B 878(2):201–206

Schneider N, Werkmeister K, Becker C-M, Pischetsrieder M (2011) Prevalence and stability of lysozyme in cheese. Food Chem 128(1):145–151

Scott D (1988) Antimicrobial enzymes. Food Biot. 2(2):119–132

Seifu E, Buys EM, Donkin E (2005) Significance of the lactoperoxidase system in the dairy industry and its potential applications: a review. Trends Food Sci Technol 16(4):137–154

Sermon J, Vanoirbeek K, Spiegeleer PD, Houdt RV, Aertsen A, Michiels CW (2005) Unique stress response to the lactoperoxidase-thiocyanate enzyme system in Escherichia coli. Res Microbiol 156(2):225–232

Shin K, Hayasawa H, Lönnerdal B (2001) Purification and quantification of lactoperoxidase in human milk with use of immunoadsorbents with antibodies against recombinant human lactoperoxidase. Am J Clin Nutr 73(5):984–989

Shin K, Tomita M, Lönnerdal B (2000) Identification of lactoperoxidase in mature human milk. J Nutr Biochem 11(2):94–102

Shuai W, Das RK, Naghdi M, Brar SK, Verma M (2017) A review on the important aspects of lipase immobilization on nanomaterials. Biotechnol Appl Biochem 64(4):496–508
Six DA, Dennis EA (2000) The expanding superfamily of phospholipase A2 enzymes: classification and characterization. Biochim Biophys Acta (BBA) 1488(1–2):1–19
Sousa SG, Delgadillo I, Saraiva JA (2014) Effect of thermal pasteurisation and high-pressure processing on immunoglobulin content and lysozyme and lactoperoxidase activity in human colostrum. Food Chem 151:79–85
Stadler P, Kovac A, Haalck L, Spener F, Paltauf F (1995) Stereoselectivity of microbial lipases: the substitution at position sn-2 of triacylglycerol analogs influences the Stereoselectivity of different microbial lipases. Eur J Biochem 227(1–2):335–343
Stafforini DM, Prescott SM, McIntyre TM (1987) Human plasma platelet-activating factor acetylhydrolase. Purification and properties. J Biol Chem 262(9):4223–4230
Stevens C, Millar T, Clinch J, Kanczler J, Bodamyali T, Blake D (2000) Antibacterial properties of xanthine oxidase in human milk. Lancet 356(9232):829–830
Stotz E (1955) [109] Liver esterase. Methods Enzymol 1:657–659
Sun X, Caplan M, Hsueh W (1994) Tumour necrosis factor and endotoxin synergistically activate intestinal phospholipase A2 in mice. Role of endogenous platelet activating factor and effect of exogenous platelet activating factor. Gut 35(2):215–219
Sweeney EL, Al-Shehri SS, Cowley DM, Liley HG, Bansal N, Charles BG, Shaw PN, Duley JA, Knox CL (2018) The effect of breastmilk and saliva combinations on the in vitro growth of oral pathogenic and commensal microorganisms. Sci Rep 8(1):15112
Takao K, Shirahata A (2011) Fluorescent substrates for polyamine catabolic enzymes and transport. In: Pegg AE, Casero JRA (eds) Polyamines: methods and protocols. Humana Press, Totowa, NJ, pp 463–474
Tayefi-Nasrabadi H, Hoseinpour-Fayzi MA, Mohasseli M (2011) Effect of heat treatment on lactoperoxidase activity in camel milk: a comparison with bovine lactoperoxidase. Small Rumin Res 99(2–3):187–190
Teixeira-da-Cunha MG, Gomes RN, Roehrs N, Bozza FA, Prescott SM, Stafforini D, Zimmerman GA, Bozza PT, Castro-Faria-Neto HC (2013) Bacterial clearance is improved in septic mice by platelet-activating factor-acetylhydrolase (PAF-AH) administration. PLoS One 8(9):e74567
Tenovuo J (1985) The peroxidase system in human secretions. In: The lactoperoxidase system: Chemistry and biological significance. Dekker, New York, pp 101–122
Thallinger B, Prasetyo EN, Nyanhongo GS, Guebitz GM (2013) Antimicrobial enzymes: an emerging strategy to fight microbes and microbial biofilms. Biotechnol J 8(1):97–109
Thormar H, Isaacs CE, Brown HR, Barshatzky MR, Pessolano T (1987) Inactivation of enveloped viruses and killing of cells by fatty acids and monoglycerides. Antimicrob Agents Chemother 31(1):27–31
Tigerstrom RGV, Stelmaschuk S (1989) The use of tween 20 in a sensitive turbidimetric assay of lipolytic enzymes. Can J Microbiol 35(4):511–514
Tjoelker LW, Eberhardt C, Unger J, Trong HL, Zimmerman GA, McIntyre TM, Stafforini DM, Prescott SM, Gray PW (1995a) Plasma platelet-activating factor Acetylhydrolase is a secreted phospholipase A2 with a catalytic triad. J Biol Chem 270(43):25481–25487
Tjoelker LW, Wilder C, Eberhardt C, Stafforinit DM, Dietsch G, Schimpf B, Hooper S, Le Trong H, Cousens LS, Zimmerman GA (1995b) Anti-inflammatory properties of a platelet-activating factor acetylhydrolase. Nature 374(6522):549–553
Touch V, Hayakawa S, Yamada S, Kaneko S (2004) Effects of a lactoperoxidase–thiocyanate–hydrogen peroxide system on Salmonella enteritidis in animal or vegetable foods. Int J Food Microbiol 93(2):175–183
Tully DB, Jones F, Tully MR (2001) Donor milk: what's in it and what's not. J Hum Lact 17(2):152–155
Vakil J, Chandan R, Parry R, Shahani K (1969) Susceptibility of several microorganisms to milk lysozymes. J Dairy Sci 52(8):1192–1197

Van Den Munckhof RJM, Vreeling-Sindelárová H, Schellens JPM, Van Noorden CJF, Frederiks WM (1995) Ultrastructural localization of xanthine oxidase activity in the digestive tract of the rat. Histochem J 27(11):897–905

Van Hooijdonk AC, Kussendrager K, Steijns J (2000) In vivo antimicrobial and antiviral activity of components in bovine milk and colostrum involved in non-specific defence. Br J Nutr 84(S1):127–134

Varaldo PE, Valisena S, Mingari MC, Satta G (1989) Lysozyme-induced inhibition of the lymphocyte response to Mitogenic lectins. Proc Soc Exp Biol Med 190(1):54–62

Viazis S, Farkas BE, Allen JC (2007) Effects of high-pressure processing on immunoglobulin a and lysozyme activity in human Milk. J Hum Lact 23(3):253–261

Vorbach C, Harrison R, Capecchi MR (2003) Xanthine oxidoreductase is central to the evolution and function of the innate immune system. Trends Immunol 24(9):512–517

Vujcic S, Diegelman P, Bacchi CJ, Kramer DL, Porter CW (2002) Identification and characterization of a novel flavin-containing spermine oxidase of mammalian cell origin. Biochem J 367(Pt 3):665–675

Vujcic S, Liang P, Diegelman P, Kramer DL, Porter CW (2003) Genomic identification and biochemical characterization of the mammalian polyamine oxidase involved in polyamine backconversion. Biochem J 370:19–28

Wang CS, Kuksis A, Manganaro F, Myher JJ, Downs D, Bass HB (1983) Studies on the substrate specificity of purified human milk bile salt-activated lipase. J Biol Chem 258(15):9197–9202

Wang Y, Devereux W, Woster PM, Stewart TM, Hacker A, Casero RA (2001) Cloning and characterization of a human polyamine oxidase that is inducible by polyamine analogue exposure. Cancer Res 61(14):5370–5373

Welk A, Meller C, Schubert R, Schwahn C, Kramer A, Below H (2009) Effect of lactoperoxidase on the antimicrobial effectiveness of the thiocyanate hydrogen peroxide combination in a quantitative suspension test. BMC Microbiol 9(1):134

Williamson S, Finucane E, Ellis H, Gamsu HR (1978) Effect of heat treatment of human milk on absorption of nitrogen, fat, sodium, calcium, and phosphorus by preterm infants. Arch Dis Child 53(7):555–563

Winkler UK, Stuckmann M (1979) Glycogen, hyaluronate, and some other polysaccharides greatly enhance the formation of exolipase by Serratia marcescens. J Bacteriol 138(3):663–670

Wolfson LM, Sumner SS (1993) Antibacterial activity of the Lactoperoxidase system: a review. J Food Prot 56(10):887–892

Yamashiro Y, Sato M, Shimizu T, Oguchi S, Maruyama K, Kitamura S (1989) Possible biological growth factors in breast Milk and postnatal development of the gastrointestinal tract. Pediatr Int 31(4):417–423

Yang B, Wang J, Tang B, Liu Y, Guo C, Yang P, Yu T, Li R, Zhao J, Zhang L (2011) Characterization of bioactive recombinant human lysozyme expressed in milk of cloned transgenic cattle. PLoS One 6(3):e17593

Yener FY, Korel F, Yemenicioğlu A (2009) Antimicrobial activity of lactoperoxidase system incorporated into cross-linked alginate films. J Food Sci 74(2):M73–M79

Yoshida A, Takagaki Y, Nishimune T (1991) Enzyme immunoassay for hen egg white lysozyme used as a food additive. J Assoc Off Anal Chem 74(3):502–505

Zou Z, Bouchereau-De Pury C, Hewavitharana AK, Al-Shehri SS, Duley JA, Cowley DM, Koorts P, Shaw PN, Bansal N (2020) A sensitive and high-throughput fluorescent method for determination of oxidase activities in human, bovine, goat and camel milk, Food Chemistry

Chapter 6
Enzymes Associated with Milk Phospholipid Membrane Structures: Milk Fat Globule Membranes and Extracellular Vesicles

Maria Stenum Hansen and Jan Trige Rasmussen

6.1 Introduction

Estimates of the total amount of phospholipids (PL) in bovine milk vary considerably, which might reflect the use of different extraction methods. Addressing this query, Barry and co-workers reported values for PL to be between 0.9 and 2.3% of the total lipid content in bovine milk; the first value was based on the Röse Gottlieb method and the latter from a Folch extraction (Barry et al. 2016).

The membrane (MFGM) enveloping and solubilizing milk fat in the form of milk fat globules accounts for 60–65% of milk PL, whereas the remaining 35–40% is found in the skim milk phase (Morton 1954; Huang and Kuksis 1967; Patton and Keenan 1971). Even though the existence of phospholipid membrane structures in skim milk has been known for a long time, knowledge about their biological origin and possible functions in milk is just now emerging. At the beginning of the 1950s Morton reported the presence of a specific type of phospholipid particles in milk (Morton 1953). These structures were named "milk microsomes", and in the following decades the origin of "milk microsomes" was discussed (e.g., Bailie and Morton 1958a, b; Wellings and Deome 1961; Plantz and Patton 1973). However, as evidence accumulated, it was generally accepted that skim milk phospholipid components originate from detached remnants of the outer layer of MFGM (e.g., reviewed by Fox and Kelly 2006; Silanikove et al. 2006). Recent results contradict this proposal, because milk from various species has been shown to contain nano-sized phospholipid structures in skimmed milk that are distinct from milk fat globules (MFGs). They are generically termed extracellular vesicles (EVs), because they are likely to encompass both exosomes and microvesicles (Admyre et al. 2007; Hata et al. 2010; Gu et al. 2012; Blans et al. 2017).

M. S. Hansen · J. T. Rasmussen (✉)
Aarhus University, Department of Molecular Biology and Genetics, Aarhus C, Denmark
e-mail: jatr@mbg.au.dk

A. L. Kelly, L. B. Larsen (eds.), *Agents of Change*, Food Engineering Series,
https://doi.org/10.1007/978-3-030-55482-8_6

Besides milk, EVs have been identified in a number of other biological fluids and rediscovered as an important part of intercellular communication, which involves genetic exchanges of RNA and microRNAs (Valadi et al. 2007; Théry et al. 2009; Pegtel et al. 2010). Importantly, although MFGs and EVs differ in intracellular origin, they are both secreted through the apical membrane in a way that results in overlapping content of constituents. The purpose of the present contribution is to provide an overview of data and views on significant indigenous membrane-associated milk enzymes. The number of enzymes identified in MFGs and EVs has increased considerably following the use of refined proteomic techniques. However, the abundance of many of these enzymes is very low, and they largely end up in these two fractions as carry over from ER- or Golgi membranes or cytosolic crescents (known to follow MFGs, especially in human milk), e.g., superoxide dismutase. Moreover, a large pool of the MFGM/EV-associated enzymes will be inactive in milk due to the lack of relevant substrates or a suitable environment, and they are therefore not considered here. Also omitted is another category of enzymes that has been reported in MFGM-preparations, i.e., enzymes from the processes involved in synthesizing the MFG core lipids, e.g., fatty acid synthase, acetyl-CoA carboxylase, and long-chain-fatty-acid CoA ligase. Thus, only enzymes with significance for mammary gland biology, milk integrity, and/or physiological function upon consumption will be considered here. Table 6.1 summarizes characteristics of the enzymes which will be discussed.

6.2 Sulfhydryl Oxidase

Sulfhydryl oxidase (EC. 1.8.3.2) catalyses the oxidation of protein thiols (SH-groups of cysteine residues) to form disulphides with simultaneous reduction of oxygen to hydrogen peroxide. This activity may contribute to disulphide bond formation in secreted proteins. The existence of both a metal- and a flavin-dependent type of sulfhydryl oxidases has been reported in milk. Swaisgood and coworkers performed a large amount of work in connection to sulfhydryl oxidase activity in bovine milk, and they attributed this to a 89-kDa protein containing 0.5 equiv of iron per monomer (Janolino and Swaisgood 1975). Investigations with purified samples resulted in data on steady-state kinetics, amino acid composition, presence of carbohydrate (~10%), and showed that the enzyme activity could be inhibited by EDTA (Janolino and Swaisgood 1975). However, a later study failed to verify the existence of significant amounts of iron-binding sulfhydryl oxidase activities in bovine milk (Jaje et al. 2007). Instead, the presence of a flavin-dependent sulfhydryl oxidase in bovine milk was methodically documented, and information from sequences of derived peptides revealed the protein to be a member of the Quiescin-sulfhydryl oxidase family, more precisely QSOX1. These authors sought for an explanation for the discrepancy between their own and the previously published observations but did not reach any. Coming research may shed light on this unsolved query.

Table 6.1 Selected list of enzymes reported to be associated with milk fat globules (MFG) or extracellular vesicles (EV)

Enzyme	Enzyme number	UniProt/ gene name	Presence by proteomics in bovine/human MFG and/or EV	Comments
Sulfhydryl oxidase	EC 1.8.3.2	F1MM32/ QSOX1	MFG[a,b]	Flavotype
Catalase[c]	EC 1.11.1.6	P00432	n.d.	From somatic cells/ microorganisms
Lactoperoxidase[c] (LPO)	EC 1.11.1.7	P80025	MFG[a,b,d,e], EV[b]	Abundant in skim milk and whey
Xanthine oxidoreductase (XOR)	EC 1.17.3.2 XO EC 1.17.1.4 XD	P80457	MFG[a,b,d,e,f], EV[b,g]	Associates with butyrophilin/ adipophilin
γ-Glutamyltransferase (GGT)	EC 2.3.2.2 (EC 3.4.19.13)	G3N2D8 (human P19440)	MFG[a,b], EV[b]	Heterodimer
5'-Nucleotidase	EC 3.1.3.5	Q05927	MFG[b,f], EV[b]	Ecto-5'-nucleotidase
Alkaline phosphatase	EC 3.1.3.1	P09487 (tissue non-sp)	MFG[b], EV[b]	Described in Chap. 4
Lipoprotein lipase		P11151	MFG[a,b,e], EV[b]	Described in Chap. 10

[a]Liao et al. (2011)
[b]Reinhardt et al. (2013)
[c]Suggested not to be classified as a "true" MFG or EV enzyme
[d]Fong et al. (2007)
[e]Ma et al. (2019)
[f]Reinhardt and Lippolis (2006)
[g]Admyre et al. (2007)

Early research assigned sulfhydryl oxidase activity to phospholipid membranes from both skim milk and cream, in a 2.7 ratio, respectively (Kitchen 1974). The firmness of this association has been questioned in proceeding investigations, implying a more loose membrane association rendering the enzyme as a more soluble protein (Jaje et al. 2007). Nevertheless, proteomic studies have clearly shown the presence of F1MM32/QSOX1 in membrane fractions of both human and bovine milk (Liao et al. 2011; Reinhardt et al. 2013).

6.2.1 Structure of Flavin-Dependent Sulfhydryl Oxidase

In 1993, Coppock et al. described identification of two novel cDNA clones that were up-regulated in a human WI38 fibroblast cell line transitioned from logarithmic growth to quiescence (Coppock et al. 1993). It later turned out that the two related mRNAs corresponded to splice variants that are now called QSOX1 long (QSOX1-L or QSOX1a) and short (QSOX1-S or QSOX1b). QSOX1-S and –L share identical sequences until the middle of exon 12. The short form misses 733 base pairs of the 12th exon and it continues in the 13th exon, terminating after what are the last two amino acids of the encoded 604 amino acid protein (63.8 kDa). The long form (747 amino acid, 79.6 kDa) represents most of exon 12 and includes a transmembrane region (reviewed by Lake and Faigel 2014). The existence of seven additional splice variants has been reported, but to what extent they are expressed remains to be investigated (Soloviev et al. 2013).

Figure 6.1b summarises structural features of the QSOX1-enzyme, which is a multi-domain protein formed by fusion of two ancient genes: a tandem pair of thioredoxin (Trx) domains, related to protein disulphide isomerase (PDI), a region predicted to be helix-rich (HRR), an Erv/ALR FAD-binding module, and finally a region of high sequence variability. In QSOX1-L, the C-terminal also encloses a membrane spanning region. Furthermore, a CxxC motif is found in three places in the enzyme. This motif is common in proteins involved in dithiol/disulphide redox

A

```
Bovine QSOX1-S - F1MM32 (edited with knowledge from Jaje et al. 2007)
MGWCGRGSGP PPSRLLMLLS LLLAVRGAGA APRSALYSSS DPLTLLRADT VRSTVLGSSS  30
AWAVEFFASW CGHCIAFAPT WKALANDVKD WRPALNLAAL DCAEETNSAV CRDFNIPGFP  90
TVRFFKAFSK TGSGTTLSVA GADVQTLRER LIDALESHSD TWPPACPPLE PARLEEITGF 150
FARNNEEYLA LIFEKEGSYL GREVTLDLSQ HQGIAVRRVL NTERDVVNRF GVTNFPSCYL 210
LSRNGSFSRV PALTESRSFY TTYLRKFSGS TRGAVQTTAA PATTSAVAPT VWKVADRSKI 270
YMADLESALH YILRIEVGKF SVLEGQRLVA LKKFMAVLAK YFRGRPLVQN FLHSMNDWLK 330
KQQRKKIPYG FFKNALDSRK EGTVIAEKVN WVGCQGSEPH FRGFPCSLWI LFHFLTVQAA 390
QEGVDHPQER AKAQEVLQAI RGYVRFFFGC RECAGHFEQM ASGSMHRVGS LNSAVLWFWS 450
SHNKVNARLA GAPSEDPQFP KVQWPPRELC SACHNELRGT PVWDLDNILK FLKTHFSPSN 510
IVLDFPSAGP GPWRGAERMA VIPKQLI                                    537
```

B

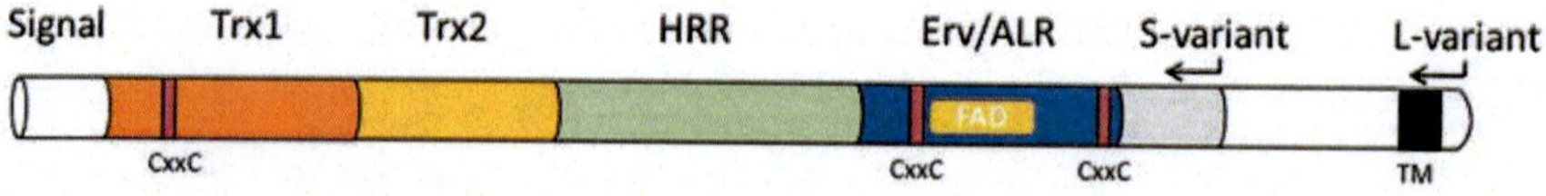

Fig. 6.1 Structure of bovine flavin-dependent sulfhydryl oxidase. (**a**) Amino acid sequence with signal sequence (underlined italics); N-glyc (bold); FAD-binding (underlined); dithiol/disulphide redox active cysteines (grey bold italics). (**b**) Schematic presentation of enzymatic domains and structural features of the short and long variant (S and L, respectively). *Trx* thioredoxin binding domain, *HRR* helix-rich region, *Erv/ALR* FAD-binding domain, *TM* transmembrane region. The amino acid sequence data is available under accession number F1MM32, where additional information can be obtained

reactions, and a model for flow of reducing equivalents has been presented for avian and human QSOX (Heckler et al. 2008). Interestingly, it was shown that the distal CxxC pair in human QSOX does not have an essential role in catalysis of disulphide bond formation in dithiothreitol or reduced RNase. This is not the case for QSOX, from an evolutionary lower level. The bovine sequence of QSOX1-S (537 amino acids) is presented in Fig. 6.1a, which is constructed from Uniprot entry F1MM32 and Jaje and coworkers (Jaje et al. 2007). A putative N-glycosylation site is found at Asn-214, and database notifications for human QSOX1 suggest that Arg-372, Trp-379, His-383, Glu-422, His-426, Lys-471, and Trp-474 are involved in the FAD-binding. The QSOX enzyme has been reported to be expressed in a wide variety of cells and tissues, and it has been suggested that there is a correlation between QSOX-overexpression and tumorigenesis (Antwi et al. 2009; Katchman et al. 2013). Furthermore, QSOX1-S is shown to be much more prevalent than the long form, which is also seen for mammary-derived cell lines (Lake and Faigel 2014). QSOX1-S has also been isolated as a soluble enzyme from mammalian blood serum (Israel et al. 2014). Finally, it has also been reported that QSOX1-L may be proteolytically modified and secreted into the extracellular matrix, as a consequence of removal of the transmembrane domain (Rudolf et al. 2013); in the same study, dimerization of QSOX1 was also demonstrated.

Jaje et al. (2007) reported isolation of a pure sample of sulfhydryl oxidase from bovine skim milk after application of five different purification steps. The separated enzyme migrated in SDS-PAGE as a 62 kDa band. After analysing the sequence information from obtained peptides, the authors reached the conclusion that the purified sample matched QSOX1-L, but it seemingly lacked a large part of C-terminus enclosing a transmembrane segment (Jaje et al. 2007). Retrospectively, it seems more likely that they purified QSOX1-S, as also achieved when blood serum was used as a source for QSOX1-activity (Israel et al. 2014).

6.2.2 Biological Role of Flavin-Dependent Sulfhydryl Oxidase and Significance in Milk

It has been shown that QSOX1 is present in cells with high secretory activity, both in association with intracellular membranes compartments, especially Golgi, and at the cell surface, but also as a secreted enzyme (Israel et al. 2014 and cited references). Recently, it has amply been documented that QSOX1 is over-expressed in breast tissue, in pancreas, and during prostate cancer (Ouyang et al. 2005; Song et al. 2009; Antwi et al. 2009; Soloviev et al. 2013; Katchman et al. 2013). This prompted the suggestion that QSOX1 has the potential to be a diagnostic cancer marker (reviewed by Lake and Faigel 2014), and also a marker for heart failure (Mebazaa et al. 2012). The more accurate knowledge about the long and short forms of QSOX1 is relatively new; however, there is still much to learn about their relative contributions at different locations. A lot of good arguments have been presented ascribing a role for QSOX1 during folding of secreted proteins; however, Fass and Thorpe (2018) have pointed out that the exact physiological substrates for intracellular and secreted QSOX1 still have to be discovered, and that the enzyme has a preference for flexible model substrates.

It is well accepted that poor consumer acceptability of UHT milk is attributed to "cooked" and "flat" flavours, which is linked to formation of a variety of sulphur-containing compounds, methyl ketones, and aliphatic aldehydes (Vazquez-Landaverde et al. 2005, 2006). Initially, this off-flavour was attributed to thermal denaturation of milk serum proteins during processing (Patrick and Swaisgood 1976; Vazquez-Landaverde et al. 2006). However, the sulphurous note usually dissipates after three days of storage, probably through oxidation (Allen and Joseph 1985). Simon and Hansen implied that QSOX might have impact on this oxidation process, as 60% of its activity remained after a typical pasteurization treatment (Swaisgood 1980; 2001). Moreover, it has been reported that QSOX immobilised on glass beads can be used as a tool to reduce the "cooked" flavour of UHT-treated milk (Swaisgood et al. 1987). The significance of QSOX in relation to other oxidation products remains unclear.

QSOX1 activity can be measured in a number of ways, e.g., using GSH or reduced RNase as substrates at pH 7, and by time-dependent titration of thiols by DTNB, forms a yellow product (which absorbs at 412 nm) (Janolino and Swaisgood 1975; Hoober et al. 1996). Recently, more high throughput assays have been published (Israel et al. 2014). A dimeric glycoprotein containing one FAD per ~80,000 Mr subunit has been isolated from chicken egg white and found to have sulfhydryl oxidase activity with a range of small molecular weight thiols. Dithiothreitol was the best substrate of those tested, with a turnover number of 1030/min, a K_m of 150 μM, and a pH optimum at about 7.5. Oxidation of thiol substrates generates hydrogen peroxide in aerobic solution. Anaerobically, the ferricenium ion is a facile alternative electron acceptor. Reduction of the oxidase with dithionite or dithiothreitol under anaerobic conditions yields a two-electron intermediate (EH2) showing a charge transfer band (λ_{max} 560 nm; ϵ_{obs} 2.5 mM^{-1} cm^{-1}). Complete bleaching of the flavin and discharge of the charge transfer complex require a total of four electrons. Borohydride and catalytic photoreduction give the same spectral changes. EH2, but not the oxidized enzyme, is inactivated by iodoacetamide, with alkylation of 2.7 cysteine residues/subunit.

These data indicate that the oxidase contains a redox-active disulphide bridge generating a thiolate to oxidized flavin charge transfer complex at the EH2 level. Sulphite treatment does not form the expected flavin adduct with the native enzyme but cleaves the active site disulphide, yielding an air-stable EH2-like species. The close functional resemblance of the oxidase to the pyridine nucleotide-dependent disulphide oxidoreductase family has been discussed (Hoober et al. 1996; Jaje et al. 2007). Investigation of the catalytic activities showed that the k_{cat}/K_m values were slightly higher for bovine milk QSOX1 than those of the egg white counterpart, when testing oxidation of DTT, glutathione, and reduced RNase (Hoober et al. 1996; Jaje et al. 2007). There are general catalytic similarities between the two enzymes, but some of the differences may reflect the fact that a different flow of reduction equivalents has been suggested for avian and human QSOX1 (Heckler et al. 2008).

6.3 Catalase

The enzyme catalase (EC 1.11.1.6) catalyses the breakdown of hydrogen peroxide to generate water and oxygen ($2H_2O_2 \rightarrow 2H_2O + O_2$). It is a common enzyme found in nearly all living organisms exposed to oxygen and is primarily found in peroxisomes. Monitoring the peroxide content is relatively simple and this might be the reason why catalase activity was among the first enzyme activities to be detected in milk. The history behind the early descriptions of catalase in milk has been outlined previously (Fox and Kelly 2006). Through time, catalase has often been classified as an indigenous MFGM enzyme; however, as indicated in Table 6.1, scrutiny of available proteomic data did not document presence of catalase in this milk fraction. This probably reflects the fact that catalase primarily stems from non-milk secreting cells, and is mostly present in milk during mastitis. This section will address this phenomenon, together with additional information related to catalase activity when it occurs in milk. In the authors' opinion, catalase should be disregarded as a true membrane-associated milk enzyme, because it rather originates from somatic milk cells or microorganisms (added or contaminating). However, as catalase historically has been considered as a membrane-associated milk enzyme, it is still reviewed in this chapter.

```
Bovine catalase - P00432
MADNRDPASD QMKHWKEQRA AQKPDVLTTG GGNPVGDKLN SLTVGPRGPL LVQDVVFTDE  60
MAHFDRERIP ERVVHAKGAG AFGYFEVTHD ITRYSKAKVF EHIGKRTPIA VRFSTVAGES 120
GSADTVRDPR GFAVKFYTED GNWDLVGNNT PIFFIRDALL FPSFIHSQKR NPQTHLKDPD 180
MVWDFWSLRP ESLHQVSFLF SDRGIPDGHR HMNGYGSHTF KLVNANGEAV YCKFHYKTDQ 240
GIKNLSVEDA ARLAHEDPDY GLRDLFNAIA TGNYPSWTLY IQVMTFSEAE IFPFNPFDLT 300
KVWPHGDYPL IPVGKLVLNR NPVNYFAEVE QLAFDPSNMP PGIEPSPDKM LQGRLFAYPD 360
THRHRLGPNY LQIPVNCPYR ARVANYQRDG PMCMMDNQGG APNYYPNSFS APEHQPSALE 420
HRTHFSGDVQ RFNSANDDNV TQVRTFYLKV LNEEQRKRLC ENIAGHLKDA QLFIQKKAVK 480
NFSDVHPEYG SRIQALLDKY NEEKPKNAVH TYVQHGSHLS AREKANL             527
```

Fig. 6.2 Amino acid sequence of bovine catalase. Modifications and structural features include removed amino acids (underlined italics); blocked mature N-terminal acetylated alanine (bold and grey); active site phosphorylated residues (bold); N6-succinyl/acetyl lysines (underlined); iron, heme axial ligand (double underlined). The amino acid sequence data is available under accession number P00432 where additional information can be obtained

6.3.1 Structure of Catalase

A reliable amino acid sequence of bovine liver catalase (526 amino acids) was published by Schroeder et al. (1982), and there are numerous reports on the crystallization of the enzyme, e.g., a 2.5 Å resolution structure was presented already in 1981 (Murthy et al. 1981). Features linked to the amino acid sequence of bovine catalase are presented in Fig. 6.2 using the Uniprot entry P00432. Ito and coworkers purified catalase from bovine milk (Ito and Akuzawa 1983a, b). There are some differences in the amino acid sequences between, for example, bovine and human catalase. However, all functionally important amino acid stretches are conserved, and they are structurally very similar. Mammalian catalase is a homotetramer of 60-kDa subunits and belongs to a group of monofunctional catalases. Each subunit contains a heme, ferriprotoporphyrin IX, at the active site. The inside of the enzyme is only accessible for substrates of small size. The fact that the gene for catalase encodes a precursor without a signal sequence and that the mature protein has a blocked N-terminal emphasizes that catalase is not a protein destined for secretion.

6.3.2 Biological Role of Catalase and Significance in Milk

Half a century ago (perhaps even earlier) it was suggested that catalase activity may be used as an indicator of mastitis (Kitchen 1976, 1981; Fitz-Gerald et al. 1981). This concept has been reported to be valid for several breeds and mammalian species (Savas et al. 2011), and the correlation between catalase activity and bacterial counts can be used during quick biosensor-scans for microbial challenged milk (Zhang et al. 2014). Even though this novel strategy has shown promise, conventional tests might be applied to reach confirmatory evidence for mastitis.

Silanikove and coworkers conducted a series of studies addressing the impact of bovine catalase activity in milk (reviewed in Silanikove et al. 2014). They advocate for the notion that bovine catalase catalyzes a hydrogen peroxidase-dependent reaction converting nitrite to nitrate. This oxidation process could be an important of feature for catalase, which is important for keeping a low concentration of free radicals and oxidation products (carbonyls, lipid peroxides, and nitrotyrosine). A site of action for catalase might be in the udder, and it thereby facilitates maintenance of good milk quality at an early stage. Furthermore, results presented showed that intact catalase activity might be important for preventing accumulation of oxidant products in milk after milking and during storage (Silanikove et al. 2009). This is supported by the fact that addition of a catalase inhibitor (3-amino-1,2,4-triazole) results in an increased presence of nitrotyrosine- and lipid peroxides. In milk, both lactoperoxidase and xanthine oxidoreductase are involved in the processes that provide substrates for catalase activity. These two enzymes will be dealt with later in this chapter.

A number of studies revealed that catalase is relatively susceptible to heat treatment, reducing the activity nearly as much as alkaline phosphatase (e.g., Hirvi et al. 1996). However, some evidence suggests that catalase activity can be restored by extending the heat treatment, making the picture a little more blurry. As an example, Griffiths obtained results showing that all catalase activity was largely destroyed after heating at 72 °C for 15 s, but it was also seen that heating a sample that was stored at 4 °C for 24 h to 80 °C for 15 s showed a 20-fold increase in catalase activity. This was suggested to be explained by release of catalase from somatic milk cells and bacteria, and not as reactivation of already heat-denatured catalase (Griffiths 1986). Pasteurization of milk is applied for the sake of public health and to ensure standardized conditions for cheese making, which should lower the impact of catalase activity in these products. However, as evident from the above, catalase might be brought in by presence of somatic cells or stem from deliberately added or unwanted microbes. Variance in catalase activity might also be among the factors making raw milk cheese ripening less easy to control (Gatti et al. 2014; Yoon et al. 2016). Finally, as catalase activity might increase after pasteurization, it does not meet the criteria for being a suitable index for this heat treatment of milk.

Catalase was not isolated from milk until relatively recently. This is most likely because the enzyme mainly comes from non-milk secreting cells and is prevalent during mastitis. However, various aspects of catalase in milk were reported by Prof. O. Ito in a series of papers during the period 1969–1983 (see Ito and Akuzawa 1983a). A procedure for the isolation of catalase from milk was published by Ito and Akuzawa (1983a); the enzyme was purified 23,000-fold and crystallized; it was found by gel permeation to have a molecular mass of 225 kDa. Ito and Akuzawa (1983b) reported that there were three isozymes in the catalase preparation isolated from cream. Ito and Akuzawa (1983b) reported that milk catalase was dissociated by SDS into five sub-units ranging in molecular mass from 11 to 55 kDa. Bovine liver catalase is a homotetramer of 60–65 kDa subunits (total molecular mass 250 kDa). It seems likely that the structure of catalase in milk is similar to that enzyme and that the heterogeneity reported by Ito and Akuzawa (1983b) was due to proteolysis during isolation.

The enzymatic activity of catalase can measured in many different ways. One of the simplest ways is to monitor the rate of disappearance of H_2O_2 by observing the absorbance at 240 nm (Beers and Sizer 1952). Suggestion for new assays are still being published (Hadwan 2018) together with more sensitive ones, which are commercially available from many vendors.

6.4 Lactoperoxidase

Lactoperoxidase (LPO) (EC 1.11.1.7) is a glycoprotein found in exocrine secretions such as saliva, tears, and of course also in milk. It belongs to the family of heme-containing peroxidase enzymes, where members with similar function also are found in plants and fungi. These peroxidases demonstrate differences in the mechanism of ligand binding. The prosthetic heme group in mammalian peroxidases is covalently bound, whereas this type of covalent bond is absent in fungal and plant peroxidases (Sharma et al. 2013). LPO catalyse the breakdown of hydrogen peroxide to generate water, with involvement of an acceptor molecule (H_2O_2 + reduced acceptor → oxidized acceptor + $2H_2O$). The acceptor can be a phenolic compound or alternatively an aromatic amine or an aromatic acid, but it can also be thiocyanate, bromide, or iodine (Kohler and Jenzer 1989). Monitoring the peroxidase activity is relatively simple and LPO was also one of the earliest enzymes to be described in milk (Arnold 1881).

Over time, many chromogenic substrates have been used to monitor peroxidase activity, but right now 2,2'-azino-bis(3-ethylbenzothiazoline-6-sulphonic acid), abbreviated ABTS, is probably the most frequently used substrate. The historical aspects of LPO activity in milk have been covered previously (Fox and Kelly 2006). Classically, LPO has been assigned to skim milk and whey, but application of more refined techniques has revealed that small amounts of this enzyme can be found in MFGM-preparations together with EVs (see Table 6.1). Appearance of LPO in MFGM and EVs most probably reflects cytosolic contamination, which aligns with the fact that more MS-spectral counts were found for LPO in EVs relative to MFGM in a study using an iTRAQ proteomic approach (Reinhardt et al. 2013). As for catalase, it is the authors' opinion that LPO should not be characterized as a true membrane-associated milk enzyme. However, as LPO still might end up in more less pure phospholipid-containing milk fractions, it is still covered in this chapter.

6.4.1 Structure of Lactoperoxidase

The entire amino acid sequence of bovine lactoperoxidase was published in 1990–1991. The isolated cDNA encoded a 712 amino acid long polypeptide, covering a 22 residue long signal peptide and a 78 amino acid propeptide, resulting in a mature protein of 612 amino acid residues (Dull et al. 1990; Cals et al. 1991). Over time, the database entry has been updated with information about residues involved in disulphide bridge pattern, N-glycosylations, phosphorylations, heme-binding and coordination (UniProt P80025), and a 2.3 Å resolution 3D-structure was published in 2009 (Singh et al. 2009). Key features is specified in Fig. 6.3. The gene for bovine lactoperoxidase encodes a precursor with a signal sequence, which underlines its nature as a secreted protein, but there is nothing from the protein structure or other information that implies that LPO should be regarded as a membrane protein.

Due to the versatile applicability of the enzyme, a huge body of methods for its purification has been published. As an example, a new report by Li et al. (2019) described the use of an integrated salting-out and ion-exchange approach on whey, which gave a product with very high purity as judged by SDS-PAGE. This publication also offers a short overview of what has been reported effective for isolation of LPO, e.g., salting-out, membrane filtration, chromatography, two-phase extraction, and immune-affinity chromatography. The purity of an LPO preparation can be

A

```
Bovine lactoperoxidase - P80025
MWVCLQLPVF LASVTLFEVA ASDTIAQAAS TTTISDAVSK VKIQVNKAFL DSRTRLKTTL
SSEAPTTQQL SEYFKHAKGR TRTAIRNGQV WEESLKRLRR DTTLTNVTDP SLDLTALSWE  20
VGCGAPVPLV KCDENSPYRT ITGDCNNRRS PALGAANRAL ARWLPAEYED GLALPFGWTQ  80
RKTRNGFRVP LAREVSNKIV GYLDEEGVLD QNRSLLFMQW GQIVDHDLDF APETELGSNE 140
HSKTQCEEYC IQGDNCFPIM FPKNDPKLKT QGKCMPFFRA GFVCPTPPYQ SLAREQINAV 200
TSFLDASLVY GSEPSLASRL RNLSSPLGLM AVNQEAWDHG LAYLPFNNKK PSPCEFINTT 260
ARVPCFLAGD FRASEQILLA TAHTLLLREH NRLARELKKL NPHWNGEKLY QEARKILGAF 320
IQIITFRDYL PIVLGSEMQK WIPPYQGYNN SVDPRISNVF TFAFRFGHME VPSTVSRLDE 380
NYQPWGPEAE LPLHTLFFNT WRIIKDGGID PLVRGLLAKK SKLMNQDKMV TSELRNKLFQ 440
PTHKIHGFDL AAINLQRCRD HGMPGYNSWR GFCGLSQPKT LKGLQTVLKN KILAKKLMDL 500
YKTPDNIDIW IGGNAEPMVE RGRVGPLLAC LLGRQFQQIR DGDRFWWENP GVFTEKQRDS 560
LQKVSFSRLI CDNTHITKVP LHAFQANNYP HDFVDCSTVD KLDLSPWASR EN         612
```

B

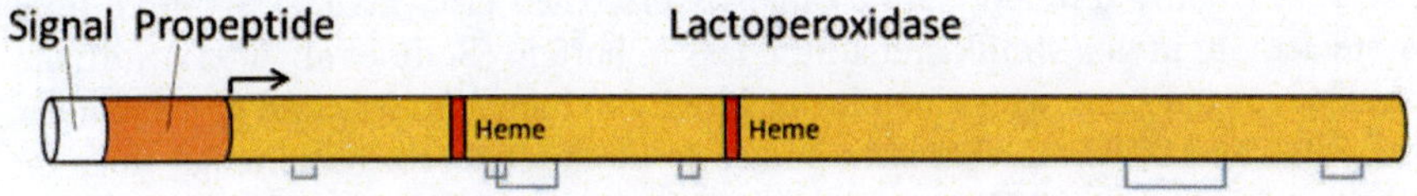

Fig. 6.3 Structure of bovine lactoperoxidase. (**a**) Amino acid sequence with signal sequence (underlined italics); propeptide (grey); N-glyc (bold); phosphorylated residues (bold italics); and heme-binding residues (underlined). The mature protein is formed by the numbered residues. (**b**) Schematic presentation of lactoperoxidase with disulphide bridges (blue lines) and heme coordination indicated. The amino acid sequence data is available under accession number P80025, where additional information can be obtained

determined by absorbance measurements, where values at 412 nm are correlated with the LPO heme concentration and related to $A_{280\ nm}$ giving values for the total content of protein (Andersson et al. 1996).

6.4.2 Biological Role of Lactoperoxidase and Significance in Milk

LPO, together with its inorganic ion substrates, hydrogen peroxide, and oxidized products form what is known as the LPO system, which constitutes an unspecific humoral immune response directed against bacteria, fungi, and viruses (Ihalin et al. 2006). The LPO system works by using hydrogen peroxide to oxidize thiocyanate ions (also iodides and bromides) to hypothiocyanite ions (hypoiodides and hypobromides). Hypothiocyanite ions oxidize amino acid thiol groups leading to impaired protein function and successive antimicrobial action (for early review see, e.g., Reiter and Härnulv 1984). It is well established that the LPO enzyme functions as a natural component of the non-immune biological defence system of mammals and contributes to the neonatal defence system against pathogenic microorganisms in the digestive system (Koksal et al. 2016). It is notable that it has been argued that the LPO-system is inactive in milk and super-physiological amounts of thiocyanate and H_2O_2 must be added for activation (Silanikove et al. 2006). However, another system is active in milk, because it is documented that nitrite reacts with lactoperoxidase, leading to production of the potent radical, nitric dioxide (NO_2^-). This system also involves the action of xanthine oxidase (described below) and catalase activity in the udder, as mentioned above (Silanikove et al. 2005).

LPO is the second most abundant enzyme in milk, after xanthine oxidoreductase, constituting ~0.5% of the total whey proteins (~0.1% of total protein; 30 mg L^{-1}). In 1988, it was reported that human milk lacked LPO, but contains myeloperoxidase instead (Hamosh 1988). However, a later study demonstrated that LPO is commonly present in mature human milk, and the enzyme was isolated from whey using immune affinity chromatography in the final step (Shin et al. 2001). Table 6.2 summarises the inter-species presence of LPO when results obtained with the ABTS are listed. A slightly different set of values is presented elsewhere, assigning a little more activity in goats milk and a little less in buffalo (Seifu et al. 2005). Still, the tendency is the same with much lower activity in human milk, close or even under the value at 0.02 U/mL that is estimated to be the limit for bactericidal function (Reiter and Härnulv 1984).

Soon after milk LPO was demonstrated to lose activity upon heat treatment (Arnold 1881), pasteurization was introduced in the Danish dairy industry in 1891, as advised by Prof N. J. Fjord, and after 1897 nearly all butter was made from pasteurized cream. Among the various tests that have been suggested to determine whether milk was adequately pasteurized, the one developed by Storch stands out (Storch 1898); however, more up to date versions are of course available. Using the principle of the Storch test, residual LPO activity is estimated, and is still used to identify super-pasteurized milk (i.e., milk heated to ≥76 °C for 15 s).

The LPO system is widely used in dairy industry to improve dairy quality, inhibiting the microbial growth, and preserving raw milk during storage and/or transportation to processing plants. The application portfolio has been broadened to other areas in food technology and functional food, e.g., oral hygiene, cold storage shelf-life extension of chicken breast fillets, degradation of various carcinogens, protection against peroxidative effects, and in fish farming (for references see Seifu et al. 2005; Li et al. 2019).

Table 6.2 Activity of lactoperoxidase in milk from different species using an assay with ABTS as chromogenic substrate

Species	Activity (Units/mL)	References
Human colostrum	0.23	Gothefors and Marklund (1975)
Human milk	0.024	Gothefors and Marklund (1975)
Bovine	3.33	Uguz and Ozdemir (2005)
Buffalo	7.37	Kumar and Bhatia (1999)
Goat	0.1	Fonteh et al. (2002)
Ewe	1.21	Medina et al. (1989)
Guinea pig	22	Stephens et al. (1979)

Data are adopted from Magacz et al. (2019)

6.5 Xanthine Oxidoreductase

Bovine xanthine oxidoreductase (XOR) is one of the longest and most extensively studied enzymes. In 1902, Schardinger reported its isolation from bovine milk as an aldehyde reductase. After a series of other discoveries, Booth documented in 1938 that the "Schardinger" enzyme was equal to the enzyme that could oxidize xanthine and hypoxanthine, with the simultaneous reduction of O_2 to H_2O_2 (Booth 1938). The enzyme was highly purified by Ball (1939). For a long time, xanthine oxidase (XO, EC 1.17.3.2) and xanthine dehydrogenase (XD, EC 1.17.1.4) were considered to be distinct enzymes; however, over time it became clear that they are interchangeable forms of the same enzyme, especially after works published by Stirpe & Della Corte (Corte and Stirpe 1968, 1972; Stirpe et al. 1969). Efforts to obtain better understanding of structural differences between the two enzymatic forms of XOR were facilitated by the publication of the complete amino acid sequences of first the human followed by the bovine enzyme (Ichida et al. 1993; Berglund et al. 1996). The molybdenum-containing XOR is very widely distributed biologically with a few exceptions. For higher animals, it is textbook knowledge that the enzyme catalyses oxidation of hypoxanthine to xanthine and xanthine to uric acid in the purine degradation pathway. A large body of the knowledge about XOR was collected by studying the enzyme from bovine milk, and frequently uses only the name xanthine oxidase or just XO. The historical aspects of tracking XOR activity in milk have been comprehensively covered previously (Fox and Kelly 2006).

Early research located XOR activity in both skim milk and cream phospholipids membranes with a 0.3 ratio (Kitchen 1974). XOR is released from MFGM during phase inversion, and large amounts of XOR can be removed from MFGM upon washing with non-ionic detergents or high amounts of salt (Briley and Eisenthal 1975; Bruder et al. 1982). Accordingly, the enzyme is thereby characterized as a peripherally associated membrane protein having both soluble- and insoluble properties. Still, proteomic studies have clearly shown the presence of XOR in EVs and MFGs in both human and bovine milk (see Table 6.1). A recent study has described the presence of XOR in subsets of EVs in bovine milk based on differential centrifugation applying different *g*-forces. It was suggested that XOR can be used as a protein marker associated with the bulk of microRNAs in commercial milk (Benmoussa et al. 2017).

6.5.1 *Structure of Xanthine Oxidoreductase*

As in the case of other animals, active bovine XOR exists as a dimer of identical catalytically independent monomers. Each monomer comprises four redox-centres: two 2Fe/2S centres (ferrodoxin type), one FAD, and one molybdopterin (Mo-pt) cofactor (Fig. 6.4). Each monomer consists of 1332 amino acid residues with an Mr of 146,600 Da and an isoelectric point of 7.7 (Berglund et al. 1996). The Mo-pt

A

```
Bovine xanthine dehydrogenase/oxidase - P80457
MTADELVFFV NGKKVVEKNA DPETTLLAYL RRKLGLRGTK LGCGEGGCGA CTVMLSKYDR   60
LQDKIIHFSA NACLAPICTL HHVAVTTVEG IGSTKTRLHP VQERIAKSHG SQCGFCTPGI  120
VMSMYTLLRN QPEPTVEEIE DAFQGNLCRC TGYRPILQGF RTFAKNGGCC GGNGNNPNCC  180
MNQKKDHTVT LSPSLFNPEE FMPLDPTQEP IFPPELLRLK DVPPKQLRFE GERVTWIQAS  240
TLKELLDLKA QHPEAKLVVG NTEIGIEMKF KNQLFPMIIC PAWIPELNAV EHGPEGISFG  300
AACALSSVEK TLLEAVAKLP TQKTEVFRGV LEQLRWFAGK QVKSVASLGG NIITASPISD  360
LNPVFMASGT KLTIVSRGTR RTVPMDHTFF PSYRKTLLGP EEILLSIEIP YSREDEFFSA  420
FKQASRREDD IAKVTCGMRV LFQPGSMQVK ELALCYGGMA DRTISALKTT QKQLSKFWNE  480
KLLQDVCAGL AEELSLSPDA PGGMIEFRRT LTLSFFFKFY LTVLKKLGKD SKDKCGKLDP  540
TYTSATLLFQ KDPPANIQLF QEVPNGQSKE DTVGRPLPHL AAAMQASGEA VYCDDIPRYE  600
NELFLRLVTS TRAHAKIKSI DVSEAQKVPG FVCFLSADDI PGSNETGLFN DETVFAKDTV  660
TCVGHIIGAV VADTPEHAER AAHVVKVTYE DLPAIITIED AIKNNSFYGS ELKIEKGDLK  720
KGFSEADNVV SGELYIGGQD HFYLETHCTI AIPKGEEGEM ELFVSTQNAM KTQSFVAKML  780
GVPVNRILVR VKRMGGGFGG KETRSTLVSV AVALAAYKTG HPVRCMLDRN EDMLITGGRH  840
PFLARYKVGF MKTGTIVALE VDHYSNAGNS RDLSHSIMER ALFHMDNCYK IPNIRGTGRL  900
CKTNLSSNTA FRGFGGPQAL FIAENWMSEV AVTCGLPAEE VRWKNMYKEG DLTHFNQRLE  960
GFSVPRCWDE CLKSSQYYAR KSEVDKFNKE NCWKKRGLCI IPTKFGISFT VPFLNQAGAL 1020
IHVYTDGSVL VSHGGTEMGQ GLHTKMVQVA SKALKIPISK IYISETSTNT VPNSSPTAAS 1080
VSTDIYGQAV YEACQTILKR LEPFKKKNPD GSWEDWVMAA YQDRVSLSTT GFYRTPNLGY 1140
SFETNSGNAF HYFTYGVACS EVEIDCLTGD HKNLRTDIVM DVGSSLNPAI DIGQVEGAFV 1200
QGLGLFTLEE LHYSPEGSLH TRGPSTYKIP AFGSIPTEFR VSLLRDCPNK KAIYASKAVG 1260
EPPLFLGASV FFAIKDAIRA ARAQHTNNNT KELFRLDSPA TPEKIRNACV DKFTTLCVTG 1320
APGNCKPWSL RV                                                     1332
```

B

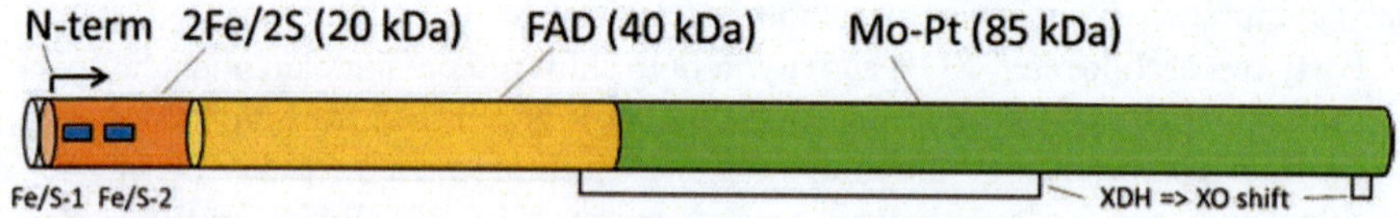

Fig. 6.4 Structure of bovine xanthine dehydrogenase/oxidase. (**a**) Amino acid sequence with N-terminal methionine removed (underlined italics); iron-sulphur binding (bold, italics, underlined); N6-succinyl/acetyl-lysines (underlined); FAD-binding (grey and underlined); molybdenum binding (bold, double underlined); substrate binding (bold, wavy underline); proton acceptor (grey italics, dotted underline); proteolytic labile sites (lysines, bold, double underlined); cysteines in XDH—XO interconversion (grey, bold, italics, double underlined). (**b**) Schematic presentation of the enzyme indicating the four redox centres 2Fe/2S (orange with blue squares), FAD (yellow), and Mo-P (molybdopterin) (green). Cysteines involved in dehydrogenase/oxidase-interconversion are connected with solid lines. *XDH* xanthine dehydrogenase, *XO* xanthine oxidase. The amino acid sequence data is available under accession number P80457, where additional information can be obtained

cofactor is located in the 85 kDa C-terminal, the 2Fe/2S centres in 20 kDa N-terminal, and FAD in intermediate 40 kDa domains (Amaya et al. 1990). The XD form is characterized by high xanthine/NAD^+-activity and low xanthine/O_2-activity, and the reverse applies for the XO form. NADH is formed in XD catalysed reactions, while in XO catalysis oxygen is converted to superoxide (O_2^-) and hydrogen peroxide (H_2O_2). The oxidation of xanthine takes place at the molybdenum centre, and electrons are rapidly transferred to FAD via the two 2F/2S-centers. The reduced enzyme is reoxidised by NAD^+ or molecular oxygen through FAD (reviewed by Hille 2013).

In the cell, XOR exists in the XD form, but during preparation it can be converted to the XO form, either irreversibly by proteolytic fragmentation or reversibly by oxidation of sulfhydryl residues. Early isolation protocols involved use of pancreatin, which facilitates release of XOR from MFGM; however, this results in a controlled nicking at specific sites in the protein chain with an irreversible conversion to the XO form as consequence (Battelli et al. 1972). Isolation of the enzyme in an oxidative environment converts the enzyme form XD to XO, but introduction of sulfhydryl reducing substances can be used to reverse this phenomenon or keep the enzyme in the XD form (Corte and Stirpe 1968; Waud and Rajagopalan 1976). Interconversion between XD and XO only comes about in mammalian species, whereas the enzyme exclusively operates as XD in chicken and lower species (Sato et al. 1995). Initially, it was reasoned that, during conversion of XOR from XD to XO by sulfhydryl reactive reagents, significant conformational changes seem to occur, leading to modifications of flavin activity and loss of NAD^+ binding (Hunt and Massey 1992). Detailed protein chemical studies pointed out that XD and XO interconversion is connected to formation of a disulphide bridge between Cys535 and Cys992, which are located in NAD^+/FAD and Mo-Pt domains respectively (Nishino and Nishino 1997; Rasmussen et al. 2000). The latter study describes the existence of a disulphide linkage between Cys1317 and Cys1325, but the significance of that was questioned. Now these discoveries are backed up by a line of publications with crystallographic, mutational, and kinetic analyses studies, which describe the structural and functional changes of XOR resulting from its conversion from the XD to the XO form. Huge contributions are a result of collaboration between the groups of Takeshi Nishino and Emil F. Pai (e.g., Enroth et al. 2000; Nishino et al. 2005, 2015). The data reveal that the Cys535-Cys992 disulphide bridge causes the disruption of a unique amino acid cluster that is central to the XD and XO interconversion. Moreover, in the XD form of XOR, the C-terminal part is shown to be inserted into the FAD-binding cavity, where it contributes to the generation of a correct NAD^+-binding site. Establishment of a bridge between Cys1316 and Cys1324 obstructs this insertion and stimulates formation of the XO form. Cys535 and Cys992 are involved in the rapid conversion from the dehydrogenase to the oxidase, whereas Cys1316 and Cys1324 are suggested to convey a slow conversion (illustrated model in Nishino et al. 2005). However, it has been shown that the an C535A/C992R/C1324S rat XDH mutant, which cannot be converted to the XO form by sulfhydryl reagents, will change form when exposed to 1.5 M guanidine hydrochloride or 6 M urea, without ruining its catalytic capabilities. This is presumably due to the disruption of the unique amino acid cluster, implying that the XD and XO forms are in a dynamic equilibrium. Routes for XD to XO interconversion are outlined in Fig. 6.5.

Early reports suggested that human milk lacks XOR; however, application of a more sensitive assay demonstrated that human milk in fact does contain XOR (Bradley and Günther 1960). Later it was demonstrated that the low XOR activity in human milk relates to the fact that 95–98% of the XOR molecules lack

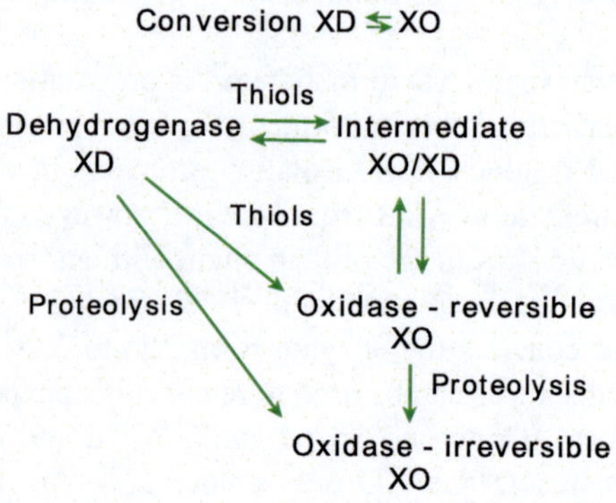

Fig. 6.5 Schematic presentation of the conversion between the xanthine dehydrogenase and oxidase forms of xanthine oxidoreductase. Reversible conversions (double arrows) include oxidation of thiols, and irreversible reactions (arrow) involve proteolysis. *XD* xanthine dehydrogenase, *XO* xanthine oxidase

molybdenum (Godber et al. 1997). Caprine and ovine milk are also shown to display a low level of XOR activity, and fortifying the diet with molybdenum leads to increased XOR activity (Atmani et al. 2004; Benboubetra et al. 2004).

Several protocols for isolation of bovine milk have been published, due to the extensive use of XOR as a model enzyme. Many of them still use elements from the original procedure (Ball 1939). Pancreatin and/or other lipase-containing substances have frequently been applied in the initial steps. Instead, a three-phase partitioning system with butanol may be used to liberate the enzyme from the cream phase (Rajagopalan 1986). Introduction of folate-affinity chromatography as the last purification step improved the purity of the bovine milk XOR preparations (Nishino et al. 1981; Berglund et al. 1996), but this step can be replaced by ion-exchange- or size-exclusion chromatography (Kristensen et al. 1996). A recent protocol used affinity chromatography based on binding to XO inhibitor p-aminobenzamidine (Beyaztaş and Arslan 2015).

6.5.2 *Biological Role of Xanthine Oxidoreductase and Significance in Milk*

XOR is a housekeeping enzyme that, in a biological context, is known for its principal role in purine catabolism (Bray 1975). In humans, it is seen that XOR mutations lead to xanthinuria, with kidney stone formation and urinary tract disorders as a consequence (Ichida et al. 1997). However, beyond this, multiple functions have been attributed to this protein, and for obvious reasons it is not possible to cover them all here (for references see (Vorbach et al. 2003). Interestingly, XOR has been shown to have an important role in the innate immune protective system, by being a direct downstream target of the Toll-like receptor NF-κB pathway and perhaps operating in the communication between innate and adaptive immunity (Vorbach et al. 2006). Therefore, XOR is one of several milk components controlling and assisting inflammatory responses.

It has long been known that XOR possesses antibacterial activity (Lipmann and Owen 1943; Green and Pauli 1943), but the explanation behind it is still not totally clear, and more than one mechanism might be relevant. In 1979, it was suggested that XOR-derived hydrogen peroxide acts as a substrate for the LPO-system in milk (Björck and Claesson 1979). XOR has been shown to catalyse the reduction of inorganic nitrite to nitric oxide (·NO) under hypoxic conditions (Godber et al. 2000b), but, if oxygen is present, superoxide is formed, which can react with NO forming peroxynitrite (Godber et al. 2000a). Consequently, XOR, LPO, and their respective substrates induce formation of both reactive oxygen species (ROS) and reactive nitrogen species (RNS) associated with bactericidal properties. Relatively recent studies in a simplified system have elegantly demonstrated the antibacterial effect of XOR in milk, and how the effect was neutralized by addition of a specific XOR inhibitor (Martin et al. 2004).

Ischemia is the medical term for decreased perfusion (hypoperfusion equals shock) resulting in reduced oxygen and nutrient supply to body tissues. A well supported model suggests that reperfusion has the potential to induce subsequent injury in ischemic tissue, and the phenomenon is termed ischemia-reperfusion injury. In an ischemic state, xanthine dehydrogenase is suggested to shift to xanthine oxidase due to a lower ATP level, oxidation of critical cysteine residues and/or limited proteolysis (see above). Upon blood flow restoration, xanthine oxidase reacts with its substrates and uses oxygen as the final electron acceptor (Carden and Granger 2000). During this process, ROS (superoxide and hydrogen peroxide) are released, increasing the oxidative stress. ROS produced by xanthine oxidoreductase lead to cytokine cascades and pathological conditions (Lee et al. 2014). Much has been written about the role of xanthine oxidoreductase during ischemia-reperfusion injury and possible ways to reduce the impact by use of enzymatic inhibitors (Granger and Kvietys 2015).

Even though there is a long record of work that supports the above described functions, there is also solid evidence that XOR plays a structural role during milk lipid secretion. It has been suggested to be a good example of gene sharing, and that the role during milk fat globule secretion might be the most important for XOR (Vorbach et al. 2002; Fox and Kelly 2006). The idea was supported by two findings: (1) that human XOR displays only little catalytic capacity as it exists primarily in a demolybdo form (Abadeh et al. 1992; Godber et al. 1997), and (2) that XOR was isolated in a stable complex with two other MFGM proteins, butyrophilin and adipophilin, in a 1:5:3 molar ratio (Heid et al. 1996). Different views have been published about implication of XOR, butyrophilin, and adipophilin in milk lipid secretion. By the use of gene knockout technology in mice, XOR was judged as indispensable for successful lactation (Vorbach et al. 2002; Wang et al. 2012). Published results imply that butyrophilin alone was responsible for docking of intracellular formed lipid droplets prior to MFG secretion (Robenek et al. 2006), and later another study reached the conclusion that adipophilin alone can mediate binding of intracellular lipid droplets (Jeong et al. 2013). However, a more recent study coming from the group of James L. McManaman nicely demonstrates that XOR is not required for apocrine secretion of milk lipids, but through interactions with butyrophilin XOR mediates docking of cytosolic lipid droplets to the apical membrane (Monks et al. 2016). Results from the same work showed that a lack of XOR delays MFG secretion, but fat globule formation and secretion are not blocked. Moreover, the absence of XOR leads to a general increase in the amount of MFG-associated intracellular proteins, which illustrates considerable modifications in the envelopment process.

Due to the relatively low activity of XOR in human milk, much focus has been on the structural function of XOR during MFG secretion. Recent results do, however, provide solid evidence for a role for breast milk XOR in oral health and the initial establishment of the microbiome during early infancy. When breastmilk is mixed in vivo with neonatal saliva that contains highly elevated concentrations of XOR substrates (xanthine and hypoxanthine), antibacterial compounds are released. Accordingly, this mechanism regulates the growth of microorganisms in the neonatal mouth and gut (Al-Shehri et al. 2015; Sweeney et al. 2018). This has prompted a new study evaluating the impact of industrial heating processes on XOR activity in bovine milk, which showed that UHT treatment is detrimental (Ozturk et al. 2019).

When considering the presence of XOR, it is important to distinguish between enzyme activity and the amount of protein, due to the dual catalytic power depending on the protein conformation. It has been reported that XOR is present at levels of about 160 mg/L bovine milk (Kitchen et al. 1970). However, due to its association with butyrophilin at the inner leaflet of the apical membrane during the secretion process, it must be expected that parts of XOR are trapped inside phospholipid structures, especially in an anaerobic setting. This will keep parts of XOR in the XD form. This is backed up by results showing that, 16 h after milking, up to 15% of XOR is capable of operating as a dehydrogenase, whereas processing to buttermilk or MFGM largely converts all XOR into the XO form (Nakamura and Yamazaki 1982). Accordingly, storage at 4 °C, heating to 70 °C, and homogenization will release XOR into the aqueous environment, where it then can be measured as XO (Greenbank and Pallansch 1962; Gudnason and Shipe 1962; Cerbulis and Farrell 1977; Griffiths 1986). The heat stability of XOR has been reviewed by Griffiths (1986), who reported that XOR is most heat-stable in cream and least stable in skim milk; this work provides detailed information about response of XOR to different temperatures and time of treatment. Notably, these authors showed that heating to 80 °C for 120 s did not completely inactivate XOR, and concluded that XOR is not suitable as an indicator for heat treatment of milk. This was contradicted by Andrews and coworkers, who proposed XOR activity as a suitable indicator for heating in the range 80–90 °C (1987). A newer study also suggested XOR activity as a marker for milk ultra-pasteurization and cream pasteurization (Vetsika et al. 2014). The amount of XOR in MFGM has been reported to vary considerably, e.g., XOR was measured to comprise 19% and 12% of MFGM proteins from Holstein and Jersey milk, respectively (Mondy and Keenan 1993). A comparison of MFGM from milk of buffalo and bovine Frisian breeds showed a six-fold higher presence of XOR in buffalo milk, which was linked to the reported higher natural shelf-life of the latter (Nguyen et al. 2017).

The presence of XOR prooxidant reaction products has been reported to affect flavor and lipid oxidation. XOR action might lead to oxidation of a range of aldehydes present in milk, including acetaldehyde, 3-methyl butanal, 2-methyl-butanal and 2-methyl propanol (Kathriarachchi et al. 2014). Off-flavour in bovine milk has been reported be linked to XOR-coupled oxidation of unsaturated fatty acids, leading to the formation of aldehydes such as butanal, but-2-enal, pent-2-enal and hex-2-enal (Aurand et al. 1967). Appearance of spontaneous oxidative rancidity in individual milk samples has been attributed to XOR activity, as it could be induced in milk upon addition of XOR (Aurand et al. 1967, 1977).

Uncontrolled growth of *Clostridium tyrobutyricum* strains causes flavour defects and late blowing of semi-hard and hard cheeses. Common preventative approaches include bactofugation or microfiltration of milk, or addition of nitrate or lysozyme (van den Berg et al. 2004). Addition of sodium nitrate leads to XOR-mediated production of ROS and RNS, which inhibits butyric acid fermentation by delaying the germination of spores (Vissers et al. 2007). Addition of nitrate is, however, controversial, due to possible production of undesirable derived products like carcinogenic nitrosamines (Beresford et al. 2001). Alternative ways to mitigate late cheese blowing have therefore been pursued, e.g., addition of lactic acid bacteria modified to produce specific antimicrobial peptides (bacteriocins) (Silva et al. 2018).

In the early 1970s, XOR was subject to substantial attention, because it was suggested that homogenization of milk facilitated entrance of the enzyme into the vascular system after consumption and in that way could stimulate atherosclerosis. This implication has never been substantiated, and arguments for discarding it has been given on several occasions (see Harrison 2006 for references).

A multitude of assays have been used to study XOR activity. The principle of the method developed by Cerbulis and Farrell (1977) is perhaps the most widely used for determining XOR activity in milk (e.g., Rasmussen et al. 2000; Aboul-Enein et al. 2003; Kathriarachchi et al. 2014). This method uses spectrophotometric measurement of xanthine to urate conversation around 295 nm. A recent published method suggests estimation of XOR activity by monitoring produced urate using reversed phase HPLC (Rashidinejad et al. 2016). Higher sensitivity can be achieved by using, for example, stable isotopes and liquid chromatography/triple quadrupole mass spectrometry (Murase et al. 2016) or fluorometric detection of hydrogen peroxide produced by XOR (Hwang et al. 2016).

6.6 γ-Glutamyltransferase

γ-Glutamyltransferase (GGT) (E.C. 2.3.2.2) catalyses the transfer of γ-glutamyl functional groups from molecules such as glutathione to an amino acid, a peptide or water (forming glutamate). Some authors continue to use the older name γ-glutamyl transpeptidase. GGT is omnipresent and is found associated with cell membranes.

It has been suggested that GGT could be involved in the supply of amino acids to the lactocyte, because inhibition of GGT led to decreased milk protein synthesis (Calamari et al. 2015). GGT activity is traditionally measured by monitoring the release of *p*-nitroanilide from a synthetic γ-glutamyl-*p*-nitroanilide at 410 nm. A number of other methods have been presented for detecting GGT levels, e.g., HPLC and fluorescence techniques (Ziobro and McElroy 2013; Hou et al. 2014). A recent study reported design of an activatable chemiluminescent probe that can be used for the sensitive real-time detection of GGT activity in vitro and in living mice (An et al. 2019). Early research located GGT activity in both skim milk and cream phospholipids membranes with a 5.8:1 ratio, respectively (Kitchen 1974). In a later study, 74.3% of the GGT activity was associated with skim milk, and only 6.6% with the MFGM (Baumrucker 1979). A more recent proteomic study imply a 8:1 ratio for GGT in EV and MFGM respectively (see G3N2D8 in supplementary material in Reinhardt et al. (2013).

A

```
Bovine Gamma-Glutamyltransferase (GGT) precursor - G3N2D8
MKKRSILLAL GAVVLVLLII GLCIWLPETSK PQSHVYPRAA VAADAKLCS EIGRDALLDG  60
GSAVDAAIAA LLCMGLLNAH SMGIGGGLFLT IYNSTMRKAE VINAREVAP RLASAGMFNS 120
SEQSEEGGLS VAVPGELRGY ELAHQRHGRLP WARLFQPSIE LARQGFPVG KALAGALQRQ 180
RETVERHPAL CEVFCRDGKV MREGDRVTMPR LADTLETLAS EGAQAFYNG SLTAQIVKDI 240
QEAGGIVTAE DLNSYRAELV EQPLSISLGDA QLYVPSAPLS GPVLALIIN ILKGYNFSRA 300
SVEMPEQKGL TYHRIVEAFR FAYAKRTLLGD PKFVNVTEVV RNMTSEFFA AQLWARISDS 360
TTHPASYYEP EFYTPDGGGT AHLSVVSEDGS AVSATSTINL YFGSKVRSR VSGILFNDEM 420
DDFSSPNIIN QFGVPPSPAN FIAPGKQPLSS MCPVIIVGDD GQVRMVVGA SGGTQITTST 480
ALAIINSLWF GYDAKQAVEE PRLHNQLLPNT TVLEKGIDQA VVAALKTRH HAIEETSTYI 540
AVVQAVVRTA GGWAAASDSR KGGEPAGY                                   568
```

B

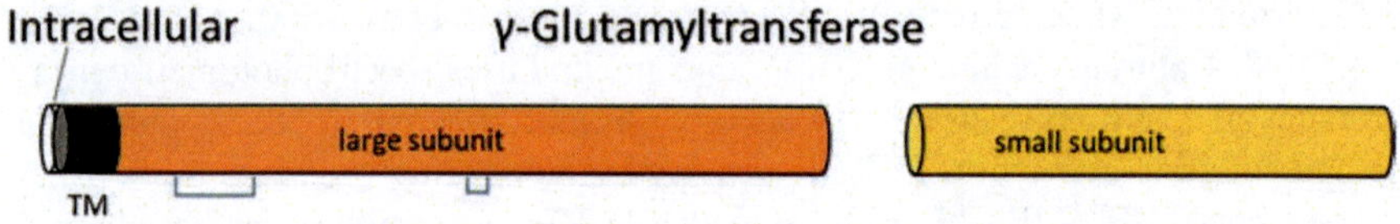

Fig. 6.6 Structure of γ-glutamyltransferase. (**a**) Amino acid sequence indicating an intracellular part (underlined); a transmembrane domain (grey); extracellular C-terminal; putative consecutive disulphide bonded cysteines (double underlined italics); putative N-glyc (bold); cross species conserved GGT motif and N-terminal of the small subunit (bold italics). (**b**) Schematic presentation of γ-glutamyltransferase with a transparent intracellular part; transmembrane domain (TM) (black); large and small subunit as result of autocleavage (orange and yellow, respectively), disulfide bonds (blue lines). The amino acid sequence data is available under accession number G3N2D8, where additional information can be obtained

6.6.1 Structure of γ-Glutamyltransferase

The amino acid sequence of bovine GGT was deduced from the whole genome and given an entry G3N2D8 at the UniProt database (Fig. 6.6). By analogy with human GGT, it is expected that bovine GGT is expressed as a 568 amino acid residue precursor that is autocleaved into a large subunit (41.0 kDa, residues 1–379) and a small subunit (19.8 kDa, residues 380–569). The heterodimer localizes to the cell surface and is bound to the plasma membrane via a single-pass transmembrane domain (residues 5–26) located at the N-terminus of the large subunit. The threonine at position 380 is responsible for the autocleavage of the propeptide and also serves as the catalytic nucleophile in the active site of the mature enzyme. Details about the structure are extractable from the released high resolution 1.67 Å crystal structure of human GGT (West et al. 2013). There are relatively few studies reporting purification of GGT, but extensive studies have been performed on the enzyme purified from rat kidneys (Castonguay et al. 2007).

Studies of human GGT have shown that it cleaves the γ-glutamyl amide bond of a wide range substrates. The catalytic mechanism is initiated by a nucleophilic attack on the γ-glutamyl amide bond by the side chain hydroxyl oxygen of active threonine. An acyl bond is formed and the rest of the substrate released. In the subsequent rate-limiting hydrolysis step glutamate is released. GGT can also catalyze a transpeptidation reaction, where the γ-glutamyl group bound to the active threonine is transferred to the α-nitrogen of a di- or tripeptide acceptor, forming a new γ-glutamyl compound (West et al. 2013).

6.6.2 Biological Role and Significance of Glutamyltransferase in Milk

GGT is a cell surface enzyme that hydrolyses extracellular γ-glutamyl compounds, including glutathione and leukotriene C4, and it is thereby a central part of the essential γ-glutamyl cycle involved in maintaining adequate intracellular glutathione levels, which is important for cysteine homeostasis, intracellular redox status and inflammation (Taniguchi 2009). Evaluation of the GGT activity in the circulation is widely used in diagnosis of liver diseases, and it is used as a marker of alcohol consumption. In that light, GGT activity has been investigated for its association with diabetes and metabolic syndrome, cancer, atherosclerosis, and cardiovascular disease (Ndrepepa et al. 2018).

The bovine enzyme has been purified from a milk membrane pool (MFGM and skim milk membranes) using a seven-step procedure involving treatment with detergents, acetone precipitation, salting out, ion-exchange-, and size exclusion chromatography. The yield was 0.74 mg of 3 g membrane protein, and the molecular mass was estimated to be 80 kDa in total, constituted by two subunits at 57 and 25 kDa, respectively. Both subunits were glycosylated (Baumrucker 1980). GGT has its optimal activity at pH 8.5–9.0 and 45 °C and it is strongly inhibited by diisopropylfluorophosphate, iodoacetamide and metals, e.g., Cu^{2+} and Fe^{3+} (Farkye 2003). For kinetic studies, 2-amino-4-[mono(4-cyanophenyl)phosphono] butanoic acid has been used as a very specific and effective inhibitor (Castonguay et al. 2007).

It is reported that GGT is more resistant to heat than alkaline phosphatase, but less than LPO. As a result, it has been recommended as an indicator for monitoring thermal processes in the range 70–80 °C, and especially >77 °C (Andrews et al. 1987; Zehetner et al. 1995; Lorenzen et al. 2010). Upon heating milk to 69 °C, 50% inactivation of GGT was observed after 3 min, whereas the activity of ALP was lost. In comparison, 50% of LPO activity was measured even after 0.9 min at 77 °C (Blel et al. 2002). In the same study, residual GGT activity could be monitored in Camembert cheese made from raw milk. A recent study investigated the effects of heat treatments on GGT activity in cows, ewes, and goats milk. Pasteurization reduced the GGT activity in all three kinds of milk to 7–12%. By comparison to XOR activity, it was suggested that GGT could be used as a marker of milk pasteurization treatments, and XOR as markers of milk ultra-pasteurization and cream pasteurization (Vetsika et al. 2014). Figure 6.7, which is redrawn from the original by Zehetner et al. (1995), illustrates the impact of heating for the activity of GGT in milk. The heat stability of other indigenous milk enzymes is treated in Chap. 12 in the present book.

6.7 5'-Nucleotidase

5'-Nucleotidase (EC 3.1.3.5) is a glycoprotein enzyme involved in the nucleotide catabolism. It is responsible for catalyzing the hydrolysis of the phosphate esterified at C5 of ribose and deoxyribose to a ribonucleoside and orthophosphate, e.g., AMP + H_2O → adenosine + phosphate. Reis was the first to describe the enzyme in muscle and heart skeletal muscle (Reis 1934). Since then, the presence of

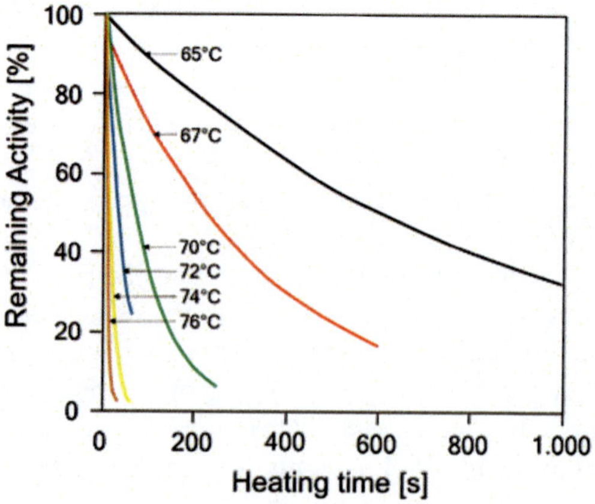

Fig. 6.7 Thermal inactivation of γ-glutamyltransferase (GGT) during heating of milk (Zehetner et al. 1995). Redrawn from the original by Zehetner et al. (1995)

5'-nucleotidase has been described in various mammalian tissues and cells well as in bacteria and plant cells. The broad appearance of 5'-nucleosidase activity is also reflected in significantly different preferences of substrate specificity and pH optima among this group of enzymes. 5'-nucleotidase is the common name for a group of enzymes with the ability to hydrolyze 5'-mononucleotides. Furthermore, different 5'-nucleotidases display hydrolytic abilities towards 5'-dinucleotides, 5'-trinucleotides, or more complex nucleotides such as FAD. It is therefore up for discussion whether the various 5'nucleotidases belong to one principal enzymatic group or originate from different forms of enzymes with overlapping substrates. The review by Zimmermann considers this matter (Zimmermann 1992).

The enzyme is not only expressed in various cells, tissues, and species, but also occurs in different subcellular locations as either membrane-bound or in soluble forms. So far, at least seven human 5'-nucleotidases have been cloned, one type being membrane-associated and the remaining in cytosolic forms with one in the mitochondrial matrix. The soluble cytosolic 5'-nucleotidases can be grouped based on their biochemical features. Some forms display high affinity for IMP and have a high K_m value towards AMP, whereas other forms prefer AMP (low K_m-AMP forms) (Bianchi and Spychala 2003). The surface-located 5'-nucleotidase goes by the name ecto-5'-nucleotidase and is attached to the membrane *via* a GPI anchor (Misumi et al. 1990). In 1992, it was noted by Zimmerman that up to 50% of the enzyme may be associated with intracellular membranes, which then are released during cell disruption (Zimmermann 1992). Gordon and coworkers described how ecto-5'-nucleotidases exhibit a high degree of stereoselectivety, characteristic for each enzyme. By using enantiomers in which the ribose moiety was exchanged from the natural D-enantiomer to the L-form, and similar with the phosphate side chain, they showed that the catabolic rate fell to one-fifth and one-fifteenth when exchanging the natural compounds (Cusack et al. 1983). This stereo-selectivity is not observed in 5'-nucleotidases isolated from bacteria. In 1954, Wang was the first to report on the presence of bacterial 5'-nucleotidase activity; the bacteria in which the nucleotidase activity was found came from soil (Wang 1954). Around a decade later, 5'-nucleotidase from *E. coli* was thoroughly isolated and the enzymatic activity described of 5'-nucleotidase from *E. coli* (Neu 1967). By using the osmotic shock technique followed by DEAE-cellulose and then hydroxylapatite chromatography, it was possible to obtain a 5000-fold purification of the enzyme; the molecular weight was determined to be 52,000 Da. Enzymatic analyses identified all 5'-ribo- and deoxy ribonucleotides as substrates but with optimal activity against 5'-AMP. This multi-substrate function is unusual, since AMP- and UDP-hydrolysis are quite unrelated, but might be a consequence of multifunctional enzymes often found in bacteria. Moreover, it was described how the enzymatic activity was stimulated by the addition of divalent cations as Co^{2+} and Ca^{2+}. This is in contrast to the mammalian 5'-nucleotidases, which often require Mg^{2+} or Mn^{2+} as activators (Neu 1967).

Activity of 5'-nucleotidase can be measured in several ways. Often the rate of appearance of one of the two products, orthophosphate or a nucleoside, is measured either spectrophotometrically or by using radioactive labels on one of the two products (Edelson and Duncan 1981; Martínez-Martínez et al. 1998). However, measuring the amount of released ortho-phosphate requires a highly purified enzyme preparation to exclude the possible action of phosphatases. Kits are commercially available to determine 5'nucleotidase activity colorimetrically, e.g., through a reaction where the enzyme converts a substrate to ammonia, which then is quantified. Detecting the presence of an individual 5'-nucleotidase in a cell lysate is challenging and might be problematic due to a co-expression of different nucleotidases in one type of tissue or cell. However, this can partly be overcome by the use of specific nucleotidase inhibitors in enzymatic assays (Bianchi and Spychala 2003).

6.7.1 Structure of 5'-Nucleotidase

As mentioned earlier, 5'-nucleotidase exists in multiple organisms in different forms with different substrate affinities. The multitude of origins is also reflected in structural differences among the enzymes. Ecto-5'-nucleotidase exists as a homodimer with a hydrophilic and an amphiphilic dimer (Bailyes et al. 1984; Buschette-Brambrink and Gutensohn 1989), whereas cytosolic 5'-nucleotidases can be found as monomers, dimers, or in tetrameric form (Bianchi and Spychala 2003; Walldén et al. 2007).

The crystal structure of a eukaryotic ecto-5'-nucleotidase was published by the group of Sträter (Knap et al. 2012). Prior to that, the crystal structure of a variant of human cytosolic 5'-nucleotidase was determined (Walldén et al. 2007) and bacterial 5'-nucleotidase structures had been determined on several occasions (e.g. Knöfel and Sträter 1999). Thereby, the structure of the dimeric human ecto-5'-nucleotidase, in both open and closed conformation to a 1.55–2.00 Å resolution was disclosed. Originally the dimers were thought to be disulphide-bonded, but this hypothesis has been disproven (Martínez-Martínez et al. 2000).

The amino acid sequence of bovine 5'-nucleotidase (ecto-form) in its unprocessed form is shown in Fig. 6.8. It has a length of 574 amino acid residues and a calculated mass on 62,966 Da in its unprocessed form. Four disulfide bonds are present between Cys51-Cys57, Cys353-Cys358, Cys365-Cys387, and Cys476-479 (Fini et al. 2000). The glycosylation pattern is species- and tissue specific, but cDNA analyses reveal four or five potential *N*-linked sites of glycosylation and no *O*-linked glycosylations (Misumi et al. 1990; Zimmermann 1992). The C-terminal contains the site for substrate binding and the position for the GPI anchor. The active site is located at the interface between the two N- and C-terminal domains and is inaccessible to solvents (Knapp et al. 2012).

A

```
Bovine                5'-nucleotidase                -                 Q05927
MNPGAARTPA LRILALGALL WPAARPWELT ILHTNDVHSR LEQTSEDSSK CVNASRCVGG  60
VARLATKVHQ IRRAEPHVLL LDAGDQYQGT IWFTVYKGTE VAHFMNALGY DAMALGNHEF 120
DNGVEGLIDP LLKEVNFPIL SANIKAKGPL ASKISGLYSP YKILTVGDEV VGIVGYTSKE 180
TPFLSNPGTN LVFEDEITAL QPEVDKLKTL NVNKIIALGH SGFEVDKLIA QKVKGVDVVV 240
GGHSNTFLYT GNPPSKEVPA GQYPFIVTSD DGRKVPVVQA YAFGKYLGYL KVEFDEKGNV 300
VTSHGNPILL NSSIPEDPNI KADINKWRVK LDNYSTQELG KTIVYLDGTA QSCRFRECNM 360
GNLICDAMIN NNLRHPDEMS WNHVSMCILN GGGIRSPIDE RNNGTITWEN LAAVLPFGGT 420
FDLVQLKGST  LKKAFEHSVH  RYGQATGEFL  QVGGIHVVYD  ISRNPGDRVV  KLEVLCTQCR  480
VPSYEPLRMD KVYKVILPSF LVSGGDGFQM IKDEKIKHDS GDQDINVVSG YISKMKVLYP 540
AVEGRIQF A GSHCCGSFSL IFLSVLAVII ILYQ                             574
```

B

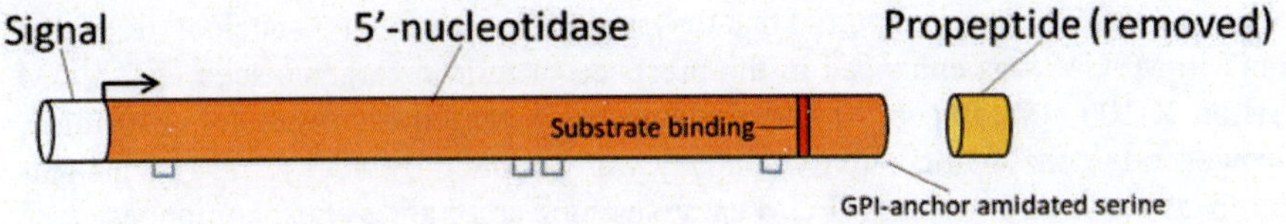

Fig. 6.8 (**a**) Amino acid sequence of bovine 5'-nucleotidase. Signal peptide (bold, double underlined); removed in mature form (italics, wave underlined); metal binding (bold, double underlined); transition state stabilizer (gray background); substrate binding site (bold); proposed glycosylations (bold underlined); GPI-anchor amidated serine (white on black). (**b**) Schematic presentation of 5'-nucleotidase with a signal peptide (white); a propeptide (yellow); enzymatic part (orange) with disulphide bridges (blue lines), substrate binding, and a C-terminally located GPI-anchor. The amino acid sequence data is available under accession number Q05927, where additional information can be obtained

6.7.2 *Biological Role and Significance of 5'-Nucleotidase in Milk*

Much of the work on 5'-nucleotidase in milk uses the enzyme as a marker of the plasma membrane. The presence of the enzyme on the apical plasma membranes of secretory lactating cells in mammary glands has, e.g., made it suitable in studying secretory mechanisms. It was hypothesized by Patton (1978) that 5'-nucleotidase plays a role in milk secretion. Injection of concavalin A, a molecule known to bind cell surface receptors such as 5'-nucleotidase, into the mammary gland of goats resulted in inhibition of milk flow. The first milk, after the drug effect was terminated, had higher levels of fat and protein, and this led to the hypothesis that 5'-nucleotidase plays a role in exocytosis of milk in the mammary gland (Snow et al. 1980).

Historically, milk has been as an easily accessible and scalable source of 5'-nucleotidase in the search to describe the membrane associated ecto-form. In 1972, Huang and Keenan were the first to isolate and describe 5'-nucleotidase from bovine MFGM. Multiple steps including addition of detergents, precipitation with ammonium sulfate, heat treatment, sonication, and chromatography on a Sepharose 4-B column were applied. This approach yields two enriched fractions of 5'-nucleotidase. The two isoforms display different preferences towards substrates and their

K_M values for AMP are 0.94 and 5 mM. 5'-nucleotidase in milk does not require a divalent cation as cofactor, and the optimal activity is at pH 7.0–7.5. However, addition of Mg^{2+} yields another activity peak at pH 10.0, which has also been described for other 5'-nucleotidases (Mennella and Jones 1977). Maximum catalytic rates of the MFGM 5'-nucleotidase are achieved at 69 °C. Both isoforms lose up to 50% activity after heat treatment at 60 °C for 60 min (Huang and Keenan 1972). 5'-nucleotidase can be extracted from MFGM preparations in a single phase system using 8% n-butanol, resulting in over 60% activity after extraction when performed at room temperature. At 0 °C, a 10% n-butanol solution is required to obtain similar activity; however, the enzyme is less susceptible to denaturation at 0 °C (Ahn and Snow 1993). It has been reported that the enzymatic activity of 5'-nucleotidase from buffalo MFGM was enhanced in the presence of mild detergents such as 2.5 mM Triton X-100, 1% Tween 20 and the stronger detergent deoxycholate (40 mM). However, the enzymatic activity contrary was inhibited by SDS (2 mM). This pattern was the same for other MFGM enzymes such as alkaline phosphomonoesterase and xanthine oxidase (Bhavadasan and Ganguli 1978). Higher concentrations of nonionic detergents have been shown to inhibit enzymatic activity of 5-nucleotidase from bovine MFGM preparations (Kanno and Yamauchi 1979).

It is not known whether 5'-nucleotidase serves any biological roles in milk. Nucleotides are naturally present in substantial amounts in human milk, whereas the content in bovine milk is lower and the composition is different. A new-born infant's intestinal system is immature at birth, and the capacity to digest nucleotides has not been established, and, since nucleosides are the preferred form for cellular absorption, it is likely that 5'-nucleotidase in milk helps to create absorbable nucleosides (Janas and Picciano 1982).

6.8 Conclusion

Six enzymes have been reviewed in this chapter, but the authors suggest that only four of them should be recognized as true components linked to milk phospholipid structures. The appearance of catalase most likely reflect presence of somatic milk cells or microorganisms, which illustrates either biological status of the mammary gland or deliberate addition. LPO lacks traditional structural features of a membrane-associated protein, and thus probably ends up in MFGM and EVs due to fraction contamination by skim milk constituents or cytosolic remnants from the milk-producing cell. Comprehensive information is given for four major phospholipid membrane attached milk enzymes (sulfhydryl oxidase, xanthine oxidoreductase, γ-glutamyltransferase, and 5'-nucleotidase). Alkaline phosphatase and lipoprotein lipase are also found bound to phospholipid membranes, but are dealt with in other chapters of the present book.

References

Abadeh S, Killacky J, Benboubetra M, Harrison R (1992) Purification and partial characterization of xanthine oxidase from human milk. Biochim Biophys Acta 1117:25–32.

Aboul-Enein HY, Refaie MOI, El-Gazzar H, El-Aziz MA (2003) Chemical modification of milk xanthine oxidase with different modifiers. Prep Biochem Biotechnol 33:173–187.

Admyre C, Johansson SM, Qazi KR, Filén J-J, Lahesmaa R, Norman M, Neve EPA, Scheynius A, Gabrielsson S (2007) Exosomes with immune modulatory features are present in human breast milk. J Immunol 179:1969–1978

Ahn YS, Snow LD (1993) Selective extraction of alkaline phosphatase and 5'-nucleotidase from milk fat globule membranes by a single phase n-butanol procedure. Prep Biochem 23:409–419.

Allen JC, Joseph G (1985) Deterioration of pasteurized milk on storage. J Dairy Res 52:469–487.

Al-Shehri SS, Knox CL, Liley HG, Cowley DM, Wright JR, Henman MG, Hewavitharana AK, Charles BG, Shaw PN, Sweeney EL, Duley JA (2015) Breastmilk-saliva interactions boost innate immunity by regulating the oral microbiome in early infancy. PLoS One 10:e0135047.

Amaya Y, Yamazaki K, Sato M, Noda K, Nishino T, Nishino T (1990) Proteolytic conversion of xanthine dehydrogenase from the NAD-dependent type to the O2-dependent type. Amino acid sequence of rat liver xanthine dehydrogenase and identification of the cleavage sites of the enzyme protein during irreversible conversion by trypsin. J Biol Chem 265:14170–14175

An R, Wei S, Huang Z, Liu F, Ye D (2019) An activatable chemiluminescent probe for sensitive detection of γ-glutamyl transpeptidase activity in vivo. Anal Chem 91:13639–13646.

Andersson LA, Bylkas SA, Wilson AE (1996) Spectral analysis of lactoperoxidase evidence for a common heme in mammalian peroxidases. J Biol Chem 271:3406–3412.

Andrews AT, Anderson M, Goodenough PW (1987) A study of the heat stabilities of a number of indigenous milk enzymes. J Dairy Res 54:237–246.

Antwi K, Hostetter G, Demeure MJ, Katchman BA, Decker GA, Ruiz Y, Sielaff TD, Koep LJ, Lake DF (2009) Analysis of the plasma peptidome from pancreas cancer patients connects a peptide in plasma to overexpression of the parent protein in tumors. J Proteome Res 8:4722–4731.

Arnold C (1881) Einige neue Reactionen der Milch. Arch Pharm 219:41–42.

Atmani D, Benboubetra M, Harrison R (2004) Goats' milk xanthine oxidoreductase is grossly deficient in molybdenum. J Dairy Res 71:7–13

Aurand LW, Chu TM, Singleton JA, Shen R (1967) Xanthine oxidase activity and development of spontaneously oxidized flavor in milk. J Dairy Sci 50:465–471.

Aurand LW, Boone NH, Giddings GG (1977) Superoxide and singlet oxygen in milk lipid peroxidation. J Dairy Sci 60:363–369.

Bailie MJ, Morton RK (1958a) Comparative properties of microsomes from cow's milk and from mammary gland. 1. Enzymic activities. Biochem J 69:35–44

Bailie MJ, Morton RK (1958b) Comparative properties of microsomes from cow's milk and from mammary gland. 2. Chemical composition. Biochem J 69:44–53

Bailyes EM, Soos M, Jackson P, Newby AC, Siddle K, Luzio JP (1984) The existence and properties of two dimers of rat liver ecto-5'-nucleotidase. Biochem J 221:369–377.

Ball EG (1939) Xanthine oxidase: purification and properties. J Biol Chem 128:51–67

Barry KM, Dinan TG, Murray BA, Kelly PM (2016) Comparison of dairy phospholipid preparative extraction protocols in combination with analysis by high performance liquid chromatography coupled to a charged aerosol detector. Int Dairy J 56:179–185.

Battelli MG, Corte ED, Stirpe F (1972) Xanthine oxidase type D (dehydrogenase) in the intestine and other organs of the rat. Biochem J 126:747–749.

Baumrucker CR (1979) γ-Glutamyl transpeptidase of bovine milk membranes: distribution and characterization. J Dairy Sci 62:253

Baumrucker CR (1980) Purification and identification of γ-glutamyl transpeptidase of milk membranes. J Dairy Sci 63:49–54.

Beers RF, Sizer IW (1952) A spectrophotometric method for measuring the breakdown of hydrogen peroxide by catalase. J Biol Chem 195:133–140

Benboubetra M, Baghiani A, Atmani D, Harrison R (2004) Physicochemical and kinetic properties of purified sheep's milk xanthine oxidoreductase. J Dairy Sci 87:1580–1584.
Benmoussa A, Ly S, Shan ST, Laugier J, Boilard E, Gilbert C, Provost P (2017) A subset of extracellular vesicles carries the bulk of microRNAs in commercial dairy cow's milk. J Extracellular Vesicles 6:1401897.
Beresford TP, Fitzsimons NA, Brennan NL, Cogan TM (2001) Recent advances in cheese microbiology. Int Dairy J 11:259–274.
Berglund L, Rasmussen JT, Andersen MD, Rasmussen MS, Petersen TE (1996) Purification of the bovine xanthine oxidoreductase from milk fat globule membranes and cloning of complementary deoxyribonucleic acid. J Dairy Sci 79:198–204.
Beyaztaş S, Arslan O (2015) Purification of xanthine oxidase from bovine milk by affinity chromatography with a novel gel. J Enzyme Inhib Med Chem 30:442–447.
Bhavadasan MK, Ganguli NC (1978) Selective solubilization of marker enzymes of buffalo milk fat globule membrane by detergents. J Dairy Sci 61:697–700.
Bianchi V, Spychala J (2003) Mammalian 5'-nucleotidases. J Biol Chem 278:46195–46198.
Björck L, Claesson O (1979) Xanthine oxidase as a source of hydrogen peroxide for the lactoperoxidase system in milk. J Dairy Sci 62:1211–1215.
Blans K, Hansen MS, Sørensen LV, Hvam ML, Howard KA, Möller A, Wiking L, Larsen LB, Rasmussen JT (2017) Pellet-free isolation of human and bovine milk extracellular vesicles by size-exclusion chromatography. J Extracell Vesicles 6:1294340.
Blel M, Guingamp M-F, Gaillard J-L, Humbert G (2002) Studies on the thermal sensitivity of ?-glutamyl transpeptidase measured with a modified test procedure and compared with that of alkaline phosphatase and lactoperoxidase in milk. Lait 82:555–566.
Booth V (1938) The specificity of xanthine oxidase. Biochem J 32:494–502
Bradley PL, Günther M (1960) The xanthine oxidase of human milk and colostrum. Biochem J 74:15P
Bray RC (1975) Molybdenum iron–sulfur flavin hydroxylases and related enzymes. In: The enzymes. Academic Press, New York, pp 299–419
Briley MS, Eisenthal R (1975) Association of xanthine oxidase with the bovine milk-fat-globule membrane. Nature of the enzyme-membrane association. Biochem J 147:417–423
Bruder G, Heid H, Jarasch E-D, Keenan TW, Mather IH (1982) Characteristics of membrane-bound and soluble forms of xanthine oxidase from milk and endothelial cells of capillaries. Biochim Biophys Acta Protein Struct Mol Enzymol 701:357–369.
Buschette-Brambrink S, Gutensohn W (1989) Human placental ecto-5'-nucleotidase: isoforms and chemical crosslinking products of the membrane-bound and isolated enzyme. Biol Chem Hoppe Seyler 370:67–74.
Calamari L, Gobbi L, Russo F, Cappelli FP (2015) Pattern of γ-glutamyl transferase activity in cow milk throughout lactation and relationships with metabolic conditions and milk composition1. J Anim Sci 93:3891–3900.
Cals M-M, Mailliart P, Brignon G, Anglade P, Dumas BR (1991) Primary structure of bovine lactoperoxidase, a fourth member of a mammalian heme peroxidase family. Eur J Biochem 198:733–739.
Carden DL, Granger DN (2000) Pathophysiology of ischaemia-reperfusion injury. J Pathol 190:255–266.
Castonguay R, Halim D, Morin M, Furtos A, Lherbet C, Bonneil E, Thibault P, Keillor JW (2007) Kinetic characterization and identification of the acylation and glycosylation sites of recombinant human γ-glutamyltranspeptidase. Biochemistry 46:12253–12262.
Cerbulis J, Farrell HM (1977) Xanthine oxidase activity in dairy products. J Dairy Sci 60:170–176.
Coppock DL, Kopman C, Scandalis S, Gilleran S (1993) Preferential gene expression in quiescent human lung fibroblasts. Cell Growth Differ 4:483–493
Corte ED, Stirpe F (1968) Regulation of xanthine oxidase in rat liver: modifications of the enzyme activity of rat liver supernatant on storage at 20 degrees. Biochem J 108:349–351.

Corte ED, Stirpe F (1972) The regulation of rat liver xanthine oxidase. Involvement of thiol groups in the conversion of the enzyme activity from dehydrogenase (type D) into oxidase (type O) and purification of the enzyme. Biochem J 126:739–745.

Cusack NJ, Pearson JD, Gordon JL (1983) Stereoselectivity of ectonucleotidases on vascular endothelial cells. Biochem J 214:975–981.

Dull TJ, Uyeda C, Strosberg AD, Nedwin G, Seilhamer JJ (1990) Molecular cloning of cDNAs encoding bovine and human lactoperoxidase. DNA Cell Biol 9:499–509.

Edelson PJ, Duncan RA (1981) 48 – 5'-nucleotidase assay. In: Adams DO, Edelson PJ, Koren HS (eds) Methods for studying mononuclear phagocytes. Academic Press, New York, pp 461–467

Enroth C, Eger BT, Okamoto K, Nishino T, Nishino T, Pai EF (2000) Crystal structures of bovine milk xanthine dehydrogenase and xanthine oxidase: structure-based mechanism of conversion. Proc Natl Acad Sci U S A 97:10723–10728.

Farkye NY (2003) Other enzymes. In: Fox PF, McSweeney PLH (eds) Advanced dairy chemistry—1. Proteins: part A/part B. Springer, Boston, pp 571–603

Fass D, Thorpe C (2018) Chemistry and enzymology of disulfide cross-linking in proteins. Chem Rev 118:1169–1198.

Fini C, Amoresano A, Andolfo A, D'auria S, Floridi A, Paolini S, Pucci P (2000) Mass spectrometry study of ecto-5'-nucleotidase from bull seminal plasma. Eur J Biochem 267:4978–4987.

Fitz-Gerald CH, Deeth HC, Kitchen BJ (1981) The relationship between the levels of free fatty acids, lipoprotein lipase, carboxylesterase, N-acetyl-β-D-glucosaminidase, somatic cell count and other mastitis indices in bovine milk. J Dairy Res 48:253–265.

Fong BY, Norris CS, MacGibbon AKH (2007) Protein and lipid composition of bovine milk-fat-globule membrane. Int Dairy J 17:275–288.

Fonteh FA, Grandison AS, Lewis MJ (2002) Variations of lactoperoxidase activity and thiocyanate content in cows' and goats' milk throughout lactation. J Dairy Res 69:401–409

Fox PF, Kelly AL (2006) Indigenous enzymes in milk: overview and historical aspects - part 1. Int Dairy J 16:500–516.

Gatti M, Bottari B, Lazzi C, Neviani E, Mucchetti G (2014) Invited review: microbial evolution in raw-milk, long-ripened cheeses produced using undefined natural whey starters. J Dairy Sci 97:573–591.

Godber B, Sanders S, Harrison R, Eisenthal R, Bray RC (1997) ≥95% of xanthine oxidase in human milk is present as the demolybdo form, lacking molybdopterin. Biochem Soc Trans 25:519S.

Godber BL, Doel JJ, Durgan J, Eisenthal R, Harrison R (2000a) A new route to peroxynitrite: a role for xanthine oxidoreductase. FEBS Lett 475:93–96.

Godber BL, Doel JJ, Sapkota GP, Blake DR, Stevens CR, Eisenthal R, Harrison R (2000b) Reduction of nitrite to nitric oxide catalyzed by xanthine oxidoreductase. J Biol Chem 275:7757–7763.

Gothefors L, Marklund S (1975) Lactoperoxidase activity in human milk and in saliva of newborn infants. Infect Immun 11:1210–1215

Granger DN, Kvietys PR (2015) Reperfusion injury and reactive oxygen species: the evolution of a concept. Redox Biol 6:524–551.

Green DE, Pauli R (1943) The anti-bacterial action of the xanthine oxidase system. Proc Soc Exp Biol Med 54:148–150.

Greenbank GR, Pallansch MJ (1962) Inactivation and reactivation of xanthine oxidase in dairy products. J Dairy Sci 45:958–961.

Griffiths MW (1986) Use of milk enzymes as indices of heat treatment. J Food Prot 49:696–705.

Gu Y, Li M, Wang T, Liang Y, Zhong Z, Wang X, Zhou Q, Chen L, Lang Q, He Z, Chen X, Gong J, Gao X, Li X, Lv X (2012) Lactation-related microRNA expression profiles of porcine breast milk exosomes. PLoS One 7:e43691.

Gudnason GV, Shipe WF (1962) Factors affecting the apparent activity and heat sensitivity of xanthine oxidase in milk. J Dairy Sci 45:1440–1448.

Hadwan MH (2018) Simple spectrophotometric assay for measuring catalase activity in biological tissues. BMC Biochem 19:7.

Hamosh M (1988) Enzymes in milk: their function in the mammary gland, in milk, and in the infant. In: Biology of human milk. Raven Press, New York, pp 45–61

Harrison R (2006) Milk xanthine oxidase: properties and physiological roles. Int Dairy J 16:546–554.

Hata T, Murakami K, Nakatani H, Yamamoto Y, Matsuda T, Aoki N (2010) Isolation of bovine milk-derived microvesicles carrying mRNAs and microRNAs. Biochem Biophys Res Commun 396:528–533.

Heckler EJ, Alon A, Fass D, Thorpe C (2008) Human quiescin-sulfhydryl oxidase, QSOX1: probing internal redox steps by mutagenesis. Biochemistry 47:4955–4963.

Heid HW, Schnölzer M, Keenan TW (1996) Adipocyte differentiation-related protein is secreted into milk as a constituent of milk lipid globule membrane. Biochem J 320(Pt 3):1025–1030.

Hille R (2013) The molybdenum oxotransferases and related enzymes. Dalton Trans 42:3029–3042.

Hirvi Y, Griffiths MW, McKellar RC, Modler HW (1996) Linear-transform and non-linear modelling of bovine milk catalase inactivation in a high-temperature short-time pasteurizer. Food Res Int 29:89–93.

Hoober KL, Joneja B, White HB, Thorpe C (1996) A sulfhydryl oxidase from chicken egg white. J Biol Chem 271:30510–30516.

Hou X, Yu Q, Zeng F, Yu C, Wu S (2014) Ratiometric fluorescence assay for γ-glutamyltranspeptidase detection based on a single fluorophore via analyte-induced variation of substitution. Chem Commun 50:3417.

Huang CM, Keenan TW (1972) Preparation and properties of 5'-nucleotidases from bovine milk fat globule membranes. Biochim Biophys Acta 274:246–257.

Huang TC, Kuksis A (1967) A comparative study of the lipids of globule membrane and fat core and of the milk serum of cows. Lipids 2:453–460.

Hunt J, Massey V (1992) Purification and properties of milk xanthine dehydrogenase. J Biol Chem 267:21479–21485

Hwang K-A, Hwang Y-J, Song J (2016) Antioxidant activities and oxidative stress inhibitory effects of ethanol extracts from Cornus officinalis on raw 264.7 cells. BMC Complement Altern Med 16:196.

Ichida K, Amaya Y, Noda K, Minoshima S, Hosoya T, Sakai O, Shimizu N, Nishino T (1993) Cloning of the cDNA encoding human xanthine dehydrogenase (oxidase): structural analysis of the protein and chromosomal location of the gene. Gene 133:279–284.

Ichida K, Amaya Y, Kamatani N, Nishino T, Hosoya T, Sakai O (1997) Identification of two mutations in human xanthine dehydrogenase gene responsible for classical type I xanthinuria. J Clin Invest 99:2391–2397.

Ihalin R, Loimaranta V, Tenovuo J (2006) Origin, structure, and biological activities of peroxidases in human saliva. Arch Biochem Biophys 445:261–268.

Israel BA, Jiang L, Gannon SA, Thorpe C (2014) Disulfide bond generation in mammalian blood serum; detection and purification of quiescin-sulfhydryl oxidase. Free Radic Biol Med 69:129–135.

Ito O, Akuzawa R (1983a) Purification, crystallization, and properties of bovine milk catalase. J Dairy Sci 66:967–973.

Ito O, Akuzawa R (1983b) Isozymes of bovine milk catalase. J Dairy Sci 66:2468–2473.

Jaje J, Wolcott HN, Fadugba O, Cripps D, Yang AJ, Mather IH, Thorpe C (2007) A flavin-dependent sulfhydryl oxidase in bovine milk. Biochemistry 46:13031–13040.

Janas LM, Picciano MF (1982) The nucleotide profile of human milk. Pediatr Res 16:659–662.

Janolino VG, Swaisgood HE (1975) Isolation and characterization of sulfhydryl oxidase from bovine milk. J Biol Chem 250:2532–2538

Jeong J, Lisinski I, Kadegowda AKG, Shin H, Wooding FBP, Daniels BR, Schaack J, Mather IH (2013) A test of current models for the mechanism of milk-lipid droplet secretion. Traffic 14:974–986.

Kanno C, Yamauchi K (1979) Selective extraction of marker enzymes of bovine milk fat globule membrane by nonionic detergents. J Biochem 85:529–534.

Katchman BA, Ocal IT, Cunliffe HE, Chang Y-H, Hostetter G, Watanabe A, LoBello J, Lake DF (2013) Expression of quiescin sulfhydryl oxidase 1 is associated with a highly invasive phenotype and correlates with a poor prognosis in Luminal B breast cancer. Breast Cancer Res 15:R28.

Kathriarachchi K, Leus M, Everett DW (2014) Oxidation of aldehydes by xanthine oxidase located on the surface of emulsions and washed milk fat globules. Int Dairy J 37:117–126.

Kitchen BJ (1974) A comparison of the properties of membranes isolated from bovine skim milk and cream. Biochim Biophys Acta 356:257–269

Kitchen BJ (1976) Enzymic methods for estimation of the somatic cell count in bovine milk. J Dairy Res 43:251–258.

Kitchen BJ (1981) Review of the progress of dairy science: bovine mastitis: milk compositional changes and related diagnostic tests. J Dairy Res 48:167–188.

Kitchen B, Taylor G, White I (1970) Milk enzymes - their distribution and activity. J Dairy Res 37:279.

Knapp K, Zebisch M, Pippel J, El-Tayeb A, Müller CE, Sträter N (2012) Crystal structure of the human ecto-5'-nucleotidase (CD73): insights into the regulation of purinergic signaling. Structure 20:2161–2173.

Knöfel T, Sträter N (1999) X-ray structure of the Escherichia coli periplasmic 5'-nucleotidase containing a dimetal catalytic site. Nat Struct Biol 6:448–453.

Kohler H, Jenzer H (1989) Interaction of lactoperoxidase with hydrogen peroxide: formation of enzyme intermediates and generation of free radicals. Free Radic Biol Med 6:323–339.

Koksal Z, Gulcin I, Ozdemir H (2016) An important milk enzyme: lactoperoxidase. In: Milk proteins - from structure to biological properties and health aspects. IntechOpen Limited, London.

Kristensen D, Nylander T, Rasmussen JT, Paulsson M, Carlsson A (1996) Bovine milk sphingomyelin at the air/water interface and its interaction with xanthine oxidase. Langmuir 12:5856–5862.

Kumar R, Bhatia KL (1999) Standardization of method for lactoperoxidase assay in milk. Lait 79:269–274.

Lake DF, Faigel DO (2014) The emerging role of QSOX1 in cancer. Antioxid Redox Signal 21:485–496.

Lee M-C-I, Velayutham M, Komatsu T, Hille R, Zweier JL (2014) Measurement and characterization of superoxide generation from xanthine dehydrogenase: a redox-regulated pathway of radical generation in ischemic tissues. Biochemistry 53:6615–6623.

Li T, Ma L, Sun D, Liu L, Qayum A, Jiang Z, Hou J (2019) Purification of lactoperoxidase from bovine milk by integrating the technique of salting-out extraction with cation exchange chromatographic separation. J Food Meas Charact 13:1400–1410.

Liao Y, Alvarado R, Phinney B, Lönnerdal B (2011) Proteomic characterization of human milk fat globule membrane proteins during a 12 month lactation period. J Proteome Res 10:3530–3541.

Lipmann F, Owen CR (1943) The antibacterial effect of enzymatic xanthine oxidation. Science 98:246–248.

Lorenzen PC, Martin D, Clawin-Rädecker I, Barth K, Knappstein K (2010) Activities of alkaline phosphatase, γ-glutamyltransferase and lactoperoxidase in cow, sheep and goat's milk in relation to heat treatment. Small Rumin Res 89:18–23.

Ma Y, Zhang L, Wu Y, Zhou P (2019) Changes in milk fat globule membrane proteome after pasteurization in human, bovine and caprine species. Food Chem 279:209–215.

Magacz M, Kędziora K, Sapa J, Krzyściak W (2019) The significance of lactoperoxidase system in oral health: application and efficacy in oral hygiene products. Int J Mol Sci 20:1443.

Martin HM, Hancock JT, Salisbury V, Harrison R (2004) Role of xanthine oxidoreductase as an antimicrobial agent. Infect Immun 72:4933–4939.

Martínez-Martínez A, Flores-Flores C, Campoy FJ, Muñoz-Delgado E, Fini C, Vidal CJ (1998) Biochemical properties of 5'-nucleotidase from mouse skeletal muscle. Biochim Biophys Acta Protein Struct Mol Enzymol 1386:16–28.

Martínez-Martínez A, Muñoz-Delgado E, Campoy FJ, Flores-Flores C, Rodríguez-López JN, Fini C, Vidal CJ (2000) The ecto-5'-nucleotidase subunits in dimers are not linked by disulfide bridges but by non-covalent bonds. Biochim Biophys Acta 1478:300–308.

Mebazaa A, Vanpoucke G, Thomas G, Verleysen K, Cohen-Solal A, Vanderheyden M, Bartunek J, Mueller C, Launay J-M, Van Landuyt N, D'hondt F, Verschuere E, Vanhaute C, Tuytten R, Vanneste L, De Cremer K, Wuyts J, Davies H, Moerman P, Logeart D, Collet C, Lortat-Jacob B, Tavares M, Laroy W, Januzzi JL, Samuel J-L, Kas K (2012) Unbiased plasma proteomics for novel diagnostic biomarkers in cardiovascular disease: identification of quiescin Q6 as a candidate biomarker of acutely decompensated heart failure. Eur Heart J 33:2317–2324.

Medina M, Gaya P, Nuñez M (1989) The lactoperoxidase system in ewes' milk: levels of lactoperoxidase and thiocyanate. Lett Appl Microbiol 8:147–149.

Mennella MR, Jones MR (1977) The activity of some nucleolytic enzymes in semen and in the secretion of the male reproductive tract. Andrologia 9:15–22.

Misumi Y, Ogata S, Ohkubo K, Hirose S, Ikehara Y (1990) Primary structure of human placental 5'-nucleotidase and identification of the glycolipid anchor in the mature form. Eur J Biochem 191:563–569.

Mondy BL, Keenan TW (1993) Butyrophilin and xanthine oxidase occur in constant molar proportions in milk lipid globule membrane but vary in amount with breed and stage of lactation. Protoplasma 177:32–36.

Monks J, Dzieciatkowska M, Bales ES, Orlicky DJ, Wright RM, McManaman JL (2016) Xanthine oxidoreductase mediates membrane docking of milk-fat droplets but is not essential for apocrine lipid secretion. J Physiol 594:5899–5921.

Morton RK (1953) Microsomal particles of normal cow's milk. Nature 171:734–735

Morton RK (1954) The lipoprotein particles in cow's milk. Biochem J 57:231–237

Murase T, Nampei M, Oka M, Miyachi A, Nakamura T (2016) A highly sensitive assay of human plasma xanthine oxidoreductase activity using stable isotope-labeled xanthine and LC/TQMS. J Chromatogr B Analyt Technol Biomed Life Sci 1039:51–58.

Murthy MR, Reid TJ, Sicignano A, Tanaka N, Rossmann MG (1981) Structure of beef liver catalase. J Mol Biol 152:465–499.

Nakamura M, Yamazaki I (1982) Preparation of bovine milk xanthine oxidase as a dehydrogenase form. J Biochem 92:1279–1286.

Ndrepepa G, Colleran R, Kastrati A (2018) Gamma-glutamyl transferase and the risk of atherosclerosis and coronary heart disease. Clin Chim Acta 476:130–138.

Neu HC (1967) The 5'-nucleotidase of Escherichia coli. II. Surface localization and purification of the Escherichia coli 5'-nucleotidase inhibitor. J Biol Chem 242:3905–3911

Nguyen HTH, Ong L, Hoque A, Kentish SE, Williamson N, Ang C-S, Gras SL (2017) A proteomic characterization shows differences in the milk fat globule membrane of buffalo and bovine milk. Food Biosci 19:7–16.

Nishino T, Nishino T (1997) The conversion from the dehydrogenase type to the oxidase type of rat liver xanthine dehydrogenase by modification of cysteine residues with fluorodinitrobenzene. J Biol Chem 272:29859–29864.

Nishino T, Nishino T, Tsushima K (1981) Purification of highly active milk xanthine oxidase by affinity chromatography on sepharose 4B/folate gel. FEBS Lett 131:369–372.

Nishino T, Okamoto K, Kawaguchi Y, Hori H, Matsumura T, Eger BT, Pai EF, Nishino T (2005) Mechanism of the conversion of xanthine dehydrogenase to xanthine oxidase - identification of the two cysteine bonds and crystal structure of a non-convertible rat liver xanthine dehydrogenase mutant. J Biol Chem 280:24888–24894.

Nishino T, Okamoto K, Kawaguchi Y, Matsumura T, Eger BT, Pai EF, Nishino T (2015) The C-terminal peptide plays a role in the formation of an intermediate form during the transition between xanthine dehydrogenase and xanthine oxidase. FEBS J 282:3075–3090.

Ouyang X, DeWeese TL, Nelson WG, Abate-Shen C (2005) Loss-of-function of Nkx3.1 promotes increased oxidative damage in prostate carcinogenesis. Cancer Res 65:6773–6779.

Ozturk G, German JB, de Moura Bell JMLN (2019) Effects of industrial heat treatments on the kinetics of inactivation of antimicrobial bovine milk xanthine oxidase. Npj Sci Food 3:1–7.

Patrick PS, Swaisgood HE (1976) Sulfhydryl and disulfide groups in skim milk as affected by direct ultra-high-temperature heating and subsequent storage. J Dairy Sci 59:594–600.

Patton S (1978) Milk secretion at the cellular level: a unique approach to the mechanism of exocytosis. J Dairy Sci 61:643–650.

Patton S, Keenan TW (1971) The relationship of milk phospholipids to membranes of the secretory cell. Lipids 6:58–61

Pegtel DM, Cosmopoulos K, Thorley-Lawson DA, van Eijndhoven MAJ, Hopmans ES, Lindenberg JL, de Gruijl TD, Würdinger T, Middeldorp JM (2010) Functional delivery of viral miRNAs via exosomes. Proc Natl Acad Sci U S A 107:6328–6333.

Plantz PE, Patton S (1973) Plasma membrane fragments in bovine and caprine skin milks. Biochim Biophys Acta 291:51–60

Rajagopalan K (1986) Purification of bovine milk xanthine oxidase. In: Greenwald REA (ed) CRC handbook of methods for oxygen radical research. CRC Press, Boca Raton, pp 21–24

Rashidinejad A, Birch EJ, Everett DW (2016) Green tea catechins suppress xanthine oxidase activity in dairy products: an improved HPLC analysis. J Food Compos Anal 48:120–127.

Rasmussen JT, Rasmussen MS, Petersen TE (2000) Cysteines involved in the interconversion between dehydrogenase and oxidase forms of bovine xanthine oxidoreductase. J Dairy Sci 83:499–506.

Reinhardt TA, Lippolis JD (2006) Bovine milk fat globule membrane proteome. J Dairy Res 73:406–416.

Reinhardt TA, Sacco RE, Nonnecke BJ, Lippolis JD (2013) Bovine milk proteome: quantitative changes in normal milk exosomes, milk fat globule membranes and whey proteomes resulting from Staphylococcus aureus mastitis. J Proteome 82:141–154.

Reis J (1934) Nucleotidase and its relation to the deamination of nucleotides in the heart and the muscles. Bull Soc Chim Biol 16:385–399

Reiter B, Härnulv G (1984) Lactoperoxidase antibacterial system: natural occurrence, biological functions and practical applications. J Food Prot 47:724–732.

Robenek H, Hofnagel O, Buers I, Lorkowski S, Schnoor M, Robenek MJ, Heid H, Troyer D, Severs NJ (2006) Butyrophilin controls milk fat globule secretion. Proc Natl Acad Sci U S A 103:10385–10390.

Rudolf J, Pringle MA, Bulleid NJ (2013) Proteolytic processing of QSOX1A ensures efficient secretion of a potent disulfide catalyst. Biochem J 454:181–190.

Sato A, Nishino T, Noda K, Amaya Y, Nishino T (1995) The structure of chicken liver xanthine dehydrogenase cDNA cloning and the domain structure. J Biol Chem 270:2818–2826.

Savas A, Huseyin E, Ertugrul K (2011) Relationship between somatic cell count and catalase activity in raw milk of Anatolian buffaloes. Sci Res Essays 6:4109–4112.

Schroeder WA, Shelton JR, Shelton JB, Robberson B, Apell G, Fang RS, Bonaventura J (1982) The complete amino acid sequence of bovine liver catalase and the partial sequence of bovine erythrocyte catalase. Arch Biochem Biophys 214:397–421.

Seifu E, Buys EM, Donkin EF (2005) Significance of the lactoperoxidase system in the dairy industry and its potential applications: a review. Trends Food Sci Technol 16:137–154.

Sharma S, Singh AK, Kaushik S, Sinha M, Singh RP, Sharma P, Sirohi H, Kaur P, Singh TP (2013) Lactoperoxidase: structural insights into the function, ligand binding and inhibition. Int J Biochem Mol Biol 4:108–128

Shin K, Hayasawa H, Lönnerdal B (2001) Purification and quantification of lactoperoxidase in human milk with use of immunoadsorbents with antibodies against recombinant human lactoperoxidase. Am J Clin Nutr 73:984–989.

Silanikove N, Shapiro F, Shamay A, Leitner G (2005) Role of xanthine oxidase, lactoperoxidase, and NO in the innate immune system of mammary secretion during active involution in dairy cows: manipulation with casein hydrolyzates. Free Radic Biol Med 38:1139–1151.

Silanikove N, Merin U, Leitner G (2006) Physiological role of indigenous milk enzymes: an overview of an evolving picture. Int Dairy J 16:533–545.
Silanikove N, Shapiro F, Silanikove M, Merin U, Leitner G (2009) Hydrogen peroxide-dependent conversion of nitrite to nitrate as a crucial feature of bovine milk catalase. J Agric Food Chem 57:8018–8025.
Silanikove N, Merin U, Leitner G (2014) Nitrite and catalase levels rule oxidative stability and safety properties of milk: a review. RSC Adv 4:26476–26486.
Silva CCG, Silva SPM, Ribeiro SC (2018) Application of bacteriocins and protective cultures in dairy food preservation. Front Microbiol 9:594.
Simon M, Hansen AP (2001) Effect of various dairy packaging materials on the shelf life and flavor of pasteurized milk. J Dairy Sci 84:767–773.
Singh AK, Singh N, Sharma S, Shin K, Takase M, Kaur P, Srinivasan A, Singh TP (2009) Inhibition of lactoperoxidase by its own catalytic product: crystal structure of the hypothiocyanate-inhibited bovine lactoperoxidase at 2.3-Å resolution. Biophys J 96:646–654.
Snow LD, Doss RC, Carraway KL (1980) Cooperativity of the concanavalin A inhibition of bovine milk fat globule membrane 5'-nucleotidase. Response to extraction of nucleotidase and of putative cytoplasmic surface coat components. Biochim Biophys Acta 611:333–341.
Soloviev M, Esteves MP, Amiri F, Crompton MR, Rider CC (2013) Elevated transcription of the gene QSOX1 encoding quiescin Q6 sulfhydryl oxidase 1 in breast cancer. PLoS One 8:e57327.
Song H, Zhang B, Watson MA, Humphrey PA, Lim H, Milbrandt J (2009) Loss of *Nkx3.1* leads to the activation of discrete downstream target genes during prostate tumorigenesis. Oncogene 28:3307–3319.
Stephens S, Harkness RA, Cockle SM (1979) Lactoperoxidase activity in guinea-pig milk and saliva: correlation in milk of lactoperoxidase with bactericidal activity against Escherichia coli. Br J Exp Pathol 60:252–258
Stirpe F, della Corte E, With the technical assistance of E. Lorenzoni (1969) The regulation of rat liver xanthine oxidase - convertion in vitro of the enzyme activity from dehydrogenase (type D) to oxidase (type O). J Biol Chem 244:3855–3863
Storch V (1898) Chemical method of ascertaining whether milk has been heated to at least 80°C. Centra Agrikulturchemie 27:711–714
Swaisgood H (1980) Sulfhydryl oxidase - properties and applications. Enzym Microb Technol 2:265–272.
Swaisgood H, Janolino V, Skudder P (1987) Continuous treatment of ultrahigh-temperature sterilized milk using immobilized sulfhydryl oxidase. Methods Enzymol 136:423–431
Sweeney EL, Al-Shehri SS, Cowley DM, Liley HG, Bansal N, Charles BG, Shaw PN, Duley JA, Knox CL (2018) The effect of breastmilk and saliva combinations on the in vitro growth of oral pathogenic and commensal microorganisms. Sci Rep 8:1–9.
Taniguchi N (2009) From the γ-glutamyl cycle to the glycan cycle: a road with many turns and pleasant surprises. J Biol Chem 284:34469–34478.
Théry C, Ostrowski M, Segura E (2009) Membrane vesicles as conveyors of immune responses. Nat Rev Immunol 9:581–593.
Uguz MT, Ozdemir H (2005) Purification of bovine milk lactoperoxidase and investigation of antibacterial properties at different thiocyanate mediated. Appl Biochem Microbiol 41:349–353.
Valadi H, Ekström K, Bossios A, Sjöstrand M, Lee JJ, Lötvall JO (2007) Exosome-mediated transfer of mRNAs and microRNAs is a novel mechanism of genetic exchange between cells. Nat Cell Biol 9:654–659.
van den Berg G, Meijer WC, Düsterhöft E-M, Smit G (2004) Gouda and related cheeses. In: Fox PF, McSweeney PLH, Cogan TM, Guinee TP (eds) Cheese: chemistry, physics and microbiology. Academic Press, Amsterdam, pp 103–140
Vazquez-Landaverde PA, Velazquez G, Torres JA, Qian MC (2005) Quantitative determination of thermally derived off-flavor compounds in milk using solid-phase microextraction and gas chromatography. J Dairy Sci 88:3764–3772.

Vazquez-Landaverde PA, Torres JA, Qian MC (2006) Quantification of trace volatile sulfur compounds in milk by solid-phase microextraction and gas chromatography-pulsed flame photometric detection. J Dairy Sci 89:2919–2927.

Vetsika F, Boukidi K, Roussis IG (2014) gamma-Glutamyl-transferase, xanthine oxidase and total free sulfhydryls as potential markers for pasteurization treatments in dairy technology. J Food Nutr Res 53:324–332

Vissers MMM, Driehuis F, Te Giffel MC, De Jong P, Lankveld JMG (2007) Minimizing the level of butyric acid bacteria spores in farm tank milk. J Dairy Sci 90:3278–3285.

Vorbach C, Scriven A, Capecchi MR (2002) The housekeeping gene xanthine oxidoreductase is necessary for milk fat droplet enveloping and secretion: gene sharing in the lactating mammary gland. Genes Dev 16:3223–3235.

Vorbach C, Harrison R, Capecchi MR (2003) Xanthine oxidoreductase is central to the evolution and function of the innate immune system. Trends Immunol 24:512–517

Vorbach C, Capecchi MR, Penninger JM (2006) Evolution of the mammary gland from the innate immune system? Bioessays 28:606–616.

Wallden K, Stenmark P, Nyman T, Flodin S, Gräslund S, Loppnau P, Bianchi V, Nordlund P (2007) Crystal structure of human cytosolic 5'-nucleotidase II: insights into allosteric regulation and substrate recognition. J Biol Chem 282:17828–17836.

Wang TP (1954) Specific 5'-nucleotidase from a soil bacterium1. J Bacteriol 68:128

Wang W, Lv N, Zhang S, Shui G, Qian H, Zhang J, Chen Y, Ye J, Xie Y, Shen Y, Wenk MR, Li P (2012) Cidea is an essential transcriptional coactivator regulating mammary gland secretion of milk lipids. Nat Med 18:235–243.

Waud WR, Rajagopalan KV (1976) The mechanism of conversion of rat liver xanthine dehydrogenase from an NAD+-dependent form (type D) to an O2-dependent form (type O). Arch Biochem Biophys 172:365–379.

Wellings SR, Deome KB (1961) Milk protein droplet formation in the Golgi apparatus of the C3H/Crgl mouse mammary epithelial cells. J Biophys Biochem Cytol 9:479–485

West MB, Chen Y, Wickham S, Heroux A, Cahill K, Hanigan MH, Mooers BHM (2013) Novel insights into eukaryotic γ-glutamyltranspeptidase 1 from the crystal structure of the glutamate-bound human enzyme. J Biol Chem 288:31902–31913.

Yoon Y, Lee S, Choi K-H (2016) Microbial benefits and risks of raw milk cheese. Food Control 63:201–215.

Zehetner G, Bareuther C, Henle T, Klostermeyer H (1995) Inactivation kinetics of y-glutamyltransferase during the heating of milk. Z Lebensm Unters Forsch 201:336

Zhang S-B, Zhai H-C, Hu Y-S, Wang L, Yu G-H, Huang S-X, Cai J-P (2014) A rapid detection method for microbial spoilage of agro-products based on catalase activity. Food Control 42:220–224.

Zimmermann H (1992) 5'-Nucleotidase: molecular structure and functional aspects. Biochem J 285:345–365

Ziobro GC, McElroy KM (2013) Fluorometric detection of active alkaline phosphatase and gamma-glutamyl transferase in fluid dairy products from multiple species. J Food Prot 76:892–898.

Chapter 7
Milk and Other Glycosidases

Anne Vuholm Sunds, Søren Drud-Heydary Nielsen, Lotte Bach Larsen, and Nina Aagaard Poulsen

7.1 Introduction

Glycosidases (EC 3.2.1) are enzymes hydrolyzing the glycosidic bonds in sugars. They are widespread in nature and, like many other biological fluids, milk also contains glycosidases. This Chapter reviews the literature on milk glycosidases mainly of bovine origin, whereas human milk glycosidases are described by Chan et al. (2019) in Chap. 9. Milk glycosidases are of interest, due to their sugar substrates, which include not only lactose, but also more complex free oligosaccharides, as well as oligosaccharides (glycans) bound to either proteins or lipids. These combined molecules are referred to as glycoconjugates, and include glycoproteins, glycolipids and proteoglycans. Despite their varied activities, glycosidases are generally present in milk at low levels and, while activity assays have shown variation in activity, little is known about their origin and effect on milk sugars during intake and processing. Within the field of milk glycosidases, only a limited number of glycosidases have been characterized in detail, and research has mainly addressed specific aspects, like the use of N-acetyl-β-glucosaminidase (NAGase) as marker for mastitis. It is thus an immature area with few published studies in relation to foods and in relation to milk. As a result, the knowledge reviewed in this Chapter is based on a relatively limited number of studies.

Glycan metabolism involves both glycosyltransferases, sugar transporters and glycosidases in the formation, modification and hydrolysis of sugars, where the glycosidases participate in trimming during glycan synthesis and turn-over. Glycosidases are also referred to as glycoside hydrolases or carbohydrases, and are thus characterized by catalyzing the hydrolysis of glycosidic linkages between monosaccharides. Like other milk enzymes, the milk relevant glycosidases can have

A. V. Sunds · S. D.-H. Nielsen · L. B. Larsen · N. A. Poulsen (✉)
Department of Food Science, Aarhus University,
Agro Food Park, Denmark
e-mail: annesunds@food.au.dk; sodn@food.au.dk; lbl@food.au.dk; Nina.Poulsen@food.au.dk

A. L. Kelly, L. B. Larsen (eds.), *Agents of Change*, Food Engineering Series,
https://doi.org/10.1007/978-3-030-55482-8_7

different origins. This includes indigenous enzymes, i.e., of bovine origin, as well as those of microbial origin. Microorganisms can be present in raw milk, e.g., due to udder infections or contamination at some point in the cold chain. Furthermore, they can be added exogenously during processing, such as the lactases (β-galactosidases) used in the manufacture of lactose-free products.

The sugar fraction in mature bovine milk is constituted of 46–48 g/L lactose, approximately 0.03–0.06 g/L free oligosaccharides (Urakami et al. 2018; Urashima et al. 2013), as well as several glycoconjugates. In comparison, the composition of human milk sugars is very different; mature human milk contains approximately 68–70 g/L lactose, 7–14 g/L free oligosaccharides (Boehm and Stahl 2007; Coppa et al. 1999), as well as several glycoconjugates. These differences are of significant interest and a challenge in the production of infant formula based on bovine milk. Since the glycan part of glycoconjugates affects digestion and is responsible for many bioactivities related to milk, these sugar compounds are highly valuable food ingredients. The bioactive functions of glycans are associated with antitoxin, antiviral, immunoregulatory and prebiotic effects (Kawasaki et al. 1993; Parker et al. 2010), for example, by stimulating the growth of beneficial microorganisms, especially *Bifidobacteria* and *Lactobacilli* (Sela 2011), and prevention of pathogen adhesion to the intestinal epithelium, e.g., of *Helicobacter pylori* (Simon et al. 1997). Furthermore, sialylated oligosaccharides are essential for brain development in newborns (Wang et al. 2001). It is thus of interest to gain more knowledge on the enzymes influencing the glycan structures in milk. Even though β-galactosidases and their use as lactases in dairy context is well-known, the field of glycosidases present in milk, that be either of bovine or microbial origin, is still far from elucidated.

7.2 Origin and Significance of Glycosidases in Milk

Overall, glycosidases can be secreted into the milk from mammary epithelial cells, directly from the blood through the paracellular pathways, be secreted by immune cells during phagocytosis or cell lysis, or arise from microorganisms from the environment (Le Maréchal et al. 2011; Pyörälä 2003). The milk glycosidases are active in degradation of free oligosaccharides and glycoconjugates, especially in the lysosomes originating from the mammary epithelial cells (Sellinger et al. 1960; Von Figura and Hasilik 1986; Fox and Kelly 2006). Somatic cells from healthy cows comprise immune cells, including a mixture of polymorphonuclear (PMN) neutrophils or leucocytes, lymphocytes and macrophages (~12%, 28% and 60% respectively), as well as a small amount of epithelial cells (Wickramasinghe et al. 2011; Burvenich et al. 1995).

The glycosidase NAGase has been found to be widely associated with somatic cell count (SCC) in milk (Hovinen et al. 2016). As the content of neutrophils in milk increases exponentially during early inflammation of the mammary gland (Saad and Östensson 1990), it has been suggested that NAGase leaks into the milk during cell lysis or phagocytosis, especially by neutrophils (Kaartinen et al. 1988), but it is also possible that some active secretion takes place.

Glycosidases can furthermore be excreted by a variety of microorganisms that may contaminate during milking, e.g., coliforms and *Streptococcus uberis* (Djabri et al. 2002), or psychrotrophic bacteria during the cold chain e.g. *Sphingobacterium* and *Rahnella* (Yuan et al. 2018). They can further be added as processing enzymes at the dairy originating from starter cultures, or by addition of β-galactosidase from the dairy yeast, *Kluyveromyces lactis* or from, e.g., *Bacillus circulans* or *Aspergillus oryzae* in the production of lactose-free products (Dekker et al. 2019).

Wickramasinghe et al. (2011) investigated genes associated with bovine milk oligosaccharide metabolism, and identified 11 genes encoding different glycosidases that may play a role, based on expression studies. These genes encoding glycosidases included fucosidase (FUC A1 and A2), galactosidase (GLA and GLB1), NAGase (HEXA and HEXB), neuraminidase (NEU1–4) and α-glucosidase (GANAB). While no expression was observed for NEU2 and NEU4, significant increases in the gene expression of the remaining nine glycosidases were observed over lactation. This observation was based on a comparison of the glycosidase transcriptome from somatic cells isolated from early lactation milk (15 days in milking) with transcripts of glycosidases in cells from late lactation milk (250 days in milking) in both Holstein and Jersey cows, which could indicate increased catabolism of glycoconjugates in later lactation (Wickramasinghe et al. 2011). Even though this study suggests that increasing levels of glycosidases are present in milk over lactation, most of these do not seem to be equivalently activated, based on activity assays of glycosidases in bovine milk from day 1–90 of lactation. Though, the activities may increase later in lactation, but there is a lack of knowledge on this in relation to bovine milk.

Overall, the highest activities of glycosidases are found in colostrum, decreasing to constant levels in mature milk (O'Riordan et al. 2014). The higher activities of most glycosidases in colostrum and early lactation milk (1–10 days) suggests that milk glycosidases are important contributors to glycan digestion in neonates until colonization of a healthy gut microflora rich in *Bifidobacteria* and *Lactobacillus*, which in calves takes approximately 7 days post-parturition (Alipour et al. 2018). This is possibly the most important biological function of glycosidases naturally present in milk from healthy cows, since milk sugar moieties bigger than lactose are indigestible by mammalian digestive enzymes.

Besides lactation, the milk glycosidase activities found are affected by several parameters, including genetics, utter health status (mastitis), cow breed, feeding system, season and processing (Chagunda et al. 2006; Jóźwik et al. 2004; Jóźwik et al. 2016). Most glycosidases increase significantly in activity during summer season. The highest effect of feeding system was, in a study by Jóźwik et al. (2016), observed for β-galactosidase and α-glucosidase activities, with significant increases in these activities during pasture.

Overall, the activity of only nine different milk glycosidases in bovine milk have been described in the literature, as summarized in Table 7.1. These nine comprise NAGase, α-fucosidase, α-galactosidase, β-galactosidase (lactase), α-glucosidase, β-glucosidase, N-acetyl-α-neuraminidase (sialidase), β-glucuronidase and α-mannosidase (O'Riordan et al. 2014; Jóźwik et al. 2016). Four of these have been

Table 7.1 Glycosidases in bovine milk and their origin, cleavage specificity, thermostability, pH optimum and activities in colostrum and mature milk

Glycosidases	Origin	Cleavage specificity	Thermostability	pH optimum	Activity in colostrum[a] (pmol/μL/min)	Activity in mature milk[a, b] (pmol/μL/min)
N-acetyl-β-glucosaminidase (NAGase) EC 3.2.1.52	Lysosomal, originating from mammary epithelial cells, from phagocytes and ~ 15% from somatic cells	Acts on N-acetylglucosides and N-acetylgalactosides, these enzymes are thus also referred to as N-acetyl-β-hexosaminidase	Completely inactivated after heating at 75 °C for 15 s in milk	pH 4.2	~5.9	~0.008–39
α-Fucosidase EC 3.2.1.51	Lysosomal[c]	Hydrolyses α-L-fucosides by cleavage of terminal L-fucose residues	3–25% activity after 75 °C for 15 s in milk	pH 5–6	~1.379	~0.342
α-Galactosidase EC 3.2.1.2	Lysosomal[c]	Hydrolyses the terminal alpha-galactosyl moieties from glycolipids and glycoproteins	Completely inactivated after heating at 65 °C for 15 s in milk	pH 6.8–7.4	~0.994	~0.096
β-Galactosidase EC 3.2.1.23	Lysosomal[c]	Hydrolyses the β-glycosidic bond formed between a galactose and its organic moiety	Completely inactivated after heating at 65 °C for 15 s in milk	pH 7.3–7.8	~0.071	~0.04
α-Glucosidase EC 3.2.1.20	Lysosomal[c]	Hydrolysis of terminal, non-reducing α-1,4 linkages of α-D-glucose residues with release of α-D-glucose	Completely inactivated after heating at 65 °C for 15 s in milk	pH 6.5	–	–
β-Glucosidase EC 3.2.1.21	Lysosomal[c]	Hydrolysis of terminal non-reducing β-D-glucose residues	Completely inactivated after heating at 65 °C for 15 s in milk	pH 5.0	~0.026	~0.086
N-acetyl-α-neuraminidase (sialidase) EC. 3.2.1.18	Lysosomal[c]	Hydrolysis of terminal sialic acid residues	No studies on milk	pH 4.5–6.0	~0.409	~0.14

Glycosidases	Origin	Cleavage specificity	Thermostability	pH optimum	Activity in colostrum[a] (pmol/μL/min)	Activity in mature milk[a, b] (pmol/μL/min)
β-Glucuronidase EC 3.2.1.31	Lysosomal[c]	Hydrolysis of β-D-glucuronosides, with cleavage of glucuronate	Completely inactivated after heating at 70 °C for 15 s in milk	pH 6–6.5	~0.095	~0.005
α-Mannosidase EC 3.2.1.24	Lysosomal[c]	Hydrolysis of terminal, non-reducing α-D-mannose residues in α-D-mannosides	Inactivated after heating at 86 °C for 25 s in milk	pH 3	–	–

Thermostability and pH optimum are based on Griffiths (1986), Mellors and Harwalkar (1968), McKellar and Piyasena (2000) and Andrews et al. (1987), and reported activities are based on O'Riordan et al. (2014) and Hovinen et al. (2016)

N-acetyl-β-glucosaminidase = β-hexosaminidase = hexosaminidase A = NAGase

N-acetyl-α-neuraminidase = acetyl-neuraminyl hydrolaselysosomal sialidase = sialidase

[a]Values for activity was recalculated for this table from reported units (μM/mL/h) in O'Riordan et al. (2014) to pmol/μL/min as used in Hovinen et al. (2016)

[b]Mature milk represents milk from day 90 after parturition, measured at 37 °C

[c]Potentially both from mammary epithelial cells and somatic cells

suggested to be biologically most relevant in bovine milk, due to their elevated activities in colostrum, including NAGase, α-fucosidase, α-galactosidase and sialidase (O'Riordan et al. 2014). These findings are consistent with studies on human milk, where NAGase and α-fucosidase are highest in activity (Wiederschain and Newburg 2001b).

7.3 Cleavage Specificity and Biological Relevance in Relation to Substrates

In nature, sugars are the most diverse and complex class of biomolecules, varying in anomeric structures, glycosylation positions and composition of monosaccharide moieties. These variations make cleavage of sugar structures complex, requiring glycosidases with the same regio- and anomeric specificity or unspecific glycosidases which are able to cleave various bonds (Fig. 7.1, Table 7.3). Glycosidases are, based on their glyco-specificity, classified into exo- and endoglycosidases. Exoglycosidases cleave off monosaccharides from the non-reducing terminal position, whereas endoglycosidases hydrolyse the inner part of the sugar chain. The group of exoglycosidases are usually known to possess a strong substrate

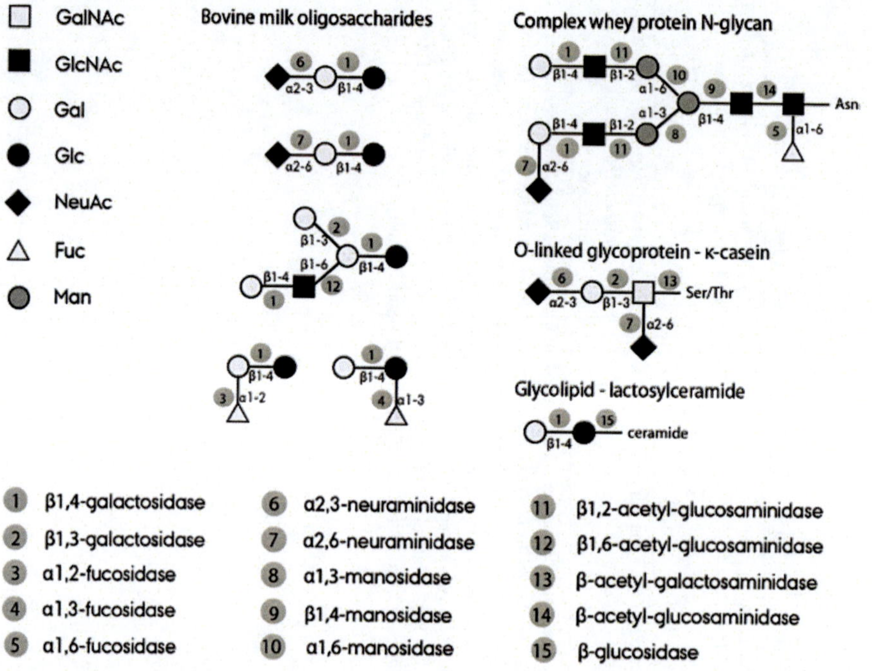

Fig. 7.1 Selected glycans in bovine milk and glycosidases with corresponding regio- and anomeric specificity. *Glc* glucose, *Gal* galactose, *GlcNAc* N-acetyl-glucosamine, *GalNAc* N-acetyl-galactosamine, *Fuc* fucose, *NeuAc* N-acetyl-neuraminic acid, *Man* mannose

specificity, both according to anomeric structure of the glycosidic bond, as well as, to the targeted monosaccharide.

The glycans in bovine milk are present as lactose, as free oligosaccharides and as bound glycans. Lactose, which is composed of a galactose and a glucose moiety linked by a β-1,4-glycosidic bond, is the major sugar in both bovine and human milk and the main osmotic component in milk. The free milk oligosaccharides are sugar structures of 3–15 monosaccharides linked by different glycosidic bonds (Boehm and Stahl 2007). The monosaccharide building blocks in bovine milk consist of glucose (Glc), galactose (Gal), N-acetyl-glucosamine (GlcNAc), N-acetyl-galactosamine (GalNAc), fucose, N-glycolyl-neuraminic acid (NeuGc), and N-acetyl-neuraminic acid (NeuAc) (Aldredge et al. 2013). At the reducing end, most oligosaccharides consist of a lactose core, and can further vary in the degree of terminal fucosylation and sialylation at the non-reducing end. Bovine colostrum is rich in acidic oligosaccharides and glycoproteins (containing NeuAc and NeuGc), whereas mature milk contains a larger fraction of neutral glycans (Wickramasinghe et al. 2011; Takimori et al. 2011). A similar decrease over lactation is observed for fucosylated glycoproteins (Takimori et al. 2011).

The highest concentrations of free oligosaccharides are found in colostrum, where levels of 0.7–1.2 g/L have been reported (Tao et al. 2008), decreasing in mature bovine milk to 0.03–0.06 g/L (Urashima et al. 2013; Urakami et al. 2018). The decreased levels of free oligosaccharides over lactation may be correlated with the decreased glycosidase activities through a possible co-regulation mechanism in the mammary gland, reflecting the physiological need for glycosidases. Breed-specific differences in oligosaccharide profiles have been observed between Holstein and Jersey milk (McJarrow and van Amelsfort-Schoonbeek 2004; Sundekilde et al. 2011; Robinson et al. 2019). Sundekilde et al. (2011) found that Holstein-Friesian milk contained a higher abundance of smaller and simpler neutral free oligosaccharides, whereas Jersey milk contained higher relative amounts of both sialylated and fucosylated oligosaccharides. Such differences may be related to breed differences in glycosidases, e.g., fucosidases.

Glycoproteins are found within the different families of milk proteins: the caseins, whey proteins and proteins of the milk fat globule membrane (MFGM) (Table 7.2). N-linked glycans are divided into three classes; complex, hybrid or high-mannose depending on their composition and how the core is modified. The O-linked glycans are less abundant in milk and consist of GalNAc elongated with a few monosaccharides. The N-linked glycans are mainly present in proteins associated with the MFGM and in whey proteins like lactoferrin, α-lactalbumin and immunoglobulins, whereas O-linked glycans are present in κ-casein and MFGM proteins. Targeted studies of N-glycans in milk have largely been done using PNGase F or, in some cases, EndoBI-1 to release protein-bound glycans before MS identification (Cao et al. 2019; Nwosu et al. 2012; Karav and Le Parc 2016). Bound glycans are difficult to isolate and the total amount present in milk is still challenging to quantify. Recently, Huang et al. (2017) used triple Q multiple reaction monitoring (MRM) for identification of glycosylations of major proteins in human milk and found high glycosylation heterogeneity at each N-glycosylation-site.

A N-glycan level of 374.9 pmol/mg dry whey protein concentrate has been reported (Sriwilaijaroen et al. 2012), and a recent study by Cao et al. (2019) identified up to 100 and 98 N-glycans in the whey fraction of bovine colostrum and mature milk, respectively. In comparison, only 68 and 58 N-glycans were identified in human colostrum and mature milk. Furthermore, up to 31% of bovine N-glycans have shown to be fucosylated (Nwosu et al. 2012), compared to less than 1% of the total free bovine milk OS (Tao et al. 2008; Aldredge et al. 2013), illustrating the biological importance of bound glycans in bovine milk.

The most abundant O-linked glycoprotein in bovine milk is κ-casein, where 30–60% of κ-casein molecules are glycosylated (Table 7.2); it holds six potential O-glycosylation sites. The five predominant glycosylation forms attached to κ-casein are: GalNAc (Mw = 203.19 Da), Galβ(1–3)GalNAc (Mw = 365.34 Da), NeuAcα(2–3)Galβ(1–3)GalNAc (Mw = 656.59 Da), Galβ(1–3)[NeuAcα(2–6)]GalNAc (Mw = 656.59 Da), and NeuAcα(2–3)Galβ(1–3)[NeuAcα(2–6)]GalNAc (Mw = 947.85 Da) (Table 7.2; Fig. 7.1). These glycan structures have previously been reported to be present in amounts of approximately 1%, 6%, 18%, 19% and 56%, respectively (Saito and Itoh 1992). More recently, an additional glycosylation form has been observed constituting GalNAc and NeuAc (Mw = 494.45 Da) (Nwosu et al. 2010).

Approximately 70% of all the glycolipids in bovine milk are associated with the MFGM (Newburg and Chaturvedi 1992). The predominant neutral glycolipids situated in the MFGM are the glycosphingolipids, glucosylceramide and lactosylceramide, whereas the predominant acidic glycolipids are the gangliosides (Table 7.2) (Sánchez-Juanes et al. 2009; Rombaut and Dewettinck 2006).

Table 7.2 Levels of major glycoconjugates in bovine milk, including degrees of glycosylation

Major glycoconjugates	Milk fraction	Concentration in mature bovine milk	Glycosylation degree (%)
Lactoferrin	Whey	0.02–0.2 g/L	100[a]
Immunoglobulin A	Whey	0.075 g/L	6–10[b]
Immunoglobulin G	Whey	0.47 g/L	2.6–3.1[b]
Immunoglobulin M	Whey	0.52 g/L	10–12[b]
α-Lactalbumin	Whey	1.2 g/L	3–10[a]
Lactoperoxidase	Whey	0.4 μM	6.4–11.5[b]
Folate binding protein	Whey	–	Asn49 = 49[a] Asn141 = 74[a]
κ-Casein	Casein	~12% of total casein	30–60[a]
Butyrophilin	MFGM	40% of the MFGM proteins	4–6[b]
Lactadherin (PAS 6/7)	MFGM	–	7.1/5.5[b]
Mucins	MFGM	5–10% of the MFGM proteins	>30[b]
Glucosylceramide	MFGM	0.006–0.011 g/L	Monomer, 100[a]
Lactosylceramide	MFGM	0.0065–0.015 g/L	Dimer, 100[a]
Gangliosides	MFGM	~0.011 g/L	Oligomer, 100[a]

MFGM milk fat globule membrane, *PAS* periodic acid-Shiff
[a]The % of protein or lipid that are glycosylated
[b]The carbohydrate content (%) of the total molecular weight

Table 7.3 presents some of the major glycosidic linkage types present in milk sugar substrates and the specific glycosidases required to hydrolyse them. At present, not all the glycosidases listed in Table 7.3 have been confirmed in bovine milk. Neuraminidase and fucosidase are particularly important glycosidases in relation to sugar substrates possessing biological functions which are important in relation to infant nutrition. Neuraminidase (EC 3.2.1.18), also referred to as sialidase, releases sialic acid from glycans. In bovine milk, sialylated glycoconjugates include both NeuAc and NeuGc, although the acetyl form (NeuAc) is significantly more abundant. As is shown in Table 7.3, the following isomeric forms are present in milk, NeuAc-α-2,3-Gal and NeuAc-α-2,6-Gal. Two of the most abundant free oligosaccharides in milk differ by these glycosidic bonds: 3′sialyllactose (3′-SL) and 6′-sialyllactose (6′-SL). The regio-specificity of the neuraminidases are still undetermined.

α-Fucosidases (EC 3.2.1.51) catalyze the release of α-fucose residues from the terminal non-reducing end of glycans. They possess very specific cleavage properties, as they only cleave α-fucopyranosides, and not the isomeric form,

Table 7.3 Glycan bonds, examples of sugar substrates in milk containing these bonds, and corresponding glycosidases able to cleave the specific bonds

Glycan bond	Examples of sugar substrate	Glycosidase required
NeuAc-α-2,3-gal	3-sialyllactose or κ-casein	α-2,3-neuraminidase
NeuAc-α-2,6-gal	6-sialyllactose	α-2,6-neuraminidase
NeuAc-α-2,6-GalNAc	κ-Casein	α-2,6-neuraminidase
Fuc-α-1,2-gal	2-fucosyllactose	α-1,2-fucosidase
Fuc-α-1,3-Glc	3-fucosyllactose	α-1,3-fucosidase
Fuc-α-1,6-GalNAc	Lactoferrin and complex-type N-glycans in whey	α-1,6-fucosidase
Gal-α-1,3-gal	α-3′-galactosyl lactose	α-1-3-galactosidase
Gal-β-1,3-gal	3-galactosyllactose	β-1,3-galactosidase
Gal-β-1,3-GalNAc	κ-Casein	β-1,3-galactosidase
Gal-β-1,4-GlcNAc	Lacto-N-neotetraose (LNnT)	β-1,4-galactosidase
Gal-β-1,4-Glc	Lactose	β-1,4-galactosidase
Gal-β-1,6-gal	6-galactosyllactose	β-1,6-galactosidase
Glc-α-gal	–	α-Glucosidase
Glc-β- gal	–	β-Glucosidase
GalNAc-β-1,4-Glc	N-acetyl-galactosaminyl glucose	β-1,4-N-acetylgalactosaminidase
GlcNAc-β-1,6-gal	GlcNAc-gal-Glu	β-1,6-N-acetylglucosaminidase
Man-α-1,3-man	Lactotransferrin	α-1,3-mannosidase
Man-α-1,6-man	Lactotransferrin	α-1,6-mannosidase
Man-β-1,4-GlcNAc	Lactotransferrin	β-1,4-mannosidase

Gal galactose, *Glc* glucose, *NeuAc* Neuraminic acid, *GalNAc* N-acetyl-galactosamine, *GlcNAc* N-acetyl-glucosamine, *Fuc* fucose, *Man* mannose

β-fucopyranosides (Kobata 2013). Furthermore, bovine milk oligosaccharides contain both α1–2- and α1–3-linked fucose residues and some glycoconjugates contain α1–6-linked fucose residues (Table 7.3) (Aldredge et al. 2013; Spik et al. 1994; van Leeuwen et al. 2012). Several fucosidases with high anomeric- and regio-specificity should thus be present in order to hydrolyse all fucosylated glycoconjugates in milk, or more unspecific fucosidases are required. The degree of specificity of α-fucosidases in bovine milk is still unclear. The presence of fucosidases in bovine milk have been suggested to explain why only few fucosylated oligosaccharides have been detected, as they might be hydrolysed. However, only 170 μg/mL of free fucose is liberated from human milk oligosaccharides and glycoconjugates after incubation at body temperature for 16 h (Wiederschain and Newburg 2001a). This equals approximately 5% of the total amount of bound fucosylated sugars in human milk, which should not be of major concern in relation to hydrolysis of sugars in the mammary gland between feedings of the neonate, or until bovine milk is refrigerated (4 °C) after milking, at which temperature no increases in fucose amount were reported (Wiederschain and Newburg 2001a).

NAGase (EC: 3.2.1.30) is an example of a less specific exoglycosidase. This glycosidase cleaves not only the glycosidic bond of GalNAc, but also GlcNAc residues (Fig. 7.1). These glycosidases are thus more unspecific and are often able to cleave both O-linked and N-linked glycans. O-linked glycans are mainly linked *via* GalNAc to Ser/Thr residues, and N-linked glycans are linked *via* GlcNAc to Asn in the sequence Asn-X-Ser/Thr (X can be any amino acid except proline). The NAGase enzymes are, due to their unspecific nature, also referred to as N-acetylhexosaminidase (Karav and Le Parc 2016).

7.4 Glycosidases as Indicators of Mastitis

Mastitis, which is an inflammation disease caused by a bacterial infection in the mammary gland, greatly affects milk quality and yield. The most common marker for detection of mastitis is an increase in SCC (Le Maréchal et al. 2011). Inflammation of the udder affects the concentration of ions, immunoglobulins, indigenous enzymes including glycosidases, lactose concentration and pH, as well as the casein to whey protein ratio (Auldist and Hubble 1998; Le Maréchal et al. 2011).

The activity of the lysosomal enzyme NAGase (Sellinger et al. 1960) has been shown to be highly affected by mastitis infection. Mellors (1968) was the first to suggest the use of NAGase as an indicator of mastitis. Since then, several studies have demonstrated the reliability of NAGase as indicator enzyme (Hovinen et al. 2016; Kitchen et al. 1978; Kitchen 1976). Where increased SCC is a defence mechanism, increased NAGase activity is a tissue damage response. NAGase activities in normal mature milk highly varies amongst studies, where mean values of 0.62–1.2 pmol/μL/min have been reported, and mean values for clinical mastitic milk of 1.9–16.7 pmol/μL/min (Hovinen et al. 2016).

Hovinen et al. (2016) found a high correlation between SCC and NAGase activity in cows with subclinical mastitis, whereas no correlation existed in milk from healthy cows. Perdigon et al. (1986) also revealed increasing activities of β-glucuronidase with increasing SCC. More recently, Larsen and Aulrich (2012) found moderate but significant correlation between β-glucuronidase activity and SCC in bovine milk, and Oliszewski et al. (2002) found similar associations in caprine milk. The pH optimum for β-glucuronidase activity was tested and, as milk samples with different strains of pathogens had the same optimum, it was suggested that the enzymes were of mammary tissue and leucocyte origin rather than from bacterial cells. Furthermore, Hurley (1987) showed β-glucuronidase activities in leucocyte fractions from bovine blood.

Only little attention has been paid to β-glucuronidase as an indicator enzyme compared to NAGase. A major reason for this is the higher activities of NAGase at infection, leading to more precise activity measurements and less time needed for incubation when conducting fluorometric assays (Larsen and Aulrich 2012).

7.5 Industrial Relevance and Effect of Milk Glycosidases during Processing

7.5.1 Effect of Processing on Milk Glycosidases

Few previous studies have investigated heat stability of milk glycosidases, as summarized in Table 7.1. Based on these studies, α-mannosidase seems to be the most heat-stable of the milk glycosidases, with inactivation after 86 °C for 25 s in milk (Andrews et al. 1987). α-Fucosidases have been shown to be very heat labile (De Noni 2006), and it has therefore been suggested to use α-fucosidase as indicator for heat treatment (McKellar and Piyasena 2000; Andrews et al. 1987). Exogenous glycosidases, originating from microbiological contamination, may also contribute to the presence of glycosidases, and thus potentially affect sugar structures in milk during storage and processing. Heat-stable glycosidases can thus affect product quality, due to degradation of valuable sugars and lead to potential changes, e.g., in perception of sweetness.

Cleavage of glycoconjugates may also alter functional properties of the final product (Svanborg et al. 2016; Manso and López-Fandiño 2004), as well as affect shelf-life and bioactivities. An example is glycosylated κ-casein; a high proportion of κ-casein isoforms contain negatively charged sialic acid residues, which influences hydrophilic and electrostatic properties. Cleavage of sialic acid by neuraminidases may therefore alter microstructure, foaming and gelling properties of this protein (Kreuß et al. 2009). Furthermore, the pH value has an impact on activity of glycosidases; different optimal conditions have been reported in literature, varying from pH 3 to 7.8 (Table 7.1).

7.5.2 *Applications of Lactases in Processing of Dairy Products*

Application of various glycosidases might also have potentials in the dairy industry. The most common use of glycosidases in dairy industry is the use of β-galactosidase (lactase) in the production of lactose-free products. Furthermore, lactase has also found application in enhancing sweetness, solubility and flavour of certain dairy products (Contesini et al. 2013). Development of lactose-hydrolyzed dairy products has its basis in the large market potential, as 65% of the world's population are lactose-intolerant (Leonardi et al. 2012). Lactase hydrolyses lactose into its two constituent monosaccharides. Lactase is produced in the small intestine of mammals, with high activity during suckling. However, as the diet changes from exclusively milk based into one which is more complex during weaning, lactase expression is down-regulated to a level where it is insufficient to hydrolyse ingested lactose from a dairy-rich diet in most of the world's population (Leonardi et al. 2012).

The lactase used in production of lactose-free milk can be extracted from bacteria, yeast or fungi, but it is mainly isolated from the dairy yeast, *Kluyveromyces lactis* (Ansari and Satar 2012). Commercially available lactase preparations differ in purity, depending on extraction procedures and purification degrees, and can thus contain different side-activities, such as residual protease activity (Nielsen et al. 2018). These side-activities affect the quality of especially long shelf-life dairy products, such as ultra-high temperature (UHT) treated lactose-free milk, while, in fresh lactose-free dairy products, the effect of lactase impurities on product deterioration may be less relevant, due to the shorter shelf-life.

Lactose-free UHT milk has been shown to have reduced shelf-life compared to conventional UHT milk and exhibits increased off-flavor development (Jensen et al. 2015). Residual proteolysis in commercial lactase enzyme preparations was observed to have a high impact on product deterioration depending on the purity of the lactase enzyme (Mittal et al. 1991). Such proteolysis leads to an increase in free amino acids, peptides with potential bitterness and reduced level of intact protein, as well as, in some cases, aggregate formation (Nielsen et al. 2017; Nielsen et al. 2018; Troise et al. 2016).

Increased formation of furosine and 2-methylbutanal in addition to a visual browning of the milk was also observed in lactose-hydrolysed UHT milk compared to conventional UHT milk, indicating more favorable conditions for the Maillard reaction (Jansson et al. 2014). This is explained by the glycan composition in the lactose-hydrolyzed milk, as glucose and galactose have a higher reactivity with amines compared to lactose, enhancing the formation of early stage Maillard reactions (Naranjo et al. 2013). This effect could be enhanced by proteolysis of milk proteins, due to the increased level of free amino groups, especially in relation to formation of free amino acids due to exopeptidase side-activities, and may result in formation in Strecker aldehydes and aroma compounds, which can affect taste and flavor (Jansson et al. 2014, 2015).

Two processes (pre-hydrolysis and post-hydrolysis) are used in production of lactose-free milk. In pre-hydrolysis, the lactase enzyme is added to a tank of raw milk and incubated for approximately 24 h at a cool temperature (4–8 °C) to avoid microbial spoilage prior to heat treatment. Pre-hydrolysis requires a higher dose of lactase than post-hydrolysis, and the enzyme should be active at the low holding temperatures. The pre-hydrolysis method is less sensitive to deterioration from lactase side activity, as heating after the addition of lactase will inactivate most enzymatic activities, but also requires the milk to be held in a tank for a day (Dekker et al. 2019). In post-hydrolysis, lactase is aseptically added to milk which has been heat-treated, but the doses used can be lower than with pre-hydrolysis. This procedure requires a high quality of the lactase enzyme, mostly free from proteolytic side-activity and arylsulfatase side-activity (Dekker et al. 2019).

7.6 Detection and Methods for Analysis of Glycosidases

Various analytical methods can be used in determination of glycosidases. Commonly used methodologies for analysis of milk glycosidases include assays targeting glycosidase activities with use of fluorescent 4-methylumbelliferyl substrates (O'Riordan et al. 2014), or colorimetric assays based on *ortho*-nitrophenyl substrates (ONPG) (McKellar and Piyasena 2000). Both assays are frequently used to assess various glycosidases. Furthermore, the latter method is commonly used to monitor lactase activities in commercial milk products. The principle of these methods is to use modified substrates which are non-fluorescent and colorless until they are hydrolysed by the glycosidases specific for each substrate. Hydrolysis causes release of either 4-methylumbelliferone or *ortho*-nitrophenole, resulting in a fluorometric or colorimetric signal, respectively. The measured intensities reflect the amount of substrate hydrolysed, and thereby the glycosidase activity. These assays are easy to apply, but they face some limitations, such as insufficient accuracy in detection of low activity levels, e.g., of fucosidase.

Alternatively, HPLC has been used to monitor fucosidase activity levels (De Noni 2006). Since most glycosidases are associated with the cellular compartments, freezing will probably release glycosidases, which needs to be considered in relation to monitoring activities (Pyörälä 2003). Besides analysis of enzyme activities, monitoring of glycosidase substrates (glycoconjugates and oligosaccharides) and cleavage products (monosaccharides) are commonly used, mostly in combination with activity assays (Wiederschain and Newburg 2001a; Gnoth et al. 2000).

To our knowledge, glycosidases have not been identified by proteomic studies of milk. The precision and sensitivity of recent mass spectrometry instruments should, however, be capable of detecting and quantifying these enzymes, even at low abundances in the complex milk matrix. One mass spectrometry approach suitable for such purpose could be MRM experiments. This mode of operation exploits the unique capabilities of triple quadrupole (QQQ) MS instruments. Because only specified transitions are recorded and no full spectra are obtained, this technique has high sensitivity and repeatability, and would therefore be ideal for quantification of low-abundance proteins in complex mixtures.

Structural elucidation of monosaccharide sequences can be conducted with application of cleavage specific exoglycosidases, often in combination with LC-MS. Hydrolysis by α-2,3-neuraminidase and α-2,6-neuraminidase can, e.g., be used to confirm the presence of positional isomers containing either NeuAc-α-2,3-Gal or NeuAc-α-2,6-Gal in the non-reducing termini (Table 7.3), or to study the functional properties of specific isomers of glycoconjugates like e.g. κ-casein (Sunds et al. 2019).

7.7 Concluding Remarks

Glycosidases play an important role in the lysosomal catabolism of milk sugars and bound glycans, as a part of the normal turnover. These glycosidases are represented in milk and is, for some types, elevated at increased somatic cell count or mastitis. In particular, NAGase and β-glucuronidase are associated with mammary infection. Other glycosidases in milk such as α-fucosidase and neuraminidase are thought to be important contributors in the digestion of oligosaccharides by neonates until colonization of a healthy gut microflora, as fucose and sialic acid are important for microbiome colonization. Thus, glycosidases seem to be present in the milk for various biological reasons and from different origins. Overall the importance of glycosidases on cleavage of valuable oligosaccharides and milk sugars as ingredients is still not fully elucidated and requires further investigation.

References

Aldredge DL, Geronimo MR, Hua S, Nwosu CC, Lebrilla CB, Barile D (2013) Annotation and structural elucidation of bovine milk oligosaccharides and determination of novel fucosylated structures. Glycobiology 23(6):664–676

Alipour MJ, Jalanka J, Pessa-Morikawa T, Kokkonen T, Satokari R, Hynonen U, Iivanainen A, Niku M (2018) The composition of the perinatal intestinal microbiota in cattle. Sci Rep 8(1):10437

Andrews AT, Anderson M, Goodenough PW (1987) A study of the heat stabilities of a number of indigenous milk enzymes. J Dairy Res 54:237–246

Ansari SA, Satar R (2012) Recombinant β-galactosidases – past, present and future: a mini review. J Mol Catal B Enzym 81:1–6

Auldist MJ, Hubble IB (1998) Effects of mastitis on raw milk and dairy products. Aust J Dairy Technol 53:28–36

Boehm G, Stahl B (2007) Oligosaccharides from Milk. J Nutr 137:847–849

Burvenich C, Guidry AJ, Pape MJ (1995) Natural defence mechanisms of the lactating and dry mammary gland. In: Proceedings of the Third IDF International Mastitis Seminar, pp 3–13

Cao X, Yang M, Yang N, Liang X, Tao D, Liu B, Wu J, Yue X (2019) Characterization and comparison of whey N-glycoproteomes from human and bovine colostrum and mature milk. Food Chem 276:266–273

Chagunda MG, Larsen T, Bjerring M, Ingvartsen KL (2006) L-lactate dehydrogenase and N-acetyl-beta-D-glucosaminidase activities in bovine milk as indicators of non-specific mastitis. J Dairy Res 73(4):431–440

Chan LE, Beverly RL, Dallas DC (2019) The enzymology of human milk. In: Dairy enzymology. Springer, Berlin

Contesini FJ, de Alencar FJ, Kawaguti HY, de Barros Fernandes PC, de Oliveira CP, da Graca Nascimento M, Sato HH (2013) Potential applications of carbohydrases immobilization in the food industry. Int J Mol Sci 14(1):1335–1369

Coppa GV, Pierani P, Zampini L, Carloni I, Carlucci A, Gabrielli O (1999) Oligosaccharides in human milk during different phases of lactation. Acta Paediatr 430:89–94

De Noni I (2006) Study on the variability of fucosidase activity in bovine milk by means of HPLC. Int Dairy J 16(1):9–17

Dekker P, Koenders D, Bruins M (2019) Lactose-free dairy products: market developments, production, nutrition and health benefits. Nutrients 11(3):551

Djabri B, Bareille N, Beaudeau F, Seegers H (2002) Quarter milk somatic cell count in infected dairy cows: a meta-analysis. Vet Res 33(4):335–357

Fox PF, Kelly AL (2006) Indigenous enzymes in milk: overview and historical aspects—part 2. Int Dairy J 16(6):517–532

Gnoth MJ, Kunz C, Kinne-Saffran E, Rudloff S (2000) Human milk oligosaccharides are minimally digested in vitro. J Nutr 130:3014–3020

Griffiths MW (1986) Use of milk enzymes as indices of heat treatment. J Food Prot 49:696–705

Hovinen M, Simojoki H, Poso R, Suolaniemi J, Kalmus P, Suojala L, Pyorala S (2016) N-acetyl beta-D-glucosaminidase activity in cow milk as an indicator of mastitis. J Dairy Res 83(2):219–227

Huang J, Kailemia MJ, Goonatilleke E, Parker EA, Hong Q, Sabia R, Smilowitz JT, German JB, Lebrilla CB (2017) Quantitation of human milk proteins and their glycoforms using multiple reaction monitoring (MRM). Anal Bioanal Chem 409(2):589–606

Hurley WL (1987) Assays and activities of glycosidic enzymes in bovine peripheral blood leucocytes. Vet Immunol Immunopathol 16:85–93

Jansson T, Clausen MR, Sundekilde UK, Eggers N, Nyegaard S, Larsen LB, Ray C, Sundgren A, Andersen HJ, Bertram HC (2014) Lactose-hydrolyzed milk is more prone to chemical changes during storage than conventional ultra-high-temperature (UHT) milk. J Agric Food Chem 62(31):7886–7896

Jensen S, Jansson T, Eggers N, Clausen MR, Larsen LB, Jensen HB, Ray C, Sundgren A, Andersen HJ, Bertram HC (2015) Storage-induced changes in the sensory characteristics and volatiles of conventional and lactose-hydrolyzed UHT processed milk. Eur Food Res Technol 240(6):1247–1257

Józwik A, Bagnicka E, Slivá-Józwik A, Strzalkowska N, Sloniewski K, Krzyzewski J, Kolataj A (2004) Activity of selected glycosidases of whole milk in cows as related to feeding season. Anim Sci Paper Rep 22:673–677

Józwik A, Strzalkowska N, Lipinska P, Markiewicz-Kęszycka M, Łysek-Gładysińska M, Wróblewska B, Stanisławska I (2016) The effect of breed and the feeding system on the activity of glycosidases in cow's milk. Anim Sci Paper Rep 34:41–52

Karav S, Le Parc A, de Moura Bell JMLN, Frese SA, Kirmiz N, Block DE, Barile D, Mills DA (2016) Oligosaccharides released from milk glycoproteins are selective growth substrates for infant-associated bifidobacteria. Appl Environ Microbiol 82(12):3622–3630

Kawasaki Y, Isoda H, Shinmoto H, Tanimoto M, Dosako SI, Idota T, Nakajima I (1993) Inhibition by κ-casein glycomacropeptide and lactoferrin of influenza virus hemagglutination. Biosci Biotechnol Biochem 57(7):1214–1215

Kitchen BJ (1976) Enzymatic methods for estimation of the somatic cell count in bovine milk. I. Development of assay techniques and a study of their usefulness in evaluating the somatic cell content of milk. J Dairy Res 43:251–258

Kitchen BJ, Mitleton G, Salmon M (1978) Bovine milk N-acetylb-D-glucosaminidase and its significance in the detection of abnormal udder secretions. J Dairy Res 45:15–20

Kobata A (2013) Exo- and endoglycosidases revisited. Proc Jpn Acad Ser B 89(3):97–117

Kreuß M, Strixner T, Kulozik U (2009) The effect of glycosylation on the interfacial properties of bovine caseinomacropeptide. Food Hydrocoll 23(7):1818–1826

Kaartinen L, Kuosa PL, Veijalainen K, Sandholm M (1988) Compartmentalization of milk N-acetyl-β-D-glucosaminidase (NAGase). J Vet Med 35:408–414

Larsen T, Aulrich K (2012) Optimizing the fluorometric beta-glucuronidase assay in ruminant milk for a more precise determination of mastitis. J Dairy Res 79(1):7–15

Le Maréchal C, Thiéry R, Vautor E, Le Loir Y (2011) Mastitis impact on technological properties of milk and quality of milk products—a review. Dairy Sci Technol 91(3):247–282

Leonardi M, Gerbault P, Thomas MG, Burger J (2012) The evolution of lactase persistence in Europe. A synthesis of archaeological and genetic evidence. Int Dairy J 22(2):88–97

Manso MA, López-Fandiño R (2004) κ-casein macropeptides from cheese whey: physicochemical, biological, nutritional, and technological features for possible uses. Food Rev Intl 20(4):329–355

McJarrow P, van Amelsfort-Schoonbeek J (2004) Bovine sialyl oligosaccharides: seasonal variations in their concentrations in milk, and a comparison of the colostrums of Jersey and Friesian cows. Int Dairy J 14(7):571–579

McKellar RC, Piyasena P (2000) Predictive modelling of inactivation of bovine milk α-L-fucosidase in a high-temperature short-time pasteurizer. Int Dairy J 10:1–6

Mellors A (1968) b-N-acetylglucosaminase in bovine milk. Can J Biochem 46:451–455

Mellors A, Harwalkar VR (1968) Glycosidases in bovine milk: α-mannosidase and its inhibition by zwitterions. Can J Biochem 46:1351–1356

Mittal SB, Newell G, Hourigan JA, Zadow JG (1991) The effect of protease contamination in lactase on the flavor of lactose-hydrolyzed milks. Aust J Dairy Technol 46:46–47

Naranjo GB, Pereyra Gonzales AS, Leiva GE, Malec LS (2013) The kinetics of Maillard reaction in lactose-hydrolysed milk powder and related systems containing carbohydrate mixtures. Food Chem 141(4):3790–3795

Newburg DS, Chaturvedi P (1992) Neutral glycolipids of human and bovine milk. Lipids 27:923–927

Nielsen SD, Jansson T, Le TT, Jensen S, Eggers N, Rauh V, Sundekilde UK, Sørensen J, Andersen HJ, Bertram HC, Larsen LB (2017) Correlation between sensory properties and peptides derived from hydrolysed-lactose UHT milk during storage. Int Dairy J 68:23–31

Nielsen SD, Zhao D, Le TT, Rauh V, Sørensen J, Andersen HJ, Larsen LB (2018) Proteolytic side-activity of lactase preparations. Int Dairy J 78:159–168

Nwosu CC, Aldredge DL, Lee H, Lerno LA, Zivkovic AM, German JB, Lebrilla CB (2012) Comparison of the human and bovine milk N-glycome via high-performance microfluidic chip liquid chromatography and tandem mass spectrometry. J Proteome Res 11(5):2912–2924

Nwosu CC, Strum JS, An HJ, Lebrilla CB (2010) Enhanced detection and identification of glycopeptides in negative ion mode mass spectrometry. Anal Chem 82:9654–9662

O'Riordan N, Kane M, Joshi L, Hickey RM (2014) Glycosidase activities in bovine milk over lactation. Int Dairy J 35(2):116–121

Oliszewski R, Núnes de Kairús MS, González de Elias SN, Oliver G (2002) Assessment of b-glucuronidase levels in goat's milk as an indicator of mastitis: comparison with other mastitis detection methods. J Food Prot 65:864–866

Parker P, Sando L, Pearson R, Kongsuwan K, Tellam RL, Smith S (2010) Bovine Muc1 inhibits binding of enteric bacteria to Caco-2 cells. Glycoconj J 27(1):89–97

Perdigon G, Medici M, Cecilia M, Nader de Macias ME, Haedo R, Oliver G, Peche de Ruiz Holgado AA (1986) Significance of the presence of bovine milk β-glucuronidase in mastitis detection. J Dairy Sci 69:27–31

Pyörälä S (2003) Indicators of inflammation in the diagnosis of mastitis. Vet Res 34(5):565–578

Robinson RC, Poulsen NA, Colet E, Duchene C, Larsen LB, Barile D (2019) Profiling of aminoxyTMT-labeled bovine milk oligosaccharides reveals substantial variation in oligosaccharide abundance between dairy cattle breeds. Sci Rep 9(1):5465

Rombaut R, Dewettinck K (2006) Properties, analysis and purification of milk polar lipids. Int Dairy J 16(11):1362–1373

Saito T, Itoh T (1992) Variations and distribution of O-glycosidically linked sugar chains in bovine κ-casein. J Dairy Sci 75:1768–1774

Sánchez-Juanes F, Alonso JM, Zancada L, Hueso P (2009) Distribution and fatty acid content of phospholipids from bovine milk and bovine milk fat globule membranes. Int Dairy J 19(5):273–278

Sela DA (2011) Bifidobacterial utilization of human milk oligosaccharides. Int J Food Microbiol 149:58–64

Sellinger OZ, Beaufay H, Jacques P, Doyen A, de Duve C (1960) Tissue fractionation studies. 15. Intracellular distribution and properties of b-N-acetylglucosaminidase and b-galactosidase in rat liver. Biochem J 74:450–456

Simon PM, Goode PL, Mobasseri A, Zopf D (1997) Inhibition of helicobacter pylori binding to gastrointestinal epithelial cells by sialic acid-containing oligosaccharides. Infect Immun 65:750–757

Spik G, Coddeville B, Mazurier J, Bourne Y, Cambillaut C (1994) Primary and three-dimensional structure of lactotransferrin (lactoferrin) glycans. Adv Exp Med Biol 357:21–32

Sriwilaijaroen N, Kondo S, Yagi H, Hiramatsu H, Nakakita S, Yamada K, Ito H, Hirabayashi J, Narimatsu H, Kato K, Suzuki Y (2012) Bovine Milk whey for preparation of natural N-glycans: structural and quantitative analysis. Open Glycosci 5:41–50

Sundekilde UK, Frederiksen PD, Clausen MR, Larsen LB, Bertram HC (2011) Relationship between the metabolite profile and technological properties of bovine milk from two dairy breeds elucidated by NMR-based metabolomics. J Agric Food Chem 59(13):7360–7367

Sunds AV, Poulsen NA, Larsen LB (2019) Short communication: application of proteomics for characterization of caseinomacropeptide isoforms before and after desialidation. J Dairy Sci 102(10):8696–8703. Accepted paper

Svanborg S, Johansen A-G, Abrahamsen RK, Schüller RB, Skeie SB (2016) Caseinomacropeptide influences the functional properties of a whey protein concentrate. Int Dairy J 60:14–23

Saad AM, Östensson K (1990) Flow cytometric studies on the alteration of leucocyte populations in blood and milk during endotoxin-induced mastitis in cows. Am J Vet Res 51:1603–1607

Takimori S, Shimaoka H, Furukawa J, Yamashita T, Amano M, Fujitani N, Takegawa Y, Hammarstrom L, Kacskovics I, Shinohara Y, Nishimura S (2011) Alteration of the N-glycome of bovine milk glycoproteins during early lactation. FEBS J 278(19):3769–3781

Tao N, DePeters EJ, Freeman S, German JB, Grimm R, Lebrilla CB (2008) Bovine milk glycome. J Dairy Sci 91(10):3768–3778

Troise AD, Bandini E, De Donno R, Meijer G, Trezzi M, Fogliano V (2016) The quality of low lactose milk is affected by the side proteolytic activity of the lactase used in the production process. Food Res Int 89:514–525

Urakami H, Saeki M, Watanabe Y, Kawamura R, Nishizawa S, Suzuki Y, Watanabe A, Ajisaka K (2018) Isolation and assessment of acidic and neutral oligosaccharides from goat milk and bovine colostrum for use as ingredients of infant formulae. Int Dairy J 83:1–9

Urashima T, Taufik E, Fukuda K, Asakuma S (2013) Recent advances in studies on milk oligosaccharides of cows and other domestic farm animals. Biosci Biotechnol Biochem 77(3):455–466

van Leeuwen SS, Schoemaker RJW, Timmer CJAM, Kamerling JP, Dijkhuizen L (2012) N- and O-glycosylation of a commercial bovine whey protein product. J Agric Food Chem 60(51):12553–12564

Von Figura K, Hasilik A (1986) Lysosomal enzymes and their receptors. Annu Rev Biochem 55:167–193

Wang B, Brand-Miller J, McVeagh P, Petocz P (2001) Concentration and distribution of sialic acid in human milk and infant formulas. Am J Clin Nutr 74:510–515

Wickramasinghe S, Hua S, Rincon G, Islas-Trejo A, German JB, Lebrilla CB, Medrano JF (2011) Transcriptome profiling of bovine milk oligosaccharide metabolism genes using RNA-sequencing. PLoS One 6(4):e18895

Wiederschain GY, Newburg DS (2001a) Glycoconjugate stability in human milk: glycosidase activities and sugar release. J Nutr Biochem 12:559–564

Wiederschain GY, Newburg DS (2001b) Glycosidase activities and sugar release in human milk. Adv Exp Med Biol 501:573–577

Yuan L, Sadiq FA, Liu TJ, Li Y, Gu JS, Yang HY, He GQ (2018) Spoilage potential of psychrotrophic bacteria isolated from raw milk and the thermo-stability of their enzymes. J Zhejiang Univ Sci B 19(8):630–642

Chapter 8
The Enzymology of Non-bovine Milk

Marzia Albenzio, Antonella Santillo, and Golfo Moatsou

8.1 Introduction

The indigenous enzymes are important milk components that have been studied since 1881, when the peroxidase isolated from milk was given the name lactoperoxidase and was the first enzyme reported to be found in milk (Arnold 1881). Over the years since, the number of enzymes identified and characterized gradually increased, and about 70 indigenous enzymes are reported in normal bovine milk (Fox and Kelly 2006). Although milk from species other than bovine shares wide homology in terms of number and function of enzyme systems, specific studies in small ruminants, buffalo, camel, horse and donkey milk are limited compared to bovine milk; however, this field of research has gained increasing importance through the years, and will be the subject of this chapter.

M. Albenzio (✉) · A. Santillo
Department of Sciences of Agriculture, Food and Environment, University of Foggia, Foggia, Italy
e-mail: marzia.albenzio@unifg.it; antonella.santillo@unifg.it

G. Moatsou
Laboratory of Dairy Research, Department of Food Science and Human Nutrition, Agricultural University of Athens, Athens, Greece
e-mail: mg@aua.gr

A. L. Kelly, L. B. Larsen (eds.), *Agents of Change*, Food Engineering Series,
https://doi.org/10.1007/978-3-030-55482-8_8

8.2 Enzymology of Ovine and Caprine Milk

8.2.1 Proteolytic Enzyme Activities

Ovine and caprine milk are mainly used for cheesemaking, and it is noteworthy that both milk protein and indigenous enzymes impact directly on the ability of milk to be processed into cheese and on the quality of dairy products. The increasing demand for ovine and caprine dairy products, therefore, necessitates increasing knowledge on the role of indigenous proteolytic and lipolytic enzymes on the cheesemaking suitability of milk and on their effects on ripening process.

Several indigenous proteolytic enzymes are associated with somatic cells (SC), which include polymorphonuclear leucocytes (PMN), macrophages, lymphocytes and perhaps mammary epithelial cells. Numerous enzymes are associated with somatic cells in milk: one of the plasminogen activators comprising the complex enzymatic plasmin-plasminogen system is associated with somatic cells. Elastase, cathepsins, and collagenases may be released by the lysosomes of somatic cells (Kelly and McSweeney 2002); alkaline phosphatase originates from myoepithelial cells of the mammary gland (Bingham et al. 1992). Two acid phosphatases are located in skim milk and originate from leucocytes (Andrews and Alichanidis 1975), while catalase and N-acetyl-b-D-glucosaminidase originate from somatic cells and from mammary gland epithelial cells (Kitchen et al. 1970; Kaartinen and Jensen 1988).

Moatsou (2010) reviewed the indigenous enzymatic activities in ovine and caprine milk, focusing on the principal enzymes found in these milks in comparison with bovine milk: plasmin (PL) and plasminogen (PG), cathepsin D, elastase, lipoproteinlipase (LPL), alkaline phosphatase (ALP), acid phosphatase (ACP), ribonuclease (RNase), lactoperoxidase (LPO), catalase, superoxide dismutase (SOD), xanthine oxidoreductase (XOR), γ-glutamyl trasferase (g-GGT), N-acetyl-b-Dglucosaminidase (NAGase), lysozyme and glutathione peroxidase (GPx).

Plasmin is the predominant proteolytic enzyme in all investigated mammalian milks and is also present at a significant level in ovine milk (Albenzio et al. 2005a; Bianchi et al. 2004; Caroprese et al. 2007; Leitner et al. 2004a, b). In caprine milk, the PG:PL ratio, representing the rate of zymogen conversion, is much lower than that observed in other ruminant species, and total PL + PG-derived activities are in general also lower (Moatsou 2010).

Due to the relevance of plasmin activity for protein hydrolysis and milk and dairy product quality, there have been a number of studies of the plasmin system of ovine and caprine milk. As shown in Table 8.1, several studies have shown that the plasmin-plasminogen system varies with SCC, as well as stage of lactation in ovine and caprine milk (Albenzio et al. 2004a, b, 2005a, 2006, 2009a, b, 2012, 2015, 2016; Albenzio and Santillo 2011; Caroprese et al. 2007; Santillo et al. 2009).

Some authors have reported effects of SCC on the activity of the plasmin system in ovine milk, with inconsistent results. Bianchi et al. (2004) observed that high SCC is not correlated with plasminogen-derived plasmin activity, and proposed that

Table 8.1 Indigenous proteolytic enzymes in ovine and caprine milk

Lactating species		Plasmin, U/mL	Cathepsin D, U/mL	Elastase, U/mL	References
Ovine	Stage of lactation				
	Early (<70 DIM)	42.05			Caroprese et al. (2007)
	Mid from (110 < DIM < 130)	31.29			
	Late (>160DIM)	28.19			
Caprine	Early (43 ± 5 DIM)	7.58			Albenzio et al. (2016)
	Mid (120 ± 10 DIM)	10.30			
	Late (205 ± 12 DIM)	18.10			
Ovine	Level of SCC				
	<300,000	6.46	4.98	0.15	Albenzio et al. (2012)
	301,000 < SCC < 500,000	8.42	5.06	0.14	
	501,000 < SCC < 1,000,000	10.44	5.13	0.13	
	1,001,000 < SCC < 2,000,000	11.71	5.61	0.14	
	>2,001,000	10.72	5.01	0.15	
Caprine	<700,000	7.98	0.42	0.10	Albenzio et al. (2015)
	700,000 < SCC < 1,500,000	11.33	0.84	0.19	
	>1,500,000	19.07	0.70	0.14	

the increase in plasminogen level is due to increased quantities of this pro-enzyme by-passing the blood-milk barrier. The same authors proposed that this mechanism of increased plasmin activity in high SCC ovine milk is similar to that found in bovine milk in the final phase of lactation, but opposite from that occurring in mastitic cows, where plasmin and plasminogen both increase with SCC (Politis et al. 1989, 1991). Contrasting results obtained by Leitner et al. (2004a), Rebucci et al. (2005), Theodorou et al. (2007) could be explained by overlapping effect of lactation stage and level of somatic cells that lead to unpredictable behaviour of PL and PG (Albenzio et al. 2004a, b, 2005a).

Albenzio et al. (2004a, b, 2005a) suggested that, in ovine milk, plasminogen activation is related to SCC and that plasmin activity probably depends on plasminogen activators released by macrophages. Indeed, Caroprese et al. (2007) reported that, in bulk ovine milk from healthy flocks, the contribution of macrophage-derived urokinase-type plasminogen activators to the total plasmin activity in milk is minimal. Albenzio and Santillo (2011) reported that plasmin activity was highest in ovine milk samples with SCC higher than 2,000,000 cells/mL, and lowest in milk samples with SCC lower than 300,000 cells/mL. A similar result was reported in caprine milk, where the same authors reported that plasmin activity increased with increasing SCC (from <700,000 cell/mL to >1,500,000 cell/mL), and suggested 700,000 cell/mL as a threshold value to discriminate the health status of the ovine mammary gland (Albenzio et al. 2015).

The study of changes in indigenous proteolytic enzymes in ovine milk during lactation conducted by Albenzio et al. (2009b) suggested that the plasmin and plasminogen activity and PL:PG ratio do not vary significantly throughout lactation;

however, an increase in the content of these enzymes with advancing lactation could be explained by the gradual involution of the mammary gland with all the enzymatic system components passing the disrupted cells *via* para-cellular routes (Caroprese et al. 2007).

In caprine milk, as lactation progresses, plasmin activity significantly increases (Albenzio et al. 2016). In addition, the health of mammary gland in combination with lactation stage magnifies the effects on the PL system at increased SCC in late lactation. In a study conducted by Albenzio et al. (2012), plasmin activity increased with increasing SCC and was always higher in ovine milk samples containing pathogenic bacteria. Udder health also affects the activities of the PL system in caprine milk. Leitner et al. (2004b) reported that PL activity from infected udders was almost double that than found in milk with low SCC, with PA activity being 25% higher (Leitner et al. 2004b). Leitner et al. (2004b) concluded that the PL system works at a lower rate in goats than in sheep and concluded that the enzymology of caprine milk was less responsive to external factors such as infection.

Theodorou et al. (2007) compared the effect of breed on the activities of PL, PG, and PA in milk of sheep of different breeds and, between the tested breeds, PL and PA activities were similar, while differences were detected in the PG:PL ratio, which was attributed to differences in the influx of PG or PL from the blood. Indeed, compared to bovine milk, the PG:PL ratio reported for the milk of various sheep breeds is in general low (from 0.80 to 5.77); for this reason, Theodorou et al. (2007) suggest that PG is converted to PL more efficiently in ovine than in bovine milk.

Housing conditions can affect milk PL activity, as studied by Albenzio et al. (2007); these authors reported that sheep subjected to low ventilation rate had higher PG level, probably due to the negative effect of poor air quality on udder health that results in increased leakage of blood proteins into milk.

Less information is available about the activity of lysosomal proteases, such as cathepsin D and elastase, in ovine and caprine milk in comparison with bovine milk. Cathepsin D and elastase, although less intensely investigated than plasmin, may contribute to the breakdown of casein, impairing coagulation properties of milk before cheese-making (Santillo et al. 2009). Total cathepsin D and elastase concentrations were measured by Santillo et al. (2009) to be 1.14 and 1.81 mg/L, respectively, in milk Garganica goats. Albenzio et al. (2009a, b) found a lower concentration of elastase (0.41 mg/L) and a higher concentration of cathepsin D (2.56 mg/L) in ovine bulk milk, respectively, than those found in caprine milk.

Cathepsin D and elastase activities in ovine milk vary throughout lactation, but their changes do not parallel each other (Albenzio et al. 2009a, b); indeed, elastase activity increases during lactation, whereas cathepsin D activity is highest at mid-lactation. It is known that, in ovine milk, cathepsin D is associated with PMN leucocytes and macrophages, whereas elastase is mainly associated with PMN (Albenzio and Santillo 2011). In caprine milk, the positive correlation between cathepsin D activity and PMN level suggests that PMN may be a main source of this protease in goat milk. Albenzio et al. (2015) reported the variation of cathepsin D and elastase activity during lactation in caprine milk, and the high relative activity of these enzymes observed in medium-SCC milk (from 701,000 to 1,500,000 cells/

mL) could reflect the first activation of the immune response at mammary gland level, consistent with the recruitment of PMN through macrophages, that act as a fast, but nonspecific, defense. In caprine milk with SCC >1,501,000 cells/mL, the same authors explained the minor contribution to the overall indigenous proteolytic activities of cathepsin D and elastase through the activation of specific immune mechanisms that represent the second line of the immune response at mammary gland level. Furthermore, the clotting time of ovine milk is positively correlated both with SCC and with cathepsin D activity (Albenzio et al. 2015). On the other hand, the increase in proportion of PMN parallels elastase content in ovine milk, suggesting that elastase concentration could be a reliable indicator of mammary gland involution in healthy sheep udders (Albenzio et al. 2009b).

8.2.2 Activities of Other Enzymes

Lipoprotein lipase (LPL) is another key enzyme in dairy technology, because it represents almost all lipolytic activity in raw milk. LPL is a glycoprotein which is electrostatically and hydrophobically associated with the casein micelles in bovine milk, and active only at the oil-water interface. In raw bovine milk, about the 80% of lipoprotein lipase is associated with the casein micelles. Chandan et al. (1968) reported that lipase activity in ovine milk is about one-tenth of that in bovine milk. The hydrolysis pattern of ovine milk fat due to ovine LPL is similar to that reported for bovine LPL, with preferential hydrolysis of medium-chain fatty acids over long-chained. In parallel with bovine milk, pasteurization of ovine milk LPL causes 73–95% inactivation of enzyme activity (Chavarri et al. 1998).

In line with the above, Chilliard et al. (2003) reported that the LPL activity also is lower in caprine than in bovine milk, although compared with bovine milk more LPL is bound to the milk fat globules in caprine milk, and the activity is better correlated with spontaneous lipolysis. Two distinct phenomena characterize the hydrolysis of milk triglycerides: lipolysis of the milk fat substrate and on the other hand the LPL activity itself. The first generally takes place at 4 °C, at natural milk pH, in the presence of biochemical factors naturally occurring in milk, and the result of it is generally lower than 1.0 mmol released FA/24 h/L in caprine milk (Chilliard 1982). On the other hand, milk LPL activity is measured as maximal potential activity on an artificial lipid emulsion, at 37 or 39 °C, at alkaline pH, in the presence of apoprotein activators from blood serum, and is generally higher than 20 μmol/h/mL of caprine milk (Chilliard 1982). Thus, potential LPL activity is about 500 times higher than spontaneous lipolysis in caprine milk (Chilliard et al. 2003).

Despite lipase activity in caprine milk being about one-third of the activity of bovine milk, caprine milk has a greater tendency to undergo spontaneous lipolysis, which contributes to the characteristic caprine milk flavor (Chandan et al. 1968). In bovine milk, spontaneous lipolysis is not correlated with LPL activity, and lipolysis remains generally low despite the very high LPL activity of this milk, because of hindered access to the substrate by the milk fat globule due to protection by the

MFGM, and maybe also due to presence of some inhibitors in bovine milk (Cartier et al. 1990). Moreover, bovine milk LPL is largely bound to casein micelles, thus diminishing enzyme-fat substrate interactions, except when milk is homogenized (and not pasteurized). Furthermore, in caprine milk, a substantial proportion of the LPL is bound to the cream layer, which could explain the higher level of lipolysis due to LPL in caprine milk.

The soluble nitrogen fraction derived from hydrolysis of β-casein, the proteose peptone fraction, was found to inhibit both spontaneous and induced lipolysis of caprine milk, as similarly observed for bovine milk (Chilliard et al. 2003). The characteristic flavor of caprine milk is related to caprine α_{s1}-casein alleles. Both spontaneous lipolysis and LPL activity are linked to α_{s1}-casein genotype; both lipolytic events are higher in milk from goats with the weak expression of the α_{s1}-casein genotype than in milk from goats with the more strongly expressing genotype. In animals with the weak α_{s1}-casein genotype, the fat content is also lower and furthermore the caprine flavor was found to be stronger (Delacroix-Buchet et al. 1996). This was explained by Chilliard et al. (2003), who reported that lipolysis increased with advancing lactation, and that this was more pronounced in caprine milk with the weak a_{s1}-casein genotypes containing "weak" variants E, F or O.

It is known that heat induces inactivation of indigenous enzymes in milk and this has been well studied in bovine milk. Alkaline phosphatase (ALP) is used as a means of monitoring the accuracy of pasteurization or the addition of raw milk to pasteurized milk (Shakeel-Ur-Rehman et al. 2003; Harding and Garry, 2005; Commission Regulation (EC) No. 1664/2006) (see also Chap. 4). γ-glutamyl transferase (GGT), is more heat-stable than ALP and it may be useful for testing milk heated between 75 and 80 °C (Farkye 2003). Lactoperoxidase (LPO) is one of the most heat-stable enzymes in bovine milk (Seifu et al. 2005) and LPO inactivation has been proposed as an index for monitoring thermal treatments above 78 °C for 15 s (Griffiths 1986).

The effects of high-pressure treatment applied at various temperatures on the activities of some indigenous proteases such as plasmin, plasminogen and cathepsin D in ovine milk were determined by Moatsou et al. (2008). Those authors concluded that high-pressure treatment decreased PL and PA activities in ovine milk and did not affect their distribution between the casein and milk serum fractions.

Raynal-Ljutovac et al. (2007) suggested that values for ALP activity in raw milk are not related to the fat content of milk, but are species-specific. This was explained by the same authors showing that ovine milk contains much more ALP activity than bovine milk, (8000–17,000 vs 1800–4800 μg phenol/mL). ALP activity of ovine milk was negatively correlated with daily milk yield and positively correlated with milk fat content, and its activity was higher in mastitic milk (Anifantakis and Rosakis 1983). The same authors found highest alkaline phosphatase activity at mid-lactation in Greek flocks, while Martini et al. (2013) reported that the maximum activity was detected within 60–120 days of the start of lactation, which was 40 times the activity at day one of lactation.

Caprine milk contains about 2–5 times lower ALP activity than bovine milk, and its activity ranges from 120 to 1300 μg phenol/mL (Merin et al. 1988; Coburn et al.

1992; Roseiro de Bivar and Barbosa 1995; Assis et al. 2000; Raynal-Ljutovac et al. 2007; Sharma et al. 2009). Ying et al. (2002), studying milk from Alpine goats, measured 128 U/L ALP activity in early lactation and 236 U/L in late lactation, and reported that ALP activity in caprine milk is negatively correlated with log SCC. Studies comparing raw milk from cow, sheep and goat breeds showed average ALP activities of 330, 2200 and 95 U/L, respectively (Assis et al. 2000; Barbosa 2005). Vamvakaki et al. (2006), analyzing milk from bovine, ovine and caprine raw milk samples in Greece, reported average ALP activities of 390, 2430 and 134 U/L, respectively. The respective average activities reported by Klotz et al. (2008) were 577, 1216 and 61 U/L or 1562, 3512 and 164 μg phenol/mL.

Medina et al. (1989) reported that LPO activity in individual morning and evening ovine milk samples at mid-lactation ranged from 0.14 to 2.38 U/mL, with a mean value of 0.77 U/mL; however, activity was not significantly correlated with milk production, but significantly varied between individual animals. For this reason, the differences in LPO activity among studies may be attributed to different ovine breeds, as suggested by Chavarri et al. (1998). LPO activity in bulk ovine milk of the Latxa breed has been found to be 4 U/mL on average at mid-lactation, which is about three times higher than that of bovine milk, to increase at mid-lactation, and to be at its lowest value at the end of lactation (Chavarri et al. 1998; Moatsou 2010). Lorenzen et al. (2010) determined that LPO activities were 2015, 2796 and 5190 U/L on average for bovine, ovine, and caprine raw milk, respectively. Values reported by Dumitraşcu et al. (2012) were 970, 1720 and 810 U/L, respectively.

Vamvakaki et al. (2006) reported that heat treatments at 59 °C at various holding times reduced the ALP activity in caprine milk to a lesser extent than in ovine milk. Wilińska et al. (2007) suggested higher stability for bovine milk ALP compared to the caprine enzyme within a temperature range of 54–69 °C. Lorenzen et al. (2010) compared the effects of isochrone (35–85 °C for 90 s) heating with different time-temperature combinations (LTLT- and HTST-heating) on residual activities of alkaline phosphatase (ALP), γ-glutamyltransferase (GGT) and lactoperoxidase (LPO) in bovine (Holstein Friesian Cow), ovine (East Friesian Dairy Sheep) and caprine (German Improved Fawn) milk, and concluded that the investigated enzymes were more heat-stable in ovine and caprine milk compared with bovine milk. Klotz et al. (2008) reported that pasteurization at 72.5 °C for 15 s decreased ALP activity from 61 to 0.223 U/L in caprine milk, from 1459 to 0.181 in ovine milk and from 744 to 0.085 U/L in bovine milk. Dumitraşcu et al. (2012) reported mean z-values of 3.38, 4.11 and 3.58 °C and activation energies of 678, 560 and 641 kJ/mol within the range of temperature from 70 to 77 °C for caprine, ovine and bovine milk LPO, respectively; they concluded that LPO in ovine milk was less heat-stable than its caprine or bovine counterparts.

Chavarri et al. (1998) studied ACP activity in bulk ovine milk from flocks of Latxa ewes and found values higher than those reported for bovine milk, which increased four-fold in the initial phase of lactation, and then remained constant till end of lactation. They also reported that pasteurization did not significantly inactivate the enzyme and that ACP was active at levels of 65–80% in skim milk and 30–34% in whey after renneting.

The activity of GGT in bovine skim milk was, on average, 72% of that found in whole milk, and only limited seasonal variation was noted in either whole or skim raw milk over a period of 300 days (McKellar et al. 1991). Anjos et al. (1998) found, on average, 2880 U/l GGT activity in raw bovine milk. Using two validated analytical procedures, Piga et al. (2009) found GGT activities of 2720–3460 U/l in raw milk of a Sardinian sheep breed (Moatsou 2010).

With regard to RNase activity in ovine and caprine milk, Chandan et al. (1968) found it was much lower than bovine milk, being in ovine milk about 1/4 and in caprine milk about 1/3 the activity in bovine milk. Furthermore, Liu and Williams (1982) report that the RNase activity in caprine milk was 1/10 the activity in bovine milk.

Benboubetra et al. (2004) purified and studied xanthine oxidoreductase (XOR) from ovine milk. The ovine enzyme is similar to the enzyme of bovine milk and was seen as a single band of 150 kDa in SDS-PAGE. The specific XOR activity of ovine milk is about 2.5 times lower than that in bovine milk and about 2.5 times higher than that of the caprine milk enzyme. The low XOR activity is related to a low molybdenum content of ovine milk XOR, which has been estimated as 0.18 atoms Mo per subunit, compared to 0.55 atoms Mo per subunit of the bovine milk enzyme (Moatsou 2010). Sharma et al. (2009) reported the XOR activity in bulk caprine to be 55% of that in bulk bovine milk, and 95% of this activity was found in the cream fraction. However, Atmani et al. (2004, 2005) reported lower activities of XOR in goat milk compared to bovine milk. Caprine milk XOR gave a single band on the SDS-PAGE gels at about 150 kDa, similarly to bovine and ovine XOR (Atmani et al. 2004; Zamora et al. 2009). Cebo et al. (2010) and Martini et al. (2013) studied caprine XOR; the oxidase activity of XOR of ovine milk changed during lactation, being maximum at day 45, and its changes were inversely proportional to the average size of milk fat globules. On the other hand, xanthine dehydrogenase activity of XOR was much lower and constant throughout lactation.

8.2.3 *Effect of Enzyme Activity on Milk and Dairy Product Quality*

Milk proteins and indigenous enzymes impact directly on the ability of milk to be processed and the quality of dairy products. The characteristics of casein in ovine and caprine milk are of particular interest, due to the high degree of variety that is related to cheese-making properties of milk. Ovine and caprine milk are mainly processed into cheese, which is in increasing demand, and therefore it is of importance in order to understand the role of indigenous proteolytic and lipolytic enzymes on milk quality and cheesemaking abilities, as well as their effects on the ripening process.

Proteolysis in milk occurs in the udder before milking (Saeman et al. 1988) and during milk storage after milking (Albenzio et al. 2004b, 2005b, 2009b; Caroprese et al. 2007). Albenzio et al. (2005a) and Santillo et al. (2009) studied indigenous proteolytic enzyme activities using natural substrates of ovine sodium caseinate and caprine milk, respectively, to investigate proteolytic activity in small ruminant milk during storage prior to cheese manufacture. In these studies, the authors concluded that the total proteolytic activity was largely dependent on plasmin activity. Several authors have suggested that, if ovine milk has an elevated SCC, its cheese-making properties will deteriorate, i.e., show longer coagulation time and weak coagulums, leading to poor syneresis and increased moisture content, lower cheese yield and lower overall fat contents in the cheeses (Albenzio et al. 2004b, 2005b; Bencini and Pulina 1997; Jaeggi et al. 2003; Pellegrini et al. 1997; Pirisi et al. 1996, 2000).

During cheese ripening, fresh curd is converted to one of the numerous cheese varieties with particular appearance, taste, flavor, texture, and functionality characteristics. All these aspects are involved in the definition of cheese quality and result from the balance and intensity of the ripening process in terms of proteolysis, lipolysis, and glycolysis.

Raw milk compositional and biochemical characteristics, milk technological treatment and renneting process, curd production, and ripening of cheese represent a multiplicity of factors that impact, directly or indirectly, on the quality of the manufactured cheese. Each factor in this complex system is affected by the activity of enzymes originating from milk (i.e., indigenous enzymes and microflora), coagulants, and starter and non-starter microflora.

Regarding indigenous proteolytic enzymes, it has been reported that plasmin acts primarily on β- and α_s-CN, leading to the formation of the polypeptides γ_1-, γ_2 and γ_3-CN and the resulting proteose-peptones, as well as α_s-CN degradation products (Albenzio et al. 2010). However, the hydrolysis of principal caseins proceeds differently along with ripening time, with α_s-CN products being released within the first phase of ripening, whereas the cleavage of β-CN proceeds more slowly as a consequence of the different specificity of proteolytic agents towards caseins (Albenzio et al. 2010; Irigoyen et al. 2000; Revilla et al. 2007).

Casein degradation and casein breakdown products represent an important index of cheese ripening and involve a complex pattern of events occurring during cheese ripening. However, all these parameters are influenced by several factors and, therefore, they are not easily directly correlated. Albenzio et al. (2004b) found that, in fresh ovine cheese curd, the decrease of PL and PG-derived activities in milk coincided with an increase in non-protein nitrogen (NPN), suggesting that NPN components derived from CN hydrolysis may be involved in the regulation of the plasmin-plasminogen enzyme system. Furthermore, as reported in previous studies (Albenzio et al. 2005b, 2010; Santillo and Albenzio 2008), changes in the formation of γ-CN mainly derived from plasmin activity were not correlated with the changes in PL activity in the cheese during ripening, indicating that factors other than the enzyme activities themselves are important for the resulting breakdown observed.

Furthermore, cheese-making technology is one of the main factors able to affect indigenous enzyme level and activity. In Canestrato Pugliese cheese made from ovine milk, the production protocol includes heating the curd in hot whey, which can promote syneresis and inactivate heat-labile inhibitors of PL and PG activators

(Albenzio et al. 2007). Examples for the effect of cheese-making conditions and cheese physicochemical composition on plasmin and PG-derived activity in various cheese varieties have been presented by Nega and Moatsou (2012).

Salting of fresh curd is a further step able to influence enzymatic activity in cheese during ripening; in fact, salting decreases the activity of most of the enzymes in cheese, including indigenous milk proteinases such as plasmin (Sutherland 2003). Fox (2003) reported that hydrolysis of β-CN is strongly inhibited by a NaCl content in cheese above 5%. In Canestrato Pugliese cheese from ovine milk, a salt concentration of about 4% was able to inhibit β-CN hydrolysis, as indicated by the limited accumulation of small peptides ascribed to γ-CNs over 45 d of ripening (Albenzio et al. 2004b, 2005b). Also, lipolytic activity appeared to be limited at NaCl levels >13 g/kg of cheese in Idiazabal cheese made from raw ovine milk using natural rennet (Nájera et al., 1994).

Flavor of cheese is determined by its taste and aroma and results from the correct balance and concentration of numerous sapid and aromatic compounds perceived during cheese consumption. Proteolysis and lipolysis are of great importance in the development of cheese flavor and are largely due to the activities of residual milk clotting enzyme, milk proteinase and lipase, proteolytic and lipolytic enzymes from starter and non-starter bacteria, and lipases associated with certain coagulants (Collins et al. 2003; Visser 1993). However, excessive proteolysis and lipolysis could result in off-flavors because of high concentrations of bitter peptides and volatile FFA, respectively, and influence cheese flavor either directly or as precursors of other compounds (Pinho et al. 2004). The contribution of indigenous proteinases to the development of cheese flavor has possible negative implications through the accumulation of bitter peptides, which are gradually formed by further degradation of β-CN compounds (Visser 1993).

Revilla et al. (2007) reported a lower overall acceptance of ovine hard cheese made with high SCC milk compared to milk of medium and low SCC, finding that the former was perceived as weakly bonded, very grainy, and crumbly. In contrast, Pirisi et al. (1996, 2000) did not find significant differences in sensory characteristics and lipolysis when comparing ovine cheeses made from milk with low and high SCC. Jaubert et al. (1996) and Morgan and Gaspard (1999) also found a minor effect of SCC of caprine milk on the 'goatish flavor' of cheese, which is instead mainly influenced by cheese-making technology, in particular by the ripening conditions.

Milk contains various components, mainly associated with protein and lipid molecules, that are recognized as biofunctional components. Peptides derived from milk proteins, which are inactive within the sequence of the precursor protein, can be released by enzymatic proteolysis, for instance during cheese ripening (Gómez-Ruiz et al. 2004). Santillo et al. (2009) investigated the effects of indigenous proteolytic enzymes on the release of bioactive peptides from caprine milk using different proteolytic enzyme inhibitors. The authors found that caprine milk incubated with or without inhibitors of acid proteases (e.g., pepstatin, which inhibits cathepsin D activity) showed several peptides with potential bioactivity deriving from β- and α_{s2}-CN, indicating that both aspartic proteases and other protease types were involved in the generation of bioactive peptides.

8.3 Enzymology of Buffalo Milk

In the last few decades, many enzyme activities have been detected in buffalo milk, some of which have been isolated and partially characterized (El-Salam and El-Shibiny 2011). Lombardi et al. (2000) established the levels of ALP, LDH, GGT, and AST in raw buffalo milk, highlighting higher concentrations of enzymes than in bovine milk, which could be related to the different composition of the milk, i.e., higher concentrations of protein, fat, and other solids than bovine milk. The activity of the following enzymes has been reported for buffalo milk: GGT (712 ± 601 IU/L), LDH (386 ± 183 IU/L), ALP (295 ± 164 IU/L) and AST (18 ± 4 IU/L); also, the patterns of heat inactivation of these enzymes suggested that ALP, LDH and GGT are potential markers for heat treatment of buffalo milk.

Concentrations and characteristics of lysozyme, lactoperoxidase, xanthine oxidase, lipase, alkaline phosphatase, ribonuclease, and protease activities in buffalo milk have been reviewed (Ahmad et al. 2013). Priyadarshini and Kansal (2002b) found the molecular weight of buffalo milk lysozyme to be 16 kDa, and determined its antibacterial activity, highlighting that buffalo colostrum contains five times more lysozyme activity than mature milk and higher specific activity than that of bovine milk. Levels of lysozyme in buffalo milk are, however, still controversial, probably due to the different methodologies used for enzyme determination. Kumari and Mathur (1981) reported lower concentration in buffalo milk in comparison to bovine milk (average 15.21 μg/100 mL and 18.00 μg/100 mL in buffalo and bovine milk, respectively). On the other hand, more recently, Priyadarshini and Kansal (2002a) reported a twofold higher lysozyme activity in buffalo milk ($60 \pm 3.9 \times 10^{-3}$ U/mL) than in bovine milk ($29.1 \pm 1.5 \times 10^{-3}$ U/mL), with the enzymatic activity in buffalo milk also being more stable compared with bovine during storage and heat treatment (Priyadarshini and Kansal 2003). Concerning the factors affecting lysozyme activity of buffalo milk, Kumari and Mathur (1981) reported an increase during lactation, while Priyadarshini and Kansal (2002a) found it to be influenced neither by parity nor stage of lactation; however, it increased during extreme weather conditions in winter and summer. The higher lysozyme activity in buffalo milk has been proposed as one of the factors responsible for lesser incidences of udder infections in buffaloes. Furthermore, lysozyme, though mainly from egg-white sources, has found application in food preservation, being already used successfully as an antimicrobial in many foods (Benkerroum 2008).

Lactoperoxidase is abundant in buffalo milk. Its activity has been reported to vary widely, possibly due to different assay methods used, and values from 0.9 U/mL (Hamulv and Kandasamy 1982) to 16.84 U/mL (Tayefi-Nasrabadi and Asdpour 2008) have been reported. Sharma et al. (2009) reported that lactoperoxidase was found free in buffalo milk and exhibited 1.76-fold higher activity (12.02 ± 0.83 U/mL) over that of bovine milk (6.83 U/mL).

The lipase activity of buffalo milk was found to increase gradually during the lactation period from 0.86 U/mL in the first month to 2.48 U/mL in sixth month of lactation (Hofi et al., 1976) The lipase activity was affected by feeding system, being higher for dry concentrate feeding than during green feeding, and it also decreased with increased number of lactations, although no differences have been reported between morning and afternoon milking (Abd El-Hamid et al. 1977). Feeding protected fat diets containing supplemental tallow had no significant effect on lipase activity of buffalo milk, compared to the control diet (Srivastava et al. 1989).

The activity of XOR measured in Indian buffalo breeds displayed average level of 0.075 ± 0.027 U/mL (Sanhotra and Dutta 1986). The major portion of its activity in buffalo milk was found in the skim milk fraction, with approximately 25% of the total activity located in the cream fraction (El-Gazzar et al. 1999). The same authors also found an increase of about 10% of XOR activity upon heating buffalo milk at 55 °C for 1 min, whereas heating at 70 °C for 1 min and 80 °C for 1 min caused a decrease in activity by 75% and 90%, respectively.

Only a limited number of studies have reported on indigenous proteinase systems and activities in buffalo milk; Manjunath and Bhat (1992) showed 55.0 and 70.2 μg/mL in buffalo and bovine milk of total indigenous proteinase, respectively. Buffalo milk contains the complete plasmin system, cathepsins and elastase, similar to milk of other ruminants, and the presence and activity of plasmin, plasminogen, activators, cathepsin D and elastase have been reported (Madkor and Fox 1991; Fantuz et al. 1998; Santillo et al. 2011). A study of indigenous proteolytic enzyme activity in Mediterranean buffalo milk highlighted, as expected, the main role of the plasmin-plasminogen system, although its activity seemed to be limited compared with milk of other species (Santillo et al. 2011). Moreover, similarly to other ruminants' milk, plasmin activity in buffalo milk was associated primarily with the casein micelles, and completely dissociated at pH <4.6 (Madkor and Fox 1991). Addition of urokinase increased the proteolytic activity by 3.5-fold in milk and fivefold in micellar suspension (Madkor and Fox 1991), while addition of NaCl (at 3–15%) decreased the plasmin activity, and this effect was even more pronounced at high NaCl concentrations (Madkor and Fox 1991).

Heating at 70 °C for 10 min increased the plasmin activity of buffalo milk, while 15% of its activity was lost after heating at 80 °C for 10 min. Most of the plasminogen activators (PA) were associated with casein micelles (Fantuz et al. 1998). Several PAs were identified in buffalo milk, i.e., major fractions with molecular masses of 75 and 120 kDa and two minor fractions of 48 and 38 kDa (Fantuz et al. 1998). The proteins at 38, 75, and 120 kDa were tissue-type PAs, and the 48 kDa fraction was urokinase-type PA (Fantuz et al. 1998).

River buffalo milk is characterized by a peculiar γ-CN defined as β-CN f(69–209) which has no counterpart in bovine milk. This fragment was previously known as "component D" by Trieu-Cuot and Addeo (1981) and was named "protein X" by Angeletti et al. (1998). It was later identified by Di Luccia et al. (2009), who also demonstrated that the formation of this fragment formed earlier with respect to that of γ1-CN, because of conformational exposition of the plasmin cleavage site Lys_{68}-Ser_{69}. For this reason, β-CN f(69–209) and its complementary proteose peptone β-CN f(1–68) have been proposed as candidate molecular markers for the assessment of buffalo milk and curd freshness.

The main indigenous proteolytic enzymes have also been detected in buffalo Mozzarella cheese (Table 8.2, Santillo et al. 2011). The curd heating step has been considered partially responsible for the higher plasmin activity in cheese compared with the corresponding milk (Gobbetti et al. 2002). The measurable activity of acid proteases, such as cathepsin D, have been also found to be higher in Mozzarella than in the cheese milk due to the low pH of the acidified cheese mass before the stretching phase, in line with some activation occurring and leading to elevated measurable cathepsin D activity at lower pH conditions of this cheese matrix. The contribution of the activity of elastase to the proteolysis pattern of buffalo mozzarella was found to be of minor relevance compared to the other enzymes.

8.4 Enzymology of Camel Milk

Raw camel milk is produced by one-humped Dromedary camels in the desert areas of Africa, Asia and Middle East and by two-humped Bactrian camel in more cool areas in Mongolia, China and Russia; therefore, it is consumed in arid and hot areas of Africa and Asia and mainly processed into fermented products (El-Agamy 2017, Al haj and Al Kanhal 2010). The mean gross composition of Dromedary camel milk is similar to that of bovine milk, except for the higher whey protein content of the former. According to El-Agamy (2017), the mean casein content in Dromedary camel milk is 2.40% vs. 2.51% of bovine milk. Bactrian camel milk contains more fat and protein—on average, 5.32 and 4.09%, respectively—and consequently more total solids than bovine milk. The whey protein:casein ratio is 0.34–0.36 in both camel milks, higher than that of 0.24 for bovine milk (El-Agamy 2017).

There are several recent reports about the enhanced biofunctionality and therapeutic potential of camel milk, e.g., low or lack of allergy effects due to the absence of β-lg, considerable amounts of bioactive peptides, better digestibility than bovine milk, higher lactoferrin content than in ruminants' milk, and possible hypoglycaemic/antihyperlipidemic effects, due to high concentrations of zinc, vitamin B3, polyunsaturated fatty acids, non-enzymatic antioxidants and an insulin-like protein that mimics the interaction of insulin with its receptor (Al haj and Al Kanhal 2010; El-Agamy 2017; Khalesi et al. 2017; Maqsood et al. 2019; Shori 2015).

Indigenous enzymatic activities have been identified in raw camel milk; Ryskaliyeva et al. (2018) reported that, among the top 70 proteins of skimmed

Table 8.2 Indigenous proteolytic enzymes in buffalo milk and mozzarella cheese produced from Mediterranean buffalos' milk (adapted from Santillo et al. 2011)

Item	Buffalo Milk U/mL	Buffalo mozzarella cheese U/g
Plasmin	3.02 ± 0.1	3.4 ± 0.45
Plasminogen	3.13 ± 0.14	5.33 ± 0.7
Cathepsin D	0.011 ± 0.01	0.22 ± 0.01
Elastase	0.07 ± 0.05	0.02 ± 0.01

camel milk, xanthine oxidoreductase (XOR), lactoperoxidase (LPO) and lipoprotein lipase (LPL) were detected, with molecular masses of 150, 87.7 and 46.5 kDa, respectively.

Lysozyme (LYZ) and lactoperoxidase (LPO) were the focus of early studies. Duhaiman (1988) estimated a level of 0.5 mg lysozyme per 100 mL of camel milk, with a specific activity lower than that of from human milk and egg white, whereas Elagamy et al. (1992) observed that camel milk and egg-white LYZ exhibited a similar antibacterial activity spectrum. Later, Elagamy et al. (1996) estimated a level of 15 μg LYZ per 100 mL camel milk, twice as much as in bovine milk, with higher lytic activity than for bovine milk but lower than egg white LYZ. They report no immunological cross-reactivity between camel and bovine and between camel and egg white LYZ. Lysozyme C (EC: 3.2.1.17) of camel milk consists of 130 AA and has a molecular mass of 14.8 kDa - www.uniprot.org/uniprot/P11376, www.uniprot.org/uniprot/P11375 and www.uniprot.org/uniprot/P37712- and, as reviewed by Hailu et al. (2016), there is about 85% sequence similarity among camel, human and ruminant LYZ. On the other hand, Kappeler (1998) did not found LYZ activity in camel milk, or a cDNA corresponding to LYZ by means of screening with oligonucleotides derived from conserved regions of the already known C-type lysozymes. He suggested either an effect of the stage of lactation or that the literature reported LYZ activity could be related to other enzymes.

Elagamy et al. (1992, 1996) observed that camel milk was bacteriostatic against Gram-positive and bactericidal against Gram-negative bacteria. LPO activity in the presence of H_2O_2 and thiocyanate (SCN^-) similar to that of bovine milk was detected, and molecular masses of 78 and 72.5 kDa for camel and bovine milk LPO, respectively, were estimated and cross-reactivity between them was detected. Similarly, Mahdi et al. (2018) reported that the molecular mass of LPO from camel colostrum was 79 kDa and its concentration was 2.35 mg/mL. Camel LPO consists of 611 amino acids (Hailu et al. 2016). The SCN content of camel milk varies considerably, i.e. from 2.4 to 33 ppm, being higher than in other ruminants' milk (Kamau et al. 2010; Younan et al. 2006).

The enhancement of the LPO system in raw camel milk has been effectively used for the extension of storage of raw camel milk, e.g., resulting in constant total bacterial counts over 6 h at 30 °C (Younan et al. 2006). Addition of 30:30 ppm NaSCN: H_2O_2 was suggested as the most effective supplementation for the extension of raw camel milk shelf-life, which was more effective at 20 and 30 °C than at 10 °C, due to higher activity of LPO at ambient temperatures (Kamau et al. 2010). Similarly, addition of 20:20 ppm SCN: H_2O_2 preserved raw camel milk for up to 18–20 h at 37 °C (Singh et al. 2017). On the other hand, Kamau and Wangoh (2010) noted greater shelf-life extension when pasteurized camel milk supplemented with 8.5 ppm H_2O_2 was stored at 10 °C compared to 20 °C.

Activities of camel LPO other than its preservation effect in milk have been also reported. According to Redwan et al. (2015), the direct interaction of camel milk LPO at a concentration of 1 mg/mLwith hepatitis C virus inhibits virus amplification, while a concentration of 1.5 mg/mL is needed for the bovine milk counterpart. In accordance with the human and bovine counterparts, camel milk LPO exhibits activity against the herpes simplex virus type 1, which is enhanced in the presence of H_2O_2 (El-Fakharany et al. 2017). LPO from camel colostrum, either alone or in combination with lactoferrin, exhibits significant antibacterial activity *in vivo* and *in vitro* against *Acinetobacter baumannii*, being more effective than the sensitive antibiotic imipenem, due to immunomodulation expressed by the activation of IL-10 and IL-4 levels in lung homogenates (Mahdi et al. 2018). Sheikh et al. (2018, 2019) performed structural studies on camel milk LPO and concluded that common antibiotics can inhibit it, in accordance with the findings of Al-Nazawi and Homeida (2015).

The average alkaline phosphatase activity in camel milk has been estimated as 21.31 ± 1.26 μg/mL, which is lower than for bovine and buffalo milks, i.e., 575.00 ± 18.42 μg/ml and 253.75 ± 8.22 μg/ml, respectively (Yoganandi et al. 2014). Increased ALP activity in milk from camels with sub-clinical mastitis has been observed (Ali et al. 2016).

Indigenous enzymes, mainly alkaline phosphatase (ALP) and lactoperoxidase (LPO), have been also studied in relation to heat treatment efficacy for camel milk. LPO and γ-glutamyltransferase (GGT) of camel milk are destroyed on heating at 74 °C for between 15 s and 5 min, and they have been suggested as suitable markers for the correct pasteurization of camel milk, because ALP exhibits residual activity after treatment at 72 °C for 5 min (Wernery et al. 2006, 2008). Indeed, Lorenzen et al. (2011) estimated mean ALP activities from 15.9 to 23.4 U/L for raw camel milk and from 16 to 10 U/L after pasteurization at 75 ± 1 °C for 15–30 s. According to their findings, mean GGT activity in raw camel milk was 293–343 U/L and residual activity of 0.1–2.2 U/L was detected after pasteurization, implying that it is not suitable as an indicator. Moreover, they estimated too low and highly variable leucine arylamidase (LAP) activity in raw camel milk, in the range of 4–6 U/L, such that, despite its inactivation after pasteurization, it cannot be used as an indicator. In the same study, high residual lipase activity was estimated after pasteurization, i.e., mean lipase activity was in the range of 57–71 U/L in raw and 12–14 U/L in pasteurized camel milk.

Finally, LPO has been proposed for the assessment of pasteurization of camel milk due its high concentration in raw milk (869–1172 U/L) and non-detectable amounts <5 U/L in pasteurized milk. The findings of Tayefi-Nasrabadi et al. (2011) showed that camel milk LPO is less heat-stable than its bovine counterpart within the temperature range 67–73 °C, due to lower values of activation energy and change in enthalpy of denaturation. They report that D-values for camel milk LPO at 71 °C and 67 °C were about 3.1 and 6.7 times lower than for bovine milk, with a higher z-value (6.42 °C vs 4.7 °C), which indicates that camel milk LPO is more sensitive to the extension of heating duration. It has been confirmed that LPO in camel milk is destroyed at 75 °C for 15 s and has been suggested as an indicator for pasteurization of such milk (Wernery et al. 2013).

Studies on other indigenous enzymatic activities in camel milk are limited. Variable plasmin and plasminogen activities were detected in individual camel milks using a fluorogenic substrate by Baer et al. (1994), with proteolytic activity on camel casein similar to the proteolytic activity of bovine plasmin. Ryskaliyeva et al.

(2018) suggested the presence of plasmin activity in camel milk based on the identification of β-casein fragments missing 946 Da, which they hypothesized to result from the cleavage of the first seven N-terminal residues by plasmin.

In the proteomic analysis of Felfoul et al. (2017), lipoprotein lipase isoform 3 was detected in raw camel milk. Xanthine oxidoreductase (XOR) has been also identified in the camel milk fat globule membrane (Saadaoui et al. 2013). The XOR content of camel milk has been found as up to 22.2 mg/L, which is comparable to that obtained for bovine and human milk. Purified camel milk XOR showed no detectable activity towards xanthine or NADH. The mean molybdenum content per XOR monomer was 0.02 ± 0.002, similar to the value of 0.03 ± 0.005 estimated for the human milk enzyme and much lower than the value of 0.56 ± 0.051 for bovine milk (Baghiani et al. 2003). The average catalase activity in camel milk has been found to be higher than in bovine or buffalo milks, being estimated at 6.20 ± 0.18 units, 0.86 ± 0.04 units and 0.62 ± 0.04 units, respectively (Yoganandi et al. 2014).

8.5 Enzymology of Non-ruminant Milk

The milk of members of the Equidae family, such as horses and donkeys, has drawn the attention of dairy and nutrition scientists due to its similarity with human milk in terms of composition, possible low allergy potential, and suggested therapeutic properties. As a result, horse (mare) milk (MM) and donkey milk (DM) have been proposed as more suitable for infant formulas compared to ruminants' milk (Alichanidis et al. 2016; Altomonte et al. 2019; Martini et al. 2018; Massouras et al. 2017; Salimei and Fantuz 2012, 2013; Salimei and Park 2017; Souroullas et al. 2018; Vincenzetti et al. 2017; Uniacke-Lowe et al. 2010).

In brief, MM and DM contain (expressed in g/kg) 110 and 95.3 total solids, 61 and 65.8 lactose, 21.4 and 16.5 N × 6.38, 14.0 and 7.6 fat and 4.5 and 4.1 ash, respectively. The composition of fat of equid milk differs substantially from ruminants' milk. Triglygerides comprise <80% of total lipids, free fatty acids approx. 10% and phospholipids >5%, cholesterol content is low, at 5.0–8.8 mg/dL, and milk fat globules are small with a mean diameter of 2–3 μm. Casein is <50% of crude protein in equid milk and lysozyme is more than 10% of whey proteins. Moreover, the number of somatic cells is low (average SCC is in general <5 log SCC/mL) in these milks, because the mammary gland status is good and mastitis is very rare. The bacterial counts are in general <10^3 cfu/mL, the presence of undesirable *Staphylococcus* spp. is rare, and the absence of *Listeria monocytogenes* and *Salmonella* spp. has been reported for raw equid milk produced under hygienic conditions (Doreau and Martin-Rosset 2011; Salimei and Fantuz 2012, 2013; Uniacke-Lowe et al. 2010).

There are a few reports on indigenous enzymatic activities of equid milk, i.e., horse (mare) milk (MM) and donkey milk (DM) and on the effect of treatments on them; the greater part of the recent literature is about DM.

The role of alkaline phosphatase (ALP) as an indicator of proper equid milk pasteurization has been studied. Marchand et al. (2009) reported that mean alkaline ALP activity, determined using the Fluorophos® substrate ranges from 3.1 to 20.8 and from 740 to 1050 U/L for different pools of raw MM and bovine milk respectively, i.e., 35–350 times lower ALP activity for equine milk compared to bovine milk. Also, those authors reported that, in contrast to bovine milk, no specific association of ALP with the fat fraction of MFGM was detected in equine milk. They estimated that the inactivation of ALP in raw MM follows first-order kinetics at 48–64 °C, with a z-value 5.31 °C, and concluded that ALP could not be a useful pasteurization indicator for this milk type, due to the low D-values in the pasteurization range and the low initial ALP activity. On the other hand, Giacometti et al. (2016) reported that ALP can be used as an indicator for proper pasteurization of DM. They estimated 36,047 ± 1009 mU/L ALP activity in raw donkey milk, which after pasteurization at 65 °C for 30 min was 106 ± 10.5, ranging from <100 to 118 mU/L. Moreover, high pressure treatment at 400 MPa at 4–6 °C for 100 s resulted in approximately one tenth of the original residual ALP activity.

The well-known antibacterial activity in equid milk is associated mainly with the rather high levels of lactoferrin and especially of lysozyme (LYZ), which are major components of the whey protein fraction compared to bovine milk. While lactoferrin concentration is intermediate between that of human and bovine milk, lysozyme level is much higher than that of human milk, especially in DM (Uniacke-Lowe et al. 2010; Salimei and Fantuz 2012, 2013). Doreau and Martin-Rosset (2011) reported that MM contains 0.8 g/kg LYZ, while Markiewicz-Kęszycka et al. (2013) found 1.03 ± 0.27 g/L LYZ at late lactation. As reviewed by Altomonte et al. (2019), LYZ concentration in DM ranges from 0.9 to 3 g/L and in human milk ranges from 0.8 to 1.1 g/L. LYZ activity in DM ranges from 4000 to 5000 U/mL and exhibits maximum values at the beginning of the lactation (Pilla et al. 2010; Vincenzetti et al. 2008), being very much higher than the 0.0292 U/mL found in bovine milk and very much lower than the 39,000 U/mL in human milk (Martini et al. 2018). Indicative concentrations or activities of lysozyme in DM milk are shown in Table 8.3.

Equid milk lysozyme binds calcium strongly, which results in a more stable configuration opposite to lysozyme of egg-white, tears and saliva (Brumini et al. 2016; Uniacke-Lowe et al. 2010). MM and DM lysozymes (LYZ), 1,4-beta-N-acetylmuramidase C (EC 3.2.17), contain 129 amino acid residues and bind one Ca^{2+} ion per subunit. Residues Glu35 and Asp52 are related to enzymatic activity and the conserved high-affinity calcium binding site 81–92 similar to that of α-lactalbumin - www.uniprot.org/uniprot/P11376, www.uniprot.org/uniprot/P11375 and www.uniprot.org/uniprot/P37712 -. Two variants of the LYZ of DM differing in three amino acids were identified by Herrouin et al. (2000). Sixty five percent of LYZ of raw and heat treated (95 °C for 1 min) MM remains mainly intact after *in vitro* gastric and duodenal digestions (Inglingstad et al. 2010). Tidona et al. (2011, 2014) reported higher residual intact lysozyme, from 75 to 85%, after *in vitro* digestion of DM with human gastric and duodenal juice.

Table 8.3 Mean lysozyme (LYZ) level or activity in raw donkey milk as reported in various studies

Quantity/activity	Analysis method	Reference
11,531 ± 923 U/mL	Fluorimetric commercial kit	Addo and Ferragut (2015)
4000 mg/L	Microbiological assay[a]	Coppola et al. (2002)
2.03 mg/mL[b]	RP-HPLC	Cosentino et al. (2015)
1090 mg/L	RP-HPLC	Cosentino et al. (2016)
1 mg/mL	RP-HPLC	Polidori and Vincenzetti (2010)
0.95 ± 0.07 mg/mL	RP-HPLC	Ozturkoglu-Budak (2018)
9500 ± 4200 U/mL	Fluorimetric commercial kit	Murgia et al. (2016)
4000–5000 U/mL	Fluorimetric commercial kit	Pilla et al. (2010)
3750 ± 250 mg/L	Microbiological assay[a]	Chiavari et al. (2005)
0.76–1.34 mg/mL[c]	RP-HPLC	Vincenzetti et al. (2008)
3986 ± 278 U/mL	Fluorimetric commercial kit	Ragona et al. (2016)
0.98–3.74 g/L	SDS-PAGE	Šarić et al. (2014)
1.17 ± 0.10 mg/mL	RP-HPLC	Vincenzetti et al. (2018)

[a]According to Lodi et al. (1984)
[b]Under similar conditions, the LYZ content in bovine milk was 0.18 mg/mL
[c]Throughout lactation

Due to its antimicrobial effect, the residual activity of equid milk LYZ after various types of treatments has been studied. Jauregui-Adell (1975) reported that the activity of LYZ in MM was not affected by treatment for 15 min at 71 °C or 2 min at 82 °C. In particular, after 30 min at 62, 71 and 82 °C, 102, 86 and 37% residual activity, respectively, was detected. Residual LYZ in MM treated at 75 °C for 10 min, which is consistent with the treatment of infant formulas, was 85% of the initial activity (Civardi et al. 2007). Under denaturing conditions, LYZ of MM forms a very stable molten globule, which is a compact state with native-like secondary structures, but with little tertiary structure compared to native proteins (Morozova-Roche 2007).

Lysozyme in DM survives heat treatments at 90 °C for 1 min and 100 °C for 5 min, but is destroyed by treatment at 121 °C for 10 min (Coppola et al. 2002), while it does not decrease significantly after pasteurization of DM at 63 °C for 30 min or 72 °C for 15 s (Giribaldi et al. 2017). According to the results of Addo and Ferragut (2015), LYZ activity of DM heat treated at 70 ± 2 and 80 ± 2 °C for 1 min or treated by ultra-high pressure homogenization at 100, 200 and 300 MPa was not significantly decreased compared to raw milk throughout 28 days of storage. Ozturkoglu-Budak (2018) reported that freezing or heating at 75 °C for 2 min did not affect the level of native lysozyme, while heating at 85 °C did.

Chiavari et al. (2005) reported that storage for 30 days did not significantly affect the antimicrobial lysozyme activity in raw and heat treated (63 °C for 30 min) DM. Polidori and Vincenzetti (2010) found that the lysozyme content of fresh, frozen and powdered DM did not differ significantly, although a 30% decrease of activity was observed in the DM powder. Those authors also reported a 50% decrease of activity at 70 °C for 5 min. Concentration of raw or pasteurized (63 °C for 30 min)

DM by rotary vacuum evaporation, to 40% and 20% of the initial value, did not affect either concentration or antimicrobial activity of lysozyme against gram-positive and gram-negative bacteria, with the exception of *Bacillus mojaventis* (Cosentino et al. 2016). Vincenzetti et al. (2018) observed that the level of lysozyme after freeze-drying or spray-drying of DM was significantly reduced compared to control raw milk, being 85% and 82% of the initial, respectively. A more severe reduction was observed for lysozyme activity, i.e., 96% and 58% respectively.

DM has been used in cheese-making as a source of lysozyme against late gas production. According to Cosentino et al. (2015), the addition of DM to bovine cheese milk at levels from 2 to 7.5% decreased the counts of *Clostridium tyrobutyricum* and bacterial spores in semi-hard cheese in a dose-dependent manner. Niro et al. (2017) reported that the supplementation of bovine cheese milk with 2% DM had effects on microbial counts, composition and proteolysis similar to the addition of 1.6 g egg-white lysozyme in cheese milk; no late blowing was observed in either case, although 32 MPN clostridia spores/L were determined in bovine cheese milk.

Information about other indigenous enzymes in equid milk is sporadic. Egito et al. (2002, 2003) suggested that small quantity of PP5-like peptides and 23-kDa γ-like caseins found in MM sodium caseinate could come from the action of indigenous plasmin. They also reported that sodium caseinate prepared from MM exhibited very low, but detectable, plasmin activity. On the other hand, Humbert et al. (2005) found high plasmin activity in fresh milk, along with intense presence of γ-caseins in MM, whereas plasminogen could not be detected following the usual activation protocols.

In contrast to bovine milk, lactoperoxidase (LPO) activity in skimmed DM was found to be very low, 4.83 ± 0.35 mU/mL. Its concentration in DM was estimated to be 0.11 mg/L, seven times lower than that in human milk, and very much lower than the 30–100 mg/L LPO content of bovine milk (Vincenzetti et al. 2017).

Chilliard and Doreau (1985) showed that MM contains high lipase activity that exhibits the characteristics of lipoprotein lipase, i.e., it was activated with 2–8 volumes of equine blood serum, inactivated by 95% after heating at 60 °C for 30 min, was sensitive to sodium chloride, and was not significantly affected by heparin.

Xanthine dehydrogenase/oxidase (XOR) with molecular mass 146 kDa and an electrophoretic profile similar to its bovine and caprine counterparts has been identified among MFGM proteins of pony mares' milk by Cebo et al. (2012). NAGase activity <50 U/mL of has been identified in raw DM for the most part of the lactation period, and it was not affected by the bacteriological status (Pilla et al. 2010).

8.6 Conclusion

Several decades of research has led to significant knowledge of indigenous enzymes in milk from small ruminants, such as buffalo, camel, horse and donkey, and several enzymes have been partially identified and characterized. However, in species with less common use for milk production, significant gaps in understanding the nature, biochemical properties and technological significance of milk enzymes remain. From this perspective, it is important that the study of the various aspects of all indigenous enzymes continues to be an active area of research in terms of milk of the species covered by this review.

Among the milk indigenous enzymes, the study of proteolytic enzymes has perhaps been most intensive in the milk of small ruminants and buffalo, due to the implications for the quality of the milk protein fraction and therefore on the processing of milk and the quality of dairy products, but deficits in our understanding remain.

As expected, the limited research on camel milk enzymatic activities has focussed on the antimicrobial lysozyme and LPO and on indicators of heat-treatment such as ALP and LPO. The findings for lysozyme are contradictory, but LPO activity in the presence of H_2O_2 and SCN^- similar to that of bovine milk has been detected and interesting antiviral and immunomodulatory behavior has been reported for the enzyme. In contrast to bovine milk, high residual ALP and LPL activities and non-detectable amounts of LPO have been found in pasteurized camel milk, indicating that different approaches to the assessment of heat treatment of this milk kind should be developed.

The profile of indigenous enzymatic activities of the milk of the members of the Equidae family differs from their ruminant counterpart; for example, very high lysozyme–and therefore antimicrobial–activity and low ALP and LPO activity are reported in such milk.

The implementation of 'omics' techniques has produced significant new information regarding the presence of indigenous enzymes or enzyme systems in non-bovine milks. However, studies on their behavior under the conditions and treatments applied in raw milk storage and processing are scarce. In particular, this research topic is of great interest for the exploitation of the biological potential of non-ruminants milk in both infant and adult nutrition.

References

Abd El-Hamid LB, Mahran GA, Shehata AE, Osman SG (1977) Lipase activity in buffaloes milk 2. Effect of feeding system, animal age and milking phase and meal. Egypt J Dairy Sci 5:7–10

Addo CNA, Ferragut V (2015) Evaluating the ultra-high pressure homogenization (UHPH) and pasteurization effects on the quality and shelf-life of donkey milk. Int J Food Stud 4:104–115

Ahmad S, Anjum FM, Huma N, Sameen A, Zahoor T (2013) Composition and physico-chemical characteristics of buffalo milk with particular emphasis on lipids, proteins, minerals, enzymes and vitamins. J Anim Plant Sci 23:62–74

Al haj OA, Al Kanhal HA (2010) Compositional, technological and nutritional aspects of dromedary camel milk. Int Dairy J 20:811–821

Albenzio M, Santillo A (2011) Biochemical characteristics of ewe and goat milk: effect on the quality of dairy products. Small Rumin Res 101:33–40

Albenzio M, Caroprese M, Santillo A, Marino R, Taibi L, Sevi A (2004a) Effects of somatic cell count and stage of lactation on the plasmin activity and cheese-making properties of ewe milk. J Dairy Sci 87:533–542

Albenzio M, Marino R, Caropresе M, Santillo A, Annicchiarico G, Sevi A (2004b) Quality of milk and Canestrato Pugliese cheese from ewes exposed to different ventilation regimens. J Dairy Res 7:434–443
Albenzio M, Caroprese M, Santillo A, Marino R, Muscio A, Sevi A (2005a) Proteolytic patterns and plasmin activity in ewes' milk as affected by somatic cell count and stage of lactation. J Dairy Sci 72:86–92
Albenzio M, Santillo A, Caroprese M, Marino R, Centoducati P, Sevi A (2005b) Effect of different ventilation regimens on ewe milk and canes-trato Pugliese cheese quality in summer. J Dairy Res 72:447–455
Albenzio M, Caroprese M, Marino R, Muscio A, Santillo A, Sevi A (2006) Characteristics of Garganica goat milk and Cacioricotta cheese. Small Rumin Res 64:35–44
Albenzio M, Santillo A, Caroprese M, Marino R, Annicchiarico G, Sevi A (2007) Effects of ventilation rate and of dietary protein level in an intensive dairy sheep system on the features of Canestrano Pugliese cheese. J Dairy Res 74:26–33
Albenzio M, Santillo A, d'Angelo F, Sevi A (2009a) Focusing on casein gene cluster and protein profile in Garganica goat milk. J Dairy Res 76:83–89
Albenzio M, Santillo A, Caroprese M, d'Angelo F, Marino R, Sevi A (2009b) Role of endogenous enzymes in proteolysis of sheep milk. J Dairy Sci 92:79–86
Albenzio M, Santillo A, Caroprese M, Marino R, Trani A, Faccia M (2010) Biochemical patterns in ovine cheese: influence of probiotic strains. J Dairy Sci 93:3487–3496
Albenzio M, Santillo A, Caroprese M, Ruggieri D, Ciliberti M, Sevi A (2012) Immune competence of the mammary gland as affected by somatic cell and pathogenic bacteria in ewes with subclinical mastitis. J Dairy Sci 95:3877–3887
Albenzio M, Santillo A, Marino R, Della Malva A, Caroprese M, Sevi A (2015) Identification of peptides in functional Scamorza ovine milk cheese. J Dairy Sci 98:8428–8432
Albenzio M, Santillo A, Avondo M, Nudda A, Chessa S, Pirisi A, Banni S (2016) Nutritional properties of small ruminant food products and their role on human health. Small Rumin Res 135:3–12
Ali F, Hussain R, Qayyum A, Gul ST, Iqbal Z, Hassan MF (2016) Milk somatic cell counts and some hemato-biochemical changes in sub-clinical mastitic dromedary she-camels (Camelus dromedarius). Pak Vet J 36:405–408
Alichanidis E, Moatsou G, Polychroniadou A (2016) The composition and the properties of non-cow milks and products. In: Tsakalidou E, Papadimitriou K (eds) Non-bovine milk and milk products. Academic, Cambridge, pp 81–116
Al-Nazawi MH, Homeida AM (2015) Effects of antibacterial drugs on camel milk lysozyme concentration and lactoperoxidase activity. J Camel Pract Res 22:97–99
Altomonte I, Salari F, Licitra R, Martini M (2019) Donkey and human milk: insights into their compositional similarities. Int Dairy J 89:111–118
Andrews AT, Alichanidis E (1975) The acid phosphatases of bovine leucocytes, plasma and the milk of healthy and mastitic cows. J Dairy Res 42:391–400
Angeletti R, Gioacchini AM, Deraglia R, Piro R, Traldi P (1998) The potential of matrix-assisted laser desorption/ionization mass spectrometry in the quality control of water buffalo mozzarella cheese. J Mass Spectrom 33:525–531
Anifantakis EM, Rosakis PS (1983) Alkaline phosphatase activity of sheep's milk and some factors affecting it Egypt. J Dairy Sci 11:173–182
Anjos FD, Machado A, Ferro C, Otto F, Bogin E (1998) Gamma-glutamyltransferase as a marker for the pasteurisation of raw milk. J Food Prot 61:1057–1059
Arnold C (1881) Einige neue Reactionen der Milch. Arch Pharm 219:41–42
Assis G, Roseiro IB, Barbosa M (2000) Determination of alkaline phosphatise activity levels in milk from indigenous Portuguese ewe and goat breeds by the fluorometric method. In: Proceedings of development strategy for the sheep and goat dairy sector, international dairy federation bulletin no 351. International Dairy Federation, Brussels, p 34

Atmani D, Benboubetra M, Harrison R (2004) Goat's milk xanthine oxidoreductase is grossly deficient in molybdenum. J Dairy Res 71:7–13
Atmani D, Banghiani A, Harrison R, Benboubetra M (2005) NADH oxidation and superoxide production by caprine milk oxidoreductase. Int Dairy J 15:1113–1121
Baer A, Ryba I, Farah Z (1994) Plasmin activity in camel milk. LWT Food Sci Technol 27:595–598
Baghiani A, Harrison R, Benboubetra M (2003) Purification and partial characterisation of camel milk xanthine oxidoreductase. Arch Physiol Biochem 111:407–414
Barbosa M (2005) Interest in controlling alkaline phosphatase activity in sheep and goat milks. The future of the sheep, goat dairy sectors (part 3, special issue 0501). IDF-FIL, Brussels, pp 117–127
Benboubetra M, Baghiani A, Atmani D, Harrison R (2004) Physicochemical and kinetic properties of purified sheep's milk xanthine oxidoreductase. J Dairy Sci 87:1580–1584
Bencini R, Pulina G (1997) The quality of sheep milk: a review. Aust J Exp Agric 37:485–504
Benkerroum N (2008) Antimicrobial activity of lysozyme with special relevance to milk. Afr J Biotechnol 7:4856–4867
Bianchi L, Bolla A, Budelli E, Caroli A, Casoli C, Pauselli M, Duranti E (2004) Effect of udder health status and lactation phase on the characteristics of Sardinian ewe milk. J Dairy Sci 87:2401–2408
Bingham EW, Garrer K, Powlem D (1992) Purification and properties of alkaline phosphatase in the lactating mammary gland. J Dairy Sci 75:3394–3401
Brumini D, Criscione A, Bordonaro S, Vegarud GE, Marletta D (2016) Whey proteins and their antimicrobial properties in donkey milk: a brief review. Dairy Sci Technol 96:1–14
Caroprese M, Marzano A, Schena L, Marino R, Santillo A, Albenzio M (2007) Contribution of macro phages to proteolysis and plasmin activity in ewe bulk milk. J Dairy Sci 90:2767–2772
Cartier P, Chillard Y, Paquet D (1990) Inhibiting and activating effects of skim milks and proteose-peptone fractions on spontaneous lipolysis and purified lipoprotein lipase activity in bovine milk. J Dairy Sci 73:1173–1177
Cebo C, Caillat H, Bouvier F, Martin P (2010) Major proteins of the goat milk fat globule membrane. J Dairy Sci 93:868–876
Cebo C, Rebours E, Henry C, Makhzami S, Cosette P, Martin P (2012) Identification of major milk fat globule membrane proteins from pony mare milk highlights the molecular diversity of lactadherin across species. J Dairy Sci 95:1085–1098
Chandan RC, Parry RM, Shahani KM (1968) Lysozyme, lipase and ribonuclease in milk of various species. J Dairy Sci 51:606–607
Chavarri F, Santistebam A, Virto M, de Renobales M (1998) Alkaline phosphatase, acid phosphatase, lactoperoxidase, and lipoprotein lipase activities in industrial ewes' milk and cheese. J Agric Food Chem 46:2926–2932
Chiavari C, Coloretti F, Nanni M, Sorrentino E, Grazia L (2005) Use of donkey's milk for a fermented beverage with lactobacilli. Lait 85:481–490
Chilliard Y (1982) Variations physiologiques desactivite´slipasiques et de la lipolyse spontane´e dans les laits de vache de chevre et de femme: revue bibliographique (suite). Lait 62:126–154
Chilliard Y, Doreau M (1985) Characterization of lipase in mare milk. J Dairy Sci 68:37–39
Chilliard Y, Ferlay A, Rouel J, Lamberet G (2003) A review of nutritional and physiological factors affecting goat milk synthesis and lypolysis. J Dairy Sci 86:1751–1770
Civardi G, Curadi MC, Orlandi M, Cattaneo TMP, Giangiacomo R (2007) Mare's milk: monitoring the effect of thermal treatments on whey proteins stability by SDS-capillary electrophoresis (CE-SDS). Milchwissenschaft 62:32–35
Coburn SP, Mahuren JD, Pauly TA, Ericson KL, Townsend DW (1992) Alkaline phosphatase activity and pyridoxal phosphate concentrations in the milk of various species. J Nutr 122:2348–2353
Collins YF, McSweeney PLH, Wilkinson MG (2003) Lipolysis and free fatty acid catabolism in cheese: a review of current knowledge. Int Dairy J 13:841–866

Coppola R, Salimei E, Succi M, Sorrentino E, Nanni M, Ranieri P, Belli Blanes R, Grazia L (2002) Behaviour of *Lactobacillus rhamnosus* strains in ass's milk. Ann Microbiol 52:55–60

Cosentino C, Paolino R, Valentini V, Musto M, Ricciardi A, Adduci F, D'Adamo C, Pecora G, Freschi P (2015) Effect of jenny milk addition on the inhibition of late blowing in semihard cheese. J Dairy Sci 98:5133–5142

Cosentino C, Labella C, Elshafie HS, Camele I, Musto M, Paolino R, D'Adamo C, Freschi P (2016) Effects of different heat treatments on lysozyme quantity and antimicrobial activity of jenny milk. J Dairy Sci 99:5173–5179

Delacroix-Buchet A, Degas C, Lamembert G, Vassal L (1996) Influence des variants AA et FF de la case´ine a_{s1} caprine sur le rendement fromager et les characte´ristiques des fromages. Lait 76:217–241

Di Luccia A, Picariello G, Trani A, Alviti G, Loizzo P, Faccia M, Addeo F (2009) Occurrence of β-casein fragments in cold-stored and curdled river buffalo (Bubalus bubalis L.) milk. J Dairy Sci 92:1319–1329

Doreau M, Martin-Rosset W (2011) Animals that produce dairy foods. Horse. In: Fuquay JW (ed) Encyclopaedia of dairy sciences. Elsevier, London, pp 358–364

Duhaiman AS (1988) Purification of camel milk lysozyme and its lytic effect on Escherichia coli and *Micrococcus lysodeikticus*. Comp Biochem Physiol 91B:793–796

Dumitraşcu L, Stănciuc N, Stanciu S, Râpeanu G (2012) Thermal inactivation of lactoperoxidase in goat, sheep and bovine milk – a comparative kinetic and thermodynamic study. J Food Eng 113:47–52

Egito AS, Miclo L, Lopez C, Adam A, Girardet J-M, Gaillard J-L (2002) Separation and characterisation of mares' milk a_{s1}-, b-, k-caseins, g-casein-like, and proteose peptone component 5-like peptides. J Dairy Sci 85:697–706

Egito AS, Girardet J-M, Poirson C, Mollé D, Humbert G, Miclo L, Gaillard J-L (2003) Action of plasmin on equine β-casein. Int Dairy J 13:813–820

El-Agamy EI (2017) Camel milk. In: Park YW, Haenlein GF, Wendorff WL (eds) Handbook of milk of non-bovine mammals. Wiley, Oxford, pp 409–480

Elagamy E, Ruppanner R, Ismail A, Champagne C, Assaf R (1992) Antibacterial and antiviral activity of camel milk protective proteins. J Dairy Res 59:169–175

Elagamy EI, Ruppanner R, Ismail A, Champagne CP, Assaf R (1996) Purification and characterization of lactoferrin, lactoperoxidase, lysozyme and immunoglobulins from camel's milk. Int Dairy J 6:129–145

El-Fakharany EM, Uversky VN, Redwan EM (2017) Comparative analysis of the antiviral activity of camel, bovine, and human lactoperoxidases against herpes simplex virus type 1. Appl Biochem Biotechnol 182:294–310

El-Gazzar H, Rafaie MOI, El-Aziz MA (1999) Activity of xanthine oxidase in milk and its products and effect of heat and folic acid upon its activity. Ann Agric Sci Ain Shams Univ Cairo 44:631–639

El-Salam MA, El-Shibiny S (2011) A comprehensive review on the composition and properties of buffalo milk. Dairy Sci Technol 91:663–699

Fantuz F, Baldi A, Dellorto V, Polidori F, Rossi CS, Pollitis I, Heegaard CW (1998) Distribution of plasminogen activator forms in different fractions of buffalo milk. J Dairy Res 65:521–527

Farkye NY (2003) Other enzymes. In: Fox PF, McSweeney PLH (eds) Advanced dairy chemistry. Elsevier Applied Science, London, pp 571–603

Felfoul I, Jardin J, Gaucheron F, Attia H, Ayadi MA (2017) Proteomic profiling of camel and cow milk proteins under heat treatment. Food Chem 216:161–169

Fox PF (2003) Significance of indigenous enzymes in milk and dairy products. In: Whitaker JR, Voragen AGJ, Wong DWS (eds) Handbook of food enzymology. Marcel Dekker, New York, pp 255–277

Fox PF, Kelly AL (2006) Review indigenous enzymes in milk: overview and historical aspects-part 2. Int Dairy J 16:517–532

Giacometti F, Bardasi L, Merialdi G, Morbarigazzi M, Federici S, Piva S, Serraino A (2016) Shelf life of donkey milk subjected to different treatment and storage conditions. J Dairy Sci 99:4291–4299

Giribaldi M, Antoniazzi S, Gariglio GM, Coscia A, Bertino E, Cavallarin L (2017) A preliminary assessment of HTST processing on donkey milk. Vet Sci 4:50

Gobbetti M, Morea M, Baruzzi F, Corbo MR, Matarante A, Considine T, Di Cagno R, Guinee T, Fox PF (2002) Microbiological, compositional, biochemical and textural characterisation of Caciocavallo Pugliese cheese during ripening. Int Dairy J 12:511–523

Gómez-Ruiz JA, Ramos M, Recio I (2004) Identification and formation of angiotensin-converting enzymes-inhibitory peptides in Manchego cheese by high-performance liquid chromatography-tandem mass spectrometry. J Chromotogr A 1054:269–277

Griffiths MW (1986) Use of milk enzymes as indices of heat treatment. J Food Prot 49:696–705

Hailu Y, Hansen EB, Seifu E, Eshetu M, Ipsen R, Kappeler S (2016) Functional and technological properties of camel milk proteins: a review. J Dairy Res 83:422–429

Hamulv G, Kandasamy C (1982) Increasing the keeping quality of milk by activation of its lacto-peroxidase system: results from Sri Lanka. Milchwissenschaft 37:454–457

Harding F, Garry E (2005) Collaborative evaluation of a fluorometric method for measuring alkaline phosphatase activity in cow's, sheep's, and goat's milk. J Food Prot 68:1047–1053

Herrouin M, Mollé D, Fauquant J, Ballestra F, Maubois JL, Léonil J (2000) New genetic variants identified in donkey's milk whey proteins. J Protein Chem 19:105–115

Hofi AA, Mahran GA, Abd El-Hamid LB, Osman SG (1976) Lipase activity in buffalo's milk. 1. Effect of lactation period and individuality. Egypt J Dairy Sci 4:111–114

Humbert G, Chang O, Gaillard JL (2005) Plasmin activity and plasminogen activation in equine milk. Milchwissenschaft 60:134–137

Inglingstad RA, Devold TG, Eriksen EK, Holm H, Jacobsen M, Liland KH, Rucke OE, Vegarud GE (2010) Comparison of the digestion of caseins and whey proteins in equine, bovine, caprine and human milks by human gastrointestinal enzymes. Dairy Sci Technol 90:549–563

Irigoyen A, Izco JM, Ibánez FC, Torre P (2000) Evaluation of the effect of rennet type on casein proteolysis in an ovine milk cheese by means of capillary electrophoresis. J Chromatogr A 881:59–67

Jaeggi JJ, Govindasamy-Lucey S, Berger YM, Johnson ME, McKusick BC, Thomas DL, Wendorff WL (2003) Hard ewes milk cheese manufactured from milk of three different groups of somatic cell count. J Dairy Sci 86:3082–3089

Jaubert G, Bodin JP, Jaubert A (1996) Flavour of goat farm bulk milk. In: Proceedings of the sixth international conference on goats, vol 1. International Academic Publishers, Beijing, pp 66–69

Jauregui-Adell J (1975) Heat stability and reactivation of mare milk lysozyme. J Dairy Sci 58:835–838

Kumari V, Mathur MP (1981) Buffalo milk lysozyme. Indian J Dairy Sci 34:385–390

Kaartinen L, Jensen NE (1988) Use of N-acetyl-beta-glu-cosaminidase to detect teat can inflammations. J Dairy Sci 55:603–607

Kamau PNM, Wangoh J (2010) Use of the lactoperoxidase system to enhance keeping quality of pasteurised camel milk. FoodReview 4:61–63

Kamau PNM, Lamuka P, Wangoh J (2010) Effect of lactoperoxidase-thiocyanate-hydrogen peroxide system and storage temperature on keeping quality of raw camel milk. Afr J Food Agric Nutr Dev 10:4186–4201

Kappeler S (1998) Compositional and structural analysis of camel milk proteins with emphasis on protective proteins. Zurich. Dissertation for the degree of Doctor of Technical Sciences, Swiss Federal Institute of Technology. Diss. ETH No. 12947

Kelly AL, McSweeney PLH (2002) Indigenous proteinases in milk. Adv Dairy Chem 1:494–519

Khalesi M, Salami M, Moslehishad M, Winterburn J, Moosavi-Movahedi AA (2017) Biomolecular content of camel milk: a traditional superfood towards future healthcare industry. Trends Food Sci Technol 62:49–58

Kitchen BJ, Taylor GC, White IC (1970) Milk enzymes – their distribution and activity. J Dairy Res 37:279–288

Klotz V, Hill A, Warriner K, Griffiths M, Odumeru J (2008) Assessment of the colorimetric and fluorometric assays for alkaline phosphatase activity in cow's, goat's, and sheep's milk. J Food Prot 71:1884–1888

Leitner G, Chaffer M, Shamay A, Shapiro F, Merin U, Erza E, Saran A, Silanikove N (2004a) Changes in milk composition as affected by subclinical mastitis in sheep. J Dairy Sci 87:46–52

Leitner G, Merin U, Silanikove N (2004b) Changes in milk composition as affected by subclinical mastitis in goats. J Dairy Sci 87:1719–1726

Liu DK, Williams GH (1982) Species differences in ribonuclease activity of milk and mammary gland. Comp Biochem Physiol Part B Comp Biochem 71:535–538

Lodi R, Todesco R, Carini S (1984) Preliminary study on the use of lysozyme in provolone cheese production. Ind Latte 20:37–50

Lombardi P, Avallone L, D'Angelo A, Mor T, Bogin E (2000) Buffalo-milk enzyme levels, their sensitivity to heat inactivation, and their possible use as markers for pasteurization. J Food Prot 63:970–973

Lorenzen PC, Martin D, Clawin-Rädecker I, Barth K, Knappstein K (2010) Activities of alkaline phosphatase, glutamyltransferase and lactoperoxidase in cow, sheep and goat's milk in relation to heat treatment. Small Rumin Res 89:18–23

Lorenzen PC, Wernery R, Johnson B, Jose S, Wernery U (2011) Evaluation of indigenous enzyme activities in raw and pasteurised camel milk. Small Rumin Res 97:79–82

Madkor SA, Fox PF (1991) Plasmin activity in buffalo milk. Food Chem 39:139–156

Mahdi L, Mahdi N, Al-kakei S, Musafer H, Al-Joofy I, Essa R, Zwain L, Salman I, Mater H, Al-Alak S, Al-Oqaili R (2018) Treatment strategy by lactoperoxidase and lactoferrin combination: Immunomodulatory and antibacterial activity against multidrug-resistant *Acinetobacter baumannii*. Microb Pathog 114:147–152

Maqsood S, Al-Dowaila A, Mudgil P, Kamal H, Jobe B, Hassan HM (2019) Comparative characterization of protein and lipid fractions from camel and cow milk, their functionality, antioxidant and antihypertensive properties upon simulated gastro-intestinal digestion. Food Chem 279:328–338

Marchand S, Merchiers M, Messens W, Coudijzer K, De Block J (2009) Thermal inactivation kinetics of alkaline phosphatase in equine milk. Int Dairy J 19:763–767

Markiewicz-Kęszycka M, Wójtowski J, Kuczyńska B, Puppel K, Czyżak-Runowska G, Bagnicka E, Strzałkowska N, Jóźwik A, Krzyżewski J (2013) Chemical composition and whey protein fraction of late lactation mares' milk. Int Dairy J 31:62–64

Martini M, Altomonte I, Pesi R, Tozzi MG, Salari F (2013) Fat globule membranes in ewes' milk: the main enzyme activities during lactation. Int Dairy J 28:36–39

Martini M, Salari F, Altomonte I, Ragona G, Piazza A, Gori R, Casati D, Brajon G (2018) Effects of pasteurization and storage conditions on donkey milk nutritional and hygienic characteristics. J Dairy Res 85:445–448

Massouras T, Triantaphyllopoulos KA, Theodossiou I (2017) Chemical composition, protein fraction and fatty acid profile of donkey milk during lactation. Int Dairy J 75:83–90

McKellar RC, Emmons DB, Farber J (1991) Gamma-glutamyl transpep-tidase in milk and butter as indicator of heat treatment. Int Dairy J 1:241–251

Medina M, Gaya P, Nunez M (1989) The lactoperoxidase system in ewe's milk: levels of lactoperoxidase and thiocyanate. Lett Appl Microbiol 8:147–149

Merin U, Rosental I, Maltz E (1988) The composition of goat milk as affected by nutritional parameters. Milchwissenschaft 43:363–365

Moatsou G (2010) Indigenous enzymatic activities in ovine and caprine milks. Int J Dairy Technol 63:16–31

Moatsou G, Katsaros G, Bakopanos C, Kandarakis I, Taoukis P, Politis I (2008) Effects of high-pressure treatment at various temperatures on activity of indigenous proteolytic enzymes and denaturation of whey proteins in ovine milk. Int Dairy J 18:1119–1125

Morgan F, Gaspard CE (1999) Influence des cellules somatiques sur les qualities technologiques du lait de chèvre et sur les caractéristiques des fromages de chèvre (Influence of somatic cells on technological properties of goat milk and on characteristics of goat cheese). Renc Rech Rumin 6:317

Morozova-Roche L (2007) Equine lysozyme: the molecular basis of folding, self-assembly and innate amyloid toxicity. FEBS Lett 581:2587–2592

Murgia A, Scano P, Contu M, Ibba I, Altea M, Bussu M, Demuru M, Porcu A, Caboni P (2016) Characterization of donkey milk and metabolite profile comparison with human milk and formula milk. LWT Food Sci Technol 74:427–433

Manjunath GM, Bhat GS (1992) Effect of processing on native proteinases in milk. J Food Sci Technol 29:195–196

Nájera AI, Barron LJR, Barcina Y (1994) Changes in free fatty acids during the ripening of Idiazabal cheese: influence of brining time and smoking. J Dairy Res 61:281–288

Nega A, Moatsou G (2012) Proteolysis and related enzymatic activities in ten Greek cheese varieties. Dairy Sci Technol 92:57–73

Niro S, Fratianni A, Colavita G, Galassi L, Zanazzi M, Salimei E (2017) Technological use of donkey milk in cheesemaking. Int J Dairy Technol 70:439–442

Ozturkoglu-Budak S (2018) Effect of different treatments on the stability of lysozyme, lactoferrin and-lactoglobulin in donkey's milk. Int J Dairy Technol 71:36–45

Pellegrini O, Remeuf F, Rivemale M, Barillet F (1997) Renneting properties of milk from individual ewes: influence of genetic and non-genetic variables, and relationship with physicochemical characteristics. J Dairy Res 64:355–366

Piga C, Urgeghe PP, Piredda G, Scintu M, Sanna G (2009) Asssessment and validation of methods for the determination of glutamyltransferase activity in sheep milk. Food Chem 115:1519–1523

Pilla R, Daprà V, Zecconi A, Piccinini R (2010) Hygienic and health characteristics of donkey milk during a follow-up study. J Dairy Res 77:392–397

Pinho O, Mendes E, Alves MM, Ferreira IM (2004) Chemical, physical, and sensorial characteristics of "Terrincho" ewe cheese: changes during ripening and intravarietal comparison. J Dairy Sci 87:249–257

Pirisi A, Piredda G, Podda F, Pintus S (1996) Effect of somatic cell count on sheep milk composition and cheesemaking properties. In: Somatic cells and Milk of small ruminants. EAAP publication no. 77. Wageningen Press, Wageningen, pp 245–251

Pirisi A, Piredda G, Corona M, Pes M, Pintus S, Ledda A (2000) Influence of somatic cell count on ewe's milk composition, cheese yield and cheese quality. In: 6th proceedings of Great Lakes dairy sheep symposium, Guelph, Canada. Ontario Dairy Sheep Association, Shelburne, ON, pp 47–59

Polidori P, Vincenzetti S (2010) Differences of protein fractions among fresh, frozen and powdered donkey milk. Recent Pat Food Nutr Agric 2:56–60

Politis I, Lachance E, Block E, Turner JD (1989) Plasmin and plasminogen in bovine milk: a relationship with involution. J Dairy Sci 72:900–906

Politis I, Zhao X, McBraid BW, Burton JH, Turner JD (1991) Plasminogen activator production by bovine milk macrophages and blood monocytes. Am J Vet Res 52:1208–1213

Priyadarshini S, Kansal VK (2002a) Lysozyme activity in buffalo milk: effect of lactation period, parity, mastitis, season in India, pH and milk processing heat treatment. Asian-Aust J Animal Sci 15:895–899

Priyadarshini S, Kansal VK (2002b) Purification, characterization, antibacterial activity and Nterminal sequencing of buffalo-milk lysozyme. J Dairy Res 69:419–431

Priyadarshini S, Kansal VK (2003) Biochemical characterization of buffalo (Bubalusbubalis)milk lysozyme. J Dairy Res 70:467–472

Ragona G, Corrias F, Benedetti M, Paladini I, Salari F, Altomonte I, Martini M (2016) Amiata donkey milk chain: animal health evaluation and milk quality. Italian J Food Saf 5:173–178

Raynal-Ljutovac K, Park YW, Gaucheron F, Bouhallab S (2007) Heat stability and enzymatic modifications of goat and sheep milk. Small Rumin Res 68:207–220

Rebucci R, Fusi E, Pecorini L, Pinotti L, Cheli F, Baldi A (2005) Evaluation of the biological activation of plasmin plasminogen system in sheep and goat milk. Italian J Anim Sci 4:330–332

Redwan EM, Almehdar HA, El-Fakharany EM, Baig A-WK, Uversky VN (2015) Potential antiviral activities of camel, bovine, and human lactoperoxidases against hepatitis C virus genotype 4. RSC Adv 5:60441–60452
Revilla I, Rodríguez-Nogales GM, Vivar-Quintana AM (2007) Proteolysis and texture of hard ewe's milk cheese during ripening as affected by somatic cell count. J Dairy Res 74:127–136
Roseiro de Bivar ML, Barbosa M (1995) Phosphatase activity levels in pasteurized goat's milk. J Soc Dairy Technol 48:9–12
Ryskaliyeva A, Henry C, Miranda G, Faye B, Konuspayeva G, Martin P (2018) Combining different proteomic approaches to resolve complexity of the milk protein fraction of dromedary, Bactrian camels and hybrids, from different regions of Kazakhstan. PLoS One 13(5):e0197026
Saadaoui B, Henry C, Khorchani T, Mars M, Martin P, Cebo C (2013) Proteomics of the milk fat globule membrane from Camelus dromedarius. Proteomics 13:1180–1184
Saeman AI, Verdi RJ, Galton DM, Barbano DM (1988) Effect of mastitis on proteolytic activity in bovine milk. J Dairy Sci 71:505–512
Salimei E, Fantuz F (2012) Equid milk for human consumption. Int Dairy J 24:130–142
Salimei E, Fantuz F (2013) Horse and donkey milk. In: Park YW, Haenlein GFW (eds) Milk and dairy products in human nutrition. Wiley-Blackwell, Oxford, pp 594–612
Sanhotra M, Dutta SM (1986) Effect of cold storage, heating and homogenization on xanthine oxidase activity in buffalo milk. Indian J Dairy Sci 39:423–425
Santillo A, Albenzio M (2008) Influence of lamb rennet paste containing probiotic on proteolysis and rheological properties of pecorino cheese. J Dairy Sci 91:1733–1742
Santillo A, Kelly AL, Palermo C, Sevi A, Albenzio M (2009) Role of indigenous enzymes in proteolysis of casein in caprine milk. Int Dairy J 19:655–660
Santillo A, Ruggieri D, Albenzio M (2011) Caratterizzazione chimica, enzimatica, e microbiologica del latte e della mozzarella di bufalo prodotti in provincia di Foggia. Sci Tec Latt Casearia 62:377–386
Šarić LC, Šarić BM, Mandić AI, Kevrešan ZS, Ikonić BB, Kravić SZ, Jambrec DJ (2014) Role of calcium content in antibacterial activity of donkeys' milk toward *E. coli*. Eur Food Res Technol 239:1031–1039
Seifu E, Buys EM, Donkin EF (2005) Significance of the lactoperoxidase system in the dairy industry and its potential applications: a review. Trends Food Sci Technol 16:137–154
Shakeel-Ur-Rehman, Fleming CM, Farkye NY, Fox PF (2003) Indigenous phosphatases in milk. In: Fox PF, McSweeney PLH (eds) Advanced dairy chemistry. Elsevier Applied Science, London, pp 523–543
Sharma R, Kaur S, Rajput YS, Kumar R (2009) Activity and thermal stability of indigenous enzymes in cow, buffalo and goat milk. Milchwissenschaft 64:173–175
Sheikh IA, Jiffri EH, Ashraf GM, Kamal MA, Beg MA (2018) Structural studies on inhibitory mechanisms of antibiotic, corticosteroid and catecholamine molecules on lactoperoxidase. Life Sci 207:412–419
Sheikh IA, Jiffri EH, Ashraf GM, Kamal MA (2019) Structural insights into the camel milk lactoperoxidase: homology modeling and molecular dynamics simulation studies. J Mol Graph Model 86:43–51
Shori AB (2015) Camel milk as a potential therapy for controlling diabetes and its complications: a review of in vivo studies. J Food Drug Anal 23:609–618
Singh R, Mal G, Kumar D, Patil NV, Pathak KML (2017) Camel milk: An important natural adjuvant. Agric Res 6:327–340
Souroullas K, Aspri M, Papademas P (2018) Donkey milk as a supplement in infant formula: benefits and technological challenges. Food Res Int 109:416–425
Srivastava A, Mathur MP, Mudgel VD (1989) Influence of feeding protected fat to buffaloes on some enzymes in milk. Indian J Dairy Sci 42:354–355
Sutherland BJ (2003) Salting of cheese. In: Roginski H, Fuquay JW, Fox PF (eds) Encyclopedia of dairy sciences. MPG Books, Bodmin

Salimei E, Park YW (2017) Mare milk. In: Park YW, Haenlein GFW, Wendorff WL (eds) Handbook of Milk of Non-Bovine Mammals: Second Edition. John Wiley & Sons Ltd, Oxford, pp 369-408

Tayefi-Nasrabadi H, Asdpour R (2008) Effect of heat treatment on buffalo (Bubalus bubalis) lactoperoxidase activity in raw milk. J Biol Sci 8:1310–1315

Tayefi-Nasrabadi H, Hoseinpour-fayzi MA, Mohasseli M (2011) Effect of heat treatment on lactoperoxidase activity in camel milk: a comparison with bovine lactoperoxidase. Small Rumin Res 99:187–190

Theodorou G, Kominakis A, Rogdakis E, Politis I (2007) Factors affecting the plasmin-plasminogen system in milk obtained from three Greek dairy sheep breeds with major differences in milk production capacity. J Dairy Sci 90:3263–3269

Tidona F, Sekse C, Criscione A, Jacobsen M, Bordonaro S, Marletta D, Vegarud GE (2011) Antimicrobial effect of donkeys' milk digested in vitro with human gastrointestinal enzymes. Int Dairy J 21:158–165

Tidona F, Criscione A, Devold TG, Bordonaro S, Marletta D, Vegarud GE (2014) Protein composition and micelle size of donkey milk with different protein patterns: effects on digestibility. Int Dairy J 35:57–62

Trieu-Cuot P, Addeo F (1981) Occurrence of gamma casein in buffalo milk. J Dairy Res 48:311–317

Uniacke-Lowe T, Huppertz T, Fox PF (2010) Equine milk proteins: chemistry, structure and nutritional significance. Int Dairy J 20:609–629

Vamvakaki AN, Zoidou E, Moatsou G, Bokari M, Anifantakis E (2006) Residual alkaline phosphatase activity after heat treatment of ovine and caprine milk. Small Rumin Res 65:237–241

Vincenzetti S, Polidori P, Mariani P, Cammertoni N, Fantuz F, Vita A (2008) Donkey's milk protein fractions characterization. Food Chem 106:640–649

Vincenzetti S, Pucciarelli S, Polzonetti V, Polidori P (2017) Role of proteins and of some bioactive peptides on the nutritional quality of donkey milk and their impact on human health. Beverages 3:34

Vincenzetti S, Cecchi T, Perinelli DR, Pucciarelli S, Polzonetti V, Bonacucina G, Ariani A, Parrocchia L, Spera DM, Ferretti E, Vallesi P, Polidori P (2018) Effects of freeze-drying and spray-drying on donkey milk volatile compounds and whey proteins stability. LWT Food Sci Technol 88:189–195

Visser S (1993) Proteolytic enzymes and their relation to cheese ripening and flavour: an overview symposium: proteolytic enzymes and cheese ripening. J Dairy Sci 76:329–350

Wernery U, Maier U, Johnson B, George RM, Braun F (2006) Comparative study on different enzymes evaluating heat treatment of dromedary milk. Milchwissenschaft 61:281–285

Wernery U, Fischbach S, Johnson B, Jose S (2008) Evaluation of alkaline phosphatase (ALP), γ-glutamyl transferase (GGT) and lactoperoxidase (LPO) activities for their suitability as markers of camel milk heat inactivation. Milchwissenschaft 63:265–267

Wernery U, Wernery R, Masko O, Johnson B, Gnanaraj B, Jose S, Nagy P, Lorenzen PC (2013) Lactoperoxidase: a suitable enzymatic marker of camel milk pasteurization. J Camel Pract Res 20:35–38

Wilińska A, Bryjak J, Illeová V, Polakovič M (2007) Kinetics of thermal inactivation of alkaline phosphatase in bovine and caprine milk and buffer. Int Dairy J 17:579–586

Ying C, Wang HT, Hsu JT (2002) Relationships of somatic cell count, physical, chemical and enzymatic properties to the bacterial standard plate count in dairy goat milk. Livestock Prod Sci 74:63–77

Yoganandi J, Jain AK, Mehta BM, Wadhwani KN, Aparnathi KD (2014) Comparative studies on selected enzymes activities of camel, cow and buffalo milk. J Camel Pract Res 21:249–252

Younan M, Farah Z, Galett V (2006) Application of the activated lactoperoxidase-thiocyanate-hydrogen peroxide system to improve the storage of raw camel milk. Milchwissenschaft 61:391–394

Zamora A, Guamis B, Trujillo AJ (2009) Protein composition of caprine milk fat globule membrane. Small Rumin Res 82:122–129

Chapter 9
The Enzymology of Human Milk

Lauren E. Chan, Robert L. Beverly, and David C. Dallas

9.1 Introduction

Modern mammalian milk has evolved over more than 200 million years to optimally nourish the infant (Oftedal 2002; Lefèvre et al. 2010). Milk provides the infant with essential nutrients, including high quality proteins, lipids and carbohydrates, as well as micronutrients and water. In addition to its nutrient content, milk also contains enzymes that both aid in the breakdown of macronutrients into smaller fragments for easier absorption and provide additional bioactivity and protective effects to the milk. Human milk proteases aid in protein digestion and release protein fragments—peptides—that have an array of bioactivities. Human milk lipases assist with infant lipid digestion, thus increasing absorption and, ultimately, growth. Human milk also contains an array of glycosidases, yet whether these glycosidases have important functions within milk or within the infant is not completely known. Additional enzymes in human milk include phosphatases, antioxidant enzymes and many others that are not well understood. Understanding the unique functions of enzymes in human milk may reveal key insights into improving neonatal health.

9.2 Proteases

Proteases are enzymes that hydrolyze the peptide bond between amino acids within proteins and peptides. Proteases exist in most tissues of the body, and many are found in biological secretions such as milk. Although protease functions in milk have not been fully elucidated, evidence of protease function within the mammary

L. E. Chan · R. L. Beverly · D. C. Dallas (✉)
School of Biological and Population Health Sciences, College of Public Health and Human Sciences, Oregon State University, Corvallis, OR, USA
e-mail: dave.dallas@oregonstate.edu

A. L. Kelly, L. B. Larsen (eds.), *Agents of Change*, Food Engineering Series,
https://doi.org/10.1007/978-3-030-55482-8_9

gland (Nielsen et al. 2017) and milk protease activation within the infant may indicate that some proteases can directly contribute to infant health, development and protection (Dallas and German 2017).

Proteases in bovine milk have been investigated to a larger extent than those in human milk, as variations in protease concentrations in bovine milk can significantly alter dairy products (Kelly et al. 2006; Ismail and Nielsen 2010). For example, high protease activity in milk can increase casein degradation, lowering cheese yields (Barbano et al. 1991), and milk proteases can impact the flavor development during cheese ripening (Farkye and Landkammer 1992). Beyond the effect of proteases on dairy product quality, little is known about how milk proteases function to benefit neonates. Proteases are hypothesized to originate from blood, immune cell secretions or mammary epithelial cells, and function at multiple locations, including within the mammary gland and within the infant gut (Dallas et al. 2015a). As milk is a biologically optimized product created by mammals to help their young survive and develop, understanding the evolutionary importance and function of proteases within the infant may reveal routes for improving neonatal health.

Either the presence, activity or concentration of proteases within human milk have been determined for plasmin, trypsin, kallikrein, cathepsins, elastase, thrombin, amino- and carboxypeptidases and matrix metalloproteinases (Picariello et al. 2012; Molinari et al. 2012; Demers-Mathieu et al. 2017; Palmer et al. 2006; Namachivayam et al. 2013). All of these proteases, except for thrombin and trypsin, have similarly been identified in bovine milk. Table 9.1 lists the activity and concentration of identified proteases in human milk. Details on specific proteases will be provided in the following sections.

9.2.1 Protease Activators

In addition to proteases, milk also contains protease activators, which help regulate total enzymatic activity and peptide release within milk. Protease activators facilitate activation of the inactive zymogen form to the active protease through cleavage of a specific peptide bond within the zymogen (Dallas et al. 2015a). Once cleaved, the protease can then function as an active enzyme. Auto-cleavage of the zymogen form, low pH environments and proteolytic cleavage by other proteases are also means by which some proteases can become activated in human milk. Protease activators within milk offer control mechanisms for proteolytic action, and paired with antiproteases, likely help maintain an appropriate balance of active proteases and inactive zymogens in milk.

Table 9.1 The activity and concentration of enzymes identified in human milk (HM)

	Activity in HM	Reference	Concentration in HM	Reference
Plasmin	*Total plasminogen*			
	12.4 ± 9.9 units H-D-Val-Leu-Lys-l-pNA cleaved/mL (5–6 days)	Korycha-Dahl et al. (1983)	836 ± 128 ng/mL (preterm, <7 week)	Demers-Mathieu et al. (2017)
	Plasmin			
	4.8 ± 4.2 units H-D-Val-Leu-Lys-l-pNA cleaved/mL (5–6 days)	Korycha-Dahl et al. (1983)		
	0.045–0.104 (term, <4 week) 0.147–0.397 (preterm, <4 week) units AMC released/mL/min	Armaforte et al. (2010)		
	5–7 (preterm, <1 month) 5.5–6.5 (term, <1 month) μmol AFC released/L	Namachivayam et al. (2013)		
	146 ± 20 ng AMC released/mL (preterm, <7 week)	Demers-Mathieu et al. (2017)		
Cathepsin D	Not detected (preterm, <7 week)	Demers-Mathieu et al. (2017)	0.8 ± 0.2 ng/mL (preterm, <7 week)	Demers-Mathieu et al. (2017)
Elastase	2.1 ± 0.3 ng methoxy-Suc-Ala-Ala-Pro-Val-l-fluorophore cleaved/mL (preterm, <7 week)	Demers-Mathieu et al. (2017)	382 ± 10 ng/mL (preterm, <7 week)	Demers-Mathieu et al. (2017)
Trypsin	Not detected (<5 week)	Monti et al. (1986)	2.9–5.6 ng/mL (<5 week)	Monti et al. (1986)
Chymotrypsin	N/A		N/A	
Thrombin	71 ± 13 ng Pro-Arg-l-Ser-Phe-fluorogen cleaved/mL (preterm, <7 week)	Demers-Mathieu et al. (2017)	25 ± 7 ng/mL (preterm, <7 week)	Demers-Mathieu et al. (2017)
Kallikrein	16,000 ± 2000 ng His-Asp-Pro-Phe-Arg-l-pNA cleaved/mL (preterm, <7 week)	Demers-Mathieu et al. (2017)	700 ± 100 ng/mL (preterm, <7 week)	Demers-Mathieu et al. (2017)
Amino- and Carboxypeptidase	*Cytosol aminopeptidase*		*Cytosol aminopeptidase*	
	146 ± 20 ng Leu-l-pNA cleaved/mL (preterm, <7 week)	Demers-Mathieu et al. (2017)	1.1 ± 0.1 ng/mL (preterm, <7 week)	Demers-Mathieu et al. (2017)
	Carboxypeptidase B2		*Carboxypeptidase B2*	
	1900 ± 500 ng *N*-benzoyl-Gly-l-Arg cleaved/mL (preterm, <7 week)	Demers-Mathieu et al. (2017)	1574 ± 3 ng/mL (preterm, <7 week)	Demers-Mathieu et al. (2017)
Matrix metalloproteinase	N/A		N/A	

	Activity in HM	Reference	Concentration in HM	Reference
Lipoprotein lipase	0.01–3.32 μmol FFAs released/mL/min (7–380 days)	Constantopoulos et al. (1980)	0.178 mg/mg milk cream powder (<2 month)	Wang et al. (1982)
	12 ± 3 (early preterm, 0.2 week) 20 ± 4 (early preterm, 1 week) 33 ± 9 (early preterm, 3 week) 102 ± 13 (early preterm, 6 week) 101 ± 58 (early preterm, 12 week) 25 ± 6 (late preterm, 0.2 week) 50 ± 11 (late preterm, 1 week) 57 ± 10 (late preterm, 3 week) 128 ± 40 (late preterm, 6 week) 17 ± 9 (term, 0.2 week) 63 ± 15 (term, 1 week) 270 ± 80 (term, 3 week) 106 ± 50 (term, 6 week) 102 ± 60 (term, 12 week) nmol FFAs released/mL/min	Mehta et al. (1981)	100–200 ng/mL (donor milk)	Zechner (1990)
	7.985 μmol FFAs released/mg milk cream powder/h (<2 month)	Wang et al. (1982)		
	414 ± 128 nmol FFAs released/mL/min (<6 month)	(Berkow et al. 1984)		
	0.15 ± 0.080 (early preterm, 1–5 days) 0.66 ± 0.079 (early preterm, 1 week) 0.66 ± 0.079 (early preterm, 3 week) 0.31 ± 0.087 (early preterm, 6 week) 0.27 ± 0.108 (early preterm, 12 week) 0.29 ±0.072 (late preterm, 1–5 days) 0.44 ±0.066 (late preterm, 1 week) 0.49 ± 0.068 (late preterm, 3 week) 0.64 ±0.068 (late preterm, 6 week) 0.72 ±0.077 (late preterm, 12 week) 1.04 ±0.086 (term, 1 week) 1.71 ±0.079 (term, 3 week) 1.68 ±0.082 (term, 6 week) 1.10 ± 0.086 (term, 12 week) μmol FFAs released/mL/min	Freed et al. (1989)		
	51.4–102.8 nmol FFAs released/mL/min (donor milk)	Zechner (1990)		
	125–711 nmol FFAs released/mL/min (<4 month)	Neville et al. (1991)		
	624 ± 146 (<1 month) 534 ± 115 (5–6 month) nmol FFA released/mL/min	Hamosh et al. (1997)		
	1.14–1.22 μmol *p*-nitrophenol released/mL/min (preterm, 1 month)	Bertino et al. (2013)		

	Activity in HM	Reference	Concentration in HM	Reference
Bile salt-stimulated lipase	20–35 μmol FFAs released/mL/min (term & preterm, 0.2–12 week)	Mehta et al. (1982)	100–200 mg/L (<3 month)	Strömqvist et al. (1995)
	5.3 (term, 2 weeks) 4.5 (term, 4 weeks) 4.9 (term, 12 weeks) 3.6 (term, 16 weeks) μmol FFAs released/mL/min	Brown et al. (1983)		
	5.7 ± 0.7 μmol FFAs released/mL/min(<6 month)	Berkow et al. (1984)		
	18.65 ± 4.4 (term, 1–5 days) 17.5 ± 5.7 (term, 6–10 days) 19.8 ± 5.6 (term, >10 days) μmol FFAs released/mL/min	Wardell et al. (1984)		
	20.1 ± 2 (preterm, 3–4 days) 12.2 ±1.4–13.6 ± 1.4 (term, 3–4 days) μmol FFAs released/mL/min	Pamblanco et al. (1987)		
	30.8 ±1.09 (preterm & term, 1–5 days) 43.2 ± 1.04 (term, 1 week) 42.6 ± 1.03 (term, 3 week) μmol FFAs released/mL/min	Freed et al. (1989)		
	9.3 μmol 4-nitrophenol released/mg protein/min (donor milk)	Swan et al. (1992)		
	41 ± 4 (<1 month) 32 ± 7 (5–6 month) μmol FFAs released/mL/min	Hamosh et al. (1997)		
	27.6 ± 13.5–38.8 ±14.3 μmol PNP-butyrate released/mL/min	Torres et al. (2001)		
	31–45 μmol *p*-nitrophenol released/mL/min (preterm, 1 month)	Bertino et al. (2013)		
α-L-Fucosidase	1055 ± 79 nmol 4-methylumbelliferone released/mL/h (52–78 days)	Wiederschain and Newburg (1995)	N/A	
	250–2000 nmol 4-methylumbelliferone released/mL/h (2–9 month)	Wiederschain and Newburg (2001)		
α-1,3/4-Fucosyltransferase	0.0038 nmol fucose transferred/mg milk/min (donor milk)	Eppenberger-Castori et al. (1989)	N/A	
	0.0022–0.0040 nmol fucose transferred/mg milk/min (donor milk)	Prieels et al. (1981)		
N-Acetyl-β-D-hexosaminidase	800–2000 nmol 4-methylumbelliferone released/mL/h (2–9 month)	Wiederschain and Newburg (2001)	N/A	
Neuraminidase	1250–6000 (preterm, <1 month) 3750–8500 (term, <1 month) mU resorufin formed/mL	Namachivayam et al. (2013)	10 μmol	Schauer et al. (1976)

	Activity in HM	Reference	Concentration in HM	Reference
α-Amylase	8973 ± 697 (early preterm, colostrum) 2007 ± 413 (early preterm, 6 weeks) 4422 ± 1000 (late preterm, colostrum) 1462 ± 465 (late preterm, 6 weeks) 3350 ± 889 (term, colostrum) 1373 ± 400 (term, 6 weeks) μmol maltose formed/L/min	Jones et al. (1982)	N/A	
	1.11 (12–91 days) 1.04 (92–182 days) 0.79 (183–274 days) 0.67 (273–365 days) 0.78 (366–456 days) 0.81 (457–547 days) μmol maltotetraose hydrolyzed/mL/min	Dewit et al. (1990)		
Phosphatases	*Alkaline phosphatase*			
	30–540 μg phenol released/mL (donor milk)	Stewart et al. (1958)	N/A	
	4.37 ± 1.37 (3–8 days) 1.13 ± 0.83 (13–50 days) μg phenol released/mL/min	Heyndrickx (1962)		
	Adenosine triphosphatase			
	4.68 ± 1.8 (3–5 days) 5.38 ± 1.94 (7–32 days) μg P released/mL/min (donor milk)	Heyndrickx (1963)		
Xanthine oxidase	0.16–9.2 nmol uric acid formed/mL/h (donor milk, 3–261 days)	Zikakis et al. (1976)	510 μg/mL (term, 3 week) >110 μg/mL (term, 112 days)	Stevens et al. (2000)
	0.02–3.47 nmol uric acid formed/mL/min (2–87 days)	Brown et al. (1995)		
	52–465 nmol NO formed/mL/min (term, 2–112 days)	Stevens et al. (2000)		

	Activity in HM	Reference	Concentration in HM	Reference
Antioxidant enzymes	*Manganese superoxide dismutase*		*Copper-zinc superoxide dismutase*	
	0.32 ± 0.05 (term, 3–5 days) 0.94 ± 0.17 (term, 3 week) 0.13 ± 0.03 (term, 4 month) 0.32 ± 0.08 (term, 7 month) U/mg milk protein	Kasapović et al. (2005)	202 ng/mL (term, 5–10 days) 245 ng/mL (preterm, 5–10 days)	Katzer et al. (2016)
	Copper-zinc superoxide dismutase			
	36.2 ± 16 (term, 1–12 week) 32.2 ± 15 (preterm, 1–12 week) U/mL	L'Abbe and Friel (2000)		
	1.47 ± 0.25 (term, 3–5 days) 3.07 ± 0.3 (term, 3 week) 0.83 ± 0.25 (term, 4 month) 1.44 ± 0.44 (term, 7 month) U/mg milk protein	Kasapović et al. (2005)		
	3.31 ± 0.67 U/mL (term, 7 days)	Ermis et al. (2005)		
	Selenium glutathione peroxidase			
	73 ± 21 (term, 1–12 week) 85.8 ± 29 (preterm, 1–12 week) mmol NADPH oxidized/mL/min	L'Abbe and Friel (2000)		
	Total glutathione peroxidase			
	82.5 ± 26 (term, 1–12 week) 96.9 ± 32 (preterm, 1–12 week) mmol NADPH oxidized/mL/min	L'Abbe and Friel (2000)		
	114.6 ± 7.1 U/L (term, 7 days)	(Ermis et al. 2005)		

Activity is measured by the rate of appearance or disappearance of a substrate, while concentration is the amount of enzyme present in milk. Birth gestational age (early preterm, late preterm or term; donor milk) and day of lactation (days, weeks, or months) of the milk is given when provided in the reference. Values are given per mL of human milk except where otherwise noted. *HM* human milk, *pNA* p-nitroaniline, *AMC* 7-amino-4-methyl coumarin, *AFC* 7-amino-4-(trifluoromethyl) coumarin, *FFA* free fatty acid, *N/A* not analysed

9.2.2 Antiproteases

Antiproteases, which bind to and directly inhibit the activity of proteases, are present in human and bovine milk. These antiproteases contribute to the regulation of milk protease activity (Lindberg et al. 1982; Lindberg 1979; McGilligan et al. 1987). Antiproteases may act on multiple proteases or target specific proteases (D'Alessandro et al. 2011). Additional inhibition of protease activity can take place through protease activator inhibitors, which indirectly inhibit protease activity by inhibiting a protease activator to prevent the conversion of the zymogen to the active form (D'Alessandro et al. 2010). Both antiproteases and protease activator inhibitors provide regulatory action to maintain an appropriate balance of active and inactive proteases in milk.

Proteomic analyses of human milk have revealed the presence therein of α_1-antitrypsin, α_1-antichymotrypsin, α_2-antiplasmin, α_2-macroglobulin, plasma serine protease inhibitor and antithrombin III (Picariello et al. 2012; Molinari et al. 2012; Palmer et al. 2006). Immunoassay-based measurements have also been performed on term and preterm human milk to measure the concentrations of α_1-antitrypsin, α_1-antichymotrypsin, α_2-antiplasmin, antithrombin and plasma serine protease inhibitor (Lindberg et al. 1982; Demers-Mathieu et al. 2017; McGilligan et al. 1987). Evidence of α_1-antitrypsin and α_1-antichymotrypsin transcription within human milk indicates they are expressed by the human mammary gland and secreted into milk (Chowanadisai and Lönnerdal 2002).

One hypothesized function of antiprotease secretion in milk is the specific protection of bioactive proteins such as lactoferrin or immunoglobulins against proteolytic degradation (Hamosh 1998; Chowanadisai and Lönnerdal 2002). Adding α_1-antitrypsin during in vitro digestion of lactoferrin by pancreatin increased retention of intact lactoferrin, indicating potential lactoferrin-protective properties of α_1-antitrypsin in milk (Chowanadisai and Lönnerdal 2002). Lactoferrin also releases fewer peptides during gastric digestion compared with less-abundant milk proteins such as the caseins, osteopontin and polymeric immunoglobulin receptor (Beverly et al. 2019; Nielsen et al. 2018). As human milk α_1-antitrypsin can survive proteolysis by infant digestive enzymes (Chowanadisai and Lönnerdal 2002; Demers-Mathieu et al. 2018), it could prevent the degradation of lactoferrin and other bioactive proteins within the infant gastrointestinal tract.

Another potential benefit of antiproteases may be the prevention of early involution of the mammary gland, or premature shrinkage of the mammary gland prior to the discontinuation of nursing. Within bovine milk, increased plasmin activity was correlated with early involution (Politis et al. 1989a). During late lactation, plasminogen activation to plasmin increases immediately before involution. In theory, the presence of antiproteases may decrease protease activity within the mammary gland and could prevent premature tissue breakdown (Heegaard et al. 1994a; Politis et al. 1989b).

9.2.3 Protease Activity in the Mammary Gland

Milk protein digestion is initiated by proteases in milk within the mammary gland (Nielsen et al. 2017). Recent investigations of human milk indicate that protease activity is targeted, degrading specific proteins (e.g., casein proteins, polymeric immunoglobulin receptor, osteopontin), while leaving others intact (e.g., lactoferrin, immunoglobulins) (Dallas et al. 2013a). Proteins that remain intact are likely more biologically beneficial to the infant as intact proteins (Khaldi et al. 2014). Plasmin, carboxypeptidase B2, cytosol aminopeptidase, cathepsin D and elastase were predicted through peptidomic analysis and bioinformatics to actively digest proteins within the mammary gland in both preterm and term human milk (Dallas et al. 2015b; Guerrero et al. 2014). In addition to some intact proteins that are bioactive, peptides released in the mammary gland can also possess a multitude of functions beneficial to the infant, such as antimicrobial, ACE-inhibitory, cell-proliferative and immunomodulatory functions (Dallas et al. 2013a; Guerrero et al. 2014; Nielsen et al. 2017). However, despite specific protein digestion and peptide release within the mammary gland, the majority of milk proteins are intact in human milk (Dallas et al. 2013a).

General protease activity within expressed human milk declines with increased lactation time. Enzymatic protein hydrolysis in milk decreases as the infant's own digestion capabilities improve over time at digesting protein and releasing amino acids (Lindberg et al. 1982; Britton and Koldovsky 1989). Further information regarding protease activity across lactation is provided in the following sections.

9.2.4 Protein Digestion in Human Neonates

Protein hydrolysis in term and preterm infants is often much lower than the digestive capacity of older children or adults due to an elevated gastric pH and limited gastric protease function. However, while protein digestion may be limited for neonates, proteases present in human milk may offer a compensatory mechanism for protein breakdown in the infant.

Within term infants, production of gastric acid is significantly lower than in adults initially after birth but increases shortly thereafter (Tayman et al. 2011). Increased gastric pH can limit protein denaturation (loss of secondary and tertiary protein structure), leading to lower protein digestion capacity (Kauzmann 1959). Investigations have indicated that only a small amount of gastric proteolytic enzymes are secreted by neonates and that these gastric-secreted enzymes are inactivated at high gastric pH levels, thus potentially reducing protein digestion in the infant (Mason 1962). For example, pepsin, the predominant gastric protease, performs optimally at a low pH and can have limited function at the more neutral gastric pH of an infant (Dallas et al. 2012). Cathepsin D from human milk is also activated within the infant stomach, hydrolyzing proteins at the gastric pH of premature infants (Demers-Mathieu et al. 2018).

The function of enterokinase, a protease that converts trypsinogen to trypsin in the small intestine (Tarlow et al. 1970), is also limited in the neonate, having approximately 20% of the enterokinase function of older children (Antonowicz and Lebenthal 1977). In newborns, pancreatic chymotrypsin activity is 10–60% of that of adults (Lebenthal and Lee 1980) and carboxypeptidase B is 10–25% of that of 2-year-olds (Lebenthal and Lee 1980). Trypsin concentration in the duodenum is also reduced in the first month postpartum (Lebenthal and Lee 1980), contributing further to the limited digestion capacity of term neonates.

Within the small intestine and colon, brush border enzymes are necessary for hydrolysis of peptides into smaller peptides or amino acids. An investigation using fetal intestinal samples determined that brush border enzymes were present in the fetus, with γ-glutamyl-transpeptidase being found at higher activity than in adults, dipeptidyl- aminopeptidase IV and carboxypeptidase at comparable activity to that seen in adults, and the only significantly depleted enzyme being aminopeptidase A (Auricchio et al. 1981).

A potential consequence of incomplete protein digestion and limited protein absorption is bacterial protein-fermentation (putrefaction) within the colon. Putrefaction in the colon may lead to microbial metabolites such as hydrogen sulfide that can increase epithelial permeability (Hughes et al. 2008; McCall et al. 2009), reduce colonic mucus layer thickness (Toden et al. 2007), lead to inflammatory reactions (Hamer et al. 2012; Windey et al. 2012), provoke colonic epithelial cell DNA damage (Attene-Ramos et al. 2006) and reduce colonic epithelial cell viability (Pedersen et al. 2002). Microbial metabolites from protein fermentation are also evident in formula-fed infant fecal samples (Chow et al. 2014). Protein fermentation in formula-fed infants may be related to a higher protein content in infant formula than in human milk (Raiten et al. 1998) or to increased amino-acid fermenting bacterial species present in formula-fed infant colons (Harmsen et al. 2000).

Preterm infants may have even more difficulty digesting protein than their term counterparts. Preterm infants produce less gastric acid than term infants (Miclat et al. 1978; Kelly et al. 1993) and have a higher gastric pH between 3.9 and 5.3 (Palla et al. 2018; Demers-Mathieu et al. 2018). Premature infants also have decreased gastrointestinal trypsin activity, with trypsin concentration decreasing throughout the duration of digestion (Borgström et al. 1960; Engberg et al. 1999; Demers-Mathieu et al. 2018). Reduced pepsin and overall protease activity are also present in preterm infants compared with term infants.

Although preterm infants experience significant limitations in protein digestion based on their altered enzymatic abilities, milk expressed by mothers of premature infants contains proteases (Demers-Mathieu et al. 2017). Peptidomic-based enzyme activity analysis also predicts that activity of proteases in human milk, such as plasmin, kallikrein and thrombin, are higher in that of mothers of preterm infants, compared with term, based on peptide abundance (Demers-Mathieu et al. 2017).

9.2.5 Proteolytic Systems and Peptides

The combination of proteases, protease activators and antiproteases contributes to the balance of proteolytic activity, and helps mediate the protein content and release of peptides within milk. Although the precise purpose of proteolytic systems remains undefined, we hypothesize that the balance of proteolytic action is important for preserving some whole proteins with bioactive functions, while breaking down other proteins to amino acids for nourishment or to function as peptides.

Recent literature has focused on the functional properties of peptides released from milk proteins. Milk peptides may offer distinct benefits to the infant, but they must first be cleaved from their parent milk proteins by the activity of milk and digestive proteases. Peptides have been identified with a multitude of bioactive functions (Clare and Swaisgood 2000), including immunomodulation (Lahov and Regelson 1996), antimicrobial activity (Lahov and Regelson 1996; Zucht et al. 1995; Wakabayashi et al. 2003), antiviral activity (Mistry et al. 2007), commensal bacterial growth promotion (Liepke et al. 2002), aiding calcium absorption (Sato et al. 1986), decreasing diarrheal effects (Daniel et al. 1990), opioid functions (Ferranti et al. 2004) and stimulating hormonal secretions (Yen et al. 1985). Milk peptides may offer systemic functions to the infant, including immunomodulation by β-casokinins, which promotes immune enhancing effects through the inactivation of bradykinin (Clare and Swaisgood 2000). Targeted gut functions of peptides are also recognized, such as the influence of casomorphins on prolonged gastrointestinal transient time through inhibition of intestinal peristalsis and motility (Tome et al. 1987).

Identification of peptides in breast milk has been achieved by separating the peptides from the intact protein, lipid and carbohydrate content, and identifying peptide sequences through mass spectrometry. Peptidomic studies have determined the peptide profile of human milk (Dallas et al. 2013a; Nielsen et al. 2017), infant digested human milk (Nielsen et al. 2018; Beverly et al. 2019; Dallas et al. 2014), bovine milk (Dallas et al. 2013b) and bovine milk products (Sforza et al. 2012; Lambers et al. 2015). Peptide bioactivity can then either be directly measured or predicted *via* database comparisons and functional prediction algorithms (Dallas et al. 2014; Holton et al. 2014).

Some researchers have postulated that the balance of proteolytic activity has evolved to maximize benefits to infants by providing whole milk proteins and peptides (Dallas et al. 2015a). Developing novel methodologies and investigating the balance of multiple proteolytic systems will reveal more about the biological benefits of human milk composition.

9.2.6 *Specific Protease Systems*

9.2.6.1 Plasmin System

The plasmin system is the most well-studied proteolytic system in milk. Plasmin is an active serine protease in both bovine and human milks (as well as that of other mammalian species) that is initially derived from its zymogen form, plasminogen (Grufferty and Fox 1988; Ferranti et al. 2004; Heegaard et al. 1997; Korycha-Dahl et al. 1983). Plasminogen, plasmin, plasmin activators and plasmin inhibitors are present in human and bovine milks. Cleavage of plasminogen at the *C*-terminal side of an arginine residue at position 560 (human) or 557 (bovine) creates active plasmin (Schaller et al. 1985). Activation of plasmin in bovine milk is initiated by two serine proteases: urokinase-type plasminogen activator (uPA) and tissue-type plasminogen activator (tPA) (Heegaard et al. 1994b; Korycha-Dahl et al. 1983; Wickramasinghe et al. 2012). Similarly, these two plasminogen activators are present in human milk (Heegaard et al. 1997). Although these activators allow for conversion of plasminogen to plasmin, plasmin inhibitors are present in milk to regulate plasmin activation.

The plasmin present in milk is the same as the enzyme in the blood, and it crosses into the milk as both active and zymogen forms through paracellular routes (Bastian and Brown 1996). Mammary epithelial cells secrete plasminogen activators (Heegard et al. 1994), which can activate plasminogen to contribute to the total plasmin activity of milk. Both plasmin and activated plasminogen activities in human milk have been quantitatively determined by protease assays with a fluorescent substrate (Armaforte et al. 2010; Demers-Mathieu et al. 2017). Plasmin concentrations have been determined by ELISA assays (Demers-Mathieu et al. 2017). Measured plasmin activities and concentrations are listed in Table 9.1. Human plasmin can digest human β-casein into peptides and amino acids, as measured by HPLC analysis (Azuma and Yamauchi 1992). Peptidomic-based bioinformatic analyses of human milk predict that plasmin degrades all human milk caseins, as well as major whey proteins such as polymeric immunoglobulin receptor and osteopontin (Khaldi et al. 2014; Nielsen et al. 2017).

The presence and activity of uPA in human milk was identified through immunoblotting and zymography analysis of extracts of milk cells (Heegaard et al. 1997). Identification and activity studies of uPA in bovine milk have also been conducted (Heegaard et al. 1994b). An additional plasmin activator, tPA, was identified in the casein micelle of human milk through immunoblotting and zymographic analyses, with distribution of uPA and tPA investigated through ligand binding assays (Heegaard et al. 1997). In human milk, mRNA for tPA was identified (Wickramasinghe et al. 2012), yet evidence of the presence and activity of tPA has not been verified through either ELISA or enzyme activity assays.

Plasmin activity in the milk is balanced by inhibition of plasminogen activation by plasma activator inhibitors (PAI) and inhibition of active plasmin by plasmin inhibitors. The primary PAI in both bovine and human milk is type-1 PAI (Heegaard et al. 1997, 1994b; Precetti et al. 1997). Type-1 PAI is found in blood, is secreted by mammary epithelial cells (Heegard et al. 1994) and inhibits both uPA and tPA. Lactoferrin, one of the most abundant whey proteins in human milk, can also block plasminogen activation on the cell surface through direct binding to human plasminogen (Zwirzitz et al. 2018).

The major plasmin inhibitors in human milk are α_1-antitrypsin (also known as α_1-antiproteinase, SERPINA1) (D'Alessandro et al. 2010) and α1-microglobulin/bikunin precursor (AMBP) (D'Alessandro et al. 2010). α_2-Antiplasmin (a major bovine plasmin inhibitor) was identified at low concentrations in human milk (<500 ng/mL) (Demers-Mathieu et al. 2017). α_1-Antitrypsin decreases across the first 2 weeks postpartum within both preterm and term human milk (Lindberg et al. 1982; McGilligan et al. 1987). Electroimmunoassay measurements indicated a similar decrease in α_1-antichymotrypsin within human milk from an average of 0.67 g/L on day 1 to 0.012 g/L after day 14 (Lindberg et al. 1982). Within bovine milk, α_1-antitrypsin was identified but its concentration was not determined (Weber and Nielsen 1991). In bovine colostrum, α_2-antiplasmin and other antiproteases that included α_2-macroglobulin, C1-inhibitor, inter-α-trypsin inhibitor, bovine plasma elastase inhibitor, bovine plasma trypsin inhibitor and antithrombin III were identified and quantified via ELISA (Christensen et al. 1995).

With an array of activators and inhibitors being present in milk, the plasmin proteolytic system maintains a potentially adaptive balance of active plasmin and zymogen forms. Throughout lactation, plasmin activity changes, potentially based on biological needs of the infant and mother. In bovine milk, plasmin activity increases across lactation (Benslimane et al. 1990). Unlike in bovine milk, human milk plasmin activity appears to decrease across lactation (Dallas et al. 2015b), with noted decreases in α_1-antitrypsin concentrations (Lindberg et al. 1982). Notably, preterm human milk displays increased plasmin activity as compared with term milk as determined by fluorescence-based protease analyses (Armaforte et al. 2010; Greenberg and Groves 1984) and peptidomic-based analyses (Dallas et al. 2015b).

Within both human and bovine milk, the majority of plasminogen is bound to casein micelles (Heegaard et al. 1994b, 1997). tPA also binds primarily to casein micelles as opposed to the whey fraction (Heegaard et al. 1997). The presence of casein micelles in milk thus increases the activation of plasmin by bringing plasminogen and its activator into close proximity (Heegaard et al. 1997; Politis et al. 1995; Markus et al. 1993). Although activity of plasminogen activators is increased around casein micelles, degradation of the micelle is mitigated through the presence of PAIs within the micelle (Heegaard et al. 1994b). Unlike tPA, uPA does not bind to casein micelles but is instead found within the somatic cell fraction of milk (White et al. 1995; Heegaard et al. 1994b; Politis et al. 2003, 2002).

The association of plasmin with the casein micelle may contribute to the predominance of β-casein-derived peptides among protein-derived peptides in human milk, despite β-casein not being the most abundant protein (Dallas et al. 2013a, b; Nielsen et al. 2017; Beverly et al. 2019; Guerrero et al. 2014). The location of β-casein in the casein micelle may increase its rate of degradation due to the higher amounts of activated plasmin around the micelle. Conversely, whey proteins with bioactivities beneficial to the infant, such as α-lactalbumin, immunoglobulins and lactoferrin, release far fewer peptides in human milk than caseins, despite being present at higher concentrations. The limited contact of whey proteins with plasmin, in addition to their strong disulfide bonds and globular structure (Permyakov and Berliner 2000; Legrand et al. 2008), may lead to higher resistance to digestion, thus increasing their likelihood of survival within the infant gastrointestinal tract.

9.2.6.2 Cathepsin System

The cathepsin family is a large group of proteases that vary widely in function. The cathepsin family contains aspartic, serine and cysteine proteases (Bergmann and Fruton 1936). Cathepsins are typically active at a low pH, and often within the lysosomes of cells (Bergmann and Fruton 1936). Within human milk, multiple cathepsins have been identified, including cathepsin D (Větvicka et al. 1993), B (Lemay et al. 2013), H (D'Alessandro et al. 2010) and S (D'Alessandro et al. 2010).

Cathepsin D is the only cathepsin known to have measurable hydrolytic activity in the infant stomach once exposed to gastric pH (Demers-Mathieu et al. 2018). The cleavage sites for cathepsin D are between two hydrophobic amino acids with preference for Leu and Phe (Pimenta et al. 2001). Cathepsin D is a soluble lysosomal aspartic endopeptidase that is synthesized as the zymogen procathepsin D (Benes et al. 2008). Cathepsin D is most predominantly identified as inactive procathepsin D (Larsen and Petersen 1995; Větvicka et al. 1993), as limited to no active cathepsin D has been identified in the whey portion of fresh human (Palmer et al. 2006; Větvicka et al. 1993; Demers-Mathieu et al. 2018) or bovine milk (Kaminogawa and Yamauchi 1972).

Both activated lymphocytes (Hasilik 1992) and mammary epithelial cells (Capony et al. 1989; Radisky 2010) can secrete procathepsin D. Separate lysosomal cysteine proteases are necessary for activation of procathepsin D to cathepsin D (Wittlin et al. 1999), but no specific inhibitors or activators have been identified that regulate the balance of procathepsin D and cathepsin D within milk. However, under acidic conditions, procathepsin D can auto-cleave into an active pseudo-cathepsin D (Hasilik et al. 1982). While it may be possible for some active cathepsin D to leak into milk from the lysosomes of dead lymphocytes, cathepsin D activity is only detectable once the milk has reached the infant stomach and the procathepsin D has been activated by gastric acid (Demers-Mathieu et al. 2018).

Cathepsin B, a cysteine protease, was identified in bovine milk (Magboul et al. 2001), and is similarly present in human milk (Lemay et al. 2013). Cathepsin B is most active at pH 6.0 and is inactivated at > pH 7.0 (Kirschke et al. 1998). When active, cathepsin B can cleave β-casein and α_{s1}-casein, with a preference for cleavage sites with Leu, Val, Gln, Pro or Ser residues (Considine et al. 2004).

9.2.6.3 Elastase System

Neutrophil elastase is a serine protease present in milk with the ability to digest many proteins (Borulf et al. 1987). The inactive zymogen form of neutrophil elastase, proelastase, has not been identified in human milk. There is some evidence that neutrophil elastase originates from mammary epithelial cells (Borulf et al. 1987), but it may also be a lysosomal protease derived from milk immune cells (Kelly et al. 2006). Neutrophil elastase has site-specificity for peptide bonds between uncharged, non-aromatic amino acids (e.g., Ala, Val, Leu, Ile, Gly, Ser) (Naughton and Sanger 1961). Protease activity assays have determined that neutrophil elastase is active in human milk (Borulf et al. 1987; Demers-Mathieu et al. 2017). Previous peptidomic and bioinformatic analyses predicted that neutrophil elastase contributes to the digestion of milk proteins within the mammary gland (Khaldi et al. 2014; Nielsen et al. 2017). No elastase activators have been found in human milk. Some elastase inhibitors have been identified, including antielastase in human milk (Lindberg 1979) and plasma elastase inhibitor in bovine milk (Christensen et al. 1995). Inter-α-trypsin inhibitor, derived from AMBP (Salier et al. 1996), is present in bovine milk and can inhibit polymorphonuclear lysosomal granulocytic elastase (Jochum and Bittner 1983). Although AMBP has been identified in human milk, inter-α-trypsin inhibitor has yet to be identified. α_1-Antitrypsin, an inhibitor of trypsin activation in both bovine (D'Alessandro et al. 2011) and human milk (McGilligan et al. 1987), also inhibits neutrophil elastase activity (Janoff 1972; Chowanadisai et al. 2003). Further exploration of the neutrophil elastase system is needed to better understand the presence and function of elastase and its proteolytic control across lactation.

9.2.6.4 Trypsin System

Trypsin is present in human milk but not bovine milk (Monti et al. 1986; Tacoma et al. 2016). Trypsin cleaves on the *C*-terminal side of Lys or Arg residues (Borulf et al. 1987; D'Alessandro et al. 2011). Trypsin was originally identified in human milk through immunodiffusion and immunoelectrophoresis (Monti et al. 1986). This version of trypsin was known as anionic trypsin until amino acid sequence comparison revealed that it was identical to trypsin 2 (Dallas et al. 2015a). Trypsin activity in milk was predicted by the demonstrated release of milk peptides with Lys and Arg *C*-termini, although some of its activity can overlap with that of plasmin and kallikrein (Dingess et al. 2017). Although trypsin has been identified in human milk through radioimmunoassay, its zymogen form, trypsinogen, has yet to be identified.

Trypsin has a wide array of inhibitors, many of which are common to both human and bovine milk (Heyndrickx 1963). Specifically, α_1-antitrypsin has been identified in human milk from mothers delivering at both preterm and term (Chowanadisai and Lönnerdal 2002; McGilligan et al. 1987; Nielsen et al. 2018), as well as in bovine milk (Mattila et al. 1985). α_1-Antitrypsin is synthesized and secreted by mammary epithelial cells into the milk (Chowanadisai and Lönnerdal 2002). α_1-Antitrypsin can block a variety of proteases, including trypsin, chymotrypsin and elastase, through direct binding (Gettins 2002; Kalsheker 1989).

Although trypsin 2 has been identified in milk, there is little evidence that it actively digests proteins (Monti et al. 1986). The large amount of trypsin inhibitors present in milk likely limits trypsin activity, and this balance could protect bioactive milk proteins from digestion until they reach the infant gastrointestinal tract. Although the level of α_1-antitrypsin itself decreases across lactation (Lindberg et al. 1982; Chowanadisai and Lönnerdal 2002), little is known about the activity of the entire trypsin system in milk across lactation.

9.2.6.5 Chymotrypsin

Neither active chymotrypsin nor its zymogen chymotrypsinogen have been identified in either human or bovine milk. While there remains no antibody-based or proteomic evidence of this protease in milk, there remains a partial chymotrypsin system including chymotrypsin inhibitors. Starting with the first day of lactation, milk from term- and preterm-delivering mothers contains the serpin antiprotease α_1-antichymotrypsin (SERPINA3), up to at least 4 months postpartum and decreasing with continued lactation (Lindberg et al. 1982; Chowanadisai and Lönnerdal 2002; Demers-Mathieu et al. 2017). As with other serpins, α_1-antichymotrypsin limits chymotrypsin activation through direct binding (Beatty et al. 1980; D'Alessandro et al. 2011, 2010). α_1-Antitrypsin also inhibits chymotrypsin, with higher specificity than α_1-antichymotrypsin (Beatty et al. 1980). As chymotrypsin has not been identified, chymotrypsin inhibitors may offer inhibitory mechanisms for other serine proteases in milk such as elastase or trypsin.

9.2.6.6 Thrombin

The presence of the zymogen form of thrombin, prothrombin, has been confirmed in human milk via several proteomics studies (Molinari et al. 2012; Palmer et al. 2006). In one study, active thrombin was detected in human milk at a low activity level (Demers-Mathieu et al. 2017). No thrombin activators have been identified through proteomic analysis of human or bovine milk. A variety of protease inhibitors are present in milk which can suppress the activity of thrombin; α_1-antitrypsin can inhibit thrombin activity (D'Alessandro et al. 2010), and inter-α-trypsin likely inhibits thrombin as well. Proteomic studies also revealed the presence of thrombin-specific antiproteases: antithrombin III (SERPINC1) and thrombin inhibitor (SERPIND1) in bovine milk (D'Alessandro et al. 2011), and antithrombin III in human milk (D'Alessandro et al. 2010; Palmer et al. 2006).

9.2.6.7 Kallikrein

Kallikrein is a serine protease that, in blood, plays a role in coagulation, fibrinolysis and kinin formation (Schapira et al. 1982). This protease is found in both bovine (Heegaard et al. 1994b) and human milk (Palmer et al. 2006). Proteomic analysis of human colostrum identified kallikrein 6 and kallikrein 11 in either their zymogen or active forms (Palmer et al. 2006).

In human milk, active kallikrein was identified but had minimal proteolytic activity in the infant stomach (Demers-Mathieu et al. 2018). Active kallikrein has not been identified in bovine milk. Bovine milk contains the protease inhibitors kallistatin (SERPINA4), plasma serine protease inhibitor (SERPINA5) and plasma protease C1 inhibitor (SERPING1), each of which inhibits kallikrein activity (D'Alessandro et al. 2011; Chen et al. 2000). Plasma protease C1 inhibitor is also present in blood and regulates kallikrein activation during the blood clot formation process (Schapira et al. 1982). Kallistatin (SERPINA4) activity was identified in human milk (Zhou et al. 1992), but limited studies have taken place to identify other kallikrein inhibitors.

9.2.6.8 Amino- and Carboxypeptidase Systems

Amino- and carboxypeptidases are proteases that cleave amino acids from the *N*- and *C*-terminals of proteins, respectively. Investigations into the presence of amino- and carboxypeptidases via proteomics and ELISA-based assays identified their presence in both human and bovine milk (Demers-Mathieu et al. 2017). Specific proteases in human milk include cytosol aminopeptidase and carboxypeptidase B2 (Molinari et al. 2012), both of which are also present in bovine milk (D'Alessandro et al. 2011). Although proteomics offers a means to identify these enzymes, such approaches are limited in their ability to distinguish active versus zymogen forms of these proteases. Thrombin has been indicated as an activator of carboxypeptidase B2 (Valnickova et al. 2007), yet only one study has identified active thrombin at low concentrations in human milk (Demers-Mathieu et al. 2017), which may be a limiting factor of carboxypeptidase B2 activation. Cleavage site analysis of peptide sequences in human milk documents some evidence of aminopeptidase and carboxypeptidase B2 activity within the mammary gland (Nielsen et al. 2017; Dallas et al. 2015b).

9.2.6.9 Matrix Metalloproteinase System

Matrix metalloproteinases (MMP), which are integral proteases for the tissue remodeling process through extracellular matrix protein degradation, have been identified in human milk. MMP-2 and MMP-9, or the "gelatinases," are known for their unique ability among MMPs to degrade both collagen and gelatin. Both MMP-2 and MMP-9 have been detected in human milk from preterm and term-delivering mothers (Cheung et al. 2001; Lubetzky et al. 2010). MMP-2 and MMP-9 have gelatinolytic activity in milk that is similar between term and preterm (Namachivayam et al. 2013). Proteomic analysis of human milk identified tissue inhibitor of metalloproteases 4 (TIMP-4), an inhibitor of MMP-4 (Picariello et al. 2012). This inhibitor is more abundant in preterm human milk than in term human milk (Lubetzky et al. 2010).

9.3 Lipases

A high percentage of the caloric content of human milk (~50%) comes in the form of milk fat (Martin et al. 2016). To absorb this energy source, the infant must digest the fat via gastrointestinal lipases. However, term neonates often have insufficient fat digestion capacity. Compared with healthy human adults with efficient fat absorption (around 95% of dietary fat), infants experience significant deficits in fat metabolism, with the most affected group being preterm infants, who experience 20–30% lower fat absorption (Lindquist and Hernell 2010; Hamosh et al. 1981). Proposed mediators of lower fat digestion include decreased activity levels of pancreatic lipase and lower concentrations of bile salts (Fredrikzon et al. 1978; Norman et al. 1972; Zoppi et al. 1972), as well as a high proportion of fat in human milk (Fredrikzon et al. 1978), leading to higher demand for enzymes needed to digest fat. With high meal frequency and expedited transit time through the gut compared with those of adults, infants have reduced opportunity for complete nutrient absorption (Lindquist and Hernell 2010).

Limited fat digestion and absorption can be detrimental to infants during their initial growth and development periods, but compensatory methods have evolved in human milk to support the infant's digestive deficiencies. In particular, two known lipases (lipoprotein lipase and bile salt-stimulated lipase) are expressed and active in human milk and the infant gastrointestinal tract, and are major contributors to the infant's fat digestion. The activity and concentration of these two lipases in human milk are listed in Table 9.1. Human milk lipases benefit the infant by improving linear (length) growth, weight gain and brain development, as a result of increased lipid metabolism (Lindquist and Hernell 2010).

9.3.1 Lipoprotein Lipase

Lipoprotein lipase (LPL) is a glycoprotein with two *N*-linked glycans that is present in both human and bovine milk (Olivecrona et al. 2003). LPL is present in the mammary gland capillaries, where it cleaves lipids from chylomicrons to release free fatty acids for synthesis into milk lipids by mammary epithelial cells (Mendelson et al. 1977). The presence of LPL in milk is likely due to paracellular transport from the bloodstream and rupture of interstitial cells during milk expression (Freed et al.

1989; McBride and Korn 1963; Deeth 2006). LPL is typically associated with the casein micelle in milk due to its hydrophobic nature (Freed et al. 1989).

LPL is present and active in human milk (Neville et al. 1991), but has little lipolytic activity compared with bile salt-stimulated lipase, likely due to the absence of apolipoprotein C2 (a required activator for LPL) in human milk and the presence of inhibitors such as proteose-peptone component 3 (Anderson 1981; Girardet et al. 1993). Within bovine milk, LPL may play a role in hydrolytic rancidity, specifically initiated when the milk fat globule membrane has been agitated or homogenized, leading to physical damage to the membrane barrier and increasing lipase exposure to the fat substrate (Deeth 2006). While LPL may increase hydrolytic rancidity within milk, LPL is unstable to heat and can be largely inactivated with high-temperature, short-time pasteurization (Deeth 2006). The role of LPL in lipid hydrolysis could explain why frozen human milk develops rancidity (Hung et al. 2018), as freezing disrupts the milk fat globule membrane and can expose lipids to LPL activity. Regardless of control mediators, LPL offers minimal biological function in unprocessed bovine or human milk.

9.3.2 Bile Salt-Stimulated Lipase

Bile salt-stimulated lipase (BSSL) is a lipase in human milk with broad substrate specificity that plays a major role in the degradation of milk fats (Fredrikzon et al. 1978). Human milk has much higher lipolytic activity than bovine milk, which is related to the presence of BSSL (Palmer 1922). BSSL has an amino acid sequence identical to that of pancreatic carboxylic ester hydrolase, but it has a different glycosylation pattern (Nilsson et al. 1990; Landberg et al. 2000) and functions in a similar manner to carboxylic ester hydrolase as a broad-specificity lipase. BSSL is active in human milk, where it is capable of hydrolyzing mono-, di- and triglycerides, cholesteryl-, retinyl- and other fat-soluble vitamin esters, and phospholipids and ceramides (Bläckberg and Hernell 1983; Andersson et al. 2011). The activity of BSSL is significantly increased by primary bile salts once milk reaches the infant duodenum (Swan et al. 1992), while phospholipids have the opposite effect by reducing BSSL activity (Venuti et al. 2017). Two active forms of BSSL have been identified in the milk of some mothers via HPLC analysis, with similar levels of activity (Swan et al. 1992). It is unknown whether the alternative form is a shortened version of BSSL, a result of alternative splicing of mRNA, or a co-dominant allele.

As neonates are not well-equipped to digest fats after birth due to limited pancreatic lipase secretion (Manson and Weaver 1997), there is an evolutionary benefit to the presence of BSSL in human milk to improve overall lipid utilization in infants (Fredrikzon et al. 1978). BSSL activity levels in human milk are similar for preterm- and term mothers (Freed et al. 1989), and are present in colostrum within the first few days postpartum, offering increased lipid hydrolytic activity (Pamblanco et al. 1987). BSSL concentration and activity declines across lactation for both preterm- and term-delivering mothers (Brown et al. 1983; Torres et al. 2001; Freed et al. 1989), concurrently with the infant's own pancreatic lipase development. Glycosylation of BSSL also decreases across lactation, but the effect of this on its overall activity is minimal (Froehlich et al. 2010; Landberg et al. 2000).

Although BSSL is a critical component of neonatal lipid digestion, it is inactivated if subjected to processing methods such as pasteurization. Feeding preterm infants pasteurized human milk with heat-inactivated BSSL decreased lipid absorption by 17% in preterm infants, and 30% in term infants (Bläckberg and Hernell 1981; Andersson et al. 2007). This finding poses a critical challenge for human milk banks, where pasteurization is a standard protocol to reduce risk of milk contamination by infectious agents (Wight 2001). Donor milk banks are a good source of human milk for infants with mothers who are unable to produce milk in sufficient amounts; however, the increased degradation of BSSL through pasteurization likely reduces overall fat metabolism in neonates and may inhibit appropriate growth and development. Due to concerns with BSSL inactivation through pasteurization, further investigations are needed to identify processing methods for donor milk that can keep BSSL intact while still reducing infection risk.

9.4 Glycosidases

Human milk contains a variety of unique glycosidases that catalyze the hydrolysis of glycosidic bonds between oligosaccharides in free glycan structures and within the oligosaccharide portion of other complex protein-linked glycan (PLG) bond structures of N- or O-linked glycoproteins. While digestive amylases do not break down human PLGs (Dewit et al. 1990), three human milk glycosidases, including α-L-fucosidase (Wiederschain and Newburg 1996), *N*-acetyl-β-D-hexosaminidase (Wiederschain and Newburg 2001; Dudzik et al. 2008) and neuraminidase (sialidase) (Schauer et al. 1976), can interact with and break down human milk oligosaccharides and PLGs. All identified human milk glycosidases and their activities and concentrations are listed in Table 9.1.

9.4.1 α-L-Fucosidase

α-L-Fucosidase is a lysosomal glycosidase that cleaves α-1,6-linked fucose from the terminal ends of fucosylsaccharides (Bathula et al. 2017). Ultracentrifugation of human milk revealed that α-L-fucosidase is only present and active within the supernatant of milk, rather than the pelleted cellular compartment (Wiederschain and Newburg 1996). Expression of α-L-fucosidase in human milk is highest at the beginning of lactation, with a reduction in production through the second week of lactation, and a marked increase from day 28 until day 370 of lactation (Wiederschain and Newburg 1995). α-L-Fucosidase is active at human milk pH but has optimal activity at pH 5.0 (Wiederschain and Newburg 1996, 2001). However, as there is a complex mixture of α-1,2-, 1,3-, 1,4- and 1,6-linked fucosylsaccharides (Spik et al. 1994; Pierce-Crétel et al. 1989; Royle et al. 2003; Mechref et al. 1999) in milk PLGs and free oligosaccharides, multiple fucosidases with unique regioselectivities, such as those released by bifidobacteria, are necessary for full cleavage of human milk PLG.

9.4.2 *α-1,3/4-Fucosyltransferase*

α-1,3/4-Fucosyltransferase (FUT3) is the major fucosyltransferase in human milk that adds α-1,3- and α-1,4-fucose linkages to the ends of PLGs. FUT3 is involved in Lewis blood group antigen synthesis but is also detectable in human milk (Prieels et al. 1981; Johnson and Watkins 1982; Eppenberger-Castori et al. 1989). Like α-L-fucosidase, FUT3 is mostly found dissolved in the aqueous portion of milk (Wiederschain and Newburg 1996). Little is known about the impact of FUT3 on glycan degradation in human milk.

9.4.3 *N-Acetyl-β-D-hexosaminidase*

N-Acetyl-β-D-hexosaminidase is an enzyme with undetermined regioselectivity that releases β-linked *N*-acetylglucosamine (GlcNAc) and *N*-acetylgalactosamine (GalNAc) from the non-reducing end of glycans. The enzyme is active within the soluble fraction of human milk (Wiederschain and Newburg 2001; Dudzik et al. 2008). *N*-Acetyl-β-D-hexosaminidase has measurable activity in human milk (Wiederschain and Newburg 2001), and may be able to act upon a wide variety of substrates, such as GlcNAc-β-1,3-Galactose, GlcNAc-β-1,2-Mannose, GlcNAc-β-1,4-GlcNAc, GlcNAc-β-1,4-Mannose, GlcNAc-β-1,6-Galactose and GlcNAc-β-1,6-GalNAc, all of which exist in human milk oligosaccharides and PLGs (Spik et al. 1988; Mechref et al. 1999; Pierce-Crétel et al. 1984; Hanisch et al. 1989).

9.4.4 *Neuraminidase*

Neuraminidases (also called sialidases) function by releasing sialic acids, such as *N*-acetylneuraminic acid (NeuAc), from glycans (Wang et al. 2004). Whereas the regioselectivity and stereoisomeric specificities are variable, neuraminidase activity has been identified in human milk (Schauer et al. 1976; Namachivayam et al. 2013). Neuraminidases present in milk can potentially degrade both NeuAc-α-2,3-Gal and NeuAc-α-2,6-Gal, which are present in human milk oligosaccharides and PLGs (Pierce-Crétel et al. 1989; Mechref et al. 1999). Neuraminidase activates transforming growth factor-β (TGF- β), which is present in human milk (Namachivayam et al. 2013). Neuraminidase removes sialic acid residues from the latent activating peptide of latent TGF- β, thus turning it into active TGF- β (Carlson et al. 2010). Activated TGF-β assists in the maturation of the infant gut barrier and immune system and has anti-inflammatory effects (Namachivayam et al. 2013). Neuraminidase activity is lower in milk from preterm-delivering mothers than in milk from term-delivering mothers, which explains in part the lower bioactivity of TGF-β in preterm infants (Namachivayam et al. 2013).

9.4.5 α-Amylase

α-Amylase primarily cleaves α-1,4 linked glucose from starch, and is present within the first week postpartum in human milk (Dewit et al. 1990). Although there is evidence of fluctuating α-amylase activity levels in human milk throughout lactation, the lack of reports indicating α-amylase activity on the α-1,4 linked glucose on human milk oligosaccharides and PLGs supports the conclusion that α-amylase plays little to no role in human milk glycan degradation.

9.4.6 Pro- and Anti-Glycosidases

Despite there being control mechanisms for the activity of the various proteolytic systems, there are no known pro- or anti-glycosidases in human milk. Anti-glycosidase compounds can be widely found in plant-based food sources (Wafaa et al. 2010; Williams et al. 2015), and thus may be relevant for nondairy milks. *Lactobacillus* (Vankudre et al. 2015) and pepsin-digested (Lacroix and Li-Chan 2013) bovine whey protein isolate includes peptides that have glycosidase-inhibitory activity, but similar peptides have not been identified from human milk sources. No proteins or peptides have been found with pro-glycosidase activity in human or bovine milk. As such, additional research on human milk pro- and anti-glycosidases is needed to determine whether they are present and active within human milk.

9.5 Other Enzymes in Human Milk

The enzymes discussed so far are associated with the degradation of complex proteins, lipids and glycans. Human milk also contains enzymes that are not involved in macronutrient degradation. The following enzymes are involved in processes that help protect the mammary gland tissue and preserve the nutrient content of human milk from oxidation or infection until ingestion by the infant. These enzymes are important contributors to the benefits of human milk to the infant, and their activities and concentrations in milk are listed in Table 9.1.

9.5.1 Phosphatases

Phosphatases are enzymes that function by removing a phosphate group from a protein. Multiple kinds of phosphatases have been identified in human and bovine milk (Heyndrickx 1963; Geneix et al. 2007; Walentin et al. 1988). Although previous research identified phosphatases in human milk, such as pyrophosphatase (Heyndrickx 1963) and acid phosphatase (Walentin et al. 1988), most recent research has focused on the presence and functions of alkaline phosphatase and adenosine triphosphatase (ATPase).

In general, alkaline phosphatases are present in many body tissues and have various functions, but their physiological function in milk remains unknown (Lowe and John 2018; Stewart et al. 1958). Human milk contains alkaline phosphatase in varying activity levels throughout lactation, although with limited consistency between individual mothers and across lactation duration (Stewart et al. 1958). Alkaline phosphatase in human milk is heat-stable up to 65 °C (Chuang 1987). However, alkaline phosphatase is present in much higher quantities in bovine milk, at approximately 40-fold greater activity levels than in human milk (Heyndrickx 1962; Stewart et al. 1958).

ATPase has also been investigated for quantity and functional properties in human milk. Unlike alkaline phosphatase, ATPase is present at 23-fold higher activity levels in human milk than in bovine milk (Heyndrickx 1963). Two specific ATPases were identified in the membrane fraction of human milk: neutral- and acid Ca^{2+}ATPase (Kanno et al. 1992). In general, ATPases vary only slightly in activity throughout stages of lactation (Heyndrickx 1963).

9.5.2 Xanthine Oxidase

Xanthine oxidase, a cellular redox enzyme present in human milk, is expressed in mammary epithelial cells and secreted into milk as part of the milk fat globule membrane. Xanthine oxidase generates nitric oxide from organic and inorganic nitrates and nitrites (Millar et al. 1998). As a reactive-oxygen species, xanthine oxidase-derived nitric oxide has bactericidal activity in milk against the pathogens *Escherichia coli* and *Salmonella enteritidis,* offering protection to the infant stomach from infectious species (Stevens et al. 2000). Xanthine oxidase concentration varies widely, with one investigation reporting an average peak between 15–20 days (Brown et al. 1995); the concentration is hypothesized to be correlated with the period of greatest susceptibility to infection in infants. Human milk nitric-oxide-generating capacity was higher in hind milk (milk expressed at the end of a feeding) (Stevens et al. 2000). Xanthine oxidase proceeds through an activation-inactivation cycle in human milk, the cause of which is unknown but is hypothesized to relate to hormonal variation (Brown et al. 1995). Whether human milk contains xanthine oxidasc inhibitors remains unknown.

9.5.3 *Antioxidant Enzymes*

Several enzymatic antioxidants are present in human milk, including superoxide dismutase (SOD), catalase and glutathione peroxidases (GPx), all of which offer cyto-protective functions to defend neonates, and perhaps the mammary gland itself, from cellular damage and oxidative stress (Zarban et al. 2009). In human milk, total antioxidant capacity and radical-scavenging activities peak in colostrum and decrease throughout lactation (Zarban et al. 2009; Marinkovic et al. 2016). Although pasteurization and cold storage do not affect non-enzymatic antioxidative capacity in human milk, these treatment methods inactivate the antioxidant enzymes and reduce their capacity to scavenge free radicals (Marinkovic et al. 2016). Fresh milk is thus preferred, to maximize antioxidative benefits to the infant.

GPx are a family of enzymes that catalyze the reduction of hydrogen peroxide and lipid hydroperoxides into water and alcohols to protect cell membranes and functions (Torres et al. 2003). Multiple investigators hypothesized that GPx plays antioxidant roles in protecting lipids and the milk fat globule membrane from peroxidation, thus conserving the membrane's structural integrity and fluidity (L'Abbe and Friel 2000; Torres et al. 2003). The content of GPx in human milk gradually decreases throughout lactation (Torres et al. 2003; Hojo 1986). The activity of GPx is higher in preterm mothers' milk than in term mothers' milk (Păduraru et al. 2018; L'Abbe and Friel 2000; Ellis et al. 1990), potentially compensating for diminished endogenous antioxidant enzyme production in preterm infants (Georgeson et al. 2002). GPx are heat-sensitive and not functional after pasteurization (Hojo 1982). Furthermore, GPx lose activity with prolonged freezing at low temperatures, both −20 °C and −80 °C (Silvestre et al. 2009).

SOD is present in human milk in two forms: copper-zinc SOD (CuZnSOD) and manganese SOD (MnSOD). SOD catalyzes the conversion of the superoxide radical into molecular oxygen or hydrogen peroxide to protect cells from harmful oxidation (Fridovich 1972). SOD concentration in human milk increases across lactation (Kasapović et al. 2005). Data are contradictory as to whether SOD is higher in milk from term-delivering mothers or preterm-delivering mothers (Katzer et al. 2016; L'Abbe and Friel 2000). When separating MnSOD and CuZn SOD measurements, one study reported significant increases in both throughout lactation, with the highest reported measures for both forms of the enzyme at 2–3 weeks of lactation (Kasapović et al. 2005). Although human milk SOD concentrations throughout lactation vary, they have also been reported to change based on maternal dietary intake during pregnancy and lactation (Kasapović et al. 2005). Furthermore, milk from mothers who smoked had significantly decreased activity levels of SOD in their milk compared with milk from non-smoking mothers (Ermis et al. 2005), although interestingly, smoking did not impact the activity of GPx.

9.5.4 *Additional Enzymes*

Human milk also contains other enzymes, including ribonucleases (Kwan et al. 1980; Dalaly et al. 1980), peroxidases (Heyndrickx 1963), glucose-6-phosphatase (Kwan et al. 1980), lactic and malic acid dehydrogenase (Kjellberg and Karlsson 1967; Kwan et al. 1980), lactose synthetase (Kwan et al. 1980) and cholinesterase (Heyndrickx 1963; Walentin et al. 1988). Although these enzymes have been reported, current evidence of their functions and concentrations in human milk and within the infant is limited at this time.

9.6 Conclusion

Human milk has evolved to provide not only the optimal balance of nutrients to promote the growth of the neonate, but also the means to access those nutrients through its wide variety of expressed enzymes. Proteases, originating from blood, immune cell secretions, or mammary epithelial cells, function within the mammary gland, the milk, and the infant stomach to break down macromolecular protein structures for the absorption and utilization of peptides and amino acids. Lipases, likely originating from the blood or interstitial cell rupture during milk expression, function in the same locations, inducing enzymatic degradation of lipids to free fatty acids. Glycosidases catalyze the degradation of complex PLGs and oligosaccharides, though the effect these have on infant health is not well understood. Other enzymes in milk provide additional bioactive benefits to the infant, either by protecting the milk and its nutritional components or by directly acting inside the mammary gland or infant gastrointestinal tract. The fact that the mother must expend her energy reserves to synthesize and excrete these enzymes into the milk lends support to the hypothesis that these enzymes are a necessary and vital part of milk to promote infant development. However, there is still much to discover with regards to the presence and activity of enzymes in milk. By learning more about these enzymes and their functions in human milk and in the infant, we can begin to develop new technologies and improvements to donor milk and formula, such as the inclusion of synthetic enzymes or new milk processing procedures that mimic the activity of the natural enzymes, to best match native human milk.

References

Anderson M (1981) Inhibition of lipolysis in bovine milk by proteose peptone. J Dairy Res 48(2):247–252

Andersson Y, Sävman K, Bläckberg L, Hernell O (2007) Pasteurization of mother's own milk reduces fat absorption and growth in preterm infants. Acta Paediatr 96(10):1445–1449

Andersson E-L, Hernell O, Bläckberg L, Fält H, Lindquist S (2011) BSSL and PLRP2: key enzymes for lipid digestion in the newborn examined using the Caco-2 cell line. J Lipid Res 52(11):1949–1956

Antonowicz I, Lebenthal E (1977) Developmental pattern of small intestinal enterokinase and disaccharidase activities in the human fetus. Gastroenterology 72(6):1299–1303

Armaforte E, Curran E, Huppertz T, Ryan CA, Caboni MF, O'Connor PM, Ross RP, Hirtz C, Sommerer N, Chevalier F, Kelly AL (2010) Proteins and proteolysis in pre-term and term human milk and possible implications for infant formulae. Int Dairy J 20(10):715–723

Attene-Ramos MS, Wagner ED, Plewa MJ, Gaskins HR (2006) Evidence that hydrogen sulfide is a genotoxic agent. Mol Cancer Res 4(1):9–14

Auricchio S, Stellato A, Vizia BD (1981) Development of brush border peptidases in human and rat small intestine during fetal and neonatal life. Pediatr Res 15(7):991–995

Azuma N, Yamauchi K (1992) Plasmin cleavage of human beta-casein. Biosci Biotechnol Biochem 56(7):1140–1141

Barbano D, Rasmussen R, Lynch J (1991) Influence of milk somatic cell count and milk age on cheese yield. J Dairy Sci 74(2):369–388

Bastian ED, Brown RJ (1996) Plasmin in milk and dairy products: an update. Int Dairy J 6(5):435–457

Bathula C, Ghosh S, Hati S, Tripathy S, Singh S, Chakrabarti S, Sen S (2017) Bioisosteric modification of known fucosidase inhibitors to discover a novel inhibitor of α-l-fucosidase. RSC Adv 7(6):3563–3572

Beatty K, Bieth J, Travis J (1980) Kinetics of association of serine proteinases with native and oxidized alpha-1-proteinase inhibitor and alpha-1-antichymotrypsin. J Biol Chem 255(9):3931–3934

Benes P, Vetvicka V, Fusek M (2008) Cathepsin D—many functions of one aspartic protease. Crit Rev Oncol Hematol 68(1):12–28

Benslimane S, Dognin-Bergeret MJ, Berdague J-L, Gaudemer Y (1990) Variation with season and lactation of plasmin and plasminogen concentrations in Montbeliard cows' milk. J Dairy Res 57(4):423–435

Bergmann M, Fruton JS (1936) Regarding the general nature of catheptic enzymes. Science 84(2169):89–90

Berkow SE, Freed LM, Hamosh M, Bitman J, Larry Wood D, Happ B, Hamosh P (1984) Lipases and lipids in human milk: effect of freeze-thawing and storage. Pediatr Res 18:1257–1262

Bertino E, Giribaldi M, Baro C, Giancotti V, Pazzi M, Peila C, Tonetto P, Arslanoglu S, Moro GE, Cavallarin L, Gastaldi D (2013) Effect of prolonged refrigeration on the lipid profile, lipase activity, and oxidative status of human milk. J Pediatr Gastroenterol Nutr 56(4):390–396

Beverly RL, Underwood MA, Dallas DC (2019) Peptidomics analysis of milk protein-derived peptides released over time in the preterm infant stomach. J Proteome Res 18(3):912–922

Bläckberg L, Hernell O (1981) Bile salt stimulated lipase in human milk purification and characterization. Eur J Biochem 116(2):221–226

Bläckberg L, Hernell O (1983) Further characterization of the bile salt-stimulated lipase in human milk. FEBS Lett 157(2):337–341

Borgström B, Lindquist B, Lundh G (1960) Enzyme concentration and absorption of protein and glucose in duodenum of premature infants. AMA J Dis Child 99(3):338–343

Borulf S, Lindberg T, Mansson M (1987) Immunoreactive anionic trypsin and anionic elastase in human milk. Acta Paediatr 76(1):11–15

Britton J, Koldovsky O (1989) Development of luminal protein digestion: implications for biologically active dietary polypeptides. J Pediatr Gastroenterol Nutr 9(2):144–162

Brown PB, Clark RM, Hundrieser KE, Ferris AM, Jensen RG (1983) Bile salt-stimulated lipase in human milk from 2 to 16 weeks postpartum. Lipids 18(5):393–396

Brown A-M, Benboubetra M, Ellison M, Powell D, Reckless JD, Harrison R (1995) Molecular activation—deactivation of xanthine oxidase in human milk. Biochim Biophys Acta Gen Subj 1245(2):248–254

Capony F, Rougeot C, Montcourrier P, Cavailles V, Salazar G, Rochefort H (1989) Increased secretion, altered processing, and glycosylation of pro-cathepsin D in human mammary cancer cells. Cancer Res 49(14):3904–3909

Carlson CM, Turpin EA, Moser LA, O'Brien KB, Cline TD, Jones JC, Tumpey TM, Katz JM, Kelley LA, Gauldie J, Schultz-Cherry S (2010) Transforming growth factor-β: activation by neuraminidase and role in highly pathogenic H5N1 influenza pathogenesis. PLoS Pathog 6(10):e1001136

Chen VC, Chao L, Chao J (2000) Reactive-site specificity of human kallistatin toward tissue kallikrein probed by site-directed mutagenesis. Biochim Biophys Acta 1479(1-2):237–246

Cheung P, Sawicki G, Gross S, Van Aerde J, Radomski M (2001) Differential expression of matrix metalloproteinases and the tissue inhibitor in human milk. Proc West Pharmacol Soc 44:97–98

Chow J, Panasevich MR, Alexander D, Vester Boler BM, Rossoni Serao MC, Faber TA, Bauer LL, Fahey GC (2014) Fecal metabolomics of healthy breast-fed versus formula-fed infants before and during in vitro batch culture fermentation. J Proteome Res 13(5):2534–2542

Chowanadisai W, Lönnerdal B (2002) α1-Antitrypsin and antichymotrypsin in human milk: origin, concentrations, and stability. Am J Clin Nutr 76(4):828–833

Chowanadisai W, Huang J, Huang N, Lönnerdal B (2003) Stability of recombinant human alpha-1-antitrypsin produced in rice in infant formula. J Nutr Biochem 14(7):386–393

Christensen S, Wiegers T, Hermansen J, Sottrup-Jensen L (1995) Plasma-derived protease inhibitors in bovine milk. Int Dairy J 5(5):439–449

Chuang N-N (1987) Alkaline phosphatase in human milk: a new heat-stable enzyme. Clin Chim Acta 169(2-3):165–174

Clare DA, Swaisgood HE (2000) Bioactive milk peptides: a prospectus. J Dairy Sci 83(6):1187–1195

Considine T, Healy A, Kelly A, McSweeney P (2004) Hydrolysis of bovine caseins by cathepsin B, a cysteine proteinase indigenous to milk. Int Dairy J 14(2):117–124

Constantopoulos A, Messaritakis J, Matsaniotis N (1980) Breast milk jaundice; the role of lipoprotein lipase and the free fatty acids. Eur J Pediatr 134(1):35–38

D'Alessandro A, Scaloni A, Zolla L (2010) Human milk proteins: an interactomics and updated functional overview. J Proteome Res 9(7):3339–3373

D'Alessandro A, Zolla L, Scaloni A (2011) The bovine milk proteome: cherishing, nourishing and fostering molecular complexity. An interactomics and functional overview. Mol BioSyst 7(3):579–597

Dalaly BK, Eitenmiller RR, Friend BA, Shahani KM (1980) Human milk ribonuclease. Biochim Biophys Acta 615(2):381–391

Dallas DC, German JB (2017) Enzymes in human milk. Nestle Nutr Inst Workshop Ser 88:129–136

Dallas DC, Underwood MA, Zivkovic AM, German JB (2012) Digestion of protein in premature and term infants. J Nutr Disord Ther 2(3):112–121

Dallas DC, Guerrero A, Khaldi N, Castillo PA, Martin WF, Smilowitz JT, Bevins CL, Barile D, German JB, Lebrilla CB (2013a) Extensive in vivo human milk peptidomics reveals specific proteolysis yielding protective antimicrobial peptides. J Proteome Res 12(5):2295–2304

Dallas DC, Guerrero A, Parker EA, Garay LA, Bhandari A, Lebrilla CB, Barile D, German JB (2013b) Peptidomic profile of milk of Holstein cows at peak lactation. J Agric Food Chem 62(1):58–65

Dallas DC, Guerrero A, Khaldi N, Borghese R, Bhandari A, Underwood MA, Lebrilla CB, German JB, Barile D (2014) A peptidomic analysis of human milk digestion in the infant stomach reveals protein-specific degradation patterns. J Nutr 144(6):815–820

Dallas DC, Murray NM, Gan J (2015a) Proteolytic systems in milk: perspectives on the evolutionary function within the mammary gland and the infant. J Mammary Gland Biol Neoplasia 20(3-4):133–147

Dallas DC, Smink CJ, Robinson RC, Tian T, Guerrero A, Parker EA, Smilowitz JT, Hettinga KA, Underwood MA, Lebrilla CB, German JB, Barile D (2015b) Endogenous human milk peptide release is greater after preterm birth than term birth. J Nutr 145(3):425–433

Daniel H, Vohwinkel M, Rehner G (1990) Effect of casein and beta-casomorphins on gastrointestinal motility in rats. J Nutr 120(3):252–257
Deeth HC (2006) Lipoprotein lipase and lipolysis in milk. Int Dairy J 16(6):555–562
Demers-Mathieu V, Nielsen SD, Underwood MA, Borghese R, Dallas DC (2017) Analysis of milk from mothers who delivered prematurely reveals few changes in proteases and protease inhibitors across gestational age at birth and infant postnatal age. J Nutr 146(6):1152–1159
Demers-Mathieu V, Nielsen SD, Underwood MA, Borghese R, Dallas DC (2018) Changes in proteases, antiproteases and bioactive proteins from mother's breast milk to the premature infant stomach. J Pediatr Gastroenterol Nutr 66(2):318–324
Dewit O, Dibba B, Prentice A (1990) Breast-milk amylase activity in English and Gambian mothers: effects of prolonged lactation, maternal parity, and individual variations. Pediatr Res 28(5):502–506
Dingess KA, de Waard M, Boeren S, Vervoort J, Lambers TT, van Goudoever JB, Hettinga K (2017) Human milk peptides differentiate between the preterm and term infant and across varying lactational stages. Food Funct 8(10):3769–3782
Dudzik D, Knas M, Gocal M, Borzym-Kluczyk M, Szajda S, Kna-Karaszewska K, Tomaszewski J, Zwierz K (2008) Activity of N-acetyl-D-hexosaminidase (HEX) and its isoenzymes A and B in human milk during the first 3 months of breastfeeding. Adv Med Sci 53(2):300–304
Ellis L, Picciano MF, Smith AM, Hamosh M, Mehta NR (1990) The impact of gestational length on human milk selenium concentration and glutathione peroxidase activity. Pediatr Res 27:32
Engberg S, Mansson M, Andersson Y, Jakobsson I, Lindberg T (1999) Trypsin and elastase activity in duodenal juice from preterm infants before and after a meal of human milk. Prenat Neonatal Med 4(6):466–471
Eppenberger-Castori S, Lötscher H, Finne J (1989) Purification of the N-acetylglucosaminide alpha (1-3/4) fucosyltransferase of human milk. Glycoconj J 6(1):101–114
Ermis B, Yildirim A, Örs R, Tastekin A, Ozkan B, Akcay F (2005) Influence of smoking on serum and milk malondialdehyde, superoxide dismutase, glutathione peroxidase, and antioxidant potential levels in mothers at the postpartum seventh day. Biol Trace Elem Res 105(1):27–36
Farkye N, Landkammer CF (1992) Milk plasmin activity influence on cheddar cheese quality during ripening. J Food Sci 57(3):622–639
Ferranti P, Traisci MV, Picariello G, Nasi A, Boschi V, Siervo M, Falconi C, Chianese L, Addeo F (2004) Casein proteolysis in human milk: tracing the pattern of casein breakdown and the formation of potential bioactive peptides. J Dairy Res 71(1):74–87
Fredrikzon B, Hernell O, Bläckberg L, Olivecrona T (1978) Bile salt-stimulated lipase in human milk: evidence of activity in vivo and of a role in the digestion of milk retinol esters. Pediatr Res 12(11):1048–1052
Freed LM, Berkow SE, Hamosh P, York CM, Mehta NR, Hamosh M (1989) Lipases in human milk: effect of gestational age and length of lactation on enzyme activity. J Am Coll Nutr 8(2):143–150
Fridovich I (1972) Superoxide radical and superoxide dismutase. Acc Chem Res 5(10):321–326
Froehlich JW, Dodds ED, Barboza M, McJimpsey EL, Seipert RR, Francis J, An HJ, Freeman S, German JB, Lebrilla CB (2010) Glycoprotein expression in human milk during lactation. J Agric Food Chem 58(10):6440–6448
Geneix N, Dufour E, Venien A, Levieux D (2007) Development of a monoclonal antibody-based immunoassay for specific quantification of bovine milk alkaline phosphatase. J Dairy Res 74(3):290–295
Georgeson GD, Szőny BJ, Streitman K, Varga IS, Kovács A, Kovács L, László A (2002) Antioxidant enzyme activities are decreased in preterm infants and in neonates born via caesarean section. Eur J Obstet Gynecol Reprod Biol 103(2):136–139
Gettins PGW (2002) Serpin structure, mechanism, and function. Chem Rev 102(12):4751–4804
Girardet J-M, Linden G, Loye S, Courthaudon J-L, Lorient D (1993) Study of mechanism of lipolysis inhibition by bovine milk proteose-peptone component 3. J Dairy Sci 76(8):2156–2163

Greenberg R, Groves ML (1984) Plasmin cleaves human β-casein. Biochem Biophys Res Commun 125(2):463–468

Grufferty MB, Fox PF (1988) Milk alkaline proteinase. J Dairy Res 55(04):609–630

Guerrero A, Dallas DC, Contreras S, Chee S, Parker EA, Sun X, Dimapasoc L, Barile D, German JB, Lebrilla CB (2014) Mechanistic peptidomics: factors that dictate specificity in the formation of endogenous peptides in human milk. Mol Cell Proteomics 13(12):3343–3351

Hamer HM, De Preter V, Windey K, Verbeke K (2012) Functional analysis of colonic bacterial metabolism: relevant to health? Am J Physiol Gastrointest Liver Physiol 302(1):G1–G9

Hamosh M (1998) Protective function of proteins and lipids in human milk. Biol Neonate 74(2):163–176

Hamosh M, Scanlon JW, Ganot D, Likel M, Scanlon KB, Hamosh P (1981) Fat digestion in the newborn. Characterization of lipase in gastric aspirates of premature and term infants. J Clin Invest 67(3):838–846

Hamosh M, Henderson TR, Ellis LA, Mao J-I, Hamosh P (1997) Digestive enzymes in human milk: stability at suboptimal storage temperatures. J Pediatr Gastroenterol Nutr 24(1):38–43

Hanisch FG, Uhlenbruck G, Peter-Katalinic J, Egge H, Dabrowski J, Dabrowski U (1989) Structures of neutral O-linked polylactosaminoglycans on human skim milk mucins. A novel type of linearly extended poly-N-acetyllactosamine backbones with Gal beta (1-4) GlcNAc beta (1-6) repeating units. J Biol Chem 264(2):872–883

Harmsen HJ, Wildeboer-Veloo AC, Raangs GC, Wagendorp AA, Klijn N, Bindels JG, Welling GW (2000) Analysis of intestinal flora development in breast-fed and formula-fed infants by using molecular identification and detection methods. J Pediatr Gastroenterol Nutr 30(1):61–67

Hasilik A (1992) The early and late processing of lysosomal enzymes: proteolysis and compartmentation. Experientia 48(2):130–151

Hasilik A, von Figura K, Conzelmann E, Nehrkorn H, Sandhoff K (1982) Lysosomal enzyme precursors in human fibroblasts. Eur J Biochem 125(2):317–321

Heegaard CW, Christensen T, Rasmussen MD, Benfeldt C, Jensen N, Sejrsen K, Petersen TE, Andreasen P (1994a) Plasminogen activators in bovine milk during mastitis, an inflammatory disease. Fibrinolysis 8(1):22–30

Heegaard CW, Rasmussen LK, Andreasen PA (1994b) The plasminogen activation system in bovine milk: differential localization of tissue-type plasminogen activator and urokinase in milk fractions is caused by binding to casein and urokinase receptor. Biochim Biophys Acta, Mol Cell Res 1222(1):45–55

Heegaard CW, Larsen LB, Rasmussen LK, Højberg KE, Petersen TE, Andreasen PA (1997) Plasminogen activation system in human milk. J Pediatr Gastroenterol Nutr 25(2):159–166

Heegard CE, White JH, Zavizion B, Turner JD, Politis I (1994) Production of various forms of plasminogen activator and plasminogen activator inhibitor by cultured mammary eplthelial cells. J Dairy Sci 77(10):2949–2958

Heyndrickx G (1962) Investigations on the enzymes in human milk. Pediatrics 198:356–362

Heyndrickx G (1963) Further investigations on the enzymes in human milk. Pediatrics 31(6):1019–1023

Hojo Y (1982) Selenium concentration and glutathione peroxidase activity in cow's milk. Biol Trace Elem Res 4(2-3):233–239

Hojo Y (1986) Sequential study on glutathione peroxidase and selenium contents of human milk. Sci Total Environ 52(1-2):83–91

Holton TA, Vijaykumar V, Dallas DC, Guerrero A, Borghese RC, Lebrilla CB, German JB, Barile D, Underwood MA, Shields DC, Khaldi N (2014) Following the digestion of milk proteins from mother to baby. J Proteome Res 13(12):5777–5783

Hughes R, Kurth MJ, McGilligan V, McGlynn H, Rowland I (2008) Effect of colonic bacterial metabolites on Caco-2 cell paracellular permeability in vitro. Nutr Cancer 60(2):259–266

Hung H-Y, Hsu Y-Y, Su P-F, Chang Y-J (2018) Variations in the rancid-flavor compounds of human breastmilk under general frozen-storage conditions. BMC Pediatr 18(1):94–94

Ismail B, Nielsen S (2010) Invited review: plasmin protease in milk: current knowledge and relevance to dairy industry. J Dairy Sci 93(11):4999–5009

Janoff A (1972) Inhibition of human granulocyte elastase by serum α1-antitrypsin. Am Rev Respir Dis 105(1):121–122

Jochum M, Bittner A (1983) Inter-alpha-trypsin inhibitor of human serum: an inhibitor of polymorphonuclear granulocyte elastase. Hoppe Seylers Z Physiol Chem 364(12):1709–1715

Johnson P, Watkins W (1982) Separation of an alpha-3-L-fucosyltransferase from the blood-group-Le-gene-specified alpha-3/4-L-fucosyltransferase in human milk. Biochem Soc Trans 10:445–446

Jones JB, Mehta NR, Hamosh M (1982) α-Amylase in preterm human milk. J Pediatr Gastroenterol Nutr 1(1):43–48

Kalsheker N (1989) Alpha 1-antitrypsin: structure, function and molecular biology of the gene. Biosci Rep 9(2):129–138

Kaminogawa S, Yamauchi K (1972) Acid protease of bovine milk. Agric Biol Chem 36(13):2351–2356

Kanno C, Takeda Y, Kook Cho J (1992) Enzymatic properties of two calcium-stimulated ATPases in the neutral and acidic pH regions found in the membrane fraction of human milk. Biosci Biotechnol Biochem 56(6):900–905

Kasapović J, Pejić S, Mladenovic M, Radlovic N, Pajović S (2005) Superoxide dismutase activity in colostrum, transitional and mature human milk. Turk J Pediatr 47:343–347

Katzer D, Pauli L, Mueller A, Reutter H, Reinsberg J, Fimmers R, Bartmann P, Bagci S (2016) Melatonin concentrations and antioxidative capacity of human breast milk according to gestational age and the time of day. J Hum Lact 32(4):105–110

Kauzmann W (1959) Some factors in the interpretation of protein denaturation. In: Anfinsen CB, Anson ML, Bailey K, Edsall JT (eds) Advances in protein chemistry, vol 14. Academic, New York, pp 1–63

Kelly EJ, Newell SJ, Brownlee KG, Primrose JN, Dear PR (1993) Gastric acid secretion in preterm infants. Early Hum Dev 35(3):215–220

Kelly AL, O'Flaherty F, Fox PF (2006) Indigenous proteolytic enzymes in milk: a brief overview of the present state of knowledge. Int Dairy J 16(6):563–572

Khaldi N, Vijayakumar V, Dallas DC, Guerrero A, Wickramasinghe S, Smilowitz JT, Medrano JF, Lebrilla CB, Shields DC, German JB (2014) Predicting the important enzymes in human breast milk digestion. J Agric Food Chem 62(29):7225–7232

Kirschke H, Barrett AJ, Rawlings ND (1998) Lysosomal cysteine proteases, vol 131. Oxford University Press, Oxford

Kjellberg B, Karlsson BW (1967) Comparative analyses of lactic and malic dehydrogenases and their multiple molecular forms in milk from various animal species and man. Comp Biochem Physiol 22(2):397–413

Korycha-Dahl M, Dumas BR, Chene N, Martal J (1983) Plasmin activity in milk. J Dairy Sci 66(4):704–711

Kwan AJ, Friend BA, Shahani KM (1980) Role and significance of enzymes in human milk. Am J Clin Nutr 33(8):1861–1868

L'Abbe MR, Friel JK (2000) Superoxide dismutase and glutathione peroxidase content of human milk from mothers of premature and full-term infants during the first 3 months of lactation. J Pediatr Gastroenterol Nutr 31(3):270–274

Lacroix IME, Li-Chan ECY (2013) Inhibition of dipeptidyl peptidase (DPP)-IV and α-glucosidase activities by pepsin-treated whey proteins. J Agric Food Chem 61(31):7500–7506

Lahov E, Regelson W (1996) Antibacterial and immunostimulating casein-derived substances from milk: casecidin, isracidin peptides. Food Chem Toxicol 34(1):131–145

Lambers TT, Gloerich J, van Hoffen E, Alkema W, Hondmann DH, van Tol EA (2015) Clustering analyses in peptidomics revealed that peptide profiles of infant formulae are descriptive. Food Sci Nutr 3(1):81–90

Landberg E, Huang Y, Strömqvist M, Mechref Y, Hansson L, Lundblad A, Novotny MV, Påhlsson P (2000) Changes in glycosylation of human bile-salt-stimulated lipase during lactation. Arch Biochem Biophys 377(2):246–254
Larsen LB, Petersen TE (1995) Identification of five molecular forms of cathepsin D in bovine milk. In: Takahashi K (ed) Aspartic proteinases, Advances in experimental medicine and biology, vol 362. Springer, Berlin, pp 279–283
Lebenthal E, Lee P (1980) Development of functional response in human exocrine pancreas. Pediatrics 66(4):556–560
Lefèvre CM, Sharp JA, Nicholas KR (2010) Evolution of lactation: ancient origin and extreme adaptations of the lactation system. Annu Rev Genomics Hum Genet 11(1):219–238
Legrand D, Pierce A, Elass E, Carpentier M, Mariller C, Mazurier J (2008) Lactoferrin structure and functions. In: Bösze Z (ed) Bioactive components of milk. Springer, New York, pp 163–194
Lemay DG, Ballard OA, Hughes MA, Morrow AL, Horseman ND, Nommsen-Rivers LA (2013) RNA sequencing of the human milk fat layer transcriptome reveals distinct gene expression profiles at three stages of lactation. PLoS One 8(7):e67531
Liepke C, Adermann K, Raida M, Magert HJ, Forssmann WG, Zucht HD (2002) Human milk provides peptides highly stimulating the growth of bifidobacteria. Eur J Biochem 269(2):712–718
Lindberg T (1979) Protease inhibitors in human-milk. Pediatr Res 13(9):969–972
Lindberg T, Ohlsson K, Westrom B (1982) Protease inhibitors and their relation to protease activity in human milk. Pediatr Res 16(6):479–483
Lindquist S, Hernell O (2010) Lipid digestion and absorption in early life: an update. Curr Opin Clin Nutr Metab Care 13(3):314–320
Lowe D, John S (2018) Alkaline phosphatase. In: StatPearls. StatPearls Publishing LLC, Treasure Island
Lubetzky R, Mandel D, Mimouni FB, Herman L, Reich R, Reif S (2010) MMP-2 and MMP-9 and their tissue inhibitor in preterm human milk. J Pediatr Gastroenterol Nutr 51(2):210–212
Magboul AA, Larsen LB, McSweeney PL, Kelly AL (2001) Cysteine protease activity in bovine milk. Int Dairy J 11(11):865–872
Manson WG, Weaver LT (1997) Fat digestion in the neonate. Arch Dis Child Fetal Neonatal Ed 76(3):F206–F211
Marinkovic V, Rankovic-Janevski M, Spasic S, Nikolic-Kokic A, Lugonja N, Djurovic D, Miletic S, Vrvic MM, Spasojevic I (2016) Antioxidative activity of colostrum and human milk: effects of pasteurization and storage. J Pediatr Gastroenterol Nutr 62(6):901–906
Markus G, Hitt S, Harvey S, Tritsch G (1993) Casein, a powerful enhancer of the rate of plasminogen activation. Fibrinolysis 7(4):229–236
Martin CR, Ling P-R, Blackburn GL (2016) Review of infant feeding: key features of breast milk and infant formula. Nutrients 8(5):279
Mason S (1962) Some aspects of gastric function in the newborn. Arch Dis Child 37(194):387–391
Mattila T, Saari S, Vartiala H, Sandholm M (1985) Milk antitrypsin as a marker of bovine mastitis—correlation with bacteriology. J Dairy Sci 68(1):114–122
McBride OW, Korn ED (1963) The lipoprotein lipase of mammary gland and the correlation of its activity to lactation. J Lipid Res 4(1):17–20
McCall IC, Betanzos A, Weber DA, Nava P, Miller GW, Parkos CA (2009) Effects of phenol on barrier function of a human intestinal epithelial cell line correlate with altered tight junction protein localization. Toxicol Appl Pharmacol 241(1):61–70
McGilligan KM, Thomas DW, Eckhert CD (1987) Alpha-1-antitrypsin concentration in human milk. Pediatr Res 22(3):268–270
Mechref Y, Chen P, Novotny M (1999) Structural characterization of the N-linked oligosaccharides in bile salt-stimulated lipase originated from human breast milk. Glycobiology 9(3):227–234
Mehta NR, Jones JB, Hamosh M, Calcagno PL (1981) 1162 lipoprotein lipase in preterm milk. Pediatr Res 15(S4):636

Mehta NR, Jones JB, Hamosh M (1982) Lipases in preterm human milk: ontogeny and physiologic significance. J Pediatr Gastroenterol Nutr 1(3):317–326

Mendelson CR, Zinder O, Blanchette-Mackie JE, Chernick SS, Scow RO (1977) Lipoprotein lipase and lipid metabolism in mammary gland. J Dairy Sci 60(4):666–676

Miclat NN, Hodgkinson R, Marx GF (1978) Neonatal gastric pH. Anesth Analg 57(1):98–101

Millar TM, Stevens CR, Benjamin N, Eisenthal R, Harrison R, Blake DR (1998) Xanthine oxidoreductase catalyses the reduction of nitrates and nitrite to nitric oxide under hypoxic conditions. FEBS Lett 427(2):225–228

Mistry N, Drobni P, Naslund J, Sunkari VG, Jenssen H, Evander M (2007) The anti-papillomavirus activity of human and bovine lactoferricin. Antivir Res 75(3):258–265

Molinari CE, Casadio YS, Hartmann BT, Livk A, Bringans S, Arthur PG, Hartmann PE (2012) Proteome mapping of human skim milk proteins in term and preterm milk. J Proteome Res 11(3):1696–1714

Monti J, Mermoud A-F, Jolles P (1986) Trypsin in human milk. Experientia 42(1):39–41

Namachivayam K, Blanco CL, Frost BL, Reeves AA, Jagadeeswaran R, MohanKumar K, Safarulla A, Mandal P, Garzon SA, Raj JU, Maheshwari A (2013) Preterm human milk contains a large pool of latent TGF-β, which can be activated by exogenous neuraminidase. Am J Physiol Gastrointest Liver Physiol 304(12):1055–1065

Naughton M, Sanger F (1961) Purification and specificity of pancreatic elastase. Biochem J 78(1):156–163

Neville MC, Waxman LJ, Jensen D, Eckel RH (1991) Lipoprotein lipase in human milk: compartmentalization and effect of fasting, insulin, and glucose. J Lipid Res 32(2):251–257

Nielsen SD, Beverly RL, Dallas DC (2017) Milk proteins are predigested within the human mammary gland. J Mamm Gland Biol 22(4):251–261

Nielsen SD, Beverly RL, Underwood MA, Dallas DC (2018) Release of functional peptides from mother's milk and fortifier proteins in the premature infant stomach. PLoS One 13(11):e0208204

Nilsson J, BlÄCkberg L, Carlsson P, EnerbÄCk S, Hernell O, Bjursell G (1990) cDNA cloning of human-milk bile-salt-stimulated lipase and evidence for its identity to pancreatic carboxylic ester hydrolase. Eur J Biochem 192(2):543–550

Norman A, Strandvik B, Ojamäe Ö (1972) Bile acids and pancreatic enzymes during absorption in the newborn. Acta Paediatr 61(5):571–576

Oftedal OT (2002) The mammary gland and its origin during synapsid evolution. J Mammary Gland Biol Neoplasia 7(3):225–252

Olivecrona T, Vilaró S, Olivecrona G (2003) Lipases in milk. In: Fox PF, McSweeney PLH (eds) Advanced dairy chemistry—1 proteins: part a/part b. Springer, Boston, pp 473–494

Păduraru L, Dimitriu DC, Avasiloaiei AL, Moscalu M, Zonda GI, Stamatin M (2018) Total antioxidant status in fresh and stored human milk from mothers of term and preterm neonates. Pediatr Neonatal 59(6):600–605

Palla MR, Harohalli S, Crawford TN, Desai N (2018) Progression of gastric acid production in preterm neonates: utilization of in-vitro method. Front Pediatr 6:211

Palmer LS (1922) Is lipase a normal constituent of cow's milk? J Dairy Sci 5(1):51–63

Palmer DJ, Kelly VC, Smit AM, Kuy S, Knight CG, Cooper GJ (2006) Human colostrum: identification of minor proteins in the aqueous phase by proteomics. Proteomics 6(7):2208–2216

Pamblanco M, Ten A, Comin J (1987) Bile salt-stimulated lipase activity in human colostrum from mothers of infants of different gestational age and birthweight. Acta Paediatr Scand 76(2):328–331

Pedersen G, Brynskov J, Saermark T (2002) Phenol toxicity and conjugation in human colonic epithelial cells. Scand J Gastroenterol 37(1):74–79

Permyakov EA, Berliner LJ (2000) α-Lactalbumin: structure and function. FEBS Lett 473(3):269–274

Picariello G, Ferranti P, Mamone G, Klouckova I, Mechref Y, Novotny MV, Addeo F (2012) Gel-free shotgun proteomic analysis of human milk. J Chromatogr A 1227:219–233

Pierce-Crétel A, Debray H, Montreuil J, Spik G, Halbeek H, Mutsaers JHGM, Vliegenthart JFG (1984) Primary structure of N glycosidically linked asialoglycans of secretory immunoglobulins from human milk. Eur J Biochem 139(2):337–349

Pierce-Crétel A, Decottignies JP, Wieruszeski JM, Strecker G, Montreuil J, Spik G (1989) Primary structure of twenty three neutral and monosialylated oligosaccharides O glycosidically linked to the human secretory immunoglobulin A hinge region determined by a combinationof permethylation analysis and 400 MHz 1H NMR spectroscopy. Eur J Biochem 182(2):457–476

Pimenta DC, Oliveira A, Juliano MA, Juliano L (2001) Substrate specificity of human cathepsin D using internally quenched fluorescent peptides derived from reactive site loop of kallistatin. Biochim Biophys Acta 1544(1):113–122

Politis I, Lachance E, Block E, Turner JD (1989a) Plasmin and plasminogen in bovine milk: a relationship with involution? J Dairy Sci 72(4):900–906

Politis I, Ng Kwai Hang K, Giroux R (1989b) Environmental factors affecting plasmin activity in milk. J Dairy Sci 72(7):1713–1718

Politis I, White JH, Zavizion B, Goldberg JJ, Guo MR, Kindstedt P (1995) Effect of individual caseins on plasminogen activation by bovine urokinase-type and tissue-type plasminogen activators. J Dairy Sci 78(3):484–490

Politis I, Zavizion B, Cheli F, Baldi A (2002) Expression of urokinase plasminogen activator receptor in resting and activated bovine neutrophils. J Dairy Res 69(02):195–204

Politis I, Voudouri A, Bizelis I, Zervas G (2003) The effect of various vitamin E derivatives on the urokinase-plasminogen activator system of ovine macrophages and neutrophils. Br J Nutr 89(02):259–265

Precetti AS, Oria MP, Nielsen SS (1997) Presence in bovine milk of two protease inhibitors of the plasmin system. J Dairy Sci 80(8):1490–1496

Prieels JP, Monnom D, Dolmans M, Beyer TA, Hill RL (1981) Co-purification of the Lewis blood group N-acetylglucosaminide alpha 1 goes to 4 fucosyltransferase and an N-acetylglucosaminide alpha 1 goes to 3 fucosyltransferase from human milk. J Biol Chem 256(20):10456–10463

Radisky ES (2010) Cathepsin D: regulation in mammary gland remodeling, misregulation in breast cancer. Cancer Biol Ther 10(5):467–470

Raiten D, Talbot JM, Waters JH (1998) Assessment of nutrient requirements for infant formulas. J Nutr 128(11 Suppl):2059–2293

Royle L, Roos A, Harvey DJ, Wormald MR, Van Gijlswijk-Janssen D, Redwan ERM, Wilson IA, Daha MR, Dwek RA, Rudd PM (2003) Secretory IgA N-and O-glycans provide a link between the innate and adaptive immune systems. J Biol Chem 278(22):20140–20153

Salier JP, Rouet P, Raguenez G, Daveau M (1996) The inter-alpha-inhibitor family: from structure to regulation. Biochem J 315(Pt 1):1–9

Sato R, Noguchi T, Naito H (1986) Casein phosphopeptide (CPP) enhances calcium absorption from the ligated segment of rat small intestine. J Nutr Sci Vitaminol 32(1):67–76

Schaller J, Moser PW, Dannegger-MÜLler GAK, Rösselet SJ, Kämpfer U, Rickli EE (1985) Complete amino acid sequence of bovine plasminogen. Eur J Biochem 149(2):267–278

Schapira M, Scott CF, Colman RW (1982) Contribution of plasma protease inhibitors to the inactivation of kallikrein in plasma. J Clin Invest 69(2):462–468

Schauer R, Veh RW, Wember M, Buscher HP (1976) Demonstration of neuraminidase activity in human blood serum and human milk using a modified, radioactively labelled 1-glycoprotein as substrate. Hoppe Seylers Z Physiol Chem 357(1):559–566

Sforza S, Cavatorta V, Lambertini F, Galaverna G, Dossena A, Marchelli R (2012) Cheese peptidomics: a detailed study on the evolution of the oligopeptide fraction in Parmigiano-Reggiano cheese from curd to 24 months of aging. J Dairy Sci 95(7):3514–3526

Silvestre D, Miranda M, Muriach M, Almansa I, Jareño E, Romero FJ (2009) Frozen breast milk at -20°C and -80°C: a longitudinal study of glutathione peroxidase activity and malondialdehyde concentration. J Hum Lact 26(1):35–41

Spik G, Coddeville B, Montreuil J (1988) Comparative study of the primary structures of sero-, lacto-and ovotransferrin glycans from different species. Biochimic 70(11):1459–1469

Spik G, Coddeville B, Mazurier J, Bourne Y, Cambillaut C, Montreuil J (1994) Primary and three-dimensional structure of lactotransferrin (lactoferrin) glycans. Adv Exp Med Biol 357:21–32

Stevens C, Millar T, Clinch J, Kanczler J, Bodamyali T, Blake D (2000) Antibacterial properties of xanthine oxidase in human milk. Lancet 356(9232):829–830

Stewart RA, Platou E, Kelly VJ (1958) The alkaline phosphatase content of human milk. J Biol Chem 232:777–784

Strömqvist M, Lindgren K, Hansson L, Juneblad K (1995) Differences in the glycosylation of recombinant and native human milk bile salt-stimulated lipase revealed by peptide mapping. J Chromatogr A 718(1):53–58

Swan JS, Hoffman MM, Lord MK, Poechmann JL (1992) Two forms of human milk bile-salt-stimulated lipase. Biochem J 283(Pt 1):119–122

Tacoma R, Fields J, Ebenstein DB, Lam Y-W, Greenwood SL (2016) Characterization of the bovine milk proteome in early-lactation Holstein and Jersey breeds of dairy cows. J Proteome 130:200–210

Tarlow MJ, Hadorn B, Arthurton MW, Lloyd JK (1970) Intestinal enterokinase deficiency. A newly-recognized disorder of protein digestion. Arch Dis Child 45(243):651–655

Tayman C, Rayyan M, Allegaert K (2011) Neonatal pharmacology: extensive interindividual variability despite limited size. J Pediatr Pharmacol Ther 16(3):170–184

Toden S, Bird AR, Topping DL, Conlon MA (2007) Differential effects of dietary whey, casein and soya on colonic DNA damage and large bowel SCFA in rats fed diets low and high in resistant starch. Br J Nutr 97(3):535–543

Tome D, Dumontier A-M, Hautefeuille M, Desjeux J-F (1987) Opiate activity and transepithelial passage of intact beta-casomorphins in rabbit ileum. Am J Physiol Gastrointest Liver Physiol 253(6):G737–G744

Torres JE, VanderJagt D, Okolo SN, Magnussen M, Bhatta SK, Glew RH (2001) Bile salt-stimulated lipase in the milk of Fulani and Kanuri women in Nigeria and native Nepalese women. J Natl Med Assoc 93(6):201–207

Torres A, Farré R, Lagarda MJ, Monleón J (2003) Determination of glutathione peroxidase activity in human milk. Food Nahrung 47(6):430–433

Valnickova Z, Thøgersen IB, Potempa J, Enghild JJ (2007) Thrombin-activable fibrinolysis inhibitor (TAFI) zymogen is an active carboxypeptidase. J Biol Chem 282(5):3066–3076

Vankudre M, Balpande A, Athale M (2015) Comparative analysis of α-amylase inhibition and antioxidant activity of whey from cow and buffalo milk fermented with Lactobacillus species. Biosci Biotech Res Commun 8(1):25–28

Venuti E, Shishmarev D, Kuchel PW, Dutt S, Blumenthal CS, Gaskin KJ (2017) Bile salt stimulated lipase: inhibition by phospholipids and relief by phospholipase A2. J Cyst Fibros 16(6):763–770

Větvicka V, Vagner J, Baudys M, Tang J, Foundling S, Fusek M (1993) Human breast milk contains procathepsin D--detection by specific antibodies. Biochem Mol Biol Int 30(5):921–928

Wafaa B, Said B, Mohamed B (2010) Antidiabetic medicinal plants as a source of alpha glucosidase inhibitors. Curr Diabetes Rev 6(4):247–254

Wakabayashi H, Takase M, Tomita M (2003) Lactoferricin derived from milk protein lactoferrin. Curr Pharm Des 9(16):1277–1287

Walentin S, Lévay G, Korányi L, Endroczi E (1988) Comparative analysis of enzyme activity in human colostrum, milk, and serum. Clin Biochem 21(2):131–133

Wang C-S, Weiser D, Alaupovic P, McConathy WJ (1982) Studies on the degradation of human very low density lipoproteins by human milk lipoprotein lipase. Arch Biochem Biophys 214(1):26–34

Wang P, Zhang J, Bian H, Wu P, Kuvelkar R, Kung TT, Crawley Y, Egan RW, Billah MM (2004) Induction of lysosomal and plasma membrane-bound sialidases in human T-cells via T-cell receptor. Biochem J 380(Pt 2):425

Wardell JM, Wright AJ, Bardsley WG, D'Souza S (1984) Bile salt-stimulated lipase and esterase activity in human milk after collection, storage, and heating: nutritional implications. Pediatr Res 18:382–386

Weber B, Nielsen S (1991) Isolation and partial characterization of a native serine-type protease inhibitor from bovine milk. J Dairy Sci 74(3):764–771

White JH, Zavizion B, O'Hare K, Gilmore J, Guo MR, Kindstedt P, Politis I (1995) Distribution of plasminogen activator in different fractions of bovine milk. J Dairy Res 62:115–115

Wickramasinghe S, Rincon G, Islas-Trejo A, Medrano JF (2012) Transcriptional profiling of bovine milk using RNA sequencing. BMC Genomics 13(1):45–58

Wiederschain GY, Newburg DS (1995) Human milk fucosyltransferase and -L-fucosidase activities change during the course of lactation. J Nutr Biochem 6(11):582–587

Wiederschain GY, Newburg DS (1996) Compartmentalization of fucosyltransferase and alpha-fucosidase in human milk. Biochem Mol Med 58(2):211–220

Wiederschain GY, Newburg DS (2001) Glycoconjugate stability in human milk: glycosidase activities and sugar release. J Nutr Biochem 12(10):559–564

Wight NE (2001) Donor human milk for preterm infants. J Perinatol 21:249

Williams LK, Zhang X, Caner S, Tysoe C, Nguyen NT, Wicki J, Williams DE, Coleman J, McNeill JH, Yuen V, Andersen RJ, Withers SG, Brayer GD (2015) The amylase inhibitor montbretin A reveals a new glycosidase inhibition motif. Nat Chem Biol 11:691

Windey K, De Preter V, Verbeke K (2012) Relevance of protein fermentation to gut health. Mol Nutr Food Res 56(1):184–196

Wittlin S, Rösel J, Hofmann F, Stover DR (1999) Mechanisms and kinetics of procathepsin D activation. Eur J Biochem 265(1):384–393

Yen SS, Quigley ME, Reid RL, Ropert JF, Cetel NS (1985) Neuroendocrinology of opioid peptides and their role in the control of gonadotropin and prolactin secretion. Am J Obstet Gynecol 152(4):485–493

Zarban A, Taheri F, Chahkandi T, Sharifzadeh G, Khorashadizadeh M (2009) Antioxidant and radical scavenging activity of human colostrum, transitional and mature milk. J Clin Biochem Nutr 45(2):150–154

Zechner R (1990) Rapid and simple isolation procedure for lipoprotein lipase from human milk. Biochim Biophys Acta 1044(1):20–25

Zhou GX, Chao L, Chao J (1992) Kallistatin: a novel human tissue kallikrein inhibitor. Purification, characterization, and reactive center sequence. J Biol Chem 267(36):25873–25880

Zikakis JP, Dougherty TM, Biasotto NO (1976) The presence and some properties of xanthine oxidase in human milk and colostrum. J Food Sci 41(6):1408–1412

Zoppi G, Andreotti G, Pajno-Ferrara F, Njai DM, Gaburro D (1972) Exocrine pancreas function in premature and full term neonates. Pediatr Res 6:880

Zucht H-D, Raida M, Adermann K, Mägert H-J, Forssmann W-G (1995) Casocidin-I: a casein-αs2 derived peptide exhibits antibacterial activity. FEBS Lett 372(2-3):185–188

Zwirzitz A, Reiter M, Skrabana R, Ohradanova-Repic A, Majdic O, Gutekova M, Cehlar O, Petrovčíková E, Kutejova E, Stanek G, Stockinger H, Leksa V (2018) Lactoferrin is a natural inhibitor of plasminogen activation. J Biol Chem 293:8600–8613

Chapter 10
Lipases from Milk and Other Sources

Hilton C. Deeth

10.1 Introduction

This chapter focusses on the lipases in bovine and caprine milk, and lipolysis caused by the main lipase, lipoprotein lipase (LPL). It also discusses the use of lipases for modifying milk fat for various applications.

Bovine milk lipase and its effects in milk were widely studied in the 1970s and 1980s (e.g. Castberg and Solberg 1974; IDF 1975, 1980, 1987), but relatively little has been published since then. There has been considerable interest in the biochemical aspects of LPL because of its similarity with other LPLs, including human LPL, and its ready availability from milk for lipid metabolic studies. These studies have been reviewed by Olivecrona and Olivecrona (1999) and Olivecrona et al. (2003).

Lipolysis caused by LPL has attracted, and continues to attract, considerable interest from the dairy industry because of its practical consequences for the quality of milk and milk products. It has been studied in detail in both bovine and caprine milk and extensively reviewed (IDF 1980, 1987; Chilliard et al. 1984, 2003, 2014; Deeth 2006; Deeth and Fitz-Gerald 2006; Ray et al. 2013).

Lipolysis caused by lipases involves the transfer of an acyl group to water with formation of a carboxylic acid. Under water-limiting conditions, lipases can cause other acyl transfer reactions, which include: interesterification, in which acyl groups are interchanged between an introduced ester and the original ester such, as a triglyceride (TG); acidolysis, in which an introduced carboxylic acid replaces one of the original esterified acids and a new free acid is formed; and alcoholysis, in which new esters are formed by transfer of an acyl group to an alcohol. The lipases involved in these biotechnological modifications of milk fat are mostly of microbial origin.

H. C. Deeth (✉)
School of Agriculture and Food Sciences, University of Queensland,
Brisbane, QLD, Australia
e-mail: h.deeth@uq.edu.au

A. L. Kelly, L. B. Larsen (eds.), *Agents of Change*, Food Engineering Series,
https://doi.org/10.1007/978-3-030-55482-8_10

Lipase-catalysed modifications now have several commercial applications (Bourlieu et al. 2009).

10.2 The Milk Enzymes

10.2.1 Lipoprotein Lipase

Lipoprotein lipase (LPL) has been established as the only lipase in bovine milk, and is present in the milk of virtually all mammals. However, in human milk it occurs with another carboxyl ester lipase, commonly known as bile-salt-stimulated lipase (Hui and Howles 2002), which is not present in bovine milk (Castberg and Solberg 1974). Milk LPL is synthesised in the mammary gland, where it hydrolyses circulating lipoproteins from the blood into free fatty acids and 2-monoglycerides (2-MGs). These products are absorbed by the mammary gland and used in the synthesis of milk glycerides, mainly TGs.

The presence of LPL in bovine milk was first reported by Quigley et al. (1958) and confirmed by Korn (1962). The former identified the enzyme in milk fat globule membrane (MFGM) material (which they called lipoprotein microsomes) from washed cream buttermilk but Korn (1962) found that, although some of the enzyme was associated with the MFGM, most was associated with the casein in the skim milk. It was not until the 1970s that it was established that LPL was responsible for virtually all of the lipase activity in normal mature bovine milk (Egelrud and Olivecrona 1973; Castberg and Solberg 1974; Castberg et al. 1975; Olivecrona et al. 1975). Before that, there had been several reports of two or more different lipases in milk, some of which had been separated by gel filtration (e.g., Downey and Andrews 1969). The early reports of these enzymes were reviewed by Schwartz and Parks (1974). Interestingly, these authors made only passing reference to LPL in the statement "*Milk also contains a lipoprotein lipase similar to that found in mammalian tissue and heart tissue* (Korn 1962)*. The enzyme is highly specific for lipoproteins and probably is not involved in rancidity development to any great extent*". As discussed below, this has turned out to be untrue.

Bovine milk LPL has been extensively studied as it has been used as a model enzyme for studies on lipoprotein metabolism, where it is the major lipase involved (Olivecrona et al. 2003). In blood, it hydrolyses TGs in chylomicrons and very-low-density lipoproteins into free fatty acids and 2-MGs. These products are absorbed by tissues and used for energy production and, in the case of the mammary gland, for the synthesis of lipids, mainly TGs.

Milk LPL has some unique features; for example, its full activity is realised only when a specific lipoprotein activator is present. Hence, LPL is assayed against TGs in systems containing blood serum, serum lipoproteins or the specific activator, apolipoprotein C-II. It is believed that the activator helps to orient the enzyme favourably towards the substrate at the oil-water interface. However, LPL is active against tributyrin in the absence of these activators, although in such cases the total activity is ≤50% of the activated LPL activity (Castberg et al. 1975).

In normal raw bovine milk, LPL is mostly associated with the casein fraction, with only a small proportion being soluble or attached to the MFGM (Deeth and Fitz-Gerald 1975; Anderson 1982). However, this does not apply in caprine milk, where a substantial proportion of the LPL is normally associated with the MFGM. Chilliard et al. (1984) reported that the percentage distribution of the enzyme between the cream, serum and caseins in caprine milk was 46, 46 and 8, respectively, while the corresponding figures for bovine milk was 6, 17 and 78 respectively. Those authors suggested that the difference may relate to different LPL secretion processes in the mammary gland, differences in casein and fat globule composition, structure or physicochemical properties, or the presence of an LPL inhibitor in the casein fraction of caprine milk.

The association with the casein micelles is mostly electrostatic, as most of it can be released by 0.75-1 M NaCl into a soluble form. The electrostatic binding is through positive charges on the enzyme to negative charges on the caseins such as κ-casein. An important feature of LPL is that it binds strongly to heparin, through what is also an electrostatic interaction as heparin is negatively charged. Thus, heparin, at ~5 μg/mL, can also be used to release LPL from casein micelles (Hoynes and Downey 1973); in fact, the method of choice for isolating LPL from milk is by affinity chromatography on immobilised heparin (Egelrud and Olivecrona 1972; Castberg et al. 1975; Kinnunen et al. 1976). Hydrophobic bonding may also be involved in the attachment of lipase to casein micelles, as the enzyme can be released by dimethylformamide.

LPL exists predominantly as a dimer of two identical monomers. According to Olivecrona et al. (1982), the molecular weight of each monomer is 41.7 kDa, of which 8% is carbohydrate. However, Senda et al. (1987) calculated the molecular weight of the unglycosylated form from the cDNA coding for it to be 50.548 kDa, and Hayashi et al. (1986) reported a molecular weight of 56 kDa as determined by SDS-PAGE. The LPL molecule has a heparin binding site in the groove between the two monomers and a surface loop known as a lid which shields the hydrophobic active site from a hydrophilic environment (van Tilbeurgh et al. 1994). The lid opens to expose the active site when the lipase adsorbs to a lipid-water interface.

The enzyme has a reactive serine in an active-site peptide which is located in a β-turn, a structure similar to that at the active sites on other serine hydrolases (Reddy et al. 1986). In addition to the active site and the heparin-binding region, LPL also has an apo-lipoprotein interaction region and a lipid-binding or interface-recognition site. The apo-lipoprotein interacting region is independent of the heparin-binding region and the active site (Matsuoka et al. 1980; Olivecrona et al. 2003).

LPL is relatively unstable to heat as well as UV light, acid and oxidising agents. In milk, it is largely inactivated by normal pasteurisation conditions, involving heating at 72 °C for 15 s (Farkye et al. 1995). This means that adequately pasteurised milk and products made therefrom do not contain active lipase. As discussed below, if this was not the case, most (homogenised) pasteurised milk would develop a rancid flavour. However, some workers have maintained that more severe heating conditions such as 79 °C for 20 s (Shipe and Senyk 1981) or 85 °C for 10 s (Driessen 1987) are required to completely inactivate LPL in milk.

LPL has an optimum pH of approximately 8.0 under ideal (activated) conditions; however, in the absence of activating cofactors, the optimum is 8.8–9.0. LPL is inactivated by acid conditions such as at pH 4.6 (Driessen 1976), and thus it is highly unlikely that it is involved in digestion of fat by the newborn, as it would be inactivated by the acidic gastric juice in the stomach. This has led researchers to suggest its presence in milk is due to spill-over from the mammary gland and has no useful function in milk.

TGs are the normal substrates for LPL. If TGs contain long-chain fatty acids, lipoproteins or apolipoproteins, especially apo-LP CII, are required to activate the enzyme (Östlund-Lindqvist and Iverius 1975). Blood serum, either bovine or human, supplies such activating factors and is used in assays of LPL using emulsified long-chain TG substrates (Egelrud and Olivecrona 1973). LPL is usually assayed in the presence of bovine serum albumin, which absorbs the released free, fatty acids and reduces product inhibition. One reason why FFAs cause inhibition is that they block the effect of activator lipoproteins (Bengtsson and Olivecrona 1980; Olivecrona et al. 2003). LPL can also be assayed against tributyrin; in this case, neither serum cofactors nor fatty acid acceptors are required (Rapp and Olivecrona 1978).

Normal unprocessed milk contains an abundance both of LPL and the ideal TG substrate. At the maximum activity in milk, of around 5 μmoles of FFA/min/mL (Castberg et al. 1975), milk would develop a very high level of FFAs in a very short time. This, however, does not happen because the LPL is physically separated from the TG substrate in the fat globules by the MFGM which it cannot penetrate; in this respect, LPL differs from bacterial lipases and pancreatic lipase which can hydrolyse the fat in intact fat globules. This difference was neatly demonstrated by Danthine and Blecker (2014) with MFGM monolayers using a Langmuir film balance. Pancreatic and bacterial lipase penetrated the monolayer but LPL did not. If blood serum, which contains activating cofactors, is added to milk, the interaction between the enzyme and the TG substrate is facilitated and lipolysis ensues (Castberg and Solberg 1974; Jellema and Schipper 1975). This serum-mediated induction of lipolysis is believed to be largely due to enhancement of binding of the LPL to the fat globule by the activating cofactors (Bengtsson and Olivecrona 1982).

In hydrolysis of mixed TGs, such as occur in milk fat, LPL does not show fatty acid specificity. However, it does have strong positional specificity, hydrolysing the fatty acids in the primary positions, *sn*-1 and *sn*-3, of TGs (Morley and Kuksis 1977). Because short-chain fatty acids such as butyric acid are mostly positioned in the primary positions, it has sometimes been interpreted that LPL preferentially releases short-chain fatty acids. When LPL hydrolyses a fatty acid from a primary position, *sn*-1,2- and *sn*-2,3-diglycerides (DGs) are formed. These are then hydrolysed to 2-MGs (see Fig. 10.1). These can only be hydrolysed to a free fatty acid and glycerol if they are converted to the *sn*-1- or *sn*-3-MGs (Nilsson-Ehle et al. 1973).

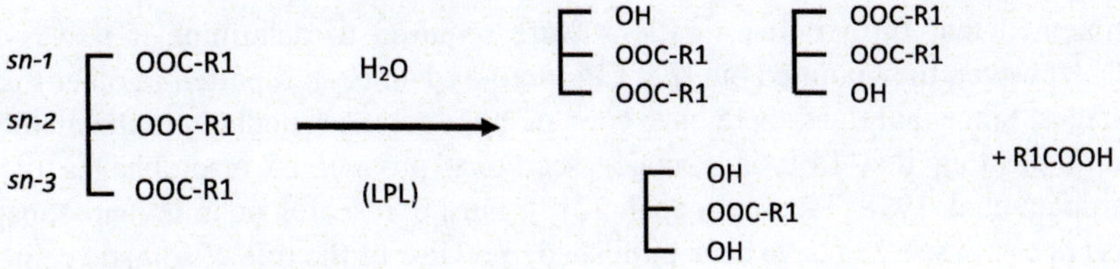

Fig. 10.1 Hydrolysis (lipolysis) of TG catalysed by LPL

LPL also hydrolyses phospholipids such as phosphatidyl choline. In this case, it hydrolyses only the primary ester bond at the *sn*-1 position to produce a 2-lysophospholipid, i.e., LPL has phospholipase A_1 activity. Interestingly, most phospholipases, such as those in snake venom, exhibit A_2 activity.

10.2.2 Somatic Cell Lipase

Chandan and Shahani (1963a, b) reported the purification and characterization of what they called "milk lipase". The enzyme, which was isolated from clarifier sediment, had a molecular weight of ~7000, determined by sedimentation-diffusion and osmotic pressure methods (Chandan et al. 1963). They considered that it was not from somatic cells as they (mistakenly) believed that leucocytes did not possess lipase activity. Gaffney and Harper (1965) isolated somatic cells from separator slime and showed that they contained a lipase which was different from the majority of the lipase in skim milk. They also showed that it was an intracellular lipase as the activity expressed was much greater after the cells had been disrupted by a ball mill. The specific lipase activity (per mg protein) against 3% homogenised milk fat at pH 8.6 was around four times greater than that of skim milk. The lipase was also active against tributyrin and dibutylfluorescein. When the somatic cell extract and skim milk were run on polyacrylamide gel electrophoresis, the band of the cell extract containing lipase activity (against dibutylfluorescein) did not match any of the four lipase-active bands in skim milk.

Azzara and Dimick (1985) isolated somatic cells from pooled raw milk with somatic cell count (SCC) of $<5 \times 10^5$/mL. The cells consisted of macrophages, polymorphonuclear leucocytes and lymphocytes. Of these, only macrophages possessed lipase activity. The total activity in the milk from the macrophages was calculated to be only about 0.1% of the total activity in skim milk. Curiously, when a concentrate of macrophages was added to milk, the total milk lipase activity after 48 h was 11.6% higher than in control milk to which cells had not been added.

The assay system used by Azzara and Dimick (1985) was one for lipoprotein lipase but the authors did not claim that the lipolytic activity present was LPL; they commented that further investigations were required to determine if it was an LPL. However, they pointed out that LPL activity had been reported in other macrophages. Since that time, there have been many reports, particularly in the medical field, indicating that LPL is a major secretory product of macrophages (e.g., Sartippour et al. 1998; Takahashi et al. 2013). Further research on milk macrophage lipase does not seem to have been published; a review of the role of somatic cells in dairy processes and products by Li et al. (2014) only lists the lipase activity reported by Azzara and Dimick (1985).

10.2.3 Colostral Lipase

Colostrum has been shown to contain a lipase which differs from lipoprotein lipase (Driessen 1976). Its activity decreases during the first four milkings, whereas LPL increases from virtually zero in the same time period. It is present in the milk serum and is stable at pH 4.6; LPL is mostly associated with casein micelles and is unstable at pH 4.6, and so clearly the colostral lipase differs from LPL.

10.2.4 Milk Esterases

There have been several reports of esterases in bovine milk. However, little detailed information on them is available. By definition, esterases differ from lipases in that they act on ester substrates in solution rather than emulsified lipid substrates (Desneulle 1972; Jaeger et al. 1994) or show a preference for esters of short-chain acids over esters of long-chain acids (Okuda and Fujii 1968).

Aryl- or A-esterase (EC 3.1.1.2) (Forster et al. 1961; Kitchen 1971), carboxyl- or B-esterase (EC 3.1.1.1) (Jensen et al. 1961; Montgomery and Forster 1961; Deeth 1978) and cholin- or C-esterase (EC 3.1.1.7; EC 3.1.1.8) (Forster et al. 1961; Kitchen 1971) have been reported. Arylesterase activity has attracted interest because it is elevated in mastitic milk; it is also present in colostrum (Forster et al. 1959; Marquardt and Forster 1965). Its activity is well correlated with other indices of mastitis (Luedecke 1964) and has been suggested as a sensitive indicator of the disease (Forster et al. 1961).

Of particular interest in this chapter is carboxylesterase (B-esterase) activity. Its presence in milk was reported (Jensen et al. 1961; Montgomery and Forster 1961) before it was established that LPL was the only lipase in normal mature milk and was responsible for lipolysis in raw milk (Castberg and Solberg 1974). It was identified as a carboxylesterase because it hydrolysed tributyrin; it appears that the enzyme concerned was LPL.

It has now been shown that a carboxylesterase which differs from LPL exists in milk. It was identified as a carboxylesterase through inhibitor studies where it was not affected by inhibitors of arylesterase (EDTA, La^{3+}) or of cholinesterase (eserine sulfate) but was almost completely inhibited by diethyl-*p*-nitrophenyl phosphate (an inhibitor of carboxylesterase). Its activity is low in normal mature milk but is elevated is colostrum and mastitic milk. In mastitic milk, it is well correlated with other indices of mastitis (Deeth 1978; Fitz-Gerald et al. 1981), and is present in both the milk serum and the cellular fractions of mastitic milk but not in the casein fraction (where most of the LPL resides). It has been assayed using 4-methylumbelliferyl heptanoate, a substrate against which LPL is not active (Deeth 1978). Its activity in non-mastitic milk was found to be around 0.004 μmoles of FFA/min/mL of milk while, in mastitic milk (SCC > 2 × 10^6/mL) and colostrum, it was up to 0.15 and 0.04 μmoles of FFA/min/mL respectively. Given that the LPL activity is around 5 μmoles of FFA/min/mL (Castberg et al. 1975), the carboxylesterase activity represents a small proportion of the carboxyl ester hydrolase activity in bovine milk, even in abnormal milks.

It seems highly likely that the carboxylesterase activity in mastitic milk corresponds to the reported lipase(s) of somatic cells (Gaffney and Harper 1965; Azzara and Dimick 1985) and colostrum (Driessen 1976). Furthermore, human milk shows very high carboxylesterase activity against 4-methylumbelliferyl heptanoate (~18 μmoles of FFA/min/mL, Deeth 1978) which is presumably due to the bile-salt-stimulated lipase (Hernell and Olivecrona 1974) which has been shown to be identical to the non-colipase-dependent pancreatic carboxylesterase/lipase (Falt et al. 2001) and cholesterol esterase (Hui and Howles 2002).

10.3 Lipolysis Due to Milk Lipase in Milk

Lipolysis by milk lipase occurs in milk before heat treatments equivalent to HTST pasteurisation. It is traditionally divided into two categories, spontaneous and induced lipolysis, as their mechanisms of action are quite different. The former is mainly related to factors associated with the cow, e.g., stage of lactation, nutritional status, while the latter is related to milk handling.

The main reason for the interest in lipolysis in milk is the fact that the free fatty acids formed, particularly the short-chain acids such as C4 (butyric) and C6 (caproic), impart an unpleasant flavour to milk. A second reason for the interest is that the lipolysis products, particularly the partial glycerides, are surface-active and reduce the foaming capacity of milk, an important consideration for making cappuccino coffee, for example.

The FFA level of milk as it leaves the cow is of the order of 0.5 mmol/L. When this level increases to around 1.5 mmol/L, many consumers detect an unpleasant taste variously described as rancid, butyric, chemical or bitter. As the FFA level increases further, more consumers will find the milk unacceptable. It should be noted that the literature varies in reports on the FFA levels deemed to make milk unacceptable. This is attributable to the different methodologies used for both analysis of the FFAs and the sensory evaluation methodology used.

Similarly, the foaming capacity declines with increasing FFA level. At around 1.5 m mol/L, the foaming capacity is usually impaired and, at 2.0 m mol/L, most milk shows little foaming. As for the sensory perception of lipolyzed milk, the reported FFA levels for impaired foaming also vary.

10.3.1 Spontaneous Lipolysis

Under normal circumstances, the milk fat globule membrane in bovine milk acts an effective barrier between the LPL in the milk serum and its TG substrate contained within the milk fat globule. In this case, when milk is extracted from the cow and cooled, no lipolysis ensues, that is, the FFA level remains the same as when the milk first leaves the cow. If the milk fat was not protected in this manner, the LPL in the milk serum would rapidly act on the fat and cause extensive lipolysis. Under ideal conditions for the enzyme action, it can be calculated that milk would become rancid within a few minutes (Olivecrona et al. 2003).

While "normal" milks do not lipolyse after leaving the cow, there are some milks in which lipolysis proceeds after they are cooled to <~10 °C; this is referred to as spontaneous lipolysis. For the sake of brevity, milks which undergo spontaneous lipolysis will be referred to as "spontaneous" milks in this chapter. There has been conjecture about the difference between non-spontaneous and spontaneous milks and the factors which influence the extent of spontaneous lipolysis. A summary of extrinsic and intrinsic factors with an assessment of their significances is given in Table 10.1. These factors are discussed below.

The situation in a herd is that the milk of some cows will always be non-spontaneous, milk of some will always be spontaneous, and the milk of others will be spontaneous under certain conditions. These conditions include late lactation and poor-quality feed. The extent of lipolysis in spontaneous milk can vary from <1 mmol/L up to 5 mmol/L or more.

Table 10.1 Factors influencing spontaneous lipolysis in raw milk

Factor	Significance
Extrinsic	
Late lactation	Medium
Poor quality feed	Medium
Individuality (genetics?)	High
Breed	Low
Mastitic infection	Medium
Intrinsic	
LPL activity	Very low
MFGM integrity	Medium
Activator-inhibitor balance	High
MFGM-associated LPL	High

A priori, the LPL activity in the milk and/or the integrity of the MFGM could be expected to account for much of the difference. Because there is ample lipase activity in all raw milk to cause extensive lipolysis under the right conditions, differences in lipase activity cannot account for the difference between the milks. Furthermore, through the use of mixing experiments where creams from spontaneous milks are mixed with skim milk from non-spontaneous milk, the nature of the MFGM does not appear to be a major factor in most cases.

What has been shown to be a major difference between non-spontaneous and spontaneous milks is the balance between inhibiting and activating factors in milks. This is indicated in the results shown in Table 10.2 of an experiment in which a non-spontaneous milk was mixed with spontaneous milks. It has long been recognised that, when non-spontaneous milk is mixed with spontaneous milk soon after milking and before cooling, the FFA level formed during low-temperature storage is less than the expected average of the FFA levels of the two milks when kept separate. In other words, the non-spontaneous milk has an inhibitory effect on the lipolysis in the spontaneous milk. This inhibition is shown in the first data row in Table 10.2. However, it can also be seen from Table 10.2 that mixing a very spontaneous milk (FFA of 4.8 mmol/L) with two other spontaneous milks (FFAs of 3.1 and 3.15 mmol/L) resulted in little change in one and a substantial (35%) activation in the other. It is generally accepted that this inhibitor-activator balance is most significant and that the nature of the MFGM is also significant in some cases (Deeth and Fitz-Gerald 1975; Sundheim and Bengtsson-Olivecrona 1987c; Cartier and Chilliard 1990).

The major activators of LPL are lipoproteins and apolipoprotein C-II, so it is logical to expect such factors to be present in spontaneous milks. In fact, milk does contain material which is immunologically cross-reactive with bovine serum lipoproteins (Anderson 1979). In support of the supposition that such lipoproteins are the activating factors is the fact that adding blood serum lipoproteins to milk initiates lipolysis. In doing so, they facilitate the movement of the LPL from the skim phase to the MFGM (Bachman and Wilcox 1990); the same occurs in spontaneous lipolysis (Sundheim and Bengtsson-Olivecrona 1985). The extent of spontaneous lipolysis correlates positively with the amount of LPL attaching to the MFGM (Ahrné and Björck 1985; Sundheim and Bengtsson-Olivecrona 1985).

Table 10.2 Inhibition and activation of spontaneous lipolysis in mixed individual bovine milk samples (data based on Deeth and Fitz-Gerald 1975)

		FFA[a] of mixed milk (1:1 with Spontaneous milk 3, FFA = 4.8)		
Added milk	FFA[a]	Observed FFA	Calculated FFA	% inhibition
Non-spontaneous	0.5	0.9	2.65	81
Spontaneous 1	3.15	4.1	4.2	2
Spontaneous 2	3.1	5.15	4.1	−35
Spontaneous 3	4.8			

[a]All FFA values were determined after overnight storage of milks at 5 °C

The attachment of LPL to the MFGM in spontaneous milk does not occur or is reduced if: the milk is not cooled; cooling is delayed; or NaCl or other inhibiting factors are added. However, it is not reduced if the milk is rewarmed after being cooled (Sundheim and Bengtsson-Olivecrona 1987a).

An effective inhibitor of spontaneous lipolysis is proteose-peptone fraction 3 (Anderson 1981; Cartier et al. 1990) which is antigenically similar to a soluble glycoprotein in the MFGM which may also act as an inhibitor (Shimizu et al. 1982). Vanbergue et al. (2018) have recently suggested that proteose-peptone fraction 5 may be an inhibitor of spontaneous lipolysis. They found a negative correlation between the level of spontaneous lipolysis and the concentration of proteose-peptone fraction 5 which appeared to be related to forage type. In addition, a caprine milk proteose peptone fraction inhibits spontaneous lipolysis in caprine milk (Chilliard et al. 1984). Interestingly, Danthine and Blecker (2014) reported that proteose peptone strengthened MFGM monolayers against penetration by both LPL and pancreatic lipase, suggesting that the mechanism of inhibition of spontaneous lipolysis may be due to strengthening of the MFGM.

The MFGM contains both activating lipoproteins (Castberg and Solberg 1974) and inhibiting proteins (Shimizu et al. 1982; Sundheim and Bengtsson-Olivecrona 1987b). In addition, LPL attaches to the MFGM of both non-spontaneous and spontaneous milk when milk fat globules are cooled in the absence of skim milk, which suggests the membrane changes its conformation to facilitate attachment of LPL, if not prevented by inhibitors (Sundheim and Bengtsson-Olivecrona 1987d). The MFGM in the milk of some cows becomes more permeable in late lactation, which may partially explain the increased susceptibility of late-lactation milk to spontaneous lipolysis.

A possible role of genetics in the susceptibility of the milk of some cows to spontaneous lipolysis has been indicated in studies on both cows and goats. Vanbergue et al. (2016) investigated the relationship between diacylglycerol acyltransferase 1 (DGAT1), an enzyme linked to milk fat synthesis, and spontaneous lipolysis through identifying cows of three different DGAT1 genotypes, KK, KA and AA. Significantly higher levels of spontaneous lipolysis were observed in the milk of the KK genotype than in the milk of either the KA or AA genotype.

In goats, the α_{S1}-casein (CSN1S1) gene polymorphism has been shown to be related to spontaneous lipolysis. More spontaneous lipolysis was observed in the milk of goats of the CSN1S1–FF genotype (sometimes referred to as the low genotype as it is associated with production of a low concentration of α_{S1}-casein in milk) than in milk from goats of the (high) AA genotype (Chilliard et al. 2003, 2013, 2014).

Breed of cow may also be a factor in spontaneous lipolysis in bovine milk. Vanbergue et al. (2017) reported that milk from Holstein cows was more susceptible than milk from Normande cows, while Bachman et al. (1988) found the milk of Jerseys to be more susceptible than that of Holsteins. However, other authors have reported no differences between breeds in relation to spontaneous lipolysis (Chilliard 1982); Chazal and Chilliard (1987) found no differences between the milk of Friesian and Montbéliarde cows.

Several authors have reported a relationship between feed and spontaneous lipolysis (reviewed by Jellema 1980). The influence of feed involves the availability, type and nutritive value. For example, Stobbs et al. (1973) found that adding 6 kg/d of a balanced concentrate to the diet of late-lactation cows fed a low quality tropical pasture significantly reduced spontaneous lipolysis in the milk of those cows (average FFA decreased from 2.09 to 1.47 mmol/L). Vanbergue et al. (2018) concluded that the higher level of spontaneous lipolysis associated with feed restriction is probably due to higher tissue lipid mobilization.

There have been several reports on the effect of mastitic infection on lipolysis in milk. Most reports show that the level of free fatty acids in raw milk increases with somatic cell count (SCC) (Gargouri et al. 2008). Mastitic milk as it leaves the udder usually has a higher FFA level than non-spontaneous milk, and this level increases during storage (Gargouri et al. 2013). However, Jurczak and Sciubisz (1981) reported that milks with different SCCs analysed immediately after milking had similar FFA levels to non-mastitic milks. Gargouri et al. (2008) analysed the bulk milk from 80 farm supplies to investigate the relationship between the total SCC, counts of different types of somatic cells (macrophages, lymphocytes and polymorphonuclear leukocytes (PMN)) and FFA level. They found a significant positive correlation between the total SCC, the PMN counts and FFA level in milk, and concluded that the PMN leucocytes could be involved in the lipolysis of mastitic milk. This suggests that the lipase in the cells, as discussed in Sect. 2.2, may play a role in elevating the FFA in mastitic milk. This is consistent with the conclusion of Jurczak and Sciubisz (1981) that lipolysis of mastitic milk was due to both spontaneous lipolysis involving LPL and lipolysis due to the activity of lipases of somatic cells, with the former being the most significant. This is despite the fact that LPL levels generally decrease with increasing somatic cell count, largely due to to impaired biosynthetic capability of the mammary tissues caused by the infection (Fitz-Gerald et al. 1981; Murphy et al. 1989). However, this decrease would not be expected to affect the level of lipolysis, as the residual LPL activity would be more than sufficient to cause considerable lipolysis.

Caprine milk differs from bovine milk in that its total LPL activity is lower. Furthermore, for spontaneous lipolysis in caprine milk, there is a reasonable correlation between LPL activity and level of spontaneous lipolysis, whereas, in bovine milk, the total LPL has little bearing on the level of spontaneous lipolysis. This may relate to the fact that a much larger proportion of the LPL is naturally associated with the MFGM in caprine milk than in bovine milk (Chilliard et al. 2003).

10.3.2 Induced Lipolysis

The major form of induced lipolysis involves the physical disruption of the MFGM allowing access of LPL to its substrate in the damaged fat globule. This type of lipolysis is equivalent to the action of the lipase on an emulsified TG substrate and does not require activating factors.

Table 10.3 Processes which cause induced lipolysis in raw milk

Location	Process
Farm	Air leaks at teat cup cluster
	Milk surging in vertical risers
	Temperature activation
	High milking frequency
	Milk transport with foaming
Factory	Homogenisation
	Mixing homogenised milk with raw milk
	Centrifugal pumping, especially with air intake
	Pumping milk over long distances, especially with sharp pipe bends
	Gas agitation of silos
	Membrane processing (UF, RO)
Laboratory	Mechanical agitation (e.g., with blenders such as Silverson and Sorvall Omni Mixer)
	Gas bubbling
	Mixing homogenised milk with raw milk

The physical disruption to induce lipolysis can take different forms. Table 10.3 summarises ways in which lipolysis can be induced in the laboratory, on the farm and in the factory. In the laboratory, agitation in a blender is an effective way of producing the effect. The temperature of the milk treated in this way has a major effect on the amount of lipolysis induced. In general, the greatest level lipolysis is induced at 37–40 °C and the least at <~5 °C. However, under certain conditions, the amount of induced lipolysis shows maxima at 12–15 °C and 30–40 °C, with a minimum at 20 °C. Furthermore, there appears to be a threshold mechanical stress intensity at each temperature above which lipolysis can be induced (Deeth and Fitz-Gerald 1977; Hisserich and Reuter 1984).

One effect of agitation is an increase in lipase activity associated with the fat portion of milk. As a result, cream separated from milk agitated at 5–10 °C or 37 °C has a several-fold higher lipase activity than cream from milk which has not been agitated (Deeth and Fitz-Gerald 1977). The lipase in the cream from the agitated milk appears to have enhanced heat stability due to protection afforded by the fat.

In industrial processing, the most effective means of inducing lipolysis is homogenisation. Homogenisation of raw milk causes rapid lipolysis, particularly at the warm temperature commonly used in commercial processing. Because of this, homogenisation is usually carried out either immediately before or immediately after pasteurisation (in which LPL is inactivated).

A related issue is the mixing of homogenised (pasteurised) milk with raw milk. In this case, rapid lipolysis occurs because the active lipase in the raw milk can act on the fine TG emulsion in the homogenised milk. This issue is of constant concern to the dairy industry, as personnel unaware of the consequences can add homogenised milk to raw milk in the misguided belief that it will prevent wastage of left-over homogenised milk.

Perhaps the most common cause of induced lipolysis in the dairy industry is through the intensive mechanical handling of raw milk, particularly if accompanied by incorporation of air with formation of foam. This can occur in milking machines at the farm and during pumping of milk where air is drawn into the milk flow through damaged or worn seals. The shear forces at the air-liquid interface disrupt the membranes of the fat globules adhering to the interface and expose the fat to the lipase.

Other industrial processes which have been implicated in inducing lipolysis in raw milk are pumping milk over long distances, especially if the pipe includes sharp bends, tanker transport, especially when the milk has a large headspace and foam is generated, and during some membrane processing where rapid changes of pressure occur (Barbano et al. 1983). It has been noted that lipolysis has increased of an issue for the dairy industry since the introduction of mechanical means of handling raw milk. A recent detailed study of on-farm factors which may affect lipolysis, as measured by FFA levels, in bulk milk supports this contention (Wiking et al. 2019); that study showed that an automated milking system contributed significantly more to FFA levels than other milking systems. Furthermore, the high milking frequency of the automated system was a major factor contributing to higher FFA levels.

Another form of induced lipolysis is what is termed "temperature activation". This involves cooling raw milk or cream to ≤5 °C, warming to 25–35 °C, and re-cooling to <10 °C. Lipolysis occurs during subsequent low temperature storage. In this way, this form of induced lipolysis resembles spontaneous lipolysis; in fact, spontaneous milks are more likely to be affected. The susceptibility of a milk to temperature-activated lipolysis is related to the MFGM which is essential for the phenomenon. The temperature manipulation causes a conformational change of the membrane which allows the lipase to attach to the membrane and gain access to the fat. The lipase–MFGM attachment is reversed on rewarming; this reversal of attachment does not occur in spontaneous or agitation-induced lipolysis (Claypool 1965; Cartier and Chillard 1989; Kon and Saito 1997).

10.4 Lipase-Catalysed Acyl-Transfer Reactions on Milk Fat

Acyl-transfer reactions occur in organic solvents containing just enough water to maintain the activity of the enzyme but minimising the hydrolysis reaction. The most common enzymic transfer reactions for modifying milk fat discussed in recent publications are, in decreasing order of number of publications, interesterification, acidolysis and alcoholysis (particularly glycerolysis) (Bourlieu et al. 2009).

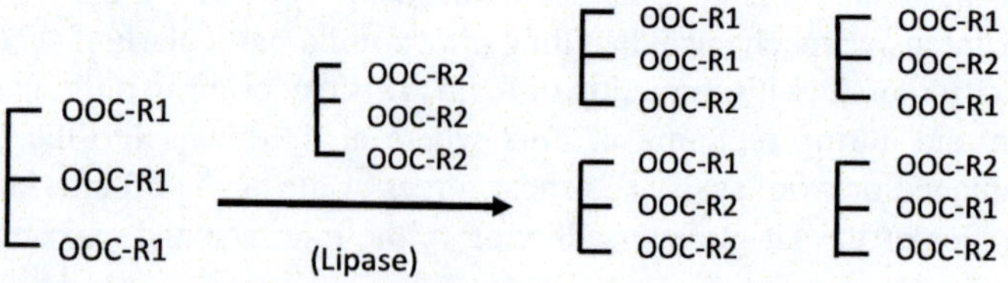

Fig. 10.2 Interesterification of two TGs catalysed by a non-sterospecific lipase

10.4.1 Interesterification

There are basically two forms of interesterification of milk fat, one in which milk fat is the only fat reacted with the lipase and another where TGs which differ from those in milk fat are incorporated (see Fig. 10.2 for generic interesterification reaction). If carried out with just milk fat, a modified milk fat is the major product, that is, a mixture of TGs with the same overall fatty acid composition as the original milk fat but with the fatty acids in different relative positions. The changes in TG structure are determined by the positional specificity of the lipase used. Therefore, to change the fatty acids at the *sn*-1 and *sn*-3 positions, a lipase which specifically attacks these positions, such as *Rhizomucor miehei* lipase, is used. If the lipase used does not have any positional specificity, a random exchange of all fatty acids present will result. If milk fat alone is reacted with a lipase, the product has different melting characteristics from the original fat. For example, interesterification of milk fat with the *sn*-1,3-specific lipase of *Rhizomucor miehei* decreased the proportion of TG molecules containing 34 to 42 carbon atoms and increased the content of those containing 44 to 52 carbons; it also resulted in an increase in hardness at 5 °C (Rousseau et al. 1996a).

The lack of solid content in milk fat at high temperatures limits its application in bakery and confectionery. One way of overcoming this limitation is to interesterify milk fat with another fat such as palm oil stearin. Nor Hayati et al. (2000) interesterified binary mixtures of milk fat and soft and hard palm oil fractions using Lipozyme IM (*Mucor miehe*i lipase) and found the interesterified blend had a higher maximum peak temperature (Tp) than the untreated mixture. Furthermore, the interesterified blends of milk fat and 25–75% soft palm stearin or milk fat and 25% hard palm stearin had similar melting behaviour to that of high melting fraction milk fat. Thus, such blends have commercial potential in pastry and confectionery.

A limitation of milk fat from a nutritional perspective is its low percentage of unsaturated fatty acids. Again, this limitation can be solved by interesterification of milk fat with unsaturated oils such as canola (Rousseau et al. 1996b) and linseed (flaxseed) (Aguedo et al. 2008: Shin et al. 2010). Another nutritional benefit to be gained by lipase-catalysed interesterification is in producing a low-*trans* fat suitable for spreads (Shin et al. 2010). Unlike chemical reactions such as hydrogenation, enzymic interesterification does not cause the production of *trans* double bonds because of the milder conditions used (Aguedo et al. 2008). For example, De et al. (2007) performed chemical inter-esterification at 180 °C for 4 h while inter-esterification using a *Mucor miehei* lipase was performed at 60 °C for 4 h.

10.4.2 Acidolysis

Another acyl-transfer approach to altering the fatty acid composition of milk fat for nutritional and functional properties is acidolysis in which a lipase is used to incorporate a fatty acid or fatty acids (see Fig. 10.3) This was used by Balcão and colleagues (Balcão and Malcata 1997, 1998a, 1998b; Balcão et al. 1998) to incorporate oleic acid into milk fat using lipases from *Mucor circinelloides* and *M. javanicus*. The acidolysis reaction increased the oleic acid content of the milk fat by 27–30% and reduced the contents of lauric and myristic acids by 8% and 2–6%, respectively. The milk fat modified by the *Mucor circinelloides* lipase contained 83% less high-melting fraction and 19% more low-melting fraction than the original milk fat.

Improved spreadability was the aim of Kim et al. (2002) who used *Rhizopus arrhizus* lipase to incorporate α-linolenic acid from perilla oil into milk fat. Their modified milk fat, which contained ~24% (w/w) α-linolenic acid, had better spreadability than the original milk fat as measured by a texture analyser after the fats had been held at 4 °C for 24 h.

In addition to increasing the content of oleic acid, in line with the current nutritional advice of increasing monounsaturated fatty acids in the diet, there has also been research on incorporating fatty acids with particular therapeutic properties, such as conjugated linoleic acid (CLA). Using *Candida antarctica* lipase in a solvent-free system at 50 °C, Garcia et al. (2000) successfully incorporated CLA into milk fat. Sehanputri and Hill (2003) achieved the same result using the same enzyme in a packed bed reactor.

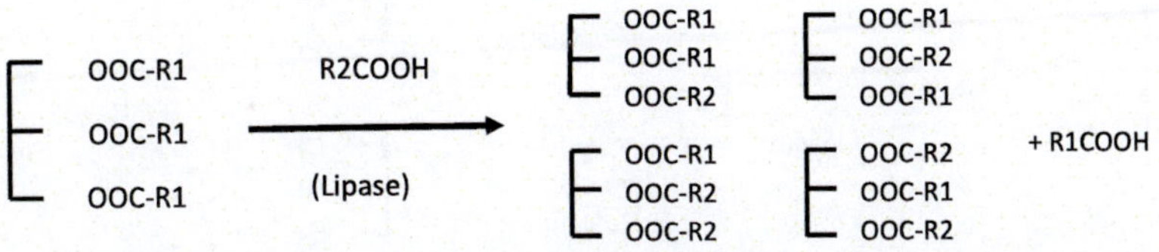

Fig. 10.3 Acidolysis of TG catalysed by a non-sterospecific lipase

The holy grail of lipase-catalysed acyl-transfer reactions on (bovine) milk fat is to modify its structure to resemble that of human milk fat for use in infant formula. To produce human milk fat substitute (HMFS), the 2 position of the TG needs to be modified to contain mostly saturated fatty acids such as palmitic acid, and the *sn*-1 and *sn*-3 positions need to be modified to contain mostly unsaturated fatty acids. This is not easy using bovine milk fat as the starting material and, for this reason, it is more common to interesterify unsaturated marine or vegetable oils with fats containing a high percentage of palmitic acid in the *sn*-2 position, even tripalmitin, to produce HMFSs (Karabulut et al. 2007; Soumanou et al. 2013). Two approaches have been used to produce HMFSs from bovine milk fat. The first is to use an *sn*-1,3-regiospecific lipase (e.g., from *Rhizomucor miehei*) to produce a fat enriched in unsaturated fatty acids in the *sn*-1 and *sn*-3 positions, and then fractionate the interesterified fat and retain a solid fraction as a HMFS. Using this approach, Sorensen et al. (2010) obtained a fat which contained 56% palmitic acid in the *sn*-2 position; human milk fat contains ~72%. Alternatively, a solid milk fat fraction can be interesterified with unsaturated oils using the same enzyme (Sorensen et al. 2010).

10.4.3 Alcoholysis

Alcoholysis of milk fat involves the transfer of an acyl group to an alcohol with the formation of an ester (see Fig. 10.4). A commercial example is glycerolysis, in which the lipase-catalysed reaction forms diglycerides (DGs) and monoglycerides (MGs) from milk fat in the presence of glycerol. These compounds are used widely as emulsifiers in the food industry (Balcão and Malcata 2002). Lipases vary in their ability to catalyse glycerolysis of milk fat; Yang et al. (1994) found two lipases from *Pseudomonas* species which produced mixtures of 55–60% of MG and 24–39% of DG.

Since DG-rich oils are physiologically superior to TG-rich oils in relation to obesity management (Rudkowska et al. 2005), ways of producing them *via* lipase catalysis have been sought. One approach is ethanolysis using a *Pseudomonas*

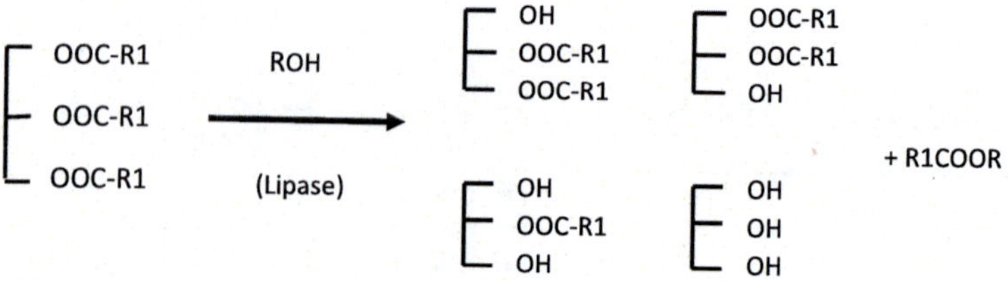

Fig. 10.4 Alcoholysis of TG catalysed by a non-sterospecific lipase

fluorescens lipase. This produces a mixture of 1,2-DGs and fatty acid ethyl esters. If pure DGs are required, the ethyl esters can be removed by supercritical carbon dioxide. Furthermore, if 1,3-DGs are required, they can be obtained by isomerisation of the 1,2-DGs by the *P. fluorescens* lipase (Lubary et al. 2011).

Lipase-catalysed ethanolysis is also used in synthesising structured TGs with palmitic acid in the 2 position, the predominant form in human milk fat. The alcoholysis reaction is carried out on tripalmitin (PPP) with anhydrous ethanol and a 1,3-regiospecific lipase such as that from *Rhizopus delemar.* The 2-monopalmitin formed can then be esterified with oleic acid to produce the TG, OPO, by acidolysis with oleic acid using the same lipase (Schmid et al. 1998).

Andrewes (2018) recently reported the use of the lipase-catalysed alcoholysis reaction for the analysis of residual bacterial lipase activity in UHT milk. Using methanol as the acyl acceptor, the amount of fatty acid methyl esters formed was quantified as a measure of the lipase activity in the milk.

10.5 Conclusion

Lipases are ubiquitous in nature and, in dairy science, are recognised for both their detrimental and beneficial effects. This chapter focuses on the detrimental lipolytic effects of the native LPL and on some beneficial acyl-transferase effects of microbial lipases in modifying milk fat. LPL is present in all raw milk in relative high activities. Under ideal circumstances, it could cause hydrolysis of the milk triglycerides to such an extent that the milk would be unacceptable in a short time. However, lipolysis only occurs under certain circumstances which can be categorised as spontaneous or induced. The reason why spontaneous lipolysis only occurs in the milk of some animals has been a puzzle for some time, but recent research indicating a genetic involvement is an exciting development. If this is confirmed in future research, it could lead to the inclusion of susceptibility of the milk to spontaneous lipolysis as a selection trait in future breeding programs. The causes of induced lipolysis are now quite well known, but knowledge of the causes and vigilance of dairy factory operatives are on-going challenges in the dairy industry. The composition and structure of milk fat do not fit every purpose in the food industry. Lipases used under restricted water conditions are now being used to effect modifications of milk fat to meet a range of needs. Commercial applications of such modifications are expected to increase and make use of new enzymes with improved or new specificities.

References

Aguedo M, Hanon E, Danthine S, Paquot M, Lognay G, Thomas A, Vandenbol M, Thonart P, Wathelet J-P, Blecker C (2008) Enrichment of anhydrous milk fat in polyunsaturated fatty acid residues from linseed and rapeseed oils through enzymatic interesterification. J Agric Food Chem 56:1757–1765

Ahrné L, Björck L (1985) Lipolysis and the distribution of lipase activity in bovine milk in relation to stage of lactation and time of milking. J Dairy Res 52:55–64

Anderson M (1979) Enzyme immunoassay for measuring lipoprotein lipase activator in milk. J Dairy Sci 62:1380–1383

Anderson M (1981) Inhibition of lipolysis in bovine milk by proteose peptone. J Dairy Res 48:247–252

Anderson M (1982) Factors affecting the distribution of lipoprotein lipase activity between serum and casein micelles in bovine milk. J Dairy Res 49:51–59

Andrewes P (2018) Indirect detection of lipase in UHT milk by measuring methyl ester formation. Int Dairy J 79:1–4

Azzara CD, Dimick PS (1985) Lipolytic enzyme activity of macrophages in bovine mammary-gland secretions. J Dairy Sci 68:1804–1812

Bachman KC, Wilcox CJ (1990) Effect of blood and high-density lipoprotein preparations upon lipase distribution and spontaneous lipolysis in bovine milk. J Dairy Sci 73:3393–3401

Bachman KC, Hayen MJ, Morse D, Wilcox CJ (1988) Effect of pregnancy, milk yield, and somatic cell count on bovine milk fat hydrolysis. J Dairy Sci 71:925–931

Balcão VM, Malcata FX (1997) Lipase-catalyzed modification of butterfat via acidolysis with oleic acid. J Mol Catal B Enzymatic 3:161–169

Balcão VM, Malcata FX (1998a) Interesterification and acidolysis of butterfat with oleic acid by *Mucor javanicus* lipase: changes in the pool of fatty acid residues. Enzyme Microb Technol 22:511–519

Balcão VM, Malcata FX (1998b) Lipase catalyzed modification of milkfat. Biotechnol Adv 16:309–341

Balcão VM, Malcata FX (2002) Enzyme-mediated modification of milkfat. In: Kuo TM, Gardner HW (eds) Lipid biotechnology. Marcel Dekker, New York, pp 479–492

Balcão VM, Kemppinen A, Malcata FX, Kalo PJ (1998) Lipase-catalyzed acidolysis of butterfat with oleic acid: characterization of process and product. Enzyme Microb Technol 23:118–128

Barbano DM, Bynum D, Senyk GF (1983) Influence of reverse osmosis on milk lipolysis. J Dairy Sci 66:2447–2451

Bengtsson G, Olivecrona T (1980) Lipoprotein lipase mechanism of product inhibition. Eur J Biochem 106:557–562

Bengtsson G, Olivecrona T (1982) Activation of lipoprotein lipase by apolipoprotein CII. Demonstration of an effect of the activator on the binding of the enzyme to milk-fat globules. FEBS Lett 147:183–187

Bourlieu C, Bouhallab S, Lopez C (2009) Biocatalyzed modifications of milk lipids: applications and potentialities. Trends Food Sci Technol 20:458–469

Cartier P, Chillard Y (1989) Lipases redistribution in cows′ milk during induced lipolysis I. activation by agitation, temperature, change, blood serum and heparin. J Dairy Res 56:699–709

Cartier P, Chilliard Y (1990) Spontaneous lipolysis in bovine milk: combined effects of nine characteristics in native milk. J Dairy Sci 73:1178–1186

Cartier P, Chilliard Y, Paquet D (1990) Inhibiting and activating effects of skim milks and proteose-peptone fractions on spontaneous lipolysis and purified lipoprotein lipase activity in bovine milk. J Dairy Sci 73:1173–1177

Castberg HB, Solberg P (1974) The lipoprotein lipase and the lipolysis in bovine milk. Meieriposten 63:961–975

Castberg HB, Egelrud T, Solberg P, Olivecrona T (1975) Lipases in bovine milk and the relationship between the lipoprotein lipase and tributyrate hydrolysing activities in cream and skimmilk. J Dairy Res 42:255–266

Chandan RC, Shahani KM (1963a) Purification and characterization of milk lipase 1 purification. J Dairy Sci 46:275–283

Chandan RC, Shahani KM (1963b) Purification and characterization of milk lipase 2 characterization of purified enzyme. J Dairy Sci 46:503–509

Chandan RC, Shahani KM, Hill RM, Scholz JJ (1963) Purification and characterization of milk lipase 3 sedimentation coefficient, diffusion coefficient and molecular weight. Enzymologia 26:87–97

Chazal M-P, Chilliard Y (1987) Effect of breed of cow (Friesian and Montbéliarde) on spontaneous and induced lipolysis in milk. J Dairy Res 54:7–11

Chilliard Y (1982) Variations physiologiques des activités lipasiques et de la lipolyse spontanée dans les laits de vache, de chèvre et de femme: revue bibliographique. Lait 62:1–31

Chilliard Y, Selselet-Attou G, Bas P, Morand-Fehr P (1984) Characteristics of lipolytic system in goat milk. J Dairy Sci 67:2216–2223

Chilliard Y, Ferlay A, Rouel J, Lamberett G (2003) A review of nutritional and physiological factors affecting goat milk lipid synthesis and lipolysis. J Dairy Sci 86(5):1751–1770

Chilliard Y, Rouel J, Guillouet P (2013) Goat alpha-s1 casein genotype interacts with the effect of extruded linseed feeding on milk fat yield, fatty acid composition and post-milking lipolysis. Anim Feed Sci Technol 185(3–4):140–149

Chilliard Y, Toral PG, Shingfield KJ, Rouel J, Leroux C, Bernard L (2014) Effects of diet and physiological factors on milk fat synthesis, milk fat composition and lipolysis in the goat: a short review. Small Rumin Res 122(1–3):31–37

Claypool LL (1965) Studies on milk lipase activation. PhD thesis, University of Minnesota

Danthine S, Blecker C (2014) Interactions of lipases with milk fat globule membrane monolayers using a Langmuir film balance. Int Dairy J 35(1):81–87

De BK, Hakimji M, Patel A, Sharma D, Desai H, Kumar T (2007) Plastic fats and margarines through fractionation, blending and interesterification of milk fat. Eur J Lipid Sci Technol 109:32–37

Deeth HC (1978) Fluorimetric detection of carboxyl-esterase activity in milk. In: Proceedings 20th international dairy congress, vol E. International Dairy Federation, Brussels, pp 364–365

Deeth HC (2006) Lipoprotein lipase and lipolysis in milk. Int Dairy J 16:555–562

Deeth HC, Fitz-Gerald CH (1975) Factors governing the susceptibility of milk to spontaneous lipolysis. IDF Bull 86:24–34

Deeth HC, Fitz-Gerald CH (1977) Some factors involved in milk lipase activation by agitation. J Dairy Res 44:569–583

Deeth HC, Fitz-Gerald CH (2006) Lipolytic enzymes and hydrolytic rancidity. In: Fox PF, McSweeney P (eds) Advanced dairy chemistry: lipids, vol 2. Springer, New York, pp 481–556

Desneulle P (1972) The lipases. In: Boyer PD (ed) The enzymes, vol 7. Academic, New York, pp 575–617

Downey WK, Andrews P (1969) Evidence for the presence of several lipases in cow's milk. Biochem J 112:559–562

Driessen FM (1976) Comparative-study of lipase in bovine colostrum and in bovine milk. Neth Milk Dairy J 30(3–4):186–196

Driessen FM (1987) Inactivation of lipases and proteinases (indigenous and bacterial). IDF Bull 238:71–93

Egelrud T, Olivecrona T (1972) The purification of lipoprotein lipase from bovine skim milk. J Biol Chem 247:6212–6217

Egelrud T, Olivecrona T (1973) Purified bovine milk (lipoprotein) lipase: activity against lipid substrates in the absence of exogenous serum factors. Biochim Biophys Acta 306:115–127

Falt H, Hernell O, Blackberg L (2001) Do human bile salt stimulated lipase and colipase-dependent pancreatic lipase share a common heparin-containing receptor? Arch Biochem Biophys 386(2):188–194

Farkye NY, Imafidon GI, Fox PF (1995) Thermal denaturation of indigenous milk enzymes. In: Heat induced changes in milk. International Dairy Federation, Brussels, pp 331–348

Fitz-Gerald CH, Deeth HC, Kitchen BJ (1981) The relationship between the levels of free fatty acids, lipoprotein lipase, carboxylesterase, *N*-acetyl-β-D-glucosaminidase, somatic cell count and other mastitis indices in bovine milk. J Dairy Res 48:253–265

Forster TL, Bendixen HA, Montgomery MW (1959) Some esterases of cows milk. J Dairy Sci 42(12):1903–1912

Forster TL, Montgomery MW, Montoure JE (1961) Some factors affecting the activity of the A- B- and C-esterases of bovine milk. J Dairy Sci 44:1420–1430

Gaffney PJ, Harper WJ (1965) Lipase activity in somatic cells from separator slime. J Dairy Sci 48:613–615

Garcia HS, Keough KJ, Arcos JA, Hill CG (2000) Interesterification (acidolysis) of butterfat with conjugated linoleic acid in a batch reactor. J Dairy Sci 83(3):371–377

Gargouri A, Hamed H, ElFeki A (2008) Total and differential bulk cow milk somatic cell counts and their relation with lipolysis. Livest Sci 113(2–3):274–279

Gargouri A, Hamed H, ElFeki A (2013) Analysis of raw milk quality at reception and during cold storage: combined effects of somatic cell counts and psychrotrophic bacteria on lipolysis. J Food Sci 78(9):M1405–M1411

Hayashi R, Tajima S, Yamamoto A (1986) Purification and characterization of lipoprotein-lipase from human postheparin plasma and its comparison with purified bovine-milk lipoprotein-lipase. J Biochem 100(2):319–331

Hernell O, Olivecrona T (1974) Human milk lipases 2 Bile salt-stimulated lipase. Biochim Biophys Acta 369(2):234–244

Hisserich D, Reuter H (1984) Grenzen der mechanischen Beanspruchung von Rohmilch für die induzierte Lipolyse. Milchwissenschaft 39:333–335

Hoynes MCT, Downey WK (1973) Relationship of the lipase and lipoprotein lipase activities of bovine milk. Biochem Soc Trans 1:256–259

Hui DY, Howles PN (2002) Carboxyl ester lipase: structure-function relationship and physiological role in lipoprotein metabolism and atherosclerosis. J Lipid Res 43(12):2017–2030

IDF (1975) Proceedings of the lipolysis symposium. Cork, Ireland, March 1975. Bull Int Dairy Fed Doc 86

IDF (1980) Flavor impairment of milk and milk products due to lipolysis. Bull Int Dairy Fed Doc 118

IDF (1987) Significance of lipolysis in the manufacture and storage of dairy products. Bull Int Dairy Fed Doc 144

Jaeger K-E, Ransac S, Dijkstra BW, Colson C, van Heuvel M, Misset O (1994) Bacterial lipases. FEMS Microbiol Rev 15:29–63

Jellema A (1980) Physiological factors associated with lipolytic activity in cow's milk. IDF Bull 118:33–40

Jellema A, Schipper CJ (1975) Influence of physiological factors on the lipolytic susceptibility of milk. IDF Bull 86:2–6

Jensen RG, Gander GW, Sampugna J (1961) Lipolysis by a B-esterase preparation from milk. J Dairy Sci 44(5):943–944

Jurczak ME, Sciubisz A (1981) Studies on the lipolytic changes in milk from cows with mastitis. Milchwissenschaft 36(4):217–219

Karabulut I, Turan S, Vural H, Kayahan M (2007) Human milk fat substitute produced by enzymatic interesterification of vegetable oil blend. Food Technol Biotechnol 45:434–438

Kim JS, Maeng IK, Lee BO, Kim CK, Kwon Y, Yj K (2002) Lipase-catalyzed acidolysis of butterfat with α-linolenic acid from perilla oil. Food Sci Biotechnol 11:66–70

Kinnunen PKJ, Huttunen JK, Ehnholm C (1976) Properties of purified bovine milk lipoprotein lipase. Biochim Biophys Acta 450:342–351

Kitchen BJ (1971) Bovine milk esterases. J Dairy Res 38(2):171–177

Kon H, Saito Z (1997) Factors causing temperature activation of lipolysis in cow's milk. Milchwissenschaft 52:435–440

Korn ED (1962) The lipoprotein lipase of cow's milk. J Lipid Res 3:246–250

Li N, Richoux R, Boutinaud M, Martin P, Gagnaire V (2014) Role of somatic cells on dairy processes and products: a review. Dairy Sci Technol 94:517–538

Lubary M, Hofland GW, ter Horst JH (2011) A process for the production of a diacylglycerol-based milk fat analogue. Eur J Lipid Sci Technol 113:459–468

Luedecke LO (1964) Relationship between California Mastitis Test reactions and leucocyte count, catalase activity and A-esterase activity of milk from opposite quarters. J Dairy Sci 47:696

Marquardt RR, Forster TL (1965) Milk A-esterase levels as influenced by stage of lactation. J Dairy Sci 48:1526–1528

Matsuoka N, Shirai K, Jackson RL (1980) Preparation and properties of immobilized lipoprotein-lipase. Biochim Biophys Acta 620(2):308–316

Montgomery MW, Forster TL (1961) Partial purification of B-esterase of bovine milk. J Dairy Sci 44(4):721

Morley N, Kuksis A (1977) Lack of fatty acid specificity in the lipolysis of oligo- and polyunsaturated triacylglycerols by milk lipoprotein lipase. Biochim Biophys Acta 487:332–342

Murphy SC, Cranker K, Senyk GF, Barbano DM, Saeman AI, Galton DM (1989) Influence of bovine mastitis on lipolysis and proteolysis in milk. J Dairy Sci 72:620–626

Nilsson-Ehle P, Egelrud T, Belfrage P, Olivecrona T, Borgström B (1973) Positional specificity of purified milk lipoprotein lipase. J Biol Chem 248:6734–6737

Nor Hayati I, Aminah A, Mamot S, Nor Aini I, Lida HMN, Sabariah S (2000) Melting characteristic and solid fat content of milk fat and palm stearin blends before and after enzymatic interesterification. J Food Lipids 7:175–193

Okuda H, Fujii S (1968) Relationship between lipase and esterase. J Biochem 64:377–385

Olivecrona T, Olivecrona G (1999) Lipoprotein and hepatic lipase in lipoprotein metabolism. In: Betteridge D, Illingworth D, Shepherd J (eds) Lipoproteins in health and disease. Arnold, London, pp 223–246

Olivecrona T, Egelrud T, Castberg HB, Solberg P (1975) Is there more than one lipase in bovine milk. IDF Bull 86:61–72

Olivecrona T, Bengtsson G, Osborne JC (1982) Molecular-properties of lipoprotein-lipase - effects of limited trypsin digestion on molecular-weight and secondary structure. Eur J Biochem 124(3):629–633

Olivecrona T, Vilaro S, Olivecrona G (2003) Lipases in milk. In: Fox PF, McSweeney PLH (eds) Advanced dairy chemistry vol 1, Proteins, 3rd edn. Kluwer Academic/Plenum Publishers, New York, pp 473–494

Östlund-Lindqvist A-M, Iverius P-H (1975) Activation of highly purified lipoprotein lipase from bovine milk. Biochem Biophys Res Commun 65:1447–1455

Quigley TW, Roe CE, Pallansch MJ (1958) A lipoprotein lipase in cows milk. Fed Proc 17(1):292–292

Rapp D, Olivecrona T (1978) Kinetics of milk lipoprotein lipase. Studies with tributyrin. Eur J Biochem 91:379–385

Ray PR, Chatterjee K, Chakrabortyand C, Ghatak P (2013) Lipolysis of milk: a review. Int J Agric Vet Med 1:58–74

Reddy MN, Maraganore JM, Meredith SC, Heinrikson RL, Kezdy FJ (1986) Isolation of an active-site peptide of lipoprotein lipase from bovine milk and determination of its amino acid sequence. J Biol Chem 261:9678–9683

Rousseau D, Forestiere K, Hill AR, Marangoni AG (1996a) Restructuring butterfat through blending and chemical interesterification.1. Melting behavior and triacylglycerol modifications. J Am Oil Chem Soc 73:963–972

Rousseau D, Hill AR, Marangoni AG (1996b) Restructuring butterfat through blending and chemical interesterification.3. Rheology. J Am Oil Chem Soc 73:983–989

Rudkowska I, Roynette C, Demonty I, Vanstone CA, Jew S, Jones PJH (2005) Diacylglycerol: efficacy and mechanism of action of an anti-obesity agent. Obes Res 13:1864–1876

Sartippour M, Lambert A, Laframboise M, St-Jacque P, Renier G (1998) Stimulatory effect of glucose on macrophage lipoprotein lipase expression and production. Diabetes 47:431–438

Schmid U, Bornscheuer UT, Soumanou MM, McNeill GP, Schmid RD (1998) Optimization of the reaction conditions in the lipase-catalyzed synthesis of structured triglycerides. J Am Oil Chem Soc 75:1527–1531

Schwartz DP, Parks OW (1974) The lipids in milk: deterioration part 1 lipolysis and rancidity. In: Webb BH, Johnson AH, Alford JA (eds) Fundamentals of dairy chemistry, 2nd edn. AVI Publishing, Westport, pp 220–272

Sehanputri PS, Hill CG (2003) Lipase-mediated acidolysis of butteroil with free conjugated linoleic acid in a packed bed reactor. Biotechnol Bioeng 83:608–617

Senda M, Oka K, Brown WV, Qasba PK (1987) Molecular cloning and sequence of a cDNA coding for bovine lipoprotein lipase. Proc Natl Acad Sci U S A 84:4369–4373

Shimizu M, Miyaji H, Yamauchi K (1982) Inhibition of lipolysis by milk fat globule membrane materials in model milk fat emulsion. Agric Biol Chem 46:795–799

Shin J-A, Akoh CC, Lee K-T (2010) Enzymatic interesterification of anhydrous butterfat with flaxseed oil and palm stearin to produce low-trans spreadable fat. Food Chem 120:1–9

Shipe WF, Senyk GF (1981) Effects of processing conditions on lipolysis in milk. J Dairy Sci 64:2146–2149

Sorensen A-DM, Xu X, Zhang L, Kristensen JB, Jacobsen C (2010) Human milk fat substitute from butterfat: production by enzymatic interesterification and evaluation of oxidative stability. J Am Oil Chem Soc 87(2):185–194

Soumanou MM, Perignon M, Villeneuve P (2013) Lipase-catalyzed interesterification reactions for human milk fat substitutes production: a review. Eur J Lipid Sci Technol 115:270–285

Stobbs TH, Deeth HC, Fitz-Gerald CH (1973) Effects of energy intake on spontaneous lipolysis in milk from cows in late lactation. Aust J Dairy Technol 28:170–172

Sundheim G, Bengtsson-Olivecrona G (1985) Lipolysis in milk induced by cooling or by heparin: comparisons of amount of lipoprotein lipase in the cream fraction and degree of lipolysis. J Dairy Sci 68:589–593

Sundheim G, Bengtsson-Olivecrona G (1987a) Isolated milk fat globules as substrate for lipoprotein lipase: study of factors relevant to spontaneous lipolysis in milk. J Dairy Sci 70:499–505

Sundheim G, Bengtsson-Olivecrona G (1987b) Hydrolysis of bovine milk fat globules by lipoprotein lipase: inhibition by proteins extracted from milk fat globule membrane. J Dairy Sci 70:815–1821

Sundheim G, Bengtsson-Olivecrona G (1987c) Methods to assess the propensity of milk fat globules toward lipolysis and the ability of skim milk to inhibit lipolysis. J Dairy Sci 70:2040–2045

Sundheim G, Bengtsson-Olivecrona G (1987d) Hydrolysis of bovine and caprine milk fat globules by lipoprotein lipase effects of heparin and of skim milk on lipase distribution and on lipolysis. J Dairy Sci 70:2467–2475

Takahashi M, Yagyu H, Tazoe F, Nagashima S, Ohshiro T, Okada K, Osuga JI, Goldberg IJ, Ishibashi S (2013) Macrophage lipoprotein lipase modulates the development of atherosclerosis but not adiposity. J Lipid Res 54:1124–1134

van Tilbeurgh H, Roussel A, Lalouel JM, Cambillau C (1994) Lipoprotein-lipase. Molecular model based on the pancreatic lipase x-ray structure - consequences for heparin-binding and catalysis. J Biol Chem 269(6):4626–4633

Vanbergue E, Peyraud JL, Guinard-Flament J, Charton C, Barbey S, Lefebvre R, Gallard Y, Hurtaud C (2016) Effects of DGAT1 K232A polymorphism and milking frequency on milk composition and spontaneous lipolysis in dairy cows. J Dairy Sci 99(7):5739–5749

Vanbergue E, Delaby L, Peyraud JL, Colette S, Gallard Y, Hurtaud C (2017) Effects of breed, feeding system, and lactation stage on milk fat characteristics and spontaneous lipolysis in dairy cows. J Dairy Sci 100(6):4623–4636

Vanbergue E, Peyraud JL, Ferlay A, Miranda G, Martin P, Hurtaud C (2018) Effects of feeding level, type of forage and milking time on milk lipolytic system in dairy cows. Livest Sci 217:116–126

Wiking L, Bjerring M, Løkke M, Løvendahl P, Kristensen T (2019) Herd factors influencing free fatty acids concentration in bulk tank milk. J Dairy Res 86(2):226–232

Yang B, Harper WJ, Parkin KL, Chen J (1994) Screening of commercial lipases for production of mono- and diacylglycerols from butteroil by enzymic glycerolysis. Int Dairy J 4:1–13

Chapter 11
Heat-Stable Microbial Peptidases Associated with the Microbiota of Raw Milk

Claudia Glück, Timo Stressler, and Lutz Fischer

11.1 The Microbiota of Raw Milk

Although milk in healthy udder cells is thought to be sterile, the microbiota of raw milk is very diverse (Glück 2017). The multiplicity of recontamination sources, including the teat apex, milking equipment, air, water, feed, grass, soil and other environments, results in a broad biodiversity of the microbiota in raw milk (Fig. 11.1; Quigley et al. 2013).

It has to be noted that there are several taxa which can be found in milk but not in the farm environment and *vice versa*. For example, *Clavibacter* and *Arcanobacterium* spp. can be detected on the teat surface but are not commonly found in milk, suggesting a lack of competitiveness (Quigley et al. 2013; Verdier-Metz et al. 2012). Nonetheless, milk is an ideal growth medium for many microorganisms, due to its high nutritional value, high water activity and nearly neutral pH (Hassan and Frank 2011). Both Gram-positive (*Bacillus*, *Clostridium*, *Corynebacterium*, *Microbacterium*, *Micrococcus*, *Streptococcus*, *Staphylococcus* and *Lactobacillus*) and Gram-negative bacteria (*Pseudomonas*, *Aeromonas*, *Serratia*, *Acinetobacter*, *Alcaligenes*, *Achromobacter*, *Enterobacter* and *Flavobacterium*) have been isolated from raw milk (Cousin 1982; Champagne et al. 1994; Munsch-Alatossava and Alatossava 2006; Hantsis-Zacharov and Halpern 2007; von Neubeck et al. 2015).

The isolation of yeasts from raw milk, such as *Candida*, *Kluyveromyces* and *Pichia*, has also been described in the literature (Cocolin et al. 2002; von Neubeck et al. 2015). The microorganisms found in milk can be either of technologically importance, such as the fermentation of milk through production of lactic acid, or

C. Glück · T. Stressler (✉) · L. Fischer (✉)
Departement of Biotechnology and Enzyme Science, Institute of Food Science and Biotechnology, University of Hohenheim, Stuttgart, Germany
e-mail: Lutz.Fischer@uni-hohenheim.de; t.stressler@uni-hohenheim.de; claudiaglueck@t-online.de

A. L. Kelly, L. B. Larsen (eds.), *Agents of Change*, Food Engineering Series,
https://doi.org/10.1007/978-3-030-55482-8_11

Fig. 11.1 Contamination sources and respective genera of raw milk (modified from Hassan and Frank 2011)

the formation of texture, flavor or organoleptic properties (Wouters et al. 2002). They can also negatively influence milk quality and cause spoilage by enzymes such as peptidases or lipases (Datta and Deeth 2003; Santos et al. 2003; Barbano et al. 2006; Baur et al. 2015a). Furthermore, they can have health-promoting effects in case of probiotic microorganisms or can cause illness in case of pathogenic microorganisms (Wouters et al. 2002; Quigley et al. 2013).

The total microbial cell count and the composition strongly depends on season, farm hygiene, feed and efficiency of cooling (Hassan and Frank 2011). Total cell counts in raw milk are typically in the range of 3–6 log10 cfu (colony forming units) mL^{-1} (Desmasures et al. 1997; Fricker et al. 2011; von Neubeck et al. 2015). The cell counts found in dairy samples tend to be higher than in farm samples. In a study which covered ten farms and ten dairies in Germany, the highest cell count of above 8 log10 cfu mL^{-1} was detected in raw milk from a dairy bulk milk tank (von Neubeck et al. 2015). In general, raw milk contains a significant lactic acid bacteria population that includes *Lactococcus*, *Streptococcus*, *Lactobacillus*, *Leuconostoc* and *Enterococcus* spp. (Quigley et al. 2013). As already mentioned, the microbiota of raw milk is very complex and diverse. For example, a total of 169 different species were identified in 20 raw milk samples using culturing and subsequent sequencing analysis (von Neubeck et al. 2015). The three dominant genera were *Pseudomonas*, *Acinetobacter* and *Lactococcus*, which were isolated from nearly all 20 samples. In another study, by using a direct high-throughput DNA sequencing approach, 256 bacterial species were identified in Danish raw milk (Masoud et al. 2012). The main bacteria detected belonged to the group of lactic acid bacteria, namely *Lactobacillus*, *Lactococcus* and *Streptococcus*. Due to this high biodiversity, raw milk is a rich source for the isolation of unknown species and even genera. Different studies showed that 15–20% of the isolates belonged to novel species (Hantsis-Zacharov and Halpern 2007; von Neubeck et al. 2015).

The refrigerated storage of raw milk is a selection criterion that promotes psychrotolerant bacteria, which can grow at temperatures ≤7 °C, although their optimal growth temperature may be higher, between 20 and 30 °C (McPhee and Griffiths 2011). Prior to refrigeration, the proportion of psychrotolerant bacteria is <10% (Cousin 1982). During storage, the biodiversity decreases and psychrotolerant bacteria are able to outgrow others and become predominant. The genera *Pseudomonas* and *Acinetobacter* are often reported to be the predominant bacteria in refrigerated raw milk (Sørhaug and Stepaniak 1997; Marchand et al. 2009a; De Jonghe et al. 2011; Raats et al. 2011; von Neubeck et al. 2015). Further genera were identified as *Staphylococcus*, *Streptococcus*, *Bacillus*, *Lactobacillus*, *Enterococcus*, *Clostridium*, *Corynebacterium*, *Micrococcus* and *Microbacterium*. These microorganisms represent 37–54% of the microbiota of cold-stored milk samples (Raats et al. 2011).

11.1.1 The Genus Pseudomonas

The genus *Pseudomonas* was described by Migula in 1894 and is well known for its metabolic and genetic flexibility. Some species are human, animal or plant pathogens (Moore et al. 2006). Species of the genus *Pseudomonas* are straight or slight-curved rods, 0.5–1.0 μm in diameter and 1.5–5.0 μm in length, Gram-negative, and motile by one or more polar flagella. They are aerobic, catalase-positive and the majority are oxidase-positive. In some cases, nitrate can be used instead of oxygen as electron acceptor, allowing anaerobic growth (Moore et al. 2006). *Pseudomonas* species are truly ubiquitous and can be found multitude of habitats ranging from soil and water environments to plant and animal tissues, from the Antarctica to the Tropics. Species of the genus *Pseudomonas* are capable to use a wide range of organic and inorganic compounds and are able to live under various environmental conditions (Moore et al. 2006). They grow in a temperature range from 4 to 42 °C and at pH values between 4 and 8. *Pseudomonas* species grow relatively rapidly and are often able to outgrow other bacteria. The requirement for oxygen is apparently the major constraining factor for their environment (Moore et al. 2006; Peix et al. 2009; McPhee and Griffiths 2011).

The genome of *Pseudomonas* species is approximately 6 Mbp in size, including approximately 5500 putative genes, and about 135 whole genomes have been recorded (Winsor et al. 2015). Almost 10% of the genomes encode for molecules which are involved in gene regulation. This fact shows the evolutionary emphasis of the genus *Pseudomonas* in monitoring and responding to environmental signals. The major control mechanisms take place on the level of transcription, and post-transcriptional control systems are also included for the fine-tuning of the expression. The regulatory machinery includes various features such as sigma factors, the LysR transcriptional regulator family (>100 members), a two-component phosphorelay systems for the sensing of nutrients and starvation and quorum sensing (Moore et al. 2006). The genome includes a large battery of stress response systems; for example, efflux systems for antibiotics, detergents, heavy metals, biocides, solvents and toxins impart several resistances (Moore et al. 2006). Species of the genus *Pseudomonas* are well known for the ability to transfer genetic material horizontally. This fact complicates the identification and taxonomic analyses of species belonging to this genus (Moore et al. 2006). In comparative genome studies, it was shown that the pathogenicity of some strains, as well as their high environmental adaptability, can be correlated to the exchange of genes from a pool of genes that interact in various combinations in different genetic backgrounds (Lee et al. 2006; Peix et al. 2009).

Especially under iron-limiting conditions, many *Pseudomonas* species (fluorescent pseudomonads) produce pigments that fluoresce at short wavelength ultraviolet light (254 nm) (Moore et al. 2006). Some of these pigments act as siderophores for the efficient uptake of traces of iron(III). The major siderophores of fluorescent *Pseudomonas* species is pyoverdine. Pyoverdines are principally composed of three domains: (1) the chromophore (quinoline derivative), (2) an acyl chain linked to the NH_2 group of the chromophore and (3) a peptide chain attached to the carboxylic group of the chromophore. Up to 50 different pyoverdines have been identified, illustrating the diversity of these biomolecules (Meyer et al. 2002; Meyer 2000). Another, non-fluorescent blue pigment, typically produced by the type strain *Pseudomonas aeruginosa,* is pyocyanin. Other pigments produced by several *Pseudomonas* species are pyorubin (*P. aeruginosa*, red color), oxychlororaphin (*P. aureofaciens*, *P. chlororaphis*, orange color), chlororaphin (*P. chlororaphis*, green color), etc. (Moore et al. 2006). The determination of pyoverdines can be used for the identification *Pseudomonas* species (Meyer et al. 2002).

Species of the genus *Pseudomonas* are able to form biofilms to effectively colonize, for example, the rhizosphere of plants. Bacterial communities attached to surfaces can dramatically increase their resistance against antibiotics and provide an ideal niche for the exchange of genetic material (Whiteley et al. 2001; Donlan 2002). The ability to form biofilms in milk environment (e.g., pipelines or bulk tanks) and the ability to form spoilage enzymes has been investigated *in vitro* by Teh et al. (2011, 2012). They showed that 52 of 153 *Pseudomonas* isolates were capable to produce thermostable enzymes and form biofilms on stainless steel chips (Teh et al. 2011).

Pseudomonas species are known to secrete various extracellular enzymes, such as proteolytic or lipolytic enzymes, alkaline phosphatase, chitinase and β-lactamase. The secreted enzymes are either responsible for the degradation of biopolymers for nutrient acquisition or acting as virulence factors (Ashdown and Koehler 1990; Tielen et al. 2010; Allison et al. 2014). In case of *Pseudomonas aeruginosa*, the extracellular protease IV is one of the virulence factors (Engel et al. 1998; Lee et al. 2006). *Pseudomonas* spp. are major organisms in spoilage of food and have been identified in vegetables, meat, dairy products and milk (Liao 1989; Rajmohan et al. 2002; Dogan and Boor 2003; Olofsson et al. 2007; Ercolini et al. 2009). Thermostable peptidases of *Pseudomonas spp*. play an important role in spoilage of milk products with long shelf life and will be discussed in the next section.

11.2 Microbial Peptidase from Raw Milk Isolates

Peptidases have the EC-number (Enzyme Commission) 3.4, belong to the enzyme group of hydrolases (EC 3) and catalyze the hydrolysis of peptides bonds. Also ester bonds can be cleaved if the structure of the ester is similar to the structure of the peptide. The terms proteases, proteinases and peptidases are commonly used synonymously for proteolytic enzymes, cleaving peptide bonds. According to the "Nomenclature Committee of The International Union of Biochemistry and Molecular Biology (NC-IUBMB)" it is recommended that the term "peptidase" is used for any enzyme that hydrolyses peptide bonds.

Besides indigenous enzymes in milk, enzymes may be secreted by psychrotolerant microorganisms during cold storage of raw milk. In particular, the secretion of heat-resistant peptidases can cause instability and spoilage in dairy products long before the expiry is reached (Sørhaug and Stepaniak 1997; Marchand et al. 2009b; von Neubeck et al. 2015; Stoeckel et al. 2016a). In a study of Glück et al. (2016), 231 isolates covering the broad range of species found in raw milk were investigated for their secretion of enzymes at 6 °C. After cultivation, peptidase activity was determined in the supernatant of 53 isolates, and it was found that the majority of isolates secreting thermostable peptidases belonged to the genus *Pseudomonas* (Table 11.1).

As seen in Table 11.1, the genus *Pseudomonas* is the most potent source for thermostable peptidases in raw milk and can, therefore, be involved in spoilage of milk products. This fact has been demonstrated by several studies (Adams et al. 1975; Fairbairn and Law 1986; Champagne et al. 1994; Sørhaug and Stepaniak 1997; Rajmohan et al. 2002; Stoeckel et al. 2016a).

11.2.1 Peptidases Secreted by Pseudomonas Species

The majority of *Pseudomonas* species secrete one type of peptidase (Fairbairn and Law 1986; Chen et al. 2003). This extracellular peptidase is typically a metallopeptidase (EC 3.4.24.) which is *aprA*/*aprX*-gene-encoded (*apr* = alkaline protease), with a molecular weight of 45–50 kDa (Liao and McCallus 1998; Dufour et al. 2008; Mu et al. 2009; Baur et al. 2015b). The genes encoding for *Pseudomonas* peptidases are designated as *aprA* and *aprX* synonymously in literature. In this chapter, the term *aprA* for the gene and AprA for the respective protein will be used. Mutation in the *aprA* gene in *Pseudomonas fluorescens* resulted in extracellular peptidase-deficient isolates (Liao and McCallus 1998; Anderson et al. 2004; Maunsell et al. 2006).

Peptidases can be classified regarding their evolutionary relationship, based on information about their primary and three-dimensional structure. Based on these information peptidases are grouped into clans and families in the *MEROPS* database (Rawlings et al. 2014). The classification into families of peptidases is based on the

Table 11.1 Comparison of the relative extracellular peptidase activity of raw milk isolates after cultivation at 6 °C in 10% (v/v) milk medium for 8 days and residual peptidase activity after heat-treatment at 70 and 95 °C for 5 min (adapted from Glück et al. 2016)

	Microorganism	Relative peptidase activity [%]	Residual peptidase activity [%] after heat treatment at 70 °C	95 °C
Bacteria	*Acinetobacter parvus*	0.37	78.5	65.6
	Chryseobacterium jejuense	0.80	2.15	14.1
	Chryseobacterium oncorhynchi	9.35	10.7	32.3
	Chryseobacterium piscium	2.11	n.d.	n.d.
	Devosia glacialis	4.87	113	90.9
	Microbacterium maritypicum	0.66	50.3	76.5
	Microbacterium oxydans	1.07	56.6	68.1
	Pseudomonas aeruginosa	3.22	76.9	4.93
	Pseudomonas brenneri	4.37	41.8	38.1
	Pseudomonas fluorescens	72.2	9.78	4.76
	Pseudomonas fragi	0.98	64.6	9.43
	Pseudomonas gessardii	49.6	13.9	20.6
	Pseudomonas lundensis	0.29	55.2	76.1
	Pseudomonas lurida	16.0	5.20	11.7
	Pseudomonas meridiana	29.1	8.69	9.71
	Pseudomonas protegens	9.98	103	91.4
	Pseudomonas proteolytica	32.4	15.8	17.3
	Pseudomonas sp.nov. (1)	66.5	10.3	17.0
	Pseudomonas sp.nov. (2)	0.35	95.2	59.9
	Pseudomonas sp.nov. (4)	3.70	95.2	76.0
	Pseudomonas sp.nov. (7)	15.4	9.37	10.7
	Pseudomonas sp.nov. (8)	0.47	93.5	56.9
	Pseudomonas sp.nov. (11)	70.7	11.5	9.28
	Pseudomonas sp.nov. (12)	12.9	88.6	69.2
	Pseudomonas sp.nov. (13)	5.52	61.5	75.6
	Pseudomonas sp.nov. (16)	100	48.9	37.3
	Pseudomonas sp.nov. (23)	10.0	3.79	0.21
	Pseudomonas sp.nov. (24)	28.2	20.4	10.9
	Pseudomonas sp.nov. (27)	14.5	73.2	79.4
	Renibacterium salmoninarum	0.45	14.8	17.0
	Serratia marcescens	3.14	99.4	77.4
	Stenotrophomonas chelatiphaga	1.28	19.6	16.2
	Stenotrophomonas rhizophila	2.05	48.1	70.7
Yeast	*Candida parapsilosis*	0.64	66.5	57.1
	Candida pseudoglaebosa	0.89	80.5	84.1
Fungi	*Geotrichum clavatum*	0.21	76.9	65.1

Enzyme activity was determined at 40 °C, pH 6.7, using azocasein as a substrate
100% relative peptidase activity: 478 $\Delta A\ h^{-1}\ mL^{-1}$
n.d. = not detectable

homology of amino acid sequences. Every member of a *MEROPS* family shows a statistically significant relationship in amino acid sequence to at least one other member of the family. Each family has a representative member, called the type example and is named with a letter denoting the catalytic type (A, C, G, M, S, T or U unknown) followed by a number. If the divergence in one family is too high, subfamilies are constructed (Barrett et al. 2013) A clan contains one or more families that appear to have arisen from a unique peptidase. The best evidence for homology at clan level is similarity of the three-dimensional structure. However, the arrangement of catalytic residues in the polypeptide chain or limited similarities of the residues around the catalytic residues can be revealing. Each clan is identified with two letters, the first representing the catalytic type of the families included in the clan followed by an arbitrary second letter. Not every peptidase can be assigned to a clan yet (Barrett et al. 2013; Rawlings et al. 2014).

The extracellular peptidases produced by *Pseudomonas* isolates belong to the subfamily M10B of the clan MA of metallopeptidases according to the MEROPS database for proteases (Baumann 2013; Rawlings et al. 2014). The type of peptidase of the subfamily is serralysin (EC 3.4.24.40), an extracellular peptidase from *Serratia marcescens*, and therefore, peptidases belonging to this subfamily are often called serralysins (Gomis-Rüth 2003). Other metallopeptidases in this family are secreted by *Pseudomonas aeruginosa*, *Proteus mirabilis* or *Erwinia chrysanthemi* and are often virulence factors (Gomis-Rüth 2003; Rawlings et al. 2014).

The peptidase gene (*aprA*) is located on an operon (*aprA-lipA* operon, Fig. 11.2), which is observed in various *Pseudomonas* spp. (Woods et al. 2001).

In case of *Pseudomonas fluorescens,* the operon spans 14 kb and contains eight genes. It includes genes for the metallopeptidase AprA, a peptidase inhibitor (Inh), a Type I secretion system (AprD, E, F), two autosecreted peptidase-homologues (PrtA, B) and an extracellular lipase (LipA) (Kawai et al. 1999; Woods et al. 2001). Variation in the operon structure for different *Pseudomonas fluorescens* strain such as the absence of the *prtA*/*B* genes or the existence of two lipase genes (*lipA* and *lipB*) are described (Johnson et al. 1992; Liao and McCallus 1998; Ahn et al. 1999; Lo et al. 2015). All genes are contiguous in the genome and a single transcription start is located upstream from *aprA* (Woods et al. 2001). The AprA inhibitor (12–14 kDa) is solely located in the periplasm and its presumed biological role is to protect periplasmic proteins from hydrolysis by AprA (Baumann et al. 1995; Bardoel et al. 2012).

The crystal structure of serralysins from *Pseudomonas aeruginosa* and *Serratia marcescens* and a psychrophilic metallopeptidase from an Antarctic *Pseudomonas* sp. have been solved (Baumann et al. 1993, 1995; Miyatake et al. 1995; Aghajari et al. 2003). Comparative studies showed that their overall structures are similar.

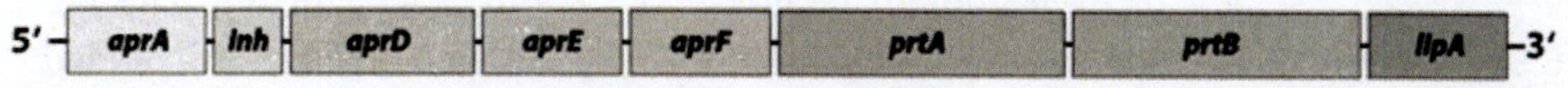

Fig. 11.2 Structure of the *apr*A-*lip*A operon. Adapted from Woods et al. (2001) and McCarthy et al. (2004)

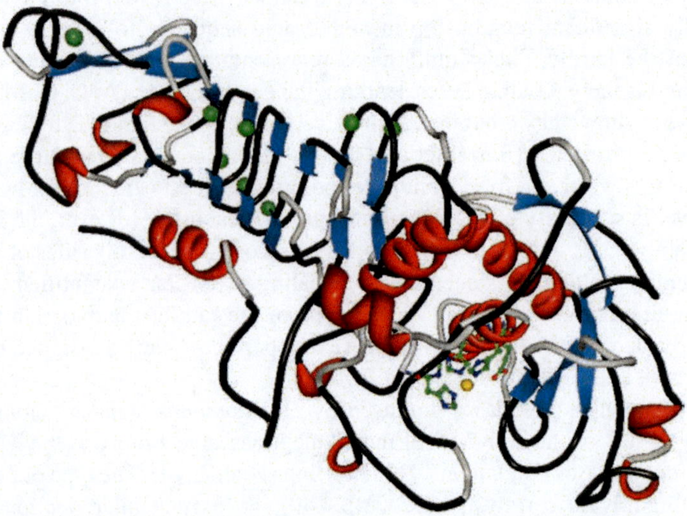

Fig. 11.3 Crystal structure of AprA from *Pseudomonas aeruginosa*. PDB accession number: 1KAP, according to Baumann et al. (1993) and Moreland et al. (2005)

The active site of the psychrophilic peptidase revealed a greater accessibility of the substrates due to deletions of two to four residues (Aghajari et al. 2003). The three-dimensional structure (PDB accession number: 1KAP) of the alkaline metallopeptidase AprA from *Pseudomonas aeruginosa* was solved to a resolution of 1.64 Å (Baumann et al. 1993). This crystal structure is shown in Fig. 11.3.

The peptidase protein is an elongated molecule and includes one Zn^{2+} ion which is required for catalysis and eight Ca^{2+} ions which are responsible for stabililty. It is built up of two structural domains, an N-terminal globular catalytic domain and a C-terminal domain that forms an extended, two-layer β-sandwich. The interface between the two domains contains serine and threonine residues and is generally hydrophilic (Baumann et al. 1993). The catalytic domain shares an open sandwich topology with other metallopeptidases. The Zn^{2+} ion is bound by three histidine residues, which are part of the active-site consensus sequence motif HEXXHXUGUXH (H = histidine; E = glutamic acid; G = glycine; X = arbitrary amino acid; and U = bulky, hydrophobic amino acid). The glutamic acid residue, which is supposed to activate the zinc-bound water molecule for nucleophilic attack, is also located in this part of the sequence (Baumann et al. 1993; Miyatake et al. 1995; Hege and Baumann 2001). The entrance to the active site is closed by a highly flexible loop when a substrate molecule is bound. An asparagine residue located in this loop forms hydrogen bonds with the bound substrate (Miyatake et al. 1995; Baumann et al. 1995). The *β*-sandwich domain includes 21 *β*-strands with mixed parallel/antiparallel topology. Seven Ca^{2+} binding sites are fully and one is partially included in this domain. The five Ca^{2+} ions in the central region of this domain are bound in a six-coordinate, approximately octahedral site, either exclusively to protein ligands or to one or two water molecules. Glycine-rich repeats are responsible for binding the Ca^{2+} ions and forming the parallel β-roll. The sequence motif is GGXGXDXUX (D = aspartic acid). The first six residues are involved in Ca^{2+} binding, while the last three form a short β-strand. Each Ca^{2+} is bound by a pair loops with the motif GGXGXD. As well Ca^{2+} and Zn^{2+} ions can be removed from the molecule by adding chelating agents such as ethylendiaminetetraacetic acid (EDTA) (Baumann 2013).

The type I secretion system (T1SS) for AprA is encoded by the genes *aprD*, *E* and *F* within the *aprA-lipA* operon (Woods et al. 2001). Secretion systems of this type are relatively simple and facilitate the translocation of proteins up to 800 kDa in a single step from the cytoplasm to the extracellular space without a stable periplasmic intermediate (Delepelaire 2004; Holland et al. 2005). In Gram-negative bacteria, TT1S consists of three proteins spanning the cell envelope, an ATP-binding cassette protein, an inner membrane protein, and an outer membrane protein (Holland et al. 2005). The ATP-binding cassette is a dimeric protein located at the cytoplasmic membrane and is responsible for substrate recognition. The inner membrane protein, which is also called membrane fusion protein, protrudes into the periplasm and connects the inner and outer membrane components. The outer membrane protein is a trimeric protein, which forms a water-filled channel in the outer membrane (Delepelaire 2004). Type I secretion signals are located at the C-terminus of the proteins and are not cleaved. Therefore, only completely synthesized molecules are recognized by the ATP-binding cassette protein and can be transported. Multiple glycine-rich repeats such as GGXGXDXXX are assumed to be secretion motifs and mediate the substrate recognition (Delepelaire 2004; Holland et al. 2005).

The family of serralysins, to which AprA belongs, posses 4–6 of these repeats. Serralysins are secreted as inactive zymogens with an N-terminal, 9–20-residues long propeptide. This propeptide is autocatalytically cleaved in the presence of divalent cations such as Ca^{2+} or Zn^{2+} (Liao and McCallus 1998; Gomis-Rüth 2003; Baumann 2013). The highest secretion of AprA is typically observed at the end of the exponential growth or at the beginning of the stationary phase (Heeb and Haas 2001; Rajmohan et al. 2002; Liu et al. 2007; Marchand et al. 2008). The peptidase production is regulated in a complex manner by different molecular mechanisms. In general, almost 10% of the genes of *Pseudomonas* ssp. genomes encodes for products that are involved regulation, due to monitoring and responding to a huge number of environmental signals. In this genus, the majority of regulation takes place at the level of transcription, but an increasing number of posttranscriptional control systems are becoming evident (Moore et al. 2006).

The production of AprA and the siderophore pseudobactin by *Pseudomonas fluorescens* was shown to be negatively regulated by iron (Adams et al. 1994). The transcriptional regulation is based on iron starvation extracytoplasmic function (ECF) alternative sigma factors PbrA. In *pbrA* knockout experiments, the requirement of PbrA for full transcription of *aprA* under low iron conditions on skim milk agar was confirmed. Furthermore, it was shown that the ferric uptake regulator (Fur) represses the transcription of *aprA* under high iron conditions (Maunsell et al. 2006). Interestingly, the iron regulation of AprA was culture-dependent, as proteolysis of skim milk agar with yeast extract by a *pbrA* deficicnt mutant was observed when supplemented with iron or pseudobactin. These results imply that yet unidentified sigma factors may be required for the PbrA-independent *aprA* expression induced by iron and siderophore (Maunsell et al. 2006).

The GacS/GacA two-component system is widely distributed in Gammaproteobacteria and regulates many physiological processes in response to environmental stimuli (Heeb and Haas 2001; Workentine et al. 2009). In *Pseudomonas* ssp., the GacS/GacA regulatory system controls the production of secondary metabolites and exoenzymes such as AprA (Lapouge et al. 2008; Loper et al. 2012; Cheng et al. 2013). Typically, the two-component system is composed of a sensor kinase (GacS) and a cognate response regulator (GacA). GacS is an unorthodox sensor histidine sensor kinase (EC 2.7.13), which is a homodimeric transmembrane protein. In response to environmental triggers (temperature, pH, signals produced by bacterial population, quorum sensing, etc.) a conformational change of GacS is induced, which favors the autophosphorylation. The phosphate group is then transferred to the response regulator GacA by a phosphorelay mechanism and activates the transcription of small RNA genes. These small regulatory RNAs are supposed to bind post-transcriptional repressor proteins, RsmA and RsmE, and therefore relieve the repression of target genes (Lapouge et al. 2008; Workentine et al. 2009). In the case of *Pseudomonas fluorescens* SBW25, transcriptome analysis revealed that 702 genes were differentially regulated in a *gacS* mutant. The *aprA* gene was downregulated 221-fold in the *gacS* mutant and a loss of extracellular peptidase activity was observed (Cheng et al. 2013). The GacS/GacA is activated during transition from the exponential to the stationary growth phase. This is in accordance to the detection of peptidase activity at this time point during cultivation of *Pseudomonas* isolates (Liao and McCallus 1998; Heeb and Haas 2001; Rajmohan et al. 2002). Furthermore, peptidase secretion is regulated by quorum sensing, which is a cell-to-cell communication system based on acylated homoserine lactones (AHL) as response to fluctuations in cell density. The requirement of GacS/GacA for AHL production was shown for several pseudomonads (Kinscherf and Willis 1999; Juhas et al. 2005).

A correlation between phase-variation and peptidase production can be recognized. In general, phase-variation is used to generate population diversity, to increase bacterial fitness, and to adapt to changing environmental conditions. In the case of *Pseudomonas brassicacearum*, one phenotypic variant (phase II) showed no peptidase and lipase activity because the *aprA-lipA* operon was not transcribed in this variant (Chabeaud et al. 2001). This phenotypic-switching occurred due to spontaneous mutations in the GacA/GacS system (Lalaouna et al. 2012).

11.2.2 Interactions of Plasmin and Microbial Peptidases in Milk

Several studies have shown that microbial peptidases such as *Pseudomonas* peptidases are able to affect the plasmin system in milk and therefore influence the quality of milk and milk products (Ismail and Nielsen 2010). Peptidases produced by *Pseudomonas fluorescens* increased plasmin activity in milk, due to an enhancement of plasminogen activator activity by conversion of the one-chain molecule into the more active two-chain protein of lower weight. The hydrolysis resulted in a lowered K_m value (Frohbieter et al. 2005). Furthermore, the *Pseudomonas fluorescens* peptidase may act as a plasminogen activator itself (Frohbieter et al. 2005). Further studies showed that microbial peptidases are able to release plasmin from casein micelles in reconstituted non-fat dry milk and fresh milk; therefore, higher plasmin activity is found in the whey fraction (Fajardo-Lira and Nielsen 1998; Fajardo-Lira et al. 2000). The resulting whey products have higher plasmin activity and can cause undesired proteolysis in food systems to which they are added as functional ingredients (Nielsen 2002). Furthermore, the reduction of plasmin in the casein fraction is detrimental in case of cheese-making, since plasmin plays a key role in ripening of certain cheeses (Nielsen 2002).

11.3 Destabilization of Milk by Peptidases

The spoilage of milk and milk products caused by thermostable enzymes of psychrotolerant microorganisms, especially the species of the genus *Pseudomonas,* is well known and has been under research for the last decades. Thermostable peptidases can withstand pasteurization (72 °C/ 2 s) and even UHT (138 °C/2 s) processes and remain active in the derived milk products (Marchand et al. 2009b; Baur et al. 2015b; Matéos et al. 2015; Glück et al. 2016). Defects occurring during storage include gelation, creaming or formation of sediment, particle and bitterness (Barach et al. 1976; Stead 1987; Kohlmann et al. 1991; Datta and Deeth 2003; Barbano et al. 2006; Teh et al. 2012; Matéos et al. 2015; Stoeckel et al. 2016a). Whey proteins are less sensitive to *Pseudomonas* peptidases and plasmin, probably due to their globular structure (Baglinière et al. 2013; Datta and Deeth 2003; Matéos et al. 2015). The order of susceptibility of the caseins in milk to hydrolysis by Pseudomonas peptidases and plasmin is suggested to be $\kappa > \beta > \alpha_{S1}$ and $\beta = \alpha_{S2} > \alpha_{S1} > \kappa$, respectively (Datta and Deeth 2001). Matéos et al. (2015) investigated the hydrolysis of sodium caseinate and micellar casein in buffered systems by a *Pseudomonas* peptidase. They found the order to be $\beta > \alpha_S > \kappa$ and $\kappa = \beta > \alpha_S$ for sodium caseinate and micellar casein, respectively.

Bitterness is often the first change observed during storage of UHT milk (Adams et al. 1975; Gebre-Egziabher et al. 1980; Stoeckel et al. 2016a). Intact proteins do not have any bitter taste, since their molecular mass does not allow interaction with bitterness receptors. The hydrolysis of proteins may result in the formation of bitter peptides. The bitterness of peptides is associated with the molecular weight and the presence of hydrophobic residues. Peptides with a molecular weight smaller than 6 kDa and a high content of Leu, Pro, Phe, Tyr, Ile and Trp residues are likely to be bitter (Ney 1971; FitzGerald and O'Cuinn 2006). Caseins are hydrophobic proteins with a high Pro content and have therefore a high potential for bitterness when cleaved into peptides (Walstra et al. 2006).

Gelation or age gelation is an irreversible process and a major factor limiting the shelf-life of UHT milk (Kelly et al. 2012). Gelation occurs due to the formation of a three-dimensional network by β-lactoglobulin and κ-casein. The ultra-high temperature treatment in the production process (140–150 °C, 4–10 s) causes denaturation of β-lactoglobulin and the formation of β-lactoglobulin-κ-casein complexes (βκ-complex) at the surface of the casein micelles. During storage of UHT milk, these complexes may be released and form a network by aggregation, causing the gelation of milk (Datta and Deeth 2001, 2003). The process can be initiated by bacterial thermostable peptidases or plasmin. Bacterial peptidases cleave preferentially κ-casein at the surface of the casein micelles, which leads to a destabilization of the micelles and aggregation of the micelles occur comparable to chymosin coagulation (Datta and Deeth 2001, 2003). Plasmin preferably acts on β-casein, which is inside the micelles, causing a disintegration of the micelle and the formation of a less stable network or sedimentation (Datta and Deeth 2003; Kelly et al. 2012). A recent study by Anema (2017) proposed an alternative mechanism for physico-chemical age gelation during storage. It was shown that the level of β-lactoglobulin-κ-casein complexes in the serum does not change in the time between UHT treatment and gelation. Furthermore the gelled phase was composed of casein micelles without κ-casein. The new hypothsis is that the UHT treatment leads to κ-casein-depleted casein micelles which are meta-stable. During storage, these κ-casein-depleted casein micelles can cause gelation when the conditions such as pH and ioni calcium level change (Anema 2017, 2019). Creaming of the milk fat is suggested to be the result of the proteolysis of proteins in the milk fat globule membrane (Stoeckel et al. 2016a).

11.4 Determination of Peptidase Activity in Milk

The quantification of heat-resistant peptidase activity in raw milk is an important point to avoid or minimize textural and organoleptic changes in milk and milk products during storage. This is due to the fact that the enzyme activity would allow a prediction of the quality of the manufactured milk and milk products during storage. In general, a variety of methods for the determination of peptidase activity using either natural or synthetic substrates are available (Bendicho et al. 2002; Chen et al. 2003).

Methods using natural substrates are based on measuring the increase of peptides. A very simple method for determining the increase of peptides is to measure the absorption at 280 nm, which depends on the presence of aromatic amino acid residues (mainly tryptophan) and therefore on the amino acid sequence of the substrate. For quantification, the absorption coefficient is necessary (Kunitz 1947). Further methods include the derivatization of new, released α-amino groups with reagents such as fluorescamine, trinitrobenzene sulfonic acid or *o*-phthaldialdehyde and their subsequent photo- or fluorometric detection (Chism et al. 1979; Church et al. 1983; Humbert et al. 1990; Chen et al. 2003). Synthetic or chromogenic substrates used are either labeled proteins such as azocasein and fluoroescein thiocarbamoyl casein or chromogenic peptides such as *p*-nitroanilide-peptides (Twining 1984; Bendicho et al. 2002; Chen et al. 2003).

The challenges when determining peptidase activity in milk are on one hand the low levels of peptidase activity, which can still result in spoilage over the storage durations up to 6 months, and on the other hand inherently high levels of protein and fat in milk. Casein micelles and fat globules are responsible for the turbidity of milk, which hinders photometric measurements and quenches fluorescence (Bendicho et al. 2002; Humbert et al. 2006).

In general, there are two approaches for the determination of enzyme activity in food samples. The activity can be quantified directly in the respective product or after isolation of the enzymes from the sample. The first direct approach does not change the matrix conditions or remove inhibitory substances and therefore reveals the effective enzyme activity in the sample. Acidic precipitation of proteins or the clarifying reagents are used to render the solution transparent for spectrometric measurements (Humbert et al. 2006; Kelly and Fox 2006; Krewinkel et al. 2016; Stoeckel et al. 2016a). For direct determination, either synthetic substrates can be added or indigenous proteins can be used as a substrate. If using synthetic substrates, the indigenous milk protein competes with the added substrate and the determined activity is maybe underestimated and therefore unreliable. The use of synthetic substrates is possible when the specificity (k_{cat}/K_m) for the added substrate is many orders of magnitude higher than for the in situ protein (Chen et al. 2003). A more reliable method would be to use the *in situ* protein as a substrate and quantify the freed amino groups with one of the aforementioned methods. But these methods are in general less convenient and not rapid enough to deliver results on the peptidase activity during commercial milk production (Chen et al. 2003).

The second possibility is the application of a sample preparation method which is designated to isolate the peptidases and allow subsequent activity determination under defined conditions, with substrates added that do not compete with indigenous substrates. Furthermore, the application of a sample preparation prior to activity determination allows a simultaneously up-concentration and therefore an increase in sensitivity. Disadvantages are a possible loss of activity during sample preparation and a loss of information about the effective activity in the respective product (Chen et al. 2003; AiF 16588 N 2014).

11.5 Thermal Inactivation of Peptidases

As mentioned in earlier sections, peptidases from psychrotolerant microorganisms, such as *Pseudomonas* ssp., can withstand heat treatments during milk processing such as UHT treatments (138 °C, 2 s) and spoil the respective products during storage (Stoeckel et al. 2016a). On the other hand to this high thermostability at high temperatures, the peptidases from *Pseudomonas* ssp. are relatively susceptible to inactivation at low temperatures betwcen 40 and 70 °C (Barach et al. 1978; Kroll

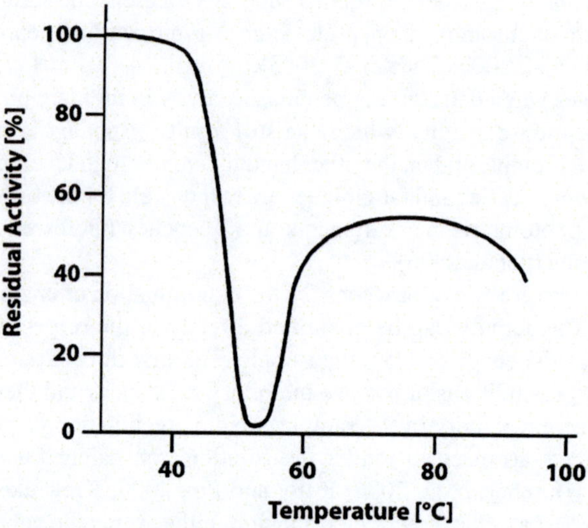

Fig. 11.4 Typical course of residual activity of *Pseudomonas* peptidase after heat treatment at various temperatures for 5 min, schematically depicted according to Glück et al. (2016)

and Klostermeyer 1984; Schokker and van Boekel 1998a; Alves et al. 2016). This inactivation behavior is schematically depicted in Fig. 11.4.

The highest inactivation is often found in a temperature range between 45 and 65 °C (Barach et al. 1976; Stepaniak et al. 1982; Owusu and Doble 1994; Baral et al. 1995; Schokker and van Boekel 1997a, 1998b; Matéos et al. 2015). This inactivation is caused by intermolecular autoproteolysis. Under certain conditions, such as temperatures near denaturation temperature (T_d), (partly) unfolded peptidase molecules can be hydrolyzed by native molecules (Schokker and van Boekel 1998a, b). Furthermore, autoproteolysis can be intramolecular; this means that a partly unfolded molecule hydrolyzes itself. Schokker and van Boekel (1998a) tried to discriminate between inter- and intramolecular autoproteolysis for a peptidase from *Pseudomonas fluorescens*, and concluded that intermolecular hydrolysis was more likely due to the occurrence of smaller fragments (Schokker and van Boekel 1998b). As seen in Fig. 11.4 at temperatures higher than 55 °C, the residual endopeptidase activity increases. In theory, reversible unfolding of the amino acid sequence is assumed and thus autoproteolysis cannot occur. However, when decreasing the temperature afterwards (<45 °C), refolding of the unfolded endopeptidases is supposed and thus residual activity increases. Kinetic parameters of thermal inactivation of peptidases of the genus *Pseudomonas* are summarized in Table 11.2.

When investigating the inactivation behavior of peptidases, the influence of heating medium has to be considered. In the context of milk spoilage, milk would be the most appropriate heating medium. In several studies, the protective influence milk or milk constituents have been proven (Mayerhofer et al. 1973; Birkeland et al. 1985; Schokker and van Boekel 1998a; Baur et al. 2015b; Glück et al. 2016). In case of a peptidase from *Pseudomonas panacis,* the *D*-values were 8–18-fold higher in

Table 11.2 Kinetic parameters for the thermal inactivation of peptidases of *Pseudomonas* ssp. (adapted from Stoeckel et al. 2016b)[a]

Species	Heating medium	T [°C]	E_a [kJ mol^{-1}]	D_ϑ [min] (ϑ [°C])	z [°C]	Reference
*P. fluorescens*22F	Demineralized water	90–110	87.6	69.2 (100)[b]	–	Schokker et al. (1997)
P. fluorescens AR-11	Phosphate buffer	120–160	92.8	–	–	Alichanidis and Andrews (1977)
Pseudomonas AFT21	Phosphate buffer	70–150	87	1.3 (130)	29.7	Stepaniak and Fox (1985)
Pseudomonas MC 60	Tris-HCl buffer	110–150	93.3[b]	–	–	Barach and Adams (1977)
P. fluorescens P38	Tris-HCl buffer	100–145	36.1[b]	–	–	Owusu and Doble (1994)
P. tolaasii	Tris-HCl buffer	80–140	82	2.6 (140)[b]	34.0	Baral et al. (1995)
P. fluorescens B52	Tris-HCl buffer + $CaCl_2$	110–150	104.3	–	–	Richardson (1981)
P. fluorescens AFT-36	Synthetic milk salt buffer	70–150	88.1	1.2 (130)	31.9	Stepaniak and Fox (1983)
Pseudomonas MC 60	SMUF	110–150	–	1.5 (149)	32.5	Adams et al. (1975)
Pseudomonas AFT21	SMUF	70–150	92	2.0 (130)	30.3	Stepaniak and Fox (1985)
P. fluorescens	SMUF	74, 140	–	0.1–5.0 (140)	–	Mitchell et al. (2009)
P. fluorescens biotype I	SMUF	80–140	100	110 (90)	–	(Diermayr et al. 1987)
P. panacis	SMUF	70–90	–	25.6 (80)	63.6[b]	Baur et al. (2015b)
P. fluorescens NCD02085	Whey	107–137	95	3.0 (127)	–	Vercet et al. (1997)
P. fluorescens	Skim milk	130–150	115.5	2.2 (140)	28.0	Kroll and Klostermeyer (1984)
P. panacis	Skim milk	70–90	–	422 (80)	29.0[b]	Baur et al. (2015a, b)
Pseudomonas sp. W15a	Whole milk	120–140	–	4.8 (130)	25.8	Stoeckel et al. (2016a)
P. proteolytica 691	Whole milk	120–140	–	4.0 (130)	28.7	Stoeckel et al. (2016a)

[a]Abbreviations are: *Ea* = activation energy, D_ϑ = Decimal reduction value at temperature ϑ, z = z-value, SMUF = simulated milk ultrafiltrate

[b]Calculated from data presented

skim milk than in SMUF (simulated milk ultrafiltrate) at temperatures from 70 to 90 °C (Baur et al. 2015b). Schokker and van Boekel (1998a) observed a temperature shift of maximal inactivation and increased residual activity when sodium caseinate was added to a purified peptidase from *Pseudomonas fluorescens*. They suggested that small amounts of caseinate stabilize the native form, which leads to an increase in the denaturation temperature. The reduction of inactivation with increasing the caseinate concentration has been explained by competitive inhibition of autoproteolysis by substrate molecules and steric hindrance by the reversible formation of peptidase-caseinate complexes (Schokker and van Boekel 1998a). Furthermore, the influence of calcium ions on the stability of *Pseudomonas* peptidases has been studied (Richardson 1981; Kroll and Klostermeyer 1984; Schokker and van Boekel 1999). A stabilizing effect of 10 or 20 mM $CaCl_2$ in buffer in a temperature range of 70–90 °C was observed by Kroll and Klostermeyer (1984) and Schokker et al. (1997). As described before for the addition of caseinate, the temperature of maximum inactivation shifted to a higher temperature, indicating stabilization against unfolding (Kroll and Klostermeyer 1984; Schokker et al. 1997). At higher temperatures, the influence of calcium is less clear. While Birkeland et al. (1985) observed a positive effect of calcium at 150 °C, Schokker and van Boekel (1999) found no significant influence on the heat stability of a *Pseudomonas* peptidase.

11.6 Conclusion and Outlook

Heat-resistant peptidases are one reason for spoilage of UHT-milk within the shelf life. While microorganisms are inactivated during ultra-high temperature processing, peptidases, which were secreted during raw milk storage, can withstand and remain active in the final products. As a result, products can deteriorate within the shelf-life, showing defects such as bitterness and coagulation. The peptidases secreted are well characterized and their inactivation behavior has been investigated in several studies. However, there is actually no detection system for peptidases in milk that is sensitive and specific enough to predict milk spoilage during the shelf life. Possibilities to overcome this issue could be the use of an enzyme-linked immunosorbent assay (ELISA), or use of specific, chromogenic peptides to detect heat-resistant peptidases in raw milk. Therefore, it would be necessary to investigate relevant peptidases regarding their substrate specificity using defined protein substrates such as β-casein. Based on this knowledge, chromogenic peptides might be designed which would serve as specific substrates for heat-resistant *Pseudomonas* peptidases. Ideally, the developed assay will be specific and sensitive enough that no elaborately sample preparation and concentration will be necessary to determine the proteolytic activity of raw milk. This could enable the application as a quality assurance test during milk processing in the future.

References

Adams DM, Barach JT, Speck ML (1975) Heat resistant proteases produced in milk by psychrotrophic bacteria of dairy origin. J Dairy Sci 58:828–834

Adams C, Dowling DN, O'Sullivan DJ et al (1994) Isolation of a gene (pbsC) required for siderophore biosynthesis in fluorescent *Pseudomonas* sp. strain M114. Mol Gen Genet MGG 243:515–524

Aghajari N, Van Petegem F, Villeret V et al (2003) Crystal structures of a psychrophilic metalloprotease reveal new insights into catalysis by cold-adapted proteases. Proteins 50:636–647

Ahn JH, Pan JG, Rhee JS (1999) Identification of the tliDEF ABC transporter specific for lipase in *Pseudomonas fluorescens* SIK W1. J Bacteriol 181:1847–1852

AiF 16588 N (2014) Schlussbericht FEI-Projekt "Hitzestabile mikrobielle Enzyme in Rohstoffen zur Milchverarbeitung - Qualitätssicherung, Entwicklung eines Testsystems und technologische Optionen"

Alichanidis E, Andrews AT (1977) Some properties of the extracellular protease produced by the psychrotrophic bacterium *Pseudomonas fluorescens* strain AR-11. Biochim Biophys Acta Enzymol 485:424–433

Allison SD, Lu L, Kent AG et al (2014) Extracellular enzyme production and cheating in *Pseudomonas fluorescens* depend on diffusion rates. Front Microbiol 5:169

Alves MP, Salgado RL, Eller MR et al (2016) Characterization of a heat-resistant extracellular protease from *Pseudomonas fluorescens* 07A shows that low temperature treatments are more effective in deactivating its proteolytic activity. J Dairy Sci 99(10):7842–7851

Anderson LM, Stockwell VO, Loper JE (2004) An extracellular protease of *Pseudomonas fluorescens* inactivates antibiotics of *Pantoea agglomerans*. Phytopathology 94:1228–1234

Anema SG (2017) Storage stability and age gelation of reconstituted ultra-high temperature skim milk. Int Dairy J 75:56–67

Anema SG (2019) Age gelation, sedimentation, and creaming in UHT milk: a review. Compr Rev Food Sci Food Saf 18:140–166

Ashdown LR, Koehler JM (1990) Production of hemolysin and other extracellular enzymes by clinical isolates of *Pseudomonas pseudomallei*. J Clin Microbiol 28:2331–2334

Baglinière F, Matéos A, Tanguy G, et al (2013) Proteolysis of ultra high temperature-treated casein micelles by AprX enzyme from Pseudomonas fluorescens F induces their destabilisation. Int Dairy J 31:55–61

Barach JT, Adams DM (1977) Thermostability at ultrahigh temperatures of thermolysin and a protease from a psychotrophic *Pseudomonas*. Biochim Biophys Acta 485:417–423

Barach JT, Adams DM, Speck ML (1976) Low temperature inactivation in milk of heat-resistant proteases from psychrotrophic bacteria. J Dairy Sci 59:391–395

Barach JTT, Adams DMM, Speck MLL (1978) Mechanism of low temperature inactivation of a heat-resistant bacterial protease in milk. J Dairy Sci 61:523–528

Baral A, Fox PF, O'Connor TP (1995) Isolation and characterization of an extracellular proteinase from *Pseudomonas tolaasii*. Phytochemistry 39:757–762

Barbano DM, Ma Y, Santos MV (2006) Influence of raw milk quality on fluid milk shelf life. J Dairy Sci 89(Supple):E15–E19

Bardoel BW, van Kessel KPM, van Strijp JAG, Milder FJ (2012) Inhibition of *Pseudomonas aeruginosa* virulence: characterization of the AprA–AprI Interface and species selectivity. J Mol Biol 415:573–583

Barrett AJ, Ito K, Nakajima Y, Yoshimoto T (2013) Handbook of Proteolytic Enzymes

Baumann U (2013) Chapter 180 – Serralysin and related enzymes. In: Rawlings ND, Salvesen G (eds) Handbook of proteolytic enzymes, 3rd edn. Academic, Cambridge, pp 864–867

Baumann U, Wu S, Flaherty K et al (1993) 3-dimensional structure of the alkaline protease of *Pseudomonas aeruginosa*: a 2-domain protein with a calcium-binding parallel-beta roll motif. EMBO J 12:3357–3364

Baumann U, Bauer M, Létoffé S et al (1995) Crystal structure of a complex between *Serratia marcescens* Metallo-protease and an inhibitor from Erwinia chrysanthemi. J Mol Biol 248:653–661
Baur C, Krewinkel M, Kranz B et al (2015a) Quantification of the proteolytic and lipolytic activity of microorganisms isolated from raw milk. Int Dairy J 49:23–29
Baur C, Krewinkel M, Kutzli I et al (2015b) Isolation and characterisation of a heat-resistant peptidase from *Pseudomonas panacis* withstanding general UHT processes. Int Dairy J 49:46–55
Bendicho S, Martí G, Hernández T et al (2002) Determination of proteolytic activity in different milk systems. Food Chem 79:245–249
Birkeland SE, Stepaniak L, Sorhaug T (1985) Quantitative studies of heat-stable proteinase from *Pseudomonas fluorescens* P1 by the enzyme-linked immunosorbent assay. Appl Environ Microbiol 49:382–387
Chabeaud P, de Groot A, Bitter W et al (2001) Phase-variable expression of an operon encoding extracellular alkaline protease, a serine protease homolog, and lipase in *Pseudomonas brassicacearum*. J Bacteriol 183:2117–2120
Champagne CP, Laing RR, Roy D et al (1994) Psychtotrophs in dairy products: their effects and their control. Crit Rev Food Sci Nutr 34:1–30
Chen L, Daniel RMM, Coolbear T (2003) Detection and impact of protease and lipase activities in milk and milk powders. Int Dairy J 13:255–275
Cheng X, de Bruijn I, van der Voort M et al (2013) The Gac regulon of *Pseudomonas fluorescens* SBW25. Environ Microbiol Rep 5:608–619
Chism GW, Huang AE, Marshall JA (1979) Sensitive assay for proteases in sterile milk. J Dairy Sci 62:1798–1800
Church FC, Swaisgood HE, Porter DH, Catignani GL (1983) Spectrophotometric assay using o-phthaldialdehyde for determination of proteolysis in milk and isolated milk proteins. J Dairy Sci 66:1219–1227
Cocolin L, Aggio D, Manzano M et al (2002) An application of PCR-DGGE analysis to profile the yeast populations in raw milk. Int Dairy J 12:407–411
Cousin MA (1982) Presence and activity of psychrotrophic microorganisms in milk and dairy products: a review. J Food Prot 45:172–207
Datta N, Deeth HC (2001) Age gelation of UHT milk—a review. Food Bioprod Process 79:197–210
Datta N, Deeth HC (2003) Diagnosing the cause of proteolysis in UHT milk. LWT Food Sci Technol 36:173–182
De Jonghe V, Coorevits A, Van Hoorde K et al (2011) Influence of storage conditions on the growth of *Pseudomonas* species in refrigerated raw milk. Appl Environ Microbiol 77:460–470
Delepelaire P (2004) Type I secretion in gram-negative bacteria. Biochim Biophys Acta, Mol Cell Res 1694:149–161
Desmasures N, Bazin F, Guéguen M (1997) Microbiological composition of raw milk from selected farms in the camembert region of Normandy. J Appl Microbiol 83:53–58
Diermayr P, Kroll S, Klostermeyer H (1987) Mechanisms of heat inactivation of a proteinase from *Pseudomonas fluorescens* biotype I*. J Dairy Res 54:51–60
Dogan B, Boor KJ (2003) Genetic diversity and spoilage potentials among *Pseudomonas* spp. isolated from fluid milk products and dairy processing plants. Appl Environ Microbiol 69:130–138
Donlan RM (2002) Biofilms: microbial life on surfaces. Emerg Infect Dis 8:881–890
Dufour D, Nicodème M, Perrin C et al (2008) Molecular typing of industrial strains of *Pseudomonas* spp. isolated from milk and genetical and biochemical characterization of an extracellular protease produced by one of them. Int J Food Microbiol 125:188–196
Engel LS, Hill JM, Caballero AR et al (1998) Protease IV, a unique extracellular protease and virulence factor from *Pseudomonas aeruginosa*. J Biol Chem 273:16792–16797
Ercolini D, Russo F, Ferrocino I, Villani F (2009) Molecular identification of mesophilic and psychrotrophic bacteria from raw cow's milk. Food Microbiol 26:228–231
Fairbairn DJ, Law BA (1986) Proteinases of psychrotrophic bacteria: their production, properties, effects and control. J Dairy Res 53:139–177

Fajardo-Lira CE, Nielsen SS (1998) Effect of psychrotrophic microorganisms on the plasmin system in milk. J Dairy Sci 81:901–908

Fajardo-Lira C, Oria M, Hayes KD et al (2000) Effect of psychrotrophic bacteria and of an isolated protease from *Pseudomonas fluorescens* M3/6 on the plasmin system of fresh milk. J Dairy Sci 83:2190–2199

FitzGerald RJ, O'Cuinn G (2006) Enzymatic debittering of food protein hydrolysates. Biotechnol Adv 24:234–237

Fricker M, Skånseng B, Rudi K et al (2011) Shift from farm to dairy tank milk microbiota revealed by a polyphasic approach is independent from geographical origin. Int J Food Microbiol 145(Suppl):S24–S30

Frohbieter KA, Ismail B, Nielsen SS et al (2005) Effects of *Pseudomonas fluorescens* M3/6 bacterial protease on plasmin system and plasminogen activation. J Dairy Sci 88:3392–3401

Gebre-Egziabher A, Humbert ES, Blankenagel G (1980) Hydrolysis of milk proteins by microbial enzymes. J Food Prot 9:672–735

Glück C (2017) Investigation of heat-stable, microbial peptidases associated to the microbiota of raw milk. Dissertation, University Hohenheim

Glück C, Rentschler E, Krewinkel M et al (2016) Thermostability of peptidases secreted by microorganisms associated with raw milk. Int Dairy J 56:186–197

Gomis-Rüth FX (2003) Structural aspects of the metzincin clan of metalloendopeptidases. Mol Biotechnol 24:157–202

Hantsis-Zacharov E, Halpern M (2007) Culturable psychrotrophic bacterial communities in raw milk and their proteolytic and lipolytic traits. Appl Environ Microbiol 73:7162–7168

Hassan AN, Frank JF (2011) Microorganisms assosciated with milk. In: Fuquay JW (ed) Encyclopedia of dairy science, 2nd edn. Academic, Cambridge, pp 447–457

Heeb S, Haas D (2001) Regulatory roles of the GacS/GacA two-component system in plant-associated and other gram-negative bacteria. Mol Plant-Microbe Interact 14:1351–1363

Hege T, Baumann U (2001) Protease C of *Erwinia chrysanthemi*: the crystal structure and role of amino acids Y228 and E189. J Mol Biol 314:187–193

Holland IB, Schmitt L, Young J (2005) Type 1 protein secretion in bacteria, the ABC-transporter dependent pathway (review). Mol Membr Biol 22:29–39

Humbert G, Guingamp M-F, Kouomegne R et al (1990) Measurement of proteolysis in milk and cheese using trinitrobenzene sulphonic acid and a new dissolving reagent. J Dairy Res 57:143

Humbert G, Guingamp M-F, Linden G et al (2006) The clarifying reagent, or how to make the analysis of milk and dairy products easier. J Dairy Res 73:464–471

Ismail B, Nielsen SS (2010) Invited review: plasmin protease in milk: current knowledge and relevance to dairy industry. J Dairy Sci 93:4999–5009

Johnson LA, Beacham IR, MacRae IC et al (1992) Degradation of triglycerides by a pseudomonad isolated from milk: molecular analysis of a lipase-encoding gene and its expression in Escherichia coli. Appl Environ Microbiol 58:1776–1779

Juhas M, Eberl L, Tummler B (2005) Quorum sensing: the power of cooperation in the world of *Pseudomonas*. Environ Microbiol 7:459–471

Kawai E, Idei A, Kumura H et al (1999) The ABC-exporter genes involved in the lipase secretion are clustered with the genes for lipase, alkaline protease, and serine protease homologues in *Pseudomonas fluorescens* no. 33. Biochim Biophys Acta Gene Struct Expr 1446:377–382

Kelly AL, Fox PF (2006) Indigenous enzymes in milk: a synopsis of future research requirements. Int Dairy J 16:707–715

Kelly AL, Datta N, Deeth H (2012) Thermal processing of dairy products. In: Sun D-W (ed) Thermal food processing. CRC Press, Boca Raton, FL, pp 273–306

Kinscherf TG, Willis DK (1999) Swarming by *Pseudomonas syringae* B728a requires gacS (lemA) and gacA but not the acyl-Homoserine lactone biosynthetic gene ahlI. J Bacteriol 181:4133–4136

Kohlmann KL, Nielsen SS, Steenson LR et al (1991) Production of proteases by psychrotrophic microorganisms. J Dairy Sci 74:3275–3283

Krewinkel M, Baur C, Kranz B et al (2016) A sensitive and robust method for direct determination of lipolytic activity in natural milk environment. Food Anal Methods 9(3):646–655

Kroll S, Klostermeyer H (1984) Heat inactivation of exogenous proteinases from *Pseudomonas fluorescens*. Z Lebensm Unters Forsch 179:288–295

Kunitz M (1947) Crystalline soybean trypsin inhibitor: II. general propertieS. J Gen Physiol 30:291–310

Lalaouna D, Fochesato S, Sanchez L et al (2012) Phenotypic switching in *Pseudomonas brassicacearum* involves GacS- and GacA-dependent Rsm small RNAs. Appl Environ Microbiol 78:1658–1665

Lapouge K, Schubert M, Allain FHT et al (2008) Gac/Rsm signal transduction pathway of γ-proteobacteria: from RNA recognition to regulation of social behaviour. Mol Microbiol 67:241–253

Lee DG, Urbach JM, Wu G et al (2006) Genomic analysis reveals that *Pseudomonas aeruginosa* virulence is combinatorial. Genome Biol 7:1–14

Liao CH (1989) Analysis of pectate lyases produced by soft rot bacteria associated with spoilage of vegetables. Appl Environ Microbiol 55:1677–1683

Liao C-H, McCallus DE (1998) Biochemical and genetic characterization of an extracellular protease from *Pseudomonas* biochemical and genetic characterization of an extracellular protease from *Pseudomonas fluorescens* CY091. Appl Environ Microbiol 64:914–921

Liu M, Wang H, Griffiths MW (2007) Regulation of alkaline metalloprotease promoter by N-acyl homoserine lactone quorum sensing in Pseudomonas fluorescens. J Appl Microbiol 103:2174–2184

Lo R, Stanton-Cook MJ, Beatson SA et al (2015) Draft genome sequence of *Pseudomonas fluorescens* SRM1, an isolate from spoiled raw milk. Genome Announc 3:e00138–e00115

Loper JE, Hassan KA, Mavrodi DV et al (2012) Comparative genomics of plant-associated *Pseudomonas* spp.: insights into diversity and inheritance of traits involved in multitrophic interactions. PLoS Genet 8:e1002784

Marchand S, Coudijzer K, Heyndrickx M et al (2008) Selective determination of the heat-resistant proteolytic activity of bacterial origin in raw milk. Int Dairy J 18:514–519

Marchand S, Heylen K, Messens W et al (2009a) Seasonal influence on heat-resistant proteolytic capacity of *Pseudomonas lundensis* and *Pseudomonas fragi*, predominant milk spoilers isolated from Belgian raw milk samples. Environ Microbiol 11:467–482

Marchand S, Vandriesche G, Coorevits A et al (2009b) Heterogeneity of heat-resistant proteases from milk *Pseudomonas* species. Int J Food Microbiol 133:68–77

Masoud W, Vogensen FK, Lillevang S et al (2012) The fate of indigenous microbiota, starter cultures, *Escherichia coli*, *Listeria innocua* and *Staphylococcus aureus* in Danish raw milk and cheeses determined by pyrosequencing and quantitative real time (qRT)-PCR. Int J Food Microbiol 153:192–202

Matéos A, Guyard-Nicodème M, Baglinière F et al (2015) Proteolysis of milk proteins by AprX, an extracellular protease identified in *Pseudomonas* LBSA1 isolated from bulk raw milk, and implications for the stability of UHT milk. Int Dairy J 49:78–88

Maunsell B, Adams C, O'Gara F (2006) Complex regulation of AprA metalloprotease in *Pseudomonas fluorescens* M114: evidence for the involvement of iron, the ECF sigma factor, PbrA and pseudobactin M114 siderophore. Microbiology 152:29–42

Mayerhofer HJ, Marshall RT, White CH et al (1973) Characterization of a heat-stable protease of *Pseudomonas fluorescens* P26. Appl Microbiol 25:44–48

McCarthy CN, Woods RG, Beacham IR (2004) Regulation of the aprX-lipA operon of *Pseudomonas fluorescens* B52: differential regulation of the proximal and distal genes, encoding protease and lipase, by ompR-envZ. FEMS Microbiol Lett 241:243–248

McPhee JD, Griffiths MW (2011) *Pseudomonas* spp. In: Fuquay JW (ed) Encyclopedia of dairy science, 2nd edn. Academic, Cambridge, pp 379–383

Meyer JM (2000) Pyoverdines: pigments, siderophores and potential taxonomic markers of fluorescent Pseudomonas species. Arch Microbiol 174:135–42

Meyer J-M, Geoffroy VA, Baida N, et al (2002) Siderophore Typing, a Powerful Tool for the Identification of Fluorescent and Nonfluorescent Pseudomonads. Appl Environ Microbiol 68:2745–2753

Mitchell GE, Ewings KN, Bartley JP (2009) Physicochemical properties of proteinases from selected psychrotrophic bacteria. J Dairy Res 53:97–115

Miyatake H, Hata Y, Fujii T et al (1995) Crystal structure of the unliganded alkaline protease from *Pseudomonas aeruginosa* IFO3080 and its conformational changes on ligand binding. J Biochem 118:474–479

Moore ERB, Tindall BJ, Martins Dos Santos VAP et al (2006) Nonmedical: *Pseudomonas*. In: Dworkin M, Falkow S, Rosenberg E et al (eds) The prokaryotes: volume 6: proteobacteria: gamma subclass. Springer, New York, pp 646–703

Moreland JL, Gramada A, Buzko OV et al (2005) The Molecular Biology Toolkit (MBT): A modular platform for developing molecular visualization applications. BMC Bioinformatics 6:21

Mu Z, Du M, Bai Y (2009) Purification and properties of a heat-stable enzyme of *Pseudomonas fluorescens* Rm12 from raw milk. Eur Food Res Technol 228:725–734

Munsch-Alatossava P, Alatossava T (2006) Phenotypic characterization of raw milk-associated psychrotrophic bacteria. Microbiol Res 161:334–346

Ney KH (1971) Voraussage der Bitterkeit von Peptiden aus deren Aminosäurezu-sammensetzung. Z Lebensm Unters Forsch 147:64–68

Nielsen SS (2002) Plasmin system and microbial proteases in milk: characteristics, roles, and relationship. J Agric Food Chem 50:6628–6634

Olofsson TC, Ahrné S, Molin G (2007) Composition of the bacterial population of refrigerated beef, identified with direct 16S rRNA gene analysis and pure culture technique. Int J Food Microbiol 118:233–240

Owusu RK, Doble C (1994) Heat-stability of a proteinase from psychrotrophic *Pseudomonas fluorescens* P38, chymotrypsin and thermolysin. Food Chem 51:137–142

Peix A, Ramírez-Bahena M-H, Velázquez E (2009) Historical evolution and current status of the taxonomy of genus *Pseudomonas*. Infect Genet Evol 9:1132–1147

Quigley L, O'Sullivan O, Stanton C et al (2013) The complex microbiota of raw milk. FEMS Microbiol Rev 37:664–698

Raats D, Offek M, Minz D, Halpern M (2011) Molecular analysis of bacterial communities in raw cow milk and the impact of refrigeration on its structure and dynamics. Food Microbiol 28:465–471

Rajmohan S, Dodd CER, Waites WM (2002) Enzymes from isolates of *Pseudomonas fluorescens* involved in food spoilage. J Appl Microbiol 93:205–213

Rawlings ND, Waller M, Barrett AJ et al (2014) MEROPS: the database of proteolytic enzymes, their substrates and inhibitors. Nucleic Acids Res 42:D503–D509

Richardson BC (1981) The purification and characterization of a heat-stable protease from *Pseudomonas fluorescens* B52. N Z J Dairy Sci Technol 16:195–207

Santos MV, Ma Y, Caplan Z et al (2003) Sensory threshold of off-flavors caused by proteolysis and lipolysis in milk. J Dairy Sci 86:1601–1607

Schokker EP, van Boekel MAJS (1997a) Kinetic modeling of enzyme inactivation: kinetics of heat inactivation at 90-110 C of extracellular proteinase from *Pseudomonas fluorescens* 22F. J Agric Food Chem 45:4740–4747

Schokker EP, van Boekel MAJS (1997b) Production, purification and partial characterization of the extracellular proteinase from *Pseudomonas fluorescens* 22F. Int Dairy J 7:265–271

Schokker EP, van Boekel MAJS (1998a) Effect of protein content on low temperature inactivation of the extracellular proteinase from *Pseudomonas fluorescens* 22F. J Dairy Res 65:347–352

Schokker EP, van Boekel MAJS (1998b) Mechanism and kinetics of inactivation at 40–70°C of the extracellular proteinase from *Pseudomonas fluorescens* 22F. J Dairy Res 65:261–272

Schokker EP, van Boekel MAJS (1999) Kinetics of thermal inactivation of the extracellular proteinase from *Pseudomonas fluorescens* 22F: influence of pH, calcium, and protein. J Agric Food Chem 47:1681–1686

Sørhaug T, Stepaniak L (1997) Psychrotrophs and their enzymes in milk and dairy products: quality aspects. Trends Food Sci Technol 8:35–41

Stead D (1987) Assay of proteinases of psychrotrophic bacteria in milk using hide powder azure and a simple procedure for clarification of milk. J Dairy Res 54:295–302

Stepaniak L, Fox PF (1983) Thermal stability of an extracellular proteinase from *Pseudomonas fluorescens* AFT 36. J Dairy Res 50:171–184

Stepaniak L, Fox PF (1985) Isolation and characterization of heat stable proteinases from *Pseudomonas* isolate AFT 21. J Dairy Res 52:77–89

Stepaniak L, Fox PF, Daly C (1982) Isolation and general characterization of a heat-stable proteinase from *Pseudomonas fluorescens* AFT 36. Biochim Biophys Acta Gen Subj 717:376–383

Stoeckel M, Lidolt M, Achberger V et al (2016a) Growth of *Pseudomonas weihenstephanensis*, *Pseudomonas proteolytica* and *Pseudomonas* sp. in raw milk: impact of residual heat-stable enzyme activity on stability of UHT milk during shelf-life. Int Dairy J 59:20–28

Stoeckel M, Lidolt M, Stressler T et al (2016b) Heat stability of indigenous milk plasmin and proteases from *Pseudomonas*: a challenge in the production of ultra-high temperature milk products. Int Dairy J 61:250–261

Teh KH, Flint S, Palmer J et al (2011) Thermo-resistant enzyme-producing bacteria isolated from the internal surfaces of raw milk tankers. Int Dairy J 21:742–747

Teh KH, Flint S, Palmer J et al (2012) Proteolysis produced within biofilms of bacterial isolates from raw milk tankers. Int J Food Microbiol 157:28–34

Tielen P, Rosenau F, Wilhelm S et al (2010) Extracellular enzymes affect biofilm formation of mucoid *Pseudomonas aeruginosa*. Microbiology 156:2239–2252

Twining SS (1984) Fluorescein isothiocyanate-labeled casein assay for proteolytic enzymes. Anal Biochem 143:30–34

Vercet A, Lopez P, Burgos J (1997) Inactivation of heat-resistant lipase and protease from *Pseudomonas fluorescens* by manothermosonication. J Dairy Sci 80:29–36

Verdier-Metz I, Gagne G, Bornes S, et al (2012) Cow teat skin, a potential source of diverse microbial populations for cheese production. Appl Environ Microbiol 78:326–33

von Neubeck M, Baur C, Krewinkel M et al (2015) Biodiversity of refrigerated raw milk microbiota and their enzymatic spoilage potential. Int J Food Microbiol 211:57–65

Walstra P, Wouters JTM, Geurst TM (2006) Dairy science and technology, 2nd edn. CRC Press Taylor & Francis Group, Boca Raton, FL

Whiteley M, Bangera MG, Bumgarner RE et al (2001) Gene expression in *Pseudomonas aeruginosa* biofilms. Nature 413:860–864

Winsor GL, Griffiths EJ, Lo R et al (2015) Enhanced annotations and features for comparing thousands of *Pseudomonas* genomes in the *Pseudomonas* genome database. Nucleic Acids Res 44:D646–D653

Woods RG, Burger M, Beven C-A et al (2001) The aprX-lipA operon of *Pseudomonas fluorescens* B52: a molecular analysis of metalloprotease and lipase production. Microbiology 147:345–354

Workentine ML, Chang L, Ceri H et al (2009) The GacS-GacA two-component regulatory system of *Pseudomonas fluorescens*: a bacterial two-hybrid analysis. FEMS Microbiol Lett 292:50–56

Wouters JT, Ayad EH, Hugenholtz J et al (2002) Microbes from raw milk for fermented dairy products. Int Dairy J 12:91–109

Chapter 12
The Heat Stability of Indigenous and Bacterial Enzymes in Milk

Britta Graf, Johannes Schäfer, Zeynep Atamer, and Jörg Hinrichs

12.1 Enzyme Sources and Problems Encountered in Milk Products

The enzymes in milk and milk products can be of indigenous and microbial origin. Figure 12.1 shows different pathways of the transfer of bacterial and indigenous enzymes during the processing of milk. Beginning with the milking process, several (indigenous) enzymes, such as cathepsin D and plasmin, which are expressed from different bovine tissues and cells (e.g., liver tissue and mammary epithelial cells) and secreted into the milk, are already present in the milk (Table 12.1). Their level depends on various factors, e.g., genetic factors and health status of animal. In cases of mastitis, the somatic cell count (SCC) increases (Larsen and Petersen 1995) and an increased cathepsin D activity can be observed, because the somatic cell count has been found to be positively correlated with cathepsin D activity (Somers et al. 2003). However, during milking, the milk is unavoidably contaminated with udder commensals from the teat canal, which are mainly lactic acid bacteria that are present in small, limited numbers only. The vast majority of contaminants (e.g., micro-, strepto-, and enterococci), are external to the udder and arise from different sources, such as the udder skin, hide and milking equipment, as well as utensils (Hassan and Frank 2011). After milking, further contamination can occur during storage, milk collection, delivery, reception, and any further processing step in the dairy (Fig. 12.1). Besides their replication, the microorganisms generate (exogenous) enzymes, such as lipases and peptidases and in some cases even activators, e.g., for plasminogen (Lähteenmäki et al. 2001). All in all, an (excessive) activity of

B. Graf · J. Schäfer · Z. Atamer · J. Hinrichs (✉)
Department of Soft Matter Science and Dairy Technology, Institute of Food Science and Biotechnology, University of Hohenheim, Stuttgart, Germany
e-mail: britta.graf@uni-hohenheim.de; Johannes.Schaefer@uni-hohenheim.de; zeynep.atamer@uni-hohenheim.de; j.hinrichs@uni-hohenheim.de

A. L. Kelly, L. B. Larsen (eds.), *Agents of Change*, Food Engineering Series,
https://doi.org/10.1007/978-3-030-55482-8_12

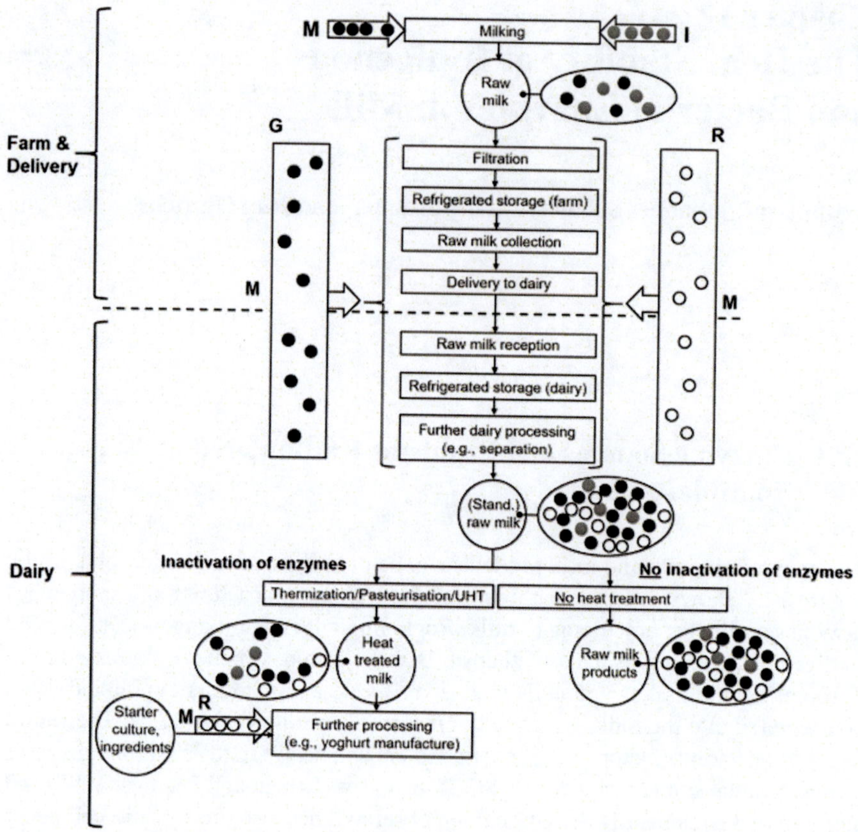

Fig. 12.1 Pathways for the transfer of microbial enzymes/microorganisms (M, filled circle) and indigenous enzymes (I, gray circle) during the processing of raw milk. Possible generation (G, filled circle) and recontamination (R, open circle) of microbial enzymes/microorganisms during these processing steps; *Stand* standardized. Figure according to Hassan and Frank (2011) and Walstra et al. (2006)

indigenous and microbial enzymes can cause various product defects, e.g., bitter taste in fresh cheese/UHT milk or rancid off-tastes, resulting in impaired sensorial and textural properties (Table 12.1).

However, an adequate treatment eliminates pathogens and reduces spoilage of microorganisms as well as the enzyme activity, depending on the technology and process parameter combination chosen. The risk of product defects is drastically lowered and the shelf-life, in particular of non-fermented, unripened dairy products, e.g., UHT milk, is prolonged.

Table 12.1 Summary of enzymes highlighted in this chapter and their importance regarding heat inactivation

Enzymes	Origin	Optimum pH (-)	Optimum temperature (°C)	Catalyzed reaction (enzyme class)	Significance	References
Indigenous enzymes						
Alkaline phosphatase	Sloughed-off myco-epithelial cells; lipid microdroplets with intracellular acquisition	9.0–10.5	37	Proteolysis	Heat indicator for pasteurization in the dairy industry, flavor profile of raw milk cheese	Fox and Kelly (2006), Meyer (1996)
γ-glutamyltransferase	Release from MEC	8.5–9	45	Transfer of γ-glutamyl residues	Indicator for heat treatments between 70–80 °C, 16 s	Claeys et al. (2002), Fox and Kelly (2006)
Lactoperoxidase	Mammary gland	5.0–7.0	37[a]	$H_2O_2 + 2\ HA \rightarrow 2\ H_2O + 2\ A$	Inhibition of microorganisms	Buys (2011), Walstra et al. (2006)
Lipoprotein lipase	Mammary gland	9.2	37	Lipolysis (hydrolase)	Rancid or soapy flavor in milk (cleavage of triglycerides); aroma profile in cheese	Bengtsson and Olivecrona (1982), Deeth (2011), Dickow et al. (2012), Meyer (1996)
Cathepsin D	Indigenous, secretion into milk via different pathways, e.g., release from MEC	~4	37	Proteolysis	Bitterness of fresh cheese and UHT milk induced by bitter peptides	Dallas et al. (2015), Ebner et al. (2016), Kaminogawa and Yamauchi (1972), Sebald et al. (2018)
Plasmin	Indigenous, expression in liver tissue, secretion into milk	7.4–8	37	Proteolysis	(Bitter) peptides, sensorial and textural defects of UHT treated products	Berglund et al. (1995), Humbert and Alais (1979), Meyer (1996), Stoeckel et al. (2016b)

(continued)

Table 12.1 (continued)

Enzymes	Origin	Optimum pH (-)	Optimum temperature (°C)	Catalyzed reaction (enzyme class)	Significance	References
Bacterial enzymes						
Lipases	*Pseudomonas* ssp.	9	30–45	Lipolysis	Rancidity, bitter flavor (e.g., in fresh cheese, UHT milk)	Fox and Kelly (2006), Nielsen (2002)
Peptidases	*Pseudomonas* ssp.	6.5–8	30–45	Proteolysis	UHT milk defects like bitterness, particle formation, gelation	Baur et al. (2015), Nielsen (2002), Stoeckel et al. (2016b)

MEC mammary epithelial cells, *UHT* ultra-high temperature
[a]not indicated but assumed to be at physiological temperature

12.2 Inactivation Methods

The inactivation of dairy enzymes is of great importance, as undesirable product changes such as bitterness or gelation are prevented, and thus the shelf life is prolonged. Moreover, the effectiveness of a heat treatment can be confirmed (e.g., low pasteurization through alkaline phosphatase) (Walstra et al. 2006). Table 12.2 provides an overview on possible inactivation methods with selected examples and process conditions. In general, the methods are divided into thermal and non-thermal treatments. In special cases, e.g., for cheese manufacture and ripening, specific enzymes are desired and the inactivation should thus be controlled. Therefore, some gentle non-thermal methods aim to retain enzymes, like lipoprotein lipase or γ-glutamyltransferase, by applying a high pressure treatment instead of utilizing thermal methods (Pandey and Ramaswamy 2004).

Non-thermal methods are primary designated to obtain a safe product while the negative effects of thermal treatments like cooked flavor development or protein denaturation should be reduced or prevented. Techniques such as pulsed electric field treatment and ultrasound are gaining popularity and their fields of application within the food industry and are currently under investigation. However, the inactivation of alkaline phosphatase in milk is below 30% for the investigated treatments and conditions (see Table 12.2) (Jaeger et al. 2009; Riener et al. 2009; Villamiel and De Jong 2000).

High pressure treatment has been intensely studied in the last decades and it was shown that treatments at 200–400 MPa for up to 180 min resulted in inactivation of alkaline phosphatase and γ-glutamyltransferase in milk of up to 15%, whereas lipoprotein lipase showed an increase of enzyme activity of 50% after the treatment. Thus, these studies showed that aroma profiles during cheese ripening can be modulated by controlling the enzymatic activity (Mussa and Ramaswamy 1997). Nevertheless, higher pressure treatments of 500–800 MPa for 128 min resulted in high enzyme inactivation rates or complete inactivation (Rademacher and Hinrichs 2006). Other non-thermal methods like ultraviolet (UV) treatment or radiation have been shown to inactivate enzymes in milk (Glew 1962; Kung et al. 1953; Lante et al. 2013). It has to be considered that alternative treatments have to be approved by each country's legal authority. For instance, UV treatment of milk has been considered as safe for human consumption by the European Food Safety Authority (EFSA) (EFSA Panel on Dietetic Products, Nutrition and Allergies 2016). The Food and Drug Administration (FDA) has recently published a legal regulation about the usage of irradiation within food production, valid for the United States of America (U.S. Food and Drug Administration 2018).

Besides conventional heat treatment (indirect through heat exchangers), the novel direct heating methods ohmic and microwave heating are included in Table 12.2. All considered references demonstrate that dairy enzymes can be inactivated by means of a thermal treatment, independently of the heating method (Jakób et al. 2010; Lin and Ramaswamy 2011; Lorenzen et al. 2010; Meyer 1996).

Table 12.2 Technologies investigated for the inactivation of dairy enzymes (for each technology, selected references have been chosen)

Method		Enzyme	Medium	Conditions	Inactivation (%)	D_{ref}-value (at ϑ_{ref}) (min)	Reference
Thermal	Heat treatment (indirect)	Alkaline phosphatase	Skim milk	60–71 °C, 1840–6 s	35–100	16.4 (60 °C) 0.1 (71.7 °C)	Meyer (1996)
		Alkaline phosphatase	Raw milk	62 °C 30 min 65 °C 32 min	99		Lorenzen et al. (2010)
		γ-glutamyltransferase	Raw milk	62 °C 30 min 65 °C 32 min	20–30 60–90		Lorenzen et al. (2010)
		Lipoprotein lipase	Raw milk	62 °C 30 min 65 °C 32 min	0		Lorenzen et al. (2010)
	Microwave heating	Alkaline phosphatase	Raw milk	60–75 °C, 1000 W, 2450 MHz	35–98	0.3 (65 °C) 0.1 (67 °C) 0.02 (70 °C)	Lin and Ramaswamy (2011)
	Ohmic heating	Alkaline phosphatase	Milk	52–64 °C, 0–750 min, 50 Hz	90–100		Jakób et al. (2010)

Method		Enzyme	Medium	Conditions	Inactivation (%)	D_{ref}-value (at ϑ_{ref}) (min)	Reference
Non-thermal	UV treatment	Protease	Buffer	15–300 min	10–100		Lante et al. (2013)
	High pressure treatment	Alkaline phosphatase	Raw milk	200–400 MPa, 5–120 min	5–15	1039–301	Mussa and Ramaswamy (1997)
		Lipoprotein lipase	Raw milk	300–400 MPa, 0–180 min	50 (increase)		Pandey and Ramaswamy (2004)
		γ-glutamyltransferase	Raw milk	300–400 MPa, 0–180 min	0–15	848.5–555.5	Pandey and Ramaswamy (2004)
		Alkaline phosphatase	Raw milk	500–800 MPa, 0–128 min, 5, 20, 40 °C	0–45 (500 MPa) 0–100 (800 MPa)		Rademacher and Hinrichs (2006)
	Irradiation	Alkaline phosphatase	Raw and heat treated milk	γ-radiation X-radiation 0–25 Mrad, 1 Mrad/min Electron radiation	5–80		Glew (1962)
		Phosphatase	Milk	γ-radiation, 8×10^5 roentgen h^{-1}, 1–12 h	1.6–9.8		Kung et al. (1953)
	Ozone treatment	Protease	Freeze concentrated milk	150 ppm, 7 L/min, 1–5 min	20–93		Hwang et al. (2007)
		Lipase	Freeze concentrated milk	150 ppm, 7 L/min, 1–5 min	40–96		Hwang et al. (2007)
	Drying (decrease of a_w-value)	Alkaline phosphatase	Spray dried condensed skim milk	Inlet feed: 12.1–29.3 g/min at outlet air temperatures: 120–80 °C	20 (80 °C)–90 (120 °C)		Etzel et al. (1996)
	Pulsed electric field	Alkaline phosphatase	Raw milk	35 kV/cm, 75 μs	29		Riener et al. (2009)
		Alkaline phosphatase	Raw milk	29–38 kV/cm, 3–8 μs	8–22		Jaeger et al. (2009)
	Ultrasound	Alkaline phosphatase	Skim milk/whole milk	150 W, 23.5 °C, 102.3 s	1.2/0		Villamiel and De Jong (2000)
		γ-glutamyltransferase	Skim milk/whole milk	150 W, 23.5 °C, 102.3 s	17.2/22.1		Villamiel and De Jong (2000)

D_{ref}-value: Decimal reduction time; the time necessary to reduce the enzyme activity to 1/10 of its original value at a specific temperature. ϑ_{ref}: Reference temperature, *UV* Ultraviolet light

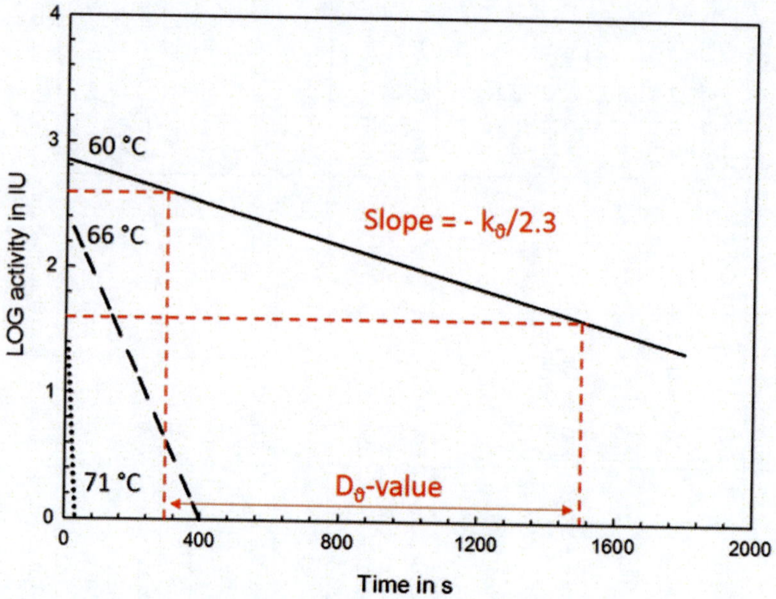

Fig. 12.2 Inactivation curves of an example of an indigenous enzyme (alkaline phosphatase) in skim milk at 60 °C (solid line), 66 °C (dashed line) and 71 °C (dotted line) according to the data of Meyer (1996). D_ϑ-value: Decimal reduction time; the time necessary to reduce the enzyme activity to 1/10 of its original value at a specific temperature (here in s); k_ϑ: Reaction rate constant for a given reference temperature ϑ (in s^{-1})

Figure 12.2 displays the heat inactivation of alkaline phosphatase in skim milk at 60, 66, and 71 °C, based on data determined by Meyer (1996). The graph represents typical enzyme inactivation curves, where the enzyme activity is plotted against the heating time. For each heating temperature, a characteristic plot is found. With increasing heating temperature, enzymatic inactivation proceeds for lower heating times, graphically recognizable by a steeper slope of the data points. Thus, an increasing heating temperature results in a decreasing enzyme activity for a similar heating time.

In order to make comparisons between different trials, experimental setups, and even enzyme types, characteristic heat inactivation parameters are therefore calculated.

12.3 Basics of Heat Inactivation Kinetics

Heating processes, which are characterized by the heating temperature and heating time, are responsible for the reactions occurring in processed media. In order to evaluate the effectiveness of a treatment (e.g., regarding the destruction of microorganisms or enzymes), the heating time and temperature as well as the concentration of the substance need to be considered. The same approach can be applied for other treatments, e.g., high-pressure processes. The decomposition of a substance (reaction rate of a substance) at a constant temperature T is described as the change in the concentration dC (g L^{-1}) with time dt (s) according to Eq. (12.1) (Hinrichs and Atamer 2011).

$$\frac{dC}{dt} = -k_T \cdot C^n \tag{12.1}$$

This has been found generally valid for decomposition reactions of the nth order (where k_T is the rate constant at reference temperature (s^{-1})). A first-order reaction (n = 1) is usually assumed and observed for the inactivation of enzymes. Equation (12.1) can be integrated and simplified as follows (Eq. 12.2):

$$C_t = C_0 \exp(-k_T \cdot t) \tag{12.2}$$

where C_t is the concentration at time t (g L^{-1}), C_0 is the initial concentration (g L^{-1}), k_T is the rate constant at reference temperature (s^{-1}), and t is the holding time (s). Utilizing the Arrhenius equation (Eq. (12.3)) to describe the inactivation rate, k_T:

$$k_T = k_{ref} \cdot \exp\left[-\frac{E_A}{R} \cdot \left(\frac{1}{T} - \frac{1}{T_{ref}}\right)\right] \tag{12.3}$$

where k_{ref} represents the rate constant at reference temperature T_{ref}, T is the absolute temperature (K), E_A is the activation energy (J mol^{-1}) and R = 8.314 J (mol K)$^{-1}$ is the universal gas constant, Eq. (12.1) can be transformed into Eq. (12.4):

$$\frac{C_t}{C_0} = \exp\left\{-\left[k_{ref} \cdot \exp\left[-\frac{E_A}{R} \cdot \left(\frac{1}{T} - \frac{1}{T_{ref}}\right)\right] \cdot t\right]\right\} \tag{12.4}$$

In order to compare or evaluate the sensitivity of enzymes regarding processing conditions, the parameters calculated using Eq. (12.5), which are given in Table 12.3 (k_{ref} and E_A), are basically used. However, in practice, often additional parameters like D_ϑ-value (defined as the time necessary at a specific temperature ϑ to reduce the population of microorganisms to 1/10 of the original value) and z_ϑ-value (represents the increase in temperature necessary to obtain the same effect of destruction or reaction in 1/10 of the time) are used, which are related to the kinetics presented above (Hinrichs and Atamer 2011). These parameters can only be calculated for the first-order reactions using Eqs. (12.5) and (12.6) (see also Fig. 12.3).

$$D_\vartheta = \frac{2.303}{k_T} \quad \text{with } T(K) = 273K + \vartheta(^\circ C) \tag{12.5}$$

Table 12.3 Summary of reported temperature-dependent kinetic data for inactivation of indigenous and bacterial enzymes

Enzymes (EC number)	Heating medium	ϑ_{ref} (°C)	D_{ref} (min)	z_{ref} (K)	T_{ref} (K)	n (-)	k_{ref} (s^{-1})	E_A (kJ mol^{-1})	References
Indigenous enzymes									
Alkaline phosphatase (3.1.3.1)	Milk	70	0.6	5.6	343	1	6.98×10^{-2}	402	Walstra et al. (2006)
Lactoperoxidase (1.11.1.7)	Skim milk	70	168	3.3	343	1.4–1.8	1.98×10^{-3}	689	Meyer (1996)
Lactoperoxidase (1.11.1.7)	Milk	70	0.3	4.2	343	1	1.16×10^{-2}	536	Horak (1980)
Plasmin (EC 3.4.21.7)[a]	Skim milk	80	6.0	n.d.	353	1	6.40×10^{-3}	n.d.	Walstra et al. (2006)
Plasmin (EC 3.4.21.7)[b]	Skim milk	120	0.5	n.d.	393	1	7.68×10^{-2}	n.d.	Walstra et al. (2006)
Acid phosphatase (3.1.3.2)	Milk	100	0.8	9.8	373	1	5.12×10^{-2}	272	Horak (1980)
Cathepsin D (EC 3.4.23.5)	Skim milk	55	387	8.0	328	1	9.92×10^{-5}	257	Hayes et al. (2001)
Lipoprotein lipase (EC 3.1.1.34)	Skim milk	70	0.3	7.9	343	1	4.61×10^{-2}	276	Meyer (1996)
γ-glutamyltransferase (EC 2.3.2.2)	Skim milk	60	66	9.9	333	1	3.40×10^{-2}	215	Stănciuc et al. (2011)
		70	8.4	10.5	343	1	2.74×10^{-1}	215	
		80	1.0	11.1	353	1	2.32	215	
γ-glutamyltransferase (EC 2.3.2.2)	Cream	60	33	12.1	333	1	7.04×10^{-2}	175	Stănciuc et al. (2011)
		70	5.0	12.9	343	1	4.62×10^{-1}	175	
		80	0.9	13.6	353	1	2.55	175	
Bacterial (exogenous) enzymes[c]									
Lipase of *Pseudomonas fluorescens*	Milk	130	8.3	88	403	1	4.61×10^{-3}	35	Walstra et al. (2006)
Lipase of *Pseudomonas fluorescens* B52	Buffer, pH 7.2	130	1.5	36	403	1	2.52×10^{-2}	81.7	Vercet et al. (2002)
Lipase of *Pseudomonas* spp.	Milk	130	12	26	403	1	3.28×10^{-3}	120	Walstra et al. (2006)
Peptidase of *Pseudomonas fluorescens*	Milk	130	11	31	403	1	3.66×10^{-3}	100	Walstra et al. (2006)
Peptidase of *Pseudomonas fluorescens* NCDO 2085	Milk whey, pH 7.2	130	1.8	31	403	1	2.14×10^{-2}	93.5	Vercet et al. (2002)
Peptidase of *Achromobacter*	Milk	130	8.5	31	403	1	4.52×10^{-3}	100	Walstra et al. (2006)
Peptidase of *Pseudomonase panacis*	Skim milk	90	132	29	363	1	2.90×10^{-4}	86.9	Baur et al. (2015)
Peptidase of *Pseudomonas sp.* W15a	Milk	130	4.8	26	403	1	7.90×10^{-3}	120	Stoeckel et al. (2016b)
Peptidase of *Pseudomonase proteolytica* 691	Milk	130	4.0	29	403	1	9.59×10^{-3}	108	Stoeckel et al. (2016b)

Symbols are: ϑ_{ref}: Temperature, D_{ref}-value: Decimal reduction time; the time necessary to reduce the enzyme activity to 1/10 of its original value at a specific temperature, z_{ref}-value: The increment in temperature (K) required to obtain the same degree of enzyme inactivation in 1/10 of the time, T_{ref}: Absolute temperature, n: Order of reaction, k_{ref}: Reaction rate constant for a given reference temperature, E_A: Activation energy, n.d.: not determine

[a]Kinetic data are calculated from experiments conducted at a temperature range of 70–91.8 °C

[b]Kinetic data are calculated from experiments conducted at a temperature range of 91.8–140 °C

[c]The values vary widely among the strains and depending on conditions during growth

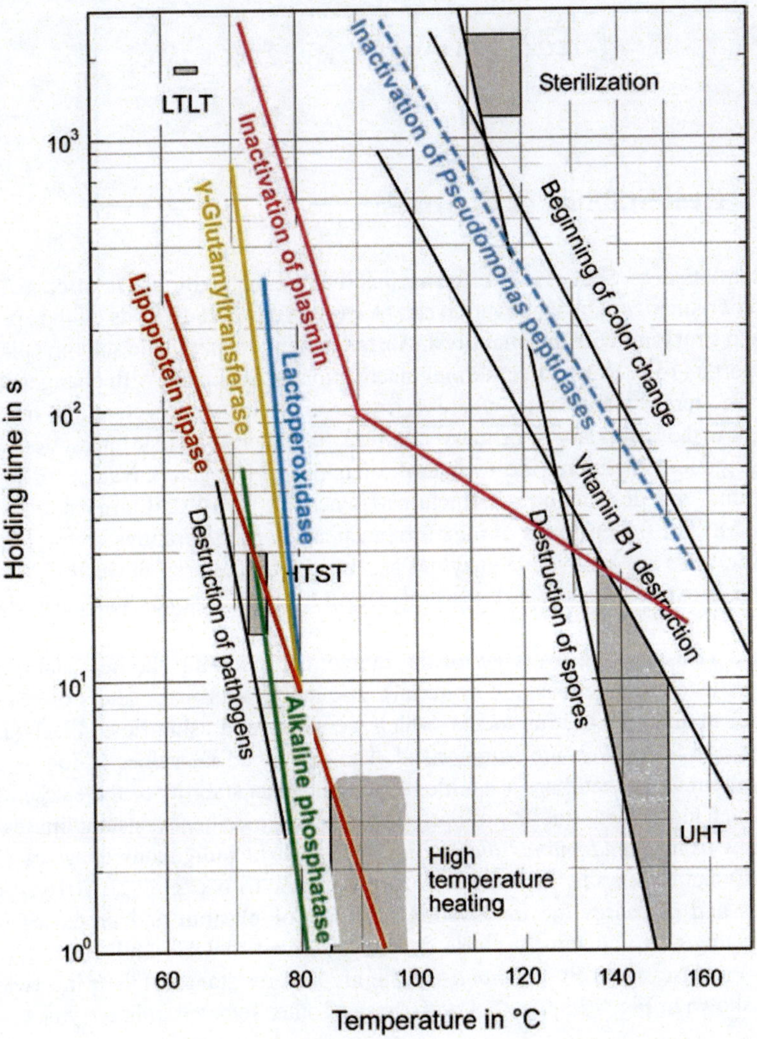

Fig. 12.3 Temperature-time graph for the heat treatment of milk showing the inactivation lines of selected endogenous enzymes (the 99% inactivation of plasmin and 90% inactivation of Pseudomonas peptidases are given, together with the inactivation lines of other endogenous milk enzymes, alkaline phosphatase, lactoperoxidase, lipoprotein lipase, and γ-glutamyltransferase). *LTLT* low temperature long time, *HTST* high temperature short time, *UHT* ultra-high temperature treatment. Figure modified according to Kessler (2002), Stoeckel et al. (2016b), Walstra et al. (2006) and Zehetner et al. (1995)

$$z_{\vartheta} = \frac{2.303 \cdot R \cdot T^2}{E_A} \quad \text{with } T(K) = 273K + \vartheta(^{\circ}C) \tag{12.6}$$

12.4 Heat Stability of Enzymes

As mentioned in Sect. 12.3, enzyme inactivation behavior most often follows first-order kinetics. The inactivation can be irreversible and depends on the enzyme type and structure, with the quantity and arrangement of enzymatic subunits playing an important role. Reversible thermal inactivation is associated with changes in the enzymes' ternary structure, whereas irreversible enzyme inactivation involves changes in the secondary or primary structure, like intermolecular interactions such as aggregation, including the formation of bonds (hydrogen, covalent, or ionic), glycations, and destruction of structures (Lencki et al. 1992; Lumry and Eyring 1954). Most of the enzymes in milk are inactivated at temperatures below 100 °C, however, there are heat-stable enzymes surviving temperatures above 100 °C, e.g., plasmin or extracellular peptidases synthesized by *Pseudomonas* subspecies (ssp.) (see Fig. 12.3).

Table 12.3 gives an overview of the literature concerning the heat stability of dairy enzymes (indigenous and bacterial). The inactivation parameters of selected enzymes in different heating media, which were obtained using Eqs. (12.1)–(12.6) presented in Sect. 12.3, are summarized. The calculated D_{ϑ}-values (1-log or 90% inactivation) show that there is a wide range in the heat stability of the enzymes.

A much higher heat stability of bacterial enzymes (e.g., heat-resistant lipases and proteases from *Pseudomonas fluorescens*) than that of indigenous enzymes (e.g., plasmin) can be seen in Table 12.3. A recent review (Stoeckel et al. 2016b) summarizes and discusses the inactivation behavior of plasmin and proteases from *Pseudomonas*, which mostly cause defects observed in UHT milk (Rauh et al. 2014). Proteases from *Pseudomonas* are highly heat resistant and their inactivation line is shown in Fig. 12.3 together with those of other selected milk enzymes.

12.4.1 Heat Stability of Indigenous Enzymes

Alkaline Phosphatase

Alkaline phosphatase (ALP) is one of the most frequently studied dairy enzymes, as it is commonly used as a marker enzyme for milk pasteurization, being more heat stable than pathogenic microorganisms present in milk. Various rapid, sensitive and to some extent automated analysis methods have turned ALP into a suitable marker enzyme for the dairy industry (Fadiloglu et al. 2006; Fox and Kelly 2006; Vamvakaki et al. 2006).

However, in comparison to other dairy enzymes (see also Fig. 12.3), ALP is very heat-sensitive, and various studies showed a rapid decrease in enzyme activity upon the thermal treatment of milk between 60 and 85 °C within minutes to seconds (Claeys et al. 2002; Dumitraşcu et al. 2014; Fadiloglu et al. 2006; Lorenzen et al. 2010; Meyer 1996; Wilińska et al. 2007) (see Table 12.3). Its activity is completely inactivated by a heat treatment at 70 °C for 15 s (Blel et al. 2002). Although the presence of ALP in milk is related to the fat globule membrane, which leads to different initial enzyme activities in whole milk and skim milk, the inactivation behavior was not shown to be influenced by the fat content of the milk (Dumitraşcu et al. 2014). Meyer (1996) hypothesized that the inactivation of ALP is mainly dependent on the opening of the enzyme structure, but also on aggregation reactions. Some studies have reported that ALP in milk was more heat-resistant than, for instance, in a buffer system. This observation was attributed to a protective effect of milk constituents, e.g., the whey protein β-lactoglobulin, on the enzyme system (Dumitraşcu et al. 2014; Fadiloglu et al. 2006; Wilińska et al. 2007).

During storage of UHT milk, ALP can be reactivated, which is associated with free sulfhydryl groups (Fox and Kelly 2006; Lorenzen et al. 2010). However, this effect has only been reported for UHT milk, but not for pasteurized milk as only high temperatures in the UHT range (130–145 °C) induce the formation of free sulfhydryl groups in milk (Fox and Kelly 2006).

γ-Glutamyltransferase

The membrane-bound milk enzyme γ-glutamyltransferase (GGT) was reported to be heat stable at 50 °C for 30 min, in contrast to many other indigenous dairy enzymes (Anjos et al. 1998). Complete inactivation was reported in several studies after a heat treatment of milk ranging between 70–80 °C for 10–1 min (Anjos et al. 1998; Blel et al. 2002; Fox 2003; Fox and Kelly 2006; Lorenzen et al. 2010; McKellar et al. 1991; Stănciuc et al. 2011; Zehetner et al. 1995). As GGT is slightly more heat-resistant than ALP, it could be suited to monitor heat treatments of milk between 75–80 °C for 15 s (Fox and Kelly 2006; Lorenzen et al. 2010; Stănciuc et al. 2011). Possible reactivation after a heat treatment of milk within 6 h to 3 days was not detected (McKellar et al. 1991; Stănciuc et al. 2011).

In milk, one-quarter of the GGT is located in the cream phase (entrapped in the fat membrane) and three quarters within the skim milk (Claeys et al. 2002; Zehetner et al. 1995). One study demonstrated GGT to be less heat-stable in cream than in skim milk (Stănciuc et al. 2011). However, another study could not confirm these findings and did not report any correlation (Claeys et al. 2002).

Lactoperoxidase

Lactoperoxidase (LPO) is more heat-stable than ALP and GGT and thus it is still active in pasteurized milk. Heat treatments at 80 °C were reported to inactivate the enzyme partly or even completely, depending on the holding time (2.5–60 s) (Blel et al. 2002; Lorenzen et al. 2010; Meyer 1996). The activation energy E_A for inactivation ranges between 500–700 kJ mol^{-1} and is the highest within all enzymes considered (see Table 12.3). This indicates that the enzyme inactivation is highly temperature-dependent (Meyer 1996).

Lipoprotein Lipase

There has not been done much research about the heat inactivation of lipoprotein lipase (LPL) in dairy matrices so far, with the exception of consideration of its significance for stability of butter. Meyer (1996) observed almost complete inactivation of LPL after a heat treatment of milk at 70 °C for >20 s (Table 12.3). That author proposed that the heat inactivation mechanisms of LPL are a mixture of aggregation reactions and the destruction of the globular protein structure rather than being pure destruction, as the specific reaction enthalpy and entropy are rather small. Moreover, a very small but highly heat-stable fraction of LPL in milk was assumed.

This suggestion is in accordance with the results of Dickow et al. (2012), where LPL activity could still be found after heat treatments between 70–90 °C, while heat treatments at >90 °C resulted in complete LPL inactivation (Dickow et al. 2012). The results of both studies are comparable, as both found residual activities of LPL of 50–70% in milk after a heat treatment at 60–64 °C for 30 s (Dickow et al. 2012; Meyer 1996).

Cathepsin D

Cathepsin D is an acid proteinase naturally present in milk, but little information about its heat inactivation is available. The onset of inactivation of cathepsin D has been reported after a heat treatment at 50 °C for 600 s in buffer (Kaminogawa and Yamauchi 1972) and at 55 °C for 1800 s in skim milk (Hayes et al. 2001). A heat treatment at 70–72 °C for 15–600 s resulted in an inactivation of 65–100%, depending on the heating medium (Fox 2003; Hayes et al. 2001; Kaminogawa and Yamauchi 1972; Larsen et al. 2000). Thus, residual activity of cathepsin D may be detectable in pasteurized milk. Moreover, it has been reported that milk constituents have a protective effect on the enzyme, possibly associated with casein (Larsen et al. 2000).

Plasmin

Plasmin, the predominant indigenous proteinase in milk, is characterized by a complex functional mechanism with activators as well as inhibitors playing an important role (see Chap. 2). Thus, the heat inactivation of the plasmin system is rather complex, as plasmin, plasminogen and plasminogen activators are heat-stable and can still be active in UHT products (Lu and Nielsen 1993; Nielsen 2002; Rauh et al. 2014; Van Asselt et al. 2008), whereas plasmin inhibitors and plasminogen activator inhibitors are heat-labile and are thus already inactivated at pasteurization temperatures (72–85 °C) (Ismail and Nielsen 2010, 2011; Nielsen 2002; Prado et al. 2006; Stoeckel et al. 2016b).

Upon heating milk at various temperatures for 30 s, plasminogen activation as well as inactivation can both occur (Stoeckel et al. 2016b). The activation takes place at temperatures ranging between 70 and 85 °C, indicating that treatments like pasteurization can induce increased plasmin levels. This results from the inactivation of the plasmin inhibitor or the activation of denatured plasminogen into plasmin (Kelly and Fox 2006; Lu et al. 2009; Nielsen 2002). Otherwise, the inactivation of plasmin starts at elevated temperatures (>80 °C) (Stoeckel et al. 2016b).

In general, the activation energy E_A for the inactivation of plasmin in milk is higher at temperatures <90 °C (160–200 kJ mol^{-1}) and lower at temperatures > 90 °C (20–50 kJ mol^{-1}) (see Table 12.3). This indicates that inactivation below 90 °C depends on temperature, while above 90 °C the heat holding time is the dominating factor, which explains the biphasic inactivation curve illustrated in Fig. 12.3 (Meyer 1996; Stoeckel et al. 2016b).

The thermal inactivation of plasmin and plasminogen in milk was thought to result from unfolding, whereupon free thiol groups (from denatured whey proteins or other plasmin molecules) react with each other (Metwalli et al. 1998; Meyer 1996; Stoeckel et al. 2016b). Such free thiol groups become also available during denaturation of β-lactoglobulin. Thus, the inactivation of plasmin is supposed to be directly linked to the denaturation of β-lactoglobulin. In contrast, the milk protein casein has been reported to have a stabilizing effect on plasmin upon heating (Metwalli et al. 1998; Stoeckel et al. 2016b), but the mechanism and causes of this have not been fully understood yet (Rauh et al. 2014).

Regarding the manufacture of UHT products, it has been recommended to include a preheating step at 90 °C for 180 s into the process, in order to inactivate plasmin to a residual activity of 0.3% and thus obtain a shelf-stable product (Stoeckel et al. 2016b).

12.5 Heat Stability of Bacterial Enzymes

Pseudomonas ssp. Peptidases

Peptidases secreted by psychrotolerant *Pseudomonas* ssp. are among the most common endogenous enzymes in milk. Various studies have monitored their thermal stability and reported that they are able to survive common UHT treatments of milk at temperatures ranging between 130 and 140 °C for 20 s (Adams et al. 1975; Barach and Adams 1977; Baur et al. 2015; Glück et al. 2016; Mu et al. 2009; Nielsen 2002; Schokker and van Boekel 1999; Stoeckel et al. 2016a; Stoeckel 2016). As summarized in Table 12.3, the D_ϑ values from different studies range between 1.5 and 11 min at 130 °C. Thus, if *Pseudomonas* spp. are present in the raw milk, these proteases remain active after a UHT treatment in the final product (with 80–90% of their residual activity). As a consequence, they can induce product defects (Baur et al. 2015; Glück et al. 2016; Stoeckel et al. 2016a).

According to Stoeckel et al. (2016b), the heat inactivation of microbial peptidases can be classified into three stages (see Fig. 12.4). The first stage is the *active enzyme stage* and occurs at temperatures from 5 to 45 °C. The proteases remain active within this stage and show their maximal enzyme activity, while unfolding of the enzyme molecule begins. The second stage, the so-called *autoproteolysis stage*, is marked by decreasing heat stability of the enzyme with increasing temperature. After a minimum heat stability is reached, the enzymatic activity of microbial proteases increases again towards the end of this stage until a temperature of 75–80 °C. The inactivation results in intermolecular autoproteolysis of the enzyme

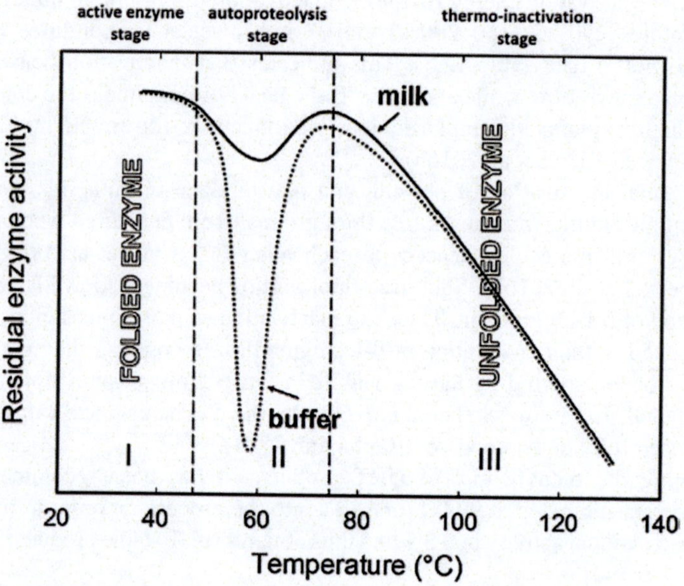

Fig. 12.4 Schematic illustration of the inactivation of proteases from *Pseudomonas* spp. as a function of temperature in milk (solid line) and in a buffer system (calcium and casein-free; dotted line). Figure modified according to Stoeckel et al. (2016b)

molecules while they are present as a mixture of folded and active molecules as well as unfolded and inactive molecules. This stage does not follow first-order inactivation kinetics and is often referred to as the low-temperature effect (Adams et al. 1975; Barach and Adams 1977; Glück et al. 2016; Schokker and Van Boekel 1999; Stoeckel et al. 2016b). The third stage (*thermo-inactivation stage*) is characterized by completely unfolded enzyme molecules with high heat stability even at UHT temperatures (Adams et al. 1975; Stoeckel et al. 2016b). It is hypothesized that, upon cooling, the protease molecules undergo refolding reactions, resulting in retained enzyme activities after heating. This stage follows first-order inactivation kinetics. The final heat inactivation is proposed to be induced by deamidation, as well as protein aggregation and reactions of disulphide groups with each other (Schokker and Van Boekel 1999; Stoeckel et al. 2016b).

The second stage is less pronounced in milk in comparison to other systems like buffers, as milk constituents are supposed to have a protective effect on the enzymes. In particular, caseins and calcium are thought to reduce the autoproteolytic effect through steric hindrance (Baur et al. 2015; Glück et al. 2016; Schokker and Van Boekel 1999; Stoeckel et al. 2016b).

Strategies to avoid spoilage of UHT milk through microbial proteases are processes with two heating steps where the UHT treatment is followed by a mild heating at 55–60 °C for 60–5 min. However, this treatment could induce the hydrolysis of milk protein, which in turn decreases the storage stability. Another strategy is to prolong the heating times at 125–130 °C to >150 s in order to inactivate 90% of the microbial proteases (see Fig. 12.3) (Stoeckel et al. 2016b).

12.6 Conclusion

Enzymes in milk can be of indigenous (expressed from different bovine tissues and cells, then secreted into milk) or exogenous (bacterial) origin. During (dairy) food processing, the activity of enzymes should be monitored and controlled as they can be either undesired or desired due to the following reasons: (1) in order to preserve products and obtain long shelf lives, enzyme inactivation by thermal (indirect or direct heating) or non-thermal methods (e.g., ultrasound, high-pressure treatment, or radiation) is favored; (2) in order to create a specific flavor and aroma profile, enzyme activity and thus low enzyme inactivation can be desired (e.g., during cheese making and ripening); (3) dairy enzymes can be used as markers, indicating, for instance, the raw milk quality by the total load of bacterial enzymes or a successful heat treatment (e.g., alkaline phosphatase for pasteurization processes).

References

Adams DM, Barach JT, Speck ML (1975) Heat resistant proteases produced in milk by psychrotrophic bacteria of dairy origin. J Dairy Sci 58:828–834

Anjos FD, Machado A, Ferro C, Otto F, Bogin E (1998) Gamma-Glutamyltransferase as a marker for the pasteurization of raw milk. J Food Prot 61:1057–1059

Barach JT, Adams DM (1977) Thermostability at ultrahigh temperatures of thermolysin and a protease from psychrotrophic Pseudomonas. Biochim Biophys Acta - Enzymol 485:417–423

Baur C, Krewinkel M, Kutzli I, Kranz B, Von Neubeck M, Huptas C, Wenning M, Scherer S, Stoeckel M, Hinrichs J, Stressler T, Fischer L (2015) Isolation and characterisation of a heat-resistant peptidase from Pseudomonas panacis withstanding general UHT processes. Int Dairy J 49:46–55

Bengtsson G, Olivecrona T (1982) On the pH dependency of lipoprotein lipase activity. Biochim Biophys Acta - Lipids Lipid Metab 712:196–199

Berglund L, Andersen MD, Petersen TE (1995) Cloning and characterization of the bovine plasminogen cDNA. Int Dairy J 5:593–603

Blel M, Guingamp M-F, Gaillard J-L, Humbert G (2002) Studies on the thermal sensitivity of gamma-glutamyl transpeptidase measured with a modified test procedure and compared with that of alkaline phosphatase and lactoperoxidase in milk. Lait 82:555–566

Buys EM (2011) Enzymes indigenous to milk | lactoperoxidase. In: Fuquay JW (ed) Encyclopedia of dairy sciences, 2nd edn. Academic Press, Amsterdam, pp 319–323

Claeys WL, van Loey AM, Hendrickx ME (2002) Intrinsic time temperature integrators for heat treatment of milk. Trends Food Sci Technol 13:293–311

Dallas DC, Murray NM, Gan J (2015) Proteolytic systems in milk: perspectives on the evolutionary function within the mammary gland and the infant. J Mammary Gland Biol Neoplasia 20:133–147

Deeth HC (2011) Enzymes indigenous to milk | lipases and esterases. In: Fuquay JW (ed) Encyclopedia of dairy sciences, 2nd edn. Academic Press, Amsterdam, pp 304–307

Dickow JA, Nielsen MT, Hammershøj M (2012) Effect of lenient steam injection (LSI) heat treatment of bovine milk on the activities of some enzymes, the milk fat globule and pH. Int J Dairy Technol 65:191–200

Dumitraşcu L, Stănciuc N, Stanciu S, Râpeanu G (2014) Inactivation kinetics of alkaline phosphatase from different species of milk using quinolyl phosphate as a substrate. Food Sci Biotechno 23:1773–1778

Ebner J, Baum F, Pischetsrieder M (2016) Identification of sixteen peptides reflecting heat and/or storage induced processes by profiling of commercial milk samples. J Proteomics 147:66–75

EFSA Panel on Dietetic Products, Nutrition and Allergies (2016) Safety of UV-treated milk as a novel food pursuant to Regulation (EC) No 258/97. EFSA J 14:4370

Etzel MR, Suen S-Y, Halverson SL, Budijono S (1996) Enzyme inactivation in a droplet forming a bubble during drying. J Food Eng 27:17–34

Fadiloglu S, Erkmen O, Sekeroglu G (2006) Thermal inactivation kinetics of alkaline phosphatase in buffer and milk. J Food Process Preserv 30:258–268

Fox PF (2003) Significance of indigenous enzymes in milk and dairy products. In: Whitaker JR (ed) Handbook of food enzymology. Dekker, New York

Fox PF, Kelly AL (2006) Indigenous enzymes in milk: overview and historical aspects—part 1. Int Dairy J 16:500–516

Glew G (1962) Some effects of ionizing radiations on the alkaline phosphatase in cow's milk. J Dairy Res 29:1

Glück C, Rentschler E, Krewinkel M, Merz M, Von Neubeck M, Wenning M, Scherer S, Stoeckel M, Hinrichs J, Stressler T, Fischer L (2016) Thermostability of peptidases secreted by microorganisms associated with raw milk. Int Dairy J 56:186–197

Hassan AN, Frank JF (2011) Microorganisms associated with milk. In: Fuquay JW (ed) Encyclopedia of dairy sciences, 2nd edn. Academic Press, Amsterdam

Hayes MG, Hurley MJ, Larsen LB, Heegaard CW, Magboul AAA, Oliveira JC, McSweeney PLH, Kelly AL (2001) Thermal inactivation kinetics of bovine cathepsin D. J Dairy Res 68:267–276

Hinrichs J, Atamer Z (2011) Heat treatment of milk | sterilization of milk and other products. In: Fuquay JW (ed) Encyclopedia of dairy sciences, 2nd edn. Academic Press, Amsterdam, pp 714–724

Horak FP (1980) Reaction kinetics of spore inactivation and chemical change during thermal treatment of milk for optimisation of heating processes (Über die Reaktionskinetik der Sporenabtötung und chemischer Veränderungen bei der thermischen Haltbarmachung von Milch zur Optimierung von Erhitzungsverfahren), Dissertation, Technical University Munich, Munich, Germany

Humbert G, Alais C (1979) The milk proteinase system. J Dairy Res 46:559

Hwang JH, Lee SJ, Park HS, Min SG, Kwak HS (2007) Comparison of physicochemical and sensory properties of freeze-concentrated milk with evaporated milk during storage. Asian-Australas J Anim Sci 20:273–282

Ismail B, Nielsen SS (2010) Invited review: plasmin protease in milk: current knowledge and relevance to dairy industry. J Dairy Sci 93:4999–5009

Ismail B, Nielsen SS (2011) Enzymes indigenous to milk | plasmin system in milk. In: Fuquay JW (ed) Encyclopedia of dairy sciences, 2nd edn. Academic Press, Amsterdam, pp 308–313

Jaeger H, Meneses N, Knorr D (2009) Impact of PEF treatment inhomogeneity such as electric field distribution, flow characteristics and temperature effects on the inactivation of E. coli and milk alkaline phosphatase. Innovative Food Sci Emerg Technol 10:470–480

Jakób A, Bryjak J, Wójtowicz H, Illeová V, Annus J, Polakovič M (2010) Inactivation kinetics of food enzymes during ohmic heating. Food Chem 123:369–376

Kaminogawa S, Yamauchi K (1972) Acid protease of bovine milk. Agric Biol Chem 36:2351–2356

Kelly AL, Fox PF (2006) Indigenous enzymes in milk: a synopsis of future research requirements. Int Dairy J 16:707–715

Kessler H-G (2002) Food and bio process engineering: dairy technology, 5th edn. Kessler, Munich

Kung H-C, Gaden EE, King CG (1953) Vitamins and enzymes in milk, effect of gamma-radiation on activity. J Agric Food Chem 1:142–144

Lähteenmäki K, Kuusela P, Korhonen TK (2001) Bacterial plasminogen activators and receptors. FEMS Microbiol Rev 25:531–552

Lante A, Tinello F, Lomolino G (2013) Effect of UV light on microbial proteases: from enzyme inactivation to antioxidant mitigation. Innovative Food Sci Emerg Technol 17:130–134

Larsen LB, Petersen ET (1995) Identification of five molecular forms of cathepsin D in bovine milk. In: Takahashi K (ed) Aspartic proteinases: structure, function, biology, and biomedical implications. Plenum Press, New York, pp 279–283

Larsen LB, Wium H, Benfeldt C, Heegaard CW, Ardö Y, Qvist KB, Petersen TE (2000) Bovine milk procathepsin D: Presence and activity in heated milk and in extracts of rennet-free UF-Feta cheese. Int Dairy J 10:67–73

Lencki RW, Arul J, Neufeld RJ (1992) Effect of subunit dissociation, denaturation, aggregation, coagulation, and decomposition on enzyme inactivation kinetics: I. First-order behaviour. Biotechnol Bioeng 40:1421–1426

Lin M, Ramaswamy HS (2011) Evaluation of phosphatase inactivation kinetics in milk under continuous flow microwave and conventional heating conditions. Int J Food Prop 14:110–123

Lorenzen PC, Martin D, Clawin-Rädecker I, Barth K, Knappstein K (2010) Activities of alkaline phosphatase, γ-glutamyltransferase and lactoperoxidase in cow, sheep and goat's milk in relation to heat treatment. Small Rumin Res 89:18–23

Lu DD, Nielsen SS (1993) Heat inactivation of native plasminogen activators in bovine milk. J Food Sci 58:1010–1012

Lu R, Stevenson CD, Guck SE, Pillsbury LA, Ismail B, Hayes KD (2009) Effect of various heat treatments on plasminogen activation in bovine milk during refrigerated storage. Int J Food Sci Technol 44:681–687

Lumry R, Eyring H (1954) Conformation changes of proteins. J Phys Chem 58:110–120

McKellar RC, Emmons DB, Farber J (1991) Gamma-glutamyl transpeptidase in milk and butter as an indicator of heat treatment. Int Dairy J 1:241–251

Metwalli AAM, de Jongh HHJ, van Boekel MAJS (1998) Heat inactivation of bovine plasmin. Int Dairy J 8:47–56

Meyer JM (1996) Vergleichende Untersuchung zur Hitzestabilität von originären Enzymen, ß-Lactoglobulin und Vitaminen in Milch und daraus hergestellten Ultrafiltrationsretentaten. Dissertation, Eidgenössische Technische Hochschule (ETH)

Mu Z, Du M, Bai Y (2009) Purification and properties of a heat-stable enzyme of pseudomonas fluorescens Rm12 from raw milk. Eur Food Res Technol 228:725–734

Mussa DM, Ramaswamy HS (1997) Ultra high pressure pasteurization of milk: kinetics of microbial destruction and changes in physico-chemical characteristics. LWT - Food Sci Technol 30:551–557

Nielsen SS (2002) Plasmin system and microbial proteases in milk: characteristics, roles, and relationship. J Agric Food Chem 50:6628–6634

Pandey PK, Ramaswamy HS (2004) Effect of high-pressure treatment of milk on lipase and gamma-glutamyl transferase activity. J Food Biochem 28:449–462

Prado BM, Sombers SE, Ismail B, Hayes KD (2006) Effect of heat treatment on the activity of inhibitors of plasmin and plasminogen activators in milk. Int Dairy J 16:593–599

Rademacher B, Hinrichs J (2006) Effects of high pressure treatment on indigenous enzymes in bovine milk: Reaction kinetics, inactivation and potential application. Int Dairy J 16:655–661

Rauh VM, Sundgren A, Bakman M, Ipsen R, Paulsson M, Larsen LB, Hammershøj M (2014) Plasmin activity as a possible cause for age gelation in UHT milk produced by direct steam infusion. Int Dairy J 38:199–207

Riener J, Noci F, Cronin DA, Morgan DJ, Lyng JG (2009) Effect of high intensity pulsed electric fields on enzymes and vitamins in bovine raw milk. Int J Dairy Technol 62:1–6

Schokker EP, van Boekel MAJS (1999) Kinetics of thermal inactivation of the extracellular proteinase from pseudomonasfluorescens 22F: influence of ph, calcium, and protein. J Agric Food Chem 47:1681–1686

Sebald K, Dunkel A, Schäfer J, Hinrichs J, Hofmann T (2018) Sensoproteomics: a new approach for the identification of taste-active peptides in fermented foods. J Agric Food Chem 66:11092–11104
Somers JM, O'Brien B, Meaney WJ, Kelly AL (2003) Heterogeneity of proteolytic enzyme activities in milk samples of different somatic cell count. J Dairy Res 70:45–50
Stănciuc N, Dumitrascu L, Râpeanu G, Stanciu S (2011) γ-glutamyl transferase inactivation in milk and cream: a comparative kinetic study. Innovative Food Sci Emerg Technol 12:56–61
Stoeckel M (2016) Shelf-stable milk products. Dissertation, Verlag Dr. Hut
Stoeckel M, Lidolt M, Achberger V, Glück C, Krewinkel M, Stressler T, von Neubeck M, Wenning M, Scherer S, Fischer L, Hinrichs J (2016a) Growth of pseudomonas weihenstephanensis, pseudomonas proteolytica and pseudomonas sp. in raw milk: impact of residual heat-stable enzyme activity on stability of UHT milk during shelf-life. Int Dairy J 59:20–28
Stoeckel M, Lidolt M, Stressler T, Fischer L, Wenning M, Hinrichs J (2016b) Heat stability of indigenous milk plasmin and proteases from pseudomonas: a challenge in the production of ultra-high temperature milk products. Int Dairy J 61:250–261
U.S. Food and Drug Administration (2018) Irradiation in the production, processing and handling of food. Code of Federal Regulations
Vamvakaki A-N, Zoidou E, Moatsou G, Bokari M, Anifantakis E (2006) Residual alkaline phosphatase activity after heat treatment of ovine and caprine milk. Small Rumin Res 65:237–241
Van Asselt AJ, Sweere APJ, Rollema HS, de Jong P (2008) Extreme high-temperature treatment of milk with respect to plasmin inactivation. Int Dairy J 18:531–538
Vercet A, Burgos J, Lopez-Buesa P (2002) Manothermosonication of heat-resistant lipase and protease from pseudomonas fluorescens: effect of pH and sonication parameters. J Dairy Res 69:243–254
Villamiel M, De Jong P (2000) Influence of high-intensity ultrasound and heat treatment in continuous flow on fat, proteins, and native enzymes of milk. J Agric Food Chem 48:472–478
Walstra P, Wouters JTM, Geurts TJ (2006) Dairy science and technology. In: Food science and technology, vol 147, 2nd edn. CRC Taylor & Francis, Boca Raton
Wilińska A, Bryjak J, Illeová V, Polakovič M (2007) Kinetics of thermal inactivation of alkaline phosphatase in bovine and caprine milk and buffer. Int Dairy J 17:579–586
Zehetner G, Bareuther C, Henle T, Klostermeyer H (1995) Inactivation kinetics of gamma-glutamyltransferase during the heating of milk. Zeitschrift fuer Lebensmittel-Untersuchung und -Forschung 201:336–338

Chapter 13
The Role of Proteases in the Stability of UHT-Treated Milk

Nivedita Datta and Alan Kelly

13.1 Introduction

Ultra-high temperature (UHT) treatment of milk is a continuous process whereby milk is heat-treated at 135–150 °C for holding periods of 0.2–20 s, to produce a commercially sterile product that may only contain viable thermophilic bacterial spores, but in which bacterial growth is highly unlikely to occur under normal storage conditions.

UHT treatment is typically combined with aseptic packaging, which involves filling the UHT-treated milk into sterile containers in a sterile atmosphere and sealing the containers in a sterile method in a continuous process which allows the UHT milk to remain bacteriologically stable at ambient temperatures for at least 6 months. According to European Commission Regulation (EC) 2074/2005, a UHT milk treatment must be "sufficient to ensure that the products remain microbiologically stable after incubating for 15 days at 30 °C in closed containers or for 7 days at 55 °C in closed containers or after any other method demonstrating that the appropriate heat treatment has been applied".

The continuous heating system and aseptic packaging of the heat-treated milk are the key elements of UHT processing and are applied with the goal of retaining good organoleptic properties and physical stability of UHT milk at ambient storage for several months. No additives or preservatives are used to achieve its long shelf-life without refrigeration. The non-refrigerated storage and transportation of UHT treated milk allows wide distribution opportunities to open in distant countries (Sunds et al. 2018) and reduces potential energy requirements, thereby reducing green-house gas emissions (Kokkinidou and Peterson 2014).

The current typical shelf-life of UHT milk is 6–9 months under ambient storage. However, the transportation and time required to satisfy the legal and customs

N. Datta · A. Kelly (✉)
School of Food and Nutritional Sciences, University College Cork, Cork, Ireland
e-mail: a.kelly@ucc.ie

A. L. Kelly, L. B. Larsen (eds.), *Agents of Change*, Food Engineering Series,
https://doi.org/10.1007/978-3-030-55482-8_13

logistics for exporting UHT milk from distant countries to local markets can take around 2–4 months, driving demands for an extension of shelf-life of UHT milk from 6–9 months to 12 months or more. In addition, the definition of ambient temperature, meaning essentially non-refrigerated storage, can vary widely from country to country, and season to season, and so UHT milk has to be stable under a potentially wide range of storage conditions.

As stated above, high-temperature treatment of milk followed by rapid cooling and aseptic packaging renders milk free from pathogens and other spoilage microorganisms and safe for human consumption, while remaining physically stable at ambient storage for prolonged times.

However, heat-resistant bacterial enzymes (e.g. proteases, lipases and amylases generated by the psychrotolerant bacteria that may grow in raw milk during cold storage) and the indigenous milk protease, i.e., plasmin, can withstand even the severity of heat treatment during UHT process (Deeth and Lewis 2017a; Stoeckel et al. 2016b; Datta and Deeth 2001). The residual activities of these enzymes may result in destabilization of UHT milk during storage (Datta and Deeth 2001) and significantly reduce the shelf-life (Stoeckel et al. 2016a).

The most common defects in UHT treated milk caused by proteases are summarised in Table 13.1. The destablized UHT milk caused by heat-resistant enzymes is safe to drink, but may show attributes which are unacceptable to consumers, such as gelation, sedimentation, fat separation, and the presence of off-odours due to proteolysis and lipolysis (Richards et al. 2016). The mechanisms responsible for gelation/phase separation of UHT milk are becoming better understood; therefore, control strategies may be emerging for significant improvement of the quality of UHT milk during storage and transportation. Table 13.2 shows causes and remedies for gelation in UHT milk.

This chapter will address enzymatic factors contributing to instability of UHT milk and outlines strategies for enhancing shelf-life of UHT milk.

13.2 Principles of UHT Processing

13.2.1 Definition of UHT Treatment

According to the Codex Alimentarius Code of Hygienic Practice for Milk and Milk Products, "UHT treatment is normally in the range of 135 to 150 °C in combination with appropriate holding times necessary to achieve commercial sterility" (Codex-Alimentarius 2004). Minimum temperature–time conditions are not indicated. In general milk is heated at 135 and 145 °C for 1–10 s in commercial plants for UHT processing (Tran et al. 2008) with exceptions for some industrial plants operating at 152 °C with holding times of 13 s (Cattaneo et al. 2008).

UHT milk is a commercially sterile product when stored at ambient temperature, which means that bacteria or their spores that survive the heat treatment are

Table 13.1 Common defects caused by proteases in UHT milk during storage

Product type	Defect type	Cause of defect	Storage temp (°C)	Storage period (months)	Reference
UHT whole milk UHT semi-skim milk UHT skim milk	Bitter flavour Bitter taste	Proteases (plasmin and bacterial)	25 20	2 >1–2	Topçu et al. (2006) and Stoeckel et al. (2016b)
UHT whole milk UHT semi-skim milk	Creaming/fat separation	Pseudomonas proteases hydrolyse casein in the secondary milk fat globule membrane leading to aggregation of fat globules	20	2–4	Stoeckel et al. (2016a)
UHT whole milk UHT semi-skim milk UHT skim milk	Sedimentation	Bacterial protease	20	3–4 2	Stoeckel et al. (2016a) Vesconsi et al. (2012)
UHT whole milk UHT semi-skim milk UHT skim milk	Gelation	Proteases (plasmin and bacterial)	20		Stoeckel et al. (2016b)
UHT whole milk UHT semi-skim milk UHT skim milk	Off flavour	Proteolysis by heat-resistant proteases	20	5	McKellar (1984)

Table 13.2 Causes and remedies for gelation in UHT milk

UHT milk destabilisation	Characteristic feature	Appearance	Storage temperature (°C)	Storage period (months)	Reasons	Remedy
Gelation (enzymatic and non-enzymatic)	Sudden increase in viscosity from 3–4 mPa s to 10 mPa s	Either looks like custard or yoghurt or gelled sediment; irreversible change	20–30 or > 40	3–9 or >12	Bacterial protease or plasmin action or non-enzymatic mechanism	Good quality raw milk and storage at <20 °C
Gel initiated by bacterial protease						
		Custard-like and compact casein particle gel	20–30	After 3–4	High levels of psychrotrophic bacteria producing heat-resistant proteases	Raw milk should be cooled to 2–4 °C and needs to be processed within 2 days from the milking time
Gel initiated by plasmin						
		Yoghurt-like and thin gel, also known as a fine-stranded gel	20–30	After 6–9	Particularly in high somatic cell count, late lactation milk	Somatic cell count should be EU standard, pre-heating of milk at 90 °C for 30 or 60 s to inactivate plasmin

thermophilic in nature, i.e., they can only grow at temperatures >~50 °C. It is noteworthy here that ambient temperature refers to 20–25 °C in most countries; however, ambient temperature in tropical countries is often over 30 °C, and infrequently raises over 50 °C. Thus, the residual thermophilic bacteria and their spores either will germinate or may show enhanced growth of the bacteria which had already germinated in UHT milk throughout transportation, distribution and storage under these conditions, and will thus influence the stability and acceptability of UHT milk (Deeth and Lewis 2017a). A similar situation may arise for activity of heat-resistant enzymes, residual proteases and lipases, which may make UHT milk unacceptable in these tropical climates. Therefore, a strict control of every step from the raw materials procurement to UHT processing, transportation, distribution and storage is of utmost importance when UHT milk is exported to tropical countries.

Another emerging challenge for UHT processors is occurrences of *Bacillus sporothermodurans* (Hammer et al. 1995) and *Paenibacillus lactis* (Scheldeman et al. 2004) in UHT milk. Both bacteria are extreme heat-resistant mesophilic and have the capacity to form spores, necessitating a higher heating regime for achieving commercial sterility in UHT processing of milk.

Heat treatment of milk very rapidly to attain temperatures as high as ≥180 °C for short times, ≤0.2 s, has been recommended for inactivation of high heat resistance bacterial spores, *Bacillus sporothermodurans* (Huijs et al. 2004). This severe treatment process, known as Innovative Steam Injection (ISI) system, has been developed by Van Asselt et al. (2008) In the ISI system, milk was treated at 150–180 °C for ≤0.2 s in order to destroy the bacterial spores at ≥5 log scale. Plasmin was not effectively inactivated, due to very low level of denatured β-lactoglobulin formation in the ISI process, and the milk developed a bitter flavour within 30 days storage at room temperature. However, there was no bitterness perceived in ISI-processed milk during storage at 7 °C for 30 days (Huijs et al. 2004) possibly due to autolysis of plasmin (Crudden et al. 2005a).

Rauh et al. (2014b) reported that heat treatment at temperatures of >150 °C and holding times of <0.2 s resulted in commercially sterile UHT milk. Preheating milk at 95 °C for 180 s before high heat treatment inactivated plasmin effectively, and resulted in little proteolysis during storage for 112 days at 20 °C.

The typical temperature–time combination used for a UHT process, for example, 140 °C for 4 s, denotes further heat-treatment of milk in the holding tube, which is located immediately after the high heat exchanger and maintains the same sterilization temperature (140 °C) as the peak temperature reached, for the set time, e.g., 4 s. However, the specification of a single temperature and time as describing any UHT process generalizes the wide range of processing conditions in UHT plants and indicates two common assumptions:

1. that the only temperatures and times in the holding tube at the highest temperature reached are relevant to inactivation of bacteria and enzymes;
2. that all particles in the milk receive identical heat treatment.

The first assumption ignores the fact that different temperature–time profiles exist in heating and cooling sections in different UHT plants which also contribute heat load for inactivation of bacteria and enzymes, while the second takes no account of the residence time distribution of the particles in milk as they pass through the holding tube, and assumes that all particles pass through the holding tube at an identical flow velocity. If a minute percentage of bacterial spores with high flow velocity travel too fast through the holding tube and are not destroyed, they may cause non-sterility in UHT milk (Cerf and Davey 2001).

Conditions Used for UHT Processing

The fundamental principle of UHT processing is based on the fact that a high-temperature, short-time treatment than a low-temperature, long-time treatment can achieve a high level of given bacterial destruction but with a lower level of chemical change or nutritional impairment compared to longer treatments at lower temperatures (i.e., retorting). The need to maximise inactivation of thermophilic bacterial spores (9-$\log_{10}$ cycles inactivation) dictates the minimum times and temperatures which can be used in UHT processing, while the need to minimize undesirable chemical changes, such as flavour impairment, brown colour formation and loss of vitamins, dictates the maximum times and temperatures.

On this basis, the minimum required temperature in UHT processing to ensure an adequate commercial sterilization is 135 °C with holding periods of several seconds. This lower limit of heat treatment for UHT sterilisation is generally considered as the minimal conditions necessary to cause a 9-$\log_{10}$ cycle inactivation of thermophilic bacterial spores. Kessler (1981) developed the bacteriological index, B*, which has a value of 1 at the conditions minimal conditions necessary to cause a 9-$\log_{10}$ cycle in thermophilic spores, which is equivalent to a heat treatment at a temperature of 135 °C for 10.1 s in the holding tube. It is noteworthy that a B* value of 1 implies a more severe heat treatment than a 9-$\log_{10}$ reduction of mesophilic spores. However, due to the emergence appearance of the extremely heat-resistant mesophilic sporeformers of *Bacillus sporothermodurans* in UHT milk, an even more severe heat treatment at 148 °C for 10 s or 150 °C for 6 s is recommended for 9 $\log_{10}$ reduction of such highly heat-resistant spore formers (Hammer et al. 1996).

The higher limit of temperature–time combinations for UHT processing of milk is governed by the minimum amount of chemical change to the constituents of milk. The content of thiamine, i.e., vitamin B1, in milk is typically used as an index of chemical damage as, being heat-sensitive, it is considered as a standard milk component for determination of the chemical changes during high-temperature heat-treatment in UHT processing. The upper temperature limit has been standardised to match a heat treatment, which causes 3% destruction of vitamin B_1. This is defined as having a chemical index, C*, of 1. A process with C* = 1 thus reduces the concentration of thiamine by 3%, assuming a Z-value for this change of 31.4, and is equivalent to heat treatment at 135 °C for a holding period of 30.5 s.

Overall, the satisfactorily operating conditions for UHT processing to achieve commercial sterilization are then: B* > 1 and C* < 1 (Von Bockelmann and Von Bockelmann 1998).

The upper temperature limit is largely determined by the physical configuration of the UHT plant and the ability to provide the short holding time necessary to restrict chemical damage to the equivalent of a maximum of 3% destruction of thiamine. The maximum temperature used in industrial plants is ~150 °C. Therefore, UHT processing of milk at temperatures ≥150 °C for very short holding periods, ≤0.2 s, (ISI system) will produce little destruction of nutrients, minimal flavour change and no detectable colour change due to Maillard browning during heat treatment and storage (Rauh et al. 2014b).

Heating Systems

There are various heating systems used in commercial UHT plants, and every system has its own merits and disadvantages; selection of a particular system depends on the desired type and quality of the finished product, as well as operational costs of the plant. In practice, three media/sources i.e., steam, superheated water, and electrical heating are used for processing milk in commercial UHT plant.

When steam is introduced directly into milk, it is classified as a direct heating mode, and results in instantaneous increase of temperature of milk, by transferring the latent heat of condensation of the steam to the milk. Concurrently, steam is condensed and the milk is diluted, and this extra water is removed after holding milk in a vacuum chamber by rapid cooling of milk in a subsequent step. In indirect heating, hot water is used as a heating source in a heat exchanger, and milk is heated by conductive and convective heat transfer from this hot water to milk, which is slower than heat transfer by direct steam addition to milk. In a direct heating system, the temperature of milk is raised almost instantaneously by transfer of latent heat of steam into the milk. Biziak et al. (1985) reported that heat transfer from heating media to milk was less than 1 s using direct steam heating compared with greater than 10 s with in-direct use of hot water in the high–heat section of the plant.

Electrical heating is used in the high–heat section of some UHT plants, whereby milk is heated in the stainless steel tube by passing of an electrical current due to Joule's law of heating. According to this law, thermal energy is generated as a result of the passage of current through a stainless steel tube which act as an electrical resistor.

The heat is transferred to milk by conduction and mixed convection (Lefebvre and Leuliet 1997). This technology is known as Current Passage Tube (Tube à Passage de Courant [TPC] in French), and has been used in several plants in Europe for sterilization of milk, sweetened milk, cream and other dairy products (Deeth 2010). Important aspects of this technology include a small constant temperature difference between the tube and milk, a linear rise in temperature of milk, and a fast heating rate for milk.

The processing steps of commercial production of UHT milk are depicted in Fig. 13.1. The preheating and final cooling steps are always operated by in-direct heat exchangers, while the heating to sterilisation temperature after preheating can be either direct or indirect. Direct plants use expansion cooling in a vacuum chamber, while in-direct plants use tubular or plate heat exchangers for cooling (Burton 1967a, b; Burton 1969). A comparison of direct and in-direct heat treatment in UHT processing of milk was reviewed by Datta et al. (2002).

Raw Milk (4-5°C)

↓

Preheating to ~70-80°C and heat recovery

↓

Homogenization (indirect heating system)

↓

Preheating to ~90-95°C and heat recovery

↓

Holding at preheat temperature

↓

Heating to 140-150°C for sterilisation (direct, indirect heating systems)

↓

Holding at sterilization temperature

↓

Cooling to 85-95°C

↓

Homogenization (direct heating system)

↓

Cooling to 20°C

↓

Aseptic packaging at 20°C

↓

Storage of packed UHT at 20°C

Fig. 13.1 Typical operational pathways used in UHT treatment of milk

13.3 Chemical Changes Occurring in Milk During UHT Processing

High temperature heat treatment of milk during UHT processing produces major chemical changes in milk due to the following reactions as:

- whey protein denaturation;
- structural changes of casein micelles;
- isomerization of lactose;
- Maillard reactions.

Each of these changes has significant practical implications for the quality and stability of UHT milk during storage. As an example, β-lactoglobulin denaturation is linked with cooked flavour development in UHT milk immediately after processing and during storage (Hutton and Patton 1952; Gaafar 1987) and fouling of UHT plants during processing (Burton 1968; Patil and Reuter 1986). The Maillard reaction is responsible for development of heated flavour in UHT milk during heat treatment, and the quantity of lactulose, which is the isomer of lactose, formed during heat treatment is used as an index of heat-induced damage of UHT milk.

The rate of denaturation of whey proteins varies depending on their relative heat-sensitivity, and the order of denaturation rates is: immunoglobulins> bovine serum albumin (BSA) > β-lactoglobulin > α-lactalbumin. Immunoglobulins and BSA are denatured almost completely by UHT treatment, while denaturation of β-lactoglobulin is more prominent than that of α-lactalbumin during UHT treatment of milk, and the extent of denaturation of β-lactoglobulin can be measured by the amount of protein remaining un-denatured (native) by reversed-phase HPLC (IDF 1999).

The changes in structure, size and composition of casein micelles during heat treatment in a UHT process play an important role in determining the storage stability of UHT milk (Burton 1984). The typical size of casein micelles in raw skim milk is 170 nm (Gaucher et al. 2008). The casein micelle size increases in UHT milk, which is linked to the interactions of κ-casein with reactive thiol groups, generated from denatured whey proteins during heat treatment, leading to the formation of a β-lactoglobulin/κ-casein complex (βκ-complex). α_{s2}-Casein, being the other cysteine-containing casein, also forms a complex with denatured β-lactoglobulin, known as the '$\beta\alpha_{s2}$-complex' (Corredig and Dalgleish 1996). These chemical reactions alter conformation of the κ-casein in the casein micelle and weaken the physico-chemical bonding between κ-casein and other caseins (e.g., α_{s1}-casein).

More severe heat treatment in a UHT process causes more incremental size increases in casein micelles, which are influenced by the pH of milk, which also affects the extent of whey protein aggregation with the casein micelles. The βκ-complex remains associated with the casein micelle at pH between 6.5–6.7 while, at pH values above 6.9, the complex dissociates from the micelle (Oldfield et al. 2000).

During UHT treatment of milk, the Maillard reactions (MR) commences at temperatures of 90 °C and above (Burton 1988), consisting of a complex series of chemical reactions, e.g., the early MR, the advanced MR with formation of galactosyl-β-pyranone (Cattaneo et al. 2008) and the brown-coloured melanoidins in the final MR products (Van Boekel 1998). The consequences of MR in UHT milk include development of a new flavour known as 'heated flavour', which is characteristic of UHT milk, loss of available lysine, as well as production of formic acid, which reduces the pH of UHT milk, and brown discoloration of final product due to MR. The reduction of pH in UHT milk as a consequence of MR produces a stable βκ-complex on the casein micelles, due to the highest extent of bonding between β-lactoglobulin and casein occurring at pH 6.48 (Oldfield et al. 2000). Thus, reduction of pH prevents the dissociation of the βκ-complex from the micelles, which results in increased stability of UHT milk.

In the early MR, the first step is the lactosylation, when €-amino groups of lysine residues in caseins (particularly α_{S1}-casein and β-casein) (Steffan et al. 2006; Johnson et al. 2011) and carbonyl group of lactose occurs with the formation of lactulosyl-lysine which is a stable intermediate. The level of lactulosyl-lysine depends on the severity of heat treatments used in UHT processing, and on the lactose content in raw milk (Evangelisti et al. 1994) and structure of the protein (Scaloni et al. 2002). Rauh et al. (2015) measured furosine content as an overall marker of lactulosyl-lysine by LC-ESI-MS in UHT milks and found that lactulosyl-lysine content was ~ 10–15 times higher in the indirectly processed UHT milk when compared to directly processed milk.

The higher heat load in indirect processing when compared with direct processing caused more extensive inactivation of plasmin activity in indirectly processed UHT milk, while 19% of the initial plasmin activity remained in directly heated (infusion) milk (Manji et al. 1986). Severe heat treatment in the indirectly processed milk modified the carboxyl site of lysine residue in the β-casein in the lactosylation stage of MR. This modification of the β-casein structure reduces the accessibility of plasmin on its target site, resulting in no or very low levels of proteolysis by plasmin (Dalsgaard et al. 2007). This is a very practical method to retard proteolysis by milk protease.

The effect of lactosylation in the early and advanced stages of the MR on the plasmin-induced hydrolysis was studied by Bhatt et al. (2014). The lactosylation reaction modified the β-casein structure in such a way that plasmin was not able to bind on its specific site, due to loss of lysine residues in both stages of the MRs, which resulted in no or very low levels of proteolysis. The lactosylation reaction plays a significant role for the stability of UHT milk, lowering plasmin-induced gelation by blocking interactions with lysine (Samel et al. 1971).

13.4 Physico-Chemical Changes in Ultra-High-Temperature (UHT) Treated Milk During Storage

In UHT processing, the high heat treatment of milk induces many chemical changes in treated milk, some of which continue during storage, resulting in physicochemical changes in UHT milk during storage.

The key changes occurring during storage include production of volatile sulfur compounds, covalent cross-linking of caseins and serum proteins, redistribution of proteins between the colloidal and serum phases, Maillard reactions, and proteolysis. The extent and magnitude of these changes are dependent on the applied heat load during processing, temperature of storage, length of storage period and type of packaging (Deeth 2010). These changes in milk can lead to various types of defects, including off-flavour, fat separation, sedimentation and gelation during storage

(Anema 2019; Karlsson et al. 2019). Out of all these defects, gelation of UHT milk is the principal one relating to enzymes (Anema 2019), and limits the shelf-life of UHT milk. For this reason, gelation of UHT milk is the focus of this chapter.

13.4.1 Gelation of UHT Milk

Gelation of UHT milk during storage period, also known as age gelation, is the formation of a custard-like (Harwalkar 1992) or curd-like (Hardham 1998) consistency in UHT milk, which commences at the bottom and gradually progresses upwards before extending in the entire package. The temperature of maximum occurrence of age gelation is at 25–28 °C (Datta and Deeth 2001). Age gelation is an irreversible process and alters milk fluidity severely, limiting the shelf-life of UHT milk. It is not a problem in countries where shelf-life of UHT milk is 3 months (McKellar et al. 1984); however, it remains as a persisting issue in those UHT milks which need to be exported to distant countries, since the shelf-life is required to be more than 6 months.

Gelation occurs more readily (3–4 weeks) in concentrated UHT milk (Datta and Deeth 2001) and the mechanism of gel formation in concentrated milk is entirely different from that of the single-strength milk (common UHT milk). This chapter focuses on age gelation of single-strength milk only.

Age gelation of UHT milk has been studied for more than 35 years, and it can be prevented using appropriate methods (Anema 2019). However, to date, there is no physical, nor bio-chemical indices nor any measuring devices available for assessing either in raw milk before UHT processing, or measuring parameters in UHT process-line that can predict whether the UHT milk will gel before use-by date.

Typically, in UHT milk, gelation occurs by changing the physical state of milk, which is manifested by a rise in viscosity (Anema 2017; Datta and Deeth 2001). The increase in viscosity occur only a few days before the visible gelation is observed (Anema 2019) and this parameter is not appropriate for early prediction of age gelation.

Gelation Induced by Enzymatic (Proteolytic) Hydrolysis

Thermostable proteases have a prominent role for stability of UHT milk. They can withstand heat loads employed in UHT processing, being not completely inactivated, and can remain partially active in UHT milk. The residual proteases in the milk cause hydrolysis of milk proteins, resulting in destabilisation of the milk protein system and leading to gelation of UHT milk during ambient storage. Two types of thermostable proteases are involved in gelation of UHT milk: plasmin, which is an indigenous milk enzyme, and proteases produced by psychrotrophic bacteria, which are endogenous proteases. The unique specificities of these enzymes causing different gelation mechanisms in UHT milk. These will be discussed in the following sections.

Bacterial-Protease Mediated Gel in UHT Milk During Storage

The bacterial flora of raw milk is acknowledged to be a key influence on the stability of UHT milk, especially those from the *Pseudomonas* (Zhang et al. 2019; Bagliniere et al. 2017a; Baglinière et al. 2013; Baur et al. 2015; Mateos et al. 2015; Mitchell et al. 1986), *Bacillus*, *Paenibacillus* (Vithanage et al. 2016; Janštová et al. 2006) and *Serratia* genera (Bagliniere et al. 2017a, b; Machado et al. 2017) due to their secretion of heat-stable proteases. Pseudomonads comprise 70–90% of the psychrotrophic microbiota in refrigerated raw milk and are regarded as the most prominent microbiota in raw milk kept under refrigerated conditions (Adams et al. 1975). Even though bacteria present in raw milk are easily destroyed by UHT treatment, their heat-resistant enzymes persist upon UHT processing of the milk (Chen et al. 2004; Chaps. 11 and 12, this book).

Some of the bacterial proteases are extremely heat-resistant to UHT heating conditions and distinctly more heat-stable than plasmin (Bagliniere et al. 2012; Machado et al. 2017; Stoeckel et al. 2016b). Heat treatment at 140 °C for 80 s was required to attain 90% reduction of the proteolytic activity of two proteases from *Pseudomonas* species (Stoeckel et al. 2016b). In another study, a bacterial protease was inactivated only by 60–80% after heat treatment at 140 °C for 5 s (Griffiths et al. 1981). The decimal reduction times at 140 °C of bacterial proteases ranges from 2 to 300 s (Mitchell and Ewings 1985).

Thus, residual protease activity of these enzymes after typical UHT processing can result in age gelation of UHT milk. It has been reported that synthesis of bacterial proteases in raw milk commences when bacterial cell counts reach 10^7–10^8 cfu/mL at the end of the log phase of growth/early stationary growth phase of psychrotrophic bacteria (Guinot-Thomas et al. 1995; Stoeckel et al. 2016b). Thus, these proteases are not expected to be synthesized in raw milk when psychrotrophic bacterial count is below 10^7–10^8 cfu/mL, suggesting that UHT milk manufactured from good quality raw milk should be less likely to show age gelation issues.

It is noteworthy that the level of psychrotrophic bacteria at which proteases are produced depends on the strains and species of the bacteria (Deeth and Lewis 2016; Anema 2019), indicating that high levels of all psychrotrophic bacteria in raw milk may not lead to proteases in raw milk samples. For example, stored raw milks containing high psychrotrophic bacterial counts such as 10^9 cfu/mL, had no sign of proteases (Haryani et al. 2003; Datta and Deeth 2003). Additionally, not all proteases produced by psychrotrophs are adequately heat-stable, and thus they cannot survive after UHT processing and will not produce any challenges to UHT processors.

Interestingly, while bacterial proteases are resistant to UHT treatment conditions, they may be inactivated at lower temperature. So-called low temperature inactivation (LTI) involves heat treatment at ~55 °C for 30–60 min (Barach et al. 1976) before or after UHT treatment. LTI is most effective when applied 24 h after UHT processing. The combination of LTI and UHT sterilization has been shown to reduce bacterial proteolysis and enhance the shelf-life (up to 2–3 times) of UHT skim milk containing residual bacterial proteases (West et al. 1978; Kocak and Zadow 1985d).

Bacterial proteases preferentially hydrolyse κ-casein, similarly to chymosin, while β- and α_{S1}-caseins are hydrolysed non-specifically at a lower rate (Law et al. 1977; Driessen and Stadhouders 1979). The order of specificity of the enzymes is κ->β->α_{S1}-casein. The products released from hydrolysis of κ-casein by bacterial proteases are *para*-κ-casein and other peptides (Miralles et al. 2003). However, a varied specificity of bacterial proteases has been reported by some authors. Bagliniere et al. (2012) observed that the specificity of proteases from five strains of *Ps. fluorescens* was in the order β->α_{S1}> κ-> α_{S2}-casein.

Proteolysis by residual bacterial proteases is one of the causative agents for gelation in UHT milk. This was demonstrated by addition of bacterial proteases to the raw milk (Zhang et al. 2018) or addition of a sterile enzyme preparation to treated milk (Button et al. 2011).

Multiple studies in recent years have pointed to the role of these enzymes in proteolysis and gelation of UHT milk. For example, Stoeckel et al. (2016a) reported that UHT-treated whole milk made from raw milk into which *Pseudomonas* proteases were added spoiled after storage for 4 months at 20 °C. The development of a bitter taste was noted by a sensory panel after 2 months of storage of UHT milk at 20 °C, which was the starting point of spoilage, and occurred before appearance of visual defects (at 3.5 months in the milk). In that study, raw milk was inoculated with *Pseudomonas sp.* W15a and *Pseudomonas proteolytica* 691 and stored at 6 °C for 4 days before UHT processing, so that the strains reached late exponential/early stationary growth phase, which is ideal for synthesising extracellular heat- stable proteases. The enzymes from both strains required heating at 140 °C for 80 s for 90% inactivation, which is much higher than the conditions typically used for UHT treatment. The spoilage pattern by these residual heat-stable enzymes in UHT whole milk was bitterness>creaming>gelation, and the UHT milk shelf-life was only 4 months at 20 °C for the lowest protease activity in the residual enzymes.

For improvement of stability of UHT milk up to 1 year transported for export, Stoeckel et al. (2016a) and Gebre-Egziabher et al. (1980) recommended:

- use of raw milk free from proteolytic activity produced by *Pseudomonas* spp.
- storage of UHT milk at temperatures lower than 20 °C.
- LTI after UHT treatment (West et al. 1978; Kocak and Zadow 1985d).

As described earlier, Pseudomonads are predominant microorganisms in raw milk; all subspecies of *Pseudomonas* secrete discrete types of specific enzymes. These enzymes have similar characteristics, being alkaline metalloproteases belonging to the serralysin family, known as AprX, or sometimes AprA (Mateos et al. 2015; Baur et al. 2015; Zhang et al. 2018). In another study, Glück et al. (2016) studied the inactivation temperatures of AprX secreted by *Pseudomonas* species which are frequently found in raw milk stored at 6 °C. Heat treatment of these enzymes at 55 °C for 5 min in skim milk inactivated 90% of activity, while only 55% of enzyme activity was inactivated by UHT treatment (138 °C for 18 s). The flexible structure and low denaturation temperature (at 50–60 °C) of AprX are responsible for the dual role in its inactivation process: (a) promoting autoproteolysis after denaturation of AprX due to low temperature treatment; and (b) facilitating reversible re-folding from its unfolded structure after denaturation during UHT treatment (Diermayr et al. 2009). Similar phenomena were also observed by Barach and Adams (1977) and Barach et al. (1978).

Baur et al. (2015) reported that a new isolate of psychrotrophic bacteria from dairy farms in Germany, which was identified as *Pseudomonas panacis,* produced AprX in raw milk stored at 6 °C. The purified AprX was inactivated only by 12% after UHT treatment of skim milk at 138 °C for 18 s. The heat-stability of the protease in skim milk during UHT treatment was found to be due to the shielding effects of skim milk matrix on the protease, which decreases its inactivation by the heat treatment. The residual enzyme activity in UHT skim milk caused complete gelation of milk during storage of UHT milk for one month at 20 °C.

Similar protective effects of skim milk on the protease were obtained when purified protease from *Ps. fluorescens* F was added to raw skim milk which was UHT treated at 140 °C for 4 s and then stored at 20 °C for 3 months (Baur et al. 2015). However, the destabilisation rate of UHT skim milk was higher with added purified protease than by inoculation of the *Ps. fluorescens* F; gelation of UHT milk was observed after only 8 days with the purified enzyme, in comparison to 90 days with the bacterial strain (Baglinière et al. 2013).

Gaucher et al. (2011) reported gelation of UHT skim milk after storage of 92 days at 20 °C following proteolysis of casein micelles by a protease from *Pseudomonas fluorescens* CNRZ 798, which was added to raw milk before UHT processing (140 °C for 4 s). The residual protease of this *Pseudomonas* strain caused instability in that milk, as measured by size, charge and hydration of casein micelles, and proteolysis of the milk.

Psychrotrophic bacteria other than *Pseudomonas* spp in raw milk also contribute to UHT milk instability. Janštová et al. (2006) reported that protease activity by *Bacillus cereus* spores showed pronounced proteolysis in semi-skimmed UHT milk during 3 weeks of storage at 24 °C. *Bacillus cereus* spores were completely inactivated by combined treatments, comprising heat treatment of spores at 100 °C for 10 min followed by UHT treatment at 135 °C for 5 s. The spores which were not initially subjected to heating at 100 °C for 10 min were not inactivated completely by UHT treatment and showed an increased level of vegetative bacteria (10^8 CFU/mL) after 3 weeks of storage at 24 °C. This indicates that an additional heat treatment at 100 °C for 10 min prior to UHT treatment was significant for complete inactivation of these bacterial spores. No proteolysis was evident in samples with combined treatments, while marked proteolysis was observed in UHT-treated samples after only 3 weeks of storage at 24 °C. Interestingly, no proteolysis was evident in samples stored at 4 °C for 4 months indicating that *Bacillus cereus* spores were not able to grow into vegetative bacterial cells (*Bacillus cereus* bacteria) at low temperatures.

Zhang et al. (2018) found a correlation between the time of gelation and the level of proteolysis by AprX in direct processed UHT milk. They observed more than 95% hydrolysis of κ-casein by AprX at the onset of gelation. Similar level of proteolysis (80–90%) (Sandra et al. 2007) occurs in cheese making, although, many authors reported that lesser extent of proteolysis occur in UHT milk gel than rennet treatment used in cheese making. The difference in reported proteolysis can be attributed to the use of different proteolysis method and/or different experimental environments, different bacterial species and strains (Nicodeme et al. 2005) and non-specific cleavage to release the glycomacropeptide (Gaucher et al. 2011).

Plasmin-Mediated Gelation of UHT Milk During Storage
The naturally occurring milk protease present in raw milk is plasmin, which is the same enzyme as found in blood (Korycka-Dahl et al. 1983). It is associated with the casein micelles and exists in raw milk in both the active form (plasmin) and its enzymatically inactive precursor form, plasminogen (Visser 1981). Raw milk normally contains approximately 0.3 mg/L plasmin and more than 2.7 mg/L plasminogen (Richardson 1983a) and both levels vary on lactation (Bastian and Brown 1996; Ismail and Nielsen 2010; Richardson 1983b). Plasmin is a heat-stable enzyme that survives UHT processing and is inactivated only by 90% by heat treatment at 140 °C for 32 s ($D_{140\ °C}$ is 32 s) which is more severe than common UHT treatment (Deharveng and Nielsen 1991). The activity of plasmin in milk is controlled by a system of plasminogen activators and inhibitors showed in Fig. 13.2 (see also Chap. 2). The inhibitors present in raw milk serum phase are heat-sensitive, and are inactivated by UHT treatment, whereas the activators are known to be heat-stable (Lu and Nielsen 1993), are present in casein micelles, and survive UHT treatment. Consequently, UHT treatment of milk alters the natural balance between the activators and inhibitors of the plasmin system and favours plasminogen activation. This results in enhanced proteolysis in UHT milk (Deharveng and Nielsen 1991) during storage when plasminogen is converted into plasmin by plasminogen activators.

Unravelling the exact nature and heat stability of plasmin activators and inhibitors in raw milk remains a real challenge to UHT processors. The amount of plasminogen surviving after UHT treatment and how much plasmin is generated by the activation of plasminogen during storage is not yet known and needs to be established in order to control gelation phenomenon induced by plasmin in UHT milk. Crudden et al. (2005a) reported that autolysis of plasmin and activation of plasminogen occurred at different temperatures during storage, which has a major

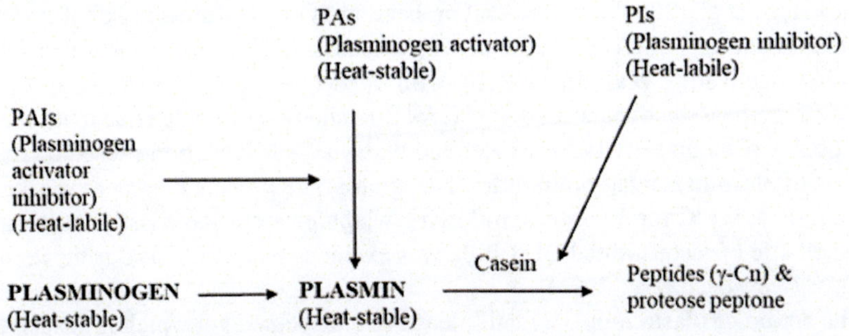

Fig. 13.2 The plasmin system

influence on plasmin activity of milk; autolysis of plasmin was significant at 5 °C, while plasminogen activation occurred mainly at 37 °C. Plasmin activity on the milk proteins can be completely inactivated or reduced by using the following strategies:

- severe heat treatment;
- pre-heating of milk;
- lactosylation of milk protein.

Plasmin hydrolyses peptide bonds in which lysine and arginine residues are in the P1' position, and the preferred substrates are β-casein and α_{S2}-casein, while α_{S1}-casein is hydrolysed at a lower rate, producing a range of products comprising mainly γ-caseins, proteose peptones and smaller peptides (Crudden et al. 2005b; Grufferty and Fox 1988).

Proteolysis induced by plasmin is one of the causative factors for gelation of UHT milk (Richardson and Newstead 1979). Although plasmin-induced proteolysis causes age gelation of UHT milk, however, there is no quantitative relationship between the onset of gelation and the degree of proteolysis (Anema 2019; Deeth and Lewis 2017b; Manji et al. 1986; Kohlmann et al. 1988).

Rauh et al. (2014b) examined the age gelation in semi-skimmed UHT milk which was treated at >150 °C for <0.2 s using a direct heated steam infusion UHT plant, as a result of which a low degree of β-lactoglobulin denaturation (~37%) occurred. Plasmin-induced proteolysis in UHT milk occurred, leading to destabilisation of the casein micelle by hydrolysing casein–casein and casein–calcium phosphate interaction sites (Rauh et al. 2014a), instead of hydrolysing κ-CN—β--lactoglobulin complex sites, which have been reported by other authors to cause age gelation in UHT milk during storage. The preferential order of proteolysis which occurred was α_{S2}-> α_{S1}-> β-caseins, and plasmin exhibited the highest affinity for the hydrophilic regions in the caseins, which are present in the serum phase. The different sites and affinity of plasmin described in previous studies could also be attributed to differences in processing conditions and levels of plasmin activity than in the study of Rauh et al. (2014b); plasmin-induced proteolysis of α_S- and β-caseins in semi-skimmed UHT milk yielded 23 bitter peptides and 43 other peptides in UHT milk stored for 14 weeks at 20 °C.

Newstead et al. (2006) reported that preheating of reconstituted milk before high heat treatment during UHT treatment will extend shelf-life of reconstituted UHT milk by inactivating plasmin activity, with subsequent lowering of proteolysis. However, preheat treatments at 80–85 °C for 30 s should be avoided due to dramatic increases in plasmin activity by accelerated plasminogen activation caused by inactivation of plasmin activator inhibitors. The appropriate preheat treatment was heating of milk at 90 °C for 30 or 60 s, following which plasmin was mostly inactivated and shelf-life of reconstituted UHT milk was extended to >8 months during storage at 30 °C (Newstead et al. 2006).

The action of plasmin in UHT milk leads to a dispersed gel which is suspended throughout the milk, by disrupting the inner sections of casein micelles, which inhibits the formation of a hard gel. De Koning et al. (1985) described a

Fig. 13.3 Plasmin-mediated gel in UHT milk during storage after 9 months; the plasmin was naturally present in the raw milk (unpublished work at University of Queensland, Australia). Proteolysis of the gel in pH 4.6 filtrate (plasmin induced proteolysis) was 0.35 mM Tyr-Leu equivalent, while the same in TCA filtrate (bacterial protease-induced proteolysis) showed a value close to zero (Le et al. 2006)

plasmin-mediated gel in UHT skim milk as having a thread-like structure, suggesting complete collapse of casein micellar structure. Figure 13.3 shows an example of a plasmin-mediated gel in UHT milk during storage. Proteolysis was measured using the fluorescamine method and quantified as 0.35 mM Tyr-Leu equivalent in pH 4.6 filtrate (Le et al. 2006) and zero value in TCA filtrate indicated plasmin-led proteolysis in the gelation of UHT milk (data not shown). A similar gel was observed by Zhang et al. (2018). However, they added plasmin to directly processed UHT milks and observed 60% hydrolysis of both β- and α_{s1}-casein, which caused gelation after 45 days.

It is known that heat-induced lactosylation of proteins can influence the activity of plasmin in UHT milk, by changing the accessibility of critical peptide bonds in the substrate (Dalsgaard et al. 2007). Bhatt et al. (2014) demonstrated an effect of lactosylation of β-casein on proteolysis during storage at 37 °C for 48 h by adding plasmin into a model system in which different extents of lactosylation had been induced. Making the lysine sites inaccessible to plasmin by modification during their lactosylation which reduced plasmin-induced proteolysis in the model system. The reduced rate of proteolysis by plasmin was proportional to the increase in the degree of lactosylation, as shown by the minimum proteolysis at 40% lactosylation of the model system during 48 h incubation at 37 °C and was independent of the stages of the Maillard reaction. These results suggest that lactosylation of casein can be a useful tool to control the plasmin-induced proteolysis of milk proteins.

Rauh et al. (2015) reported that the degree of lactosylation in direct steam infusion (>150 °C for <0.2 s) UHT-treated semi-skimmed (1.5%) milk was approximately 10–15 times lower compared to indirectly heated (140 °C for 4 s in a tubular heat exchanger) UHT milk. This lower level of lactosylation in direct heat-treated UHT milk was linked to increased plasmin-induced hydrolysis of milk protein due to more available lysine residues in comparison to in-direct heat-treated UHT milk (Bhatt et al. 2014).

Plasmin-induced proteolysis of UHT milk is influenced by the length of holding time at a particular heating temperature during manufacturing of UHT milk (Anema

2017). According to experiments carried out by Anema (2017), gelation occurred after 4 months in UHT milk which was reconstituted from low-heat skim milk powder and directly processed at 143 °C/3 s, while milk which was processed with extended holding period for 30 s (10 times) at the same heating temperature (143 °C) showed no gelation within storage period of 10 months. More heat-load was applied during processing by extending the holding period, which actually increased the level of denaturation of β-lactoglobulin from 55% (at 143 °C/3 s) to 95% (143 °C/30 s). Plasmin inactivation is directly correlated with level of denatured β-lactoglobulin; the more denaturation of the whey protein, the greater inactivation of plasmin (Bastian et al. 1993). Thus, a higher level (95%) of denaturation of β-lactoglobulin inactivated plasmin totally and resulted in a very low level of proteolysis over an extended holding period, leading the samples that did not gel for more than 10 months.

Diagnosing the Enzyme Causing Gelation of UHT Milk
The limitation of shelf-life of UHT milk by gelation prompted UHT processors to determine the causative agents in the gelling mechanism in order to improve the stability of UHT milk. Proteolysis involved in gelling processes in UHT milk may be quantified by measuring peptides which are generated after proteolysis. Interestingly, peptides released after proteolysis by plasmin and bacterial proteases have different physical and chemical properties. Bacterial proteases release smaller sized peptides and amino acids, that are hydrophilic in nature and are soluble at pH 4.6 and in trichloroacetic acid (TCA) solutions (4–12% w/v). On the other hand, plasmin produces larger sized peptides, which are hydrophobic and are soluble at pH 4.6, but precipitated in TCA solutions (4–12% w/v) (Chen and Ledford 1971; Driessen 1981).

These differences in properties of the peptides are utilised as basic principles of diagnostic methods for distinguishing between the two types of proteolysis (Datta and Deeth 2003). The study employed reversed-phase HPLC and the fluorescamine method to analyse the peptides of UHT milk soluble in 4–12% TCA and those soluble at pH 4.6. The results showed the TCA filtrate had substantial peptide peaks in the RP-HPLC chromatogram only if the milk was contaminated by bacterial protease, while the pH 4.6 filtrate showed peptide peaks in the chromatogram when either or both bacterial and native milk proteases caused the proteolysis. The proteolysis results from the fluorescamine test were in agreement with the HPLC results, whereby the TCA filtrate exhibited significant proteolysis values only when bacterial proteases were present, but the pH 4.6-soluble filtrates showed significant values when the milk contained either milk protease or both types of proteases. Table 13.3 shows the interpretation of the data generated by RP-HPLC and fluorescamine methods on the proteolysed products of UHT milk.

Visual Analysis of Gels
The consistency of gels mediated by bacterial protease and milk protease differ significantly due to the difference in the cleavage site on the casein micelles. Bacterial proteases produce a thick, firm and compact gel, which persists either throughout the whole sample or is strongly held at the bottom of the package

Table 13.3 Diagnosing the source of proteolysis in UHT milk by classifying peptides (Datta and Deeth 2003)

UHT milk proteolysis value using fluorescamine assay		UHT milk proteolysis pattern by RP-HPLC[a]		
TCA-soluble peptides	pH 4.6-soluble peptides	TCA-soluble peptides	pH 4.6-soluble peptides	Inference
Zero or close to zero value	Low values	No peaks in the chromatogram	Only few small peaks	Good quality UHT milk; will have a long shelf-life
Zero or close to zero value	Significant values	No peaks in the chromatogram	Major peaks after 20 min	Proteolysis caused by plasmin only
Significant values (slightly lower than pH 4.6 soluble peptides)	Significant values	Major peaks before 20 min in the chromatogram	Major peaks before 20 min but few if any after 20 min	Proteolysis caused by bacterial protease only
Significant values (lower than pH 4.6 soluble peptides)	Significant values	Major peaks before 20 min in the chromatogram	Major peaks both before and after 20 min	Proteolysis caused by both proteases

[a]HPLC retention data according to the HPLC method described by Datta and Deeth (2003)

(Fig. 13.4) while plasmin causes a soft, dispersed or fine-stranded gel. As described earlier, bacterial proteases preferentially hydrolyse the hydrophilic section (glycomacropeptide) of the κ-casein which is located on the outside of the micelle. Thus, the release of glycomacropeptide from the κ-casein leaves the casein micelle largely intact and reduces steric repulsion between micelles, allowing the formation of a more compact gel. In contrast, plasmin attacks β- and αs2-caseins located inside the micelle, thereby disrupting the micelle, which inhibits the formation of a strong gel (De Koning et al. 1985; Zhang et al. 2018). However, visual distinction of gelation is not possible when the gelled samples are old or without knowing the gel history of the samples. It is suitable for UHT milk samples which are monitored for gelation during storage so that on-set of gelation and cause of the gelation can be determined. Figure 13.5 shows a schematic diagram of enzymatic destabilisation of milk proteins in UHT milk.

Extensive proteolysis of UHT skim milk either by bacterial protease or plasmin completely clarifies skim milk, which is seen by lowering of viscosity due to the breakage of gel matrix, separating a clear serum layer and protein curd (Datta and Deeth 2003; Visser 1981) while in whole milk fat separation occurs. Similar phenomena occur in UHT milk when stored for several months (greater than 6 months) at higher than typical temperature (e.g., 40 °C),

It is important to acknowledge that the level of heat-stable protease required for gelation of UHT milk and the degree of proteolysis necessary for gelation to occur are not yet known; thus, the prediction of gelation of UHT milk is still a great challenge to UHT processors, partly due to the diverse nature and wide variation in biochemical characteristics of the heat-stable bacterial proteases. Moreover, new

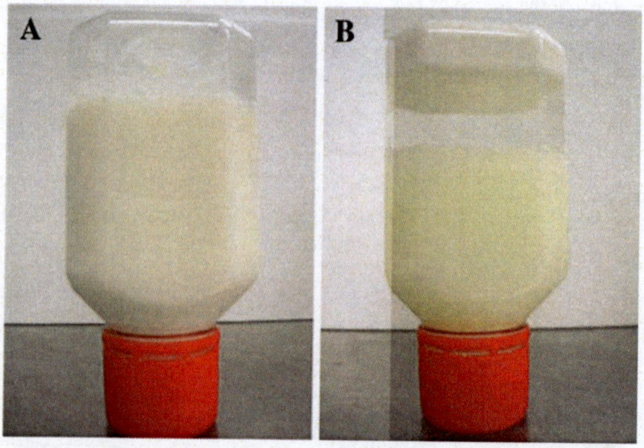

Fig. 13.4 (**a**) Control UHT milk after 90 days of storage and (**b**) gelled UHT milk after 90 days of storage containing bacterial enzyme which was added to the raw milk before UHT processing (Source: Baglinière et al. 2013)

isolates of psychrotrophic bacteria in different parts of the world producing high-heat-resistant enzymes in raw milk have been reported (see Chap. 12), which make controlling age gelation of UHT milk more difficult. Thus, the mapping and characterization of these enzymes in raw milk, and determination of their activity after UHT processing, as well as the level required for gelation of UHT milk, are key current priorities.

The diagnosing method provides a powerful tool for determining the type of proteolysis and subsequent gelation of UHT milk during storage and is also a useful tool of practical significance to UHT processors. For example, if the proteolysis and subsequent gelation that limits the shelf-life of UHT milk is caused by bacterial proteases, then the quality of the raw milk is implicated. The most common cause of high levels of psychrotrophic bacteria in the raw milk is inadequate cooling of raw milk in the farms and during transportation or keeping raw milk for excessive time periods in refrigerated conditions before UHT processing.

If the destabilisation is caused by plasmin, this indicates that the UHT processing conditions are too mild and are not adequate enough to denature β-lactoglobulin sufficiently, resulting in less whey protein–casein interaction, a low level of lactosylation in MR (Bhatt et al. 2014) and less inhibition of plasmin action on the casein. Therefore, changing a process, e.g., by optimising the pre- and high- heat treatment regime, may be required when proteolysis and subsequent gelation of UHT milk occur due to plasmin action.

Storage Temperature

Temperature is one of the significant factors for development of UHT milk gel (Anema 2019; Karlsson et al. 2019; Malmgren et al. 2017; Deeth and Lewis 2016; Datta and Deeth 2001), indicating that routine monitoring of storage temperature of UHT milk is mandatory for improvement of UHT milk stability. As described

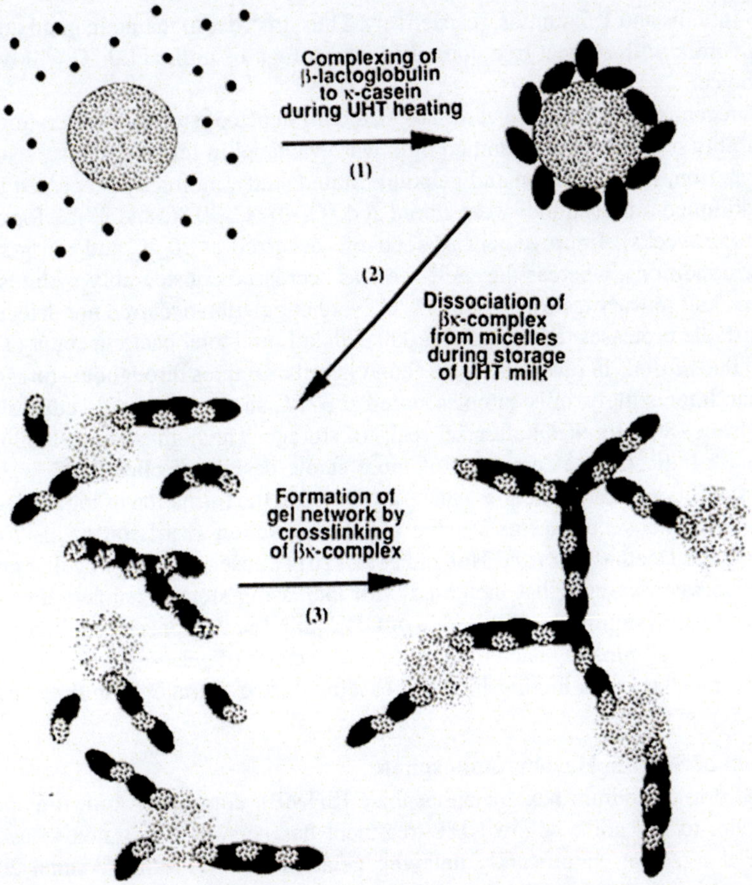

Fig. 13.5 Age gelation of UHT milk by proteases during storage (Source: McMahon 1996). (1) UHT treatment; (2) proteolysis at anchor points and dissociation of βκ-complex; (3) gel formation by cross-linking of βκ-complex

earlier, UHT milks stored at high temperature, i.e., at >~30 °C (summer in tropical countries) or at low temperatures (~4–5 °C) exhibit delayed gelation and gelation even hindered at temperature >~30 °C (Kocak and Zadow 1985a; Manji et al. 1986; Malmgren et al. 2017). Importantly, the optimum temperature for development of gelation is at ~25–28 °C (Deeth and Lewis 2017b) suggesting that normal room temperature storage of UHT milk leads to vulnerability to gel formation.

Sunds et al. (2018) studied the effect of storage temperature on quality parameters related to Maillard reaction in skimmed and full fat UHT milk stored for 6-months at 10–50 °C at 10 °C intervals. They incorporated temperature cycles to simulate variations in temperature during shipping and storage of UHT milk. The authors observed that milk stored at 20 °C had insignificant colour changes in the entire 6-months, while remarkable changes in colour occurred at 30 °C and 40 °C after 6-months and 1.5 months, respectively. Thus, in order to maintain good quality of UHT milk with respect to colour changes, storage of milk at 20 °C will be the best choice.

In a recent study (Karlsson et al. 2019), the influence of storage temperatures on the stability of UHT semi-skimmed milk was evaluated in relation to taste, colour, fat separation, sedimentation and gelation. In that study, indirectly processed UHT semi-skimmed milk samples were stored at 4 °C, 20 °C, 30 °C and 37 °C for up to 1 year (52 weeks). Improvement of shelf-life occurred at 20 °C and refrigerated storage conditions, whereas the shelf-life was decreased considerably with respect to colour and sensory taste at 30 and 37 °C. Neither gelation occurred nor detectable thermostable proteases (SCC was 182,000 cells/mL and total bacteria count (TBC) was 29,000 cfu/mL in raw milk) were found in milk samples throughout storage for one year. Interestingly, milk samples stored at 37 °C showed increased ethanol stability (from ~80% to~90%) after 52 weeks of storage. The high ethanol stability of UHT milk indicates the formation of more stable casein micelles at 37 °C. The higher stability of casein micelles can be attributed to the formation of intra-micellar cross-linking between caseins via the Maillard reaction (Karlsson et al. 2019). Al-Saadi and Deeth (2008) and Holland et al. (2011) observed intra-micellar cross-linking between caseins that increased with increasing storage temperature. This indicates that plasmin-induced gelation of UHT milk (made from raw milk in which TBC was low, eliminating bacterial proteases as a cause of gelation) can be retarded by intra-micellar cross-linking between caseins during storage of milk at higher temperatures.

Addition of Sodium Hexametaphosphate

The addition of sodium hexametaphosphate (SHMP), commonly known as polyphosphate, to raw milk before UHT treatment has been demonstrated to have a beneficial effect by significantly delaying gelation of UHT milk (Anema 2019; Deeth and Lewis 2017b; Chavan et al. 2011; Datta and Deeth 2001; Kocak and Zadow 1985b, c). Kocak and Zadow (1985b) observed that addition of ~0.1% (w/w) SHMP to milk prior to UHT processing delayed gelation from 49–70 days to 500 days at 25 °C. Some UHT processors add it to raw milk as a regular practice for enhancing storage stability of the milk (Deeth and Lewis 2017b). It does not, however, retard or inhibit proteolysis of casein rather increases the proteolysis. This phenomenon indicates that SHMP may interact with hydrolysed caseins and may alter the charge of casein micelles, possible by increasing the net negative charge of the casein micelles. The negative charge provides stronger electrostatic repulsion between micelles, causing delayed aggregation, which results increased stability of UHT milk (Deeth and Lewis 2017b). However, the efficacies of SHMP varies and the cause of the variation in its ability to control gelation depends on the processing conditions used during manufacturing of SHMP. Thus, careful selection of SHMP by determining its characteristics is required before its use in UHT processing (Kocak and Zadow 1985c).

Mechanism of Gelation

A two-stage mechanistic model of age gelation in UHT milk has been proposed by most authors (Manji and Kakuda 1988; Kohlmann et al. 1991; Kocak and Zadow 1985a; McMahon 1996). The residual proteases in UHT milk hydrolyse casein micelles, releasing peptides into the serum and causing destabilisation of the casein micelles during storage. Proteolysis is an integral part of gel formation and constitutes the first stage of the proposed model. Formation of aggregates involves interactions between the hydrolysed casein particles and the residual casein micelles. The physico-chemical changes lead to these aggregates forming a gel, which is the second stage of the model and known as "physico-chemical" or "nonenzymatic" age gelation. Proteolysis of caseins had the main role in this proposed mechanism.

However, several authors did not find a direct correlation between the time of onset of gelation and the level of proteolysis (Manji and Kakuda 1988; Kohlmann et al. 1991). The model does not account for the facts of: (1) inhibition of gelation in milk samples at temperatures >30 °C while proteolysis increases with increasing temperature; (2) occurrence of gelation (sometimes) without any sign of proteolysis. McMahon (1996) proposed a modification of the two-stage mechanism which can explain inhibition of gelation in milk samples at temperatures >30 °C. In his model, β-lactoglobulin-κ-casein-complexes (βκ-complex) had a pivotal role in the mechanistic pathway. The UHT process denatures β-lactoglobulin and forms the βκ complex on the casein micelles by interactions between denatured β-lactoglobulin and κ-casein of the casein micelles (as shown in Fig. 13.6). Then, this βκ-complex dissociates from the casein micelle via breaking the κ-casein anchor points in the micelle occurs during proteolysis. This causes partial or complete release of the complex from the casein micelle into the serum phase. The level of this complex in the serum increases during hydrolysis of the casein in stored UHT milk. When the complex attains a certain critical level in the serum, the cluster of the βκ-complex incorporates into other casein micelles (McMahon 1996) and casein micelles lose colloidal stability and form an interconnected gel network of proteins molecules.

The release of the complex has a significant influence on the stability of the casein micelles in UHT milk. The method or treatment which makes a stronger attachment of βκ-complex with casein micelles after UHT treatment will reduce the risk of release of the complex from the casein micelles, which will result delayed gelation of milk during storage. McMahon (1996) suggested the cause for delayed gelation of milk at higher storage temperatures was covalent cross-linking of the casein micelles. The intra-micellar cross linking of the casein micelles at higher storage temperatures produced a more stable and integrated casein micelles which prevented the release of βκ-complex from the micelles, delaying gelation of milk during storage.

As mentioned, the above two models do not account for the occurrence of gelation of unconcentrated milk without proteolysis. Some researchers found striking observations during manufacturing and storage of reconstituted UHT milk which was prepared from low-heat milk powder. These observations facilitated for development of an alternate model which is consistent with gelation of UHT milk in the absence of proteolysis.

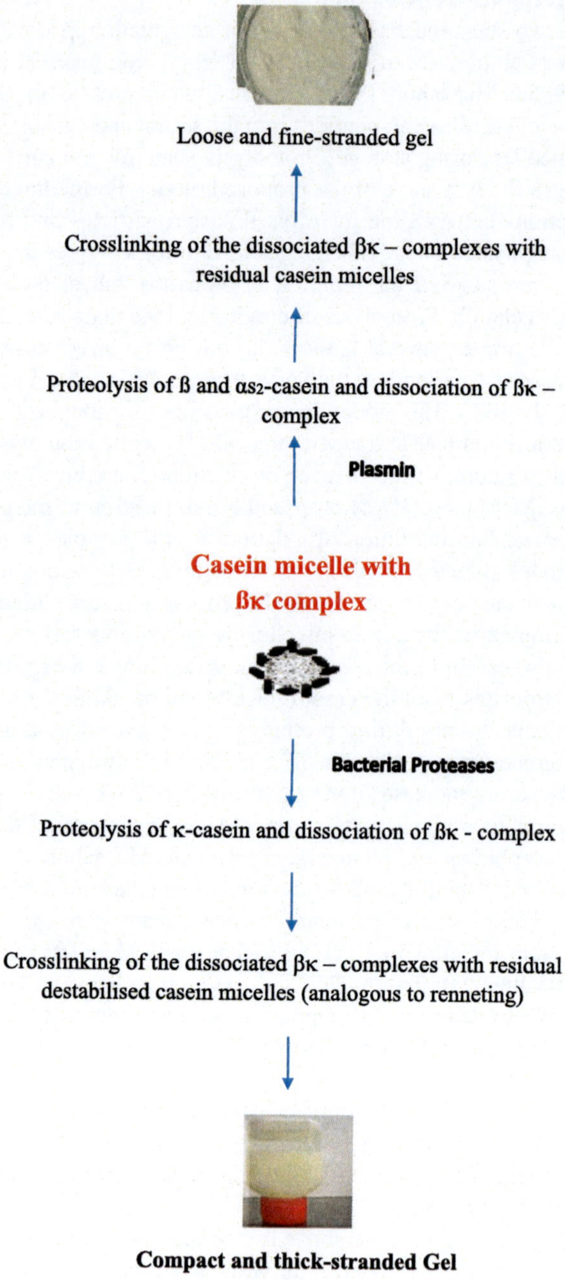

Fig. 13.6 Schematic diagram of destabilisation of milk proteins in UHT milk during storage (adapted from McMahon 1996)

Plasmin-induced proteolysis can be prevented in reconstituted UHT milk and age gelation may be delayed by introducing more heat load either in the pre-heating stage or extending the holding period. This was demonstrated by Anema (2017) to increase stability of reconstituted UHT milk; he observed that, although proteolysis is prevented, some samples still gelled after long storage. Analysis of gelled material showed the presence of κ-casein-depleted casein micelles as a major component and building block of the gel, while κ-casein and whey proteins were present as a trace level. He also noted that UHT treatment of milk released substantial amount of κ-casein from the casein micelles to serum and that κ-casein reacted with β-lactoglobulin in serum. The concentration of κ-casein and β-lactoglobulin in serum did not increase further on storage.

These findings prompted Anema (2017) to propose a non-enzymatic model known as physico-chemical age gelation which can explain gelation of milk samples without proteolysis. In this model, κ-casein-depleted casein micelles had a key role in the mechanistic pathway. These micelles are produced by UHT treatment of milk and are meta-stable due to loss of hairy κ-casein; they slowly move to the bottom of the container as a result of gravitational settling during storage of milk. Consequently, the top section of the UHT milk container has low level of κ-casein-depleted micelles, while the bottom section contains high levels of these micelles. When a critical concentration of the micelles is reached, crosslinking between κ-casein depleted micelles results gelation at the bottom of the container. This model can account for delayed gelation (>12-months) of fresh UHT milk after long storage at ambient temperature (Anema 2017).

13.5 Strategies to Control Destabilisation of UHT Treated Milk During Storage

The principal enzymatic factors, which have significant roles in destabilisation of UHT milk during storage have been extensively researched and causative factors have been identified. Attempts to formulate strategies for controlling destabilisation of UHT milk have focussed on raw milk quality, use of emerging technologies and accelerated stability testing of UHT milk.

13.5.1 Raw Milk Quality

The quality of UHT milk and its shelf-life are unquestionably correlated with the microbial status of raw milk from which it is manufactured. Keeping raw milk at high grade by maintenance of adequate cooling during milking, immediately after milking on dairy farms, and during transportation is hugely important in reducing psychrotrophic bacterial growth in raw milk. O'Connell et al. (2016) reported that raw milk in the bulk tank can be stored up to 4-days at 2 °C or 4 °C with least deterioration of quality if the initial bacterial level in raw milk is low (total bacteria count <4000 cfu/mL; SCC <290,000 cells/mL).

Karlsson et al. (2017) investigated the effect of compositional variation of Swedish raw milk throughout the year on the stability and quality of UHT milk. They observed minimum variation in raw milk composition due to good hygiene management in the farms. The UHT milk showed uniform stability for 8 months while stability was different for July and the winter months (December, January, and February). The higher number of psychrotrophic bacteria in raw milk during the outdoor period of cows (month of July) may be the reason for different stability of UHT milk in July. The variations of stability in UHT milk for winter months were not known.

The heat-stable proteases and lipases in raw milk arise from the growth of psychrotrophic bacteria in late exponential log phase or stationary phase. The concentration and activity of these heat-resistant enzymes can be lowered using the following strategies, which can thereby improve the quality of raw milk for UHT processing.

Three main causative agents that arise through raw milk are the key factors for assessing quality and stability of UHT milk: (1) gram-negative psychotropics aerobic microorganisms; (2) somatic cells and (3) heat-resistant bacterial spores from psychrotrophic bacteria.

The criteria of a good quality raw milk for UHT processing was specified by Burton (1988):

- pH 6.65–6.8;
- >74% alcohol stability;
- titratable acidity of <0.17% lactic acid.

The exact conditions used for UHT treatment may also be relevant in considering the impact of raw milk quality on product stability. UHT processing by steam injection at a higher temperature, e.g., 150 °C, for a holding period of 4 s rather than 145 °C for 4 s reduces levels of proteolysis during storage at 25 °C for 6-months. UHT milk produced from low- (SCC ~ 621,000 cells/ml, psychrotrophic bacteria count ~ 4872 cfu/mL) and medium-quality raw milk (SCC~315,000 cells/ml, psychrotrophic bacteria count~5100 cfu/mL) at the lower UHT temperature (145 °C) showed gelation after 5 and 6 months of storage, respectively, while bitterness was perceived after 2 and 4 months in those milk samples. No gelation occurred in UHT milk processed at 145 °C from high-quality raw milk (SCC ~212,000 cells/ml, psychrotrophic bacteria count ~2121 cfu/ml) and bitterness appeared after 4 months of storage. No gelation occurred in any UHT milk processed at 150 °C regardless of the quality of raw milk, and no bitterness was perceived in these UHT milk produced from high-quality raw milk; however, very low levels of bitterness were found after 180 days in UHT milk produced from low and medium-quality raw milk which was processed at 150 °C. Thus, UHT processing at 150 °C for 4 s can extend the shelf-life of the UHT milk produced from high-quality raw milk by reducing proteolysis, gelation and bitterness (Topçu et al. 2006).

Improvement of Raw Milk Quality

During refrigerated storage of raw milk, psychrotrophic bacteria can undergo a period of intensive growth, producing high concentrations of enzymes, particularly lipases and proteinases. Even low levels of enzymatic activity of lipases and/or proteases may cause defects in UHT products stored at ambient temperature for long time. These heat-stable enzyme activities are significant when the growth of psychrotrophic bacteria reaches levels of 10^6–10^8 cfu/mL, in the late log or stationary phase (Sørhaug and Stepaniak 1997).

Any process which causes extension of the lag phase of psychrotrophic bacteria or reduces the growth of these bacteria will thus be effective for lowering the concentration of heat-stable enzymes they can produce. Activation of the milk lactoperoxidase system and addition of CO_2 to raw milk for reduction of growth of psychrotrophic bacteria will be discussed below as possible approaches in this regard.

It has been suggested (Prof McMahon, personal communication) that UHT milk which is destined to export for long distances should be UHT processed from raw milk within 24 h of milking and that, before UHT processing, such raw milk should be kept at 2–4 °C; the net effect of such measures is to greatly suppress the growth of psychrotrophic bacteria, and hence production of their enzymes.

Lactoperoxidase System (LPS)

The activation of the lactoperoxidase system (LPS) in raw milk by addition of sodium thiocyanate and sodium percarbonate to raw milk, to induce bacteriostatic effects against gram negative bacteria, has been reported as a method for the preservation of raw milk. The application of LPS in raw milk before UHT processing proved beneficial where refrigeration or lowering temperature during milking and transportation of milk is not possible, particularly in developing countries and where there are poor hygienic conditions on farms (Haddadin et al. 1996).

Moussa et al. (2013) applied activation of the lactoperoxidase (LP) system in raw milk to determine the efficiency of LP system on the proteolysis and lipolysis induced by bacterial enzymes in UHT milk stored at 30 °C for 6 months. The raw milk with and without LPS was kept at 5 °C for 48 h, raw milk was centrifuged and standardised as semi-skimmed milk before in-direct heated UHT treatment. A UHT process of semi-skimmed milk at 137 °C for 4 s and stored UHT milk for 6 months at 30 °C for investigation on efficacy of LP system. The UHT milk which was processed from raw milk with LPS activation showed lower proteolysis caused by both bacterial and milk proteases and bacterially induced lipolysis during 6 months of storage.

Addition of Carbon Dioxide to Raw Milk Before UHT Processing

Vianna et al. (2012) reported that addition of carbon dioxide (CO_2) to raw milk at refrigerated temperature before UHT processing enhanced the lag phase of psychrotrophic bacteria, reducing heat-stable protease production in raw milk stored at 4 °C for 6 days. Thus, low residual levels of heat-stable proteases in UHT whole milk processed from carbon dioxide treated raw milk resulted in lower proteolysis (1.4 times lower than untreated) in UHT whole milk during storage at 30 °C for 6 months. The proteolysis in CO_2 treated milk caused by bacterial proteases is lowered by 1.4 times compared with untreated milk. The method involved the addition of CO_2 to raw milk in bulk tank during refrigerated storage for 20 min until a pH 6.20 was reached; the treated milk was kept in a closed bulk tank for 6 days at 4 °C before UHT treatment. The removal of (CO_2) from raw milk before UHT processing was performed by preheating of treated milk at 80 °C under stirring until original raw milk pH was restored. This restoration of pH was also important in order to lower fouling in the UHT plant. The CO_2-treated milk was cooled to 4 °C before UHT processing to simulate the commercial UHT processing and 0.1% sodium citrate was added to the treated milk for preventing destabilization of the caseins during UHT treatment. A lower rate of proteolysis in treated UHT milk was seen, indicating the inhibitory effect of CO_2 on the psychrotrophic bacteria count in the raw milk. Thus, the addition of CO_2 is a good option for enhancing the shelf-life of UHT milk when bacterial proteases limit the shelf-life of UHT milk.

Microfiltration of Raw Milk Before UHT Processing

Microfiltration (MF) of milk reduces bacterial load by physical separation of bacteria which are larger than the membrane pores. On the other hand, heat treatment kills the bacterial cells and keeps the dead cells with their endogenous enzymes in milk. MF remove bacteria without retaining any dead cells in the microfiltered milk (Saboya et al. 2000) which enhances the organoleptic taste of microfiltered milk, compared with heat treated milk.

A combined processing system comprised of MF and UHT treatment has been suggested by for enhancing storage stability of UHT milk. Zhang et al. (2016) observed lower plasmin activity in the microfiltered UHT milk. Raw skim milk with high or low somatic cell count (SCC) was microfiltered with a 1.4 μm ceramic membrane and the resultant microfiltered UHT milk had no residual plasmin activity, while control UHT milk showed ~4% residual plasmin activity. The removal of somatic cells by MF was more effective for high somatic SCC content of milk than the low level of SCC. The shelf-lives of microfiltered UHT milks prepared from low SCC and high SCC raw milk were enhanced by 21 and 63 days, respectively (Zhang et al. 2016).

In a recent study, the combined treatments (MF + UHT) were used for improving storage stability of whole, semi-skimmed and skim milks (D'Incecco et al. 2018). In that study, three types of raw milks were microfiltered with 1.4 μm ceramic membranes for lowering bacterial loads and SCC prior to indirect UHT processing at 137 °C for 3 s. The control milk without MF was heated at 147 °C for 5 s. The microfiltered whole and semi-skimmed UHT milks were not gelled after up to 12 months of storage at room temperature while control whole and semi-skimmed samples (without MF) were gelled after 8 and 10 months of storage, respectively. The microfiltered skim UHT milk gelled after 11 months of storage, while the control skim milk showed gelation after 6-months of storage.

Modification of Casein Structure to Inhibit Plasmin Activity

The casein structure of UHT milk can be modified by either changing process parameters or increasing storage temperature and length of storage periods. Plasmin showed decreased proteolytic activity on modified caseins, which decreased proteolysis, leading to delayed gelation (Bhatt et al. 2014; Rauh et al. 2015; Dalsgaard et al. 2007). The chemical reactions induced at higher heat loads or at higher storage temperature modify the casein structure by blocking ε-NH_2 groups of the lysine residues in casein. This blockage inhibits access of the plasmin to the Lys-X residues on the casein and results in low proteolysis of casein by plasmin, leading to delayed gelation of UHT milk during storage (Bhatt et al. 2014; Dalsgaard et al. 2007; Samel et al. 1971). The major chemical reactions, which produce stable caseins, resulting in lower proteolytic activity of plasmin, include:

- cross-linking of caseins;
- lactosylation of milk through the Maillard Reaction (as discussed in Sect. 13.3; Steffan et al. 2006; Johnson et al. 2011).

Plasmin can also be inactivated to a greater extent by increasing applied heat-load on milk during UHT treatment. The treatments with higher heat-load includes:

- pre-heat treatment (Newstead et al. 2006);
- extended holding time in high temperature heating (Anema 2017).

In summary, the above strategies can be used in order to reduce proteolysis caused by plasmin for prevention of gelation in UHT milk during storage at ambient conditions.

13.5.2 *Accelerated Shelf-Life Testing of UHT Milk*

UHT processors would like to know how long their product will remain stable from the day of production and the probable causes of any destabilisation (e.g., gelation), before the use-by date. One option is to accelerate the formation of gel of UHT milk in laboratories, to determine the practical shelf-life of the milk. A Lumisizer (L.U.M., Berlin, Germany) instrument, an analytical centrifuge, can simulate gelation of UHT milk by centrifugal forces and could be a useful tool for determination of stability of the milk.

To examine this possibility, the authors studied formation of an accelerated gel in UHT milk by chymosin (to simulate the hydrolysis of κ-casein by bacterial protease) at 37 °C inside the Lumisizer tube under centrifugal forces (unpublished work). Three sets of UHT milk samples were prepared in order to create accelerated gel in milks. One sample contained only UHT skim milk which was incubated at 37 °C inside the Lumisizer tube and was centrifuged at 6 g for 50 min, at 145 g for 20 min and at 2325 g for 30 min. A second sample was prepared from the same UHT milk but had a low level of chymosin added lower than in cheese-making and was treated inside the Lumisizer tube as per the first set. The third sample also had same quantity of chymosin as per the second set and was stored at 37 °C for 100 min (without centrifugal force).

There was no gelation in the first type of milk sample due to absence of rennet or residual proteases, confirming that it was stable UHT milk. There was also no visual gelation showed in the third milk sample, which was renneted without the influence of centrifugal force (viscosities of control milk and the third sample at 37 °C were 2.9 mPa.S and 3.0 mPa.S, respectively) confirming the lack of gelation. The second sample, however, gelled, as the combination of rennet, incubation temperature and the application of centrifugal forces appeared to induce early gelation in milk.

The intensity of transmission of NIR light through the formed gel was very different than that through an un-gelled sample; the slope of the average transmission of light with time graph represents the clarity of milk, and higher values of the slope indicate more transparency in the upper regions of the tubes containing gelled samples due to the enzymatic reaction and subsequent separation of a gel.

The potential destabilising effect of enzymes may thus be assessed qualitatively in UHT milk samples by measuring the particle movements in milks through NIR light transmission profiles using the Lumisizer. Low particle movement in UHT milk samples indicate an opaque sample (due to absence of any destabilisation reaction), resulting in low transmission of light and confirming the stability of the milk. Thus, a stable milk is characterized by no gap between the first and last transmission profiles and a compact pattern of the profile. In contrast, when more particle movement occurs due to enzymatic reaction on caseins, there are more notable differences between the transmission profiles. The protein aggregates which are formed by enzymatic action gel at an accelerated rate due to centrifugal forces of the Lumisizer and then settle as a sediment at the bottom of the sample tube, which creates a large transparent region above the sediment, through which more light will pass. Thus, a gel (and hence an unstable UHT milk sample) is characterised by wide gaps between profiles, a higher transmission of light in the later profiles (as shown in Fig. 13.7b), and a large region at the bottom of the tube with a very low transmission of light, as the light does not transmit through the sediment which is formed at the bottom of the tube.

Two UHT milk profiles are presented in Fig. 13.7a, b. Figure 13.7a depicts centrifugation of a stable UHT skim milk, showing a very small gap between the first and last transmission profiles due to low particle movements as there were no proteolysis and subsequent gelation due to absence of rennet or any residual proteases in UHT milk. This results in a compact transmission profile in milk, indicates a homogenous and storage-stable UHT milk.

Figure 13.7b indicates a gelled UHT skim milk with lower stability, due to gelation of the milk inside the Lumisizer tube. The lower stability was also demonstrated from its higher clarification rate than the sample A. The clarification rates of gelled and stable samples were 3.19 (%/h) and 0.21 (%/h) respectively.

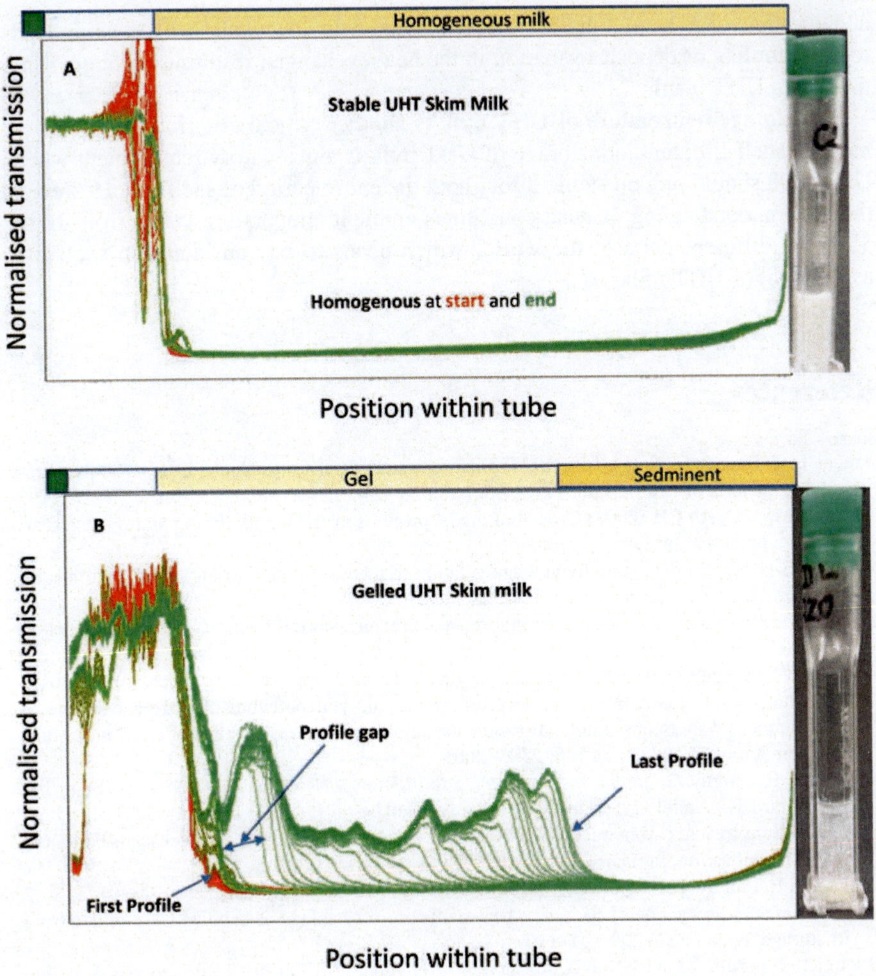

Fig. 13.7 Transmission profiles of UHT skim mik sample: (**a**) stable UHT skim milk and (**b**) gelled UHT skim milk

13.6 Conclusions

A key consideration for production of UHT milk with the longest possible shelf-life is the use of high-quality raw milk with low bacteria and somatic cell counts. UHT processing within 24 h of milking and keeping milk at ~4 °C (psychrotrophic bacterial counts ~1000 CFU/mL) before UHT processing has been shown to facilitate production of UHT milk with shelf-life > 2 years. Adequate pre-heating of raw milk, e.g., at 90 °C for 30 or 60 s, before the final high heat treatment is also critically important for inactivation of plasmin, with subsequently less gelation of UHT milk during storage. Pre-heating of milk at 90 °C for 30 or 60 s produces controlled denaturation of β-lactoglobulin, which is linked to inactivation of plasmin and reduces fouling or deposit formation in the heat exchanger, resulting in longer run-times in a UHT plant.

The storage temperature of UHT milk is another critical criterion for enhancement of shelf-life; ambient storage of UHT milk is a mandatory requirement when UHT milk should remain stable throughout the entire period of shelf life. However, the local understanding of what constitutes ambient temperature varies from 10 to 55 °C in different parts of the world, which needs to be considered in shelf-life assessment of UHT milk.

References

Adams DM, Barach JT, Speck ML (1975) Heat resistant proteases produced in milk by psychrotrophic bacteria of dairy origin. J Dairy Sci 58:828–834

Al-Saadi MSJ, Deeth CH (2008) Cross-linking of proteins in UHT milk during storage and their spoilage potential. Int Dairy J 57:80–90

Anema SG (2017) Storage stability and age gelation of reconstituted ultra-high temperature skim milk. Int Dairy J 75:56–67

Anema SG (2019) Age gelation, sedimentation, and creaming in UHT milk: a review. Compr Rev Food Sci Food Saf 18(1):140–166

Bagliniére F, Tanguy G, Jardin J, Matéos A, Briard-Bion V, Rousseau F, Gaucheron F (2012) Quantitative and qualitative variability of the caseinolytic potential of different strains of Pseudomonas fluorescens: Implications for the stability of casein micelles of UHT milks during their storage. Food Chem 135: 2593–2603

Baglinière F, Tanguy G, Jardin J, Matéos A, Briard-Bion V, Rousseau F, Robert B, Beaucher E, Gaillard JL, Amiel C, Humbert G, Dary A, Gaucheron F (2013) Proteolysis of ultra high temperature-treated casein micelles by AprX enzyme from Pseudomonas fluorescens F induces their destabilisation. Int Dairy J 31:55–61

Bagliniere F, Tanguy G, Salgado RL, Jardin J, Rousseau F, Robert B, Gaucheron F (2017a) Ser2 from *Serratia liquefaciens* L53: a new heat stable protease able to destabilize UHT milk during its storage. Food Chem 229:104–110

Bagliniere F, Jardin J, Gaucheron F, de Carvalho AF, Vanetti MCD (2017b) Proteolysis of casein micelles by heat-stable protease secreted by *Serratia liquefaciens* leads to the destabilisation of UHT milk during its storage. Int Dairy J 68:38–45

Barach JT, Adams DM (1977) Thermostability at ultrahigh temperatures of thermolysin and a protease from a psychotrophic Pseudomonas. BBA 485:417–423

Barach JT, Adams DM, Speck ML (1976) Low temperature inactivation in milk of heat-resistant proteases from psychrotrophic bacteria. J Dairy Sci 59(3):391–395

Barach JT, Adams DM, Speck ML (1978) Mechanism of low temperature inactivation of a heat-resistant bacterial protease in milk. J Dairy Sci 61:523–528

Bastian ED, Brown RJ (1996) Plasmin in milk and dairy products: an update. Int Dairy J 6(5):435–457

Bastian ED, Hanse KG, Brown RJ (1993) Inhibition of plasmin by β-lactoglobulin using casein and a synthetic substrate. J Dairy Sci 76(11):3354–3361

Baur C, Krewinkel M, Kutzli I, Kranz B, Von Neubeck M, Huptas C, Fischer L (2015) Isolation and characterisation of a heat-resistant peptidase from *Pseudomonas panacis* withstanding general UHT processes. Int Dairy J 49:46–55

Bhatt H, Cucheval A, Coker C, Patel H, Carr A, Bennett R (2014) Effect of lactosylation on plasmin-induced hydrolysis of β-casein. Int Dairy J 38:213–218

Biziak RB, Swartzel KR, Jones VA (1985) Energy use for continuous thermal processing of milk. J Food Sci 50:1607–1610

Burton H (1967a) Use of alfa-laval equipment for production of sterile milk. Amer Dairy Rev 9(32–34):131

Burton H (1967b) Ahlborn and Sordi UHT plants for processing sterile milk. Amer Dairy Rev 12(28-30):71

Burton H (1968) Section G deposits from whole milk in heat treatment plant—a review and discussion. J Dairy Res 35:317–330

Burton H (1969) Ultra-high-temperature processed milk. Dairy Sci Abstr 31:287–297

Burton H (1984) Reviews of the progress of dairy science: the bacterio-logical, chemical, biochemical and physical changes that occur in milk at temperatures of 100–1500C. J Dairy Res 51:341–363

Burton H (1988) Chemical and physical changes in milk at high temperatures. In: Ultra high temperature processing of milk and milk products. Elsevier, New York, pp 44–76

Button PD, Roginski H, Deeth HC, Craven HM (2011) Improved shelf-life estimation of UHT milk by prediction of proteolysis. J Food Qual 34:229–235

Cattaneo S, Masotti F, Pellegrino L (2008) Effects of overprocessing on heat damage of UHT milk. Eur Food Res Technol 226:1099–1106

Cerf O, Davey KR (2001) An explanation of non-sterile (leaky) milk packs in well-operated UHT plant. Food Bioprod Process 79:219–222

Chavan RS, Chavan SR, Khedkar CD, Jana AH (2011) UHT milk processing and effect of plasmin activity on shelf life: a review. Compr. Rev. Food Sci. Food Saf. 10:251–268

Chen JH, Ledford RA (1971) Purification and characterization of milk proteinase. J Dairy Sci 54:763

Chen L, Coolbear T, Daniel RM (2004) Characteristics of proteinases and lipases produced by seven Bacillus sp. isolated from milk powder production lines. Int Dairy J 14:495–504

Codex-Alimentarius (2004) Code of hygienic practice for milk and milk products (CAC/RCP 57): FAO/WHO

Corredig M, Dalgleish DG (1996) Effect of different heat treatments on the strong binding interactions between whey proteins and milk fat globules in whole milk. J Dairy Res 63:441–449

Crudden A, Fox PF, Kelly AL (2005a) Factors affecting the hydrolytic action of plasmin in milk. Int Dairy J 15(4):305–313

Crudden A, Afoufa-Bastien D, Fox PF, Brisson G, Kelly AL (2005b) Effect of hydrolysis of casein by plasmin on the heat stability of milk. Int Dairy J 15(10):1017–1025

D'Incecco P, Rosi V, Cabassi G, Hogenboom JA, Pellegrino L (2018) Microfiltration and ultra-high-pressure homogenization for extending the shelf-storage stability of UHT milk. Food Res Int 107:477–485

Dalsgaard TK, Nielsen JH, Larsen LB (2007) Proteolysis of milk proteins lactosylated in model systems. Mol Nutr Food Res 51(4):404–414

Datta N, Deeth HC (2001) Age gelation of UHT milk—a review. Food Bioprod Process 79:197–210

Datta N, Deeth HC (2003) Diagnosing the cause of proteolysis in UHT milk. LWT-Food Sci Technol 36:173–182

Datta N, Elliott AJ, Perkins ML, Deeth HC (2002) Ultra-high-temperature (UHT) treatment of milk: Comparison of direct and indirect modes of heating. Aust J Dairy Technol 57:211–227

De Koning PJ, Kaper J, Roellama HS, Driessen FM (1985) Age thinning and gelation in unconcentrated and concentrated UHT-sterilized skim milk. Effect of native milk proteinase. Neth Milk Dairy J 39:71–87

Deeth HC (2010) Improving UHT processing and UHT milk products. In: Griffiths MW (ed) Improving the safety and quality of milk. Woodhead Publishing, Cambridge, pp 302–329

Deeth HC, Lewis MJ (2016) Protein stability in sterilized milk and milk products. In: McSweeney PLH, O'Mahony JA (eds) Advanced dairy chemistry, vol 1: proteins, parts B: applied aspects, 4th edn. Springer, New York, NY, pp 247–286

Deeth HC, Lewis MJ (2017a) Microbiological aspects. In: High temperature processing of milk and milk products. Wiley-Blackwell, Chichester, pp 65–101
Deeth HC, Lewis MJ (2017b) Changes during storage of UHT milk. In: High temperature processing of milk and milk products. Wiley-Blackwell, Chichester, pp 261–319
Deharveng G, Nielsen SS (1991) Partial purification and characterization of native plasminogen activators from bovine milk. J Dairy Sci 74:2060–2072
Diermayr P, Kroll S, Klostermeyer H (2009) Mechanisms of heat inactivation of a proteinase from Pseudomonas fluorescens biotype I*. J Dairy Res 54:51–60
Driessen FM (1981) Enzymic proteolysis in UHT milk products. II. Relationship between heating intensity, inactivation of native milk protease and keeping quality. Zuivelzicht 73(33/34):688–689
Driessen FM, Stadhouders J (1979) Enzymic spoilage in pasteurized and sterilized milk products. I. Proteolysis. Voedingsmid-delentechnol 12: 35–37
Evangelisti F, Calcagno C, Zunin P (1994) Relationship between blocked lysine and carbohydrate composition of infant formulas. J Food Sci 59(2):335–337
Gaafar AMM (1987) Investigation into the cooked flavour in heat-treated milk. PhD Thesis, University of Reading
Gaucher I, Mollé D, Gagnaire V, Gaucheron F (2008) Effects of storage temperature on physico-chemical characteristics of semi-skimmed UHT milk. Food Hydrocoll 22:130–143
Gaucher I, Tanguy G, Fauquant J, Jardin J, Rousseau F, Robert B, Madec MN, Gaucheron F (2011) Proteolysis of casein micelles by Pseudomonas fluorescens CNRZ 798 contributes to the destabilisation of UHT milk during its storage. Dairy Sci Technol 91:413–429
Gebre-Egziabher A, Humbert ES, Blankenagel G (1980) Hydrolysis of milk proteins by microbial enzymes. J Food Prot 43:709–712
Glück C, Rentschler E, Krewinkel M, Merz M, Von Neubeck M, Wenning M, Scherer S, Stoeckel M, Hinrichs J, Stressler T, Fischer L (2016) Thermostability of peptidases secreted by microorganisms associated with raw milk. Int Dairy J 56:186–197
Griffiths MW, Phillips JD, Muir DD (1981) Thermostability of proteases and lipases from a number of species of psychrotrophic bacteria of dairy origin. J Appl Microbiol 50(2):289–303
Grufferty MB, Fox PF (1988) Milk alkaline proteinase. J Dairy Res 55(4):609–630
Guinot-Thomas P, Al Ammoury M, Le Roux Y, Laurent F (1995) Study of proteolysis during storage of raw milk at 4^0 C: Effect of plasmin and microbial proteinases. Int Dairy J 5:685–697
Haddadin MS, Ibrahim SA, Robinson RK (1996) Preservation of raw milk by activation of the natural lactoperoxidase systems. Food Control 7:149–152
Hammer P, Lembke F, Suhren G, Heeschen W (1995) Characterization of a heat resistant mesophilic Bacillus species affecting quality of UHT-milk: a preliminary report. Kieler Milchw Forsch 47:297–305
Hammer P, Lembke F, Suhren G, Heesschen W (1996) Characterization of a heat–resistant mesophilic Bacillus species affecting the quality of UHT milk. In: Heat treatments and alternative methods, Document 9602. International Dairy Federation, Brussels, pp 9–16
Hardham JF (1998) Effect of protein standardisation of milk by addition of UF milk permeate on the composition and storage stability of UHT processed milk. Aust J Dairy Technol 53(1):22–27
Haryani S, Datta N, Elliott AJ, Deeth HC (2003) Production of proteases by psychrotrophic bacteria in raw milk stored at low temperature. Aust J Dairy Technol 58:16–20
Harwalkar VR (1992) Age gelation of sterilised milks. In: Advanced dairy chemistry, volume 1 proteins, 2nd ed, (ed P F Fox). Elsevier Applied Science Publishers, London, pp 691–734
Holland JW, Gupta R, Deeth HC, Alewood PF (2011) Proteomic analysis of temperature-dependent changes in stored UHT milk. J Agric Food Chem 59(5):1837–1846
Huijs G, Van Asselt AJ, REM V, De Jong P (2004) High speed milk. Dairy Ind Int 69:30–32
Hutton JT, Patton S (1952) The origin of sulfhydryl groups in milk proteins and their contributions to "cooked" flavor. J Dairy Sci 35:699–705
IDF (1999) Determination of acid soluble β-lactoglobulin content - reversed-phase HPLC method. In "Standard 178A". International Dairy Federation, Brussels

Ismail B, Nielsen SS (2010) Plasmin protease in milk: Current knowledge and relevance to dairy industry. J Dairy Sci 93(11):4999–5009
Janštová B, Dračková M, Vorlová L (2006) Effect of Bacillus cereus enzymes on milk quality following ultra high temperature processing. Acta Vet Brno 75:601–609
Johnson P, Philo M, Watson A, Mills ENC (2011) Rapid fingerprinting of milk thermal processing history by intact protein mass spectrometry with nondenaturing chromatography. J Agric Food Chem 59:12420–12427
Karlsson MA, Langton M, Innings F, Wikström M, Lundh Å (2017) Short communication: variation in the composition and properties of Swedish raw milk for ultra-high-temperature processing. J Dairy Sci 100(4):2582–2590
Karlsson MA, Langton M, Innings F, Malmgren B, Höjer A, Wikström M, Lundh Å (2019) Changes in stability and shelf-life of ultra-high temperature treated milk during long term storage at different temperatures. Heliyon 5(9):e02431
Kessler HG (1981) Food engineering and dairy technology. Int J Diary Technol 34(4):171
Kocak HR, Zadow JG (1985a) Age gelation of UHT whole milk as influenced by storage temperature. Aust J Dairy Technol 40:14–21
Kocak HR, Zadow JG (1985b) Controlling age gelation of UHT milk with additives. Aust J Dairy Technol 40:58–64
Kocak HR, Zadow JG (1985c) Polyphosphate variability in the control of age gelation in UHT milk. Aust J Dairy Technol 40:65–68
Kocak HR, Zadow JG (1985d) The effect of low temperature inactivation treatment on age gelation of UHT whole milk. Aust J Dairy Technol 40:53–58
Kohlmann KL, Nielsen SS, Ladisch MR (1988) Effect of serine proteolytic-enzymes (trypsin and plasmin), trypsin inhibitor, and plasminogen activator addition to ultra-high temperature processed milk. J Dairy Sci 71(7):1728–1739
Kohlmann KL, Nielsen SS, Ladisch MR (1991) Effects of a low concentration of added plasmin on ultra-high temperature processed milk. J Dairy Sci 74(4):1151–1156
Kokkinidou S, Peterson D (2014) Control of maillard-type off-flavour development in ultrahigh-temperature-processed bovine milk by phenolic chemistry. J Agric Food Chem 62:8023–8033
Korycka-Dahl M, Dumas BR, Chene N, Martal J (1983) Plasmin activity in milk. J Dairy Sci 66:704–711
Law BA, Andrews AT, Sharpe ME (1977) Gelation of ultra-high-temperature-sterilized milk by proteases from a strain of Pseudomonas fluorescens isolated from raw milk. J Dairy Res 4(1):145–148
Le TX, Datta N, Deeth HC (2006) A sensitive method for measuring bacterial proteolysis and proteinase activity in UHT milk. Food Res Int 39:823–830
Lefebvre S, Leuliet JC (1997) Laminar mixed convection in constant wall heat flux tubular heat exchangers. In: Lowitt R (ed) Engineering and food at the seventh international congress of engineering and food ICEF7, Part 1. Academic Press, Sheffield, pp C97–C100
Lu DD, Nielsen SS (1993) Heat inactivation of native plasminogen activators in bovine milk. J Food Sci 58:1010–1012
Machado SG, Baglinière F, Marchand S, Van Coillie E, MCD V, De Block J, Heyndrickx M (2017) The biodiversity of the microbiota producing heat-resistant enzymes responsible for spoilage in processed bovine milk and dairy products. Front Microbiol 8:302
Malmgren B, Ardö Y, Langton M, Altskärd A, MGEG B, Dejmek P, Paulsson M (2017) Changes in proteins, physical stability and structure in directly heated UHT milk during storage at different temperatures. Int Dairy J 71:60–75
Manji B, Kakuda Y (1988) The role of protein denaturation, extent of proteolysis, and storage-temperature on the mechanism of age gelation in a model system. J Dairy Sci 71(6):1455–1463
Manji B, Kakuda Y, Arnott DR (1986) Effect of storage temperature on age gelation of ultra-high temperature milk processed by direct and indirect heating systems. J Dairy Sci 69:2994–3001
Mateos A, Guyard-Nicodeme M, Bagliniere F, Jardin J, Gaucheron F, Dary A, Gaillard JL (2015) Proteolysis of milk proteins by AprX, an extracellular protease identified in Pseudomonas

LBSA1 isolated from bulk raw milk, and implications for the stability of UHT milk. Int Dairy J 49:78–88

McKellar RC, Froehlich DA, Butler G, Cholette H, Campbell C (1984) The effect of uncooled storage on proteolysis bitterness & apparent viscosity in ultra-high temperature milk. Can Inst Food Sci 17(1):14–17

McMahon DJ (1996) Age-gelation of UHT milk: Changes that occur during storage on shelf-life and the mechanism by which age-gelation occurs. In: Heat treatments and alternative methods, (International Dairy Federation—Ref. S.I. 9602, Vienna, Austria). International Dairy Federation, Brussels, pp 315–325

Miralles B, Amigo L, Ramos M, Recio I (2003) Analysing para -kappa-casein and related peptides as indicators of milk proteolysis. Milchwissenschaft 58:412–415

Mitchell GE, Ewings KN (1985) Quantification of bacterial proteolysis causing gelation in UHT-treated milk. New Zealand J Dairy Sci Technol 20(1):65–76

Mitchell GE, Ewings KN, Bartley JP (1986) Physicochemical properties of proteinases from selected psychrotrophic bacteria. J Dairy Res 53(1):97–115

Moussa OB, Mankai M, Fekih AB, Hassouna M (2013) Effect of the lactoperoxidase system on proteolysis and physicochemical changes in ultra high temperature milk during storage. Afr J Biotechnol 12:2041–2050

Newstead DF, Paterson G, Anema SG, Coker CJ, Wewala AR (2006) Plasmin activity in direct-steam-injection UHT-processed reconstituted milk: effects of preheat treatment. Int Dairy J 16:573–579

Nicodeme M, Grill JP, Gaillard JL (2005) Extracellular proteinase activity of different *Pseudomonas* strains: Dependence of proteolytic activity on culture conditions. J Appl Microbiol 99:641–648

O'Connell A, Ruegg P, Jordan K, O'Brien B, Gleeson D (2016) The effect of storage temperature and duration on the microbial quality of bulk tank milk. J Dairy Sci 99:3367–3374

Oldfield DJ, Singh H, Taylor MW, Pearce KN (2000) Heat-induced interactions of β-lactoglobulin and α-lactalbumin with the casein micelle in pH-adjusted skim milk. Int Dairy J 10:509–518

Patil GR, Reuter H (1986) Deposit formation in UHT plants. II Effect of forwarming treatments in directly heated plants. Milchwissenschaft 41:411–413

Rauh V, Johansen L, Ipsen R, Paulsson M, Larsen L, Hammershøj M (2014a) Plasmin activity in UHT milk: relationship between proteolysis, age gelation, and bitterness. J Agric Food Chem 62:6852–6860

Rauh V, Sundgren A, Bakman M, Ipsen R, Paulsson M, Larsen L, Hammershøj M (2014b) Plasmin activity as a possible cause for age gelation in UHT milk produced by direct steam infusion. Int Dairy J 38:199–207

Rauh VM, Johansen LB, Bakman M, Ipsen R, Paulsson M, Larsen LB, Hammershøj M (2015) Protein lactosylation in UHT milk during storage measured by liquid chromatography–mass spectrometry and quantification of furosine. Int J Dairy Technol 68:486–494

Richards M, Buys EM, De Kock HL (2016) Survival analysis, consumer perception and physico-chemical analysis of low fat UHT milk stored for different time periods. Int Dairy J 57:56–61

Richardson BC (1983a) The proteases of bovine milk and the effect of pasteurisation on their activity. New Zealand J Dairy Sci Technol 18:233–245

Richardson BC (1983b) Variation of the concentration of plasmin and plasminogen in bovine-milk with lactation. New Zealand J Dairy Sci Technol 18(3):247–252

Richardson BC, Newstead DF (1979) Effect of heat-stable proteases on the storage life of UHT milk. N Z J Dairy Sci Technol 14:273–279

Saboya LV, Aboya LV, Maubios JL (2000) Current developments of microfiltration technology in the dairy industry. Lait 80:541–553

Samel R, Weaver RWV, Gammack DB (1971) Changes on storage in milk processed by ultra-high-temperature sterilization. J Dairy Res 38:323–332

Sandra S, Alexander M, Dalgleish DG (2007) The rennet coagulation mechanism of skim milk as observed by transmission diffusing wave spectroscopy. J. Colloid Interface Sci 308(2):364–373

Scaloni A, Perillo V, Franco P, Fedele E, Froio R, Ferrara L, Bergamo P (2002) Characterization of heat-induced lactosylation products in caseins by immunoenzymatic and mass spectrometric methodologies. Biochimica et Biophysica Acta (BBA) - Proteins and Proteomics 1598:30–39

Scheldeman P, Goossens K, Rodriguez-Diaz M, Pil A, Goris J, Herman L, De Vos P, Logan NA, Heyndrickx M (2004) Paenibacillus lactis sp. nov., isolated from raw and heat-treated milk. Int J Syst Evol Microbiol 54:885–891

Sørhaug T, Stepaniak L (1997) Psychrotrophs and their enzymes in milk and dairy products: Quality aspects. Trends Food Sci Technol 8:35–41

Steffan W, Balzer HH, Lippert F, Sambor BC, Bradbury AGW, Henle T (2006) Characterization of casein lactosylation by capillary electrophoresis and mass spectrometry. Eur Food Res Technol 222:467–471

Stoeckel M, Lidolt M, Achberger V, Glück C, Krewinkel M, Stressler T, Von Neubeck M, Wenning M, Scherer S, Fischer L, Hinrichs J (2016a) Growth of *Pseudomonas weihenstephanensis, Pseudomonas proteolytica* and Pseudomonas sp. in raw milk: Impact of residual heat-stable enzyme activity on stability of UHT milk during shelf-life. Int Dairy J 59:20–28

Stoeckel M, Lidolt M, Stressler T, Fischer L, Wenning M, Hinrichs J (2016b) Heat stability of indigenous milk plasmin and proteases from Pseudomonas: a challenge in the production of ultra-high temperature milk products. Int Dairy J 61:250–261

Sunds AV, Rauh VM, Sørensen J, Larsen LB (2018) Maillard reaction progress in UHT milk during storage at different temperature levels and cycles. Int Dairy J 77:56–64

Topçu A, Numanoğlu E, Saldamlı İ (2006) Proteolysis and storage stability of UHT milk produced in Turkey. Int Dairy J 16:633–638

Tran H, Datta N, Lewis MJ, Deeth HC (2008) Predictions of some product parameters based on the processing conditions of ultra-high-temperature milk plants. Int Dairy J 18:939–944

Van Asselt AJ, Sweere APJ, Rollema HS, De Jong P (2008) Extreme high-temperature treatment of milk with respect to plasmin inactivation. Int Dairy J 18:531–538

Van Boekel MAJS (1998) Effect of heating on Maillard reactions in milk. Food Chem 62:403–414

Vesconsi CN, Valduga AT, Cichoski AJ (2012) Particle sedimentation in semi-skimmed, skimmed on whole milk UHT, during storage. Cienc Rural 42:730–736

Vianna PCB, Walter EHM, Dias MEF, Faria JAF, Netto FM, Gigante ML (2012) Effect of addition of CO2 to raw milk on quality of UHT-treated milk. J Dairy Sci 95:4258–4262

Visser S (1981) Proteolytic enzymes and their action on milk proteins. In: Heat treatments and alternative methods. International Dairy Federation, Brussels, pp 315–325

Vithanage NR, Dissanayake M, Bolge G, Palombo EA, Yeager TR, Datta N (2016) Biodiversity of culturable psychrotrophic microbiota in raw milk attributable to refrigeration conditions, seasonality and their spoilage potential. Int Dairy J 57:80–90

Von Bockelmann B, Von Bockelmann IA (1998) Long-life products: heat-treated, aseptically packed: a guide to quality. Publisher, Åkarp

West FB, Adams DM, Speck ML (1978) Inactivation of heat-resistant proteases in normal ultra-high temperature sterilized skim milk by a low temperature treatment. J Dairy Sci 61(8):1078–1084

Zhang S, Liu L, Pang X, Lu J, Kong F, Lu J (2016) Use of microfiltration to improve quality and shelf life of ultra-high temperature milk. J Food Process Pres 40:707–714

Zhang C, Bijl E, Hetting K (2018) Destabilization of UHT milk by protease AprX from *Pseudomonas fluorescens* and plasmin. Food Chem 263:127–134

Zhang C, Bijl E, Svensson B, Hettinga K (2019) The extracellular protease AprX from Pseudomonas and its spoilage potential for UHT Milk: a review. Compr Rev Food Sci Food Saf 18:834–852

Chapter 14
Milk-Clotting Enzymes

Anders Andrén

14.1 Historical Background

The use of milk-clotting enzymes in cheese making is one of the oldest enzymatic applications used by man and dates back at least 8000 years. There are different sources of milk-clotting enzymes (animal, microbial, plant), but the main origin has long been an extract of suckling-calf abomasa (rennet). Rennet contains both chymosin and pepsin, and the ratio of these in the abomasal mucosa is very much influenced by both the feeding regime and the age of the ruminant, where milk-feeding and young age stimulate chymosin production. Today, milk-clotting enzymes (i.e., chymosin) are also produced *via* fermentation by use of recombinant DNA techniques. The most specific milk-clotting enzyme is chymosin, which has a high milk-clotting activity and a low general proteolytic activity. Chymosin of different species vary regarding the specific milk-clotting activity, which also is true for all the other milk-clotting enzymes (coagulants). Since calf rennet (which is 80–90% chymosin) from the beginning has been used in the development and production of most cheese varieties, the use of other coagulants will give different flavours of the cheeses compared using a preparations dominated by chymosin. For the analysis of the total milk-clotting activity of rennets and coagulants, there are today international standard methods and units, replacing old methods and units such as Soxhlet. These topics are all reviewed in this chapter.

Cheese was, until the nineteenth century, produced on farms using fresh extracts from dried abomasa (or, in some cultures, using plant extracts from the thistle *Cynara cardunculus*). Around the 1850s, however, small cooperative dairies started to collect bigger amounts of milk for cheese-making, which required larger volumes of rennet (by definition an extract of ruminant abomasa, in contrast to coagulants with origin from other sources). This led to industrial production of calf rennet by

A. Andrén (✉)
Swedish University of Agricultural Sciences, Uppsala, Sweden

A. L. Kelly, L. B. Larsen (eds.), *Agents of Change*, Food Engineering Series,
https://doi.org/10.1007/978-3-030-55482-8_14

the end of the nineteenth century and, in 1874, rennet was the first industrial enzyme preparation sold with a standardized enzyme activity, by Chr. Hansen A/S, Denmark (Harboe and Budtz 1999). An German agricultural chemist, Von Soxhlet (1877) defined the strength (the total milk-clotting activity) of a rennet in terms of the volume of milk which can be clotted by one unit of a rennet in 40 min at 35 °C. However, due to the daily variation in the clotting ability of raw milk, the definition is not entirely satisfactory. To improve the precision of milk-clotting activity tests, the English scientist Berridge (1952) introduced the use of a standardized milk powder, reconstituted in 0.01 M $CaCl_2$, instead of raw milk in activity tests.

The ideas of Soxhlet and Berridge influenced almost all the different national definitions of rennet strength and standard methods used until late twentieth century. However, the biggest drawback at that time was that the interpretation of rennet strength was different in different countries and by different rennet producers, resulting in a great variety of national milk-clotting units. In addition to all these different milk-clotting units, there are also other factors which make the characterization of rennets and coagulants very complex. Due to an increased level of cheesemaking and a parallel shortage of suckling calf abomasa worldwide in the last century, coagulants of a variety of sources were introduced. These included abomasal extracts of adult cattle, extracts of pig stomachs, different microbial enzyme preparations and, the most important of them all, rDNA-products with the general name "Fermentation-Produced Chymosin" (FPC).

To bring order regarding the characterization of rennets and coagulants for the sake of users and producers, the International Dairy Federation appointed an expert group for the development of international standard methods of content and total milk-clotting activity of such preparations. The first international standard method for the composition of bovine rennets was published in 1987 and now there are four different international standards covering the composition and strength of bovine, ovine and caprine rennets, as well as microbial coagulants. This will be further elaborated in Sect. 14.6 below.

14.2 Different Types of Rennets and Coagulants

14.2.1 Bovine Rennets

Calf rennet produced from the abomasa of young suckling calves was the main coagulant worldwide from the start of commercial production in the middle of nineteenth century until the middle of the twentieth century. The abomasum of the young milk-fed calf is rich in the specific milk-clotting enzyme chymosin (EC 3.4.23.4) because ruminants get their maternal immunoglobulins from the colostrum, i.e., the milk secreted in the first 2–3 days of lactation. Since chymosin has a specific proteolytic activity to coagulate the milk, but too low general proteolytic activity to extensively damage the immunoglobulins, chymosin secretion has been favoured secretion in the stomachs of young ruminants during evolution (Foltmann

1981; Lopes-Marques et al. 2017). Chymosin, first named by the French pharmacist Jean-Baptist Deschamps as early as 1840 (Deschampes 1840), has also for a long time been named rennin, derived from rennet, but this term is today not used in scientific papers.

Ruminants also secrete the milk-clotting enzymes pepsin (EC 3.4.23.1) and gastricsin (EC 3.4.23.3) in addition to chymosin. Gastricsin is a minor enzyme in ruminants and will not be discussed further. In addition to its milk-clotting activity, pepsin has a strong general proteolytic activity under acidic conditions in the abomasum/stomach. Chymosin and pepsin are both produced and secreted by chief cells, and to some extent also by the mucous neck cells, in the glands of the abomasal mucosa (Andrén et al. 1982). The granules of the Golgi apparatus of these cells contain both chymosin and pepsin (Yamada et al. 1988). Both enzymes are produced as inactive zymogens, i.e., prochymosin and pepsinogen, and the zymogens are activated to chymosin and pepsin, respectively, by the low pH due to the HCl secreted by parietal cells of the abomasal glands.

The ratio of chymosin and pepsin in the abomasal mucosa is very much influenced by both the feeding regime and the age of the ruminant. By between the 10th to the 20th week of gestation (Miura et al. 1988), through the birth and suckling period, chymosin is the dominating proteinase in the abomasa of calves. As long as the calf is exclusively suckling or milk-fed, the proportion of chymosin, measured by milk-clotting activity assay at pH 6.5, is around 90% (Andrén et al. 1982). A high proportion ($\approx$75%) of chymosin activity remains in the abomasal mucosa even in 6-month-old pasture-fed calves if they are allowed to suckle their dams (Andrén and Björck 1986). On the other hand, if the calves are weaned and fed non-milk concentrates and roughage, the proportion of chymosin decreases to around 30% at the age of 6 months. At the age of about a year, the chymosin content is about 5% and, in adult cattle, only traces of chymosin are found in the abomasal mucosa (Andrén et al. 1982).

In conclusion, both milk feeding and age of the cattle are highly related to the proportion of chymosin *versus* pepsin in the abomasal mucosa and thereby to the composition of milk-clotting enzymes in bovine rennets. This is further elucidated below in Sect. 14.3.

14.2.2 Ovine and Caprine Rennets

Abomasa from lambs and kids are also extracted into ovine and caprine rennets, but to a much lesser extent. The amount of chymosin in these rennets is also dependent on age and milk feeding of the animals from which the abomasa are obtained. These rennets are mainly produced and used in the Mediterranean region and are sometimes produced as rennet paste, containing pregastric esterase, a lipase from the gullet tissues of lambs and kids (Birschbach 1992). These pastes give a sharp and peppery flavour to cheese (Harboe 1994) and are used for certain cheese varieties such as Pecorino Romano in Italy and Roncal in Spain.

14.2.3 Porcine and Chicken Pepsin

During a limited time around the 1960s, stomachs of pigs were extracted into a preparation containing porcine pepsin. This preparation was commercially sold as a 50/50-mixture with calf rennet but was withdrawn from the market in favour for adult bovine rennet. Extracts of the glandular stomach of chickens, containing chicken pepsin, have also been used for cheese-making due to religious reasons, mainly only in Israel. However, this enzyme is too proteolytic for most cheese varieties.

14.2.4 Microbial Coagulants

In around the 1950s, a shortage of calf rennet was foreseen due to an increased milk production from individual cows, leading to a decrease of the number of milking cows worldwide and thereby also the number of calves. Simultaneously, beef consumption increased, which led to postponed slaughter of calves until they were grown to adults. The foreseen shortage of young calf abomasa thus led to search for other milk-clotting enzymes, and among these were some of microbial origin. In this context, it should be mentioned that the word "rennet" by definition is an extract of the abomasa of ruminants and hence milk-clotting preparations of other origins should be named "coagulants" according to the International Dairy Federation (IDF).

The microbial (fungal) coagulants most widely known are produced by use of *Rhizomucor miehei*, *Rhizomucor pusillus* and *Cryphonectria parasitica*. The proteinase from the former dominates the market for microbial coagulants and there are two different heat-labile forms of this milk-clotting enzyme, which are less proteolytic than the native form (Harboe and Budtz 1999). The Parasitica coagulant has a very high proteolytic activity and normally only used for cheeses where scalding (the curd heat-treated at about 55 °C) is used during the manufacture. At that high temperature, the Parasitica proteinase is inactivated and will not influence the ripening process of the cheese.

Preparations with milk-clotting enzymes of bacterial origin have also been tested for cheese making. The driving force to still search for such substitutes for rennet is that some consumers, for religious or dietary reasons, ask for dairy products produced with enzymes of non-animal origin. One of the most investigated bacteria in this respect is different strains of *Bacillus* (Lemes et al. 2016).

14.2.5 Plant Coagulants

In the search for substitutes to calf rennet a lot of different milk-clotting enzymes of plant origin have been tested over the years, among them proteolytic enzymes from different fruits (papaya, kiwi, fig, etc), flowers and seeds (Amira et al. 2017). However, their practical use is very low due to their very high proteolytic activity, causing low yield of cheese and obvious bitterness. Plant proteinases in cheese-making are mainly promoted by vegetarians, who through these products might improve their nutritional intake. The only plant coagulant used in any notable extent is an aqueous extract of the flowers of *Cynara cardunculus*, a thistle (Verissimo et al. 1995). In Portuguese artisan cheese-making, extracts of these flowers are sometimes used to clot the milk in the production of certain local cheeses.

14.2.6 Fermentation Produced Chymosin (FPC)

In the 1980s, three different companies developed fermentation-produced chymosin, FPC, by use of recombinant DNA-techniques. A DNA sequence for chymosin from a calf abomasal chief cell was introduced into the genome of a host organism. As host organisms, a bacterium (non-patogenic *Escherichia coli* var K-12), a mucor (*Aspergillus niger* var *awamori*) and a yeast (*Kluyveromyces lactis*) were used (Teuber 1990; Harboe 1992; Praaning-van Dalen 1992). Today only the two latter FPCs are on the market producing bovine chymosin, since the patent of the one produced via *E. coli* has been bought by a competitor which closed the production. In addition to fermentation-produced bovine chymosin, there is also fermentation-produced camel chymosin on the market since 2009. The market shares of the FPCs have increased significantly, both due to the shortage of suckling calf abomasa and also due to concerns about issues such as mad cow disease (BSE, bovine spongiform encephalopathy). However, there are also some negative concerns regarding FPCs from some consumers, who consider cheeses produced by FPCs to be genetically engineered. In this respect, it has to be remembered that only a small amount of rennet is added to the cheese milk (≈30 mL to 100 kg of milk, i.e. 0.03%) and normally only about 20% of that amount remains in the cheese (i.e. 6 mg kg^{-1} of cheese).

Fermentation-produced bovine chymosin is identical to natural bovine chymosin (same amino acid sequence) and its properties are in principle the very same as those for natural bovine chymosin. However, there are some residues added to the chymosin molecule by the producing organism (about 10% of the chymosin molecule is glycolysated by fungi), but these modifications have not been observed to change the properties of "calf" FPC significantly compared to natural bovine chymosin. Actually, partial or complete glycosylation is a fairly common phenomenon among the milk-clotting enzymes (Harboe 1992). In terms of analysis, FPCs should be handled as a calf rennet containing 100% chymosin (see Sect. 14.6).

14.2.7 General Considerations Around Rennets and Coagulants

Since calf rennet (i.e., 80–90% chymosin) has been used from the beginning in the development and production of most cheese varieties, the use of other coagulants will give different flavours of the cheeses compared using preparations dominated by chymosin. However, in short-ripened cheeses, the impact of chymosin is less and the difference compared to other coagulants is not significant. Furthermore, high chymosin content of the rennet is also correlated with a higher yield of cheese, which is a very important aspect of the choice of coagulant. Sometimes there are also regulations regarding the production of certain cheeses such as for example the French AOC cheeses (l'Appelation d'Origine Contrôlée) and Italian DOC cheeses (Denominazione Originato Controlato), which must be produced with calf rennet only.

14.3 Production of Rennets and Coagulants

Bovine rennets (calf and adult bovine rennet) are today mainly produced from frozen abomasa (earlier, dried abomasa were used), which are cut in grinders and mixed with a 3–10% NaCl brine solution. The milk-clotting zymogens are thereby extracted from the grinded tissue into the solution, where they are activated to chymosin and pepsin by lowering the pH to about 2 for 1 h and then adjustment to pH 5.5. The pieces of abomasal tissue are then filtered away and the remaining solution concentrated by different UF-techniques. The extract is further filtered through a low cut-off filter to remove bacteria and then the concentration of NaCl is increased to about 20%. The ratio of chymosin to pepsin in the abomasal extract varies due to the age and feeding regimes of the animals used, as elaborated above, and therefore different batches of extracts are mixed to give a desired proportion of chymosin and pepsin. Finally, the bovine rennet is diluted to the desired strength (total milk-clotting activity), which varies between producers and countries (see Sect. 14.6). Production of rennets from older cattle causes difficulties during the filtration step due to a high concentration of mucins in the extract. This problem has been solved in different ways by the manufacturers of adult bovine rennets. The production of ovine and caprine rennets follows in principle the same steps as for bovine rennets. Rennets are usually distributed as liquids, but sometimes they also are sold as powders.

FPCs are, as the name indicates, produced by fermentation. After introducing the DNA coding for calf or camel prochymosin into the genome of the host organism, the latter is submerged fermented in food-grade substrates for a time fitted to the production capacity of the host organism (Teuber 1990; Harboe 1992). There is then an acid treatment to activate prochymosin to chymosin and also to kill the host organism and destroy any residual DNA and RNA. The cell debris from the host is separated by filtration from the liquid containing chymosin, before a chromatographic purification of the enzyme is done. Lastly, the FPC is formulated, standardized and sterile filtrated. The host organisms used today (*Aspergillus niger* and *Kluyveromyces lactis*) both excrete the prochymosin produced, in contrast to the fermentation with *E. coli* K12, where the prochymosin is stored in intracellular inclusion bodies. *A. niger* also produce prochymosin as a fusion protein together with glucoamylase and is converted into active chymosin automatically, which means that the acid treatment is only for killing the *A. niger* and destroying DNA/RNA (Harboe 1992).

Microbial coagulants are all produced via fermentation of the certain fungus named above. The production steps is similar to the one described for FPCs. The enzymes are produced as inactive precursors, which are activated into native enzymes during the fermentation due to slightly acidic pH of the substrate (Harboe and Budtz 1999).

14.4 Market Shares of Different Rennets and Coagulants

As indicated above, the world market for rennets and coagulants was until the 1950s totally dominated by calf rennet, with a small exception by the use of lamb and kid rennet around the Mediterranean sea and the use of thistle coagulant in Portugal. The microbial (fungal) coagulants were introduced after 1950s and are still quite common on the market, 25–30% (estimation by commercial manager M.F. Jensen, Chr. Hansen A/S), due to a lower price than animal rennets and FPC. Another reason for continued use of microbial coagulants is that the whey from cheese-making nowadays almost without exception is used by the food industry for different products. Any trace of animal enzymes in the whey would than reduce the area of use due to vegetarian or religious (halal, kosher) claims of food.

The most used coagulants today are the FPCs, 55–60% (estimation by commercial manager M.F. Jensen, Chr. Hansen A/S), and animal rennets (calf and adult bovine rennet). The latter decreases slowly due to decreased access to abomasa and is more and more replaced by fermentation produced bovine chymosin. The new fermentation-produced camel chymosin is slowly taking market share of bovine FPC.

14.5 Molecular and Catalytic Properties of Milk-Clotting Enzymes

Most milk-clotting enzymes belong to the family of aspartic proteinases (EC 3.4.23) with a tertiary structure of high homology. The structure contains an extended cleft, the binding site, which can accommodate at least seven amino acid residues of the substrate (κ-casein; Foltmann 1988). The cleft has two aspartic acid residues, which form the active catalytic site, and also give the enzymes their family name (Tang et al. 1973). The catalytic reaction when the milk clots is hydrolysis of the Phe_{105}-Met_{106} bond of κ-casein on the surface of the casein micelle. This destabilizes the casein micelles, which then aggregate with each other in the presence of Ca^{2+}. Chymosin has a strong affinity for this region of κ-casein and has the highest specific milk-clotting activity of the aspartic proteinases. The specific milk-clotting activity of the other milk-clotting aspartic proteinases such as pepsin (EC 3.4.23.1), *Rhizomucor miehei* proteinase (EC 3.4.23.6), *Rhizomucor pusillus* proteinase (EC 3.4.23.6) and *Cryphonectria parasitica* proteinase (EC 3.4.23.6) is lower than that of chymosin and varies among them. For example, to replace the milk-clotting activity of 1 mg of chymosin at pH 6.7, about 5 mg of bovine pepsin is needed (Andrén, unpublished results). In addition to the varied specific activity of the enzymes, the milk-clotting conditions are strongly influenced by pH, temperature and calcium content, among other factors. The milk-clotting activity of pepsin is especially highly pH-dependent, for porcine more than bovine, compared to chymosin. This will be further elucidated below in Sect. 14.7.

The inactive precursors produced in the gastric glands, prochymosin and pepsinogen, are converted to active enzymes by autocatalytic cleavage of the N-terminal of the molecule during acidic conditions, the faster the lower pH. The molecular weight of most of the milk-clotting enzymes is in the range 32–39 kDa and bovine chymosin has a molecular mass of 35.6 kDa according to its primary structure (323 amino acid residues; Foltmann et al. 1979). The pH optima and the isoelectric points of the milk-clotting enzymes are acidic, with pepsin as the most acidic (≈pH 2), although they all have a high milk-clotting activity at almost neutral pH (6.6). Chymosin has an isoelectric point around pH 4.5, where it also loses its activity rapidly due to auto-degradation. In the pH interval 5.3–6.3, however, chymosin is most stable (Foltmann 1966).

Bovine chymosin exists in two major genetic variants, A and B, differing in only one amino acid residue. The specific milk-clotting activity is higher for chymosin A compared to B, but the latter has on the other hand more stable activity during storage, which might be the reason for producing chymosin B in FPCs. Ovine and caprine chymosins have been shown to have a higher specificity for ovine and caprine milk, respectively, than for bovine milk. The opposite is the case with bovine chymosin.

However, very interestingly, recombinant camel chymosin has been shown to have a 70% higher specific milk-clotting activity for bovine milk than bovine chymosin and only 20% of the general proteolytic activity of the latter, although they have 85% identity regarding the sequence of the primary structure (Kappeler et al. 2006). The interest in camel chymosin arose from the findings that bovine rennet could not clot camel milk and therefore studies on camel chymosin were initiated. Regarding the low general proteolytic activity of camel chymosin, one very positive effect in producing cheddar cheese with recombinant camel chymosin is that a hydrophobic and bitter peptide from β-casein is not found in those cheeses (Bansal et al. 2009).

Langholm Jensen (2013) explained the difference between the milk-clotting activities between camel and bovine chymosin as follows. Some more positively charged patches of the camel chymosin, compared to bovine, give a higher affinity towards the negatively charged C-terminal part of the κ-casein. Secondly, the disordered N-terminus of camel chymosin leads to a greater flexibility of the binding cleft and thereby an improved binding of the substrate. A comprehensive study of the structure and function of bovine and camel chymosin is found in the thesis of Jesper Langholm Jensen (2013).

The *C. parasitica* proteinase is characterized by having a very high proteolytic activity, low pH-dependency and giving a good curd formation, but also being very heat-labile, which make it suitable for scalded cheeses such as Emmental. The *R. miehei* and *R. pusillus* proteinases are also very proteolytic, Although not as severe as *C. parasitica*. The *R. pusillus* proteinase is similar to *R. miehei* proteinase in action, but it is more pH-dependent and does not exist in any heat-labile form (Harboe and Budtz 1999).

The cardosins A and B from the thistle *Cynara cardunculus* are also aspartic proteinases with milk-clotting activities. The specific activity to clot bovine milk is about 6 times higher for cardosin B compared to the A variant, but still more than 20 times lower than that of chymosin (Silva et al. 2003). The cardosins both contains two chains with the molecular masses 31 and 15 kDa for the A variant and 34 and 14 kDa for B. Their proteolytic activities are within the acidic range and peptide bonds between residues with hydrophobic side chains are preferable cleaved (Faro et al. 1998). Due to their high proteolytic activity, the use of these coagulants is limited to certain local Portuguese cheeses like Serra and Serpa. Generally, among plant proteinases with milk-clotting activity, there are a lot of both cysteine and serine proteinases, in addition to the aspartic proteinases (Amira et al. 2017).

The influence of the catalytic activities of the milk-clotting enzymes on the cheese ripening process is comprehensively elucidated in a special chapter (see Chap. 15).

14.6 Analysis of Rennets and Coagulants

The International Dairy Federation has supported the work to developed international standards for the characterization of the composition and total milk-clotting activity of rennets and coagulants. The international standards are published as joint standards with ISO (International Organization for Standardization) and should be reviewed for revision or confirmation every fifth year.

As long as calf rennet was dominated by the milk-clotting activity of chymosin, the old activity units of Soxhlet and Berridge were sufficient to characterize the rennet, although the low pH (6.3) of the reconstituted Berridge skim milk substrate containing 0.01 mol L^{-1} $CaCl_2$ favoured pepsin activity within bovine rennets. One of the aims in the work with the international standard methods as therefore to develop a method to measure the milk-clotting activity at pH 6.5, a pH which is close to the pH of most cheese milks (pH $\approx$6.6) after the starter culture has been added. Other aims were to find a reliable standard reference for the milk-clotting activity which was constant over time and to develop a method for analysis of the composition of rennets. The latter was important because both the composition as well as the strength is needed to fully characterize the properties of renncts.

By decreasing the concentration of calcium chloride to 0.0045 mol L^{-1} (0.05%) in the standardized milk substrate (reconstituted low-heat, low-fat spray-dried milk powder) used in the milk-clotting test, a pH of 6.5 was achieved. Regarding reliable references for the milk-clotting activity, two different rennet powders were produced in the beginning of the 1990s, one calf rennet powder containing >98% chymosin activity measured at pH 6.5 and 32 °C and one adult bovine rennet powder with >98% bovine pepsin activity under the same conditions. A large quantity of these powders were, after appointment by IDF produced by AMAFE (Association of Manufacturers of Animal-Derived Food Enzymes), and the two batches were adjusted to have exactly the same milk-clotting activity on the standardized milk substrate in 0.05% $CaCl_2$ (pH 6.5) at 32 °C. The total milk-clotting activity of these powders were then once and for all set to 1000 International Milk-Clotting Units (IMCU) per g. When the stocks of the original batches are finished, the total milk-clotting activity of the new reference standard batches should be adjusted close to 1000 IMCU g^{-1} and labelled with the exact strength. The reference powders are sold and distributed by AMAFE and their address given in the corresponding International standards.

An international standard method to determine the content of chymosin and pepsin in bovine rennets (calf and adult) was based on a method where the enzymes are chromatographically separated before their individual activities are measured (Garnot et al. 1972; Collin et al. 1981). The principle of the method is that the milk-clotting activity of the two fractions from the anion-exchange column, chymosin and pepsin, is determined relative to the international calf and adult bovine rennet reference powders, respectively, tested on the standardized milk substrate at 32 °C. The percentage of milk-clotting activity by chymosin and pepsin is calculated from the results of the activity tests.

It is important to note, however, that the total milk-clotting activity should not be calculated by adding the activity of both fractions owing to some enzyme losses on the column and the difficulty of knowing exactly the amount of rennet introduced to column. The method is described in detail in International Standard: ISO 15163, IDF 110 (International Dairy Federation 2012) and can be obtained from IDF in Brussels or ISO in Geneva. In addition to expressing the results in percentage of chymosin and pepsin activity, the results could be expressed in milligrams of active chymosin and pepsin per litre. However, it is preferable to use the expression percentage of activity, due to a better repeatability and reproducibility of the analysis. This standard is also useful for the composition of chymosin and pepsin in ovine and caprine rennets and an annex to the standard describes a qualitative determination of milk-clotting enzymes, i.e., if there are any microbial or porcine milk-clotting enzymes in the rennet. If rennets are adulterated with non-ruminant milk-clotting enzymes, the results of the chymosin and pepsin content of the rennet are not reliable, since the other coagulants could have influenced the composition of the fractions from the column.

The strength of bovine rennets (calf and adult) can be analysed using the International Standard: ISO 11815, IDF 157 (International Dairy Federation 2007). The principle of the method is that the milk-clotting activity of the rennet sample is determined relative to the activity of both the international reference rennet powders (described above) on the standardized milk substrate of pH 6.5 at 32 °C. The relative activity results are interpolated with respect to the composition of the rennet sample, i.e., the higher the proportion of chymosin, the greater the relative milk-clotting activity against the standard calf rennet reference is taken into account and vice versa. The results are given in IMCU (International Milk-Clotting Units) per mL or g. Regarding FPCs (both bovine and camel), the total milk-clotting activity is measured relative to the standard calf rennet reference only, since the FPCs contain 100% chymosin activity.

To analyse the strength of ovine and caprine rennets, including rennet paste, International Standard: ISO 23058, IDF 199 (International Dairy Federation 2006) is suitable. The standard rennet reference powders and the principle of the method are the very same as for bovine rennets (ISO 11815, IDF 157).

The total milk-clotting activity of the most widely used microbial coagulants (Miehei, Pusillus and Parasitica) can be analysed using the International Standard: ISO 15174, IDF 176 (International Dairy Federation 2002). The standard reference powder for this method is of *Rhizomucor miehei* proteinase origin and is defined according to the international calf rennet standard powder, i.e., set at a fixed value relative to the strength of the calf rennet powder. The principle of the method is also the same as for bovine rennets (ISO 11815, IDF 157) and the microbial reference powder can be obtained from the producer given in the standard (ISO 15174, IDF 176).

14.7 Interpretation of the Strength of Rennets

By running the international standards described above, a reliable and good characterization of the rennets and coagulants are achieved. The strength given in IMCU mL^{-1} can be used to compare different products and, regarding FPCs (100% chymosin) and microbial coagulants (100% single enzyme), the comparison is easily done by just comparing the figures for the strength.

However, in case of bovine, ovine and caprine rennets, the enzymatic composition also plays an important role for the interpretation of IMCU mL^{-1}. The higher the proportion of pepsin in the rennet, "the higher IMCU mL^{-1} is needed" compared to a rennet rich in chymosin. The explanation to this is as follows, as illustrated schematically in Fig. 14.1. The milk-clotting activity of pepsin is much more pH-dependent than that of chymosin and the relative activity increases faster when lowering the pH. Since the standard method for the total strength of a rennet is measuring the activity at pH 6.5, which is a little bit lower (0.1 pH-units) than the pH of most cheese milks, the total strength of adult bovine rennets are somewhat overestimated compared to their activity in cheese milk (pH ≈6.6). Therefore, when comparing different rennets, calf *versus* adult bovine, for the use in cheese-making, the total strength of the adult bovine rennet has to be higher than in a calf rennet. How much

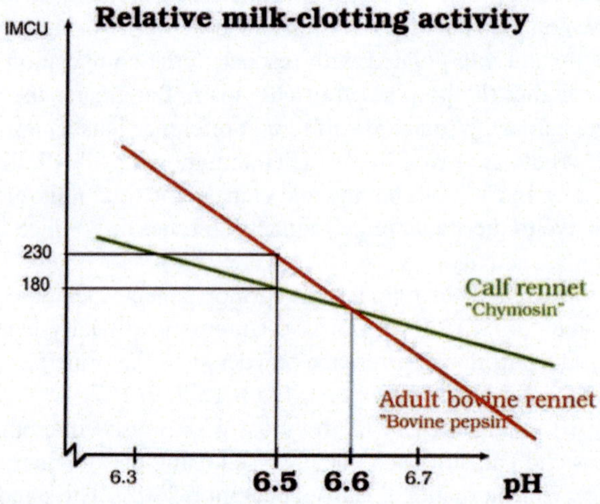

Fig. 14.1 The interpretation of the total strength of different rennets measured at pH 6.5 compared to the total strength at pH 6.6 of the cheese milk (Andrén, unpublished results)

depends on the compositions of the rennets to be compared and no exact difference in IMCU ml^{-1} can easily be given.

To illustrate this point, the difference of 50 (230 minus 180) IMCU mL^{-1} in Fig. 14.1 represents the difference between an adult bovine rennet with the composition 25% chymosin/75% pepsin and a calf rennet with the composition 75% chymosin/25% pepsin, when they are supposed to have the very same clotting activity at pH 6.6. However, the example given refers to those batches of rennets compared and should only be a guideline for the difference in IMCU mL^{-1} between a calf and an adult bovine rennet that could occur.

In this context, it is very important to mention that, in studies where the coagulation properties of milk samples from different cows are to be compared, it is necessary to use FPC (100% chymosin). If a rennet containing both chymosin and pepsin is used, small variations in milk sample pH will influence the results due to the higher pH-dependence of pepsin and other parameters influencing the coagulation properties could therefore be hidden.

To conclude, by use of the international standards for the characterization of rennets and coagulants, both producers and users will have benefits in their dialogue with each other for a better understanding. Only rennets where there is a mix of chymosin and pepsin need some extra effort for the interpretation of the total milk-clotting activity of the product.

References

Amira AB, Besbes S, Attia H, Blecker C (2017) Milk-clotting properties of plant rennets and their enzymatic, rheological, and sensory role in cheese making: a review. Int J Food Prop 20:576–593

Andrén A, Björck L (1986) Milk-feeding maintains the prochymosin production in cells of bovine abomasal mucosa. Acta Physiol Scand 126:419–427

Andrén A, Björck L, Claesson O (1982) Immunohistochemical studies on the development of prochymosin- and pepsinogen-containing cells in bovine abomasal mucosa. J Physiol 327:247–254

Bansal N, Drake MA, Piraino P, Broe ML, Harboe M, Fox PF, McSweeney PLH (2009) Suitability of recombinant camel (*Camelus dromedarius*) chymosin as a coagulant for Cheddar cheese. Int Dairy J 19:510–517

Berridge NJ (1952) Some observations on the determination of activity of rennet. Analyst 77:57–62

Birschbach P (1992) Pregastric lipases. Bull Int Dairy Fed 269:36–39

Collin JC, Martin P, Garnot P, Ribadeau-Dumas B, Mocquot G (1981) Determination of chymosin and bovine pepsin A in commercial bovine rennets and pepsins. Milchwissenschaft 36:32–35

Deschampes JP (1840) De la pressure. J Pharm 26:412–420

Faro C, Ramalho-Santos M, Verissimo P, Pissarra J, Frazào C, Costa J, Lin X-L, Tang J, Pires E (1998) Structural and functional aspects of cardosins. In: James MNG (ed) Aspartic proteinases. Plenum Press, New York

Foltmann B (1966) A review on prorennin and rennin. C R Travaux Lab Carlsberg 35:143–231

Foltmann B (1981) Gastric proteinases; structure, function, evolution and mechanism of action. Essays Biochem 17:52–84

Foltmann B (1988) Structure and function of proparts in zymogens for aspartic proteases. Biol Chem Hoppe-Seyler 369:311–314

Foltmann B, Pedersen VB, Kauffman D, Wybrandt G (1979) The primary structure of calf chymosin. J Biol Chem 254:8447–8456

Garnot P, Thapon JL, Mathieu CM, Maubois JL, Ribadeau-Dumas B (1972) Determination of rennin and bovine pepsins in commercial rennets and abomasal juices. J Dairy Sci 55:1641–1650

Harboe MK (1992) Chymogen, a chymosin rennet manufactured by fermentation of *Aspergillus niger*. Bull Int Dairy Fed 269:3–7

Harboe MK (1994) Use of lipases in cheesemaking. Bull Int Dairy Fed 294:11–16

Harboe M, Budtz P (1999) The production, action and application of rennet and coagulants. In: Law BA (ed) Technology of cheesemaking. Sheffield University Press, Sheffield

International Dairy Federation (2002) Milk and milk products – microbial coagulants – determination of total milk-clotting activity. International Standard: ISO 15174, IDF 176. ISO, Geneva

International Dairy Federation (2006) Milk and milk products – ovine and caprine rennets – determination of total milk-clotting activity. International Standard: ISO 23058, IDF 199. ISO, Geneva

International Dairy Federation (2007) Milk – Determination of total milk-clotting activity of bovine rennets. International Standard: ISO 11815, IDF 157. ISO, Geneva

International Dairy Federation (2012) Milk and milk products – Calf rennet and adult bovine rennet – Determination by chromatography of chymosin and bovine pepsin. International Standard: ISO 15163, IDF 110. ISO, Geneva

Kappeler S, Van den Brink JM, Rahbek-Nielsen H, Farah Z, Puhan Z, Hansen E, Johansen E (2006) Characterization of recombinant camel chymosin reveals superior properties for the coagulation of bovine and camel milk. Biochem Biophys Res Com 342:647–654

Langholm Jensen, J (2013) Structure and function of bovine and camel chymosin: a protein chemical study of the differences in milk-clotting abilities. Ph. D. thesis at University of Copenhagen, Copenhagen

Lemes AC, Pavón Y, Lazzaroni S, Rozycki S, Brandelli A, Kalil SJ (2016) A new milk-clotting enzyme produced by *Bacillus* sp.P45 applied in cream cheese development. LWT – Food Sci Technol 66:217–224

Lopes-Marques M, Ruivo R, Fonseca E, Teixeira A, Castro LFC (2017) Unusual loss of chymosin in mammalian lineages parallels neo-natal immune transfer strategies. Mol Phylogen Evol 116:78–86

Miura H, Yamada J, Kitamura N, Yamashita T, Andrén A (1988) An immunohistochemical study of prochymosin- and pepsinogen-immunreactive cells in the abomasal mucosa of cattle fetuses. Z Mikrosk Anat Forsch Leipzig 102:101–110

Praaning-Van Dalen DP (1992) Application and regulatory position of Maxiren®. Bull Int Dairy Fed 269:8–12

Silva SV, Allmere T, Malcata FX, Andrén A (2003) Comparative study on the gelling properties of cardosins extracted from Cynara cardunculus and chymosin on cow's skim milk. Int Dairy J 13:559–564

Tang J, Sepulveda P, Marciniszyn J, Chen KCS, Huang W-Y, Tao N, Liu D, Lanier JP (1973) Amino acid sequence of porcine pepsin. Proc Natl Acad Sci USA 70:3437–3439

Teuber M (1990) Production of chymosin (EC 3.4.23.4) by microorganisms and its use for cheesemaking. Bull Int Dairy Fed 251:3–15

Verissimo P, Esteves C, Faro C, Pires E (1995) The vegetable rennet of *Cynara cardunculus* L. contains two proteinases with chymosin and pepsin-like specificities. Biotechnol Lett 17:621–626

Von Soxhlet F (1877) Milchzeitung 6:495–513

Yamada J, Andrén A, Kitamura N, Yamashita T (1988) Electron immunocytochemical co-localization of prochymosin and pepsinogen in chief cells, mucous neck cells and transitional mucous neck/chief cells of the fundic glands. Acta Anat 132:246–252

Chapter 15
Enzymes in Cheese Ripening

Ylva Ardö

15.1 Introduction

Cheese ripening involves microbial, enzymatic, chemical and physical processes, and enzymes are main actors in the development of cheese characteristics through this process. Their activities may either be controlled by microorganisms during growth, surviving or starving, or they work uncontrolled in the cheese matrix after disruption of cell membranes and eventual lysis, as do enzymes from the milk and free enzymes added during cheese-making. The enzymology of cheese ripening is complicated and involves enzymes of all the three groups of indigenous, endogenous and exogenous milk enzymes (Table 15.1). This chapter describes enzymatic activities in cheese during ripening, with a focus on enzymes that could be possible for the cheesemaker to influence, with purposes of, e.g., controlling development of characteristic properties, accelerating ripening, intensifying flavour or increasing yield.

Indigenous milk enzymes are involved at different degrees in cheese ripening depending on type of milk animal, race and feeding, as well as cheese making technology. By far the most important technological factor influencing milk enzymes is heat treatments of milk and curd. Long ripening times may, however, allow also enzymes at low residual activities in cheese made from pasteurised milk to influence the cheese quality. Indigenous milk enzymes are described in other chapters in this book, and here the focus is on their role in cheese ripening, such as the following:

- The milk lipase, lipoprotein lipase (LPL) contributes to a rich characteristic flavour of raw milk cheeses. Only about 5% of its activity remains after

Y. Ardö (✉)
Department of Food Science, University of Copenhagen,
Rolighedsvej, Frederiksberg, Denmark
e-mail: ya@food.ku.dk

A. L. Kelly, L. B. Larsen (eds.), *Agents of Change*, Food Engineering Series,
https://doi.org/10.1007/978-3-030-55482-8_15

Table 15.1 Enzymes with potential to influence cheese ripening

Indigenous milk enzymes	Exogenous enzymes
Plasmin	Milk coagulants
Cathepsin D	Enzymes of starter bacteria
Lipoprotein lipase, LPL	Enzymes of ripening microorganisms
Alkaline phosphatases	Proteases/lipases to accelerate ripening
Acid phosphatases	Activators of plasminogen
Ribonuclease	Aminopeptidase preparations
γ-Glutamyl tranferase, GGT	Transglutaminase, TGA, for structure
Glutathione peroxidase	Phospholipases to increase yield
Lactoperoxidse	Lactose oxidase to prevent browning
Xanthine oxidase	Lysozyme to replace nitrate in cheese
Lysozyme	β-Galactosidase to stimulate starter
Endogenous enzymes	
Thermoresistent enzymes of psychrotrophic bacteria	
Enzymes of raw milk microflora developing in cheese during ripening	

pasteurisation; however, this residual activity may still be significant for flavour in long-ripened cheese;

- Plasmin is the most important milk protease involved in cheese ripening, and it survives pasteurisation very well;
- Milk contains alkaline and acidic phosphatases, of which the alkaline is inactivated by pasteurisation, while the acid phosphatase is less heat-sensitive and active at the pH of cheese;
- The milk enzyme γ-glutamyltranspeptidase (GGT) may have a role in synthesis of glutamyl-peptides during cheese ripening, which have a significant influence on flavour of long-ripened cheeses. It is partly inactivated during pasteurisation;

Other indigenous enzymes, which are active in milk but not in cheese, may also have an impact on cheese ripening. For example, milk oxidases act when oxygen is present and thereby increase the level of oxidised groups that may contribute to flavour formation in the cheese. Lactoperoxidase may oxidise amino acids in casein prior to coagulation, and oxidised Tyr residues that, after being released in the cheese, may be converted by microorganisms to dopachrom and melanin that cause pink and brown colour defects in the cheese body. Glutathione peroxidase oxidises reduced glutathione that acts as a cofactor in several enzyme activities. Indigenous enzymes may also protect the cheese from detrimental activity of *Costridium* bacteria, e.g. xanthine oxidase catalyses reduction of nitrate to nitrite that activates the anticlostridial effect of the nitrate added to some cheeses. Xanthine oxidase is inactivated by a somewhat higher heat load than pasteurisation, and nitrate will not be useful in such cases. The milk enzyme lysozyme may inhibit *Costridium tyrobutyricum*; however, its concentration in milk is low, and exogenous egg white lysozyme is used in some hard cheeses to replace nitrate. These enzymes are not further treated in this chapter.

Endogenous milk enzymes of importance for cheese originate from the microflora of the raw milk and the dairy environment. They contribute to similar ripening activities as enzymes of starter and added ripening bacteria; however, under an extended time period during ripening. Some of these bacteria may also contribute to new cheese characteristics, off-flavours or detrimental gas formation. The spontaneous microflora of cheese milk depends on conditions during milk production and very much on milk cleaning procedures at the farm, during transportation and at the dairy. Psychrotrophic bacteria, mainly *Pseudomonads,* are able to grow in milk during cold storage and are stimulated by repeated periods at higher temperatures during and after milking. The bacteria are killed by pasteurization, but their proteases and lipases remain active and may contribute to a decreased cheese yield and introduce off-flavours. *Clostridia* spores survive pasteurization and cause bad flavour and texture if they grow out in the cheese. A high proportion of the spores may be removed from the milk by bactofugation or microfiltration (Waagner Nielsen 2004). The retentate from microfiltration and the bactofugate from bactofugation is typically treated at ultra-high-temperature (UHT) at about 120 °C for a few seconds together with the cream. Raw, thermised, pasteurised, bactofuged and microfiltrated milk contain very different microflora; however, bacteria that survive heat treatments at low number and perhaps in bad condition may develop well in the cheese if the environment and nutrient supply is suitable for them. The long-time survivors in cheese consist mainly of facultative heterofermentative *Lactobacillus* such as *Lb. casei*, *Lb. plantarum* and *Lb. curvatus* but also obligate heterofermentative *Lactobacillus* such as *Lb. brevis*, *Lb. buchneri* and *Lb. parabuchneri*.

A new taxonomy for the family *Lactobacillus* based on whole genome sequencing of 261 recognized *Lactobacillus* species was published during preparation of this book, and several of the species got new names (Zheng et al. 2020; http://lactobacillus.uantwerpen.be). The three groups, homofermentative, facultative heterofermentative and obligate heterofermentative Lactobacillaceae, are still valid. The name of the species used in this chapter of the homofermentative group have not been changed, i.e., *Lactobacillus delbrueckii*, *helevticus* and *acidophilus*. The other *Lactobacillus* species used here have changed their family names to *Lacticaseibacillus casei, paracasei* and *rhamnosus, Latilactobacillus sakei* and *curvatus, Lactiplantibacillus plantarum, Pausilactobacillus danicus* and *wasatchensis, Levilactobacillus brevis, Lentilactobacillus buchneri* and *parabuchneri*. In this chapter, the old nomenclature is used.

Exogenous enzymes added during cheese making with significant activity during ripening originate from the coagulant, starter and ripening bacteria as well as specific enzyme preparations added to influence cheese making and ripening. Milk coagulants are described in Chap. 14 and their role in ripening is discussed below. Enzymes of starter bacteria and ripening microorganisms, non-starter lactic acid bacteria (LAB), non-LAB bacteria, yeast and mould of importance to cheese ripening are described below, as well as enzymes added to accelerate ripening. Other enzymes added to influence cheese structure, e.g., transglutaminase to increase cheese yield and phospholipases to prevent browning of pizza cheese, are commonly inactivated in the cheese vat and will not directly influence cheese ripening.

Cheese is made by coagulation of milk with the action of rennet enzymes in combination with acidification by lactic acid bacteria. The coagulum undergoes spontaneous contraction and expelling of whey, which is accelerated by cutting the gel into small cubes followed by heating and stirring. The resulting curd grains are then pressed and formed into cheeses. Low water activity obtained by concentration, acidification and salting prevent detrimental microbial activities and stimulate desired ripening processes. To prevent detrimental microbial growth on the surfaces, the cheeses may be dried and cleaned, immersed in brine and coated with wax, paraffin or synthetic film material. Alternatively, a controlled microflora may be stimulated to grow on the surfaces, which introduces specific flavour notes. Ripening takes place during storage in controlled environments, which may last for a couple of weeks, several months, or up to a year or more. The organic solids of the cheese will gradually be hydrolysed and metabolized, and thereby the characteristic texture, appearance, consistency and flavour of the cheese develop. Primarily lactose, citrate, milk fat and casein are catabolised to different extents and in different ways depending on cheese variety. This chapter describes enzymatic activities involved in the ripening processes as overviewed in Fig. 15.1.

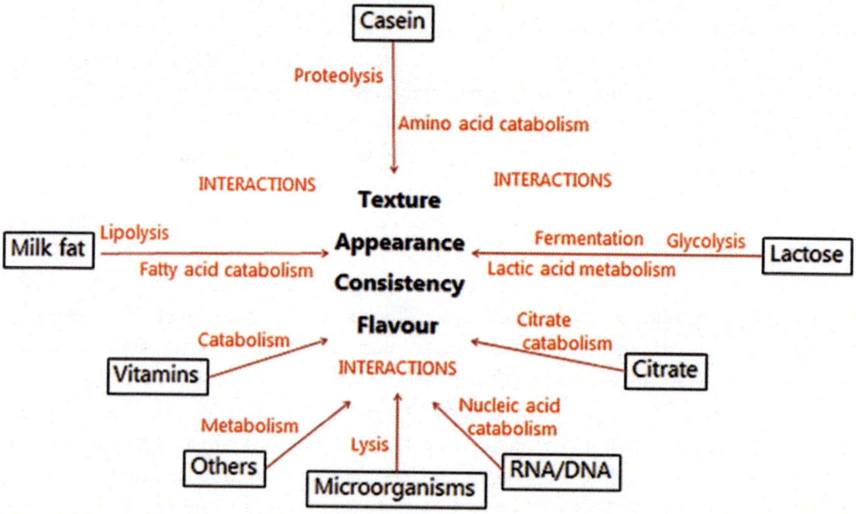

Fig. 15.1 Overview of enzymatic processes in cheese during ripening

15.2 Lactose Metabolism

15.2.1 Starter Cultures

Starter cultures are added to cheese milk to support acidification, and lactose is their main energy source. The lactose metabolism starts with glycolysis and is, in an anaerobic environment, supported by fermentation of pyruvic to lactic acid that regenerates the reduced coenzyme nicotinamide adenine dinucleotide (NADH to NAD^+). The thereby formed lactic acid lower cheese pH and may undergo further metabolism in specific cheese varieties. The lactose is used as primary energy source mainly by starter bacteria, which dominate the microflora during production and at least the first few weeks of ripening in most cheese varieties. The microbial composition of the starter culture varies with cheese type (Table 15.2), as reviewed by Cogan and Hill (1993).

The mesophilic O-starters contain lactic acid bacteria species that do not produce gas. They are commonly made of a few well-defined strains completing each other by being sensitive to different kinds of bacteriophages, and they perform

Table 15.2 Composition of commonly used starter cultures for different ripened cheeses

Starter bacteria	Lactose pathway	Use galactose	Lactate formed	Cooking temperatures	Examples of cheeses
Mesophilic O-starter *Lactococcus lactis* ssp. *cremoris* *Lc. lactis* subsp. *lactis*	EMG EMG	+ +	L(+) L(+)	36–38 °C	Closed texture Cheddar
Mesophilic DL-starter *Lc. lactis* subsp. *cremoris* *Lc. lactis* ssp. *lactis* *Lc. lactis* ssp. *lactis* biovar. *Diacetylactis* *Lecuonostoc* sp.	EMG EMG EMG PKP	+ + + +	L(+) L(+) L(+) D(−)	36–42 °C	Round-eyed or open texture semi-hard cheese Gouda, Herrgård, Danbo, Edamer, Havarti, Västerbottensost, Prästost
Thermophilic starter *Streptococcus thermophilus* *Lactobacillus helveticus or Lb. delbrüchii*	EMG EMG EMG	– + –	L(+) L(+)/D(−) D(−)	48–55 °C	Hard cheese Parmigiano Reggiano, Grana, Emmental

EMG Embden-Meyerhof glycolytic pathway, PKP phosphoketolase pathway
L(+) and D(−), names of the lactic acid enantiomers

acidification with a high repeatability over time. This is the traditional starter used in Cheddar production worldwide. Cheddar made with O-starters ferment most lactose within about 24 h. Both monomers of lactose, glucose and galactose, are consumed. In industrial production, *Streptococcus thermophilus* is sometimes used to speed up acidification and, because these bacteria only ferment glucose, galactose may be used by other microorganisms or accumulate in the cheese.

Undefined mesophilic DL-starters are used in the production of semi-hard and hard cheese varieties with eyes. They contain several strains of different species of lactic acid bacteria introducing a dynamically changing microbial population during manufacture and ripening. The bacterial strains belong to four different groups, namely, the two main lactic acid producers, *Lactococcus lactis* subsp. *cremoris* and *Lc. lactis* subsp. *lactis,* and the citrate-positive *Lc. lactis* subsp. *lactis* biovar. *diacetylactis* (D) and *Leuconostoc* sp. (L). The citrate-metabolising bacteria are responsible for eye formation by production of CO_2 from citrate with simultaneous production of flavour compounds such as diacetyl. The growth and development in cheese of the different groups of bacteria depend on the cheese manufacturing procedure. *Lc. lactis* subsp. *cremoris* and *lactis* grow to a higher number in cheeses with lower cooking temperatures, which are closer to their optimal temperatures (Ardö 1993). *Lc. lactis* subsp. *lactis* biovar. *diacetylactis* grow before *Leuconostoc* during cheese production, while *Leuconostoc* are able to grow at lower pH. The maximal number of starter bacteria is reached after a day or two, when the lactose is exhausted, and a pH minimum is reached.

Thermophilic starters are used in hard and extra-hard cheeses that are manufactured at rather high temperatures. In the Alps and southern Europe, these cheeses are commonly made from raw milk, and high cooking temperatures during production inactivate some unwanted microbial activity. They contain *Str. thermophilus* and *Lb. delbruckii* that use only the glucose part of the lactose and leave galactose in cheese for other microorganisms to use. *Lb. helveticus* that is used in some of the starters, on the other hand, convert both glucose and galactose into lactic acid. Starters used in the production of mould cheeses depend on variety and comprise mixtures of mesophilic and thermophilic bacteria.

Lactose metabolism is influenced mainly by choice of starter. Replacement of a part of the whey with water during production of some semi-hard cheeses with eye formation is used to limit acidification by lowering the concentration of lactose (Waagner Nielsen 2004). Slower gas production is obtained resulting in smooth eyes, and a higher pH in young cheese stimulates development of propionic acid bacteria (Ardö 2004).

15.2.2 Glycolysis and Regeneration of NADH

Glycolysis, the biochemical pathway of *Lactococcus* that converts the monomers of lactose to pyruvate was reviewed by Cogan and Hill (1993), and may be summarised as:

$$\text{Glucose} + 2\,\text{ADP} + 2\,\text{P} + 2\,\text{NAD}^+ \rightarrow 2\,\text{Pyruvate} + 2\,\text{ATP} + 2\,\text{NADH} + 2\,\text{H}^+ + 2\,\text{H}_2\text{O}$$

Digestion of each glucose monomer yields energy by producing two molecules of adenosine triphosphate (ATP) from adenosine diphosphate (ADP). Two molecules of NAD^+ are reduced to NADH, which needs to be regenerated back to NAD^+ for the energy production from lactose to continue. The same amount of ATP is produced from galactose in species that use both monomers. The final product of glycolysis, pyruvic acid, is fermented into lactic acid in an anaerobic environment by lactic acid bacteria to regenerate the NAD^+ from the NADH formed during the digestion of lactose. Lactic acid bacteria produce two different lactate dehydrogenases, which form two enantiomers (called L(+) and D(−), respectively) of lactic acid (Table 15.2). From each lactose molecule, four molecules of lactic acid are formed if both glucose and galactose are digested, and in homofermentative lactic acid bacteria this is the only pathway to regenerate NADH during growth under anaerobic conditions.

Leuconostoc and obligate heterofermentative *Lactobacillus* decaboxylate hexoses into pentoses that are converted into acetaldehyde, and ethanol is produced instead of lactate. Facultative heterofermentative *Lactobacillus* also produce ethanol from pentoses; however, they are not able to decarboxylate hexoses. *Leuconostoc* gains extra energy by co-metabolising lactose and citrate and produce acetate. This pathway, however, needs to be completed by parallel formation of ethanol to regenerate NADH, resulting in only one molecule of ATP instead of two from each glucose molecule. The proportions of ethanol and acetate vary, and often only ethanol is formed. CO_2 is formed either way and contributes to eye formation.

Some lactic acid bacteria, e.g., *Lc. lactis* and *Leuconostoc* from the starter, as well as *Lb. brevis* from the environment, produce NADH oxidase that regenerates the NADH if oxygen is present. Electrons from NADH are used to form superoxide in the reaction with oxygen. This activity contributes to a decrease in the redox potential, and a consequence is that acidification slows down because no acid is produced.

15.2.3 Lactic Acid Metabolism

Lactic acid is not further metabolised in cheese by the starter bacteria, nor by the common non-starter strains of the *Lb. casei* group. However, several other microorganisms may convert lactic acid into formic, acetic, propionic or butyric acid, or oxidise it to CO_2 and water as reviewed by McSweeney and Fox (2004) (Table 15.3).

Propionic acid bacteria use lactate to produce propionate, acetate and CO_2, which is responsible for the large eyes in cheeses such as Emmental, Maasdamer, Jarlsberg and Grevé. The lactate may also be used by detrimental bacteria such as *Clostridium,* resulting in late blowing, cracks and destroyed cheese texture.

Oxidation processes are limited in cheeses, especially those without a microflora growing on the surfaces in contact with the oxygen in the air. In surface-ripened

Table 15.3 Metabolism of lactic acid in some cheese-related bacteria

Reaction	Microorganisms
3 lactic acid → 2 propionic acid + acetic acid + $CO_2 + H_2O$	Propionic acid bacteria
2 lactic acid → butyric acid +2 $CO_2 + H_2O + 2\ H_2$	*Clostridium*
Lactic acid +3 O_2 → 3 CO_2 + 3 H_2O	Yeast, *Pencillium*
Lactic acid + O_2 → acetic acid + $CO_2 + H_2O$	*Pediococcus*
Lactic acid + O_2 → formic acid + X	*Pediococcus, Lactobacillus*
L-lactic acid → DL -lactic acid	*Pediococcus, Lactobacillus*

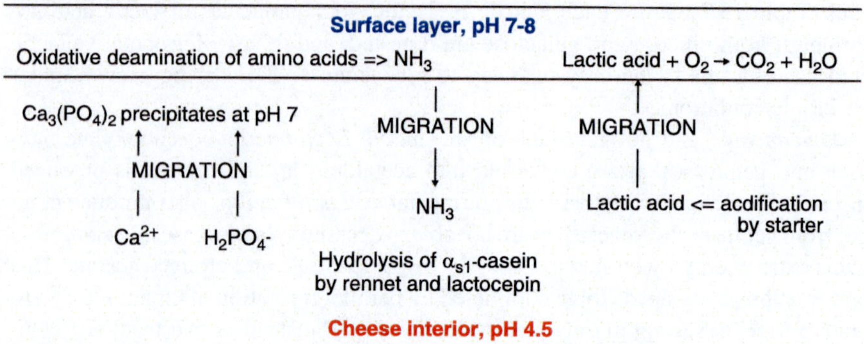

Fig. 15.2 Metabolism of lactic acid and amino acids in a white mould cheese

cheeses, such as white mould cheese, a pH gradient is formed as a result of acidification by starter bacteria in the interior of the cheese and oxidative processes at the surface (Fig. 15.2). The mould growing on the surface produces NH_3 *via* oxidative deamination of amino acids. The lactic acid of the interior of the cheese migrates to the surface, where it may be oxidised into CO_2 and H_2O by yeast and *Pencillium camemberti*. The low internal pH releases Ca^{2+} and PO_4^{3-} from the casein, and these precipitate as calcium phosphate at the higher pH close to the surface. The delicious smooth and soft interior of a white mould cheese is the result of a high pH, a low Ca^{2+} content and partly hydrolysed casein. At pH 4.5, the casein is no longer soluble as sometime can be seen as a white precipitate in the centre of the cheese, which will, however, disappear after a couple of hours in room temperature due to increased pH performed by a faster migration of ammonium into the interior and lactate to the surface.

Pediococcus oxidise lactic acid if O_2 is available, which may be the situation around cracks in old cheeses. This activity keeps the redox potential at a beneficial low level in cheese and protects thereby the cheese from detrimental oxidative activities.

Racemisation of L-lactate into DL-lactate occurs because of microbial activities and may cause white crystals in the cheese because of precipitation of DL-lactate that is less soluble in the cheese matrix than L-lactate. Commonly, this racemisation is a result of the presence of both D- and L-lactose dehydrogenase (LDH) and a need to convert NAD^+ to NADH and *vice versa* (Fig. 15.3).

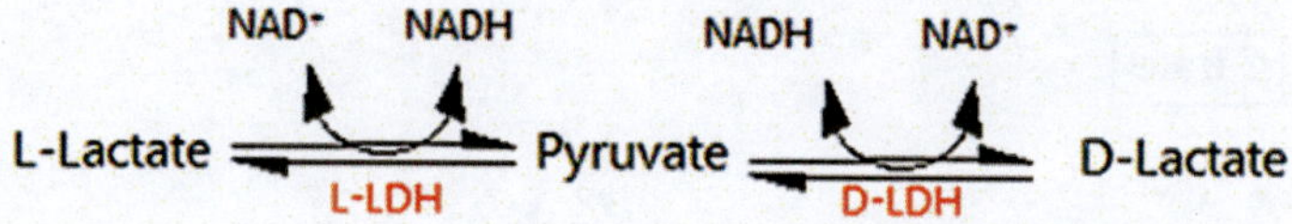

Fig. 15.3 Racemisation of lactic acid. LDH, lactic acid dehydrogenase

15.3 Citrate Catabolism

Leuconostoc and *Lc. lactis* subsp. *lactis* biovar. *diacetylactis* present in DL-starter will rapidly convert most of or all the citric acid in the cheese to acetic acid, diacetyl/acetoin and CO_2 (reviewed by Cogan and Hill 2004) (Fig. 15.4). The genetic difference of the biovariant *diacetylactis* of *Lc. lactis* is a plasmid coding for the enzymes citrate permease and lyase, allowing transport of citrate into the cell and conversion of it to oxaloacetate. The other enzymes of the metabolic pathways are present in all *Lc. lactis*, and they may convert excess of oxaloacetate and pyruvate from other sources, e.g., transamination of the amino acids Asp and Ala, to acetoin, which then may be oxidised to diacetyl or reduced to 2,3-dibutandiol (Ardö 2006). These metabolic pathways introduce alternative possibilities to regenerate the cofactors NAD^+/NADH that may be reduced or oxidised by these reactions. The CO_2 produced from citrate metabolism causes eye formation in semi-hard cheese, while diacetyl has an impact on cheese flavour.

15.4 Milk Fat Catabolism

15.4.1 Milk Fat Hydrolysis

Milk fat may undergo oxidative and hydrolytic breakdown; however, in cheese ripening, a low redox potential limits significant oxidation processes from occurring. Free fatty acids are released from fat in cheese by lipase enzymes. Short chain fatty acids have sharp and pungent flavours that are characteristic components in mould ripened cheeses, cheeses made with rennet paste, and cheeses from raw milk. A limited amount is important also for a balanced flavour in semi-hard and hard cheeses made from pasteurised milk. Hydrolysis of milk fat in cheese was recently reviewed by Thierry et al. (2017).

Lipase enzymes may be classified as lipases or esterases mainly distinguished by the size of substrate and localisation in cheese during hydrolysis. Lipases typically release larger fatty acids from triglycerides and work at interfaces between water and fat, while esterases release shorter fatty acids and work exclusively in aqueous solutions. Lipases and esterases in cheese originate mainly from milk, rennet paste, blue and white mould, but also from starter, non-starter and ripening bacteria, as well as exogenous lipase preparations added to accelerate ripening.

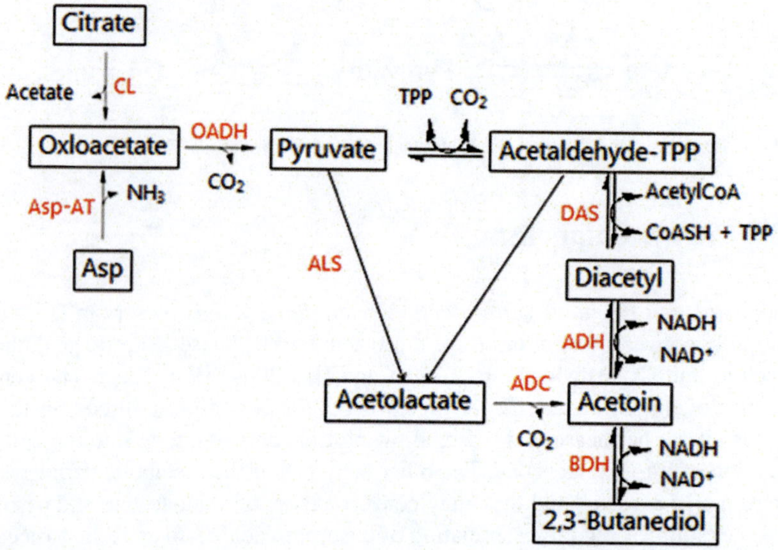

Fig. 15.4 Citrate metabolism of lactic acid bacteria. TPP triphenyl phosphate, CoA coenzyme A, CoASH reduced coenzyme A, ADC acetolactate decarboxylase, ADH acetoin dehydrogenase, ALS acetolactate synthetase, Asp-AT Asp aminotransferase, BDH, butanediol dehydrogenase, CL citrate lyase, DAS diacetyl synthetase, OADC oxaloacetate decarboxylase

The indigenous lipase of milk, lipoprotein lipase (LPL; Chap. 10), contributes to significant lipolysis in cheese made from raw milk. LPL releases fatty acids from the end positions of glycerol molecules, which in milk fat typically have short chains. Pasteurization of milk inactivates about 95% of the milk lipase; however, its activity in long-ripened cheeses may still be significant (Waagner Nielsen 2004).

Traditional rennet paste that is used mainly in Mediterrian cheese varieties is extracted from lamb and kid stomachs and includes lipases of the pregastric region; these lipases have a high specificity for releasing piquant tasting short-chain fatty acids (Birschbach 1994).

Yeast and bacteria of the smear microflora on surface-ripened cheeses produce lipases that release fatty acids, which may diffuse into the cheese and contribute to flavour (Stadhouders and Mulder 1957). Blue mould produces very active lipases. In mould-ripened cheeses with high final pH; however, the contribution of free fatty acids to the flavour is reduced because of pH-dependent dissociation of the acids.

Lactic acid bacteria possess esterases that hydrolyse di- and mono-glycerides, and their activities have been suggested to be important in cheese from pasteurised milk after disruption of their cell membranes. Esterases have been demonstrated in *Lc. lactis, Leuconostoc, Lb. helveticus, Lb delbrückii*, facultative heterofermentative *Lactobacillus, Lb danicus, Lb. parabuchnerii, Str. thermophilus* and propionic acid bacteria (Pedersen et al. 2013; Thierry et al. 2017; author's unpublished results).

In raw milk that has been cold stored for some time, psychrotrophic bacteria (commonly *Pseudomonads*) may produce thermoresistant lipases. These bacteria do not survive pasteurization, but their lipases may, and a high number of the bacteria prior to pasteurization means a risk for development of unclean rancid off-flavours in cheese (Law 1979).

15.4.2 Fatty Acid Metabolism

Free fatty acids are precursors of many important flavour compounds in cheese, including methyl ketones, secondary alcohols, lactones, esters and thioesters (Fig. 15.5). Esters and thioesters may also be formed via re-esterfication, i.e. reactions of the alcohols and methanethiol with fatty acids still attach to mono- or diglycerides (for review see Thierry et al. 2017).

Methyl ketones, especially 2-heptanone and 2-nonanone, contribute to typical flavours of mould ripened cheese. They are produced by *Penicillium roqueforti* and *Penicillium camemberti* in a series of enzyme activities starting with lipolysis to release fatty acids, which are directly oxidised to β-ketoacids and decaboxylated to 2-ketones (methylketones) that are sometimes further metabolised to secondary alcohols. While 2-heptanone is significant for the blue cheese flavour, 2-nonanone dominates in white mould cheeses (author's unpublished results).

Esters are produced by lactic acid bacteria from short fatty acids and the alcohols present in the cheese, e.g., ethanol, methanol, propanol and butanol; most commonly, ethyl esters are formed. The esters have fruity flavours and are characteristic components especially in rather dry long-ripened cheeses, e.g., Parmigiano Reggiano. A low water activity of the ripened cheese stimulates esterases to form esters and thereby release water instead of releasing fatty acids from fat that binds

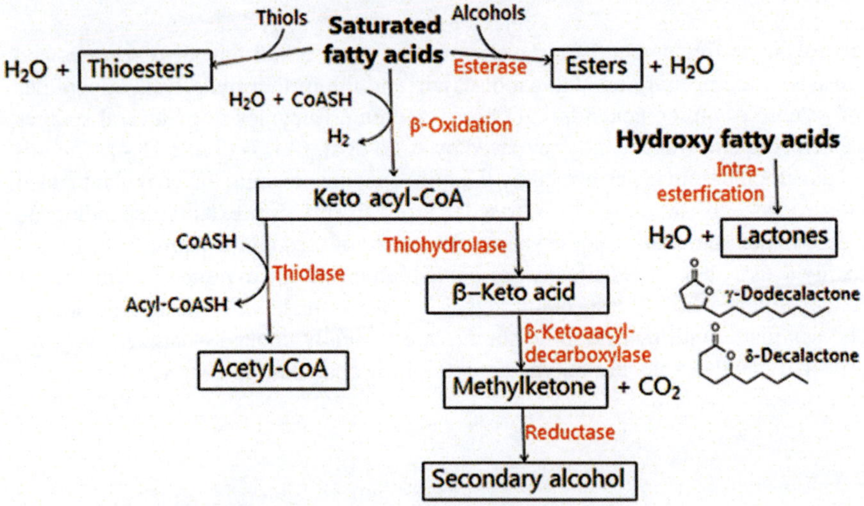

Fig. 15.5 Catabolism of fatty acids in chcese. CoASH Coenzyme A

water. Thioesters are formed when fatty acids react with thiols, and contribute to the cooked cabbage and cauliflower flavours of surface ripened cheese.

Lactones are formed by an intramolecular esterification of hydroxy fatty acids, which releases water. The stable γ- and δ-lactones have been found in cheese (Fig. 15.5), and contribute to caramel and coconut flavours.

15.4.3 Methods to Accelerate Lipolysis

Using raw milk with active LPL increases the level of lipolysis in cheese significantly. Thermisation of the cheese milk at a somewhat lower temperature than is used for pasteurisation, results in a higher content of active LPL and thereby accelerated lipolysis. Homogenization of cheese milk also accelerates lipolysis dramatically because of higher access of lipases to the fat. This technology is used for production of the blue cheese Danablu, and lipolysis, catalyzed by the milk lipase as well as mould lipases during ripening is increased significantly (Waagner Nielsen and Edelsten 1974). Yeast strains selected for high lipolytic activity, especially *Yarrowia lipolytica,* are used in production of white and blue mould cheese (Fickers et al. 2011). Pregastric animal lipases are added to increase lipolysis in parmesan cheeses made from pasteurised milk (reviewed by Kilcawley et al. 2002).

15.5 Proteolysis

15.5.1 Proteolytic Activities in Cheese

Proteolysis is fundamental for development of texture and flavour of cheese, and consists of the primary breakdown of casein and the further hydrolysis of peptides and release of amino acids during cheese ripening. Enzymes from several sources are involved and was recently reviewed by Ardö et al. (2017) (Table 15.4). The initial attack on casein is performed by the milk enzyme plasmin, the coagulant added and proteases of thermophilic starter bacteria, to different extents depending on cheese manufacture and cheese variety. Proteases of mesophilic starter bacteria are mainly active on peptides already present at high levels due to proteolysis mainly of the coagulant, which is especially active in cheese made with mesophilic starter. Cheeses made with thermophilic starters are typically cooked at higher temperatures that inactivate the proteolytic activity of the coagulant in cheese.

Table 15.4 Enzymes involved in cheese proteolysis

Enzyme	Origin	Preferred substrate	Primary products in cheese
Plasmin	Milk	β-CN	β-CN(f29–207 (γ_1), 105–209(γ_2), 107–209(γ_3)); β-CN(f1–28, 1–105, 1–107, 29–105, 29–107)
		α_{s2}-CN	α_{s2}-CN(f1–21; f22–207)
Chymosin	Rennet	α_{s1}-CN β-CN	α_{s1}-CN(f1–23, f24–199, f24–101) β-CN(f193–209)
Lactocepin	LAB cell wall	α_{s1}(f1–23)	α_{s1}-CN(f1–16, 1–14, 1–13, 1–9, 1–8, 1–7. 1–6); genetic variation
		β-CN(f1–28)	β-CN(f1–6, 7–28)
Transport enzymes	Cell membranes	Peptides outside bacterial membrane	Peptides inside bacterial membrane
Peptidases	Inside LAB cells	Peptides	Amino acids and small peptides
Lyases	Microbes	Cell membrane CH and peptides	CH, peptides, amino acids
Phosphatases	Milk and microbes	Phosphopeptides	Free phosphate and peptides

CH carbohydrate, *CN* casein, *LAB* lactic acid bacteria

15.5.2 Indigenous Milk Proteases

Activity of only two of the several proteases of milk origin has been detected in cheese, of which plasmin is the most important. The somatic cells lysosomes, which increase in number in milk during udder infection, contain several proteases and protease activators; limited activity of only one of those enzymes, cathepsin D has been shown in cheese, i.e., in rennet-free cheese from ultrafiltered milk (Larsen et al. 2000). The specificity of cathepsin D is similar to that of calf rennet and it has no significant influence in cheeses with active rennet.

Plasmin activity contributes to a shorter cheese structure and also to a deeper ripening if the peptides that are released become available for microbial aminopeptidases, which increase the amount of free amino acids. Free amino acids bind water molecules and thereby decrease water activity, and they are precursors to flavour compounds. Excessive plasmin activity in cheese milk, however, may decrease cheese yield (van Boekel and Crijns 1994).

The plasmin system comprises three components, i.e., the inactive precursor plasminogen, a plasminogen activator (PA) and a plasminogen activator inhibitor (PAI). PA is a proteolytic enzyme that splits a part off from the plasminogen peptide chain to form active plasmin, and the PAI inhibit this reaction to prevent detrimental proteolysis. Several inhibitors of plasmin also exist in milk. Cheese-making influences the plasmin system in significant ways. PAI is the most heat-sensitive of the three components and is inactivated at moderate to high cooking temperatures, which initiate the activation of plasminogen to plasmin. The plasmin inhibitors of the milk are mainly excluded with the whey, while the components of the plasmin system are highly associated with the casein micelles and retained in the cheese curd. Plasmin has long been known to be important in ripening of cheeses made with high cooking temperatures, which inactivate rennet during cheese manufacture, e.g., Emmental, Gruyère, Fontina, Parmigiano Reggiano and Grana Padano (Collin et al. 1987; Ollikainen and Kivelä 1989; Battistotti and Corradini 1993). Plasmin activity at lower rate has also been demonstrated in Cheddar (Creamer 1975; Farkye and Fox 1990), Danbo (Madsen and Ardö 2001), Herrgård (Antonsson et al. 2003a; Ardö et al. 2007) and Västerbottensost (Rehn et al. 2010).

Plasmin has activity optimum at 37 °C and pH 7.5 and is active in cheese milk and curd during the first day before temperature has decreased and a minimum pH at about 5.0–5.3 is reached. Below pH 5.2, hardly any plasmin activity has been shown to occur (Larsson et al. 2006). During ripening, the pH increases as a result of proteolysis, and plasmin becomes active again. In cheese varieties with specific ripening bacteria, e.g., surface microflora that consume lactic acid or produce ammonia, pH may increase to over 7.0. Increases in the level of amino acids released have been related to increased plasmin activity in semi-hard cheese (Barrett et al. 1999; Antonsson et al. 2003a).

Plasmin is an alkaline serine protease with preference for bonds of the type Lys-X and acts especially on β- and α_{s2} -casein, while κ-casein appears to be resistant (Bastian and Brown 1996). β-Casein has several potential plasmin-susceptible bonds, but only three are hydrolysed at significant rates in cheese, namely Lys_{28}-Lys_{29}, Lys_{105}-His_{106} and Lys_{107}-Glu_{108}. Hydrolysis of those bonds results in the release of γ-caseins and proteose-peptones (Table 15.4). Cleavage of the bond Lys_{21}-Gln_{22} in α_{s2}-casein has also been confirmed in cheese (Rehn et al. 2010; Møller et al. 2012b); however, cleavage of other sites in α_{s2}-casein has so far not been observed in cheese. α_{s1}-Casein is less susceptible to hydrolysis by plasmin, and only Lys_{79}-His_{80} has been confirmed to be cleaved in cheeses with relatively high plasmin activity; this cleavage is involved in formation of the accumulating peptide α_{s1}-CN (f80–199) (Masotti et al. 2010).

15.5.3 Coagulants

Choice of coagulant has significant impact on cheese proteolysis, and influences development of structure and flavour. The enzymes traditionally used to coagulate milk are extracted from stomachs of young milk animals, and they are specific in splitting off the hydrophilic carboxylic end of κ-casein to release the glucomacropeptide (GMP) and form *para*-κ-casein. They cleave the Phe_{105}-Met_{106} bond in κ-casein at a many times higher rate than any other bond in casein. Other coagulants are less specific and the consequences for cheese ripening have only poorly been investigated. New coagulants are being evaluated and developed because of shortage of the traditional ones or attempts to change their properties, e.g., make them more efficient as coagulants, increasing cheese yield or decreasing enzyme cost (Table 15.5).

Table 15.5 Overview of commercially used coagulants

Active protease	Origin	Advantages
Bovine chymosin and pepsin	Calf rennet	Traditional; specific to cut off GMP
Bovine chymosin B	FPC, *Aspergillus niger*	Cheap, clean, halal, kosher
Goat, lamb, buffalo chymosin	Goat, lamb, buffalo rennet	Traditional
Camel chymosin	FPC, *Aspergillus niger*	Efficient, high cheese yield
Cardosin A and B	*Cynara cardunculus*	Plant rennet, artesian cheese
Endothiapepsin	*Cryphonectria parasitica*	Cheap, vegetarian
Mucorpepsin	*Rhizomucor miehei*	Cheap, vegetarian

FPC Fermented-produced chymosin, *GMP* glucomacropeptide, κ-casein (f106–169)

The specific activities of the coagulants on casein bonds in cheese differ as described below and reviewed by Ardö et al. (2017).

Chymosin has its preferred primary attack on *para*-casein in cheese at the α_{s1}-casein bond Phe_{23}-Phe_{24}. This cleavage contributes to the initial softening of cheese texture, as does further breakdown of α_{s1}-CN (f24–199), as well as changes in calcium equilibration occurring at the same time. The bond Phe_{23}-Phe_{24} is cleaved at a considerably higher rate than any other bond in α_{s1}-casein, and the next bond to be cleaved in cheese at an initial rate about 1000 times less is Leu_{101}-Lys_{102} (Møller et al. 2012a). The rate at which the different peptide bonds are cleaved depends on the ionic strength and pH, particularly valid for Leu_{101}-Lys_{102}, which is cleaved faster at lower cheese pH (Mulvihill and Fox 1977). α_{s2}-Casein is relatively resistant to proteolysis by chymosin in cheese.

Chymosin cleaves β-casein most rapidly at the peptide bond Leu_{192}-Tyr_{193}, and the resulting bitter tasting peptide β-casein (f193–209) has been identified in cheese, especially those with a low salt content. NaCl inhibits hydrolysis to an extent that depends on pH, and hydrolysis is strongly inhibited by 5% and completely by 10% NaCl. β-Casein is also cleaved at a slower and NaCl-dependent rate at Ala_{189}-Phe_{190}, Leu_{165}-Ser_{166} and Gln_{167}-Ser_{168} (Møller et al. 2012b).

Para-κ-casein has several potential chymosin cleavage sites; however, its hydrolysis is limited in cheese. Only the κ-casein peptide (f96–105) has been isolated from Cheddar, Feta and a cheese model made to study chymosin activity. The amount of this peptide increased with decreasing salt content (Michaelidou et al. 1998; Møller et al. 2012b; Risborg and Witt 2013). The results indicate that the influence of salt on chymosin activity in cheese is due to differences in the substrate conformation.

Rennets of other young milk animals than calves are also used to coagulate milk. Some of their properties differ somewhat from those of bovine rennet, which introduces possibilities to manipulate cheese manufacture and ripening. Coagulants that are typically extracted from the stomachs of the young goats or sheep are mainly used in artisanal dairies in the Mediterranean area. Their role in cheese proteolysis is generally less understood than it is for bovine rennets. Goat chymosin is more heat stable and has a somewhat higher proteolytic activity in white-mould cheese than bovine chymosin. Recombinant chymosin has been made of lamb, kid, buffalo and camel chymosin, and some are available on the market. Recombinant ovine chymosin has been shown to be similar to bovine chymosin when it comes to proteolytic specificity and dependence on pH; however, it has lower temperature optimum and lower stability at 45 °C (Rogelj et al. 2001). Using an artificial casein substrate (fluorescein isothiocyanate-labelled casein) to estimate general proteolytic activity, Vallejo et al. (2012) found temperature optima of recombinant buffalo, cow, camel and goat chymosin to be at 37, 40, 40 and 45 °C, respectively. Camel chymosin was shown to be the least sensitive to temperatures used during cheesemaking.

Camel chymosin has a 20% lower general proteolytic activity than bovine chymosin at a 70% higher milk-clotting activity on bovine milk, which results in limited activity in cheese during ripening (Kappeler et al. 2006). Camel chymosin prefers the same cleavage sites as bovine chymosin in cheese with highest activity at the α_{s1}-casein bond Phe_{23}-Phe_{24}. It is hydrolysed at a 100 times higher rate than the next bond to be cleaved, Leu_{101}-Lys_{102} (Møller et al. 2013). As for bovine chymosin, the preferred cleavage site in β-casein is Leu_{192}-Tyr_{193}; however, this is hydrolysed at a much lower rate (Møller et al. 2012a). Cleavage rates of the bonds in the hydrophobic carboxylic end of β-casein were around 15 times higher for bovine than camel chymosin in buffer solution at pH 5.2, while sensitivity to salt (NaCl) in the range 0–5% was similar, with only 25 and 6% of the activity at 2 and 5% salt, respectively, as compared to activity without salt. These results confirm that the influence of salt on chymosin activity in cheese depends on differences in the substrate conformation rather than enzyme activity. Only 25% of the amount of the bitter peptide β-CN (f193–209) was formed in Cheddar made with camel chymosin as compared to cheese made with bovine chymosin (Børsting et al. 2012). Replacing bovine with camel chymosin, consequently, decreases the risk for bitterness. A low rate of primary proteolysis, though, may be detrimental to the cheese body, which typically becomes harder.

Bovine pepsin is present in high quality calf rennet at a concentration of about 10%; however, several bovine rennets contain as much as about 50 to 80%. Bovine pepsin has been shown to be 55 times more proteolytic on a synthetic hexapeptide than chymosin at equal clotting activity (Martin et al. 1981). Bovine pepsin is also more sensitive to pH variation in milk during gelling and more active than chymosin at the low pH of cheese during ripening (Andrén and von Reedtz 1990). Casein is hydrolysed in the same areas by bovine pepsin and chymosin; however, the pepsin is much less specific (Mulvihill and Fox 1979). Risborg and Witt (2013) identified peptides released by bovine chymosin and by calf rennet with a pepsin/chymosin ratio of 70/30 using a cheese model. They found that both coagulants hydrolysed casein in hydrophobic parts especially in the carboxylic half of the β-casein chain; however, pepsin cleaved at the double amount of sites than chymosin did.

Cardosin, which is extracted from the flowers of the thistle *Cynara cardunculus,* is the most common plant coagulant which is used mainly in the manufacture of cheeses from ovine milk in Portugal and Spain (Sousa et al. 2001). Ovine milk contains a higher concentration of casein than bovine milk, which sets limits for the release of bitter peptides. Cardosins may therefore possibly be useful in production of cheese from concentrated bovine milk. Extracts from *C. cardunculus* contain two aspartic proteases, cardosin A and cardosin B. Specificies and kinetics for cardosin A are similar to that of chymosin, and for cardosin B to pepsin, and enzymes from extracts of the dried flowers cleave the Phe_{105}-Met_{106} bond of κ-casein. The primary cleavage site in bovine α_{s1}-casein is also similar to that of bovine chymosin, i.e., Phe_{23}-Phe_{24}. *C. cardunculus* extract cleaves all the bonds in the hydrophobic bovine β-casein sequence Ala_{189}-Phe-Leu-Leu-Tyr_{193}, and at least three of the bonds were also cleaved in cheese by purified cardosin A, as well as several other bonds, which increases the risk for development of bitterness (Macedo et al. 1996; Hedlund and Ludvigsen 2005). Use of plant rennet in cheese-making has a history back to ancient times; however, their use today is limited because of a low ratio of milk clotting to proteolytic activity, low cheese yield and expensive extraction procedures.

Microbial proteases have been investigated as alternatives to bovine coagulants, and those from the moulds *Rhizomucor pusillus*, *R. miehei* (mucorpepsin) and *Cryphonectria parasitica* (endothiapepsin) are commercially available under various trade names. The activity in cheese, however, has the potential to release bitter-tasting peptides that accumulate during ripening. The gene for mucopepsin has been cloned into *Aspergillus oryzae*, resulting in coagulants with less detrimental proteolytic side activities, which has been found to be acceptable for use in Cheddar (Chen et al. 1994). Endothiapepsin has considerably more general proteolytic activity than bovine chymosin, especially on β-casein (Rea 1997). It is only used successfully for high-temperature cooked cheeses, in which it is denatured during cheese-making.

The specificities of chymosin and the fungal proteases from *R. miehei* and *C. parasitica* on caseins are quite different. The primary cleavage of κ-casein mediated by endothiapepsin is at Ser_{104}-Phe_{105}, rather than Phe_{105}-Met_{106} (Drohse and Foltmann 1989). During experiments with casein hydrolysis in milk at 30 °C for 24 h, the number of peptides released by mucopepsin, endothiapepsin and chymosin were 38, 95 and 9, respectively, while in a cheese model the numbers were 22, 15 and 21, i.e., decreased number for the microbial coagulants and increased for chymosin (Risborg and Witt 2013). The highest number of peptides released was from β-casein for endothiapepsin, while for mucopepsin and chymosin it was from α_{s1}- and α_{s2}-casein.

15.5.4 *Lactic Acid Bacterial Protease: Lactocepin*

After a long-time existence in milk and dairy environment, lactic acid bacteria have lost the ability to synthesize several amino acids. Their proteolytic system, consisting of the protease lactocepin, peptide transport enzymes and intracellular peptidases, releases amino acids from casein to be used in their metabolism. Lactocepin is a multi-domain, cell-wall anchored protease that hydrolyses casein and large peptides. It prefers peptides released by plasmin or the coagulant, of which there is abundance in most cheeses during ripening. Only in high-temperature-cooked cheeses with the coagulant inhibited during production has lactocepin of the thermophilic starter bacteria *Lactobacillus* been shown to attack casein. Lactocepin has been characterized biochemically and genetically in a number of strains of *Lc. lactis* (PrtP), *Lactobacillus* (PrtP, PrtH or PrtB) and *St. thermophilus* (PrtS) (Tan et al. 1993; Siezen 1999; Fernandez-Espla et al. 2000; Smeianov et al. 2007).

Lactocepin is homologous with the subtilisin family of serine proteases and has a catalytic domain consisting of the active site triad of Asp, His and Ser. A region of 107 amino acids including the active site and the substrate-binding region is highly conserved. Lactocepins of *Lactococcus* were initially classified into two broad groups, P_I- and P_{III}-type proteases. P_I-type enzymes degrade β-casein rapidly but act only slowly on α_{s1}-casein, whereas P_{III}-type proteases hydrolyse β-casein differently and hydrolyse also α_{s1}- and κ-caseins efficiently. Alteration of a few amino acid residues in the peptide chain of lactocepin, however, changes its specificity. Exterkate et al. (1993) and Broadbent et al. (2006) therefore proposed a classification scheme for lactocepin with eight groups (a-h) based on their specificity mainly on the peptide α_{s1}-CN (f1–23). Commonly, some of the bonds Leu_{16}-Gln_{17}, Gln_{13}-Glu_{14}, Gln_9-Glu_{10} and His_8-Gln_9, are hydrolysed. *Lc. lactis* strains of group d are commonly used in cheese starters, and this group is further divided into three sub-groups from differences in their amino acid sequences with influence on their specific hydrolysis patterns of α_{S1}- and β-casein in milk (Børsting et al. 2015).

Cell envelope-associated proteases (CEP) with properties similar to the *Lactococcus* protease lactocepin have been isolated from a number of strains of *Lactobacillus* (reviewed by Ardö et al. 2017). Only one lactocepin gene has been found in each *Lc. lactis* strain; however, a higher variation has been found in *Lb. helveticus* with up to four different genes for lactocepin (PrtH, PrtH2, PrtH3 and PrtH4) in the same strain (Broadbent et al. 2011). A large degree of strain variation has been shown in the activity of whole *Lb. helveticus* cells on intact α_{s1}-casein (Jensen et al. 2009). Only three of six dairy strains had high proteolytic activity, and they were shown to have genes for PrtH2 or PrtH4, as well as PrtH3, a gene which was present in all the strains tested and has been found to be the most common gene among *Lb. helveticus* strains, often in combinations with some of the other known lactocepin genes (Broadbent et al. 2011). The first peptide bonds cleaved in the peptide α_{s1}-casein (f1–23) was shown for different strains of *Lb. helveticus* to be Leu_{16}-Gln_{17}, Gln_{13}-Glu_{14} and Gln_9-Glu_{10} or His_8-Gln_9, while all the strains tested also cleaved Ile_6-Lys_7 at a slower rate (Oberg et al. 2002).

Mesophilic *Lactobacillus* species that commonly enter cheese uncontrolled *via* milk or production environment such as *Lb. casei*, *Lb. plantarum* and *Lb. curvatus* generally do not grow well in milk, due to lack of cell-wall protease activity. In cheese they use the pool of peptides already prepared by starter bacteria.

Leuconostoc lactis and *Leu. mesenteroides* used in DL-starters to produce CO_2 and butter-like aroma are characterized by slow growth in milk, and they lack general protease activity. A serine-like protease gene has, however, been found in *Leu. pseudomesenteroides*, which shares some homology with CEP proteases of *Lactococcus* and *Lactobacillus* (Frantzen et al. 2017). Any role of this *Leuconostoc* protease in cheese ripening is not known.

Variation in the activity of lactocepin may be obtained in a cheese by choice of starter. So called bitter starters have an unsuitable balance between activity of lactocepin to release bitter peptides and activity of intracellular aminopeptidases with the potential to break down these peptides. Bitter starters have generally been omitted from cheese production.

15.5.5 *Bacterial Cell Membrane Transport Systems*

The peptides released by coagulant, plasmin and lactocepin outside the bacteria cell membranes need to be transported into the cell and the intracellular peptidases for amino acids to be released. Different transport systems exist for amino acids, di- and tripeptides and oligopeptides. The oligopeptide transport system is essential for growth and takes up generic peptides with 4–18 amino acids (Juillard et al. 1995). The *pepO* gene that expresses an intracellular endopeptidase is located immediately downstream of the genes for the oligopeptide transport system, indicating that the two systems are physiologically linked (Tynkkynen et al. 1993).

15.5.6 *Peptidases*

Several intracellular peptidases hydrolyse peptides that have been released from casein by the coagulant, plasmin and lactocepin in cooperation and transported across the cell membrane. Amino acids have very different properties, and a series of more or less specific peptidases is needed to release them all. Lactic acid bacteria of different species have different combinations of peptidases and the variation between strains of the same species is high (Table 15.6).

Two aminopeptidases with a broad specificity, PepN and PepC, are common in lactic acid bacteria. The amino acid Pro has a secondary amino acid amino group that makes a bend in the peptide chain, which increases flavour intensity of bitter peptides, and it is important in cheese ripening to hydrolyse such peptides. Casein has a high density of Pro, and to get access to also other amino acids lactic acid bacteria need to get rid of Pro from the peptide chain. Different strategies are chosen: *Lactococcus* have several Pro-specific di- and aminopeptidases, while thermophilic *Lactobacillus* typically have high activity of X-prolyl-dipeptidyl-aminopeptidase (PepX) that releases a dipeptide from the amino end of a peptide with the second amino acid being Pro. Several aminopeptidases have high activity on Lys, which typically accumulates in cheese during ripening; however, the same enzymes release also the similar amino acid Arg that for some bacteria is useful as energy source (Ardö et al. 2002).

A high strain variation has been shown in the specificity and activity of aminopeptidases in lactic acid bacteria. In *Lb. helvetcus* multiple genes for several aminopeptidases have also been shown to exist (Jensen and Ardö 2010; Broadbent et al. 2011).

Table 15.6 Aminopeptidases involved in cheese ripening (summarized from Ardö et al. 2017)

Enzyme	Used substrate	Class	Microrganisms
Endopeptidases			
PepO	α_{s1}-CN (f1–23)	M	*1, 2, 3, 5, 7*
PepE	α_{s1}-CN (f1–23), β-CN (f193–209)	T	*7*
PepF	Phe-Ser bond in Bradykinin; 7–17 aa-peptides	M	*7, 10*
Tripeptidases			
PepT	Broad specificity, Met-Gly-Gly, Leu-Leu-Leu, Leu-Gly-Gly	M	*1, 2, 5, 7, 13*
Dipeptidases			
PepD	Gly-Tyr	T	*7*
PepQ, prolidase	Leu-Pro, X-Pro	M	*2, 5, 6, 10*
PepR, prolinase	Pro-Leu, Pro-βNA, Leu-βNA, Phe-βNA	T/S	*7, 9*
PepV	Broad specificity, Leu-Leu, Leu-Gly, β-Ala-Ala	M	*1, 2, 5, 6, 7, 10,12, 13*
X-Prolyl dipeptidyl aminopeptidases			
PepX	Gly/Ala-Pro-pNA, Arg/Lys-Pro-pNA	S	*1, 2, 4, 5, 6, 7, 8, 10, 12*
Proline iminopeptidase			
PepI	Pro-Gly-Gly Pro-pNA Pro-AMC	M S T	*2* *5* *6, 7, 10*
Aminopeptidases			
PepA	Glu-pNA, Asp-pNA	M	*1, 2, 3*
PepC	Broad specificity, Leu-Gly-Gly, Leu-pNA, His-βNA	T	*1, 2, 3, 7, 10*
PepL	Leu-βNA, Leu-pNA	–	*5, 12, 13*
PepN	Broad specificity, Lys-pNA, Leu-pNA	M	*1, 2, 3, 5, 6, 7, 8, 11, 12*
PepP	X-Pro-Pro-, X-Pro-X-, Leu-Pro-Pro	M	*1, 2*
PepS	Arg-pNA	M	*3*
PCP	Pyr-Glu-pNA	S	*2*

Class of enzyme: *M* metallo-peptidase, *T* thiol-peptidase, *S* serine-peptidase. *PCP* pyrrolidone carboxyl peptidase

Substrate: *βNA* β-nitroanilide, *pNA p*-nitroanilide, *AMC* Aminomethylcoumarin, *aa* amino acids, *pyr-Glu* pyroglutamyl

Microorganism: *1 Lactococcus. lactis* ssp. *lactis*, 2 *Lc. lactis* ssp. *cremoris*, *3 Streptococcus. thermophilus*, *4 St. macedonicus*, *5 Lactobacillus delbrueckii* spp. *bulgaricus*, *6 Lb. delbrueckii* ssp. *lactis*, *7 Lb. helveticus*, *8 Lb. acidophilus*, *9 Lb. rhamnosus*, *10 Lb. casei/paracasei*, *11 Lb. plantarum*, *12 Lb. curvatus*, *13 Lb. sake*

Specific peptides accumulate typically in different cheese varieties during ripening as a result of the presence or absence of specific peptidase activities (for review see Ardö et al. 2017).

15.5.7 Phosphatases

Phosphatases present in cheese may remove phosphate groups from phosphopeptides and facilitate their further breakdown. One alkaline and one acid phosphatase originate from milk, and the first is mainly important in cheeses made from raw milk because of heat inactivation during pasteurisation. In cheeses made from pasteurised milk, peptides with the phosphate groups from casein still intact accumulate during ripening, while they are hydrolysed to a larger extent in long-ripened cheese from raw milk (Lund and Ardö 2004; Ardö et al. 2009). Acid phosphatase activity in 12 months old Cheddar has been shown to mainly be of milk origin (Andrews and Alichanidis 1975). Microbial phosphatases may also be involved in dephosphorylation of peptides, especially in blue cheese.

Taivosalo et al. (2017) claimed that they found peptides derived from casein but with phosphate groups on serine residues that are not known to be phosphorylated in casein. Phosphatases may possibly have acted in a similar way in cheese as some esterases that reverse their activity in environments with very low water activity and contribute to ester formation instead of release of fatty acids in long-time ripened cheeses.

15.5.8 Synthesis of Specific Glu-Peptides

Several glutamyl peptides, which have an amino acid sequence without similarity to any part of casein, have been identified in ripened cheese. Roudot-Algaron et al. (1994) identified γ-Glu-Phe, γ-Glu-Tyr and γ-Glu-Leu in Comté. These peptides were also synthesised and tasted; γ-Glu-Tyr was sour and salty, whereas γ-Glu-Phe had a more intense and complex taste described as brothy, slightly sour, salty and metallic. Andersen (2009) found Glu-Phe and Glu-Tyr in Comté and Parmigiano Reggiano as well as Glu-Phe and Glu-His in Dubliner and Västerbottensost. Sforza et al. (2009) found accumulation of γ-Glu-Phe, γ-Glu-Tyr, γ-Glu-Met, γ-Glu-Ile, γ-Glu-Leu and γ-Glu-Val during ripening of Parmegiano Reggiano. The milk enzyme γ-glutamyl transferase or γ-glutamyl transpeptidase (GGT) may have a role in the synthesis of γ-glutamyl-peptides during cheese ripening. It is only partly inactivated at pasteurisation temperatures. Of its activity, 67% has been shown to remain after pasteurisation at 70.5 °C for 24 min, and 35% on slightly increasing temperature to 74 °C (Ardö et al. 1999). Quantitation of γ-glutamyl peptides in several cheese varieties and analysis of enzyme activities proposed milk GGT to be the main enzyme that catalyzes the generation of these taste-enhancing peptides

(Hillmann et al. 2016). N-lactosyl-amino acids have been suggested to be synthesized by the same GGT enzyme as the γ-Glu-peptides (Sforza et al. 2009).

The non-native amino acid pyroglutamic acid has been reported present in Parmigiano-Reggiano and other hard cheeses and is suggested to be derived from an enzymatic reaction that catalyses the internal cyclization of glutamic acid (Mucchetti et al. 2000). Pyroglutamic acid may be formed in peptides with Glu in the amino end, and pyro-γ_3-casein has been shown to accumulate in extra-hard cheese during long ripening times.

15.5.9 Acceleration of Proteolysis in Cheese

Several attempts to accelerate proteolysis and decrease expensive long storage times by adding proteolytic enzymes to cheese milk have been made over time; however, with success only for short-time ripened cheese targeted towards production of processed cheeses. The simple explanation is that the characteristic proteolysis for each cheese variety involves several different enzymes, which need to be balanced to reach the desired quality. Activity of exogenous proteases may be difficult to terminate in cheese, and commonly proteolysis is increased too much, which results in a soft and weak cheese structure and accumulation of bitter peptides.

Cheddar and semi-hard cheese have successfully been made with addition of the protease Neutrase in combination with aminopeptidase preparations (Law and Wigmore 1983; Ardö and Pettersson 1988). Peptidase preparations are extracted from microbial cell cultures and purchased on the market for addition to cheese milk or curd. These preparations are mainly made by moulds that produce aminopeptidases extracellularly, e.g., *Aspergillus* but also occasionally from *Lc. lactis* (Kilcawley et al. 2002). They are sold as general and debittering aminopeptidase preparations, and they were shown to also have a variation in side activities of proteases, phosphatases and glycosidases. A few also showed esterase and lipase activity. Exogenous enzymes are not easy to add without large losses to the whey. Encapsulation techniques that are used in medical applications have been tested with some success; however, they are quite expensive, and a considerable part of the enzymes may be lost during the encapsulation process (Skeie et al. 1997).

A more successful way to increase enzyme activities in cheese is to use attenuated bacterial cells. Heat- and freeze-shocking techniques were first tested to inhibit acidification during cheese-making while keeping ripening enzymes intact in cells added in high numbers (Bie and Sjöström 1975a, b; Pettersson and Sjöström 1975). The technique was further developed by making a well-defined heat treatment that inactivates acidification as well as general proteolytic activities while keeping the aminopeptidase activities intact (Ardö and Pettersson 1988; Ardö et al. 1989). A *Lb. helveticus* preparation was successfully made, because of a suitable interval between the heat load needed to inactivate proteases and aminopeptidases, respectively. Following this success, different *Lb. helveticus* cultures, which are selected for high aminopeptidolytic activities and low or no caseinolytic activity, have entered the market and are used successfully in some cheeses.

Different strategies to increase plasmin activity in cheese have been tested. Experiments with plasmin producing recombinant starter bacteria failed, because a procedure to prevent detrimental plasmin activity within the bacteria cells was not developed (Arnau et al. 1997). Addition of plasminogen activators into the cheese body has been shown to be difficult to perform, and the precursor plasminogen and its activator may not be close enough in the cheese body to interact (Benfeldt 1997). There is typically seven to ten times more plasminogen than active plasmin in fresh milk, and the amount is sufficient also in pasteurised milk to activate what is needed in cheese ripening. It is most efficient to perform the activation in the cheese milk, and it may be performed with bovine urokinase or the cheaper enzyme streptokinase produced by *Streptococcus uberis* or by recombinant *Lc. lactis* with the streptokinase gene incorporated (Barrett et al. 1999; Upadhyay et al. 2006). Care must be taken, however, to find the right time point for the activation of plasminogen and inactivation of the added enzymes to avoid losses of nitrogen compounds to the whey and thereby decreased cheese yield.

Acceleration of microbial activities in cheese during ripening obtained by adding ripening cultures with defined properties is beyond the scope of this chapter.

15.6 Amino Acid Catabolism

Cheese bacteria release amino acids from casein not only for their protein synthesis, but also as energy source and for synthesis of basic metabolites. Cheese flavour is highly dependent on catabolism of amino acids to characteristic aroma compounds. Both pleasant and unpleasant aroma compounds are formed (Urbach 1993). The first attack on an amino acid is performed by one of the enzymes of five groups, as reviewed by Ardö (2006) and illustrated in Fig. 15.6. Methods to influence amino acid catabolism involve use of selected ripening cultures or attenuated bacteria cells.

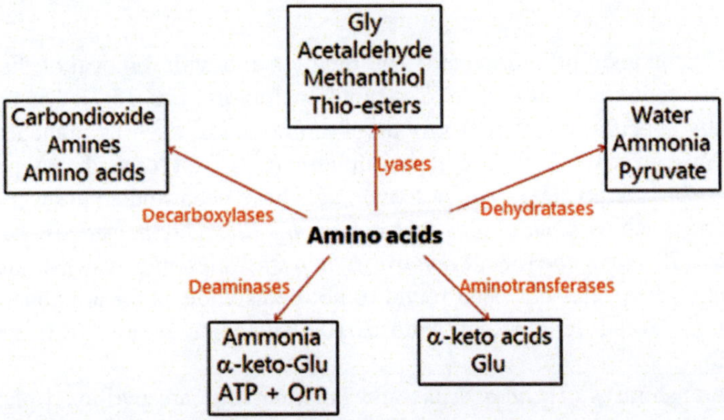

Fig. 15.6 Enzymes mediating the first step of amino acid catabolism in cheese ripening. Gly glycine, Glu glutamic acid, Orn Ornithine, ATP adenosine triphosphate

15.6.1 Lyases

Threoninaldolase catabolises Thr into Gly and acetaldehyde, which is a typical component in yoghurt flavour. This enzyme is active in *Lb. delbrüchi* subsp. *bulgaricus* and *Str. thermophilus*. Cystathionine-β-lyase, cystathione-γ-lyase and methionine-γ-lyase are active in *Brevibacterium* and involved in formation of volatile sulphur compounds with a high impact on cheese flavour. Methantiol, CH_3-SH, is released from Met and may be further metabolised to different flavour-intense sulphur compounds, chemically or enzymatically. Methanthiol also react with carboxylic acids and form thioesters mediated by acyltransferases or esterases (Weimer et al. 1999).

15.6.2 Dehydratases

Dehydratases deaminate amino acids that have an OH- or SH- group, i.e., Ser, Thr and Cys. The OH- or SH- groups make oxidation to α-ketoacids possible without using other oxidised compounds, which are highly limited in the anaerobic environment of the interior of cheese during ripening. First formed from Ser are water and aminoacrylate, which is unstable and reacts spontaneously with water to form the central metabolite pyrvate and NH_3. The α-ketoacid formed from Thr is α-ketobutyrate and from Cys pyruvate and, in this latter reaction, H_2S is also released. The free ammonium formed increases pH and influences cheese flavour. Examples of bacteria using these pathways in cheese are *Brevibacterium linens* (Weimer et al. 1999) and *Lb. plantarum* that use serine dehydratases (Antonsson et al. 2003b).

15.6.3 Aminotransferases

In cheese, catabolism of amino acids commonly starts with removal of the amino group by the redox neutral activity of aminotransferases (Fig. 15.7). The enzymes are typically specific to different groups of amino acids, such as branched-chain (Leu, Ile, Val), aromatic (Phe, Tyr, Trp), sulphuric (Cys, Met) or acidic (Asp) (Yvon and Rijnen 2001; Ardö 2006). Aminotransferases have been studied in several lactic acid bacteria, such as *Lactococcus lactis*, *Lactobacillus* (*Lb. helveticus, Lb. casei, Lb. danicus*), *Leuconostoc* (*Leu. cremoris, Leu. pseudomesenteroides*); strain variation within each species has been found to not be as wide as for aminopeptidases (Thage et al. 2004; Jensen and Ardö 2010; Broadbent et al. 2011; Pedersen et al. 2013).

Aminotransferases of *Lactococcus* and *Lactobacillus* are pyridoxal phosphate (PLP)-dependent and use α-ketoglutaric acid as amino group acceptor to produce Glu. The amino acids are converted into α-keto acids that may be further catabolized into several aroma compounds, including carboxylic acids, aldehydes and

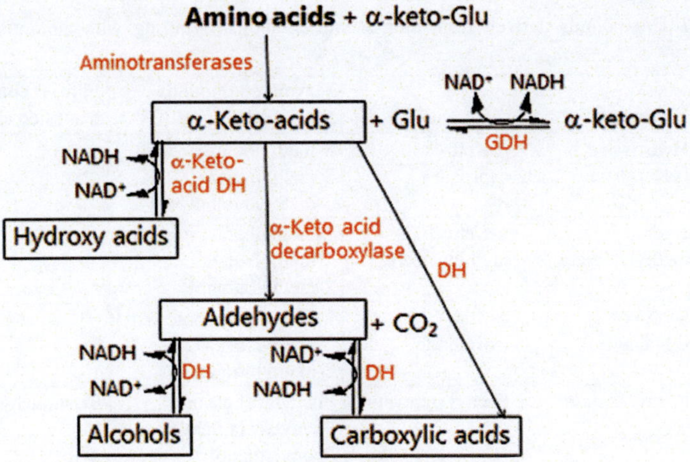

Fig. 15.7 Catabolism of amino acids starting with aminotransferase activity. DH dehydrogenase, GDH Glu dehydrogenase (revised from Ardö 2006)

alcohols. They may also be converted into hydroxy acids that have not been shown to contribute to flavour. The cheese flavour compounds are derived from branched-chain and aromatic amino acids, Met and Asp (Table 15.7).

The aldehydes formed by activity of α-keto acid dehydrogenase may be oxidised to carboxylic acids or reduced to alcohols, which give the microorganism opportunities to reduce or oxidise the cofactor NAD^+/NADH, whatever is needed in their metabolism. Activity of α-keto acid dehydrogenase is limited in many *Lactococcus* strains of dairy starters, which could be explained by the history of these cheese starters that once were selected to avoid malty flavour originating from amino acid catabolism in butter. It has been suggested that a dehydrogenase complex may convert α-keto acids to the corresponding carboxylic acids, which then are converted into aldehydes and alcohols.

The compounds derived from branched-chain α-keto acids typically contribute to flavours such as rotten fruit, sweat, pig sty, green, malt and chocolate, while those from aromatic amino acids contribute to floral and fruity flavour notes. Catabolism of sulphuric amino acids contributes to flavours like boiled potatoes, cabbage, cauliflower and garlic. Catabolism of side chains of aromatic amino acids form compounds tasting bitter almond, chemical, putrid and faecal.

Activity of α-keto acid dehydrogenases or hydroxy acid dehydrogenases has been studied in *Lb. helveticus*, *Lb. casei* and *Lb. parabuchneri*. *Lb. helveticus* had highest activity on oxaloacetate, about 10 times less on α-keto-butyrate and 100 times less on aromatic α-keto acids and none or only trace activity on branched-chain α-keto acid (Kananen 2013). *Lb. parabuchneri* showed similar high activity on oxaloacetate and α-keto-butyrate and about 20 times less activity on branched-chain α-keto-acids, while *Lb. casei* differed from the others, with 10 times less activity on α-keto-butyrate, less than half of that activity on branched-chain

Table 15.7 Compounds derived from amino acid catabolism starting with aminotransferase activity

Amino acid	Hydroxy acid	α-Keto acid	Flavour compounds from α-keto acids	Flavour compounds from side chains
Leu	Hydroxy-isocaproate	α-Keto-iso-caproate	3-methylbutanoic acid 3-methylbutanal 3-methylbutanol	
Ile	Hydroxy-methylvalerate	α-Keto-β-methylvaleate	2-methylbutanoic acid 2-methylbutanal 2-methylbutanol	
Val	Hydroxy-isovalerate	α-Keto-isovalerate	2-methylpropanoic acid 2-methylpropanal 2-methylpropanol	
Phe	Phenyl-lactate	Phenyl pyruvate	Phenylacetic acid Phenylacetaldehyde Phenylethanol	Benzaldehyde
Tyr	p-OH-phenyl-lactate	*p*-OH-phenyl pyruvate	*p*-OH-phenylacetic acid *p*-OH-phenylacetaldehyde *p*-OH-phenyl-ethanol	p-cresol Phenol
Trp	Indole-lactate	Indole pyruvate	Indol-3-acetic acid Indole-3-acetaldehyde Tryptophol	Skatole Indol
Met	Hydroxy-methyl-thio-butyrate	α-Keto-γ-methylthio-butyrate α-Keto-butyrate[a]	3-methylthio-propionic acid 3-methylthio-propanal 3-methylthio-propanol	Methanethiol DMDS DMTS DMS
Asp	Malate	Oxaloacetate	Diacetyl Acetion	

DMDS dimehtyldisulphid, *DMTS* dimethyltrisulphide, *DMS* dimethylsulphide
[a]Formed from simultaneous removal of the amino group and methanethiol

α-keto-acids and even less activity on oxaloacetate (author's unpublished results). Variation between strains tested of the same species was limited.

15.6.4 Deaminases

Capacity for transamination is dependent on the amount of α-ketoglutaric acid present in the cheese. Lactic acid bacteria that produce Glu dehydrogenase are able to release ammonium, oxidise Glu back to α-ketoglutaric acid, and increase the potential for aminotransferase activity in the cheese. This reaction is, however, dependent on NAD^+ for its oxidation. In cheeses without surface ripening, the oxidation capacity is especially limited.

Lc. lactis ssp. *lactis* produce deiminase and uses Arg to produce ATP, carbamyl phosphate and NH_3 (Christensen et al. 1999). Ammonia increase internal pH and contributes to cheese flavour; the ATP and carbamyl phosphate are used in bacterial metabolism.

15.6.5 Decarboxylases

Cheese bacteria use decarboxylase activity to regulate their internal pH in the acid environment that is created by the lactic acid they produce during growth. The products of carboxylation are mainly amines, and also the amino acids Ala, α-amino butyric acid (AABA) and γ-amino butyric acid (GABA) (Table 15.8). GABA is produced from Glu by glutamate decarboxylase activity, mainly from the surface microflora (Ardö et al. 2002). GABA-producing activity is also known in *Str. thermophilus* and the non-starter lactic acid bacteria *Lb. brevis* and *Lb. parabuchneri*. Decarboxylase activity of *Lc. lactis* strains of starter cultures is occationally found but only activated at low pH. In cheese making experiments that were performed to evaluate the influence of different milk treatments on cheese ripening, a relation was found between the use of raw milk and decarboxylating activities on Glu to GABA, as well as on Tyr to tyramine and His to histamine (Skeie and Ardö 2000). Decarboxylation of Glu is likely to decrease the savoury umami background flavour of cheese to which this amino acid highly contributes. Tyramine contributes to a sharp, numbing sensation that is characteristic for some specific ripened cheese varieties. Amines from decarboxylation of other amino acids contribute to faecal and putrid off-flavours, e.g., cadaverine and putrescine.

Table 15.8 Products of decarboxylase activity on different amino acids

Substrate	Product	Possible sensory effect
Glu	GABA, AABA	Decreased umami
Asp	Ala	Increased sweetness
Tyr	Tyramine	Sharp and numbing
His	Histamine	Sharp and numbing
Orn	Putrescine	Fecal
Lys	Cadaverine	Rotten meat
Trp	Tryptamine	–
Phe	Phenylethylamine	Fishy

GABA γ-aminobutyric acid, *AABA* α-aminobutyric acid

15.7 Breakdown of Microbial Cell Membranes and Cell Lysis

Only limited amounts of amino acids are released by viable bacteria cells, which is not sufficient for ripening of several cheese varieties. The cell membranes need to be more or less disrupted for larger amounts of peptides to migrate uncontrolled by the bacteria and interact with the intracellular peptidases. Autoysis is performed by lytic enzymes expressed by the bacteria in question, or lysis may be induced by bacteriocins formed by other bacteria in the cheese.

Robust starter cultures, which are selected for use in efficient modern cheese production because of a fast and reliable acid-production, do not commonly lyse easily, and amino acid release as well as flavour development may be retarded. This problem may be overcome by completing the starter with addition of ripening cultures with an especially good complement of several different peptidases or with antibacterial properties.

15.8 Nucleic Acid Catabolism

Bacteria cells produce new deoxyribonucleic acid (DNA) as well as ribonucleic acid (RNA) while they are growing and metabolising. The nucleotides are commonly reused and, for this purpose, the cells produce RNA hydrolases that will continue to breakdown RNA also after the cells are lysed and metabolism is no longer under control. Nucleotides that accumulate in cheese are likely contributing to umami flavour. They may also be further catabolized into the carbohydrate ribose that is one of the preferred energy source of some heterofermentative *Lactobacillus* species (e.g., *Lb. danicus*, *Lb. wasatchensis*), which live in close association with starter lactic acid bacteria. They are not always antimicrobial, but they lyse and digest already dead cells (Adamberg et al. 2005; Oberg et al. 2016).

15.9 Catabolism of Vitamins and Other Compounds

Volatile flavour compounds in cheese are successfully analysed with gas chromatography and olfactometry using the human nose as sensor. The compounds that the panellists smell originate mainly from fat and amino acid catabolism, but also from breakdown of lactose, citrate, ribonucleic acids and vitamins. Quite a lot of compounds are noticed by the human nose at so low levels that they are so far impossible to analyse chemically and, consequently, it has not been possible to establish the enzymatic pathways generating them.

15.10 Conclusions

Several enzymes of importance to cheese ripening are today identified and characterised, which generates possibilities to influence ripening processes by changes in the production. The enzymes that originate from different sources influence each other's activities, and they work together in the creation of an incredible large number of cheese varieties. The characteristic enzymes for each cheese variety need to be present in significant amounts and active at the right time points during ripening, but also to be limited and in balance with other enzyme activities. This complexity needs to be kept in mind during any development of cheese making procedures, and each step should be taken carefully and involve skilled and experienced cheese makers. Among several others, especially two gaps in the knowledge of cheese ripening enzymology are crucial to fill in. One is deaper understanding of how enzymatic activities in the milk influence ripening processes in the cheese. The other is to identify and find the metabolic pathways behind odour-active compounds that are noticed by the human nose, but have not yet been possible to analyse.

References

Adamberg K, Antonsson M, Vogensen FK, Nielsen EW, Kask S, Møller PL, Ardö Y (2005) Fermentation of carbohydrates from cheese sources by non-starter lactic acid bacteria isolated from semi-hard Danish cheese. Int Dairy J 15:873–882

Andersen LT (2009) A chemical and sensory study of water-soluble compounds in mature Cheddar cheese with savoury flavour. Ph.D. thesis, University of Copenhagen

Andrén A, von Reedtz C (1990) Effects of chromatographically pure bovine chymosin and pepsin a on cheese curd firmness. J Dairy Res 57:109–117

Andrews AT, Alichanidis E (1975) Acid phosphatase activity in cheese and starters. J Dairy Res 42:327–339

Antonsson M, Molin G, Ardö Y (2003a) Proteolysis of the semi-hard cheese Herrgård made at different dairies. Exploratory study. Milchwissenshaft 58:145–148

Antonsson M, Molin G, Ardö Y (2003b) *Lactobacillus* strains isolated from Danbo cheese and their function as adjunct cultures in a cheese model system. Int J Food Microbiol 85:159–169

Ardö Y (1993) Characterizing ripening in low-fat, semi-hard round-eyed cheese made with undefined mesophilic DL-starter. Int Dairy J 3:343–357

Ardö Y (2004) Semi-hard Scandinavian cheeses made with mesophilic DL-starter. In: Handbook of food and beverage fermentation technology. Marcel Dekker, Inc., New York, pp 277–290

Ardö Y (2006) Flavour formation by amino acid catabolism. Biotechnol Adv 24:238–242

Ardö Y, Pettersson H-E (1988) Accelerated cheese ripening with heat treated cells of *Lactobacillus helveticus* and a commercial proteolytic enzyme. J Dairy Res 55:239–245

Ardö Y, Larsson P-O, Lindmark Månsson H, Hedenberg A (1989) Studies of peptidolysis during early maturation and its influence on low-fat cheese quality. Milchwissenschaft 44:485–490

Ardö Y, Lindblad O, Qvist KB (1999) Study of methods to monitor heat load to cheese milk. Int Dairy J 9:547–552

Ardö Y, Thage BV, Madsen JS (2002) Dynamics of free amino acid composition in cheese ripening. Aust J Dairy Technol 57:109–115

Ardö Y, Lilbæk H, Kristiansen KR, Zakora M, Otte J (2007) Identification of large phosphopeptides from β-casein that characteristically accumulate during ripening of the semi-hard cheese Herrgård. Int Dairy J 17:513–524

Ardö Y, Pripp AH, Lillevang SK (2009) Impact of heat-treated *Lactobacillus helveticus* on bioactive peptides in low-fat, semi-hard cheese. Aust J Dairy Technol 64:58–62

Ardö Y, McSweeney PLH, Magboul AAA, Upadhyay VK, Fox PF (2017) Biochemistry of cheese ripening: proteolysis. In: McSweeney PLH, Fox PF, Cotter PD, Everett DW (eds) Cheese: chemistry, physics and microbiology, 4th edn. Academic, London, pp 446–482

Arnau J, Hjerl-Hansen E, Israelsen H (1997) Hetreologous gene expression of bovine plasmin in *Lactcoccus lactis*. Appl Microbiol Biotechnol 48:331–338

Barrett FM, Kelly AL, McSweeney PLH, Fox PF (1999) Use of exogenous urokinase to accelerate proteolysis in Cheddar cheese during ripening. Int Dairy J 9:421–427

Bastian ED, Brown RJ (1996) Plasmin in milk and dairy products. Int Dairy J 6:435–457

Battistotti B, Corradini C (1993) Italian cheese. In: Fox PF (ed) Cheese: chemistry, physics and microbiology, vol 2, 2nd edn. Chapman and Hall, London, pp 221–243

Benfeldt C (1997) Proteolyse og ostemodning. Ph.D. thesis, Institut for Molekylær og Strukturel Biologi, Århus universitet

Bie R, Sjöström G (1975a) Autolytic properties of some lactic acid bacteria used in cheese production. Part I. Material and methods. Milchwissenschaft 30:653–657

Bie R, Sjöström G (1975b) Autolytic properties of some lactic acid bacteria used in cheese production. Part II. Experiments with fluid substrate and cheese. Milchwissenschaft 30:739–747

Birschbach P (1994) Origins of lipases and their characteristics. Bulletin of the International Dairy Federation No. 294, pp 7–10

Børsting MW, Qvist KB, Rasmussen M, Vindeløv J, Vogensen FK, Ardö Y (2012) Influence of proteolysis and amino acid release on bitterness and structure of reduced-fat Cheddar. Dairy Sci Technol 92:593–612

Børsting MW, Qvist KB, Brockmann E, Vindeløv J, Pedersen TL, Vogensen FK, Ardö Y (2015) Classification of *Lactococcus lactis* cell envelope proteinase based on gene sequencing, peptides formed after hydrolysis of milk, and computer modelling. J Dairy Sci 98:68–77

Broadbent JR, Rodriguez BT, Joseph P, Smith EA, Steele JL (2006) Conversion of *Lactococcus lactis* cell envelope proteinase specificity by partial allele exchange. J Appl Microbiol 100:1307–1317

Broadbent J, Cai H, Larsen RL, Hughes JE, Welker DL, de Carvalho VG, Tompkins TA, Ardö Y, Vogensen FK, de Lorentiis A, Gatti M, Neviani E, Steele JL (2011) Genetic diversity in proteolytic enzymes and amino acid metabolism among *Lactobacillus helveticus* strains. J Dairy Sci 94:4313–4328

Chen CM, Jaeggi JJ, Johnson ME (1994) Comparative study of protease from *Rhizomucor miehei* expressed in *Aspergillus oryzae* and other milk coagulants in Cheddar cheese making, yield and quality. J Dairy Sci 77(Suppl 1):14 (abstr)

Christensen JE, Dudley EG, Pederson JA, Steele JL (1999) Peptidases and amino acid catabolism in lactic acid bacteria. Antonie Van Leeuwenhoek 76:217–246

Cogan TM, Hill C (1993) Cheese starter cultures. In: Fox PF (ed) Cheese: chemistry, physics and microbiology, 2nd edn, vol 1. Chapman and Hall, London, pp 193–256

Collin JC, Berdagué JL, Dognin-Bergeret M, Grappin R (1987) Affinage et qualité du Gruyère de Comté. IV. Etude de la protéolyse. Lait 67:299–318

Creamer LK (1975) β-Casein degradation in Gouda and Cheddar cheese. J Dairy Sci 58:287–292

Drohse HB, Foltmann B (1989) Specificity of milk-clotting enzymes towards bovine κ-casein. Biochim Biophys Acta 995:221–224

Exterkate FA, Alting AC, Bruinenberg PG (1993) Diversity of cell envelope proteinase specificity among strains of *Lactococcus lactis* and its relationship to charge characteristics of the substrate-binding region. Appl Environ Microbiol 59:3640–3647

Farkye N, Fox PF (1990) Observations on plasmin activity in cheese. J Dairy Res 57:443–448

Fernandez-Espla MD, Garault P, Monnet V, Rul F (2000) *Streptococcus thermophilus* cell-wall-anchored proteinase: release, purification and biochemical and genetic characterization. Appl Environ Microbiol 66:4772–4778

Fickers P, Marty A, Nicaud JM (2011) The lipases from *Yarrowia lipolytica*: genetics, production, regulation, biochemical characterisation and biotechnological applications. Biotechnol Adv 29:632–644

Frantzen CA, Kot W, Pedersen TB, Ardö Y, Broadbent JR, Neve H, Hansen LH, Dal Bello F, Vogensen FK, Holo H (2017) Genomic characterization of dairy associated *Leuconostoc* species and diversity of leuconostocs in undefined mixed mesophilic starter cultures. Front Microbiol 8:Article 132

Hedlund U, Ludvigsen R (2005) Effect of proteolytic activities of cardosin A and B on texture and flavour of low-fat Cheddar cheese. M.Sc. Thesis, University of Copenhagen

Hillmann H, Behr J, Ehrmann MA, Vogel RF, Hofmann T (2016) Formation of kokumi-enhancing γ-glutamyl dipeptides in parmesan cheese by means of γ-glutamyltransferase activity and stable isotope double-labeling studies. J Agric Food Chem 64:1784–1793

Jensen MP, Ardö Y (2010) Variation in aminopeptidase and aminotransferase activities of six cheese related *Lactobacillus helveticus* strains. Int Dairy J 20:149–155

Jensen MP, Vogensen FK, Ardö Y (2009) Variation in caseinolytic properties of six cheese related *Lactobacillus helveticus* strains. Int Dairy J 19:661–668

Juillard V, Le Bars D, Kunji ERS, Konings WN, Gripon J-C, Richard J (1995) Oligopeptides are the main source of nitrogen for *Lactococcus lactis* during growth in milk. Appl Environ Microbiol 61:3024–3030

Kananen S (2013) Amino acid catabolism by *Lactobacilus helveticus*. Understanding the influence of strain variation on flavor formation. Ph.D. thesis, University of Copenhagen

Kappeler SR, van den Brink HJM, Rahbek-Nielsen H, Farah Z, Hansen EB, Johansen E (2006) Characerisation of recombinant camel chymosin reveals superior properties for the coagulation of bovine and camel milk. Int Dairy J 16:623–632

Kilcawley KN, Wilkinson MG, Fox PF (2002) Determination of key enzyme activities in commercial peptidase and lipase preparations from microbial or animal sources. Enzym Microb Technol 31:310–320

Larsen LB, Wium H, Benfeldt C, Heegaard CW, Ardö Y, Qvist KB, Petersen TE (2000) Bovine milk procathepsin D: presence and activity in heated milk and in extracts of rennet-free UF-feta cheese. Int Dairy J 10:67–73

Larsson M, Zakora M, Dejmek P, Ardö Y (2006) Primary proteolysis studied in a cast cheese from microfiltered milk. Int Dairy J 16:623–632

Law BA (1979) Reviews of the progress of dairy science: enzymes of psychrotrophic bacteria and their effects on milk and milk products. J Dairy Res 46:573–588

Law BA, Wigmore AS (1983) Accelerated ripening of Cheddar cheese with a commercial proteinase and intracellular enzymes from starter streptococci. J Dairy Res 50:519–525

Lund M, Ardö Y (2004) Purification and identification of water soluble phosphopeptides from cheese using Fe(III) affinity chromatography and mass spectrometry. J Agric Food Chem 52:6616–6622

Macedo IQ, Faro CJ, Pires EV (1996) Caseinolytic specificity of cardosin, an aspartic protease from the cardoon *Cynara cardunculus* L: action on bovine α_s- and β-casein and comparison with chymosin. J Agric Food Chem 44:42–47

Madsen JS, Ardö Y (2001) Exploratory study of proteolysis, rheology and sensory properties of Danbo cheese with different fat contents. Int Dairy J 11:423–431

Martin P, Collin J-C, Garnot P, Dumas BR, Mocquot G (1981) Evaluation of bovine rennets in terms of absolute concentrations of chymosin and pepsin A. J Dairy Res 48:447–456

Masotti F, Hogenboom JA, Rosi V, de Noni I, Pellegrino L (2010) Proteolysis indices related to cheese ripening and typicalness in PDO Grana Padano cheese. Int Dairy J 20:352–359

McSweeney PMS, Fox PF (2004) Metabolism of residual lactose and of lactate and citrate. In: Fox PF (ed) Cheese: chemistry, physics and microbiology, General aspects, vol 1, 3rd edn. Chapman & Hall, London, pp 361–371

Michaelidou A, Alichanidis E, Urlaub H, Polychroniadou A, Zerfiridis GK (1998) Isolation and identification of some major water-soluble peptides in feta cheese. J Dairy Sci 81:3109–3116

Møller KK, Rattray FP, Ardö Y (2012a) Camel and bovine chymosin hydrolysis of bovine α_{S1}- and β-caseins studied by comparative peptide mapping. J Agric Food Chem 60:11421–11432

Møller KK, Rattray FP, Høier E, Ardö Y (2012b) Manufacture and biochemical characteristics during ripening of Cheddar cheese with variable NaCl and equal moisture content. Dairy Sci Technol 92:543–568

Møller KK, Rattray FP, Ardö Y (2013) Application of selected lactic acid bacteria and coagulants for improving the quality of low-salt Cheddar cheese: chemical, microbiological and rheological evaluation. Int Dairy J 33:163–174

Mucchetti G, Locci F, Gatti M, Neviani E, Addeo F, Dossena A, March R (2000) Pyroglutamic acid in cheese: presence, origin, and correlation with ripening time of Grana Padano cheese. J Dairy Sci 83:659–665

Mulvihill DM, Fox PF (1977) Proteolysis of α_{s1}-casein by chymosin: influence of pH and urea. J Dairy Res 44:533–540

Mulvihill DM, Fox PF (1979) Proteolytic specificity of chymosins and pepsins on β-casein. Milchwissenschaft 34:680–683

Oberg CJ, Broadbent JR, Strickland M, McMahon DJ (2002) Diversity in specificity of extracellular *Lactobacillus helveticus* and *Lactobacillus delbruckii* subsp. *bulgaricus*. Lett Appl Microbiol 34:455–560

Oberg CJ, Oberg TS, Culumber MD, Ortakci F, Broadbent JR, McMahon DJ (2016) *Lactobacillus wasatchensis* sp. Nov., a non-starter lactic acid bacteria isolated from aged Cheddar cheese. Int J Syst Evol Microbiol 66:158–164

Ollikainen P, Kivelä T (1989) The importance of plasmin in Swiss-type cheese ripening. Milchwissenschaft 44:204–206

Pedersen TB, Ristagno D, McSweeney PLH, Vogensen FK, Ardö Y (2013) Potential impact on cheese flavour of heterofermentative bacteria from starter cultures. Int Dairy J 33:112–119

Pettersson H-E, Sjöström G (1975) Accelerated cheese ripening: a method for increasing the number of lactic starter bacteria in cheese without detrimental effect to the cheese making process, and its effect on the cheese ripening. J Dairy Res 42:313–326

Rea DM (1997). Comparison of Cheddar cheese made with chymosin, *Rhizomucor miehei* proteinase or *Cryphonectria parasitica* proteinase. M.Sc. thesis, National University of Ireland, Cork

Rehn U, Petersen MA, Hallin Saedén K, Ardö Y (2010) Ripening of extra hard cheese made with mesophilic DL-starter. Int Dairy J 20:844–851

Risborg SS and Witt SM (2013) Investigation of the proteolytic activity of five commercially used coagulants by CE and RP-HPLC-MS/MS. M.Sc. thesis, University of Copenhagen

Rogelj I, Perko B, Francky A, Penca V, Pungercar J (2001) Recombinant lamb chymosin as an alternative coagulating enzyme in cheese production. J Dairy Sci 84:1020–1026

Roudot-Algaron F, Kerhoas I, Le Bars D, Einhorn J, Gripon J-C (1994) Isolation of γ-glutamyl peptides from Comté cheese. J Dairy Sci 77:1161–1166

Sforza S, Cavatorta V, Galaverna G, Dossena A, Marchelli R (2009) Accumulation of non-proteolytic aminoacyl derivatives in Parmigiano Reggiano cheese during ripening. Int Dairy J 19:582–587

Siezen R (1999) Multi-domain, cell-envelope proteinases of lactic acid bacteria. Antonie Van Leeuwenhoek 76:139–155

Skeie S, Ardö Y (2000) Influence from raw milk flora on cheese ripening studied by different treatments of milk to model cheese. Lebensmittel -Wissenshaft und -Technology 33:499–505

Skeie S, Narvhus JA, Ardö Y, Thorvaldsen K, Abrahamsen R (1997) Influence of reduced salt content on the function of liposome-encapsulated Neutrase and heat-treated lactobacilli in rindless low-fat cheese. Lait 77:575–585

Smeianov VV, Wechter P, Broadbent JR, Hughes JE, Rodryguez BT, Christensen TK, Ardö Y, Steele JL (2007) Comparative high-density microarray analysis of gene expression during growth of *Lactobacillus helveticus* in milk versus rich culture medium. Appl Environ Microbiol 73:2661–2672

Sousa MJ, Ardö Y, McSweeney PLH (2001) Advances in the study of proteolysis during cheese ripening. Int Dairy J 11:327–345

Stadhouders J, Mulder H (1957) Fat hydrolysis and cheese flavour. Neth Milk Dairy J 11:164–183

Taivosalo A, Kriščiunaite T, Seiman A, Part N, Stulova I, Vilu R (2017) Comprehensive analysis of proteolysis during 8 months of ripening of high-cooked old Saare cheese. J Dairy Sci 101:944–967

Tan PST, Poolman B, Konings WN (1993) Proteolytic enzymes of *Lactococcus lactis*. J Dairy Res 60:269–286

Thage BV, Houlberg U, Ardö Y (2004) Amino acid transamination in permeabilised cells of *Lactobacillus helveticus, Lb. paracasei* and *Lb. danicus*. J Dairy Res 71:461–470

Thierry A, Collins YF, Mukdsi MCA, McSweeney PLH, Wilkinson MG, Spinnler HE (2017) Lipolysis and metabolism of fatty acids in cheese. In: McSweeney PLH, Fox PF, Cotter PD, Everett DW (eds) Cheese: chemistry, physics and microbiology, 4th edn. Academic, London, pp 423–444

Tynkkynen S, Buist G, Kunji ERS, Kok J, Poolman B, Venema G, Haandrikman AJ (1993) Genetic and biochemical characterization of the oligopeptide transport system of *Lactococcus lactis*. J Bacteriol 175:7529–7532

Upadhyay VK, Ravn P, Israelsen H, Sousa MJ, Kelly AL, McSweeney PLH (2006) Acceleration of proteolysis during ripening of Cheddar type cheese using a streptokinase-producing strain of *Lactococcus*. J Dairy Res 73:70–73

Urbach G (1993) Reletions between cheese flavour and chemical composition. Int Dairy J 3:389–422

Vallejo JA, Ageitos JM, Poza M, Villa TG (2012) A comparative analysis of recombinant chymosins. J Dairy Sci 95:609–613

van Boekel MAJS, Crijns CL (1994) Behaviour of the proteose-peptone fraction during renneting of milk. Neth Milk Dairy J 48:117–126

Waagner Nielsen E (2004) Principles of cheese productions. In: Handbook of food and beverage fermentation technology. Marcel Dekker, Inc., New York, pp 219–239

Waagner Nielsen E, Edelsten D (1974) The hydrolysis of fat in blue cheese. XIX Int Dairy Congress 1E:252

Weimer B, Seefeldt K, Dias B (1999) Sulphur metabolism in bacteria associated with cheese. Antonie Van Leeuwenhoek 76:247–261

Yvon M, Rijnen L (2001) Cheese flavour formation by amino acid catabolism. Int Dairy J 11:185–201

Zheng J, Wittouck S, Salvetti E, Franz CMAP, Harris HMB, Mattarell P, O'Toole PW, Pot B, Vandamme P, Walter J, Watanabe K, Wuyts S, Felis GE, Gänzle MG, Lebeer S (2020) A taxonomic note on the genus *Lactobacillus*: description of 23 novel genera, emended description of the genus *Lactobacillus* Beijerinck 1901, and union of Lactobacillaceae and Leuconostocaceae. Int J Syst Evol Microbiol

Chapter 16
Enzyme Modified Cheese

Zafer Erbay, Pelin Salum, and Kieran N. Kilcawley

16.1 Introduction

Cheese has been produced for more than 5000 years, with more than 1000 cheese varieties known to exist (Fox 2011). It is the most important dairy product, with nearly one-third of the total world milk production being used for cheese production (IDF 2017). While cheese as a product is widely consumed, increasing amounts undergo further processing or are used as a food ingredient (Fox et al. 2017; Guinee and Kilcawley 2004; Guinee 2011a; West 2007). Although the use of cheese as an ingredient can be for a variety of reasons, the most important is for its sensory appeal. Cheese flavor is diverse and results from the interaction of several volatile and nonvolatile compounds originating from the raw material, as well as from proteolysis, lipolysis, glycolysis and some chemical reactions. The extent of biochemical and chemical reactions that occur vary considerably between varieties, resulting in a wide range of cheese flavors. Mature or strongly flavored cheeses are most widely used to impart cheese flavor in foods. However, there are considerable limitations of using natural cheeses as a flavor ingredient, such as non-standard flavor profile, lack of flavor, potentially excessive lactose and/or fat levels in the final product, and difficulties adding high volumes, high costs, and potential adverse sensory textural issues due to the low solubility of some components. A significant portion of cheese is also used to produce processed cheese, which are prepared by heating and blending natural cheeses, with added emulsifiers and usually also added

Z. Erbay · P. Salum
Department of Food Engineering, Faculty of Engineering, Adana Science and Technology University, Adana, Turkey

K. N. Kilcawley (✉)
Food Quality and Sensory Science Department, Teagasc Food Research Centre, Fermoy, Co. Cork, Ireland
e-mail: kieran.kilcawley@teagasc.ie

A. L. Kelly, L. B. Larsen (eds.), *Agents of Change*, Food Engineering Series,
https://doi.org/10.1007/978-3-030-55482-8_16

colors and flavors. Analog cheese also falls under this category but are mainly produced from caseinates and vegetable oil.

Therefore, alternatives to natural cheese are widely used in food products. Originally, most cheese flavors were classified as artificial, nature-identical or natural. Even though complex synthetic cheese flavors have been developed, a significant global movement towards natural cheese flavor driven by labeling legislation to comply with consumers concerns around food safety and wellbeing has resulted in the creation of natural cheese flavors through biotransformation becoming the main source of cheese flavors. There are two main methods widely used to add natural cheese flavor to food products: natural cheese powder or enzyme-modified cheese (EMC). EMC can be in paste or powder form and is the focus of this chapter. Natural cheese powders are produced by spray-drying a cheese slurry (heating and shearing) produced with added water and emulsifying salts (Fox et al. 2017). Converting the cheese matrix into powder form overcomes some of the difficulties arising from the direct use of natural cheese in the industrial formulations. Although the ease of production and improved shelf-life are advantages, high production costs and losses of flavor during spray drying remain unresolved issues. Even though advances to improve cheese powder quality (da Silva et al. 2017, 2018a, b; Erbay et al. 2015; Erbay and Koca 2015; Koca et al. 2015; Varming et al. 2011) have been undertaken, the requirement for mature cheese as the substrate inevitably causes issues with uniformity of the final product and high costs.

EMC is a form of natural cheese concentrate and seems to have developed as the ubiquitous term for this type of product, although no true standard of identity exits. As EMC can alleviate issues with variability, and as application volumes are low, economic advantages over natural cheese or natural cheese powders exist and thus usage continues to increase. This chapter aims to provide up to date information on the procedures used to generate EMC and review the diversity of flavors available and potential uses.

16.2 General Principles

EMC is a natural flavor ingredient produced from fresh cheese or curd to obtain an intensified cheese flavor in a much shorter time than natural cheese production, primarily using enzymes under controlled conditions. It is in effect a form of accelerated natural cheese ripening producing a more intensely flavored product, but which is generally comparable in flavor to its natural equivalent. Accelerating flavor development occurs by optimizing conditions for the biochemical (enzymatic) reactions known to produce cheese flavor (Moskowitz and Noelck 1987). Using such processes, flavor intensities of 15–30 times greater than natural (target) cheese have been achieved over short periods (1–7 days), as opposed to months for natural cheeses (Kilcawley et al. 1998; West 2007). Unlike cheese, where texture and visual appeal are key quality components, the only function of an EMC is flavor, as they are not produced for direct consumption, but rather added at low levels to other products.

EMCs in the paste form are more suitable for use in products such as processed cheeses and cheese sauces, whereas EMC powders have more widespread use because of their longer shelf life and high applicability (West 2007; Wilkinson et al. 2011). Semi-hard or hard cheeses (60–65% solid content) are generally used as the substrates in EMC production, while hydrolysed butter or cream are often added to augment flavor intensity. Blue cheese EMC is a good example where cream is used as the sole substrate (Salum et al. 2019); however, production processes for Blue cheese EMC varies considerably to all other EMC products (see Sect. 16.4.4). Although there are no specific regulations, the moisture contents of paste and powder EMCs generally range from 40–60% and 5–10%, respectively (Kilcawley et al. 1998).

While EMCs can be used as the unique source of cheese flavor in a food product, they can also be used to enhance cheese flavor strength, provide a specific cheese flavor character to a product, or to harmonize other flavors. There are many factors affecting the flavor properties of EMCs. Survey studies have highlighted that commercial EMCs mimic specific natural cheese flavors to an extent rather than as a total replacement (Hulin-Bertaud et al. 2000; Kilcawley et al. 2007). According to these studies, there are significant differences between the flavor of EMCs and their natural equivalents, and even considerable differences within the same EMC variety. This is not surprising as the increased flavor intensity results from enhancing the concentration of one or more volatile components or biochemical pathways, which differentiates EMC from its natural cheese equivalent. One of the most diverse EMC types is Cheddar. A myriad of different Cheddar-type EMCs exist commercially which is in part due to the fact that natural Cheddar cheese is commercially available from 2 months to >24 months, produced across the English speaking world using bovine milk from different breeds and different diets, using milk that is raw, thermized or pasteurized, with different added starter cultures, adjunct cultures, added chymosins, as well as slightly different production procedures. The only compositional points of identity for Cheddar cheese are a moisture content below 39% and a fat-in-dry-matter content above 50% (Kilcawley et al. 2007).

One of the most important factors affecting the flavor of EMC is the type of raw material used as the substrate. It was previously reported that most of the lactate and acetate in Cheddar type EMCs originate from immature natural Cheddar curd (Kilcawley et al. 2001), but these authors also found that some commercial Cheddar EMCs contained propionate, an important flavor compound in Swiss-type cheese. It is likely that propionate was added as a preservative in this EMC (Kilcawley et al. 2001), rather than to specifically mimic Cheddar flavor, as it is not a volatile compound associated with Cheddar cheese flavor. Inter-batch variations within EMC products are low, indicating effective process control during manufacture (Kilcawley et al. 2000, 2001). One study found good correlations between specific flavor attributes, high-fat content and low pH in commercial Cheddar EMCs (Hulin-Bertaud et al. 2000). In general, compositional parameters of different commercial EMCs vary widely (Kilcawley et al. 2000), highlighting that manipulation of composition to attain specific flavor attributes is an accepted practice.

16.3 The Application and Advantages of EMC as a Cheese Flavor Ingredient

EMCs, especially in powder form, have been commonly used as a flavor ingredients in a wide range of products such as soups, salad dressings, sauces, dips, fillings, quiches, gratins, snacks, crisps, crackers, biscuits, noodles and pasta products, cake mixes, patisseries, deserts, frozen foods, ready-to-serve foods/meals, processed cheeses, cheese spreads, low-fat/no-fat cheese products, and cheese analogues/substitutes/imitations (Hannon et al. 2006; Kilcawley et al. 1998; Soccol et al. 2007). Very little detailed published material is available on the use of EMC as an ingredient, except for its usage in low- or reduced-fat cheese or in imitation cheese production (Table 16.1).

EMCs with a variety of specific cheese flavors are commercially available, such as Cheddar, Colby, Provolone, Romano, Mozzarella, Parmesan, Brick, Feta, Ezine, Blue, Edam, Gouda, Swiss, Emmental, Gruyere, goat, and sheep cheese (Kilcawley et al. 1998; Moskowitz and Noelck 1987; Salum et al. 2019; Soccol et al. 2007). The advantages of the use of EMCs can be summarized as follows:

- Intense flavor at low cost (Hannon et al. 2006; West 2007; Wilkinson et al. 2011; Kilcawley et al. 1998) which can reduces the requirement for expensive mature cheeses. Cheese off-cuts, downgrades or cheeses with damaged packaging are often used/recycled in the production process. As the flavor strength of EMCs is greater than the equivalent mature natural cheese, much lower volumes are required. Production times are typically days. It is possible to produce multiple cheese flavors from one process. Storage costs (space and conditions) are reduced, especially for EMC powders.
- As the process conditions can be strictly controlled, the effect of external factors can be minimized, enabling the production of more consistent EMC products (West 2007, Kilcawley et al. 1998).
- Different EMC flavors can be produced from the same raw material by varying the processing conditions (Hannon et al. 2006; Kilcawley et al. 1998).
- The use of EMCs can aid product development, due to low volumes, high flavor strengths and a broad range of cheese flavors (Amelia et al. 2013; Hassan et al. 2007; Kilcawley et al. 1998; Soccol et al. 2007).
- The proteins in EMCs are hydrolyzed into more soluble compounds (small peptides and amino acids) and therefore, can be used in high-quality frozen products without adversely impacting on texture and mouth feel (Kilcawley et al. 1998; Missel 1996; Soccol et al. 2007).

Table 16.1 Studies of EMC addition as an ingredient in foods

Food material	Variables	Aim of the study	EMC addition	Conclusions
Low-fat Cheddar cheese (Farkye & Arnold, 2008)	–	Production of low-fat Cheddar	Added to milled curd before salting	Acceptable products with characteristic flavor of ripened cheese were obtained
Low-fat Cheddar cheese (Amelia et al. 2013)	–	Production of low-fat Cheddar	–	Use of EMC in low-fat Cheddar production (as a preliminary experiment) enhanced the aged Cheddar flavors and masked the grapey-tortilla flavor
Cheddar cheese (Hannon et al. 2006)	Use of EMC at different amounts	Accelerate Cheddar cheese ripening	Added to milled curd before salting	Inclusion of EMC enhanced flavor development by 2 months. Levels of non-starter lactic acid bacteria (NSLAB) were greater in the EMC added cheeses compared with the control and the addition of EMC enhanced secondary proteolysis, primarily because of the increase in the available substrate for peptidases originated from NSLAB. The low flavor threshold catabolic products of lipolysis predominated in EMC added cheeses
Reduced fat processed cheese (Hassan et al. 2007)	Production with different formulations	Substitute ripened Cheddar with EMC	During processing	There were no differences in flavor between cheeses made with EMC and those made from aged cheeses
Imitation cheese (Noronha et al. 2008a)	Product with different pH; Use of EMC with different lipolysis degree	Determination of the effect of short-chain fatty acids in EMCs on the sensory properties of imitation cheese	During processing	There was a good correlation between flavor intensity and degree of lipolysis of the EMC used in the imitation cheese. Optimum flavor quality obtained by using the EMC with the intermediate level of total FFAs. Lowering pH caused a "smoother", "creamier" texture and "more cheesy" with more "balanced cheesy notes"

(continued)

Table 16.1 (continued)

Food material	Variables	Aim of the study	EMC addition	Conclusions
Reduced fat high fiber imitation cheese (Noronha et al. 2008b)	Product with different fat contents and pH; Use of EMC with different lipolysis degree	Production of reduced fat high fiber imitation cheese	During processing	Main factor was fat content on the flavor perception of imitation cheese. Level of lipolysis affected and increased the cheese-like sensory characters. Low pH increased the content of short-chain FFA and intensity of the flavor
Low and high-fat cream cheeses (Miri & Habibi-Najafi, 2011)	Fat content of the product; Use of EMC at different amounts	Production of low and high-fat cream cheeses	During processing	EMC increased sensory properties of low-fat cream cheeses but decreased for high-fat cream cheeses. Optimum EMC addition was found to be less than 1% and more than 5% for high-fat and low- fat cream cheeses, respectively
Process cheese (Habibi-Najafi et al. 2010)	Use of EMC at different amounts	Formulation optimization of process cheese	During processing	EMC addition has no significant effect on process cheese texture. Increase in EMC addition caused an increase in overall acceptance, aroma and taste intensity

- As EMCs are used in low volumes they are very useful to impart flavor in reduced and low- /reduced-fat products (Kilcawley et al. 1998; Miri et al. 2011; Sibeijn and Wouters 2009; Soccol et al. 2007; West 2007; Wilkinson et al. 2011).
- EMCs are potentially a rich source of bioactive peptides such as angiotensin-converting enzyme (ACE) inhibitory peptides, which have health-promoting properties (Haileselassie et al. 1999; Isobe et al. 2008; Kilara and Chandan 2011; Tanabe et al. 2006; Wilkinson et al. 2011; Yamasaki and Mitsuyama 2014).

16.4 Production Methods of EMC

Multiple potential processing steps and parameters exist, but the key aspect of EMC production is the creation of mature cheese flavors *via* proteolysis and lipolysis using exogenous enzymes. In general, young or fresh cheese/curd (often the target cheese flavor variety) is used as the main substrate. Production typically consists of the following steps: (a) creation of cheese slurry or emulsion; (b) inclusion of exogenous enzymes and incubation under controlled conditions, and finally; (c) termination of enzyme activity and, if necessary, standardization of the final product.

16.4.1 Emulsification of the Substrate

The effective and homogenous distribution of enzymes during incubation is important for optimal processing and is attained by first converting the cheese curd substrate into an homogenous slurry. This part of the process is analogous to processed cheese production, where curd is mixed with added water and emulsifying salts using heat and high shear. Shear requirements vary depending upon the age of curd, with young fresh curd requiring considerably more energy and emulsifying salts to create a homogenous emulsion. Efficient emulsification is critically important for optimal enzyme/substrate interaction. For this reason, selecting the type and concentration of emulsifying salts is important. Emulsifying agents are typically calcium sequestrants and these salts solubilize protein by chelating Ca^{2+} and the solubilized proteins then act as emulsifiers for the fat. The emulsified fat also facilitates optimal lipase action. The fat fraction is freely emulsified in EMC, increasing the area of the oil/water interface and providing improved conditions for the action of lipases (Villeneuve and Foglia 1997). A high heat-treatment is often undertaken to pasteurize the mix prior to enzyme addition, the main purpose of which is to reduce variability but also to eliminate potential pathogenic bacteria. A wide range of emulsifying salts are available, the most common of which are sodium phosphates. The age of the cheese curd and pH will also impact on the effectiveness of emulsifying salts, and combinations are often used to ensure that protein solubility is increased sufficiently. Emulsification salts also have antibacterial properties and some can also be used to modify pH. Moreover, the addition of nitrate, potassium sorbate or nisin can be used to inhibit microbial activities (Kilcawley et al. 1998; Moskowitz and Noelck 1987; Wilkinson et al. 2011). The cheese slurry is subsequently incubated with exogenous enzymes (protease, lipase and/or esterase) under controlled conditions (see later).

Optimization of process parameters, especially incubation time, temperature and agitation rate is important. pH is usually adjusted prior to enzyme addition but can also be modified at the end of the incubation period (controlling pH during the incubation is usually impractical). Heat-treatment is applied to inactivate the enzymes and stop the reaction. This is usually after a set time, but may also be undertaken once a specific target is reached, for example a specific degree of proteolysis or lipolysis. In situations where lipolysis is a major factor, a titratable acidity index is often used as the end point, as it is a relatively cheap and easy test to perform, requiring limited training. Insufficient heat treatment can lead to residual enzyme activity that may be problematic in the final application; however, excessively high heat treatment can also result in the formation of off-flavors and other quality defects (Kilcawley et al. 1998; Wilkinson et al. 2011). At this point, the biological activity in the cheese curd is eliminated and the dispersed hydrated *para*-caseinate contributes to emulsification of the free fat by coating the surface of dispersed free fat droplets Guinee (2011b).

16.4.2 Enzyme Addition and Incubation

EMCs have been produced by three main methods. Targetting flavor formation is highly dependent upon the type and composition of the raw material, choice and concentrations of enzymes and/or cultures added, incubation conditions (time, temperature, agitation, and pH) as well as the use of processing aids and/or additives (emulsifier, antimicrobial agents, flavor ingredients, etc.) (Kilcawley et al. 1998).

Method 1: EMCs are widely produced by a one stage process (Fig.16.1). This process is based on the simultaneous proteolysis and lipolysis of the cheese curd under controlled conditions. The main advantage of this method is the minimum amount of equipment required and relatively simple procedure. However, the process is inefficient as, for example, conditions for optimal proteolysis can be quite different to those suitable for optimal lipolysis. This is mainly due to differences in pH, temperature requirements for optimal enzyme performance, but enzyme/substrate access can also be adversely impacted, as the emulsion properties may change with hydrolysis and also proteolytic enzymes may inactivate lipases (Kilcawley et al. 1998; Moskowitz and Noelck 1987; Wilkinson et al. 2011). To overcome these problems, alternative methods have been developed (Kilcawley et al. 2006; West 2007).

Method 2: A component approach is often used, in which proteolysis and lipolysis are carried out independently with different substrates, such that emulsified cheese curd is used for proteolysis and butterfat or cream for the lipolysis (Fig. 16.2). The lipolyzed and proteolyzed slurries are blended in different ratios depending on the target flavor. This approach provides greater optimization and flexibility, enabling a greater diversity of products to be generated. However, this method is not suitable as a continuous process (Kilcawley et al. 1998; Wilkinson et al. 2011). Lipolyzed butterfat or oils are also independent commercial products that are incorporated with EMCs under the concentrated natural dairy flavor umbrella (Regado et al. 2007).

Method 3: Another approach is known as the two-stage process (Kilcawley et al. 2006) (Fig. 16.3) where the cheese slurry is proteolyzed and subsequently the proteolytic activity stopped by a pasteurization process. Thereafter, the proteolyzed

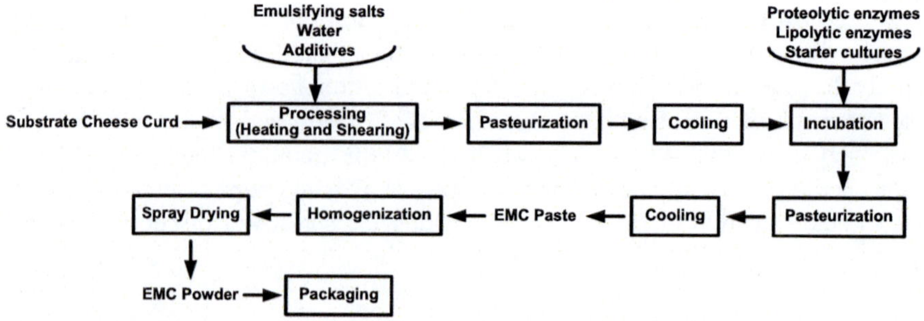

Fig. 16.1 Schematic illustration of the single-step operation for EMC production

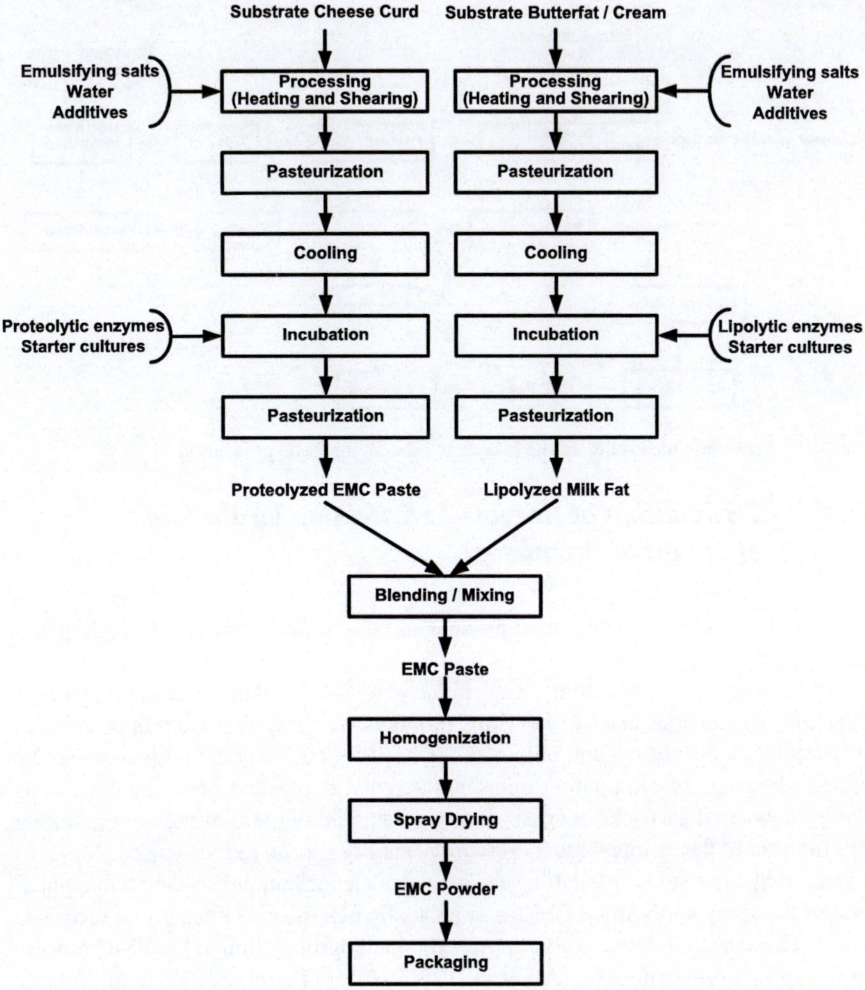

Fig. 16.2 Schematic illustration of the component approach process for EMC production

slurry is lipolyzed and EMC paste is manufactured after application of a final pasteurization process. The method enables greater diversification in terms of product flavors by differentiation of lipolytic and proteolytic profiles from a single substrate, but also enables conditions for each stage to be optimized independently. A very similar approach to produce an EMC from a mixture of soy milk and bovine milk has previously been reported (Ali et al. 2017a).

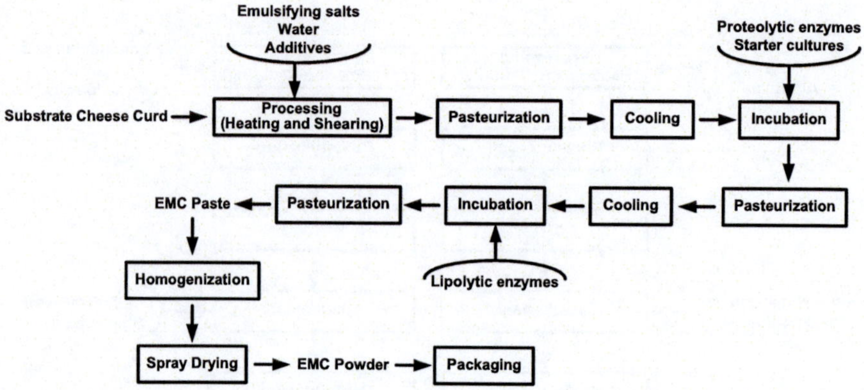

Fig. 16.3 Schematic illustration of the two-stage process for EMC production

16.4.3 Termination of Enzyme Activity/Standardization of the Final Product

It is necessary to stabilize the final product and this is best achieved through inactivation of residual enzyme activity by heat treatment. This has the additional benefit in eliminating microbial activity. The choice of heat-treatment can depend upon the nature of the enzymes used. Many EMC products are spray-dried to reduce volume, increase storage stability, and eliminate the need for cold storage. This has also the added advantage of eliminating microbial activity, depending upon the conditions used. The main drawbacks of spray-drying are potential losses of volatiles resulting in a dilution of flavor intensity. Increasing pH at drying can reduce volatile losses to an extent by creating non-volatile salts of organic acids but, unless the final application of the spray spray-dried EMC is at an acidic pH, the true intensity of the EMC will not be achieved. Some studies have examined optimization of the EMC process or the spray drying of EMC (Ali et al. 2017b; Amighi et al. 2013, 2016). Volatile compounds are often lost, due to the processing conditions, leading to a lack of characteristic aromas in the final product. Sometimes, additional flavors are added to make up for losses during processing. For example, a butter flavor may be made from a combination of a starter distillate and enzyme-treated butterfat. Starter distillates typically contain mainly water and a mixture of volatile, butter-like flavor compounds, or so-called "top notes". An example of a starter distillate procedure prepared by (steam) distilling a fermentation milk-like medium in combination with a lipase was previously reported (Ruttloff 1982).

Monosodium glutamate, yeast or starter culture extract, salt, organic acids can also be used to augment and diversify an EMC flavor with respect to the regulations (Kilcawley et al. 1998; Wilkinson et al. 2011). These additives can be included during or after production, or after termination of enzyme activity. Previous studies have shown that exogenous lactate and acetate were added in the production of Cheddar-type EMCs (Kilcawley et al. 2001), whereas exogenous acetate also appeared to be used in Parmesan EMC production (Salum et al. 2019). Moreover, emulsifiers and stabilizers, such as mono/diglycerides, phosphates, citric acid or xanthan gum, as well as antioxidants in the form of plant oil and/or fat-soluble vitamins such as tocopherols, have been utilized in EMC production. The addition of glutamate, possibly in the form of mono-sodium glutamate, was also evident in some commercial EMC (Kilcawley et al. 2000). Additionally, the key volatiles of the target cheese may sometimes be topped up by using food-grade nature-identical aroma compounds or refined by fermentation with starter cultures and moulds (Law 2010; Salum et al. 2019). Salum et al. (2019) noted that mould (likely *P. roqueforti*) must have been used to generate Blue cheese flavor in a commercial EMC as the ketone content constituted ~30.8% of total volatile content. These authors also found that 2-heptanone, 2-nonanone, and 8-nonen-2-one were present at very high concentrations and likely best represented the key sensory character of Blue cheese. The use of heat treatment in generating specific flavor notes in EMC production (goat cheese EMC) has also been reported (Salum et al. 2019).

16.4.4 Blue Cheese Flavor

Blue cheese flavor is an important component within the range of EMC flavors, because it is typically produced by a submerged culture process which is very different to the procedures previously described. As the flavor profile differs considerably from other EMC products, its applications can be more specific, yet Blue cheese remains one of the most desired flavor types. Spores and mycelia of *Penicillium roqueforti* are used to generate methyl ketones, mainly from dairy fat, to give a characteristic sharp, pungent, piquant aroma of Blue cheese (Dwivedi and Kinsella 1974; Taylor 1995). Spores are favored because they are more resistant to fatty acid toxicity and do not reduce methyl ketones to secondary alcohols as readily as mycelia. Methyl ketones, such as 2-heptanone, 2-octanone, and 2-undecanone, which are formed from free fatty acids by β-oxidation are the key aroma compounds for Blue cheese flavor. The fat content of the substrate is typically ~20–40% and is best extensively lipolyzed to give a high content of free fatty acids. High fat cream is a good substrate and other nutrients can be added if *P. roqueforti* is grown *in situ*. Exogenous lipase/esterases are added to accelerate lipolysis. The substrate is best pre-hydrolyzed or hydrolyzed *in-situ*, with the *P. roqueforti* spores added afterwards. Pre-hydrolysis ensures reproducibility and optimizes conditions for methyl ketone generation, which is the most important aspect of Blue cheese flavor production (Jolly and Kosikowski 1975). Optimum temperatures and pH for sub-cultured fermentation are ~25–40 °C and 5.8–6.2, respectively (Jolly and Kosikowski 1975; Nelson 1970). Aeration is necessary to ensure *P. roqueforti* growth and typically sufficient flavor is developed after 48 h. Other studies have highlighted that Blue cheese flavor can be developed using UHT cream or butteroil and that commercial lipases from *P. roqueforti* gave the best Blue cheesc flavor (Tomasini et al. 1993).

Other studies used a curd slurry approach and added *P. roqueforti* spores to generate Blue cheese flavor (King and Clegg 1979).

16.5 Enzymes in EMC Production

While the flavor profiles in EMCs are influenced by the processing techniques and conditions, the variety and amount of flavor compounds in EMCs are highly related to the degree and balance of proteolysis and lipolysis. The ratio and extent of proteolysis and lipolysis are in general unique for each natural cheese type and it is this aspect that is exploited in EMC manufacture. In basic terms, proteolysis tends to produce non-volatile taste compounds, such as peptides and amino acids, while lipolysis tends to produce aromatic volatile short-chain fatty acids. Thus proteolysis often provides the background notes, while lipolysis is used to enhance flavor intensity. This is clearly noted in studies of commercial EMC where, for some products, the degrees of proteolysis were identical but the degree of lipolysis corresponded with marketing information associated with intensity of flavor (Kilcawley et al. 2000, 2001). The role of glycolysis is less important in EMC, as the extent of glycolysis in most strongly flavored natural or mature cheeses is quite similar because lactose and citrate are metabolized quite quickly in the process. Many compounds arising from glycolysis are important in the flavor of short ripened soft natural cheeses, but these compounds tend to be masked in more strongly flavored mature cheese by products arising from proteolysis and lipolysis. The main exception is in Swiss-type EMC, where propionic acid is a key characteristic aroma compound. Propionic acid is produced from lactate by propionic acid bacteria in natural Swiss cheese; however, in most first generation Swiss-type EMC, it was added in the form of a starter distillate. Acetate, mainly a product of glycolysis, also plays a significant role in the flavor of many natural cheeses and is an important compound in EMC in general. It is also likely added to some commercial EMC products to adjust pH and flavor profile during standardization in the final step of the process (Kilcawley et al. 2001).

16.5.1 Proteolytic Enzymes

Microbial proteases, are mostly derived from *Bacillus, Aspergillus*, and *Rhizomucor* species and are widely used in EMC production (Kilcawley et al. 2002). They have both proteinase and peptidase activity, which aids in the formation of free amino acids and small peptides, both of which directly contribute to cheese flavor. Moreover both also provide substrates for a wide variety of aroma compounds *via* transamination, dehydrogenation, decarboxylation and reduction reactions (Marilley and Casey 2004). Previous studies have shown that indices of primary proteolysis such as water-soluble nitrogen contents were 2–4 times higher in Cheddar EMCs than that found in mature Cheddar cheese, and that indices of secondary proteolysis such as phosphotungstic acid-soluble nitrogen contents were 3–10 times higher (Kilcawley et al. 2000). Similar results were obtained from different types of EMCs such as Parmesan, Ezine, Goat, and Sheep cheese EMCs (Salum et al. 2019).

A high degree of proteolysis can lead to the formation of bitterness, which is mainly caused by the accumulation of hydrophobic peptides containing 3–15 amino acids and derived from β-casein. These peptides are characterized by the presence of leucine, isoleucine, valine, phenylalanine, tyrosine, and tryptophan, and proline. Bitterness can be reduced by further degradation of these peptides into smaller peptides and free amino acids. The removal of proline from the peptide chain has particular importance in the reduction of bitterness, as it changes all structural and functional properties of peptides and increases the susceptibility of the peptide to further hydrolysis (Wilkinson et al. 2011). Even though their processing conditions and properties vary greatly, exopeptidases cleave amino acids from ends of the polypeptide chain, either from the N-terminal or the C-terminal, whereas endopeptidases split a protein to yield two peptides from the peptide linkage between two amino acid residues in the primary protein chain (Kilara and Chandan 2011). For this reason, the balance between the use of endo- and exopeptidases in the EMC production is important.

A key aspect of many EMC products is attaining high levels of proteolysis without the development of bitterness. Previous studies have shown that two hydrophobic peptides formed by the commercial proteinase, Neutrase caused distinct bitterness in EMC (Park et al. 1995). The authors found that crude cellular extracts from *Lactobacillus casei* ssp. *casei* LLG reduced bitterness and postulated that this was due to a combination of general aminopeptidase and proline-specific peptidases in the cellular extract (Haileselassie 1999; Haileselassie and Lee 1997; Park et al. 1995). Azarnia et al. (2010) highlighted the effectiveness of non-specific general aminopeptidases from strains such as *Lb. rhamnosus* S93, *Aspergillus oryzae* and *Aspergillus niger* in enhancing secondary proteolysis in EMC. Other studies have shown the potential to mask bitterness by addition of mature natural cheese or glutamic acid (Habibi-Najafi and Lee 2007; Wilkinson et al. 2011).

16.5.2 Lipolytic Enzymes

Enzymatic modifications of milk fat are carried out with lipases and esterases producing free fatty acids, mono-glycerides, di-glycerides and glycerol. Both enzymes catalyze the same reaction, i.e., hydrolysis of an ester bond of a glyceride, yielding a fatty acid and an alcohol (glycerol), but operate best in different polar environments and have different specificities. Lipolysis contributes directly to the formation of aroma compounds by the release of free fatty acids, and also indirectly through the metabolism of free fatty acids. Due their low detection threshold values, short-chain and even-numbered fatty acids (C4:0-C10:0) contribute directly to cheese flavor (Brennand et al. 1989). Usually, the most abundant volatile compounds in natural cheese are free fatty acids. In addition, these compounds are pre-cursors of methyl ketones, aldehydes, alcohols, lactones, and esters.

The activity and specificity of esterases and lipases are very source-dependent (Kilcawley et al. 2001). Plant lipases are not commercially available, and thus animal and microbial lipases/esterases are extensively used in the food industry. Caprine pre-gastric, bovine and porcine pancreatic lipases are widely used in EMC production. These enzymes have a high specificity for fatty acids on the Sn-3 position of the glycerol, and therefore preferentially cleave the very odor active short-chain fatty acids, such as butyrate that reside at this position. However, animal enzymes cannot be used in vegetarian or kosher products, and thus alternative microbial sources of similar specificity are actively sought. Many microbial sources of lipolytic activity are available; however, identifying microbial sources of esterases/lipases with a similar specificity to animal lipases remains a keen target for the enzyme industry (Bauman et al. 2011; Kilara and Chandan 2011; Wilkinson et al. 2011).

In general the extent of lipolysis in EMC can be 10–100 times higher than that for natural cheese (Salum et al. 2019). A problem with excessive lipolysis can be the generation of soapiness or rancidity due to an imbalance between the ratio of short- and long-chain free fatty acids, and possibly the generation of off-odors based on high amounts of short-chain FFAs, and methyl ketones due to their low threshold values. The rate of lipolysis can be adversely impacted by product inhibition, which is caused by the accumulation of mono- and di-glycerides at the enzyme/substrate interface, preventing access to the substrate (Kilara and Chandan 2011). This can be alleviated to an extent by creating salts of fatty acids, but this reduces odor activity. Excessive lipolysis can also negate the selective action of esterases/lipases through a process called acyl migration. Thus, longer hydrolysis periods can result in a generic fatty acid profile and a similar aromatic profile (Kilara and Chandan 2011).

16.5.3 Use of Starter Cultures in EMC Production

Even though lactic acid bacteria are very good source of peptidases, proteinases, and esterases, they are not readily suitable in EMC production due to excess acidification. Therefore, strains used as adjunct cultures in natural cheese production to accelerate or augment flavor development can alleviate this issue, as most are not active acid producers because the strains have either been selected on this basis or attenuated so that acid production is retarded. However, little information is available on the use of cultures in EMC production.

A process was described which used full fat milk or reconstituted whole milk powder as the main substrate to produce a savory flavor EMC type product (Delabre and Bendall 2008). The substrate was sterilized and lactic acid bacteria added under controlled conditions (temperature and pH) for 48 hrs. This was subsequently heat-inactivated to terminate the fermentation. Unfortunately, no details relating to the type, dose rates or properties of the strains were provided. The final product could be evaporated to increase solids to produce a paste or spray dried (Delabre and Bendall 2008).

Some studies have used heat- and freeze-shocked lactobacilli (*Lactobacillus helveticus* and *Lactobacillus casei*) in EMC production, and it was reported that the use of freeze-shocked cultures enhanced proteolysis as a result of lysis and gave positive sensorial results (Kumar et al. 2013; Sudhir et al. 2010). Other studies used heat-treatment for *Lactobacillus helveticus* DPC 4571 to inactivate lactose fermentation ability, without obtaining significant losses of proteolytic activity (Lee et al. 2007). The addition of these heat-shocked cells enhanced secondary proteolysis, increased volatile compounds, and generated a strong savory cheesy flavor after only 16 h. However, prolonged incubation (>16 h) resulted in the advantageous growth of putative enterococci, which adversely influenced sensory perception and appeared to increase acetic acid levels and viscosity and decrease free amino acids and key volatile flavor compounds (Lee et al. 2007). In general, the utilization of lactic acid bacteria in EMC production generally requires some degree of prehydrolysis, of at least the protein content of the substrate, prior to addition. The most suitable lactic acid bacteria are those used as commercial adjuncts in semihard or hard long ripening cheeses, as these have been preselected based on high enzyme activity and poor acidification properties. Otherwise, strains can be attenuated to reduce or eliminate acidification activity, maintain intracellular or extracellular proteolytic or lipolytic enzymes, and alter the physical structure of the cell so that intracellular enzymes can be released more rapidly.

16.6 Health Effects and Functional Properties of EMCs

EMC contain high levels of peptides, free amino acids and free fatty acids in comparison to most dairy products. Some of these components appear to have functional and health-promoting characteristics, such as antihypertensive, antimicrobial, antioxidative, antithrombotic, opioid, anti-appetizing, immunomodulatory and mineral-binding activities (Nguyen et al. 2015; Park and Nam 2015).

One of the proteolytic digestion products of β-casein is β-casomorphins, which are a group of peptides with opioid properties. β-casomorphins were detected in EMC but were also found to be further degraded by peptidases resulting in the formation of antihypertensive peptides (Haileselassie 1999). The biggest risk factor in cardiovascular disease is hypertension, and some peptides have been identified as having antihypertensive properties (De Leo et al. 2009). The antihypertensive potential of EMC has been investigated *in vivo* (Suzuki et al. 2007); peptides in the EMC were found to lower the blood pressure of hypertensive rats (Suzuki et al. 2007; Tonouchi et al. 2008).

The potential of peptides in EMC to inhibit allergen permeation in the intestine was also investigated (Isobe et al. 2008; Tanabe et al. 2006; Yamasaki and Mitsuyama 2014). An effective way to prevent food allergy and inflammatory bowel disease is the enhancement of the intestinal barrier function. A peptide (GPIVLNPWDQ) derived from α_{s2}-casein in EMC was found to inhibit ovalbumin (as a representative food allergen) permeation in a Caco-2 cell model (Tanabe et al. 2006). The results of these *in vivo* studies highlighted that some EMCs are potentially useful for the prevention of food allergy by inhibiting allergen permeation (Isobe et al. 2008; Yamasaki and Mitsuyama 2014).

Another functional property of EMCs is their antibotulinal activity. One of the most important microbial safety problems in processed cheeses is the toxin produced by the heat-stable, spore-forming pathogen *Clostridium botulinum*. Studies were carried out to assess the antibotulinal activity of EMC and they found that EMC delayed toxin production in fat-free processed cheeses. While the fat content of processed cheese inversely affected the antibotulinal effect, the water activity and pH of the processed cheese did not exclusively influence this effect (Glass and Johnson 2004a, b, c).

16.7 Conclusion and Future Trends

Even though EMC-type products have been available for decades, most information still resides within the domain of individual companies. Some studies exist on production of EMC with exogenous enzymes alone, but very little information is available on the use of starter cultures in EMC production and hopefully this will be addressed in the future. The following aspects of EMC production still require investigation.

- Ensuring optimal emulsion stability during the process to maximize enzyme activity, as both fat and protein are actively hydrolyzed;
- The use of model systems (curd slurries, dairy proteins or dairy lipid substrates) to screen lactic acid bacteria (LAB) or yeasts for cheese flavor potential;
- Utilizing molecular biology techniques to screen the genotypic and phenotypic flavor potential of LAB and yeasts;
- Identification of key volatiles associated with specific cheese types using these model systems;
- Using tailor-made substrates to optimize LAB or yeasts to utilize specific pathways to generate target flavor compounds;
- The impact of shearing/heating, homogenization, mixing, incubation, spray-drying etc. on product quality;
- Impact of fat globule size on the rate of lipolysis. There is also very little information on the impact of the homogenization methods (high pressure homogenization, homogenization with rotor-stators, ultrasound application) and agitation during EMC production on enzyme activity;
- The impact of heating on volatile compound generation and loss. Specific classes of volatiles are formed while other increase and decrease;

- More in-depth studies on the losses of volatiles during spray-drying, including product composition, carriers, spray-drying conditions;
- Identification of specific peptides in EMC with bioactivity that can be produced at high concentrations and not easily further hydrolyzed into smaller peptides without bioactivity;
- Optimization of starter culture performance by attenuation, including chemical and physical methods and ultrasound;
- Application of yeasts in EMC production (with or without LAB);
- Process optimization. Two-stage production has advantages in creating a more optimized process for EMC production, whereas one-stage production has economic and practical advantages.

References

Ali B, Khan KY, Majeed H et al (2017a) Imitation of soymilk–cow's milk mixed enzyme modified cheese: their composition, proteolysis, lipolysis and sensory properties. J Food Sci Technol 54:1273–1285

Ali B, Khan KY, Majeed H et al (2017b) Soymilk-cow's milk ACE-inhibiting enzyme modified cheese. Food Chem 237:1083–1091

Amelia I, Drake MA, Nelson B et al (2013) A new method for the production of low-fat Cheddar cheese. J Dairy Sci 96:4870–4884

Amighi F, Emam-Djomeh Z, Madadlou A (2013) Spray drying of ACE-inhibitory enzyme-modified white cheese. Int J Food Sci Technol 48:2276–2282

Amighi F, Emam-Djomeh Z, Madadlou A (2016) Optimised production and spray drying of ACE-inhibitory enzyme-modified cheese. J Dairy Res 83:125–134

Azarnia S, Lee BH, Yaylayan V, Kilcawley KN (2010) Proteolysis development in enzyme-modified Cheddar cheese using natural and recombinant enzymes of Lactobacillus rhamnosus S93. Food Chem 120:174–178

Bauman DE, McGuire MA, Harvatine KJ (2011) Mammary gland, milk biosynthesis and secretion: milk fat. In: Fuquay JW, Fox PF, McSweeney PLH (eds) Encyclopedia of dairy sciences, vol 3. Elsevier, London, pp 352–358

Brennand CP, Ha JK, Lindsay RC (1989) Aroma properties and thresholds of some branched-chain and other minor volatile fatty acids occuring in milkfat and meat lipids. J Sens Stud 4:105–120

da Silva DF, Larsen FH, Hougaard AB et al (2017) The influence of raw material, added emulsifying salt and spray drying on cheese powder structure and hydration properties. Int Dairy J 74:27–38

da Silva DF, Ahrné L, Larsen FH et al (2018a) Physical and functional properties of cheese powders affected by sweet whey powder addition before or after spray drying. Powder Technol 323:139–148

da Silva DF, Hirschberg C, Ahrné L et al (2018b) Cheese feed to powder: effects of cheese age, added dairy ingredients and spray drying temperature on properties of cheese powders. J Food Eng 237:215–225

Delabre M-L, Bendall JG (2008) Flavour ingredients from fermented dairy streams. Expression of multidisciplinary flavour science. In: Blank I, Wust M, Yeretzian C (eds) Proceedings from the 12th weurman symosium. Interlaken, Switzerland, pp 375–378

De Leo F, Panarese S, Gallerani R et al (2009) Angiotensin converting enzyme (ACE) inhibitory peptides: production and implementation of functional food. Curr Pharm Des 15:3622–3643

Dwivedi BK, Kinsella JE (1974) Continuous production of Blue-type cheese flavor by submerged fermentation of Penicillium roqueforti. J Food Sci 39:620–622

Erbay Z, Koca N (2015) Effects of whey or maltodextrin addition during production on physical quality of white cheese powder during storage. J Dairy Sci 98:8391–8404

Erbay Z, Koca N, Kaymak-Ertekin F et al (2015) Optimization of spray drying process in cheese powder production. Food Bioprod Process 93:156–165

Farkye NY, Arnold M (2008) A novel technology for making lowfat cheese. In: Joint annual meeting abs, no:140, Dairy Foods Division, p 154

Fox PF (2011) CHEESE: overview. In: Fuquay JW, Fox PF, McSweeney PLH (eds) Encyclopedia of dairy sciences, vol 1. Elsevier, London, pp 535–543

Fox PF, Guinee TP, Cogan TM, McSweeney PLH (2017) Cheese as an ingredient. In: Fox PF, Guinee TP, Cogan TM, McSweeney PLH (eds) Fundamentals of cheese science, 2nd edn. Springer, New York, pp 629–679

Glass KA, Johnson EA (2004a) Antagonistic effect of fat on the antibotulinal activity of food preservatives and fatty acids. Food Microbiol 21:675–682

Glass KA, Johnson EA (2004b) Antibotulinal activity of process cheese ingredients. J Food Prot 67:1765–1769

Glass KA, Johnson EA (2004c) Factors that contribute to the botulinal safety of reduced-fat and fat-free process chesse products. J Food Prot 67:1687–1693

Guinee TP (2011a) Cheese as a food ingredient. In: Fuquay JW, Fox PF, McSweeney PLH (eds) Encyclopedia of dairy sciences, vol 1. Elsevier, London, pp 822–832

Guinee TP (2011b) Pasteurized processed cheese products. In: Fuquay JW, Fox PF, McSweeney PLH (eds) Encyclopedia of dairy sciences, vol 1. Elsevier, London, pp 805–813

Guinee TP, Kilcawley KN (2004) Cheese as an ingredient. In: Fox PF, McSweeney PLH, Cogan TM, Guinee TP (eds) Cheese: chemistry, physics and microbiology, 3rd edn. Elsevier, London, pp 395–428

Habibi-Najafi MB, Lee BH (2007) Debittering of tryptic digests from beta-casein and enzyme modified cheese by x-prolyl dipeptidylpeptidase from Lactobacillus casei ssp. casei. LLG. Iran J Sci Technol Trans A Sci 31:263–270

Habibi-Najafi MB, Sabouri S, Mahallati MN (2010) Formulation optimization of process cheese using enzyme modified cheese. In: Proceedings of International Conference of Food Science and Technology, 22nd-24th March, 2010, vol L34. University of Chester, Chester, p 43

Haileselassie SSB (1999) Production of enzyme-modified cheese and bioactive peptides by Lactobacillus and commercial enzymes. Master of Science, Department of Food Science and Agricultural Chemistry, McGill University, Montreal

Haileselassie SS, Lee BH (1997) Enzyme-modified cheese (EMC) prepared by commercial and Lactobacillus casei enzymes. In: Proceedings of the American dairy science association annual meeting abstracts, no: D39G, Dairy Foods Division, p 107

Haileselassie SS, Lee BH, Gibbs BF (1999) Purification and identification of potentially bioactive peptides from enzyme-modified cheese. J Dairy Sci 82:1612–1617

Hassan AN, Awad S, Mistry VV (2007) Reduced fat process cheese made from young reduced fat Cheddar cheese manufactured with exopolysaccharide-producing cultures. J Dairy Sci 90:3604–3612

Hannon JA, Kilcawley KN, Wilkinson MG et al (2006) Production of ingredient-type Cheddar cheese with accelerated flavor development by addition of enzyme-modified cheese powder. J Dairy Sci 89:3749–3762

Hulin-Bertaud S, Kilcawley KN, Wilkinson MG et al (2000) Sensory and compositional relationships between commercial Cheddar-flavored enzyme-modified cheeses and natural Cheddar. J Food Sci 65:1076–1082

IDF (2017) The world dairy situation 2017. Bulletin of the international dairy federation. IDF, Brussels

Isobe N, Suzuki M, Oda M et al (2008) Enzyme-modified cheese exerts inhibitory effects on allergen permeation in rats suffering from indomethacin-induced intestinal inflammation. Biosci Biotechnol Biochem 72:1740–1745

Jolly R, Kosikowski FV (1975) Blue cheese flavor by microbial lipases and mold spores utilizing whey powder, butter and coconut fats. J Food Sci 40:285–287

Kilara A, Chandan RC (2011) Enzyme-modified dairy ingredients. In: Chandan RC, Kilara A (eds) Dairy ingredients for food processing. Wiley, West Sussex, pp 317–334

Kilcawley KN, O'Connell PB, Hickey DK et al (2007) Influence of composition on the biochemical and sensory characteristics of commercial Cheddar cheese of variable quality and fat content. Int J Dairy Technol 60:81–88

Kilcawley KN, Wilkinson MG, Fox PF (1998) Review: enzyme-modified cheese. Int Dairy J 8:1–10

Kilcawley KN, Wilkinson MG, Fox PF (2000) A survey of the composition and proteolytic indices of commercial enzyme-modified Cheddar cheese. Int Dairy J 10:181–190

Kilcawley KN, Wilkinson MG, Fox PF (2001) A survey of lipolytic and glycolytic end-products in commercial Cheddar enzyme-modified cheese. J Dairy Sci 84:66–73

Kilcawley KN, Wilkinson MG, Fox PF (2002) Properties of commercial microbial proteinase preparations. Food Biotechnol 16:29–55

Kilcawley KN, Wilkinson MG, Fox PF (2006) A novel two-stage process for the production of enzyme-modified cheese. Food Res Int 39:619–627

King RD, Clegg GH (1979) The metabolism of fatty acids, methyl ketones and secondary alcohols by Penicillium roqueforti in blue cheese slurries. J Sci Food Agric 30:197–202

Koca N, Erbay Z, Kaymak-Ertekin F (2015) Effects of spray-drying conditions on the chemical, physical, and sensory properties of cheese powder. J Dairy Sci 98:2934–2943

Kumar S, Jha YK, Singh P (2013) Influence of adjuncts as debittering aids on the sensory and biochemical properties of enzyme modified cheese-base. Proc Natl Acad Sci India B Biol Sci 83:347–355

Law BA (2010) Enzymes in dairy product manufacture. In: Whitehurst RJ, van Oort M (eds) Enzymes in food technology, 2nd edn Wiley, New Jersey, p 88–102

Lee BH, Kilcawley KN, Hannon JA et al (2007) The use of viable and heat-shocked Lactobacillus helveticus DPC 4571 in enzyme-modified cheese production. Food Biotechnol 21:129–143

Marilley L, Casey MG (2004) Flavours of cheese products: metabolic pathways, analytical tools and identification of producing strains. Int J Food Microbiol 90:139–159

Miri MA, Habibi Najafi MB (2011) The effect of adding enzyme-modified cheese on sensory and texture properties of low- and high-fat cream cheeses. Int J Dairy Technol 64:92–98

Missel D (1996) Selecting the right ingredients for adding the flavor of cheese. Food Prod Des 6:51–61

Moskowitz GJ, Noelck SS (1987) Enzyme-modified cheese technology. J Dairy Sci 70:1761–1769

Nguyen DD, Johnson SK, Busetti F et al (2015) Formation and degradation of peta-casomorphins in dairy processing. Crit Rev Food Sci Nutr 55:1955–1967

Nelson JH (1970) Production of blue cheese flavor via submerged fermentation by Penicillium roqueforti. J Agric Food Chem 18:1970

Noronha N, Cronin DA, O'Riordan ED et al (2008a) Flavouring of imitation cheese with enzyme-modified cheeses (EMCs): sensory impact and measurement of aroma active short chain fatty acids (SCFAs). Food Chem 106:905–913

Noronha N, Cronin D, O'Riordan D et al (2008b) Flavouring reduced fat high fibre cheese products with enzyme modified cheeses (EMCs). Food Chem 110:973–978

Park SY, Gibbs BF, Lee BH (1995) Effects of crude enzyme of Lactobacillus casei LLG on water-soluble peptides of enzyme-modified cheese. Food Res Int 28:43–49

Park YW, Nam MS (2015) Bioactive peptides in milk and dairy products: a review. Korean J Food Sci Anim Resour 35:831–840

Regado MA, Cristóvão BM, Moutinho CG et al (2007) Flavour development via lipolysis of milk-fats: changes in free fatty acid pool. Int J Food Sci Technol 42:961–968

Ruttloff H (1982) Biotechnology and aroma production. Nahrung 26:575–589

Salum P, Erbay Z, Selli S (2019) The compositional properties, proteolytic-lipolytic ripening parameters, and volatile compositions of commercial enzyme-modified cheeses (EMCs) with different cheese flavours. Int J Dairy Technol 72:416–426

Sibeijn M, Wouters JA (2009) Production of dairy aromas and flavours: new directions. In: Corredig M (ed) Dairy-derived ingredients: food and nutraceutical uses. Woodhead Publishing, Boca Raton, pp 470–481

Soccol CR, Medeiros ABP, Vandenberghe LPS, Woiciechowski AL (2007) Flavor compounds produced by fungi, yeasts, and bacteria. In: Hui YH (ed) Handbook of food products manufacturing. Wiley, West Sussex, pp 179–191

Sudhir K, Jha YK, Pratibha S (2010) Influence of adjuncts as a debittering aids in encountering the bitterness developed in cheese slurry during accelerated ripening. Int J Food Sci Technol 45:1403–1409

Suzuki M, Uchida M, Tonouchi H et al (2007) Inhibition of angiotensin I converting enzyme and hypotensive effect in spontaneously hypertensive rats by enzyme-modified cheese. Nippon Shokuhin Kagaku Kogaku Kaishi 54:181–186

Tanabe S, Isobe N, Miyauchi E et al (2006) Identification of a peptide in enzymatic hydrolyzate of cheese that inhibits ovalbumin permeation in Caco-2 cells. J Agric Food Chem 54:6904–6908

Taylor F (1995) Bleu cheese flavor production by submerged fermentation with on-line monitoring by gas chromatography. J Ind Microbiol 15:71–74

Tomasini A, Bustillo G, Lebeault JM (1993) Fat lipolyzed with a commercial lipase for the production of Blue cheese flavour. Int Dairy J 3:117–127

Tonouchi H, Suzuki M, Uchida M et al (2008) Antihypertensive effect of an angiotensin converting enzyme inhibitory peptide from enzyme modified cheese. J Dairy Res 75:284–290

Varming C, Beck TK, Petersen MA et al (2011) Impact of processing steps on the composition of volatile compounds in cheese powders. Int J Dairy Technol 64:197–206

Villeneuve P, Foglia TA (1997) Lipase specificities: potential application in lipid bioconversions. Inform 8:640–650

West S (2007) Production of flavours, flavour enhancers and other protein-based speciality products. In: Rastall R (ed) Novel enzyme technology for food applications. CRC Press, Boca Raton, pp 183–204

Wilkinson MG, Doolan IA, Kilcawley KN (2011) Enzyme-modified cheese. In: Fuquay JW, Fox PF, McSweeney PLH (eds) Encyclopedia of dairy sciences, vol 1. Elsevier, London, pp 799–804

Yamasaki H, Mitsuyama K (2014) Enzyme-modified cheese suppresses intestinal inflammation in rats. In: Japan digestive disease week, international session workshop-1, October 24, 2014

Chapter 17
Enzymatic Protein Cross-Linking in Dairy Science and Technology

Norbert Raak, Harald Rohm, and Doris Jaros

17.1 Introduction

Proteins determine sensory and technological properties of foods to a large extent. Besides protein degradation, e.g., proteolysis during cheese ripening, their polymerisation is also of interest with respect to structure modification. Covalent cross-linking of food proteins has been extensively studied in the past and is still a topic of scientific and industrial interest. While non-enzymatic cross-linking reactions are rather undesired and occur mainly in an uncontrolled manner during storage and processing of foods, cross-linking by enzymes has a huge potential to modify techno-functional properties of proteins because of their high substrate specificity and the mild reaction conditions required. This chapter provides an overview of the most prominent enzymes used for covalent cross-linking of food proteins and analytical approaches for the analysis of cross-linking reactions, as well as a discussion of their application in dairy science and technology.

17.2 Chemical Aspects and Production of Enzymes for Protein Cross-Linking

In the recent years, four cross-linking enzymes have mainly been applied in dairy research and technology, namely transglutaminase (TGase), laccase, tyrosinase and peroxidase. Table 17.1 gives an overview of the main characteristics of these enzymes.

N. Raak (✉) · H. Rohm · D. Jaros
Chair of Food Engineering, Institute of Natural Materials Technology, Technische Universität Dresden, Dresden, Germany
e-mail: norbert@food.au.dk; harald.rohm@tu-dresden.de; doris.jaros@tu-dresden.de

A. L. Kelly, L. B. Larsen (eds.), *Agents of Change*, Food Engineering Series,
https://doi.org/10.1007/978-3-030-55482-8_17

Table 17.1 Main characteristics of relevant cross-linking enzymes for application in dairy science and technology

	Transglutaminase (TGase)	Laccase	Tyrosinase	Peroxidases
Enzyme type and EC number	Acyltransferase (protein-glutamine γ-glutamyltransferase) EC 2.3.2.13	Oxidoreductase (benzenediol:oxygen oxidoreductase) EC 1.10.3.2	Oxidoreductase (monophenol, *o*-diphenol:oxygen oxidoreductase) EC 1.14.18.1	Oxidoreductase EC 1.11.1.x
Molar mass	Monomeric; ~38 kg/mol (microbial TGase)	Monomeric; 40–100 kg/mol	Multimeric; monomers of ~13.4 and ~ 43 kg/mol (fungal tyrosinases)	Depending on the origin; for instance ~45 kg/mol (horseradish peroxidase)
Cross-linking mechanism	Acyl transfer between one γ-carboxamide group of glutamine (acyl acceptor) and one ε-amino group of lysine (acyl donor), release of one mole ammonia (Fig. 18.1b).	(1) Oxidation of protein-bound tyrosine residues to free phenolic radicals, reduction of molecular oxygen to water; (2) spontaneous radical coupling to other phenolic radicals or free aromatic and amino groups (Fig. 18.2a). Cross-linking is facilitated in the presence of small phenolic mediators (Fig. 18.2b)	(1) Oxidation of protein-bound tyrosine residues to *o*-diphenols, reduction of molecular oxygen to water; (2) oxidation of protein-bound *o*-diphenolic residues to *o*-quinones, reduction of molecular oxygen to water; (3) spontaneous 1,4-addition or radical coupling to other *o*-quinones or free hydroxyl, amine or sulfhydryl groups (Fig. 18.3a). Cross-linking is facilitated in the presence of small phenolic mediators (Fig. 18.3b)	(1) Oxidation of protein-bound functional groups to free radicals, reduction of hydrogen peroxide to water; (2) spontaneous radical coupling to other molecules or protein-bound functional groups. Substrate specificity depends on the origin, e.g., horseradish peroxidase acts on tyrosine (Fig. 18.2a). Cross-linking is facilitated in the presence of small phenolic mediators (Fig. 18.2b)
Origins	Animal tissues, plants, fungi, bacteria, yeasts	Fungi, plants, insects, bacteria	Animals, plants, fungi, bacteria	Animals, plants, fungi, bacteria
Main biological functions	Protein modification (e.g., formation of blood clots, cell wall organisation)	Lignin degradation and biosynthesis	Biosynthesis of melanin pigments (e.g., browning of fruits and vegetables), formation of fungal spores	Lignin degradation and biosynthesis, oxidation of various compounds

Table 17.1 (continued)

	Transglutaminase (TGase)	Laccase	Tyrosinase	Peroxidases
Sources of commercially available enzymes	*Streptomyces mobaraensis* (bacterial)	*Trametes* sp. (fungal; frequently referred to as Laccase C)	*Agaricus bisporus* (fungal)	Horseradish (plant-based)
Other sources (examples)	Other *Streptomyces* sp. (bacterial), Guinea pig liver (tissue TGase)	*Agaricus bisporus* (fungal), *Aspergillus* sp. (fungal)	*Bacillus megaterium* (bacterial), *Botryosphaeria obtuse* (fungal), *Trichoderma reseei* (fungal)	
Further reading	Buchert et al. (2010); Giosafatto et al. (2018); Heck et al. (2013); Jaros et al. (2006a); Zeeb et al. (2017)	Buchert et al. (2010); Heck et al. (2013); Isaschar-Ovdat and Fishman (2018); Zeeb et al. (2017)	Buchert et al. (2010); Faccio et al. (2012); Heck et al. (2013); Huppertz (2009); Isaschar-Ovdat and Fishman (2018)	Buchert et al. (2010); Heck et al. (2013); Isaschar-Ovdat and Fishman (2018); Veitch (2004)

17.2.1 Transglutaminase

TGase is the most prominent cross-linking enzyme in food processing and is well-known for its application as "meat glue" in reformed meat and seafood products. Enzymes of the TGase family are ubiquitous in nature and contribute to numerous biological functions in humans, animals, plants, and microorganisms, e.g., blood clot formation (Factor XIII) or cell wall organisation (Giosafatto et al. 2018). TGases catalyse acyl transfer reactions between γ-carboxyamide groups of protein-bound glutamine (acyl donor) and primary amines (acyl acceptor), which can be amino groups of small, functional molecules (Fig. 17.1a), but also the ε-amino group of protein-bound lysine (Fig. 17.1b). The latter will result in protein cross-linking if moieties from different protein molecules participate in the reaction. In the absence of primary amines, deamidation of glutamine to glutamic acid is catalysed using H_2O as co-substrate (Fig. 17.1c). Each reaction leads to the release of one molecule of ammonia.

At present, the major organisms for industrial production of TGase are *Streptomyces* sp., in particular *S. mobaraensis*. A review of typical fermentation conditions for TGase production by *Streptomyces* sp. was provided by Gharibzahedi and Chronakis (2018). In contrast to most tissue TGases, microbial TGases are particularly suitable for food processing, as their activity is independent of Ca^{2+}, which otherwise might cause aggregation or precipitation of proteins.

TGase activity can be determined using the hydroxamate method (Folk and Cole 1965, 1966). In this assay, TGase catalyses an acyl transfer between one mole of N-carboxybenzyl-L-glutaminyl-glycine (Z-Gln-Gly; acyl donor) and one mole of hydroxylamine (acyl acceptor), resulting in one mole hydroxamate and one mole ammonia. The reaction is stopped by adding an iron(III)-chloride containing reagent, which induces a colour reaction of hydroxamate. Hydroxamate

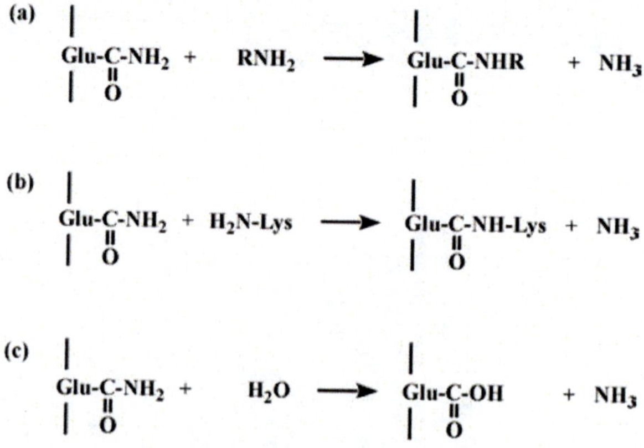

Fig. 17.1 Reactions catalysed by transglutaminase. (**a**) Acyl transfer; (**b**) Cross-linking of proteins; (**c**) Deamidation. Reprinted from Yokoyama et al. (2004) with permission from Springer Nature

concentration can then be determined photometrically at 525 nm, e.g., using L-glutamic acid γ-monohydroxamate for calibration. Enzyme activity is commonly expressed as Units (U) per g TGase powder, where 1 U corresponds to 0.5 μmol hydroxamate per min (or sometimes 1 μmol per min).

The optimum conditions for microbial TGase are pH 6–7 and 40–50 °C (Ando et al. 1989). However, the enzyme gradually loses its activity during incubation at ≥40 °C (Eissa et al. 2004; Raak et al. 2019a). Furthermore, Kütemeyer et al. (2005) reported an activity loss in the presence of Mg^{2+} and Ca^{2+}. TGase can be inactivated by heating above 70 °C (Buchert et al. 2010) as well as chemically, e.g., using N-ethylmaleimide (NEM; Jaros et al. 2010).

It has been reported that an inhibitor of microbial TGase exists in milk of different mammalian species including human milk, which considerably hampers cross-linking of milk proteins (de Jong et al. 2003). This inhibitor can be inactivated by UHT heat treatment (Bönisch et al. 2004, 2006) or through the addition of glutathione (Bönisch et al. 2007a). Bönisch et al. (2007a) suggested that glutathione only reduces interactions between inhibitor and TGase since the effective TGase inhibition could not be reversed by glutathione. However, Jacob et al. (2011) and Jaros et al. (2010) reported on successful TGase-catalysed cross-linking of raw milk without heating or glutathione addition. This indicates that inhibitor occurrence is not a general phenomenon.

17.2.2 Oxidoreductases

Laccase, tyrosinase and peroxidase belong to the group of oxidoreductases, which have a completely different mechanism than TGase. Here, cross-linking is the result of non-enzymatic reactions subsequent to enzymatic oxidation of protein-bound amino acid side chains, predominantly tyrosine (Figs. 17.2a and 17.3a). Oxidoreductases require co-substrates, namely oxygen in the case of laccase and

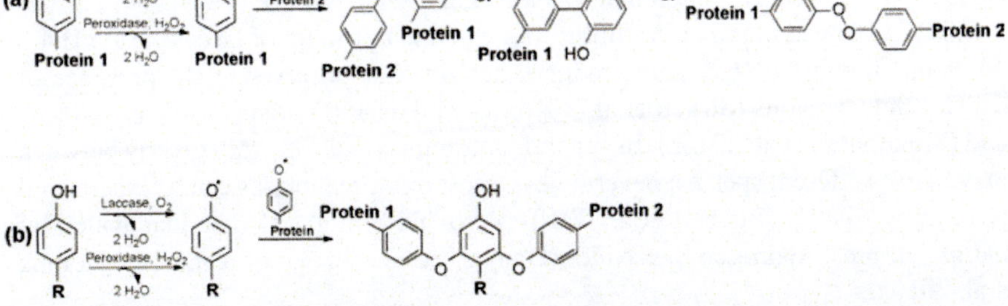

Fig. 17.2 Proposed cross-linking mechanisms of laccase and peroxidase in (**a**) the absence or (**b**) the presence of a small phenolic mediator. Reprinted from Isaschar-Ovdat and Fishman (2018) with permission from Elsevier

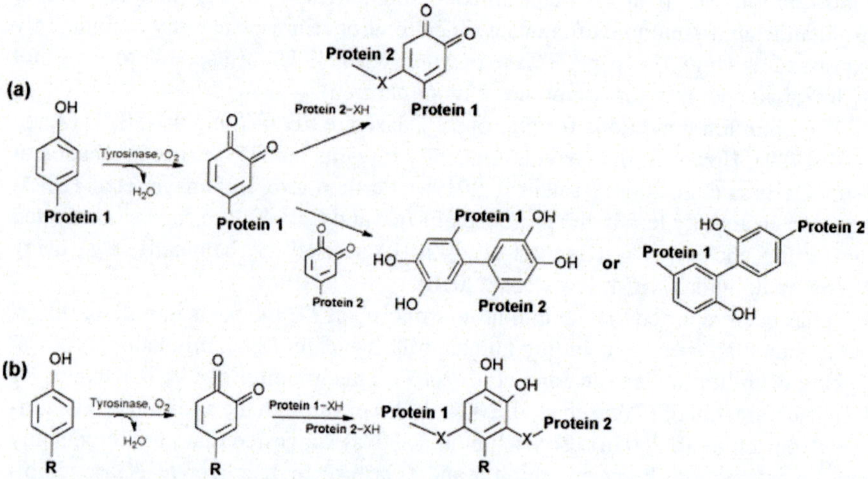

Fig. 17.3 Proposed cross-linking mechanisms of tyrosinase in (**a**) the absence or (**b**) the presence of a small phenolic mediator. X represents a hydroxyl, amine or sulfhydryl group. Reprinted from Isaschar-Ovdat and Fishman (2018) with permission from Elsevier

tyrosinase, and hydrogen peroxide in the case of peroxidases. While, in aqueous systems, oxygen is frequently dissolved, hydrogen peroxide has to be added and is therefore a possibility to control the enzymatic reaction. Protein cross-linking is facilitated by adding small phenolic molecules such as vanillin, caffeic acid or ferulic acid that are oxidised more rapidly and therefore act as mediators of the reaction (Figs. 17.2b and 17.3b).

Laccase and peroxidase activity can be determined photometrically by monitoring oxidation of 2,2′-azino-bis(3-ethylbenzthiazoline-6-sulfonic acid) (ABTS) at 405 or 420 nm (Heijnis et al. 2010a, b; Mokoonlall et al. 2016a). When considering the molar absorption coefficient of oxidised ABTS (ε_{420} = 36,000 mol/(g cm)), enzyme activity can be estimated from the increase in absorbance: 1 U corresponds to 1 μmol of oxidised substrate per min. The activity of tyrosinase is determined similarly using either tyrosine or L-3,4-dihydroxyphenylalanine (L-DOPA), and substrate oxidation is monitored at 475 nm (Labus et al. 2015).

When dissolved in milk ultrafiltrate, the reaction optimum of laccase is at pH 4.5 (Mokoonlall et al. 2016a), and a reduced activity was observed in the presence of K^+, Cl^- and Ca^{2+} ions (Struch et al. 2016) as well as antioxidants such as ascorbic acid (Mokoonlall et al. 2016b). In general, optimum conditions vary greatly between enzymes of different species; nevertheless, most oxidoreductases can be inactivated by heating ≥70 °C (Buchert et al. 2010). Peroxidase might also be inactivated by adding surplus hydrogen peroxide (Dhayal et al. 2014) or catalase (Heijnis et al. 2010a).

17.3 Analysis of Cross-Linking Reactions

17.3.1 Quantification of Enzymatically Induced Covalent Cross-Links

17.3.1.1 N-ε-(γ-Glutamyl)-Lysine Isopeptide Content

The reaction products of TGase-catalysed cross-linking can be quantified by chromatographic techniques. While N-ε-(γ-glutamyl)-lysine is not resisting acid hydrolysis usually carried out for amino acid analysis, proteases and peptidases such as pepsin, pronase (a protease from *Streptomyces griseus*), prolidase and leucine aminopeptidase are not able to cleave isopeptide bonds and were thus frequently applied in multi-step hydrolysis of cross-linked proteins (Lauber et al. 2000; Schäfer et al. 2005). As the N-ε-(γ-glutamyl)-lysine dipeptide remains intact after enzymatic hydrolysis (Nemes et al. 2005), it can be quantified by amino acid analysis based on cation exchange chromatography (Jaros et al. 2014; Lauber et al. 2000) or reverse-phase high performance liquid chromatography (RP-HPLC) after precolumn derivatisation with, e.g., *o*-phthaldialdehyde or phenylisothiocyanate (Miller and Johnson 1999; Sakamoto et al. 1995; Stachel et al. 2010).

A rough estimation of the amount of isopeptide bonds is possible from the extent of lysine modification. Free amino groups such as unmodified lysine residues are derivatised, and the reaction products are detected photometrically. Typical reagents for this approach are *o*-phthaldialdehyde (Dinnella et al. 2002) or 2,4,6-trinitrobenzenesulphonic acid (Snyder and Sobocinski 1975). Although such assays do not distinguish between different types of amino group modification, they provide a good estimate of the amount of isopeptide bonds formed, since N-ε-(γ-glutamyl)-lysine is the most abundant lysine modification caused by TGase. Djoullah et al. (2015) presented a method for rapid quantification of N-ε-(γ-glutamyl)-lysine isopeptide bonds in model substrates which is based on ^{1}H NMR spectroscopy. This, however, still needs to be validated for complex matrices.

17.3.1.2 Cross-Links from Other Enzymatic Reactions

Cross-links induced by oxidoreductases are more difficult to quantify since these cross-linking reactions can result in a broad variety of covalent bonds (Buchert et al. 2010). Monogioudi et al. (2011) applied ^{31}P NMR spectroscopy after phosphorylation of hydroxyl groups to quantify cross-links formed after oxidation of tyrosine residues by tyrosinase. Heijnis et al. (2010a) monitored peroxidase-induced formation of di-tyrosine cross-links in α-lactalbumin by UV absorbance at 318 nm and related the signal to the reduction of monomeric protein as determined by size-exclusion chromatography (SEC).

17.3.2 Static and Dynamic Light Scattering

Light scattering enables absolute size and/or molar mass determination and is therefore suitable to analyse changes in milk protein conformation as induced by enzymatic cross-linking. Two major techniques can be distinguished, namely static and dynamic light scattering (SLS and DLS). In SLS, the sample is subjected to monochromatic laser light, and the intensity of the light scattered by the analyte is directly related to its molar mass, concentration and refractive index increment and its distribution over different angles (Wyatt 1993). Light scattering detection at multiple angles, referred to as multi angle light scattering (MALS), thus allows accurate determination of molar mass and radius of gyration (R_g) and conclusions to be drawn on structural conformation of the analyte, e.g., spherical *vs.* non-spherical (Raak et al. 2018).

Size determination by DLS is based on the Stokes-Einstein equation, which describes the relationship between diffusion coefficient (D) and hydrodynamic radius (R_h), i.e., the radius of a sphere with identical diffusion coefficient:

$$D = \frac{k_B T}{6\pi R_h \eta_S} \tag{17.1}$$

Here, k_B is the Boltzmann constant, T the absolute temperature and η_S the solvent viscosity (Alexander and Dalgleish 2006). D is related to Brownian motion of the analyte in solution, because of which the scattered light intensity fluctuates over time. Brownian motion is fast for small analytes and slow for larger analytes; therefore, small analytes cause fast and large analytes slow fluctuations. These fluctuations are analysed by the autocorrelation function $g_{(1)}$, which is, for monodisperse samples:

$$g_{(1)}(t) = e^{-q^2 D t} \tag{17.2}$$

where q, the scattering vector, depends on laser wave length and scattering angle, which are both constant and known. The autocorrelation function of DLS measurements hence allows the determination of D and R_h of the analyte (Pecora 2000).

Light scattering can be performed in batch measurements or online coupled to preceding separations. SLS in batch-mode provides average values for molar mass and R_g, while approximate distributions of R_h can be obtained by batch-DLS. As light scattering intensity increases with analyte size, the estimated proportions of different fractions in polydisperse samples are not reliable. To increase the accuracy of size distribution analysis, light scattering should therefore be coupled to size separation techniques for on-line detection of eluting fractions.

17.3.3 Size Separation of Polymerised Milk Proteins

Size separation is the most common way to analyse enzymatically cross-linked milk proteins, since it provides a verification of intermolecular cross-linking as well as an estimate of the size distribution of different polymeric fractions independent of the cross-linking reaction. Gel electrophoresis and SEC were frequently applied under denaturing (e.g., urea or sodium dodecyl sulphate (SDS) containing buffers) and reducing (e.g., addition of dithiothreitol or β-mercaptoethanol) conditions to separate only according to the cross-linking reaction by suppressing non-covalent interactions and cleaving disulphide bonds (Raak et al. 2018).

17.3.3.1 Gel Electrophoresis

In gel electrophoresis, samples are injected to a polymeric gel matrix (polyacrylamide or agarose) which is subjected to an electric current that generates a migration of the analyte, usually from cathode to anode. Migration velocity depends on molar mass and net charge of the analyte, as well as the pore size of the gel matrix, facilitating separation in a particular matrix according to electrophoretic mobility. In the presence of SDS, the net charge is masked and a constant charge per mass ratio is obtained for protein-SDS-complexes, allowing a separation only according to molar mass. After separation, staining with, e.g., Coomassie brilliant blue, silver or fluorescent dyes facilitates visualisation and semi-quantitative analysis of the different fractions (O'Mahony et al. 2002; Westermeier 2011). Gel electrophoresis enables separation of the individual milk protein monomers and therefore the evaluation of their susceptibility to distinct cross-linking reactions. The separation range is limited by the pore size of the gel, and most commercial molar mass markers are not suitable as calibration substances for cross-linked caseins (Raak et al. 2018).

17.3.3.2 Size-Exclusion Chromatography

In SEC, samples are injected into a column packed with porous polymer beads. The analyte diffuses into the pores and out again during transportation through the column in liquid buffers, and smaller molecules are retained longer as they can access a higher number of pores. SEC separation range is generally limited by the largest and the smallest pores, which vary between different column materials. The eluted fractions are usually detected by concentration-sensitive detectors such as UV/Vis or refractive index, which also allows evaluation of their relative amounts (Hagel 2011). Online coupling of MALS and/or DLS detectors can further provide distributions of molar mass, R_g and R_h, which are of particular relevance for cross-linked caseins where no suitable standards for molar mass calibration are available (Raak et al. 2018).

17.3.3.3 Field Flow Fractionation

Field flow fractionation has been applied only few times to analyse milk proteins (e.g., Cragnell et al. 2017; Dhayal et al. 2014; Glantz et al. 2010). The analyte is carried in a liquid buffer through an empty channel in laminar flow while an orthogonally superimposed field causes analyte accumulation at the bottom of the channel. The concentration gradient along the channel leads to the diffusion of the analyte to the centre. This is more effective for smaller molecules with higher diffusion coefficients, which are therefore transported more rapidly to the outlet. The most abundant technique in food macromolecule analysis is asymmetrical flow field flow fractionation (AF4), where channel flow is superimposed by a cross-flow through an ultrafiltration membrane in the channel bottom (Fig. 17.4). In case of a constant cross-flow, the diffusion coefficient and hence R_h can be directly determined from elution times (Nilsson 2013; Wahlund and Nilsson 2012). Otherwise, online MALS and/or DLS can be used for conformational analyses. The major advantage of AF4 is the sufficiently wide separation range to separate casein monomers and casein micelles at native conditions (Raak et al. 2018). However, the complexity of the equipment and of the measurements is much higher compared to gel electrophoresis and SEC.

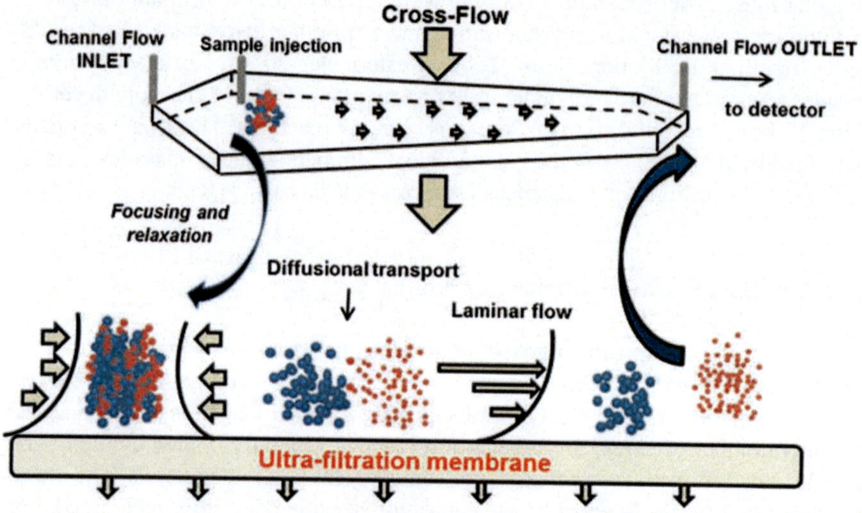

Fig. 17.4 Schematic illustration of separation channel (top) and separation principle (bottom) of asymmetric flow field flow fractionation. Particle populations with two different sizes are separated according to their diffusivity. Reprinted from Raak et al. (2018); copyright by the authors

Table 17.2 Overview of review articles and book chapters specifically referring to the application of cross-linking enzymes in dairy science and technology (since 2005; in chronological order)

Authors and type of publication	Enzyme(s)	Reviewed dairy product(s)	Main conclusions concerning the effects of protein cross-linking	Additional information
Jaros et al. (2006a) (Review article)	TGase[a]	• Acid-induced milk gels • Milk protein ingredients	• Increased firmness of non-fat, reduced-fat and full-fat yoghurts • Increased thermal stability and emulsifying activity of sodium caseinate	• Historical background of TGase • Chemical aspects, catalytic mechanism and substrate specificity of TGase • Quantification of N-ε-(γ-glutamyl)-lysine isopeptide bonds • Use for non-dairy foods
Özrenk (2006) (Review article)	TGase	• Various	• Increased gel strength and decreased permeability and syneresis of yoghurt	• Chemical aspects of TGase • Use for non-dairy foods
Huppertz (2009) (Book chapter)	TGase Laccase Tyrosinase	• Acid-induced milk gels • Rennet-induced milk gels • Heated milk • Milk proteins as emulsifiers	• Improved acid-induced gelation • Impaired rennet-induced gelation • Increased heat stability of milk at pH >6.6 • Increased stability of milk protein-stabilised emulsions	• Chemical aspects and substrate specificity of cross-linking enzymes • Techno-functional and physico-chemical properties of uncross-linked and cross-linked milk proteins

(continued)

Table 17.2 (continued)

Authors and type of publication	Enzyme(s)	Reviewed dairy product(s)	Main conclusions concerning the effects of protein cross-linking	Additional information
Jaros and Rohm (2016) (Book chapter)	TGase	• Fermented milk (yoghurt) • Cheese • Milk proteins as emulsifiers	• Increased firmness and reduced syneresis of yoghurt allows a reduction of dry matter enrichment • Delayed coagulation in cheese production because of reduced κ-casein hydrolysis • Increased surface load at oil-water interfaces, reduced coalescence and improved mechanical stability of casein-stabilised emulsions	• Chemical aspects and substrate specificity of microbial TGase • Analysis of cross-linked proteins • Application of TGase in non-dairy foods and non-food products
Romeih and Walker (2017) (Review article)	TGase	• Fermented milk (yoghurt) • Cheese • Milk protein ingredients	• Increased viscosity/gel strength and water holding capacity of yoghurt • Increased yield and water holding capacity of soft cheeses; delayed coagulation and retarded ripening in hard cheese production • Little information on the production and application of ingredients from cross-linked milk proteins (e.g., milk powder, caseinates) are available	• Chemical aspects and substrate specificity of microbial TGase • Bioavailability of lysine in cross-linked proteins • Overview of patents on TGase application in dairy products • Safety and legal aspects of TGase application

(continued)

Table 17.2 (continued)

Authors and type of publication	Enzyme(s)	Reviewed dairy product(s)	Main conclusions concerning the effects of protein cross-linking	Additional information
Akal et al. (2018) (Book chapter)	TGase	• Fermented milk (yoghurt) • Cheese • Dairy-based edible films	• Increased water holding capacity and improved texture of non- and reduced-fat yoghurt • Increased cheese yield and increased gel strength and water holding capacity of cheese curds • Reduced vapour permeability of protein films	• General aspects of milk protein cross-linking by TGase
Gharibzahedi and Chronakis (2018) (Review article)	TGase	• Fermented milk (yoghurt) • Yoghurt drinks • Encapsulated probiotics	• Increased viscosity, decreased syneresis and more homogeneous microstructure of yoghurt • Reduced phase separation of yoghurt drinks • Increased viability of probiotics encapsulated in protein matrices using transglutaminase	• Production of TGase from microbial origins
Gharibzahedi et al. (2018a) (Review article)	TGase	• Cheese • Ice cream	• Increased moisture content and yield and enhanced texture and sensory properties of cheese products • Increased consistency and reduced firmness and melting rate of ice cream	

[a]TGase = transglutaminase

17.4 Application of Cross-Linking Enzymes in Dairy Science and Technology

17.4.1 Overview of Review Articles and Book Chapters

Enzymatic cross-linking of milk proteins has been studied for years, and, as a consequence, a number of reviews and book chapters are available. From the overview in Table 17.2, it is evident that cross-linking with TGase has received by far the most attention, probably because of its commercial availability. Loveday et al. (2013) reviewed the application of TGase in the context of emerging technologies such as high pressure treatment and ultrasound processing in yoghurt production. The majority of the research was conducted on TGase application in yoghurt manufacture. There is a general consensus that cross-linking increases gel strength (Fig. 17.5) and water-holding capacity as a result of a denser, more homogeneous gel network. Increased moisture content and thus increased yield are obtained for cheese curd, but cross-linking of κ-casein on the casein micelle surface may delay rennet-induced coagulation.

Table 17.3 summarises reviews that, while not specifically focussed on dairy products, provide additional information on the enzymes and general aspects concerning cross-linked food proteins. Topics of interest are legal aspects of enzyme application, and effects of cross-linking on nutritional properties of food proteins.

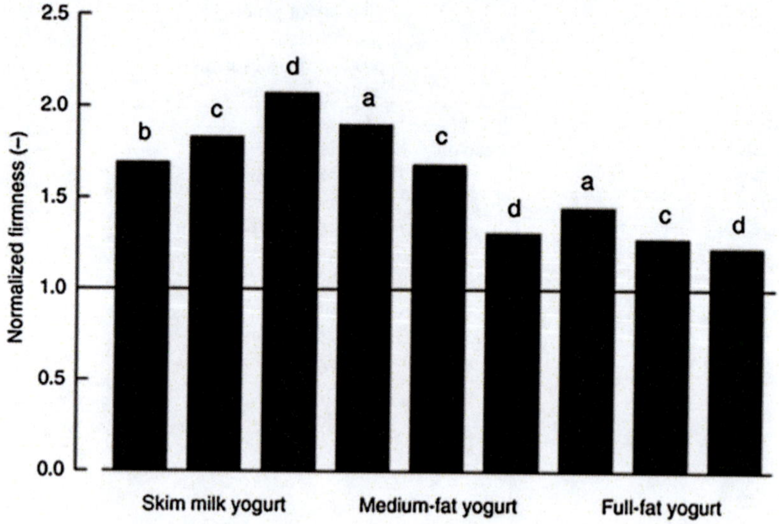

Fig. 17.5 Effect of cross-linking with microbial transglutaminase on the firmness of set yoghurt. Firmness was measured by penetration and is normalised to the firmness of the reference yoghurt (1.0). Data were compiled from different studies: (a) Færgemand et al. (1999), (b) Lauber et al. (2000), (c) Lorenzen et al. (2002) and (d) Mautner et al. (1999). Reprinted from Jaros et al. (2006a) with permission of John Wiley and Sons

Table 17.3 Overview of review articles and book chapters on the application of cross-linking enzymes in food processing with minor reference to milk proteins and dairy products (since 2005; in chronological order)

Authors and type of publication	Enzyme(s)	Main contents	Main conclusions
Buchert et al. (2010) (Review article)	TGase[a] Laccase Tyrosinase Peroxidase Others	• Sources, biological functions and reaction mechanisms of cross-linking enzymes • Application of cross-linking enzymes in cereal, dairy, meat and fish products • Digestibility and allergenicity of cross-linked food proteins	• Enzymes improved texture properties of foods • Different reaction mechanisms of the individual enzymes result in different technological properties
Heck et al. (2013) (Review article)	TGase Laccase Tyrosinase Peroxidase Others	• Biological functions and reaction mechanisms of cross-linking enzymes • Application of cross-linking enzymes in food processing, food packaging, biomimetic materials and textile and leather manufacturing	• Protein cross-linking affects long-term stability and sensory properties of foods • Cross-linked food proteins in edible films can decrease the gas and vapour permeability
Kieliszek and Misiewicz (2014) (Review article)	TGase	• Chemical aspects, reaction mechanism and sources of TGase • Production of TGase from microbial origins and available TGase preparations • Application of microbial TGase in meat, fish, dairy, and cereal products	• TGase improved texture properties of restructured meat and fish, dairy products and baked goods
Zeeb et al. (2014) (Review article)	TGase Laccase Tyrosinase Peroxidase Others	• Reaction mechanisms of cross-linking enzymes • Cross-linking enzymes for stabilisation of food dispersions (e.g., emulsions, liposomes, biopolymer dispersions)	• TGase increased pH-, salt- and temperature stability of biopolymers used as stabilisers or emulsifiers

(continued)

Table 17.3 (continued)

Authors and type of publication	Enzyme(s)	Main contents	Main conclusions
Gaspar and de Góes-Favoni (2015) (Review article)	TGase	• Reaction mechanism of microbial TGase • Effect of cross-linking of food proteins on techno-functional properties (solubility, gelation, emulsification, foam formation, viscosity)	• Cross-linking and deamidation by TGase change the hydrophobicity of food proteins and thus affect their solubility • In turn, the solubility is closely related to emulsification, foam formation, gelation and viscosity
Gharibzahedi et al. (2018b) (Review article)	TGase	• Production of TGase from microbial origins • Combination of TGase-treatment with emerging food processing techniques (e.g., high pressure, ultrasound, microwave, and UV irradiation)	• Emerging food processing techniques can increase the efficiency of protein cross-linking by TGase because of changes in secondary and tertiary structure
Gharibzahedi et al. (2018c) (Review article)	TGase	• TGase-aided encapsulation of unsaturated fatty acid rich oils in protein matrices	• Enzymatic cross-linking of microcapsule wall material increases encapsulation efficiency and yield and delays the lipid oxidation and core release of oils
Giosafatto et al. (2018) (Book chapter)	TGase	• Biological function of TGase in animals, and production of TGase from microbial origins • Application of TGase for processing of milk, egg, meat, fish, wheat and legume proteins • Digestibility and allergenicity of cross-linked food proteins • Commercially available microbial TGase preparations • Legal aspects of TGase application	• Stronger texture of homogenised pork, beef and poultry sausages • Increased hardness of fish products • Improved quality of dairy products • Cross-linked proteins might be digested to a lesser extent and could facilitate a controlled energy intake

(continued)

Table 17.3 (continued)

Authors and type of publication	Enzyme(s)	Main contents	Main conclusions
Isaschar-Ovdat and Fishman (2018) (Review article)	Laccase Tyrosinase Peroxidase	• Origins, biological functions and reaction mechanisms of oxidative cross-linking enzymes • Application of oxidative cross-linking enzymes for processing of milk, meat, cereal and soy proteins • Digestibility and allergenicity of cross-linked food proteins	• Oxidative cross-linking enzymes can contribute to protein network formation, hydrogel formation, bread making, emulsion stabilisation and production of protein nanoparticles • Commercialisation of oxidative cross-linking enzymes will be necessary for their application on the industrial level

[a]TGase = transglutaminase

Microbial TGase is "generally recognised as safe" (GRAS) without need for labelling in the ingredients list because of its classification as a processing aid (Giosafatto et al. 2018; Romeih and Walker 2017). Cross-linking can increase the protein resistance against digestive enzymes, while the bioavailability of lysine cross-linked to glutamine is hardly affected. Most studies pointed out that the allergenic potential of proteins is barely affected or even reduced by enzymatic cross-linking (Buchert et al. 2010; Giosafatto et al. 2018; Isaschar-Ovdat and Fishman 2018; Romeih and Walker 2017).

17.4.2 Limitations of Transglutaminase Application for Acid-Induced Milk Gels

Many studies indicated that TGase-catalysed protein cross-linking improves the formation of acid-induced milk gels (Table 17.2), but limitations of such approaches were also reported. For instance, Jaros et al. (2006b) reported that cross-linking had no effect or even resulted in a decreased gel stiffness depending on incubation time and enzyme dosage. The group later found out that the acidification rate has a significant impact; while cross-linked samples showed a higher gel stiffness at slow acidification (i.e., low amounts of glucono-δ-lactone) compared with the reference, the opposite was observed at higher acidification rates (Jacob et al. 2011).

Although negative effects of milk protein cross-linking by TGase on acid-induced gelation are mainly observed under "laboratory conditions", i.e., long incubation time or fast acidification, they can also become relevant for dairy products. For stirred yoghurt fortified with sodium caseinate, Bönisch et al. (2007b) reported an increase in viscosity and coarseness of the gel structure during cold storage without detecting further cross-linking. The authors suggested an alteration of non-covalent interactions in the gel network because of the covalent cross-links. The underlying mechanisms of structure deterioration of acid-induced milk protein gels caused by enzymatic cross-linking is still not fully explored. Recent research with model substrates instead of milk, however, indicated changes in the potential for non-covalent interactions due to a reduced flexibility of the protein particles (see Sect. 17.4.4).

17.4.3 Application of Laccase in Stirred Yoghurt Manufacture

Since review articles concerning enzymatic protein cross-linking in dairy products mainly focus on TGase, this section provides a brief summary of a recent, industrially targeted research project on laccase application in stirred yoghurt production.

Stirred yoghurt is fermented milk subjected to a defined breakdown of the gel network after fermentation to obtain a homogeneous, viscous product. It is commonly referred to as microgel suspension, because the dispersed particles show similar microstructure and properties as the macroscopic gel network (Mokoonlall et al. 2016c; Weidendorfer and Hinrichs 2010). The project aimed at reducing shear-induced destruction of microgel particles during processing and packaging, which causes a significant decrease of product viscosity. Here, laccase was more promising for particle stabilisation than TGase, because of its higher activity at acidic pH, which is favourable for post-processing application and storage.

In a first step, the application of a commercial laccase from *Trametes* sp., so-called Laccase C, resulted in a complex viscosity increase of commercial stirred yoghurt from skim milk, which was even more pronounced when cross-linking took place in the presence of phenolic mediators such as caffeic acid, vanillin or vanillic acid (Struch et al. 2015). This could, however, not be validated in lab- and pilot-plant-scale yoghurt manufacture, where constant or even reduced viscosities and storage moduli, as well as a more porous microstructure, were observed for laccase-treated products (Mokoonlall et al. 2016a, d). The authors attributed this to protein degradation occurring simultaneously to or instead of protein cross-linking, due to the reactivity of the radicals that are generated by the oxidation of tyrosine residues by laccase.

The activity of laccase as affected by milk components has been evaluated. Mokoonlall et al. (2016a) observed that Laccase C exhibits considerably lower activity in milk ultrafiltrate compared to sodium tartrate buffer, the standard solvent for the assay. Different milk ions were assessed for their inhibitory effect, which showed that laccase activity was most decreased in the presence of K^+, Cl^- or Ca^{2+} (Struch et al. 2016). However, the main goal of this study was to evaluate laccases from 20 fungal species to identify milk-ion-tolerant enzymes. The authors found laccases from *Funalia trogii*, *Pleurotus flabellatus*, *Pleurotus eryngii*, *Pleurotus lampas* and *Trametes versicolor* to be similarly or more tolerant against milk ions as Laccase C. Laccase from *P. eryngii* was selected for further investigations, and a critical enzyme concentration was found above which the enzyme caused more degradation than cross-linking of proteins, accompanied by decreased viscosity of commercial stirred yoghurt. Mokoonlall et al. (2016b) concluded that antioxidant activity of fermented milks may suppress oxidation of tyrosine residues by laccase and thus inhibit cross-linking reactions even in the presence of mediators.

Up to now, this has been the only comprehensive research on the application of an oxidoreductase at a scale relevant for the dairy industry. Although earlier studies demonstrated the potential of laccase and tyrosinase for structure improvement of casein gels at research level (e.g., Ercili Cura et al. 2009, 2010), this project clearly outlines difficulties of transfer to industrial level. Among these difficulties, unfavourable side reactions leading to protein degradation are probably the most critical ones. Mokoonlall et al. (2016c) therefore concluded that suitable enzymes with specific actions on target proteins are required for controlled enzymatic modification of dairy products such as stirred yoghurt.

17.4.4 Enzymatic Protein Cross-Linking in Cheese Making

A recent review on TGase application in cheese making (and ice cream production) (Gharibzahedi et al. 2018a) identified studies on all cheese types (fresh, cottage, soft, semi-hard, hard, low-fat, full-fat), including regional specialties such as Edam, Paneer, Feta or Iranian white cheese. From the available literature, they deduced five different ways for adding TGase during cheese making: (1) prior to renneting; (2) simultaneously with rennet; (3) after rennet addition but before coagulation; (4) after curd cutting; (5) after cooking. TGase first acts on κ-casein, which is located on the casein micelle surface. This can delay the rennet-induced coagulation because of reduced κ-casein hydrolysis and maintained steric repulsion between casein micelles (Bönisch et al. 2008; Huppertz et al. 2007). Therefore, adding TGase simultaneously to or after rennet might be favourable.

Recent studies on the potential of laccase were limited to rather simple systems. Loi et al. (2018) observed increased protein contents and antioxidant capacities of cheese curds from laccase-treated whole milk using chlorogenic acid as phenolic mediator. Mokoonlall et al. (2016a) applied laccase in a post-processing step to improve viscosity and shear resistance of fresh cheeses (concentrated fermented milk gels) with varying protein contents. However, the treatment increased the hysteresis loop area (i.e., higher structural loss during shearing), while the viscosity was increased for only half of the samples.

In principle, cross-linking can increase the moisture content and thus the yield of cheese curds while maintaining or improving curd firmness and cheese texture (Gharibzahedi et al. 2018a). Furthermore, cross-linking with TGase was shown to be suitable for incorporation of whey proteins to the cheese matrix (Cozzolino et al. 2003). However, process parameters have to be carefully adjusted.

Table 17.4 Overview of studies reporting on increased stability of casein micelles cross-linked with transglutaminase against different destabilisation mechanisms

Mechanism of destabilisation	Experimental approach	References
Dissociation of colloidal calcium phosphate	Addition of citrate	Huppertz et al. (2007); Nogueira et al. (2019); Smiddy et al. (2006)
	Addition of ethylenediaminetetraacetic acid (EDTA)	Heber et al. (2012); Lam et al. (2017); Partschefeld et al. (2007)
	High pressure treatment	Huppertz and de Kruif (2007a); Partschefeld et al. (2007); Smiddy et al. (2006)
Disruption of hydrophobic interactions	Addition of urea	Huppertz et al. (2007); Nogueira et al. (2019); Smiddy et al. (2006)
	Addition of sodium dodecyl sulphate (SDS)	Smiddy et al. (2006)
Loss of steric repulsion	Heat treatment ≥120 °C (dissociation of κ-casein)	Huppertz (2014); Huppertz and de Kruif (2008); Mounsey et al. (2005); Nogueira et al. (2019)
	Addition of ethanol; possibly combined with heat treatment ≤100 °C (collapse of the κ-casein layer)	Huppertz and de Kruif (2007b); Huppertz and de Kruif (2008); Nogueira et al. (2019); Partschefeld et al. (2007); Smiddy et al. (2006)
	Addition of rennet (hydrolysis of κ-casein)	Bönisch et al. (2008); Huppertz and de Kruif (2007c)
Alkaline swelling and disruption	pH increase	Duerasch et al. (2018); Lam et al. (2018)

17.4.5 *Creation of Well-Defined Micro- and Nanostructures*

17.4.5.1 Preparation of Microgel Particles by Cross-Linking of Casein Micelles

Mounsey et al. (2005) provided evidence that TGase preferentially cross-links molecules located within the same casein micelle, as their size was not affected by the treatment. When TGase acts on casein micelles, κ-casein is cross-linked first because of its good accessibility on the micelle surface, followed by β-casein, and the less accessible α_S-caseins which are located in the interior (Duerasch et al. 2018; Hinz et al. 2012; Lam et al. 2018). In this regard, the amount of cross-linked κ-casein relative to the total amount of cross-linked caseins decreased with decreasing temperature, increasing pH, decreasing $CaCl_2$ concentration and increasing amount of trisodium citrate, because of the improved accessibility of the other casein types at these conditions (Hinz et al. 2012; Moon et al. 2009). Duerasch et al. (2018) further reported that micelle-bound α_{S2}-casein was cross-linked to a larger extent on increasing pH from 6.8 to 7.9, suggesting increased accessibility of interior micelle regions due to alkaline swelling.

Internal cross-linking of casein micelles by TGase was found to increase the stability of casein micelles against destabilisation mechanisms such as dissociation of colloidal calcium phoshate (CCP) or heat-induced coagulation (Table 17.4). Modelling of light scattering and small angle neutron scattering data obtained from entirely cross-linked casein micelles in the presence of urea and/or trisodium citrate indicated a covalently bound, completely stable protein network (Huppertz et al. 2007). Thom Huppertz and colleagues were the first who referred to internally cross-linked casein micelles as microgel particles (or nanogel particles in one study), and suggested potential applications as carriers of biologically active compounds (e.g., minerals, vitamins, pharmaceuticals), or the replacement of sodium caseinate in food formulations by cross-linked, calcium-depleted casein micelles (Huppertz and de Kruif 2008; Huppertz et al. 2007). The approach of producing microgel particles from casein micelles has been pushed forward in recent years, not only by using TGase (Anema and de Kruif 2012; Bahri et al. 2018; Nogueira et al. 2019) but also *via* non-enzymatic cross-linking, e.g., using genipin (Casanova et al. 2017; Nogueira Silva et al. 2014). Nogueira Silva et al. (2014) suggested the application of casein microgel particles as carriers for hydrophobic substances or, after CCP removal, charged molecules that could be bound and released as controlled by pH. Up to now, little has been reported about the utilisation of microgel particles prepared by cross-linking casein micelles. Anema and de Kruif (2012) observed a comparable adsorption behaviour of lactoferrin to native casein micelles and casein microgel particles. However, the latter remained stable whereas lactoferrin caused rapid disintegration of native casein micelles.

17.4.5.2 Cross-Linking of Casein Nanoparticles: Molecular Aspects and Gelation Properties

For many years, research on the effect of TGase-induced cross-linking on the gelation properties of casein has been conducted using model substrates such as sodium caseinate (NaCn) to eliminate effects from whey proteins as well as from the complex features of native casein micelles. It was reported that molecules in aqueous NaCn solution self-associate to hydrated, elongated nanoparticles with $R_h \sim 10$ nm (Huppertz et al. 2017; Lucey et al. 2000), and that molar mass and R_h depend on ionic strength, pH and temperature (HadjSadok et al. 2008). As shown by AF4 under native conditions, TGase-treatment resulted in internal cross-linking of casein nanoparticles and increased their density and sphericity (Abbate et al. 2019). Denaturing and reducing SEC revealed a maximum in polymer size after moderate TGase-treatment although further cross-links were formed within the particles during prolonged incubation (Raak et al. 2019a, b). At casein concentrations relevant for acid-induced gelation (~10–50 g/kg), this maximum polymer size did not depend on incubation temperature (Raak et al. 2019a). In contrast, increasing ionic strength or adding bivalent cations increased the maximum polymer size (Raak et al. 2019b).

During acidification (e.g., using glucono-δ-lactone), casein nanoparticles aggregate because of the loss of electrostatic repulsion and form a gel network. Gelation can be monitored in time-based small amplitude oscillatory shear rheology, with the maximum storage modulus (G'_{max}) providing an indicator for gel stiffness. In the case of NaCn gels, G'_{max} increased continuously with ongoing cross-linking by TGase (Raak et al. 2019a) because of the direct contribution of N-ε-(γ-glutamyl)-lysine isopeptide bonds (Raak et al. 2017). Steady and oscillatory shear rheology of concentrated NaCn suspensions showed that internal cross-linking of casein nanoparticles increased their resistance against deformation, suggesting that the increase in gel stiffness was the result of an increased particle stiffness (Raak et al. 2020). However, when attractive electrostatic interactions were screened by the presence of ions, the highest G'_{max} was observed when maximum polymer size was just reached (Raak et al. 2019b). This indicated that excessive internal cross-linking of casein nanoparticles reduced molecular flexibility and hampered formation of non-covalent bonds by mechanisms other than electrostatic attraction.

A relationship between molecular characteristics of cross-linked caseins and rheological properties of acid-induced gels was found: the loss factor tan δ of the gels decreased with increasing maximum polymer size (Raak et al. 2019b). This indicated a lower density of particles because of a more loose and open structure (Saricay et al. 2016), as well as reduced mobility in the gel network because of steric reasons and/or greater interactions. After maximum polymer size was reached, tan δ was increased by further internal cross-linking, suggesting an increasing compactness of casein particles (Raak et al. 2019a, b).

A potential application of cross-linked casein nanoparticles might be, e.g., Pickering stabilisation of O/W emulsions, as was recently shown by Wang et al. (2018) for NaCn that was non-enzymatically cross-linked using genipin.

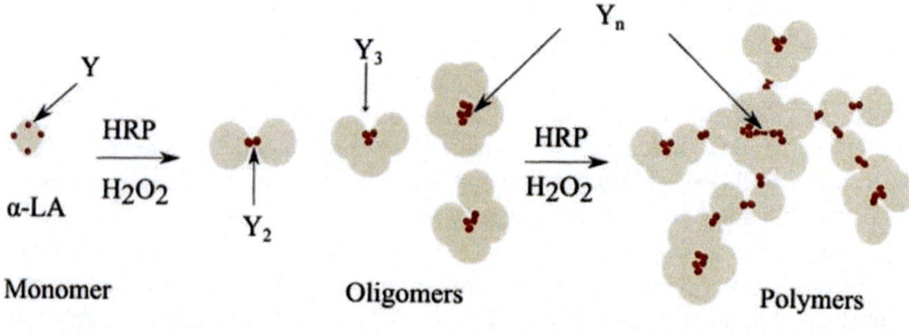

Fig. 17.6 Proposed two-stage process of nanoparticle formation through polymerisation of α-lactalbumin (α-LA; light grey circles) by horseradish peroxidase (HRP). The polymerisation is indicated by highlighting the di-tyrosine (Y_2), tri-tyrosine (Y_3) and oligo-tyrosine (Y_n) cross-links as dark red dots. Reprinted from Dhayal et al. (2015a) with permission from Elsevier

17.4.5.3 Formation and Physical Properties of Nanoparticles from Cross-Linked α-Lactalbumin

A group at Wageningen University investigated nanoparticles generated by horseradish peroxidase cross-linking of calcium-depleted α-lactalbumin, which contains four tyrosine residues that could potentially be oxidised by peroxidase. However, only cross-links between Tyr_{18} and Tyr_{50} were seen during the initial stage of the reaction, which was favourable because of a short distance between the residues and good charge compensation (Heijnis et al. 2011). This was followed by a gradual formation of different oligo-tyrosine cross-links, which possibly involved more than eight tyrosine residues per cross-link (Fig. 17.6; Dhayal et al. 2015a). The cross-linking reaction was also associated with a considerable loss of secondary and tertiary structure of α-lactalbumin, which might allow formation of cross-links between other tyrosine residues such as Tyr_{36} and Tyr_{103} (Saricay et al. 2014), but this was not studied in detail.

R_h remained constant on the particle scale during the initial stage of cross-linking, as only dimers and smaller oligomers were formed, followed by a substantial increase in R_h when oligomers were further cross-linked to larger polymers (Fig. 17.6; Dhayal et al. 2014, 2015a; Saricay et al. 2012, 2013). The final size of the nanoparticles increased from ~25 to ~100 nm with increasing protein concentration from 10 to 30 g/L, and gelation was observed at ≥40 g/L (Dhayal et al. 2014; Saricay et al. 2012, 2016). Larger nanoparticles were formed at higher ionic strength and lower pH as a consequence of conformational changes (Dhayal et al. 2014; Heijnis et al. 2010b). Independent of their size, the nanoparticles had a similar conformation and a fractal (self-similar) structure, but particle density decreased with increasing size (Dhayal et al. 2014; Saricay et al. 2013). The nanoparticles had a rather diluted and open-structured protein network because of the preferred formation of Tyr_{18}-Tyr_{50} cross-links (Saricay et al. 2012, 2013). In contrast, α-lactalbumin nanoparticles formed by TGase-induced cross-linking were twice as dense (Dhayal et al. 2015b).

At a particular protein concentration, viscosity increased with increasing R_h of the nanoparticles, and suspensions of larger nanoparticles showed more pronounced yet reversible shear thinning as well as more dominant elastic behaviour at ~50–100 g/L, indicating greater deformation and higher interpenetration of jammed particles due to the lower density (Saricay et al. 2016). Furthermore, α-lactalbumin nanoparticles increased the stability of foams compared to monomeric α-lactalbumin, whereas the foam volume was decreased (Dhayal et al. 2015b). Since Pickering stabilisation was assumed rather than drainage delay, particles with higher density would be even more favourable to increase foam stability as was shown for α-lactalbumin nanoparticles created by TGase. In fact, the nanoparticles underwent considerable flattening due to interfacial adsorption (Dhayal et al. 2015b).

In view of the structure and physical properties of α-lactalbumin nanoparticles, the research group suggested possible applications as protein-based thickeners or nanocarriers (Saricay et al. 2012, 2014). In this regard, it is worth noting that nanoparticles had a higher thermal stability compared to monomeric α-lactalbumin in terms of irreversible aggregation (Saricay et al. 2017).

17.5 Conclusions and Outlook

Enzymatic cross-linking of milk proteins has been studied extensively as a tool to modify physical and texture properties of (fermented) dairy products, but is still barely implemented on the industrial level. This is probably because it can be challenging to fully control enzymatic reactions when heat inactivation is undesired and/or unwanted side reactions occur. However, this could be managed in the future by the selection of enzymes which are suitable for the respective process and by gaining a better understanding of the effects of enzymatic cross-linking on the techno-functional behaviour of milk proteins in complex systems. Furthermore, cross-linking enzymes have a huge potential for the preparation of micro- and nanogel particles which could serve as techno-functional additives in foods or as nanocarriers for molecules with nutritional benefits. To this end, however, more research is needed.

References

Abbate RA, Raak N, Boye S et al (2019) Asymmetric flow field flow fractionation for the investigation of caseins cross-linked by microbial transglutaminase. Food Hydrocoll 92:117–124

Akal HC, Koçak C, Özer HB (2018) Transglutaminase applications in dairy technology. In: Öztürkoğlu Budak Ş, Akal HC (eds) Microbial cultures and enzymes in dairy technology, 1st edn. IGI Global, Hershey, pp 152–181

Alexander M, Dalgleish DG (2006) Dynamic light scattering techniques and their applications in food science. Food Biophys 1:2–13

Ando H, Adachi M, Umeda K et al (1989) Purification and characteristics of a novel transglutaminase derived from microorganisms. Agric Biol Chem 53:2613–2617

Anema SG, de Kruif CG (2012) Lactoferrin binding to transglutaminase cross-linked casein micelles. Int Dairy J 26:83–87

Bahri A, Martin M, Gergely C et al (2018) Topographical and nanomechanical characterization of casein nanogel particles using atomic force microscopy. Food Hydrocoll 83:53–60

Bönisch MP, Lauber S, Kulozik U (2004) Effect of ultra-high temperature treatment on the enzymatic cross-linking of micellar casein and sodium caseinate by transglutaminase. J Food Sci 69:E398–E404

Bönisch MP, Tolkach A, Kulozik U (2006) Inactivation of an indigenous transglutaminase inhibitor in milk serum by means of UHT-treatment and membrane separation techniques. Int Dairy J 16:669–678

Bönisch MP, Lauber S, Kulozik U (2007a) Improvement of enzymatic cross-linking of casein micelles with transglutaminase by glutathione addition. Int Dairy J 17:3–11

Bönisch MP, Huss M, Lauber S et al (2007b) Yoghurt gel formation by means of enzymatic protein cross-linking during microbial fermentation. Food Hydrocoll 21:585–595

Bönisch M, Heidebach T, Kulozik U (2008) Influence of transglutaminase protein cross-linking on the rennet coagulation of casein. Food Hydrocoll 22:288–297

Buchert J, Ercili Cura D, Ma H et al (2010) Crosslinking food proteins for improved functionality. Annu Rev Food Sci Technol 1:113–138

Casanova F, Nogueira Silva NF, Gaucheron F et al (2017) Stability of casein micelles cross-linked with genipin: a physicochemical study as a function of pH. Int Dairy J 68:70–74

Cozzolino A, Di Pierro P, Mariniello L et al (2003) Incorporation of whey proteins into cheese curd by using transglutaminase. Biotechnol Appl Biochem 38:289–295

Cragnell C, Choi J, Segad M et al (2017) Bovine β-casein has a polydisperse distribution of equilibrium micelles. Food Hydrocoll 70:65–68

de Jong GAH, Wijngaards G, Koppelman SJ (2003) Transglutaminase inhibitor from milk. J Food Sci 68:820–825

Dhayal SK, Gruppen H, de Vries R et al (2014) Controlled formation of protein nanoparticles by enzymatic cross-linking of α-lactalbumin with horseradish peroxidase. Food Hydrocoll 36:53–59

Dhayal SK, Sforza S, Wierenga PA et al (2015a) Peroxidase induced oligo-tyrosine cross-links during polymerization of α-lactalbumin. Biochim Biophys Acta 1854:1898–1905

Dhayal SK, Delahaije RJBM, de Vries RJ et al (2015b) Enzymatic cross-linking of α-lactalbumin to produce nanoparticles with increased foam stability. Soft Matter 11:7888–7898

Dinnella C, Gargaro MT, Rossano R et al (2002) Spectrophotometric assay using *o*-phtaldialdehyde for the determination of transglutaminase activity on casein. Food Chem 78:363–368

Djoullah A, Sok N, Djemaoune Y et al (2015) Monitoring of transglutaminase crosslinking reaction by ^{1}H NMR spectroscopy on model substrates. Colloid Surf A 475:69–74

Duerasch A, Wissel J, Henle T (2018) Reassembling of alkali-treated casein micelles by microbial transglutaminase. J Agric Food Chem 66:11748–11756

Eissa AS, Bisram S, Khan SA (2004) Polymerization and gelation of whey protein isolates at low pH using transglutaminase enzyme. J Agric Food Chem 52:4456–4464

Ercili Cura D, Lantto R, Lille M et al (2009) Laccase-aided protein modification: effects on the structural properties of acidified sodium caseinate gels. Int Dairy J 19:737–745

Ercili Cura D, Lille M, Partanen R et al (2010) Effect of *Trichoderma reesei* tyrosinase on rheology and microstructure of acidified milk gels. Int Dairy J 20:830–837

Faccio G, Kruus K, Saloheimo M et al (2012) Bacterial tyrosinases and their applications. Process Biochem 47:1749–1760

Færgemand M, Sørensen MV, Jørgensen U et al (1999) Transglutaminase: effect on instrumental and sensory texture of set-style yoghurt. Milchwissenschaft 54:563–566

Folk JE, Cole PW (1965) Structural requirements of specific substrates for guinea pig liver transglutaminase. J Biol Chem 240:2951–2960

Folk JE, Cole PW (1966) Mechanism of action of guinea pig liver transglutaminase. I. Purification and properties of the enzyme: identification of a functional cysteine essential for activity. J Biol Chem 241:5518–5525

Gaspar ALC, de Góes-Favoni SP (2015) Action of microbial transglutaminase (MTGase) in the modification of food proteins: a review. Food Chem 171:315–322

Gharibzahedi SMT, Chronakis IS (2018) Crosslinking of milk proteins by microbial transglutaminase: utilization in functional yogurt products. Food Chem 245:620–632

Gharibzahedi SMT, Koubaa M, Barba FJ et al (2018a) Recent advances in the application of microbial transglutaminase crosslinking in cheese and ice cream products: a review. Int J Biol Macromol 107:2364–2374

Gharibzahedi SMT, Roohinejad S, George S et al (2018b) Innovative food processing technologies on the transglutaminase functionality in protein-based food products: trends, opportunities and drawbacks. Trends Food Sci Technol 75:194–205

Gharibzahedi SMT, George S, Greiner R et al (2018c) New trends in the microencapsulation of functional fatty acid-rich oils using transglutaminase catalyzed crosslinking. Compr Rev Food Sci Food Saf 17:274–289

Giosafatto CVL, Al-Asmar A, Mariniello L (2018) Transglutaminase protein substrates of food interest. In: Kuddus M (ed) Enzymes in food technology, 1st edn. Springer, Singapore, pp 293–317

Glantz M, Håkansson A, Lindmark Månsson H et al (2010) Revealing the size, conformation, and shape of casein micelles and aggregates with asymmetrical flow field-flow fractionation and multiangle light scattering. Langmuir 26:12585–12591

HadjSadok A, Pitkowski A, Nicolai T et al (2008) Characterisation of sodium caseinate as a function of ionic strength, pH and temperature using static and dynamic light scattering. Food Hydrocoll 22:1460–1466

Hagel L (2011) Gel filtration: size exclusion chromatography. In: Janson J-C (ed) Protein purification: principles, high resolution methods, and applications, 3rd edn. Wiley, Hoboken, pp 51–91

Heber A, Paasch S, Partschefeld C et al (2012) ^{31}P NMR spectroscopic investigations of caseins treated with microbial transglutaminase. Food Hydrocoll 28:36–45

Heck T, Faccio G, Richter M et al (2013) Enzyme-catalyzed protein crosslinking. Appl Microbiol Biotechnol 97:461–475

Heijnis WH, Wierenga PA, Janssen AEM et al (2010a) In-line quantification of peroxidase-catalyzed cross-linking of α-lactalbumin in a microreactor. Chem Eng J 157:189–193

Heijnis WH, Wierenga PA, van Berkel WJH et al (2010b) Directing the oligomer size distribution of peroxidase-mediated cross-linked bovine α-lactalbumin. J Agric Food Chem 58:5692–5697

Heijnis WH, Dekker HL, de Koning LJ et al (2011) Identification of the peroxidase-generated intermolecular dityrosine cross-link in bovine α-lactalbumin. J Agric Food Chem 59:444–449

Hinz K, Huppertz T, Kelly AL (2012) Susceptibility of the individual caseins in reconstituted skim milk to cross-linking by transglutaminase: influence of temperature, pH and mineral equilibria. J Dairy Res 79:414–421

Huppertz T (2009) Novel applications of enzymes in the dairy sector: optimizing functional properties of milk proteins by enzymatic cross-linking. In: Corredig M (ed) Dairy-derived ingredients: food and nutraceutical uses, 1st edn. Woodhead, Cambridge, pp 394–416

Huppertz T (2014) Heat stability of transglutaminase-treated milk. Int Dairy J 38:183–186

Huppertz T, de Kruif CG (2007a) High pressure-induced solubilisation of micellar calcium phosphate from cross-linked casein micelles. Colloid Surf A 295:264–268

Huppertz T, de Kruif CG (2007b) Ethanol stability of casein micelles cross-linked with transglutaminase. Int Dairy J 17:436–441

Huppertz T, de Kruif CG (2007c) Rennet-induced coagulation of enzymatically cross-linked casein micelles. Int Dairy J 17:442–447

Huppertz T, de Kruif CG (2008) Structure and stability of nanogel particles prepared by internal cross-linking of casein micelles. Int Dairy J 18:556–565

Huppertz T, Smiddy MA, de Kruif CG (2007) Biocompatible micro-gel particles from cross-linked casein micelles. Biomacromolecules 8:1300–1305

Huppertz T, Gazi I, Luyten H et al (2017) Hydration of casein micelles and caseinates: implications for casein micelle structure. Int Dairy J 74:1–11

Isaschar-Ovdat S, Fishman A (2018) Crosslinking of food proteins mediated by oxidative enzymes – a review. Trends Food Sci Technol 72:134–143

Jacob M, Nöbel S, Jaros D et al (2011) Physical properties of acid milk gels: acidification rate significantly interacts with cross-linking and heat treatment of milk. Food Hydrocoll 25:928–934

Jaros D, Rohm H (2016) Enzymes exogenous to milk in dairy technology: transglutaminase. In: Smithers GW (ed) Reference module in food science. Elsevier, Amstersam

Jaros D, Partschefeld C, Henle T et al (2006a) Transglutaminase in dairy products: chemistry, physics, applications. J Texture Stud 37:113–155

Jaros D, Pätzold J, Schwarzenbolz U et al (2006b) Small and large deformation rheology of acid gels from transglutaminase treated milks. Food Biophys 1:124–132

Jaros D, Jacob M, Otto C et al (2010) Excessive cross-linking of caseins by microbial transglutaminase and its impact on physical properties of acidified milk gels. Int Dairy J 20:321–327

Jaros D, Schwarzenbolz U, Raak N et al (2014) Cross-linking with microbial transglutaminase: relationship between polymerisation degree and stiffness of acid casein gels. Int Dairy J 38:174–178

Kieliszek M, Misiewicz A (2014) Microbial transglutaminase and its application in the food industry. A review. Folia Microbiol 59:241–250

Kütemeyer C, Froeck M, Werlein H-D et al (2005) The influence of salts and temperature on enzymatic activity of microbial transglutaminase. Food Control 16:735–737

Labus K, Bryjak J, Polakovič M (2015) Kinetics of thermal inactivation of immobilized *Agaricus bisporus* tyrosinase. J Mol Catal B 120:136–140

Lam E, Holt C, Edwards P et al (2017) The effect of transglutaminase treatment on the physicochemical properties of skim milk with added ethylenediaminetetraacetic acid. Food Hydrocoll 69:329–340

Lam E, McKinnon I, Marchesseau S et al (2018) The effect of transglutaminase on reconstituted skim milks at alkaline pH. Food Hydrocoll 85:10–20

Lauber S, Henle T, Klostermeyer H (2000) Relationship between the crosslinking of caseins by transglutaminase and the gel strength of yoghurt. Eur Food Res Technol 210:305–309

Loi M, Quintieri L, Fanelli F et al (2018) Application of a recombinant laccase-chlorogenic acid system in protein crosslink and antioxidant properties of the curd. Food Res Int 106:763–770

Lorenzen PC, Neve H, Mautner A et al (2002) Effect of enzymatic cross-linking of milk proteins on functional properties of set-style yoghurt. Int J Dairy Technol 55:152–157

Loveday SM, Sarkar A, Singh H (2013) Innovative yoghurts: novel processing technologies for improving acid milk gel texture. Trends Food Sci Technol 33:5–20

Lucey JA, Srinivasan M, Singh H et al (2000) Characterization of commercial and experimental sodium caseinates by multiangle laser light scattering and size-exclusion chromatography. J Agric Food Chem 48:1610–1616

Mautner A, Meisel H, Lorenzen PC et al (1999) Destimmung des Dipeptids ε-(γ-glutamyl)lysin aus Transglutaminase-vernetzten Proteinen mittels Aminosäureanalyse. Kieler Milchw Forsch 51:155–163

Miller ML, Johnson GVW (1999) Rapid, single-step procedure for the identification of transglutaminase-mediated isopeptide crosslinks in amino acid digests. J Chromatogr B 732:65–72

Mokoonlall A, Pfannstiel J, Struch M et al (2016a) Structure modification of stirred fermented milk gel due to laccase-catalysed protein crosslinking in a post-processing step. Innov Food Sci Emerg Technol 33:563–570

Mokoonlall A, Hippich M, Struch M et al (2016b) Antioxidant activity of milk suppresses laccase induced radicals and the subsequent modification of acidified milk protein gels. Int Dairy J 60:24–31

Mokoonlall A, Nöbel S, Hinrichs J (2016c) Post-processing of fermented milk to stirred products: reviewing the effects on gel structure. Trends Food Sci Technol 54:26–36

Mokoonlall A, Sykora L, Pfannstiel J et al (2016d) A feasibility study on the application of a laccase-mediator system in stirred yoghurt at the pilot scale. Food Hydrocoll 60:119–127

Monogioudi E, Permi P, Filpponen I et al (2011) Protein analysis by ^{31}P NMR spectroscopy in ionic liquid: quantitative determination of enzymatically created cross-links. J Agric Food Chem 59:1352–1362

Moon J-H, Hong Y-H, Huppertz T et al (2009) Properties of casein micelles cross-linked by transglutaminase. Int J Dairy Technol 62:27–32

Mounsey JS, O'Kennedy BT, Kelly PM (2005) Influence of transglutaminase treatment on properties of micellar casein and products made therefrom. Lait 85:405–418

Nemes Z, Petrovski G, Fésüs L (2005) Tools for the detection and quantitation of protein transglutamination. Anal Biochem 342:1–10

Nilsson L (2013) Separation and characterization of food macromolecules using field-flow fractionation: a review. Food Hydrocoll 30:1–11

Nogueira Silva NF, Saint-Jalmes A, de Carvalho AF et al (2014) Development of casein microgels from cross-linking of casein micelles by genipin. Langmuir 30:10167–10175

Nogueira MH, Tavares GM, Nogueira Silva NF et al (2019) Physico-chemical stability of casein micelles cross-linked by transglutaminase as a function of acidic pH. Food Struct 19:100103

O'Mahony JA, Ardö Y, McSweeney PLH (2002) Analysis | electrophoresis. In: Roginski H, Fuquay JQ, Fox PF (eds) Encyclopedia of dairy sciences, 1st edn. Academic, New York, pp 67–74

Özrenk E (2006) The use of transglutaminase in dairy products. Int J Dairy Technol 59:1–7

Partschefeld C, Schwarzenbolz U, Richter S et al (2007) Crosslinking of casein by microbial transglutaminase and its resulting influence on the stability of micelle structure. Biotechnol J 2:456–461

Pecora R (2000) Dynamic light scattering measurement of nanometer particles in liquids. J Nanopart Res 2:123–131

Raak N, Abbate RA, Alkhalaf M et al (2020) Concentration-triggered liquid-to-solid transition of sodium caseinate suspensions as a function of temperature and enzymatic cross-linking. Food Hydrocoll 101:105464

Raak N, Rohm H, Jaros D (2017) Cross-linking with microbial transglutaminase: Isopeptide bonds and polymer size as drivers for acid casein gel stiffness. Int Dairy J 66:49–55

Raak N, Abbate RA, Lederer A et al (2018) Size separation techniques for the characterisation of cross-linked casein: a review of methods and their applications. Separations 5:14

Raak N, Brehm L, Abbate RA et al (2019a) Self-association of casein studied using enzymatic cross-linking at different temperatures. Food Biosci 28:89–98

Raak N, Schöne C, Rohm H et al (2019b) Acid-induced gelation of enzymatically cross-linked caseinate in different ionic milieus. Food Hydrocoll 86:43–49

Romeih E, Walker G (2017) Recent advances on microbial transglutaminase and dairy application. Trends Food Sci Technol 62:133–140

Sakamoto H, Kumazawa Y, Kawajiri H et al (1995) ε-(γ-glutamyl)lysine crosslink distribution in foods as determined by improved method. J Food Sci 60:416–420

Saricay Y, Dhayal SK, Wierenga PA et al (2012) Protein cluster formation during enzymatic cross-linking of globular proteins. Faraday Discuss 158:51–63

Saricay Y, Wierenga P, de Vries R (2013) Nanostructure development during peroxidase catalysed cross-linking of α-lactalbumin. Food Hydrocoll 33:280–288

Saricay Y, Wierenga PA, de Vries R (2014) Changes in protein conformation and surface hydrophobicity upon peroxidase-catalyzed cross-linking of apo-α-lactalbumin. J Agric Food Chem 62:9345–9352

Saricay Y, Wierenga PA, de Vries R (2016) Rheological properties of dispersions of enzymatically cross-linked apo-α-lactalbumin. Food Hydrocoll 56:344–351

Saricay Y, Wierenga PA, de Vries R (2017) Limited changes in physical and rheological properties of peroxidase-cross-linked apo-α-lactalbumin after heat treatment. Food Hydrocoll 66:326–333

Schäfer C, Schott M, Brandl F et al (2005) Identification and quantification of ε-(γ-glutamyl)lysine in digests of enzymatically cross-linked leguminous proteins by high-performance liquid chromatography−electrospray ionization mass spectrometry (HPLC-ESI-MS). J Agric Food Chem 53:2830–2837

Smiddy MA, Martin J-EGH, Kelly AL et al (2006) Stability of casein micelles cross-linked by transglutaminase. J Dairy Sci 89:1906–1914

Snyder SL, Sobocinski PZ (1975) An improved 2,4,6-trinitrobenzenesulfonic acid method for the determination of amines. Anal Biochem 64:284–288

Stachel I, Schwarzenbolz U, Henle T et al (2010) Cross-linking of type I collagen with microbial transglutaminase: identification of cross-linking sites. Biomacromolecules 11:698–705

Struch M, Linke D, Mokoonlall A et al (2015) Laccase-catalysed cross-linking of a yoghurt-like model system made from skimmed milk with added food-grade mediators. Int Dairy J 49:89–94

Struch M, Krahe N-K, Linke D et al (2016) Dose dependent effects of a milk ion tolerant laccase on yoghurt gel structure. LWT Food Sci Technol 65:1144–1152

Veitch NC (2004) Horseradish peroxidase: a modern view of a classic enzyme. Phytochemistry 65:249–259

Wahlund K-G, Nilsson L (2012) Flow FFF – basics and key applications. In: Williams SKR, Caldwell KD (eds) Field-flow fractionation in biopolymer analysis, 1st edn. Springer, Vienna, pp 1–21

Wang P, Chen C, Guo H et al (2018) Casein gel particles as novel soft Pickering stabilizers: the emulsifying property and packing behaviour at the oil-water interface. Food Hydrocoll 77:689–698

Weidendorfer K, Hinrichs K (2010) Online particle size measurement in microgel particle suspensions: principles and data analysis. Chem Ing Tech 82:1685–1691

Westermeier R (2011) Electrophoresis in gels. In: Janson J-C (ed) Protein purification: principles, high resolution methods, and applications, 3rd edn. Wiley, Hoboken, pp 365–377

Wyatt PJ (1993) Light scattering and the absolute characterization of macromolecules. Anal Chim Acta 272:1–40

Yokoyama K, Nio N, Kikuchi Y (2004) Properties and applications of microbial transglutaminase. Appl Microbiol Biotechnol 64:447–454

Zeeb B, Fischer L, Weiss J (2014) Stabilization of food dispersions by enzymes. Food Funct 5:198–213

Zeeb B, McClements DJ, Weiss J (2017) Enzyme-based strategies for structuring foods for improved functionality. Annu Rev Food Sci Technol 8:21–34

Chapter 18
The Production of Bioactive Peptides from Milk Proteins

Thanyaporn Kleekayai, Maria Cermeño, and Richard J. FitzGerald

18.1 Introduction: Milk Proteins and Bioactive Peptides

Bioactive peptides (BAPs) represent sequences of amino acids entrapped within protein molecules which, when released, can exert different bioactivities. To date, milk has been one of the main protein sources used for the generation of BAPs (Korhonen 2009; Park and Nam 2015). Peptides from dairy products have been widely studied in the last two decades and are known to possess a range of bioactivities associated with the modulation of various processes, e.g., the cardiovascular, digestive, immune, bone and nervous systems. Section 18.1 depicts the generation of BAPs and their impact on different bodily systems. These BAPs may therefore help in the prevention and management of different disease conditions, e.g., cardiovascular disease (CVD) and type II diabetes mellitus (T2DM). BAPs have been mainly isolated from milk and dairy products derived from bovine, ovine and caprine milk. However, there are also reports on BAPs from the milk/dairy products of camel, mare, buffalo, yak and donkey (Abd El-Salam and El-Shibiny 2017; Zenezini Chiozzi et al. 2016).

Although BAPs may not have the potency of synthetic drugs, they usually display no side-effects, and therefore represent natural alternative health-enhancing agents which are currently in demand by consumers globally (Li-Chan 2015). Being less potent than organically synthesised drugs, it is generally necessary to consume relatively high quantities in order to observe a significant physiological effect. In addition, since BAPs are generally consumed as mixtures of peptides, there is significant potential for synergistic physiological effects to be observed when consuming peptide mixtures.

BAPs from dietary proteins are naturally released *in vivo* by the enzymes in the gastrointestinal system. Processing techniques such as fermentation and enzymatic

T. Kleekayai · M. Cermeño · R. J. FitzGerald (✉)
Department of Biological Sciences, University of Limerick, Limerick, Ireland
e-mail: thanyaporn.kleekayai@ul.ie; maria.cermeno@ul.ie; dick.fitzgerald@ul.ie

A. L. Kelly, L. B. Larsen (eds.), *Agents of Change*, Food Engineering Series,
https://doi.org/10.1007/978-3-030-55482-8_18

hydrolysis have also been employed in order to generate BAPs (Fig. 18.1). Moreover, in recent years, *in silico* approaches are being increasingly used to aid in the generation of BAPs. This chapter provides an overview of the main approaches used during the production of BAPs from milk proteins.

18.2 Production of Milk Protein-Derived BAPs

18.2.1 In Vitro *Enzymatic Hydrolysis*

Enzymatic hydrolysis of proteins is considered a sustainable technology due to its relatively low environmental impact and energy consumption (Tavano 2013). During enzymatic hydrolysis, a protease/peptidase cleaves peptide bonds within proteins, resulting in the formation of protein hydrolysates (Liceaga and Hall 2019; de Castro et al. 2015). These hydrolysates are composed of multiple short chains of amino acids called peptides. The characteristics of the peptides formed will depend on the protein substrate and the specificity of the protease/peptidase activity used during enzymatic hydrolysis. Therefore, the choice of protein substrate and protease/peptidase is key to the generation of protein hydrolysates containing potent BAP sequences. Proteases and peptidases are a complex group of enzymes with several classifications, e.g., they have different specificity, optimum pH, active sites

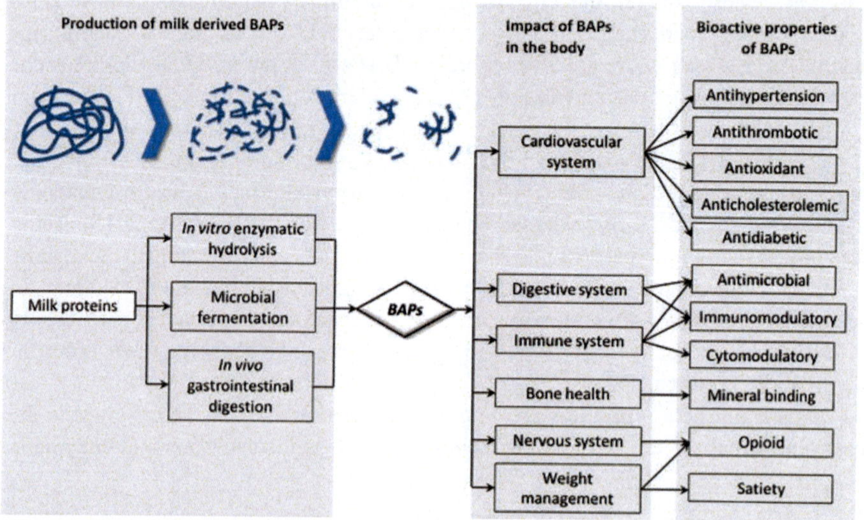

Fig. 18.1 Schematic of different approaches used for the production of milk-derived bioactive peptides (BAPs) and the impact of these BAPs on different bodily systems (adapted from Korhonen and Pihlanto (2006) and Mohanty et al. (2016b))

and modes of action (de Castro et al. 2015; Polaina and MacCabe 2007; Gurumallesh et al. 2019).

Food-grade proteolytic enzymes can be classified by their source into bacterial (dos Santos Aguilar and Sato 2018; Toldrá et al. 2018), animal (Nongonierma and FitzGerald 2011), plant (González-Rábade et al. 2011) and fungal (Martínez-Medina et al. 2019) derived enzymes. Some enzymes have broad specificity and are able to hydrolyse numerous peptide bonds while others are highly specific. Therefore, these hydrolytic enzymes can also be classified as specific and non-specific activities. Furthermore, based on their optimum operating pH conditions, proteases can be classified into acid, neutral and alkaline activities. The maximum activity for acid proteases is exerted when the pH of the reaction is between pH 2.0 and 6.0.

For instance, pepsin, which is released from mammalian gastric mucosa, has an optimum pH of 2.0. Proteases can also be classified by their mode of action into two different groups, i.e., endo- and exoproteases. Endoproteases (also called proteinases) cleave proteins internally, whereas exopeptidases cleave amino- or carboxy-terminal residues from proteins, i.e., amino- and carboxypeptidases, respectively. Neutral proteases (e.g., Neutrase®) work optimally at values close to pH 7.0. On the other hand, alkaline proteases have optimum activity between pH 8.0–13.0. This group includes the majority of the bacterial enzymes (Rao et al. 1998). Proteases can also be classified depending on the chemical group involved in the catalytic site. Cysteine proteases have a reactive cysteine residue in their active site and they include plant proteases such as papain and bromelain. Serine proteases, which possess a serine in their active site, include chymotrypsin, trypsin and elastase, as well as the subtilisin family. Aspartic proteases have two essential aspartic acid residues and include the digestive enzymes pepsin and chymosin. Threonine and glutamic acid proteases have an active-site threonine or glutamic acid residue, respectively. Finally, metalloproteases (e.g., thermolysin) require the presence of a divalent cation, e.g., zinc, manganese, calcium or copper to facilitate hydrolysis (Polaina and MacCabe 2007).

Different enzyme preparations are commercially available for protein hydrolysis (Nongonierma et al. 2016b). Examples of such proteolytic preparations are summarised in Table 18.1. Due to the different specificity of enzymes, enzyme suppliers often provide their products as a mix of proteolytic activities. Hence, the preparations contain a number of enzymes, which are active over a relatively broad range of pH and temperature values, and can promote different hydrolysis reactions depending on the result the end user wishes to achieve. These enzyme combinations can have a beneficial synergistic effect on the overall hydrolysis process, in comparison to situations where individual enzymes are used. The type and quantity of BAPs released during hydrolysis of specific protein substrates are not only dependent on the specificity of the enzyme used but is also dependent on pH, temperature, substrate concentration, substrate:enzyme ratio and the targeted bioactivity (Cheison and Kulozik 2017). The interactions and effects of these multiple parameters, e.g., pH and temperature on BAP release, can be studied by the use of design of experiments (DOE) and response surface methodology (RSM) approaches. This has been

Table 18.1 Examples of commercially available proteolytic enzyme preparations used for the generation of milk protein hydrolysates/bioactive peptides

Name	Source	Activity	pH and temperature (T) range	Manufacturer
Alcalase 2.4 L®	*Bacillus licheniformis*	Serine endopeptidase (mainly subtilisin A) broad substrate specificity	pH: 6.5–8.5 T: 60 °C	Novozymes
Protex6L® (now Multifect PR 6 L)	*Bacillus licheniformis*	Serine alkaline protease	pH: 7.0–9.0 T: 25–70 °C	DuPont Industrial Biosciences
Protamex®	*Bacillus spp.*	Alkaline endoprotease	pH: 5.0–11.0 T: 60 °C	Novozymes
Prolyve® 1000	*Bacillus licheniformis*	Serine alkaline protease	pH: 7.0–10.0 T: 50–60 °C	Lyven Enzymes
Flavourzyme® 500 L	*Aspergillus oryzae*	Exopeptidase. Aminopeptidase	pH: 5.0–7.0 T: 50 °C	Novozyme
Corolase PP®	Porcine pancreatic preparation	Trypsin, chymotrypsin and elastase. Amino and carboxypeptidase	pH: 6.0–8.5 T: 37 °C	AB enzymes
Promod® 144 MG	Papaya (papain)	Cysteine proteinase	pH: 6.0–7.0 T: 55–70 °C	Biocatalyst
Neutrase® 0.8 L	*Bacillus amyloliquefaciens*	Metalloproteinase	pH: 7.0 T: 40–50°C	Novozymes
Debitrase HYW20	*Bacillus spp., Aspergillus oryzae*	Exopeptidase, aminopeptidase and dipeptidase	pH 6.0–8.0 T: 45–55 °C	DuPont Industrial Biosciences
Thermolysin	*Bacillus thermoproteolyticus*	Metalloproteinase	pH: 5.0–8.5 T: 65–85 °C	Several
Umamizyme™	*Aspergillus oryzae*	Neutral proteinase	pH: 7.0–8.0 T: 45–50 °C	Amano
Brewer's Clarex™	*Aspergillus niger*	Prolyl endoprotease (an-PEP)	pH: 3.5–5.5 T: 40–60 °C	DSM

demonstrated, for example, during optimisation of the release of dipeptidyl peptidase IV (DPP-IV) inhibitory peptides during the hydrolysis of bovine casein with Protamex™ (Nongonierma et al. 2017a).

18.2.1.1 Enzymatic Hydrolysis for the Generation of Anti-Diabetic Peptides

Diabetes is considered to be one of the major causes of mortality worldwide and it is estimated to affect about 330 million people by 2025 (IDF 2015). While type I diabetes is an autoimmune disease, i.e., the body is unable to secrete insulin due to the inactivation or destruction of pancreatic beta cells, type II diabetes mellitus

(T2DM), which accounts for approx. 90% of all diabetes incidences, is associated with insulin resistance in peripheral tissues which evolves into insulin secretion deficiencies and, therefore, hyperglycaemia (Kahn 1998). Current treatments for T2DM include drugs which inhibit the activity of enzymes such as DPP-IV, α-amylase and α-glucosidase (Yan et al. 2019a). However, it is increasingly becoming evident that BAPs may act as an alternative to synthetic drugs for the management of T2DM (Horner et al. 2016; Nongonierma and FitzGerald 2019, 2018; Nongonierma et al. 2016b; El-Sayed and Awad 2019).

Regarding dairy-protein-derived antidiabetic BAPs, the most studied approach to date has focused on their ability to inhibit DPP-IV. DPP-IV is involved in inactivation of the incretins, glucagon like peptide 1 (GLP-1) and gastric inhibitory polypeptide (GIP), which, in turn, regulates the secretion of insulin. Therefore, when DPP-IV is inhibited, the half-life of the incretins increases, thereby attenuating an insulinotropic response (Power et al. 2014b). Table 18.2 provides details of milk protein-derived DPP-IV inhibitory peptide sequences. It has been reported that low molecular mass peptides having W, P, A or hydrophobic residues in their sequences act as good DPP-IV inhibitors (Nongonierma and FitzGerald 2014b; Lacroix and Li-Chan 2012b; Nongonierma and FitzGerald 2019). Hence, enzymes which specifically cleave proteins after P or A residues can lead to the generation of DPP-IV inhibitory protein hydrolysates (Brandelli et al. 2015). Lacroix and Li-Chan (2012a) used 11 commercial food-grade proteolytic preparations to obtain hydrolysates from sodium caseinate and whey protein isolate (WPI). In general, sodium caseinate hydrolysates displayed higher overall DPP-IV inhibitory activity than WPI hydrolysates, with the exception of the peptic digest of WPI, which had the highest inhibitory activity, with a DPP-IV IC_{50} value of 0.075 mg mL^{-1}. The antidiabetic properties of milk-derived components has been particularly attributed to whey protein-derived peptides (Adams and Broughton 2016; Turner et al. 2015). The sequence LKPTPEGDLE was found in a peptic hydrolysate of β-lactoglobulin (β-Lg) with a DPP-IV IC_{50} value of 45 μM, whereas WAGTWY (DPP-IV IC_{50} = 74 μM) and IPAVF (DPP-IV IC_{50} = 44 μM) were released from β-Lg following hydrolysis with trypsin (Lacroix and Li-Chan 2014; Silveira et al. 2013; Power et al. 2014a). IPI (DPP-IV IC_{50} = 3 μM), the most potent DPP-IV inhibiting food-derived peptide reported to date, can be found in bovine casein (Nongonierma et al. 2018a; Lacroix and Li-Chan 2014).

Hydrolysis of camel milk proteins with trypsin released LPVP and MPVQA with DPP-IV IC_{50} values of 87 and 93.3 μM, respectively. However, when the same treatment was applied to bovine milk protein, LPVP and MPVQA were not found in the hydrolysate. This was attributed to differences in the primary sequences of the milk proteins from these two different species (Nongonierma et al. 2018b).

Inhibition of α-glucosidase and α-amylase are also targeted in the management of T2DM (Yan et al. 2019b). However, this approach has been studied to a lesser extent for milk protein-derived BAPs. α-Glucosidase is a membrane-bound enzyme located in the small intestine that cleaves oligosaccharides into absorbable di- and monosaccharides. α-Amylase converts carbohydrates such as starch into glucose and maltose and is secreted by salivary glands and the pancreas. The release of

Table 18.2 Examples of milk protein derived bioactive peptides with oxygen radical absorbance capacity (ORAC), angiotensin converting enzyme (ACE) and dipeptidyl peptidase IV (DPP-IV) inhibitory activities

Species	Protein	Enzymes	Peptide Sequence[a]	Bioactivities			Reference
				ORAC value (μmol TE μmol^{-1} peptide)	ACE IC_{50} (μM)	DPP-IV IC_{50} (μM)	
Bovine	β-Casein/κ-Casein	Several (e.g., *Aspergillus oryzae*, *Lactobacillus spp.*protease)	IPP	–	5 μM	–	Nakamura et al. (1995a); Ishida et al. (2011); Stressler et al. (2013)
	β-Casein		VPP	–	9 μM	–	
Bovine	α_{S2}-Casein	Pepsin and Corolase PP®	YQK	1.819	659 μM	–	Contreras et al. (2013)
			FPQY	1.741	542 μM	–	
	α_{S1}-Casein	Pepsin and Corolase PP®	YLGY	1.476	80 μM	–	
			YLG	1.378	–	–	
			RYLGY	2.829	1 μM	–	
			RYLG	1.669	441 μM	–	
			RYL	1.746	235 μM	–	
			RY	1.944	160 μM	–	
			LGY	2.306	60 μM	–	
			AYFYPEL	3.216	7 μM	–	
			AYFYPE	4.160	330 μM	–	
			FYPEL	–	120 μM	–	
	β-Lactoglobulin	Corolase PP®	WYSLAMAASDI	2.621	–	–	Hernández-Ledesma et al. (2005)
	β-Casein	Pepsin and Corolase PP®	YPFPGPI	–	633 μM	–	Hernández-Ledesma et al. (2004)
	β-Lactoglobulin	Thermolysin	LQKW	–	59 μM	–	Hernández-Ledesma et al. (2002)

Species	Protein	Enzymes	Peptide Sequence[a]	Bioactivities			Reference
				ORAC value (μmol TE μmol^{-1} peptide)	ACE IC_{50} (μM)	DPP-IV IC_{50} (μM)	
			LLF	–	202 μM	–	
	κ-Casein	Porcine pepsin A	YVL	0.960	–	–	López-Expósito et al. (2007)
			FSDKIAK	–	113 μM	–	
			IQY	0.550	70 μM	–	
	α_{S2}-Casein	Porcine pepsin A	PYVRYL	1.820		–	
	α_{S2}-Casein	Trypsin	FALPQYLK	1.630	11 μM	–	Power et al. (2014a)
	β-Lactoglobulin	Trypsin	IPAVFK	–	144 μM	–	
			IIAEK	–	63 μM	–	
			VAGTWY	–	–	74 μM	
	Serum albumin	Prolyl endopeptidase	VR	–	–	826 μM	Nongonierma and FitzGerald (2013b)
			LP	–	–	712 μM	
			IP	–	–	149 μM	
	κ-Casein	Prolyl endopeptidase	IPI	–	–	3 μM	
	β-Casein	Prolyl endopeptidase	FLQP	–	–	65 μM	
	α_{S2}-Casein	Prolyl endopeptidase	WIQP	–	–	237 μM	
	Several	Pancreatic protease	WI	–	–	138 μM	Le Maux et al. (2015)
			WL	–	–	43 μM	
			YL	–	–	940 μM	
	β-Lactoglobulin	Trypsin	VLVLDTDYK	–	–	424 μM	Silveira et al. (2013)
			IPAVFK	–	–	143 μM	
			IPAVF	–	–	44 μM	

(continued)

Table 18.2 (continued)

Species	Protein	Enzymes	Peptide Sequence[a]	Bioactivities			Reference
				ORAC value (µmol TE μmol^{-1} peptide)	ACE IC_{50} (µM)	DPP-IV IC_{50} (µM)	
			TPEVDDEALEK	–	–	319 µM	
	β-Lactoglobulin	Thermoase PC10F	VLDTDY	–	128 µM	471 µM	Lacroix et al. (2016)
			LKALPMH	–	11 µM	193 µM	
			IQKVAGTW	–	51 µM	329 µM	
	α-Lactalbumin	Thermoase PC10F	WLAHKAL	–	29 µM	286 µM	
			LKGYGGVSLPE	–	–	486 µM	
Buffalo	β-Lactoglobulin	Thermoase PC10F	LDIQKVAGTW	–	21 µM	–	
Bovine	β-Lactoglobulin	Thermoase PC10F	LSFNPTQ	–	106 µM	–	
			LDTDY	–	121 µM	–	
	α-Lactalbumin	Thermoase PC10F	GYGGVSLPEW	–	2 µM	–	
	α-Lactalbumin	Pepsin	LCSEKLDQ	–	–	186 µM	Lacroix and Li-Chan (2014)
			LAHKALCSEKL	–	–	165 µM	
			IVQNNDSTEYGLF	–	–	337 µM	
			ILDKVGINY	–	–	263 µM	
			WLAHKALCSEKLDQ	–	–	414 µM	
			TKCEVFRE	–	–	166 µM	
	β-Lactoglobulin	Pepsin	LKPTPEGDLEIL	–	–	57 µM	
			LKPTPEGDLE	–	–	45 µM	
			LKPTPEGDL	–	–	45 µM	
			IPAVFKIDA	–	–	191 µM	
Camel	β-Casein	Trypsin	LPVP	–	–	87 µM	Nongonierma et al. (2018b)
			SPVVPF	–	–	214 µM	

Species	Protein	Enzymes	Peptide Sequence[a]	Bioactivities			Reference
				ORAC value (μmol TE μmol^{-1} peptide)	ACE IC_{50} (μM)	DPP-IV IC_{50} (μM)	
			MPVQA	–	–	93 μM	
	α_{S1}-Casein	Trypsin	LLQLEAIR	–	–	177 μM	
			ILELA	–	–	721 μM	
	α-Lactalbumin	Trypsin	ILDKEGIDY	–	–	347 μM	
Caprine	κ-Casein	Trypsin and chymotrypsin	INNQFLPYPY	–	–	40 μM	Zhang et al. (2015)
	β-Casein	Trypsin and chymotrypsin	SPTVMFPPQSVL	–	–	676 μM	
			MHQPPQPL	–	–	350 μM	
	β-Casein	Trypsin	GPFPILV	–	–	163 μM	Zhang et al. (2016)
	α_{S1}-Casein	Trypsin	HPINHR	–	–	452 μM	
Bovine	Serum albumin	Pepsin, chymotrypsin, and trypsin	VR	–	52 μM	–	Gómez-Ruiz et al. (2007)
	κ-Casein	Pepsin, chymotrypsin, and trypsin	LPYPY	–	28 μM	–	
Ovine	κ-Casein	Pepsin, chymotrypsin, and trypsin	YQQRPVA	–	14 μM	–	
			YIPIQY	–	10 μM	–	
			KDERF	–	848 μM	–	
			EKDERF	–	14 μM	–	
Bovine	Hemoglobin subunit α	Pepsin	KLLSHSL	–	42 μM	–	Adje et al. (2011)
Donkey	β-Casein	Pepsin and Corolase PP®	VAPFPQPVVP	–	46 μM	–	Bidasolo et al. (2012)

(continued)

Table 18.2 (continued)

Species	Protein	Enzymes	Peptide Sequence[a]	Bioactivities			Reference
				ORAC value (µmol TE µmol^{-1} peptide)	ACE IC_{50} (µM)	DPP-IV IC_{50} (µM)	
Bovine	β-Casein	*Bacillus lentus* alkaline peptidase	VYPFPGPIPN	–	325 µM	–	Eisele et al. (2013)
Caprine	β-Casein	Alcalase	TGPIPN	–	316 µM	–	Geerlings et al. (2006)
			SQPK	–	354 µM	–	
			SLPQ	–	330 µM	–	
Bovine	κ-Casein	AS1.398 neutral protease	RYPSYG	–	88 µM	–	Jiang et al. (2010)
			DERF	–	65 µM	–	
	Lactoferrin	Pepsin	LRPVAA	–	4 µM	–	Lee et al. (2006)
	β-Casein	*Aspergillus niger* derived prolyl endoprotease	VP	–	420 µM	–	Norris et al. (2014)
			VLGP	–	153 µM	–	
			VEP	–	63 µM	–	
			NSLP	–	755 µM	–	
			LPP	–	9 µM	–	
			KYP	–	363 µM	–	
			IQA	–	32 µM	–	
			IP	–	130 µM	–	
			HQP	–	134 µM	–	
			GP	–	252 µM	–	
			FLQP	–	68 µM	–	
			FA	–	487 µM	–	
			MAP	–	0.4 µM	–	

Species	Protein	Enzymes	Peptide Sequence[a]	Bioactivities			Reference
				ORAC value (μmol TE μmol^{-1} peptide)	ACE IC_{50} (μM)	DPP-IV IC_{50} (μM)	
	α_{S2}-Casein	*Aspergillus niger* derived prolyl endoprotease	WIQP	–	14 μM	–	Norris et al. (2015)
			ITP	–	14 μM	–	
	α-Lactalbumin	Trypsin	WLAHK	–	77 μM	–	Pihlanto-Leppälä et al. (2000)
			VGINYWLAHK	–	327 μM	–	
	α-Lactalbumin	Pepsin, trypsin and chymotrypsin	YGL	–	409 μM	–	
	β-Lactoglobulin	Trypsin	PFK	–	1029 μM	–	
			LAMA	–	1062 μM	–	
			LDAQSAPLR	–	635 μM	–	
	β-Lactoglobulin	Pepsin, trypsin and chymotrypsin	CMENSA	–	788 μM	–	
			ALPMH	–	521 μM	–	
			VLDTDYK	–	946 μM	–	
	Lactoferrin	Pepsin	RPYL	–	56 μM	–	Ruiz-Giménez et al. (2012)
			LNNSRAP	–	105 μM	–	
			LIWKL	–	0.47 μM	–	
	β-Casein	Thermolysin	VYPFPGPIPNSLPQNIPP	–	87 μM	–	Otte et al. (2007)
			LVYPFPGP	–	147 μM	–	
	α-Lactalbumin	Thermolysin	YGGVSLPEW	–	16 μM	–	
			VSLPEW	–	57 μM	–	
			LKGYGGVSLPEW	–	83 μM	–	

(continued)

Table 18.2 (continued)

Species	Protein	Enzymes	Peptide Sequence[a]	Bioactivities			Reference
				ORAC value (μmol TE μmol^{-1} peptide)	ACE IC_{50} (μM)	DPP-IV IC_{50} (μM)	
			GVSLPEW	–	30 μM	–	
	α_{S2}-Casein	Trypsin	TVY	–	15 μM	–	Tauzin et al. (2002)
			NMAINPSK	–	60 μM	–	
			FPQYLQY	–	14 μM	–	
			FALPQY	–	4 μM	–	
			ALNEINQFYQK	–	264 μM	–	
			ALNEINQFY	–	219 μM	–	
			NMAINPSKENLCSTFCK	–	129 μM	–	Tu et al. (2018)
	β-Lactoglobulin	Proteases from *Cynara cardunculus*	DAQSAPLRVY	–	12.2 μM	–	Tavares et al. (2011)
	α-Lactalbumin	Proteases from *Cynara cardunculus*	KGYGGVSLPEW	–	0.7 μM	–	
			KGYGGVSL	–	300 μM	–	
			DKVGINY	–	100 μM	–	
			DKVGINYW	–	25 μM	–	
	β-Lactoglobulin	Trypsin	ALPMHIR	–	42 μM	–	Mullally et al. (1997)
	α_{S2}-Casein	Subtilisin, bacillolysin, and trypsin	MKP	–	0.3 μM	–	Yamada et al. (2013)

[a]Peptide sequence with the one letter code. *IC_{50}* concentration of peptide resulting in 50% inhibition of ACE activity

sugars as a result of the action of these enzymes on dietary carbohydrates can lead to an increase in blood glucose levels (Yan et al. 2019a). A study by Lacroix and Li-Chan (2013) reported that peptic hydrolysates of bovine WPI, β-Lg and α-lactalbumin (α-La) were able to inhibit α-glucosidase by 36, 33, and 24%, respectively, whereas the non-hydrolysed whey protein displayed no enzyme inhibition. However, identification of α-glucosidase inhibitory peptides has yet to be reported. No studies to date appear to have reported on the α-amylase inhibitory activity in bovine milk protein derived peptides. However, a recent study has shown that the hydrolysis of camel milk protein with Alcalase® and Bromelain released KDLWDDFKGL and MPSKPPL, respectively, with potent α-amylase inhibitory activity. However, the protein origin of these peptides was not reported (Mudgil et al. 2018). Peptides with α-glucosidase and α-amylase inhibitory activity can interact with the active site of the enzyme (competitive inhibitors) or at other non-specific sites (non-competitive inhibitors), resulting in the inhibition or reduction of the enzyme activity (Yan et al. 2019a). However, the structure-activity correlation of peptides with potential to inhibit α-glucosidase and α-amylase has not yet been reported in detail.

18.2.1.2 Enzymatic Hydrolysis for the Generation of Antihypertensive Peptides

Approximately 14% of the global population displays hypertension (Frieden and Jaffe 2018). Lowering blood pressure (BP) is therefore one of the targets of the World Health Organisation (WHO), since high BP is a controllable risk factor associated with the development of cardiovascular disease (WHO 2013). Different metabolic pathways are responsible for the regulation of blood pressure, e.g., the angiotensin-renin-aldosterone (RAS), kinin-nitric oxide (KNOS), neutral endopeptidase (NEP) and endothelin-converting enzyme (ECE) systems (Norris and FitzGerald 2013; FitzGerald et al. 2004). Extensive research has been carried out on the angiotensin converting enzyme (ACE) inhibitory activity of milk-protein derived hydrolysates and peptides (for reviews see de Oliveira et al. 2018; Brandelli et al. 2015). The ACE-inhibitory activity of BAPs has been attributed to hydrophobic amino acids in the C-terminal position, e.g., P, V, L and I (Wu et al. 2006a, b). Table 18.2 provides details of bioactive peptides from milk proteins with ACE inhibitory activity. IPP and VPP, with ACE IC_{50} values of 5 and 9 μM, respectively, were considered to be the most potent ACE-inhibitory peptides derived from milk protein for many years. These peptides can be obtained by hydrolysis of casein with *Lactobacillus spp.* proteases (Stressler et al. 2013). A single dose of IPP and VPP was reported to reduce systolic blood pressure (SBP) in spontaneously hypertensive rats (SHR) by −28.3 and −32.1 mm Hg, respectively (in individual doses) (Nakamura et al. 1995a; Nakamura et al. 1995b). Subsequent studies revealed that the long-term intake of IPP and VPP attenuated the development of hypertension in SHR. This effect in lowering blood pressure was mainly attributed to IPP and VPP inhibiting ACE (Sipola et al. 2001). However, Hirota et al. (2011) reported that VPP and IPP had a vasodilatory effect based on their ability to induce nitric oxide release in endothelial cells grown in culture. A number of meta-analyses have been conducted assessing the published literature on the effect of IPP and VPP in hypertensive human subjects that result in effective BP reduction (Cicero et al. 2013; Cicero et al. 2010; Pripp 2008; Xu et al. 2008; Boelsma and Kloek 2010).

As seen from Table 18.2, a large number of studies have reported on ACE-inhibitory peptides derived from milk proteins. For instance, RYLGY (ACE $IC_{50} = 0.71$ μg mL^{-1}) was obtained from α_{s1}-casein following incubation with pepsin and Corolase PP® (Contreras et al. 2013). Incubation of α_{s1}-casein with a combination of subtilisin, bacillolysin and trypsin released a more potent peptide, MKP (ACE $IC_{50} = 0.12$ μg mL^{-1}) (Yamada et al. 2013). MAP (ACE $IC_{50} = 0.4$ μM) was released from β-casein on incubation with a prolyl endopeptidase from *Aspergillus niger* (Norris et al. 2014). Peptides originating from whey proteins with potent ACE inhibitory activity have also been reported. KGYGGVSLPEW (ACE $IC_{50} = 0.70$ μM) from α-La was identified in an hydrolysate generated using proteases from *Cynara cardunculus* (Tavares et al. 2011). Ultrafiltration permeates (<5 kDa) of whey protein hydrolysates generated using Neutrase® and Alcalase® with potent ACE-inhibitory activity (~71% inhibition) were incubated with human umbilical vein endothelial cells (HUVECs), resulting in the expression of genes involved in blood pressure regulation. The permeates induced the upregulation of endothelial nitric oxide synthase, as well as the downregulation of endothelin-1 (O'Keeffe and FitzGerald 2018). A peptic hydrolysate of bovine lactoferrin released LIWKL with an ACE IC_{50} of 0.47 μM (Ruiz-Giménez et al. 2012). The hydrolysis of ovine casein with pepsin and Corolase PP led to the identification of the pentapeptide LPYPY with an ACE IC_{50} of 28.9 μM (Gómez-Ruiz et al. 2007). This peptide was also found in hydrolysates of milks from other species such as bovine and caprine, but this sequence is not conserved in donkey, camel and pig caseins (Lammi et al. 2019) which further highlights the importance of the initial substrate as a source of BAPs.

ACE-inhibitory peptides have also been isolated from mare (Sames et al. 2018), camel (Jrad et al. 2014), and donkey (Zenezini Chiozzi et al. 2016) milk. While inhibition of renin has also been associated with hypotensive effects, this does not appear to have been extensively studied for milk-protein derived peptides (Aluko 2019). The Antihypertensive Inhibiting peptide database (AHTPDB) (http://crdd.osdd.net/raghava/ahtpdb/) includes information of BAPs with antihypertensive activity, including milk protein-derived peptides.

18.2.1.3 Enzymatic Hydrolysis for the Generation of Antioxidant Peptides

The antioxidant activity of milk-derived BAPs has been previously reviewed (El-Sayed and Awad 2019; Power et al. 2013; Sah et al. 2018). Antioxidant compounds are mainly grouped into two types; (1) primary antioxidants which neutralise free radicals (e.g., hydroxyl radicals ($^{\bullet}OH$), superoxide anion radicals ($O_2^{\bullet-}$), hydrogen peroxide (H_2O_2) and peroxynitrite ($ONOO^-$)) and (2) secondary antioxidants which inhibit the formation of these free radicals (Power et al. 2013). Some of the proposed mechanisms by which milk-protein-derived BAPs exert their antioxidant activity were reviewed by Power et al. (2013). The existence of these mechanistic differences require the use of more than one methodology to characterise the antioxidant activity of BAPs. Therefore, the assessment of antioxidant activities involves assays such as the *in vitro* oxygen radical absorbance capacity (ORAC), ferric reducing antioxidant power (FRAP), 2,2-diphenyl-1-picrylhydrazyl (DPPH) and 2,2′-azino-bis(3-ethylbenzothiazoline-6-sulphonic acid) (ABTS) assays. In addition, antioxidant activity can be assessed by quantification of the cellular levels of catalase (CAT), glutathione (GSH) and superoxide dismutase (SOD) (Lorenzo et al. 2018).

A recent study showed that the radical-scavenging activity of hydrolysates from different casein sources (cow, buffalo, goat and camel) was hydrolytic enzyme dependent. In particular, tryptic hydrolysates from camel milk protein had highest antioxidant activity (Assem et al. 2018). Hydrolysates produced with Alcalase® showed higher ORAC, FRAP and ABTS values than those produced by papain from buffalo and bovine caseins (Shazly et al. 2019). This highlights the importance of the substrate/enzyme combination during hydrolysis. Milk peptides produced with microbial-derived proteolytic enzymes have shown high antioxidant activities.

For instance, Hogan et al. (2009) employed proteases from *Aspergillus oryzae, Bacillus licheniformis* and *Bacillus subtilis* in the hydrolysis of milk protein isolate (MPI) for the production of 12 different hydrolysates. The hydrolysate generated with Validase® (from *Aspergillus oryzae*) had the highest ORAC activity. Furthermore, the addition of this hydrolysate to a beef patty significantly reduced lipid oxidation in the meat. Incubation of HUVECs with whey protein hydrolysates resulted in an increase in cellular GSH level and CAT activity (O'Keeffe and FitzGerald 2014). On the other hand, incubation with sodium caseinate hydrolysed with Prolyve® showed no effect in GSH release in U937 cells (Cermeño et al. 2016).

A detailed description of the antioxidant effect of whey proteins and peptides at a cellular level has been reported by Corrochano et al. (2018). Milk protein-derived antioxidant peptides usually contain between 5 and 11 amino acids and they may contain bulky and hydrophobic amino acid residues such as W, L, Y, P, Q and M (Li and Li 2013). However, the structure-activity criteria for antioxidant peptides is not fully understood. Some of the sequences identified in milk protein-derived peptides with ORAC antioxidant activity are summarised in Table 18.2. Bovine casein hydrolysis with trypsin released PYPQ, EMPFPK, YFYPE and PQSV with potent ABTS scavenging activity (Shazly et al. 2017). AYFYPE, AYFYPEL, RYLGY and LGY from the hydrolysis of α_{s1}-casein with pepsin and Corolase PP® had ORAC values of 4.160, 3.216, 2.829 and 2.306 μmol TE μmol^{-1} peptide, respectively (Contreras et al. 2013). Similarly, digestion of β-lactoglobulin with Corolase PP® released WYSLAMAASDI, with an ORAC value of 2.621 μmol TE μmol^{-1} peptide (Hernández-Ledesma et al. 2005). To the best of our knowledge, no studies appear to have reported antioxidant activity *in vivo* following ingestion of milk protein-derived hydrolysates.

18.2.1.4 Enzymatic Hydrolysis for the Generation of Immunomodulatory Peptides

Immunomodulatory peptides function by stimulating or supressing the immune system, i.e., *via* regulation of cytokine expression, antibody production and lymphocyte proliferation (Chalamaiah et al. 2018). Ma et al. (2014) reported that microencapsulated whey protein hydrolysed with Alcalase® significantly enhanced the proliferation of murine splenocytes. Furthermore, whey protein digested with trypsin and chymosin improved the secretion of IL-2 and IFN-γ cytokines in murine splenocytes (Saint-Sauveur et al. 2008). Hydrolysis of β-casein with a porcine trypsin preparation released peptides (<5 kDa) with anti-inflammatory activity in TNF-α treated $HEK^{nfkb\text{-}RE}$ cells. This effect was not observed when β-casein was hydrolysed with other enzyme preparations, i.e., cod trypsin, PTN 6.0 or Cryotin (Altmann et al. 2016). Sodium caseinate hydrolysates showed different cytokine secretion responses, depending on the enzyme preparation used for hydrolysis (O'Sullivan et al. 2013; Phelan et al. 2009). In particular, hydrolysates generated with Prolyve® (derived from *Bacillus licheniformis*) were associated with a decrease in the release of the pro-inflammatory cytokine, IL-6 (Cermeño et al. 2016). The tryptic digestion of bovine β-Lg was reported to increase the production of IFN-γ and TNF-α in human blood cells (Rodríguez-Carrio et al. 2014). Yak milk treated with Alcalase® downregulated the production of pro-inflammatory cytokines, i.e., IL-6, TNF-α and IL-1β (Mao et al. 2011). The tripeptide LLY was reported to stimulate the phagocytic activity of mouse peritoneal macrophages (Berthou et al. 1987), while AVPYPQRDMPIQAFLLYQ, identified after chymotryptic hydrolysis of bovine β-casein, had a stimulatory effect in lymph node cells (Coste et al. 1992). Furthermore, YGG and YG from bovine β-casein were reported to significantly enhance the synthesis of peripherial blood lymphocytes (Kayser and Meisel 1996).

18.2.1.5 Enzymatic Hydrolysis for the Generation of Anticancer Peptides

Bovine milk proteins have potential as a source of anticancer BAPs (Pepe et al. 2013; Sah et al. 2015; Chalamaiah et al. 2018). Those peptides are usually tested in cancer cells, such as endothelial intestinal Caco-2 and human blood Jurkat T cells, in order to assess their anticancer properties, i.e., cell growth inhibition and apoptosis induction. Tryptic hydrolysis of β-casein released INKKI, which induced the release of caspase-3, resulting in apoptosis in B16F10 melanoma cells (Azevedo et al. 2012). Caseinophosphopeptides (CPPs) obtained by *in vitro* enzymatic digestion showed a reduction in proliferation of HT-29 and Caco-2 cells due to the induction of apoptosis (Perego et al. 2012; Perego et al. 2013). The hexapeptide PGPIPN obtained from hydrolysis of bovine β-casein inhibited metastasis in ovarian cancer cells (Zhao et al. 2016). Lactoferrin has been the most studied of the milk proteins for its anticancer properties. Bovine lactoferrin hydrolysed with pepsin released the peptides FKCRRWQWRMKKLGAPSITCVR and

FKCRRWQWRMKKLGAPSITCVRRAF, which induced apoptosis in the Jurkat T and HL-60 cell lines (Mader et al. 2005; Roy et al. 2002).

18.2.1.6 Enzymatic Hydrolysis for the Generation of Antimicrobial Peptides

The application of milk protein-derived peptides as antimicrobial agents in food products has been extensively reviewed (Khan et al. 2018; Mohanty et al. 2016a). Antimicrobial peptides, i.e., those with activity against Gram-positive and Gram-negative bacteria, yeast and fungi, generally have low molecular mass (~2 kDa) and, are positively charged and contain hydrophobic amino acids (K, L, V and P) within their sequences (Dziuba and Dziuba 2014). Several mechanisms have been attributed to the antibacterial activity. For instance, lactoferrin chelates iron, resulting in a bacteriostatic effect since iron is necessary for bacterial growth. On the other hand, lactoferricin (obtained following peptic hydrolysis of lactoferrin) has a bacteriocidal effect by disrupting and permeabilising the bacterial cell membrane, resulting in cellular damage (Sah et al. 2018). Trypsin and chymotrypsin digests of α-La and β-Lg have shown antiproliferative effects against Gram-positive bacteria. Peptides from α_{s1}-casein (isracidin, RPKHPIKHQGLPQEVLNENLLRF, following chymosin digestion) and α_{s2}-casein (casecidin-I, KTKLTEEEKNRLNFLKKISQRYQKFALPQYLKTVYQHQK, following trypsin digestion) have shown growth inhibition in both Gram-positive and Gram-negative pathogenic bacteria (e.g., *Staphylococcus aureus, Escherichia coli*) (Zucht et al. 1995). Recently, a study reported on the antifungal and antibacterial activity of milk protein-derived BAPs obtained using fungal proteases (isolated from *Aspergillus oryzae* and *Aspergillus flavipes*) (Zanutto-Elgui et al. 2019). Furthermore, the combination of lactoferricin with other milk proteins (e.g., lactoferrin) showed a synergistic antibacterial effect (López-Expósito et al. 2008). The MilkAMP database (http://milkampdb.org/home.php) collates information for milk protein-derived antimicrobial peptides (Théolier et al. 2014).

18.2.1.7 Enzymatic Hydrolysis for the Generation of Opioid-Like Peptides

Opioid peptides have the ability to exert a morphine-like effect by binding to μ-, δ- and κ-opioid receptors, which are associated with the control of intestinal motility, emotional behaviour and satiety, respectively (Nongonierma et al. 2016b; Aslam et al. 2019; Liu and Udenigwe 2019). Opioid BAPs usually consist of 4–8 amino acids with an N-terminal Y residue (Kaur et al. 2019). Milk protein has been reported to be a good source of opioid peptides known as casomorphins (from casein) and lactorphins (from whey) (Sharma et al. 2011; Meisel and FitzGerald 2000). Opioid peptides can act as opioid agonists or antagonists. For instance, β-casomorphins (from β-casein) have opioid agonist effects, while casoxins (from κ-casein) and lactoferroxin (from lactoferrin) have an opioid antagonist effect (Chiba et al. 1989; Loukas et al. 1983). The hydrolysis of β-Lg and α-La with pepsin, trypsin, and a combination of trypsin and chymotrypsin led to the release of the opioid peptides YLLP and YGLP (β- and α-lactorphin, respectively) (Antila et al. 1991). YFYPEL and YFYPE from α_{s1}-casein showed an opioid agonist effect in guinea pig ileum (Fernández-Tomé et al. 2016). Furthermore, some opioid peptides, e.g., YLLP, YGLP and HIRL also display ACE-inhibitory activity and, therefore, are considered as multifunctional bioactive peptides (Pihlanto-Leppälä 2000; FitzGerald and Meisel 1999; Mullally et al. 1996).

18.2.1.8 Enzymatic Hydrolysis for the Generation of Other Bioactive Peptides

Milk protein-derived BAPs with mineral binding, hypocholesterolaemic and anti-thrombotic activities have been studied to a lesser extent. Milk protein-derived peptides can bind to minerals such as calcium and iron. In this regard, CPPs are associated with an enhancement in calcium bioavailability (FitzGerald 1998; Guo et al. 2014; Nongonierma and FitzGerald 2012; Ahn and Je 2019). CPPs derived from β-casein had higher iron-binding activity than CPPs from α_{s1}-casein (Kibangou et al. 2005).

The hypocholesterolaemic activity of milk peptides has been previously reported. Hydrolysates from whey protein treated with trypsin showed values of 47.67% inhibition of cholesterol micellar solubility (Jiang et al. 2018). Lactostatin (IIAEK) and β-lactotensin (HIRL) were released on incubation of β-Lg with trypsin and chymotrypsin, respectively. The administration of these peptides decreased serum levels of cholesterol and low density lipoprotein while they increased high density lipoproteins in rats fed a diet rich in cholesterol (Nagaoka et al. 2001; Yamauchi et al. 2003).

Milk protein-derived BAPs can also inhibit platelet aggregation. For instance, tryptic digestion of β-casein released YQEPVLGPVRGPFPIIV, which showed anti-coagulant and anti-thrombotic activity *in vitro* by inhibiting fibrin cross-linking and thrombin, respectively (Liu et al. 2019; Rojas-Ronquillo et al. 2012). A recent meta-analysis concluded that the ingestion of dairy products had a satiating effect (Onvani et al. 2017). Peptides from milk proteins can stimulate the secretion of specific regulatory peptides such cholecystokinin (CCK), peptide YY (PYY) and GLP-1 which are involved in the regulation of appetite (Geraedts et al. 2011; Santos-Hernández et al. 2018). Additionally, satiety can be regulated by the activation of serotonin (5-hydroxytryptamine, 5-HT) 2 receptor (5-HT_{2C}). The incubation of peptide fractions (<1 kDa), obtained from hydrolysis of sodium caseinate and whey protein using pepsin, with human embryonic kidney (Hek) cells resulted in activation of 5-HT_{2C}. However, these results were not seen by *in vivo* studies with male mice (C57BL/6), probably due to the instability of peptides during ingestion (Schellekens et al. 2014). In a recent study, a hydrolysate of sodium caseinate, named LFC25, was reported to promote the release of GLP-1. LFC25 led to a reduction in food intake in male C57BL/6 rats (O'Halloran et al. 2018). The antiviral properties of milk BAPs has been reviewed by Pan et al. (2006). For instance, Wong et al. (2014) demonstrated the effect of lactoferricin (MFKCRRWQWRMKKLGAPSITCV-RRAF) and lactoferrampin (DLIWKLLSKAQEKFGKNKSR) in the inhibition of three enzymes (human immunodeficiency virus-1 (HIV-1) protease, integrase and transcriptase) of HIV-1 indicating that lactoferrin-derived peptides could serve as beneficial agents in managing HIV infection.

18.2.2 *Microbial Fermentation*

The manufacture of fermented dairy products involves the utilisation of proteolytic and/or peptidolytic activities which may originate from the coagulant enzyme(s), the starter, non-starter and secondary microorganisms, and the indigenous enzymes. Therefore, casein/whey protein-derived BAPs can be released from milk proteins during fermentation processes through different proteolytic systems as follows: (1) coagulants, e.g., chymosin, pepsin, microbial or plant acid proteases; (2) starter and non-starter culture proteases/peptidases; (3) indigenous milk proteases, e.g., plasmin, cathepsin D and other somatic cell proteases; and (4) the proteolytic activities of secondary cultures such as *Penicillium camemberti, Penicillium roqueforti, Propionibacterium* sp., *Brevibacterium linens* and *Corynebacterium*, which are mainly utilised in mould-ripened cheeses (Gobbetti et al. 2002; Shori and Baba 2015). In this section, we focus on the contribution of bacteria, yeast and mould to the generation of BAPs during the fermentation of milk.

18.2.2.1 Bacteria

Lactic acid bacteria (LAB) such as *Lactococcus lactis, Lactobacillus helveticus* and *Lactobacillus delbrueckii* var. *bulgaricus* are the main organisms used as starter cultures in the manufacture of yoghurt, sour fermented milk and cheese. The proteolytic system of LAB is well characterised, and consists of 3 major components: (1) cell wall-bound protease(s), also referred to as the cell envelope proteinase (CEP), which is responsible for the initial hydrolysis of milk proteins to peptides containing 4–30 amino acid residues; (2) a peptide transport system which consists of an oligopeptide binding protein, two permeases and two ATPases associated with pore formation and provision of energy, respectively, and (3) intracellular peptidases which specifically cleave peptides, including endopeptidases, and aminopeptidases along with dipeptidases and tripeptidases (with potential to cleave from the N- or C-terminus), which further degrade peptides to shorter fragments and free amino acids (Hafeez et al. 2014). In addition, extracellular peptidases may play an important role in the further hydrolysis of the peptides generated by the CEPs, in the absence of bacterial lysis.

To date, a number of extracellular peptidases have been identified in LAB, e.g., aminopeptidase N (found in e.g., *L. lactis* subsp. *cremoris* AM-2, *L. helveticus*

LHE-511 and *Lactobacillus acidophilus* R-26), a tripeptidase (in e.g., *L. lactis* subsp. *cremoris* IMN-C12), an X-prolyl dipeptidyl aminopeptidase (in e.g., *L. lactis* subsp. *cremoris* P8–2-47 and *Streptococcus thermophilus*) and an aminopeptidase (in e.g., *L. delbrueckii* subsp. *bulgaricus* CNRZ397) (Hafeez et al. 2013; Hafeez et al. 2014). The short peptides and amino acids generated by these aminopeptidases are utilised by LAB to promote their growth and metabolism.

Furthermore, many of the peptides generated can act as a source of BAPs. The CEPs are considered as the major proteolytic activities for BAP generation in fermented dairy products. The catalytic site of CEPs contains aspartic acid, serine and histidine residues. All CEPs are classified as subtilisin-like serine proteases. The expression of CEP genes is influenced by several parameters, i.e., medium composition, carbon to nitrogen ratio, pH, temperature and heat treatment, which consequently impact on the level of CEP activity and the extent of formation of BAPs (Hafeez et al. 2014; Korhonen and Pihlanto 2006). On the other hand, intracellular peptidases can be released during autolysis of bacteria starters. This process has a major impact on cheese ripening and has been associated with texture changes and flavour development (Kenny et al. 2006). Different autolytic properties have been reported in different *Lactococcus* strains and, therefore, this influences the type and extent of BAPs generated during cheese ripening (Otte et al. 2011). Worsztynowicz et al. (2019) recently identified the key proteolytic system of the non-starter LAB enterococci *Enterococcus faecalis* isolated from an artisanal Polish cheese, which was shown to contain coccolysin and glutamyl endopeptidase activity.

To date, most of the studies in the literature report on the production of ACE-inhibitory or antihypertensive peptides along with immunomodulatory, antioxidant and antimicrobial peptides during the microbial fermentation of milk proteins, as outlined in Table 18.3. The most potent ACE inhibitory peptides reported to date are the casein fragments, i.e., IPP (β-CN f(74–76) and κ-CN f(108–110)), and VPP (β-CN f(84–86)), LHLPLP (β-CN f(133–138)), having IC_{50} values of 5.0, 9.0 and 5.5 μM, respectively (FitzGerald and Murray 2006; Aluko 2015). These peptides were originally identified in a commercial sour milk product, Ameal S'™, which is produced by co-fermentation of skim milk with *L. helveticus* and *Saccharomyces cerevisiae* (Nakamura et al. 1995a).

Furthermore, many studies have focused on the antihypertensive properties of the tripeptides (IPP and VPP) within fermented dairy products. Human studies have shown significant reductions in systolic blood pressure (SBP), ranging from −3.7 to −6.9 mm Hg, after administration of doses containing on average 9.4 mg IPP and VPP per day over periods up to 21 weeks (for more information on meta-analysis see Cicero et al. (2010)). Other potent ACE inhibitory peptides, LHLPLP (β-CN f(133–138)) and LVYPFPGPIPNSLPQNIPP (β-CN f(58–76)), were identified in skim milk fermented with *Enterococcus faecalis* CECT 5727 originally isolated from raw milk. These two peptides displayed potent ACE inhibitory activity *in vitro,* with IC_{50} values of 5.5 and 5.2 μM, respectively. These peptides also demonstrated antihypertensive activity in SHR. In particular, LHLPLP showed a significant effect in decreasing SBP and diastolic blood pressure (DBP) after oral administration at

Table 18.3 Examples of some bioactive peptides identified in cheese and fermented dairy products

Test samples	Microorganisms	Protein fragment	Sequence[a]	Activities	Values	References
Gouda	*L. lactis* subsp. *cremoris* or *L. lactis* subsp. *lactis*	α_{s1}-CN f(1–9)	RPKHPIKHQ	ACE inhibitory and antihypertensive	IC_{50} 13.4 μM; SBP -9.3 mmHg	Saito et al. (2000)
		β-CN f(60–68)	YPFPGPIPN	ACE inhibitory and antihypertensive	IC_{50} 14.8 μM; SBP -7.0 mmHg	
		β-CN f(70–77; 70–72)	LPQNIPPL; LPQ	DPP-IV inhibitory	IC_{50} 46 μM; 82 μM	Uenishi et al. (2012)
Cheddar	*Lactobacillus casei* ssp. *casei* 300	β-CN f(98–105)	VKEAMAPK	Antioxidant and ACE inhibitory	DPPH EC_{50} 66.7 μg mL^{-1}; ACE IC_{50} 4.21 mg mL^{-1}	Gupta et al. (2010)
		αs1-CN f(95–105)	HIQKEDVPSER	Antioxidant and ACE inhibitory	DPPH EC_{50} 74.7 μg mL^{-1}; ACE IC_{50} 3.26 mg mL^{-1}	
Manchego	*L. lactis* subsp. *lactis and L. lactis* subsp. *cremoris*	α_{s1}-CN f(102–109)	KKYNVPQL	ACE inhibitory and antihypertensive	IC_{50} 77.1 mM; SBP -11.5 ± 3.4 mmHg; DBP -9.7 ± 1.1 mmHg	Miguel et al. (2010)
Fermented skim milk	*L. helveticus* CPN4	α_{s1}-CN f(146–147)	YP	ACE inhibitory and antihypertensive	IC_{50} 720 μM; SBP -32.1 mmHg	Yamamoto et al. (1999)
Kefir from caprine milk	Co-culture of LABs and yeast from kefir grain	β-CN f(58–68)	LVYPFPGPIPN	ACE inhibitory and antihypertensive	IC_{50} 27.9 μM; SBP -28.0 mmHg	Miguel et al. (2010)
Fermented skim milk	*S. thermophilus, L. bulgaricus and K. marxianus*	α_{s1}-CN f(30–37)	VLNENLLRF	Antimicrobial against *E. coli* DPC6053	MIC 0.22 mM	Zhang et al. (2017)
AmealS® (Calpis sour milk)	*L. helveticus* and *S. cerevisiae*	β-CN f(74–76)	IPP	ACE inhibitory and antihypertensive	IC_{50} 5.0 μM; SBP -28.3 mmHg (−10.1 mm ha in human)	Nakamura et al. (1995a); Nakamura et al. (1995b)
		β-CN f(84–86)	VPP	ACE inhibitory and antihypertensive	IC_{50} 9.0 μM; SBP -32.1 mmHg (−10.1 mmHg in human)	
Fermented skim milk	*E. faecalis*	β-CN f(58–76)	LVYPFPGPIPNSLPQNIPP	ACE inhibitory and antihypertensive	IC_{50} 5.2 μM; SBP -15.0 mmHg	Quirós et al. (2007)
		β-CN f(133–138)	LHLPLP	ACE inhibitory and antihypertensive	IC_{50} 5.5 μM; SBP -21.9 mmHg	
		β-CN f(133–139)	LHLPLPL	ACE inhibitory and antihypertensive	IC_{50} 425 μM; SBP -7.7 mmHg	
		β-CN f(197–206)	VLGPVRGPFP	ACE inhibitory and antihypertensive	IC_{50} 137 μM; SBP -16.2 mmHg	
		β-CN f(201–209)	VRGPFPIIV	ACE inhibitory and antihypertensive	IC_{50} 599 μM; SBP -16.1 mmHg	

Test samples	Microorganisms	Protein fragment	Sequence[a]	Activities	Values	References
Fermented milk	*L.casei* ATCC393	Different β-CN fragments	DKIHPF f(47–52); VVPPFLQPE f(83–91); LLYQEPVLGPVRGPFPIIV f(191–209); LLYQEPVLGPVRGPFPII f(191–208); ELQDKIHPF f(44–52); LYQEPVLGPVRGPFPIIV f(192–209); LHLPLPLLQSW f(133–143); NLHLPLPLLQSW f(132–143); DVENLHLPLPLLQSW f(129–143); LYQEPVLGPVRGPFP f(192–206); MPFPKYPVEPF f(102–119); YQEPVLGPVRGPFPIIV f(193–209); LLYQEPVLGPVRGPFP f(191–206); NLHLPLPLL f(132–140)	ACE inhibitory	IC_{50} of 2 kDa permeate fraction 12.7 μg mL^{-1}	Abdel-Hamid et al. (2019)
		Different α_{s1}-CN fragments	FVAPFPEVFGKE f(39–53); ENLLRFF f(18–24); RFFVAPFPEVFGKE f(22–35)			
		κ-CN f(49–66)	ALINNQFLPYPYYAKPA			
		Different β-CN fragments	DKIHPF f(47–52); LLYQEPVLGPVRGPFPIIV f(191–209); LYQEPVLGPVRGPFP f(192–206); YQEPVLGPVRGPFPIIV f(193–209); LLYQEPVLGPVRGPFP f(191–206)	Antioxidant	EC_{50} of 2 kDa permeate 2.3 μg mL^{-1} for $ABTS^{•+}$ and 12.92 μg mL^{-1} Fe^{2+} chelating	
		Different β-CN fragments	LLYQEPVLGPVRGPFPIIV f(191–209); YQEPVLGPVRGPFPIIV f(193–209)	Anticancer (antiproliferation)	87.1% on Caco-2; 78.9% on MCF-7	

Test samples	Microorganisms	Protein fragment	Sequence[a]	Activities	Values	References
Fermented skim milk	*K. marxianus* Z17	κ-CN f(31–36)	VLSRYP	ACE inhibitory	IC_{50} 36.7 μM	Li et al. (2015)
		α_{s1}-CN f(21–24)	LRFF	ACE inhibitory	IC_{50} 116.9 μM	
Fermented bovine lactoferrin	*K. marxianus* Km2	LF f(70–76)	DPYKLRP	ACE inhibitory	IC_{50} 30.5 μM	García-Tejedor et al. (2014)
				Antihypertensive	SBP -26.8 mmHg	
				ACE and angiotensin II reduction in SHR blood	−48.1%; −27.1%	
		LF f(71–76)	PYKLRP	ACE inhibitory and antihypertensive	IC_{50} 10.2 μM; SBP -22 mmHg	
		LF f(72–76)	YKLRP	ACE inhibitory and antihypertensive	IC_{50} 16.5 μM; SBP -21 mmHg	
		LF f(130–134)	GILRP	ACE inhibitory and antihypertensive	IC_{50} 90.7 μM; SBP -20 mmHg	
Fermented sodium caseinate	*L. acidophilus* DPC6026	α_{s1}-CN f(21–29; 30–37; 195–208)	IKHQGLPQE; VLNENLLR; SDIPNPIGSENSEK	Antimicrobial against *Escherichia coli* DPC6053	MIC 0.05; 0.22; 1.0 mM	Hayes et al. (2006)
Commercial kefir; fermented casein; fermented skim milk; commercial yoghurt	Kefir grain; *L. casei* Shirota and *S. thermophilus; S. thermophilus, L. bulgaricus, and K. marxianu; S. thermophilus, L bulgaricus*	β-CN f(193–209)	YQEPVLGPVRGPFPIIV	ACE inhibitor, antithrombotic, antimicrobial against *E. coli* DPC6053	ACE of 0.14% per peptide concentration (μg mL^{-1}); thrombin inhibition efficiency ratio of 4.6% per peptide concentration (μg mL^{-1}); MIC 0.4 mg mL^{-1}	Liu and Pischetsrieder (2017); Rojas-Ronquillo et al. (2012); Birkemo et al. (2009); Zhang et al. (2017); Jin et al. (2016)

CN casein, *LF* lactoferrin, *ACE* angiotensin converting enzyme, *SHR* spontaneously hypertensive rats, IC_{50} concentration of peptide resulting in 50% inhibition of ACE activity, *SBP* systolic blood pressure, *DBP* diastolic blood pressure, *MIC* minimum inhibition concentration

[a]Peptide sequence with the one letter code

2 mg kg^{-1} BW for 2–4 h (Miguel et al. 2006; Quirós et al. 2007; Hernández-Ledesma et al. 2011).

Abdel-Hamid et al. (2019) identified BAPs possessing *in vitro* antioxidant, ACE, inhibitory and anticancer activities during storage of bovine milk fermented with *L. casei* ATCC 393. The bioactive properties of the water-soluble fractions increased with increasing proteolytic activity during storage over 21 days, especially in the 2 kDa permeate fraction which had an ACE inhibitory IC_{50} of 12.7 μg mL^{-1}, an $ABTS^{•+}$ scavenging EC_{50} of 2.3 μg mL^{-1}, an Fe^{2+} chelating EC_{50} of 12.9 μg mL^{-1} and antiproliferative activity against Caco-2 and MCF-7 cell lines of 87.1 and 78.9%. The BAPs identified containing 6–19 amino acid residues which were mainly from β-casein (14 peptides), α-casein (5 peptides) and κ-casein (2 peptides), as shown in Table 18.3.

During cheese manufacture, the starter cultures added are predominantly LAB, while other bacteria, yeast and fungi may also be involved in the later stages of the cheese making process. These organisms play a role in the development of flavour and texture during cheese ripening and maturation. The primary role of LAB is in the production of lactic acid from lactose, resulting in a pH reduction and coagulation of the milk proteins. In addition, the indigenous milk proteinase, plasmin with an alkaline optimum pH, also plays a role in the cheese ripening process (Parente et al. 2017). Various BAPs have been identified during the cheese ripening process. These BAPs have been characterised in different cheese varieties such as in Cheddar, Edam, Gouda, Camembert, Havarti, Blue cheeses, Manchego, Crescenza, and other Italian, Spanish and Finnish cheeses (FitzGerald and Murray 2006; Hernández-Ledesma et al. 2011; Korhonen and Pihlanto 2006; Gupta et al. 2010; Miguel et al. 2010; Uenishi et al. 2012).

18.2.2.2 Yeasts and Moulds

Yeasts and moulds are widely used during the manufacture of fermented dairy products. These organisms are mainly involved in the cheese ripening process and contribute to the development of texture and aroma compounds, e.g., volatile acids and carbonyl compounds (Ferreira and Viljoen 2003). In comparison to the proteolytic activity in LAB strains, yeasts showed higher caseinolytic activity, as well as possessing high peptidase activity (Bintsis et al. 2003; Klein et al. 2002). Yeasts and moulds also possess both intra- and extra-cellular proteinases. The extracellular enzymes in *Candida, Cryptococcus, Rhodotorula, Pichia* and *Yarrowia* genera have been well characterised (Ferreira and Viljoen 2003). Yeast peptidases, i.e., aminopeptidases and carboxypeptidases, play key roles in the degradation of milk proteins releasing peptides and amino acids, which are subsequently utilised as growth nutrients. Several factors affect yeast growth; these include their fermentation of lactose and galactose, and assimilation of lactate, along with their lipolytic and proteolytic activities (Chaves-López et al. 2014; Chaves-López et al. 2012). Therefore, alteration in the expression of yeast proteolytic activities could subsequently influence the type and quantity of the BAPs released.

Two main approaches, i.e., selection of the microorganisms used and optimisation of the fermentation process, are employed during the release of BAPs in

Table 18.4 Bioactive properties of peptides derived during the *in vivo* digestion of milk proteins as detected in the human gastrointestinal (GI) system and in blood (adapted from Boutrou et al. (2015) and Nongonierma and FitzGerald (2015))

Food sample	Precursor protein	Peptide fragment	Peptide sequence[a]	Bioactive properties	Detection site	References
Casein suspension	β-CN	60–66	YPFPGPI	Opioid agonist	Jejunum	Sanchón et al. (2018)
		60-68	YPFPGPIPN	Antihypertensive		
		114–119	YPVEPF	Opioid agonist and antihypertensive		
		134–138	HLPLP	Antihypertensive		
	α_{s1}-CN	143–149	AYFYPEL	Opioid agonist and antihypertensive		
Casein drink	α_{S1}-CN	24–31	FVAPFPEV	ACE inhibitor	Jejunum	Boutrou et al. (2013)
		24–32	FVAPFPEVF			
		24–33	FVAPFPEVFG			
		25–32	VAPFPEVF			
		104–109	YKVPQL			
	β-CN	59–66	VYPFPGPI	Opioid agonist	Jejunum	
		59–68	VYPFPGPIPN			
		63–68	PGPIPN	Immunomodulatory		
		114–119	YPVEPF	Opioid agonist		
		170–176	VLPVPQK	Antioxidant		
	κ-CN	106–111	MAIPPK	Antithrombotic	Jejunum	
		112–116	KNQDK	Antithrombotic		
Casein drink; skim milk	κ-CN	106–116	MAIPPKKNQDK	Antithrombotic	Jejunum; stomach	Boutrou et al. (2013); Chabance et al. (1998)

(continued)

Table 18.4 (continued)

Food sample	Precursor protein	Peptide fragment	Peptide sequence[a]	Bioactive properties	Detection site	References
Casein drink; raw bovine milk	β-CN	60–65	YPFPGP	Opioid agonist	Jejunum; duodenum	Boutrou et al. (2013); Svedberg et al. (1985)
		60–66	YPFPGPI			
		60–67	YPFPGPIP			
		60–68	YPFPGPIPN			
Raw bovine milk	β-CN	60–63	YPFP	Opioid agonist	Duodenum	Svedberg et al. (1985)
Skim milk	α_{S1}-CN	1–23	RPKHPIKHQGLPQEVLNENLLRF	Antibacterial	Plasma	Chabance et al. (1998)
		143–149	AYFYPEL	ACE inhibitor and antihypertensive	Stomach	
		144–149	YFYPEL	Antioxidant	Duodenum	
Breast milk/bovine milk-based infant formula	κ-CN	106–169	Glycomacropeptide	Antithrombotic	Plasma	Chabance et al. (1995)
Albumin enriched bovine milk	β-CN	60–66	YPFPGPI	Opioid agonist	Plasma	Kost et al. (2009)
Lactotripeptide (AmealPeptide powder, Calpis) enriched yoghurt drink	LF	81–82, 399–400	IY	ACE inhibitor	Serum	Foltz et al. (2007)
	CN, whey	Diverse	LW, FY, IW			
	α-la, BSA	Diverse	AW, VY, IPP			
	β-, κ-CN	Diverse	LPP, IL, VL, LL			

CN casein, *LF* lactoferrin, *α-la* α-lactalbumin, *BSA* bovine serum albumin, *ACE* angiotensin I-converting enzyme

[a]Peptide sequence represented by the one letter code

fermented dairy products. Li et al. (2015) reported that fermentation temperature, inoculum level and agitation intensity had a major impact on ACE inhibitory peptide release from skim milk fermented with *Kluyveromyces marxianus* Z17. This, in turn, was associated with the extent of carboxypeptidase and aminopeptidase activities. They also reported two novel ACE inhibitory peptides, VLSRYP (κ-CN f(31–36)) and LRFF (α_{s1}-CN f(21–24)) having IC_{50} values of 36.7 and 116.9 μM, respectively. Furthermore, co-culture of LABs and yeasts have been extensively used to generate BAPs.

Chaves-López et al. (2011) and Chaves-López et al. (2012) reported on selected combined LAB and yeast strains giving high ACE-inhibitory activity during milk fermentation. They also investigated the impact of co-culture of selected LAB and yeast for the production of Kumis, a Colombian fermented bovine milk product having high ACE inhibitory activity (Chaves-López et al. 2014). The most effective combination was obtained using *Pichia kudriavzevii* KL84A, *L. plantarum* LAT3 and *Enterococcus faecalis* KL06, giving a product with high ACE inhibitory activity (IC_{50} 30.63 ± 1.11 μg mL^{-1}), while having a relatively low bitterness profile (equivalent to 0–0.7 mM caffeine) after 7 days of fermentation.

Another co-cultured probiotic fermented milk beverage, kefir, which originated in the former Soviet Union has become popular due to its perceived health benefits. Kefir can be produced using bovine, caprine, ovine or buffalo milk incubated with kefir grains, which is a mixture of various polysaccharides containing different microbial species such as LAB, acetic acid bacteria, yeast and fungi (Amorim et al. 2019). The distribution of the different symbiotic microorganisms in kefir grains can vary depending on the origin of the kefir. However, several kefirs have been reported to contain *Lactobacillus brevis, L. helveticus, Lactobacillus kefir, Leuconostoc mesenteroides, Kluyveromyces lactis, K. marxianus* and *Pichia fermentans* (Ferreira et al. 2010). Kefir is a source of various BAPs, which can have an impact on different bodily systems, i.e., the cardiovascular, digestive, immune and nervous systems, as summarised in Fig. 18.2. Amorim et al. (2019) and Ebner et al.

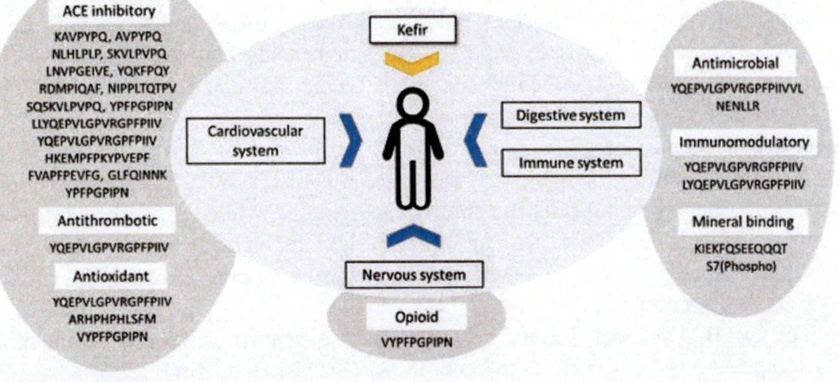

Fig. 18.2 Milk protein-derived bioactive peptides identified in kefir and their impact on different bodily systems (adapted from Ebner et al. (2015) and Abdel-Hamid et al. (2019)). Peptide sequences are displayed using the one letter amino acid code

(2015) characterised the BAPs generated during kefir fermentation; these peptides were associated with ACE inhibitory, antithrombotic, antioxidant, antimicrobial, immunomodulatory, mineral binding and opioid properties (Fig. 18.2).

18.2.3 Production of BAPs During Gastrointestinal Digestion

A wide range of dairy components that may act as sources of BAPs are consumed in the format of individual ingredients to complex dairy matrices ranging from lactoferrin, casein, whey, caseinomacropeptide, phosphopeptides, skim milk, whole milk to a range of fermented dairy products (Boutrou et al. 2015). However, while there are a limited number of studies reporting on the bioactive effects of milk-derived BAPs produced *in vivo* in humans, milk-derived peptides released *in vivo* in the human gastrointestinal (GI) system have been reported to possess numerous bioactive properties, as outlined in Table 18.4. Among the peptides detected in the human GI system to date, 11 peptides were associated with opioid agonist activity (i.e., 10 peptides derived from β-CN and 1 from α_{s1}-CN). In addition, 17 ACE inhibitory peptides have been identified, including 4 peptide fragments from α_{s1}-CN within the region 24–33 and another 11 di- and tri-peptides which are commonly found in casein and whey proteins. The other 12 BAPs have been reported to possess diverse properties such as antihypertensive, immunomodulatory, antioxidant and antithrombotic activity.

BAPs can be released from milk proteins during GI digestion by a combination of gastric and pancreatic enzymes, i.e., pepsin, trypsin and chymotrypsin (Nongonierma et al. 2016b). Pepsin represents the main digestive enzyme in the stomach where the initial hydrolysis of dietary proteins takes place. Trypsin and chymotrypsin, located in the duodenum, are involved in the further degradation of proteins to shorter peptides. Furthermore, smaller peptide fragments are digested by elastases, also secreted in pancreatic juices. In addition, pancreatic brush border enzymes, consisting of carboxypeptidases, amino peptidases and dipeptidases, contribute to the formation of shorter peptides and free amino acids in fermented dairy products (Karaś 2019). These peptides and amino acids may be absorbed in the jejunum to the blood stream *via* mucosal cells before they can act systemically. The absorption process of BAPs in the small intestine is influenced by various physiological and chemical factors, which have been outlined by Karaś (2019). These include their stability to gastric and intestinal enzymes, mucus barrier, tight junctions, intestinal epithelial cells and epithelial peptidases, while unabsorbed nutrients can pass into the large intestine and can be used for growth by gut microflora. Microbial-derived enzymes in the ileum and colon also play a role in the further hydrolysis of peptides.

The first BAPs detected in the human small intestine following raw bovine milk consumption was a group β-casomorphins (BCMs) including BCM-4 (β-CN f(60–63)), BCM-5 (β-CN f(60–64)), BCM-6 (β-CN f(60–65)), BCM-7 (β-CN f(60–66)), which are associated with immunomodulatory and opioid effects

Table 18.5 Examples of the application of *in silico* approaches during bioactive peptide generation and identification

Source	Product/Protein	Enzymes	Results/objectives achieved	References
Camel and horse	Casein	Pepsin, trypsin and combined enzymes	*In silico* prediction of ACE inhibitory peptides within hydrolysates through database search	Ugwu et al. (2019)
Yak	Casein	Chymotrypsin, trypsin, papain, thermolysin, Alcalase and protease K individually and in combination	Selection of protease to obtain maximum number of ACE inhibitory peptides	Lin et al. (2018b)
Caprine	Fermented milk	Trypsin	*In silico* prediction of ACE inhibitory peptides within hydrolysates through database search	Parmar et al. (2018)
Camel	Fermented milk	*Lactobacillus bulgaricus* NCDC(09) and *Lactobacillus fermentum* TDS030603 (LBF)	*In silico* confirmation of ACE inhibitory peptides through database	Solanki et al. (2017)
Bovine	β-Lactoglobulin	Elastase	Selection of optimum conditions (E:S, DH) to obtain DPP-IV inhibitory peptides	Le Maux et al. (2017)
Camel	Milk	Trypsin	*In silico* prediction of DPP-IV inhibitory peptides and confirmation by *in vitro* enzymatic hydrolysis	Nongonierma et al. (2017b)
Bovine and caprine	Casein	Trypsin	*In silico* prediction and comparison between DPP-IV inhibitory peptides from cow and goat caseins	Zhang et al. (2016)
Bovine	α-Lactalbumin	Elastase	*In silico* prediction of DPP-IV inhibitory peptides and confirmation by *in vitro* enzymatic hydrolysis	Nongonierma et al. (2016a)
Bovine	Whey protein (α-lactalbumin, β-lactoglobulin)	Gastrointestinal enzymes (pepsin, trypsin and chymotrypsin)	*In silico* prediction of DPP-IV inhibitory peptides and confirmation by *in vitro* enzymatic hydrolysis	Tulipano et al. (2015)

(continued)

Table 18.5 (continued)

Source	Product/Protein	Enzymes	Results/objectives achieved	References
Bovine and buffalo	Whey protein (α-lactalbumin, β-lactoglobulin)	Trypsin	*In silico* prediction of ACE inhibitory peptides and confirmation by *in vitro* enzymatic hydrolysis	Chatterjee et al. (2015a)
Bovine	Casein	Thermolysin and thermolysin-like activities	*In silico* prediction for the release of the antimicrobial peptide IKHQGLPQE by different enzymes	Guinane et al. (2015)
Bovine	α_s-casein	Prolyl endoprotease (an-PEP)	*In silico* prediction of ACE inhibitory peptides released during hydrolysis	Norris et al. (2015)
Bovine	Milk	–	Protein prediction of DPP-IV peptides based on sequence	Nongonierma and FitzGerald (2014a)
Bovine	Milk	Gastrointestinal enzymes	*In silico* prediction of ACE inhibitory peptides and confirmation by *in vitro* enzymatic hydrolysis	Nongonierma et al. (2014)
Bovine	Milk	28 different enzymes (various sources)	Prediction of antimicrobial peptides based on protein sequence	Dziuba and Dziuba (2014)
Bovine	β-Casein	Prolyl endoprotease (an-PEP)	*In silico* prediction of ACE inhibitory peptides released from hydrolysis	Norris et al. (2014)
Bovine	Milk	Gastrointestinal enzymes	*In silico* prediction of ACE inhibitory peptides released after gastrointestinal digestion	Pripp (2005)
Bovine	β-Lactoglobulin	Gastrointestinal enzymes	*In silico* prediction of ACE inhibitory peptides released from hydrolysis with gastrointestinal enzymes and comparison with those from pea protein	Vermeirssen et al. (2004)

ACE angiotensin converting enzyme, *E:S* enzyme to substrate, *DH* degree of hydrolysis, *DPP-IV* dipeptidyl peptidase-IV

(Svedberg et al. 1985; Boutrou et al. 2015). Meisel et al. (2003) were the first to report on the presence of CPPs in human ileostomy fluid. Furthermore, other BAPs derived from α_{s1}- and κ-casein and whey proteins released during transit through the human GI system have been detected in the digestive contents of the jejunum, stomach, ileum, duodenum and small intestine (Boutrou et al. 2013; Boutrou et al. 2015; Chabance et al. 1998; Meisel et al. 2003; Svedberg et al. 1985; FitzGerald and Murray 2006; Sanchón et al. 2018), as outlined in Table 18.4. Sanchón et al. (2018) reported on BAPs released in the human jejunum after consumption of casein- and whey protein-containing drinks which possessed antihypertensive (i.e., YPFPGPIPN (β-CN f(60–68)), YPFPGPI (β-CN f(60–66)), HLPLP (β-CN f(134–138)), AYFYPEL (α_{s1}-CN f(143–149)) and opioid activity (i.e., YPFPGPI (β-CN f(60–66)), YPVEPF (β-CN f(114–119)), AYFYPEL (α_{s1}-CN f(143–149)) (Table 18.4). In addition, they reported that some of these peptides were detected at different sampling times, which indicated differences in the resistance of these peptides to release in the GI system.

18.2.4 In silico *Approaches for Peptide Generation*

In silico approaches are computer-based techniques, which, e.g., can help to predict the peptides that may be released from a protein considering a range of factors such as proteolytic activity and starting material (Dziuba and Dziuba 2010). Protein and peptides databases are required for knowledge of the primary structure of the protein (e.g., Swissprot, Uniprot) and to identify the bioactive sequences encrypted within the protein, e.g., BIOPEP-UWM (University of Warmia and Mazury; http://www.uwm.edu.pl/biochemia/index.php/en/biopep), Pepbank (http://pepbank.mgh.harvard.edu/) and the Milk Bioactive Peptides DataBase (MBPDB; http://mbpdb.nws.oregonstate.edu/). Additional tools are then required in order to select the proteolytic activities which can be employed to release peptides, e.g., "enzyme action" from BIOPEP-UWM or "PeptideCutter" from ExPASy (Udenigwe 2014), along with the new online tool PeptidePredictor (www.peptidepredictor.com) also.

The DOE approach has been used to study the release of milk protein hydrolysates/peptides. During enzymatic hydrolysis, a DOE can take into account the effect of various factors, e.g., time of hydrolysis, enzyme to substrate (E:S), pH and temperature, which are combined in a design model, i.e., in factorial or response surface designs. Experimental values are then used to construct mathematical models which can be applied to predict values and optimise hydrolysis conditions to maximise the release of BAPs (Li-Chan 2015).

For instance, Nongonierma et al. (2017a) used a multifactorial design with three factors at three levels, i.e., pH, temperature and hydrolysis time to investigate hydrolysates of sodium caseinate with Protamex® with potent antioxidant and DPP-IV inhibitory activities. A total of 15 different experiments arose from the DOE. From which the optimum hydrolysis conditions were reported to be pH 8.0, 40 °C and 5 h incubation. Furthermore, knowledge of the structure-activity criteria

Table 18.6 Examples of the separation of bioactive-peptide-enriched fractions using membrane processing (adapted from Abd El-Salam and El-Shibiny 2017)

Bioactivity	Substrate	Membrane separation condition	Properties of Separated Fraction	References
Insulinotropic	Casein hydrolysate	UF membrane with 3 kDa MWCO followed by flash RP-HPLC	Successfully enriched the insulinotropic peptide fraction from 0.1% to 4% of total peptide abundance	Murray et al. (2018)
ACE inhibitory	Tryptic MCN and β-CN hydrolysates	Cross-flow UF membrane with 5 kDa MWCO, pH 5, 10 °C, 0.24 MPa transmembrane pressure	<5 kDa fraction showed significantly increased in ACE inhibitory activity with IC_{50} of 109 and 98 µg/mL for MCN and β-CN hydrolysates, respectively	Holder et al. (2013)
	Heated WPH using Corolase PP	Cascade of UF membranes 10, 5, 1 kDa followed by isoelectric focusing	1 kDa retentate exhibited highest ACE inhibitory activity (IC_{50} 0.17 g L^{-1})	O'Loughlin et al. (2014a)
	Whey concentrate (6% protein) hydrolysed with combinations of thermolysin, protease K, Alcalase, Neutrase	Two-step UF membranes using 10 kDa membrane followed by 1 kDa	1 kDa permeate exhibited 70% inhibition of ACE	Ramos et al. (2012)
	Tryptic hydrolysate of whey proteins	UF membrane 6 and 10 kDa MWCO followed by gel chromatography on Sephadex G-25 and G-10 FPLC column and RP-HPLC	The <6 kDa fraction exhibited the highest ACE inhibitory activity (at 64%), fraction with the lowest MW was the most potent fraction which contain dipeptide LL[a]	Pan et al. (2012)

(continued)

Table 18.6 (continued)

Bioactivity	Substrate	Membrane separation condition	Properties of Separated Fraction	References
Antimicrobial	Tryptic hydrolysate of WPI	UF membrane 10 kDa MWCO followed by NF membrane 2.5 kDa MWCO (pH 7.0, 1.7 MPa, 50 °C)	NF retentate was the most effective antimicrobial fraction against *Listeria monocytogenes* and *E. coli* which was rich in anionic peptide >8 amino acid residues	Demers-Mathieu et al. (2013)
DPP-IV inhibitory	Tryptic WPH	UF membrane 5 and 2 kDa followed by solid phase extraction (SPE)	Successfully enriched potent DPP-IV inhibitory peptide fraction in 5 and 2 kDa permeates	Nongonierma and FitzGerald (2013a)
	Tryptic WPH	NF membrane with 200 Da MWCO at 50 °C and 22 bar inlet operating pressure	Permeate fraction exhibited the highest DPP-IV inhibitory activity at an IC_{50} 0.66 mg protein mL^{-1}. This fraction contained short peptides, e.g., VA, VL, WL, WI[a].	Le Maux et al. (2015)
Antioxidant	β-Lg enriched WPH using Corolase PP and thermolysin	UF of thermolysin hydrolysate using 3 kDa hydrophilic membrane at 5 °C	The highest antioxidant activity was obtained in the permeate from the hydrolysate after 8 h hydrolysis at 80 °C	Contreras et al. (2011)
	Tryptic bovine casein hydrolysate	UF MWCO 10 kDa and 1 kDa (55 psi with nitrogen gas)	The highest antioxidant activity in <1 kDa permeate fraction	Irshad et al. (2015)
Immunomodulatory	Tryptic β-lg hydrolysate	UF membrane cassettes with 5 and 1 kDa MWCO and a stabilized 2 kDa cellulose membrane at pH 8, 37 °C and 7.5 × 10^4 Pa transmembrane pressure	Fraction enriched in large and acid peptides induce Th1 responses. The shorter peptides stimulated monocyte and increased TNFα secretion	Rodríguez-Carrio et al. (2014)

(continued)

Table 18.6 (continued)

Bioactivity	Substrate	Membrane separation condition	Properties of Separated Fraction	References
Mixed bioactive peptides	Tryptic β-lg hydrolysate	UF membrane cassettes with 5 and 1 kDa MWCO and 2 kDa cellulose membrane at pH 8, 37 °C and 7.5 × 10^4 Pa transmembrane pressure	<1 kDa fraction increased antioxidant, DPP-IV and ACE inhibitory activities by 1.7-, 3- and two-fold, respectively	Power et al. (2014a)
	WPH (heated, unheated) using Corolase PP (5, 10% DH)	Cascade membrane fractionation using MF membrane 0.14 μm, 30 and 10 kDa (50 °C, 4.2 bar membrane inlet pressure) and UF membrane 5 and 1 kDa MWCO membrane (at 5 bar)	Presence of multifunctional peptides; <1 kDa fraction exhibited the highest ACE-inhibitory activity from 10% DH hydrolysate of heated whey proteins; higher Fe^{2+} chelating ability found in small peptide fraction	O'Loughlin et al. (2014b)
	Camel casein hydrolysate using protenase K	10, 5 and 3 kDa UF membrane followed by semi-preparative RP-HPLC	<5 kDa fraction exhibited higher ACE inhibitory activity and <10 kDa fraction exhibited higher ABTS radical scavenging activity when compared to the hydrolysate	Rahimi et al. (2016)
	Fermented milk with *L. casei* ATCC 393	2 kDa UF membrane	<2 kDa fraction showed the highest ACE inhibitory, antioxidant (ABTS radical scavenging and Fe^{2+} chelating) and anticancer activity (against Caco-2 and MCF-7 cell lines) when compared to the water soluble extract	Abdel-Hamid et al. (2019)

DPP-IV Dipeptidyl peptidase IV, *MCN* micellar casein, *CN* casein, *WPH* whey protein hydrolysate, *β-Lg* β-lactoglobulin, *DH* degree of hydrolysis, *ATCC* the American Type Culture Collection, *UF* ultrafiltration, *NF* nanofiltration, *ACE* angiotensin I-converting enzyme, IC_{50} concentration of peptides resulting in 50% inhibition of ACE activity, *MWCO* molecular weight cut-off, *ABTS* 2,2′-azino-bis(3-ethylbenzothiazoline-6-sulphonic acid), Fe^{2+} ferrous iron, *DH* degree of hydrolysis

[a]Peptide sequence with the one letter code

by which peptides contribute to bioactivity can be exploited using *in silico* methodologies as an approach to identifying specific protein substrates as sources of potent BAPs. For instance, it is well known that peptides containing specific amino acids such as P or L in N or C terminal positions can exert specific bioactive properties such as ACE or DPP-IV inhibitory activity (Nongonierma and FitzGerald 2016b). Quantitative structure activity relationship (QSAR) studies are being used for the discovery of milk-protein-derived BAPs. QSAR studies are performed following a series of steps: (1) identification of peptides with a desired activity; (2) selection of similar peptides, based on molecular descriptors; (3) construction and application of the QSAR model to allow a ranking of peptides on the basis of their potential bioactivity; and (4) confirmatory studies with synthetic peptides (Nongonierma and FitzGerald 2016a).

In some instances, QSAR methodology involves complex mathematical formulas that can be solved using artificial neural networks (ANN) (He et al. 2012). QSAR studies often involve additional *in silico* approaches such as molecular docking. Molecular docking is a computer-based tool in which peptides and enzyme structures are represented *in silico* in order to investigate their possible interactions, e.g., as ligands and receptors, respectively (Fukunishi et al. 2017; Iwaniak et al. 2015; Lin et al. 2018a). Molecular docking approaches involve the use of specific software, e.g., AutoDock Vina, Ligplot, Dimglot (Norris et al. 2012; Singh et al. 2017; Nongonierma et al. 2014). Some examples of the discovery of milk protein-derived BAPs which have been aided using *in silico* approaches are summarised in Table 18.5. While *in silico* approaches have many benefits, the results obtained experimentally may not always reflect the predicted results (Chatterjee et al. 2015b). Therefore, confirmatory studies are required, in all instances, to validate the *in silico* predictions.

18.3 Fractionation and Enrichment of BAPs

In general, BAPs are generated from complex milk protein samples (as opposed to individual milk proteins) which results in hydrolysates with a complex mixture of residual intact proteins, peptides and free amino acids. BAPs can be separated from inactive peptides based on differences in their physicochemical properties, including molecular mass, hydrophobicity and net charge. Various techniques can be used to enrich BAPs for further characterisation or application as bioactive ingredients. These include approaches such as isoelectric precipitation, membrane filtration and chromatography (Dullius et al. 2018; Nongonierma et al. 2016b).

18.3.1 Membrane Filtration

Membrane filtration is the most versatile approach for BAP enrichment at large-scale. Membrane processing generally fractionates BAPs on the basis of differences in molecular mass using pressure as a driving force, most frequently in either ultrafiltration (UF) or nanofiltration (NF) mode. UF is one of the most commonly used membrane processing techniques for the separation and enrichment of milk protein-derived BAPs. It offers a wide range of separation capability with molecular weight cut off (MWCO) values ranging from 1–10^3 kDa, while NF separates molecules <1 kDa (Fernández et al. 2014).

Reports on the application of membrane processing for fractionation and enrichment of milk derived BAPs from different samples, e.g., from enzymatic hydrolysates of dairy proteins and fermented dairy products, are summarised in Table 18.6. Overall, a series of different membrane processing and other separation techniques, including isoelectric focusing, reversed-phase and gel permeation chromatography, have been applied to separate and enrich milk-derived BAPs possess ACE inhibitory, antimicrobial, DPP-IV, antioxidant, immunomodulatory as well as multifunctional bioactivities.

Laurent and Loubna (2013) reviewed the membrane processing techniques and devices used for BAP enrichment/separation of dairy and plant protein sources. Membrane filtration systems can be used in batch or continuous processing mode (so called membrane reactor systems). The limitations of batch separation are related to membrane fouling, which is associated with interaction between peptides and the membrane. The net charge, charge distribution and size of the peptides influences the cross-flow between retentate and permeate and thus the extent of fouling observed (Arrutia et al. 2016). An enzymatic membrane reactor has been developed to integrate enzymatic hydrolysis, peptide separation and enzyme recovery within a single operation (Bhat et al. 2015). O'Halloran et al. (2019) generated whey-protein-derived DPP-IV inhibitory peptides using enzymatic membrane reactor technology. The enzymatic membrane reactor process increased the yield of DPP-IV inhibitory peptides by 7.2–8.7% when compared to conventional enzymatic hydrolysis approaches. Hoog Antink et al. (2019) demonstrated that different membrane surface functionalisation (such as use of carboxylated and aminated membranes which bring about enzyme immobilisation *via* non-covalent and carbodiimide-activated binding) affected the interaction between the enzyme and the membrane surface. These differences in enzyme-surface interactions consequently resulted in different degrees of hydrolysis and the types of BAPs generated, in the case of enzymes immobilised to tubular ceramic membranes.

Moreover, membrane processing has been combined with other technologies to improve BAP yield. This includes the use of electrodialysis-ultrafiltration which may be applied to separate BAPs based on differences in their molecular mass and net charge under an applied electric field. For instance, Doyen et al. (2013) generated β-Lg derived BAPs using simultaneous enzymatic hydrolysis and a single step electrodialysis-ultrafiltration separation process with a reported peptide yield ranging from 5.5–66%. In total, 19 anionic and cationic peptides were separated, and 6 of the peptides were previously reported to exhibit hypocholesterolemic, antihypertensive orantibacterial activities. In particular, one of the identified cationic peptides was lactokinin, ALPMHIR (β-Lg f(142–148)), which was previously reported to exert potent ACE-inhibitory activity having an IC_{50} of 42.6 μM (Mullally et al. 1997). Suwal et al. (2017) performed a comparative study on enzymatic hydrolysis using an electrodialysis-ultrafiltration system (*in situ*) and a conventional batch hydrolysis procedure (*ex situ*). They focused on the separation parameters, and on peptide yield and composition specifically on anionic and cationic peptide fractions. They reported that the mode of hydrolysis (*in situ* vs *ex situ*) and separation had a major impact on peptide migration through the electrodialysis-ultrafiltration system.

18.3.2 Chromatography

Liquid chromatography (LC) is extensively used for BAP separation to yield high purity products. However, this technique is still limited for large scale production due to its high cost and relatively low throughput. Chromatographic separations can be classified based on the pressure applied. The two most common LC systems used for BAP separation are medium pressure, known as fast protein liquid chromatography (FPLC), and high-pressure liquid chromatography (HPLC). The chromatographic columns used in FPLC systems can separate BAPs on the basis of molecular mass, i.e., *via* gel permeation, and *via* net charge with cationic or anionic matrices, i.e., using ion-exchange separation. In general, FPLC systems are used for initial peptide separations prior to further purification using HPLC columns, especially using semi-preparative HPLC columns for larger scale BAP separations.

The use of reversed-phase columns (RP-HPLC) has been described as the most efficient approach for BAP separation (Akhavan et al. 2010). However, RP-HPLC can be time consuming and costly due to the relatively high flow rate of the mobile phase, which results in large fraction volumes that subsequently need to be dried. In addition, multiple injections of the active fraction may be required in order to obtain sufficient quantities of pure peptide (Aluko 2015). On the other hand, an understanding of the physicochemical properties of the target BAP can allow the selection of more appropriate approaches for peptide separation (Nongonierma et al. 2016b). The use of *in silico* bioprospecting models have been proposed for the selection of suitable BAP purification techniques (for reviews see Agyei et al. (2016); Dullius et al. (2018)).

Table 18.6 shows some examples of applications of membrane processing for separation and enrichment of milk protein-derived BAPs. Murray et al. (2018) fractionated insulinotropic, peptides (which stimulate insulin secretion) derived from a sodium caseinate hydrolysate using membrane filtration with a 3 kDa MWCO membrane followed by RP flash chromatography. The peptides obtained were then identified using LC-MS/MS. It was demonstrated that insulinotropic peptides were enriched from 0.1% of total peptide abundance (TPA) in the <3 kDa fraction to 4% TPA following elution with 30% ethanol by RP flash chromatography.

Rahimi et al. (2016) also used UF and RP-HPLC to fractionate ACE inhibitory and antioxidant peptides from camel casein hydrolysed with proteinase K. The hydrolysate was first fractionated using a sequential series of UF membranes with 10, 5 and 3 kDa MWCO values followed by semi-preparative RP-HPLC. The most potent peptide fractions exhibited ACE IC_{50} and $ABTS^{\bullet+}$ scavenging EC_{50} values of 36 and 3.3 μg mL^{-1}, respectively, while the intact camel casein had ACE IC_{50} and $ABTS^{\bullet+}$ scavenging EC_{50} values of >600 and 35 μg mL^{-1}, respectively. The application of UF for enrichment of BAPs derived from milk products fermented with *L. casei* ATCC 393 was demonstrated by Abdel-Hamid et al. (2019). The fractions obtained possessed multi-functional properties, including *in vitro* ACE inhibitory and antioxidant activities and *in situ* anticancer activity against Caco-2 and MCF-7 cell-lines (Table 18.6). Various short peptides (having MWs <1.5 kDa) derived from casein hydrolysed by proteolytic enzymes extracted from *L. helveticus* exhibited multi-functional activity including antioxidative, opioid agonist, hypotensive, immunomodulatory, antibacterial and antithrombotic activities were fractioned using gel permeation chromatography (Skrzypczak et al. 2017).

One of the challenges to be considered during fractionation of BAPs is the potential removal of peptides that may have synergistic effects on overall bioactivity. Removal of these peptides during fractionation may lead to a reduction, or in some instances, a complete loss in bioactivity of the fractions. Moreover, high purity peptide-based ingredients, which usually come with high production costs, may not be economically feasible for the food industry. Nevertheless, Winkelnkemper and Schembecker (2010) introduced parameters to be considered during downstream process development (including a purification performance index and a separation cost indicator) when embarking on the production of BAP-enriched preparations. These indices take into account the purification rate, product yield and overall process cost. However, more detailed studies are required to validate the cost effectiveness of large-scale BAP production for food use.

18.4 Conclusion

Several approaches can be exploited, including *in vitro* enzymatic hydrolysis, microbial fermentation and *in vivo* gastrointestinal digestion to generate milk-derived BAPs. A combination of these approaches may also be used to obtain milk-derived BAPs possessing various bioactive properties associated with beneficial effects on human bodily systems. The enzymatic hydrolysis and microbial fermentation conditions need to be carefully chosen given their influence on the release of BAP sequences. Furthermore, *in silico* approaches are being increasingly employed for process optimisation and the discovery of milk-derived BAPs with high potency. The isolation of pure BAP molecules for incorporation into foods may not always be necessary due to cost-related concerns. Another challenge associated with BAP enrichment and fractionation is the selection of appropriate techniques which maintain BAP potency. Research in the area of the generation and identification of milk-derived BAPs can lead to a better understanding the nutritional benefits of dairy product consumption, as well as in the development of functional foods and ingredients for other applications, such as food for special medical purposes or nutraceutical products.

Acknowledgements The authors acknowledge funding under the National Development Plan 2007–2013, through the Food Institutional Research Measure (FIRM), administered by the Department of Agriculture, Food and Marine, Ireland under grant issue 15/F/647 and the Dairy Processing Technology Centre (DPTC), Enterprise Ireland (Grant Agreement Number: TC 2014 0016).

References

Abd El-Salam MH, El-Shibiny S (2017). Separation of Bioactive Whey Proteins and Peptides. In Ingredients extraction by physicochemical methods in food (pp. 463–494). Academic Press

Abdel-Hamid M, Romeih E, Gamba RR, Nagai E, Suzuki T, Koyanagi T, Enomoto T (2019) The biological activity of fermented milk produced by lactobacillus casei ATCC 393 during cold storage. Int Dairy J 91:1–8

Adams RL, Broughton KS (2016) Insulinotropic effects of whey: mechanisms of action, recent clinical trials, and clinical applications. Ann Nutr Metab 69(1):56–63

Adje EY, Balti R, Kouach M, Guillochon D, Nedjar-Arroume N (2011) α 67-106 of bovine hemoglobin: a new family of antimicrobial and angiotensin I-converting enzyme inhibitory peptides. Eur Food Res Technol 232(4):637–646

Agyei D, Ongkudon CM, Wei CY, Chan AS, Danquah MK (2016) Bioprocess challenges to the isolation and purification of bioactive peptides. Food Bioprod Process 98:244–256

Ahn C-B, Je J-Y (2019) Bone health-promoting bioactive peptides. J Food Biochem 43(1):e12529

Akhavan T, Luhovyy BL, Brown PH, Cho CE, Anderson GH (2010) Effect of premeal consumption of whey protein and its hydrolysate on food intake and postmeal glycemia and insulin responses in young adults. Am J Clin Nutr 91(4):966–975

Altmann K, Wutkowski A, Klempt M, Clawin-Rädecker I, Meisel H, Lorenzen PC (2016) Generation and identification of anti-inflammatory peptides from bovine β-casein using enzyme preparations from cod and hog. J Sci Food Agric 96(3):868–877

Aluko RE (2015) Antihypertensive peptides from food proteins. Annu Rev Food Sci Technol 6:235–262

Aluko RE (2019) Food protein-derived renin-inhibitory peptides: in vitro and in vivo properties. J Food Biochem 43(1):e12648

Amorim FG, Coitinho LB, Dias AT, Friques AGF, Monteiro BL, de LCD R, TMC P, Campagnaro BP, De Pauw E, Vasquez EC, Quinton L (2019) Identification of new bioactive peptides from kefir milk through proteopeptidomics: bioprospection of antihypertensive molecules. Food Chem 282:109–119

Antila P, Paakkari I, Järvinen A, Mattila MJ, Laukkanen M, Pihlanto-Leppälä A, Mäntsälä P, Hellman J (1991) Opioid peptides derived from in-vitro proteolysis of bovine whey proteins. Int Dairy J 1(4):215–229

Arrutia F, Rubio R, Riera FA (2016) Production and membrane fractionation of bioactive peptides from a whey protein concentrate. J Food Eng 184:1–9

Aslam H, Ruusunen A, Berk M, Loughman A, Rivera L, Pasco JA, Jacka FN (2019) Unravelled facets of milk derived opioid peptides: a focus on gut physiology, fractures and obesity. Int J Food Sci Nutr:1–14

Assem FM, Abd El-Gawad MAM, Kassem JM, Abd El-Salam MH (2018) Proteolysis and antioxidant activity of peptic, tryptic and chymotryptic hydrolysates of cow, buffalo, goat and camel caseins. Int J Dairy Technol 71(1):236–242

Azevedo RA, Ferreira AK, Auada AVV, Pasqualoto KFM, Marques-Porto R, Maria DA, Lebrun I (2012) Antitumor effect of cationic INKKI peptide from bovine β-casein on melanoma B16F10. J Cancer Ther 3(04):237

Berthou J, Migliore-Samour D, Lifchitz A, Delettré J, Floc'h F, Jollès P (1987) Immunostimulating properties and three-dimensional structure of two tripeptides from human and cow caseins. FEBS Lett 218(1):55–58

Bhat ZF, Kumar S, Bhat HF (2015) Bioactive peptides of animal origin: a review. J Food Sci Technol 52(9):5377–5392

Bidasolo IB, Ramos M, Gomez-Ruiz JA (2012) In vitro simulated gastrointestinal digestion of donkeys' milk. Peptide characterization by high performance liquid chromatography–tandem mass spectrometry. Int Dairy J 24(2):146–152

Bintsis T, Vafopoulou-Mastrojiannaki A, Litopoulou-Tzanetaki E, Robinson RK (2003) Protease, peptidase and esterase activities by lactobacilli and yeast isolates from feta cheese brine. J Appl Microbiol 95(1):68–77

Birkemo GA, O'Sullivan O, Ross RP, Hill C (2009) Antimicrobial activity of two peptides casecidin 15 and 17, found naturally in bovine colostrum. J Appl Microbiol 106(1):233–240

Boelsma E, Kloek J (2010) IPP-rich milk protein hydrolysate lowers blood pressure in subjects with stage 1 hypertension, a randomized controlled trial. Nutr J 9(1):52

Boutrou R, Gaudichon C, Dupont D, Jardin J, Airinei G, Marsset-Baglieri A, Benamouzig R, Tomé D, Leonil J (2013) Sequential release of milk protein–derived bioactive peptides in the jejunum in healthy humans. Am J Clin Nutr 97(6):1314–1323

Boutrou R, Henry G, Sanchez-Rivera L (2015) On the trail of milk bioactive peptides in human and animal intestinal tracts during digestion: a review. Dairy Sci Technol 95(6):815–829

Brandelli A, Daroit DJ, Corrêa APF (2015) Whey as a source of peptides with remarkable biological activities. Food Res Int 73:149–161

Cermeño M, FitzGerald RJ, O'Brien NM (2016) In vitro antioxidant and immunomodulatory activity of transglutaminase-treated sodium caseinate hydrolysates. Int Dairy J 63:107–114

Chabance B, Jollès P, Izquierdo C, Mazoyer E, Francoual C, Drouet L, Fiat A-M (1995) Characterization of an antithrombotic peptide from α-casein in newborn plasma after milk ingestion. Br J Nutr 73(4):583–590

Chabance B, Marteau P, Rambaud JC, Migliore-Samour D, Boynard M, Perrotin P, Guillet R, Jollès P, Fiat AM (1998) Casein peptide release and passage to the blood in humans during digestion of milk or yogurt. Biochimie 80(2):155–165

Chalamaiah M, Yu W, Wu J (2018) Immunomodulatory and anticancer protein hydrolysates (peptides) from food proteins: a review. Food Chem 245:205–222

Chatterjee A, Kanawjia SK, Khetra Y, Saini P (2015a) Discordance between in silico & in vitro analyses of ACE inhibitory & antioxidative peptides from mixed milk tryptic whey protein hydrolysate. J Food Sci Technol 52(9):5621–5630

Chatterjee A, Kanawjia SK, Khetra Y, Saini P (2015b) Discordance between in silico & in vitro analyses of ACE inhibitory & antioxidative peptides from mixed milk tryptic whey protein hydrolysate. J Food Sci Technol 52(9):5621–5630

Chaves-López C, Serio A, Martuscelli M, Paparella A, Osorio-Cadavid E, Suzzi G (2011) Microbiological characteristics of kumis, a traditional fermented Colombian milk, with particular emphasis on enterococci population. Food Microbiol 28(5):1041–1047

Chaves-López C, Tofalo R, Serio A, Paparella A, Sacchetti G, Suzzi G (2012) Yeasts from Colombian Kumis as source of peptides with angiotensin I converting enzyme (ACE) inhibitory activity in milk. Int J Food Microbiol 159(1):39–46

Chaves-López C, Serio A, Paparella A, Martuscelli M, Corsetti A, Tofalo R, Suzzi G (2014) Impact of microbial cultures on proteolysis and release of bioactive peptides in fermented milk. Food Microbiol 42:117–121

Cheison SC, Kulozik U (2017) Impact of the environmental conditions and substrate pre-treatment on whey protein hydrolysis: a review. Crit Rev Food Sci Nutr 57(2):418–453

Chiba H, Tani F, Yoshikawa M (1989) Opioid antagonist peptides derived from kappa-casein. J Dairy Res 56(3):363–366

Cicero AFG, Gerocarni B, Laghi L, Borghi C (2010) Blood pressure lowering effect of lactotripeptides assumed as functional foods: a meta-analysis of current available clinical trials. J Hum Hypertens 25:425

Cicero AFG, Aubin F, Azais-Braesco V, Borghi C (2013) Do the Lactotripeptides isoleucine–Proline–Proline and Valine–Proline–Proline reduce systolic blood pressure in European subjects? A meta-analysis of randomized controlled trials. Am J Hypertens 26(3):442–449

Contreras MM, Hernández-Ledesma B, Amigo L, Martín-Álvarez PJ, Recio I (2011) Production of antioxidant hydrolyzates from a whey protein concentrate with thermolysin: optimization by response surface methodology. LWT Food Sci Technol 44(1):9–15

Contreras MM, Sanchez D, Sevilla MÁ, Recio I, Amigo L (2013) Resistance of casein-derived bioactive peptides to simulated gastrointestinal digestion. Int Dairy J 32(2):71–78

Corrochano AR, Buckin V, Kelly PM, Giblin L (2018) Invited review: whey proteins as antioxidants and promoters of cellular antioxidant pathways. J Dairy Sci 101(6):4747–4761

Coste M, Rochet V, Léonil J, Mollé D, Bouhallab S, Tomé D (1992) Identification of C-terminal peptides of bovine β-casein that enhance proliferation of rat lymphocytes. Immunol Lett 33(1):41–46

de Castro RJS, Bagagli MP, Sato HH (2015) Improving the functional properties of milk proteins: focus on the specificities of proteolytic enzymes. Curr Opin Food Sci 1:64–69

de Oliveira MR, Silva TJ, Barros E, Guimarães VM, Baracat-Pereira MC, Eller MR, dos Reis Coimbra JS, de Oliveira EB (2018) Anti-hypertensive peptides derived from caseins: mechanism of physiological action, production bioprocesses, and challenges for food applications. Appl Biochem Biotechnol 185(4):884–908

Demers-Mathieu V, Gauthier SF, Britten M, Fliss I, Robitaille G, Jean J (2013) Antibacterial activity of peptides extracted from tryptic hydrolyzate of whey protein by nanofiltration. Int Dairy J 28(2):94–101

dos Santos Aguilar JG, Sato HH (2018) Microbial proteases: production and application in obtaining protein hydrolysates. Food Res Int 103:253–262

Doyen A, Husson E, Bazinet L (2013) Use of an electrodialytic reactor for the simultaneous β-lactoglobulin enzymatic hydrolysis and fractionation of generated bioactive peptides. Food Chem 136(3):1193–1202

Dullius A, Goettert MI, de Souza CFV (2018) Whey protein hydrolysates as a source of bioactive peptides for functional foods – biotechnological facilitation of industrial scale-up. J Funct Foods 42:58–74

Dziuba M, Dziuba B (2010) In silico analysis of bioactive peptides. In: Bioactive proteins and peptides as functional foods and nutraceuticals, In, pp 325–340

Dziuba B, Dziuba M (2014) New milk protein-derived peptides with potential antimicrobial activity: an approach based on bioinformatic studies. Int J Mol Sci 15(8):14531–14545

Ebner J, Aşçı Arslan A, Fedorova M, Hoffmann R, Küçükçetin A, Pischetsrieder M (2015) Peptside profiling of bovine kefir reveals 236 unique peptides released from caseins during its production by starter culture or kefir grains. J Proteome 117:41–57

Eisele T, Stressler T, Kranz B, Fischer L (2013) Bioactive peptides generated in an enzyme membrane reactor using Bacillus lentus alkaline peptidase. Eur Food Res Technol 236(3):483–490

El-Sayed M, Awad S (2019) Milk bioactive peptides: antioxidant, antimicrobial and anti-diabetic activities. Biochemistry 7(1):22–23

Fernández A, Zhu Y, FitzGerald RJ, Riera FA (2014) Membrane fractionation of a β-lactoglobulin tryptic digest: effect of the membrane characteristics. J Chem Technol Biotechnol 89(4):508–515

Fernández-Tomé S, Martínez-Maqueda D, Girón R, Goicoechea C, Miralles B, Recio I (2016) Novel peptides derived from αs1-casein with opioid activity and mucin stimulatory effect on HT29-MTX cells. J Funct Foods 25:466–476
Ferreira AD, Viljoen BC (2003) Yeasts as adjunct starters in matured Cheddar cheese. Int J Food Microbiol 86(1):131–140
Ferreira IMPLVO, Pinho O, Monteiro D, Faria S, Cruz S, Perreira A, Roque AC, Tavares P (2010) Short communication: effect of kefir grains on proteolysis of major milk proteins. J Dairy Sci 93(1):27–31
FitzGerald RJ (1998) Potential uses of Caseinophosphopeptides. Int Dairy J 8(5):451–457
FitzGerald RJ, Meisel H (1999) Lactokinins: whey protein-derived ACE inhibitory peptides. Food Nahrung 43(3):165–167
FitzGerald RJ, Murray BA (2006) Bioactive peptides and lactic fermentations. Int J Dairy Technol 59(2):118–125
FitzGerald RJ, Murray BA, Walsh DJ (2004) Hypotensive peptides from milk proteins. J Nutr 134(4):980S–988S
Foltz M, Meynen EE, Bianco V, Van Platerink C, Koning TMMG, Kloek J (2007) Angiotensin converting enzyme inhibitory peptides from a lactotripeptide-enriched milk beverage are absorbed intact into the circulation. J Nutr 137(4):953–958
Frieden TR, Jaffe MG (2018) Saving 100 million lives by improving global treatment of hypertension and reducing cardiovascular disease risk factors. J Clin Hypertens 20(2):208–211
Fukunishi Y, Yamasaki S, Yasumatsu I, Takeuchi K, Kurosawa T, Nakamura H (2017) Quantitative structure-activity relationship (QSAR) models for docking score correction. Mol Inf 36(1–2):1600013
García-Tejedor A, Sánchez-Rivera L, Castelló-Ruiz M, Recio I, Salom JB, Manzanares P (2014) Novel antihypertensive Lactoferrin-derived peptides produced by Kluyveromyces marxianus: gastrointestinal stability profile and in vivo angiotensin I-converting enzyme (ACE) inhibition. J Agric Food Chem 62(7):1609–1616
Geerlings A, Villar IC, Hidalgo Zarco F, Sánchez M, Vera R, Zafra Gomez A, Boza J, Duarte J (2006) Identification and characterization of novel angiotensin-converting enzyme inhibitors obtained from goat milk. J Dairy Sci 89(9):3326–3335
Geraedts MCP, Troost FJ, Fischer MAJG, Edens L, Saris WHM (2011) Direct induction of CCK and GLP-1 release from murine endocrine cells by intact dietary proteins. Mol Nutr Food Res 55(3):476–484
Gobbetti M, Stepaniak L, De Angelis M, Corsetti A, Di Cagno R (2002) Latent bioactive peptides in milk proteins: proteolytic activation and significance in dairy processing. Crit Rev Food Sci Nutr 42(3):223–239
Gómez-Ruiz JÁ, Ramos M, Recio I (2007) Identification of novel angiotensin-converting enzyme-inhibitory peptides from ovine milk proteins by CE-MS and chromatographic techniques. Electrophoresis 28(22):4202–4211
González-Rábade N, Badillo-Corona JA, Aranda-Barradas JS, Oliver-Salvador MC (2011) Production of plant proteases in vivo and in vitro — a review. Biotechnol Adv 29(6):983–996
Guinane CM, Kent RM, Norberg S, O'Connor PM, Cotter PD, Hill C, Fitzgerald GF, Stanton C, Ross RP (2015) Generation of the antimicrobial peptide caseicin a from casein by hydrolysis with thermolysin enzymes. Int Dairy J 49:1–7
Guo L, Harnedy PA, Li B, Hou H, Zhang Z, Zhao X, FitzGerald RJ (2014) Food protein-derived chelating peptides: biofunctional ingredients for dietary mineral bioavailability enhancement. Trends Food Sci Technol 37(2):92–105
Gupta A, Mann B, Kumar R, Ram Bhagat S (2010) Identification of antioxidant peptides in cheddar cheese made with adjunct culture lactobacillus casei ssp. casei 300. Milchwissenschaft 65(4):396–399
Gurumallesh P, Alagu K, Ramakrishnan B, Muthusamy S (2019) A systematic reconsideration on proteases. Int J Biol Macromol 128:254–267

Hafeez Z, Cakir-Kiefer C, Girardet J-M, Jardin J, Perrin C, Dary A, Miclo L (2013) Hydrolysis of milk-derived bioactive peptides by cell-associated extracellular peptidases of Streptococcus thermophilus. Appl Microbiol Biotechnol 97(22):9787–9799

Hafeez Z, Cakir-Kiefer C, Roux E, Perrin C, Miclo L, Dary-Mourot A (2014) Strategies of producing bioactive peptides from milk proteins to functionalize fermented milk products. Food Res Int 63:71–80

Hayes M, Ross RP, Fitzgerald GF, Hill C, Stanton C (2006) Casein-derived antimicrobial peptides generated by lactobacillus acidophilus DPC6026. Appl Environ Microbiol 72(3):2260

He R, Ma H, Zhao W, Qu W, Zhao J, Luo L, Zhu W (2012, 2012) Modeling the QSAR of ACE-inhibitory peptides with ANN and its applied illustration. Int J Pept

Hernández-Ledesma B, Recio I, Ramos M, Amigo L (2002) Preparation of ovine and caprine β-lactoglobulin hydrolysates with ACE-inhibitory activity. Identification of active peptides from caprine β-lactoglobulin hydrolysed with thermolysin. Int Dairy J 12(10):805–812

Hernández-Ledesma B, Amigo L, Ramos M, Recio I (2004) Release of angiotensin converting enzyme-inhibitory peptides by simulated gastrointestinal digestion of infant formulas. Int Dairy J 14(10):889–898

Hernández-Ledesma B, Dávalos A, Bartolomé B, Amigo L (2005) Preparation of antioxidant enzymatic Hydrolysates from α-Lactalbumin and β-Lactoglobulin. Identification of active peptides by HPLC-MS/MS. J Agric Food Chem 53(3):588–593

Hernández-Ledesma B, del Mar CM, Recio I (2011) Antihypertensive peptides: production, bioavailability and incorporation into foods. Adv Colloid Interf Sci 165(1):23–35

Hirota T, Nonaka A, Matsushita A, Uchida N, Ohki K, Asakura M, Kitakaze M (2011) Milk casein-derived tripeptides, VPP and IPP induced NO production in cultured endothelial cells and endothelium-dependent relaxation of isolated aortic rings. Heart Vessel 26(5):549–556

Hogan S, Zhang L, Li J, Wang H, Zhou K (2009) Development of antioxidant rich peptides from milk protein by microbial proteases and analysis of their effects on lipid peroxidation in cooked beef. Food Chem 117(3):438–443

Holder A, Birke A, Eisele T, Klaiber I, Fischer L, Hinrichs J (2013) Selective isolation of angiotensin-I-converting enzyme-inhibitory peptides from micellar casein and β-casein hydrolysates via ultrafiltration. Int Dairy J 31(1):34–40

Hoog Antink MM, Sewczyk T, Kroll S, Árki P, Beutel S, Rezwan K, Maas M (2019) Proteolytic ceramic capillary membranes for the production of peptides under flow. Biochem Eng J 147:89–99

Horner K, Drummond E, Brennan L (2016) Bioavailability of milk protein-derived bioactive peptides: a glycaemic management perspective. Nutr Res Rev 29(1):91–101

IDF (2015) IDF Diabetes Atlas, 7th edn. International Diabetes Federation, Brussels

Irshad I, Kanekanian A, Peters A, Masud T (2015) Antioxidant activity of bioactive peptides derived from bovine casein hydrolysate fractions. J Food Sci Technol 52(1):231–239

Ishida Y, Shibata Y, Fukuhara I, Yano Y, Takehara I, Kaneko K (2011) Effect of an excess intake of casein hydrolysate containing val-pro-pro and ile-pro-pro in subjects with normal blood pressure, high-normal blood pressure, or mild hypertension. Biosci Biotechnol Biochem 75(3):427–433

Iwaniak A, Minkiewicz P, Darewicz M, Protasiewicz M, Mogut D (2015) Chemometrics and cheminformatics in the analysis of biologically active peptides from food sources. J Funct Foods 16:334–351

Jiang Z, Tian B, Brodkorb A, Huo G (2010) Production, analysis and in vivo evaluation of novel angiotensin-I-converting enzyme inhibitory peptides from bovine casein. Food Chem 123(3):779–786

Jiang C, Liu L, Li X, Ma L, Du L, Zhao Y, Li D, Zhao W (2018) Separation and purification of hypocholesterolaemic peptides from whey protein and their stability under simulated gastrointestinal digestion. Int J Dairy Technol 71(2):460–468

Jin Y, Yu Y, Qi Y, Wang F, Yan J, Zou H (2016) Peptide profiling and the bioactivity character of yogurt in the simulated gastrointestinal digestion. J Proteome 141:24–46

Jrad Z, El Hatmi H, Adt I, Girardet J-M, Cakir-Kiefer C, Jardin J, Degraeve P, Khorchani T, Oulahal N (2014) Effect of digestive enzymes on antimicrobial, radical scavenging and angiotensin I-converting enzyme inhibitory activities of camel colostrum and milk proteins. Dairy Sci Technol 94(3):205–224

Kahn BB (1998) Type 2 diabetes: when insulin secretion fails to compensate for insulin resistance. Cell 92(5):593–596

Karaś M (2019) Influence of physiological and chemical factors on the absorption of bioactive peptides. Int J Food Sci Technol 54(5):1486–1496

Kaur J, Kumar V, Sharma K, Kaur S, Gat Y, Goyal A, Tanwar B (2019) Opioid peptides: an overview of functional significance. Int J Pept Res Ther

Kayser H, Meisel H (1996) Stimulation of human peripheral blood lymphocytes by bioactive peptides derived from bovine milk proteins. FEBS Lett 383(1–2):18–20

Kenny O, FitzGerald RJ, O'Cuinn G, Beresford T, Jordan K (2006) Autolysis of selected lactobacillus helveticus adjunct strains during Cheddar cheese ripening. Int Dairy J 16(7):797–804

Khan MU, Pirzadeh M, Förster CY, Shityakov S, Shariati MA (2018) Role of milk-derived antibacterial peptides in modern food biotechnology: their synthesis, applications and future perspectives. Biomol Ther 8(4):110

Kibangou IB, Bouhallab S, Henry G, Bureau F, Allouche S, Blais A, Guérin P, Arhan P, Bouglé DL (2005) Milk proteins and Iron absorption: contrasting effects of different caseinophosphopeptides. Pediatr Res 58(4):731–734

Klein N, Zourari A, Lortal S (2002) Peptidase activity of four yeast species frequently encountered in dairy products—comparison with several dairy bacteria. Int Dairy J 12(10):853–861

Korhonen H (2009) Milk-derived bioactive peptides: from science to applications. J Funct Foods 1(2):177–187

Korhonen H, Pihlanto A (2006) Bioactive peptides: production and functionality. Int Dairy J 16(9):945–960

Kost NV, Sokolov OY, Kurasova OB, Dmitriev AD, Tarakanova JN, Gabaeva MV, Zolotarev YA, Dadayan AK, Grachev SA, Korneeva EV, Mikheeva IG, Zozulya AA (2009) β-Casomorphins-7 in infants on different type of feeding and different levels of psychomotor development. Peptides 30(10):1854–1860

Lacroix IME, Li-Chan ECY (2012a) Dipeptidyl peptidase-IV inhibitory activity of dairy protein hydrolysates. Int Dairy J 25(2):97–102

Lacroix IME, Li-Chan ECY (2012b) Evaluation of the potential of dietary proteins as precursors of dipeptidyl peptidase (DPP)-IV inhibitors by an in silico approach. J Funct Foods 4(2):403–422

Lacroix IME, Li-Chan ECY (2013) Inhibition of dipeptidyl peptidase (DPP)-IV and α-glucosidase activities by pepsin-treated whey proteins. J Agric Food Chem 61(31):7500–7506

Lacroix IME, Li-Chan ECY (2014) Isolation and characterization of peptides with dipeptidyl peptidase-IV inhibitory activity from pepsin-treated bovine whey proteins. Peptides 54:39–48

Lacroix IME, Meng G, Cheung IWY, Li-Chan ECY (2016) Do whey protein-derived peptides have dual dipeptidyl-peptidase IV and angiotensin I-converting enzyme inhibitory activities? J Funct Foods 21:87–96

Lammi C, Aiello G, Boschin G, Arnoldi A (2019) Multifunctional peptides for the prevention of cardiovascular disease: a new concept in the area of bioactive food-derived peptides. J Funct Foods 55:135–145

Laurent B, Loubna F (2013) Separation of bioactive peptides by membrane processes: technologies and devices. Recent Pat Biotechnol 7(1):9–27

Le Maux S, Nongonierma AB, Murray B, Kelly PM, FitzGerald RJ (2015) Identification of short peptide sequences in the nanofiltration permeate of a bioactive whey protein hydrolysate. Food Res Int 77:534–539

Le Maux S, Nongonierma AB, FitzGerald RJ (2017) Peptide composition and dipeptidyl peptidase IV inhibitory properties of β-lactoglobulin hydrolysates having similar extents of hydrolysis while generated using different enzyme-to-substrate ratios. Food Res Int 99:84–90

Lee N, Cheng J, Enomoto T, Nakamura I (2006) The antihypertensive activity of angiotensin-converting enzyme inhibitory peptide containing in bovine lactoferrin. Chin J Physiol 49(2):67

Li Y-W, Li B (2013) Characterization of structure–antioxidant activity relationship of peptides in free radical systems using QSAR models: key sequence positions and their amino acid properties. J Theor Biol 318:29–43

Li Y, Sadiq FA, Liu T, Chen J, He G (2015) Purification and identification of novel peptides with inhibitory effect against angiotensin I-converting enzyme and optimization of process conditions in milk fermented with the yeast Kluyveromyces marxianus. J Funct Foods 16:278–288

Liceaga AM, Hall F (2019) Nutritional, functional and bioactive protein Hydrolysates. In: Melton L, Shahidi F, Varelis P (eds) Encyclopedia of food chemistry. Academic, Oxford, pp 456–464

Li-Chan ECY (2015) Bioactive peptides and protein hydrolysates: research trends and challenges for application as nutraceuticals and functional food ingredients. Curr Opin Food Sci 1:28–37

Lin K, Zhang L, Han X, Meng Z, Zhang J, Wu Y, Cheng D (2018a) Quantitative structure-activity relationship modeling coupled with molecular docking analysis in screening of angiotensin I-converting enzyme inhibitory peptides from qula casein hydrolysates obtained by two-enzyme combination hydrolysis. J Agric Food Chem 66(12):3221–3228

Lin K, Zhang LW, Han X, Xin L, Meng ZX, Gong PM, Cheng DY (2018b) Yak milk casein as potential precursor of angiotensin I-converting enzyme inhibitory peptides based on in silico proteolysis. Food Chem 254:340–347

Liu Y, Pischetsrieder M (2017) Identification and relative quantification of bioactive peptides sequentially released during simulated gastrointestinal digestion of commercial kefir. J Agric Food Chem 65(9):1865–1873

Liu Z, Udenigwe CC (2019) Role of food-derived opioid peptides in the central nervous and gastrointestinal systems. J Food Biochem 43(1):e12629

Liu H, Tu M, Cheng S, Chen H, Wang Z, Du M (2019) An anticoagulant peptide from beta-casein: identification, structure and molecular mechanism. Food Funct 10(2):886–892

López-Expósito I, Quirós A, Amigo L, Recio I (2007) Casein hydrolysates as a source of antimicrobial, antioxidant and antihypertensive peptides. Lait 87(4–5):241–249

López-Expósito I, Pellegrini A, Amigo L, Recio I (2008) Synergistic effect between different milk-derived peptides and proteins. J Dairy Sci 91(6):2184–2189

Lorenzo JM, Munekata PES, Gómez B, Barba FJ, Mora L, Pérez-Santaescolástica C, Toldrá F (2018) Bioactive peptides as natural antioxidants in food products – a review. Trends Food Sci Technol 79:136–147

Loukas S, Varoucha D, Zioudrou C, Streaty RA, Klee WA (1983) Opioid activities and structures of alpha-casein-derived exorphins. Biochemistry 22(19):4567–4573

Ma J-J, Mao X-Y, Wang Q, Yang S, Zhang D, Chen S-W, Li Y-H (2014) Effect of spray drying and freeze drying on the immunomodulatory activity, bitter taste and hygroscopicity of hydrolysate derived from whey protein concentrate. LWT Food Sci Technol 56(2):296–302

Mader JS, Salsman J, Conrad DM, Hoskin DW (2005) Bovine lactoferricin selectively induces apoptosis in human leukemia and carcinoma cell lines. Mol Cancer Ther 4(4):612–624

Mao X-Y, Cheng X, Wang X, Wu S (2011) Free-radical-scavenging and anti-inflammatory effect of yak milk casein before and after enzymatic hydrolysis. Food Chem 126(2):484–490

Martínez-Medina GA, Barragán AP, Ruiz HA, Ilyina A, Martínez Hernández JL, Rodríguez-Jasso RM, Hoyos-Concha JL, Aguilar-González CN (2019) Chapter 14 - fungal proteases and production of bioactive peptides for the food industry. In: Kuddus M (ed) Enzymes in food biotechnology. Academic, San Diego, pp 221–246

Meisel H, FitzGerald RJ (2000) Opioid peptides encrypted in intact milk protein sequences. Br J Nutr 84(Suppl 1):S27–S31

Meisel H, Bernard H, Fairweather-Tait S, FitzGerald RJ, Hartmann R, Lane CN, McDonagh D, Teucher B, Wal JM (2003) Detection of caseinophosphopeptides in the distal ileostomy fluid of human subjects. Br J Nutr 89(3):351–358

Miguel M, Recio I, Ramos M, Delgado MA, Aleixandre MA (2006) Antihypertensive effect of peptides obtained from Enterococcus faecalis-fermented milk in rats. J Dairy Sci 89(9):3352–3359

Miguel M, Gómez-Ruiz JÁ, Recio I, Aleixandre A (2010) Changes in arterial blood pressure after single oral administration of milk-casein-derived peptides in spontaneously hypertensive rats. Mol Nutr Food Res 54(10):1422–1427

Mohanty D, Jena R, Choudhury PK, Pattnaik R, Mohapatra S, Saini MR (2016a) Milk derived antimicrobial bioactive peptides: a review. Int J Food Prop 19(4):837–846

Mohanty DP, Mohapatra S, Misra S, Sahu PS (2016b) Milk derived bioactive peptides and their impact on human health – a review. Saudi J Biol Sci 23(5):577–583

Mudgil P, Kamal H, Yuen GC, Maqsood S (2018) Characterization and identification of novel antidiabetic and anti-obesity peptides from camel milk protein hydrolysates. Food Chem 259:46–54

Mullally MM, Meisel H, FitzGerald RJ (1996) Synthetic peptides corresponding to a-lactalbumin and b-lactoglobulin sequences with angiotensin-1-converting enzyme inhibitory activity. Biol Chem Hoppe Seyler 377(4):259–260

Mullally MM, Meisel H, FitzGerald RJ (1997) Identification of a novel angiotensin-I-converting enzyme inhibitory peptide corresponding to a tryptic fragment of bovine β-lactoglobulin. FEBS Lett 402(2–3):99–101

Murray NM, O'Riordan D, Jacquier JC, O'Sullivan M, Holton TA, Wynne K, Robinson RC, Barile D, Nielsen SD, Dallas DC (2018) Peptidomic screening of bitter and nonbitter casein hydrolysate fractions for insulinogenic peptides. J Dairy Sci 101(4):2826–2837

Nagaoka S, Futamura Y, Miwa K, Awano T, Yamauchi K, Kanamaru Y, Tadashi K, Kuwata T (2001) Identification of novel hypocholesterolemic peptides derived from bovine milk β-lactoglobulin. Biochem Biophys Res Commun 281(1):11–17

Nakamura Y, Yamamoto N, Sakai K, Okubo A, Yamazaki S, Takano T (1995a) Purification and characterization of angiotensin I-converting enzyme inhibitors from sour milk. J Dairy Sci 78(4):777–783

Nakamura Y, Yamamoto N, Sakai K, Takano T (1995b) Antihypertensive effect of sour milk and peptides isolated from it that are inhibitors to angiotensin I-converting enzyme. J Dairy Sci 78(6):1253–1257

Nongonierma AB, FitzGerald RJ (2011) Enzymes exogenous to milk in dairy technology | proteinases. In: Fuquay JW (ed) Encyclopedia of dairy sciences, 2nd edn. Academic, San Diego, pp 289–296

Nongonierma AB, FitzGerald RJ (2012) Biofunctional properties of Caseinophosphopeptides in the Oral cavity. Caries Res 46(3):234–267

Nongonierma AB, FitzGerald RJ (2013a) Dipeptidyl peptidase IV inhibitory properties of a whey protein hydrolysate: influence of fractionation, stability to simulated gastrointestinal digestion and food–drug interaction. Int Dairy J 32(1):33–39

Nongonierma AB, FitzGerald RJ (2013b) Inhibition of dipeptidyl peptidase IV (DPP-IV) by proline containing casein-derived peptides. J Funct Foods 5(4):1909–1917

Nongonierma AB, FitzGerald RJ (2014a) An in silico model to predict the potential of dietary proteins as sources of dipeptidyl peptidase IV (DPP-IV) inhibitory peptides. Food Chem 165:489–498

Nongonierma AB, FitzGerald RJ (2014b) Susceptibility of milk protein-derived peptides to dipeptidyl peptidase IV (DPP-IV) hydrolysis. Food Chem 145:845–852

Nongonierma AB, FitzGerald RJ (2015) The scientific evidence for the role of milk protein-derived bioactive peptides in humans: a review. J Funct Foods 17:640–656

Nongonierma AB, FitzGerald RJ (2016a) Learnings from quantitative structure-activity relationship (QSAR) studies with respect to food protein-derived bioactive peptides: a review. RSC Adv 6(79):75400–75413

Nongonierma AB, FitzGerald RJ (2016b) Structure activity relationship modelling of milk protein-derived peptides with dipeptidyl peptidase IV (DPP-IV) inhibitory activity. Peptides 79:1–7

Nongonierma AB, FitzGerald RJ (2018) Enhancing bioactive peptide release and identification using targeted enzymatic hydrolysis of milk proteins. Anal Bioanal Chem 410(15):3407–3423

Nongonierma AB, FitzGerald RJ (2019) Features of dipeptidyl peptidase IV (DPP-IV) inhibitory peptides from dietary proteins. J Food Biochem 43(1):e12451

Nongonierma AB, Mooney C, Shields DC, FitzGerald RJ (2014) In silico approaches to predict the potential of milk protein-derived peptides as dipeptidyl peptidase IV (DPP-IV) inhibitors. Peptides 57:43–51

Nongonierma AB, Le Maux S, Hamayon J, FitzGerald RJ (2016a) Strategies for the release of dipeptidyl peptidase IV (DPP-IV) inhibitory peptides in an enzymatic hydrolyzate of α-lactalbumin. Food Funct 7(8):3437–3443

Nongonierma AB, O'Keeffe MB, FitzGerald RJ (2016b) Milk protein hydrolysates and bioactive peptides. In: McSweeney PLH, O'Mahony JA (eds) Advanced dairy chemistry, Proteins: applied aspects, vol 1B. Springer, New York, pp 417–482

Nongonierma AB, Maux SL, Esteveny C, FitzGerald RJ (2017a) Response surface methodology applied to the generation of casein hydrolysates with antioxidant and dipeptidyl peptidase IV inhibitory properties. J Sci Food Agric 97(4):1093–1101

Nongonierma AB, Paolella S, Mudgil P, Maqsood S, FitzGerald RJ (2017b) Dipeptidyl peptidase IV (DPP-IV) inhibitory properties of camel milk protein hydrolysates generated with trypsin. J Funct Foods 34:49–58

Nongonierma AB, Dellafiora L, Paolella S, Galaverna G, Cozzini P, FitzGerald RJ (2018a) In silico approaches applied to the study of peptide analogs of Ile-Pro-Ile in relation to their dipeptidyl peptidase IV inhibitory properties. Front Endocrinol 9

Nongonierma AB, Paolella S, Mudgil P, Maqsood S, FitzGerald RJ (2018b) Identification of novel dipeptidyl peptidase IV (DPP-IV) inhibitory peptides in camel milk protein hydrolysates. Food Chem 244:340–348

Norris R, FitzGerald RJ (2013) Antihypertensive peptides from food proteins

Norris R, Casey F, FitzGerald RJ, Shields D, Mooney C (2012) Predictive modelling of angiotensin converting enzyme inhibitory dipeptides. Food Chem 133(4):1349–1354

Norris R, Poyarkov A, O'Keeffe MB, FitzGerald RJ (2014) Characterisation of the hydrolytic specificity of Aspergillus niger derived prolyl endoproteinase on bovine β-casein and determination of ACE inhibitory activity. Food Chem 156:29–36

Norris R, O'Keeffe MB, Poyarkov A, FitzGerald RJ (2015) Peptide identification and angiotensin converting enzyme (ACE) inhibitory activity in prolyl endoproteinase digests of bovine αs-casein. Food Chem 188:210–217

O'Halloran J, O'Sullivan M, Casey E (2019) Production of whey-derived DPP-IV inhibitory peptides using an enzymatic membrane reactor. Food Bioprocess Technol 12(5):799–808

O'Sullivan D, Lahart N, O'Callaghan Y, O'Brien NM, FitzGerald RJ (2013) Characterisation of the physicochemical, residual antigenicity and cell activity properties of transglutaminase cross-linked sodium caseinate hydrolysates. Int Dairy J 33(1):49–54

O'Halloran F, Bruen C, McGrath B, Schellekens H, Murray B, Cryan JF, Kelly AL, McSweeney PLH, Giblin L (2018) A casein hydrolysate increases GLP-1 secretion and reduces food intake. Food Chem 252:303–310

O'Keeffe MB, FitzGerald RJ (2014) Antioxidant effects of enzymatic hydrolysates of whey protein concentrate on cultured human endothelial cells. Int Dairy J 36(2):128–135

O'Keeffe MB, FitzGerald RJ (2018) Whey protein hydrolysate induced modulation of endothelial cell gene expression. J Funct Foods 40:102–109

O'Loughlin IB, Murray BA, Brodkorb A, FitzGerald RJ, Kelly PM (2014a) Production of whey protein isolate hydrolysate fractions with enriched ACE-inhibitory activity. Int Dairy J 38(2):101–103

O'Loughlin IB, Murray BA, FitzGerald RJ, Brodkorb A, Kelly PM (2014b) Pilot-scale production of hydrolysates with altered bio-functionalities based on thermally-denatured whey protein isolate. Int Dairy J 34(1):146–152

Onvani S, Haghighatdoost F, Surkan PJ, Azadbakht L (2017) Dairy products, satiety and food intake: a meta-analysis of clinical trials. Clin Nutr 36(2):389–398

Otte J, Shalaby SMA, Zakora M, Nielsen MS (2007) Fractionation and identification of ACE-inhibitory peptides from α-lactalbumin and β-casein produced by thermolysin-catalysed hydrolysis. Int Dairy J 17(12):1460–1472

Otte J, Lenhard T, Flambard B, Sørensen KI (2011) Influence of fermentation temperature and autolysis on ACE-inhibitory activity and peptide profiles of milk fermented by selected strains of lactobacillus helveticus and Lactococcus lactis. Int Dairy J 21(4):229–238

Pan Y, Lee A, Wan J, Coventry MJ, Michalski WP, Shiell B, Roginski H (2006) Antiviral properties of milk proteins and peptides. Int Dairy J 16(11):1252–1261

Pan D, Cao J, Guo H, Zhao B (2012) Studies on purification and the molecular mechanism of a novel ACE inhibitory peptide from whey protein hydrolysate. Food Chem 130(1):121–126

Parente E, Cogan TM, Powell IB (2017) Chapter 8 - Starter cultures: general aspects. In: McSweeney PLH, Fox PF, Cotter PD, Everett DW (eds) Cheese, 4th edn. Academic, San Diego, pp 201–226

Park YW, Nam MS (2015) Bioactive peptides in milk and dairy products: a review. Korean J Food Sci Anim Resour 35(6):831–840

Parmar H, Hati S, Sakure A (2018) In vitro and in silico analysis of novel ACE-inhibitory bioactive peptides derived from fermented goat milk. Int J Pept Res Ther 24(3):441–453

Pepe G, Tenore GC, Mastrocinque R, Stusio P, Campiglia P (2013) Potential anticarcinogenic peptides from bovine milk. J Amino Acids 2013

Perego S, Cosentino S, Fiorilli A, Tettamanti G, Ferraretto A (2012) Casein phosphopeptides modulate proliferation and apoptosis in HT-29 cell line through their interaction with voltage-operated L-type calcium channels. J Nutr Biochem 23(7):808–816

Perego S, Zabeo A, Marasco E, Giussani P, Fiorilli A, Tettamanti G, Ferraretto A (2013) Casein phosphopeptides modulate calcium uptake and apoptosis in Caco2 cells through their interaction with the TRPV6 calcium channel. J Funct Foods 5(2):847–857

Phelan M, Aherne-Bruce SA, O'Sullivan D, FitzGerald RJ, O'Brien NM (2009) Potential bioactive effects of casein hydrolysates on human cultured cells. Int Dairy J 19(5):279–285

Pihlanto-Leppälä A (2000) Bioactive peptides derived from bovine whey proteins: opioid and ace-inhibitory peptides. Trends Food Sci Technol 11(9):347–356

Pihlanto-Leppälä A, Koskinen P, Piilola K, Tupasela T, Korhonen H (2000) Angiotensin I-converting enzyme inhibitory properties of whey protein digests: concentration and characterization of active peptides. J Dairy Res 67(1):53–64

Polaina J, MacCabe AP (2007) Industrial enzymes. Springer, New York

Power O, Jakeman P, FitzGerald RJ (2013) Antioxidative peptides: enzymatic production, in vitro and in vivo antioxidant activity and potential applications of milk-derived antioxidative peptides. Amino Acids 44(3):797–820

Power O, Fernández A, Norris R, Riera FA, FitzGerald RJ (2014a) Selective enrichment of bioactive properties during ultrafiltration of a tryptic digest of β-lactoglobulin. J Funct Foods 9:38–47

Power O, Nongonierma AB, Jakeman P, FitzGerald RJ (2014b) Food protein hydrolysates as a source of dipeptidyl peptidase IV inhibitory peptides for the management of type 2 diabetes. Proc Nutr Soc 73(1):34–46

Pripp AH (2005) Initial proteolysis of milk proteins and its effect on formation of ACE-inhibitory peptides during gastrointestinal proteolysis: a bioinformatic, in silico, approach. Eur Food Res Technol 221(5):712–716

Pripp AH (2008) Effect of peptides derived from food proteins on blood pressure: a meta-analysis of randomized controlled trials. Food Nutr Res 52

Quirós A, Ramos M, Muguerza B, Delgado MA, Miguel M, Aleixandre A, Recio I (2007) Identification of novel antihypertensive peptides in milk fermented with Enterococcus faecalis. Int Dairy J 17(1):33–41

Rahimi M, Ghaffari SM, Salami M, Mousavy SJ, Niasari-Naslaji A, Jahanbani R, Yousefinejad S, Khalesi M, Moosavi-Movahedi AA (2016) ACE- inhibitory and radical scavenging activities of bioactive peptides obtained from camel milk casein hydrolysis with proteinase K. Dairy Sci Technol 96(4):489–499

Ramos BP, Rodriguez AP, Telle NE, Gonzalez JPF, Rodriguez MLR, Castro LP (2012) Optimized method for obtaining ace activity inhibitory peptides from whey, ace inhibitory peptides and food comprising them

Rao MB, Tanksale AM, Ghatge MS, Deshpande VV (1998) Molecular and biotechnological aspects of microbial proteases. Microbiol Mol Biol Rev 62(3):597–635

Rodríguez-Carrio J, Fernández A, Riera FA, Suárez A (2014) Immunomodulatory activities of whey β-lactoglobulin tryptic-digested fractions. Int Dairy J 34(1):65–73

Rojas-Ronquillo R, Cruz-Guerrero A, Flores-Nájera A, Rodríguez-Serrano G, Gómez-Ruiz L, Reyes-Grajeda JP, Jiménez-Guzmán J, García-Garibay M (2012) Antithrombotic and angiotensin-converting enzyme inhibitory properties of peptides released from bovine casein by lactobacillus casei Shirota. Int Dairy J 26(2):147–154

Roy MK, Kuwabara Y, Hara K, Watanabe Y, Tamai Y (2002) Peptides from the N-terminal end of bovine lactoferrin induce apoptosis in human leukemic (HL-60) cells. J Dairy Sci 85(9):2065–2074

Ruiz-Giménez P, Salom JB, Marcos JF, Vallés S, Martínez-Maqueda D, Recio I, Torregrosa G, Alborch E, Manzanares P (2012) Antihypertensive effect of a bovine lactoferrin pepsin hydrolysate: identification of novel active peptides. Food Chem 131(1):266–273

Sah BNP, Vasiljevic T, McKechnie S, Donkor ON (2015) Identification of anticancer peptides from bovine milk proteins and their potential roles in management of cancer: a critical review. Compr Rev Food Sci Food Saf 14(2):123–138

Sah BNP, Vasiljevic T, McKechnie S, Donkor ON (2018) Antioxidative and antibacterial peptides derived from bovine milk proteins. Crit Rev Food Sci Nutr 58(5):726–740

Saint-Sauveur D, Gauthier SF, Boutin Y, Montoni A (2008) Immunomodulating properties of a whey protein isolate, its enzymatic digest and peptide fractions. Int Dairy J 18(3):260–270

Saito T, Nakamura T, Kitazawa H, Kawai Y, Itoh T (2000) Isolation and structural analysis of antihypertensive peptides that exist naturally in gouda cheese. J Dairy Sci 83(7):1434–1440

Sames L, Weimann C, Erhardt G (2018) Mare's milk as a source of ACE-inhibitory peptides for human nutrition. Zuchtungskunde 90(3):168–178

Sanchón J, Fernández-Tomé S, Miralles B, Hernández-Ledesma B, Tomé D, Gaudichon C, Recio I (2018) Protein degradation and peptide release from milk proteins in human jejunum. Comparison with in vitro gastrointestinal simulation. Food Chem 239:486–494

Santos-Hernández M, Tomé D, Gaudichon C, Recio I (2018) Stimulation of CCK and GLP-1 secretion and expression in STC-1 cells by human jejunal contents and in vitro gastrointestinal digests from casein and whey proteins. Food Funct 9(9):4702–4713

Schellekens H, Nongonierma AB, Clarke G, van Oeffelen WEPA, FitzGerald RJ, Dinan TG, Cryan JF (2014) Milk protein-derived peptides induce 5-HT2C-mediated satiety in vivo. Int Dairy J 38(1):55–64

Sharma S, Singh R, Rana S (2011) Bioactive peptides: a review. Int J Bioautomation 15(4):223–250

Shazly AB, He Z, El-Aziz MA, Zeng M, Zhang S, Qin F, Chen J (2017) Fractionation and identification of novel antioxidant peptides from buffalo and bovine casein hydrolysates. Food Chem 232:753–762

Shazly AB, Mu H, Liu Z, El-Aziz MA, Zeng M, Qin F, Zhang S, He Z, Chen J (2019) Release of antioxidant peptides from buffalo and bovine caseins: influence of proteases on antioxidant capacities. Food Chem 274:261–267

Shori AB, Baba AS (2015) Fermented milk derives bioactive peptides with antihypertensive effects. Integr Food Nutr Metab 2(3):180–183

Silveira ST, Martínez-Maqueda D, Recio I, Hernández-Ledesma B (2013) Dipeptidyl peptidase-IV inhibitory peptides generated by tryptic hydrolysis of a whey protein concentrate rich in β-lactoglobulin. Food Chem 141(2):1072–1077

Singh S, Srivastava HK, Kishor G, Singh H, Agrawal P, Raghava GPS (2017) Evaluation of protein-ligand docking methods on peptide-ligand complexes for docking small ligands to peptides. bioRxiv:212514

Sipola M, Finckenberg P, Santisteban J, Korpela R, Vapaatalo H, Nurminen M-L (2001) Long-term intake of milk peptides attenuates development of hypertension in spontaneously hypertensive rats. J Physiol Pharmacol 52(4 Pt 2):745–754

Skrzypczak K, Gustaw W, Szwajgier D, Fornal E, Waśko A (2017) κ-Casein as a source of short-chain bioactive peptides generated by lactobacillus helveticus. J Food Sci Technol 54(11):3679–3688

Solanki D, Hati S, Sakure A (2017) In Silico and in vitro analysis of novel angiotensin I-converting enzyme (ACE) inhibitory bioactive peptides derived from fermented camel milk (Camelus dromedarius). Int J Pept Res Ther 23(4):441–459

Stressler T, Eisele T, Fischer L (2013) Simultaneous monitoring of twelve angiotensin I converting enzyme inhibitory peptides during enzymatic β-casein hydrolysis using lactobacillus peptidases. Int Dairy J 30(2):96–102

Suwal S, Rozoy É, Manenda M, Doyen A, Bazinet L (2017) Comparative study of in situ and ex situ enzymatic hydrolysis of milk protein and separation of bioactive peptides in an electromembrane reactor. ACS Sustain Chem Eng 5(6):5330–5340

Svedberg J, de Haas J, Leimenstoll G, Paul F, Teschemacher H (1985) Demonstration of β-casomorphin immunoreactive materials in in vitro digests of bovine milk and in small intestine contents after bovine milk ingestion in adult humans. Peptides 6(5):825–830

Tauzin J, Miclo L, Gaillard J-L (2002) Angiotensin-I-converting enzyme inhibitory peptides from tryptic hydrolysate of bovine αS2-casein. FEBS Lett 531(2):369–374

Tavano OL (2013) Protein hydrolysis using proteases: an important tool for food biotechnology. J Mol Catal B Enzym 90:1–11

Tavares T, Contreras MM, Amorim M, Pintado M, Recio I, Malcata FX (2011) Novel whey-derived peptides with inhibitory effect against angiotensin-converting enzyme: in vitro effect and stability to gastrointestinal enzymes. Peptides 32(5):1013–1019

Théolier J, Fliss I, Jean J, Hammami R (2014) MilkAMP: a comprehensive database of antimicrobial peptides of dairy origin. Dairy Sci Technol 94(2):181–193

Toldrá F, Reig M, Aristoy MC, Mora L (2018) Generation of bioactive peptides during food processing. Food Chem 267:395–404

Tu M, Wang C, Chen C, Zhang R, Liu H, Lu W, Jiang L, Du M (2018) Identification of a novel ACE-inhibitory peptide from casein and evaluation of the inhibitory mechanisms. Food Chem 256:98–104

Tulipano G, Faggi L, Nardone A, Cocchi D, Caroli AM (2015) Characterisation of the potential of β-lactoglobulin and α-lactalbumin as sources of bioactive peptides affecting incretin function: in silico and in vitro comparative studies. Int Dairy J 48:66–72

Turner KM, Keogh JB, Clifton PM (2015) Dairy consumption and insulin sensitivity: a systematic review of short- and long-term intervention studies. Nutr Metab Cardiovasc Dis 25(1):3–8

Udenigwe CC (2014) Bioinformatics approaches, prospects and challenges of food bioactive peptide research. Trends Food Sci Technol 36(2):137–143

Uenishi H, Kabuki T, Seto Y, Serizawa A, Nakajima H (2012) Isolation and identification of casein-derived dipeptidyl-peptidase 4 (DPP-4)-inhibitory peptide LPQNIPPL from gouda-type cheese and its effect on plasma glucose in rats. Int Dairy J 22(1):24–30

Ugwu CP, Abarshi MM, Mada SB, Sanusi B, Nzelibe HC (2019) Camel and horse milk casein hydrolysates exhibit angiotensin converting enzyme inhibitory and antioxidative effects in vitro and in silico. Int J Pept Res Ther

Vermeirssen V, Van Der Bent A, Van Camp J, Van Amerongen A, Verstraete W (2004) A quantitative in silico analysis calculates the angiotensin I converting enzyme (ACE) inhibitory activity in pea and whey protein digests. Biochimie 86(3):231–239

WHO (2013) Global action plan for the prevention and control of noncommunicable diseases 2013–2020

Winkelnkemper T, Schembecker G (2010) Purification performance index and separation cost indicator for experimentally based systematic downstream process development. Sep Purif Technol 72(1):34–39

Wong JH, Liu Z, Law KWK, Liu F, Xia L, Wan DCC, Ng TB (2014) A study of effects of peptide fragments of bovine and human lactoferrins on activities of three key HIV-1 enzymes. Peptides 62:183–188

Worsztynowicz P, Schmidt AO, Białas W, Grajek W (2019) Identification and partial characterization of proteolytic activity of Enterococcus faecalis relevant to their application in dairy industry. Acta Biochim Pol

Wu J, Aluko RE, Nakai S (2006a) Structural requirements of angiotensin I-converting enzyme inhibitory peptides: quantitative structure-activity relationship modeling of peptides containing 4-10 amino acid residues. QSAR Comb Sci 25(10):873–880

Wu J, Aluko RE, Nakai S (2006b) Structural requirements of angiotensin I-converting enzyme inhibitory peptides: quantitative structure-activity relationship study of Di- and tripeptides. J Agric Food Chem 54(3):732–738

Xu J-Y, Qin L-Q, Wang P-Y, Li W, Chang C (2008) Effect of milk tripeptides on blood pressure: a meta-analysis of randomized controlled trials. Nutrition 24(10):933–940

Yamada A, Sakurai T, Ochi D, Mitsuyama E, Yamauchi K, Abe F (2013) Novel angiotensin I-converting enzyme inhibitory peptide derived from bovine casein. Food Chem 141(4):3781–3789

Yamamoto N, Maeno M, Takano T (1999) Purification and characterization of an antihypertensive peptide from a yogurt-like product fermented by lactobacillus helveticus CPN4. J Dairy Sci 82(7):1388–1393

Yamauchi R, Ohinata K, Yoshikawa M (2003) β-Lactotensin and neurotensin rapidly reduce serum cholesterol via NT2 receptor. Peptides 24(12):1955–1961

Yan J, Zhao J, Yang R, Zhao W (2019a) Bioactive peptides with antidiabetic properties: a review. Int J Food Sci Technol

Yan J, Zhao J, Yang R, Zhao W (2019b) Bioactive peptides with antidiabetic properties: a review. Int J Food Sci Technol 54(6):1909–1919

Zanutto-Elgui MR, Vieira JCS, DZD P, MAR B, Padilha PM, Elgui de Oliveira D, Fleuri LF (2019) Production of milk peptides with antimicrobial and antioxidant properties through fungal proteases. Food Chem 278:823–831

Zenezini Chiozzi R, Capriotti AL, Cavaliere C, La Barbera G, Piovesana S, Samperi R, Laganà A (2016) Purification and identification of endogenous antioxidant and ACE-inhibitory peptides from donkey milk by multidimensional liquid chromatography and nanoHPLC-high resolution mass spectrometry. Anal Bioanal Chem 408(20):5657–5666

Zhang Y, Chen R, Ma H, Chen S (2015) Isolation and identification of dipeptidyl peptidase IV-inhibitory peptides from trypsin/chymotrypsin-treated goat milk casein hydrolysates by 2D-TLC and LC–MS/MS. J Agric Food Chem 63(40):8819–8828

Zhang Y, Chen R, Zuo F, Ma H, Zhang Y, Chen S (2016) Comparison of dipeptidyl peptidase IV-inhibitory activity of peptides from bovine and caprine milk casein by in silico and in vitro analyses. Int Dairy J 53:37–44

Zhang D-D, Liu J-L, Jiang T-M, Li L, Fang G-Z, Liu Y-P, Chen L-J (2017) Influence of Kluyveromyces marxianus on proteins, peptides, and amino acids in lactobacillus-fermented milk. Food Sci Biotechnol 26(3):739–748

Zhao M, Wei C, Yang X, Zhou J, Wang J, Gu F, Lei T, Qin Y (2016) The milk-derived hexapeptide PGPIPN inhibits the invasion and migration of human ovarian cancer cells by regulating the expression of MTA1 and NM23H1 genes. Int J Oncol 48(4):1721–1729

Zucht H-D, Raida M, Adermann K, Mägert H-J, Forssmann W-G (1995) Casocidin-I: a casein-αs2 derived peptide exhibits antibacterial activity. FEBS Lett 372(2–3):185–188

Chapter 19
Reducing Allergenicity by Proteolysis

Katrine Lindholm Bøgh and Jeppe Madura Larsen

19.1 Cow's Milk Allergy

Food allergy is defined as an immune-mediated adverse reaction to otherwise harmless proteins in the food. It is caused by a failure of the immune system to develop tolerance after exposure to these proteins or due to break down of an already established oral tolerance. This process is termed "allergic sensitisation" leading to the development of food protein-specific IgE antibodies in allergic individuals. These antibodies and type-2 immune cells mediate the clinical symptoms of food allergy occurring after exposure to the sensitising food in sensitised individuals, a process termed "elicitation of food allergy" (Yu et al. 2016). Food allergy is a major health problem in Western countries, where it affects around 5–8% of young children and 2–4% of adults (Sicherer and Sampson 2014), and appears to be an increasing problem (Prescott et al. 2013; Sicherer and Sampson 2014).

Cow's milk allergy (CMA) is the most common cause of food allergy in infants and young children, affecting around 2.5% of children below the age of three years (Sampson 2004; Kattan et al. 2011). Fortunately, most children outgrow their allergy at an early age, and less than 0.5% of adults suffer from CMA (Fiocchi et al. 2010). However, the prevalence and persistency of CMA seems to be increasing in recent years (Skripak et al. 2007; Hochwallner et al. 2014). IgE-mediated CMA is caused by an adverse reaction to proteins present in the milk and is responsible for approximately 60% of all CMA-induced reactions. Symptoms appear immediately, within 2 min to 2 h, after ingestion of a dairy product in sensitised individuals. CMA may lead to symptoms in the gastrointestinal tract, the skin, the respiratory system, and, in severe cases, causes anaphylaxis, which can be fatal (Fiocchi et al. 2010; Kattan et al. 2011).

K. L. Bøgh (✉) · J. M. Larsen
Research Group for Food Allergy, National Food Institute, Technical University of Denmark, Kgs. Lyngby, Denmark
e-mail: kalb@food.dtu.dk; jeml@food.dtu.dk

A. L. Kelly, L. B. Larsen (eds.), *Agents of Change*, Food Engineering Series,
https://doi.org/10.1007/978-3-030-55482-8_19

Table 19.1 Cow's milk allergens and their characteristics

Cow's milk fraction	Protein	Allergen name	Content (%)	Size (kDa)	No of amino acids	Major/minor allergen	Natural occurring modification	Protein family	S-S bridges
Casein (80%)	α_{S1}-Casein	Bos d 9	32	23.6	199	Major	S phosphorylations	Casein family	0
	α_{S2}-Casein	Bos d 10	10	25.2	207	Major	S phosphorylations	Casein family	0
	β-Casein	Bos d 11	28	24	209	Major	S phoshorylations	Casein family	0
	K-Casein	Bos d 12	10	19	169	Major	Q pyrrolidone carboxyl acid/S and T glycosylations/S and T phophorylation	Casein	1
Whey (20%)	α-Lactalbumin	Bos d 4	5	14.2	123	Major	N glycosylation	Glycosyl Hydrolase 22 family	4
	β-Lactoglobulin	Bos d 5	10	18.3	162	Major		Lipocalin family	2 + 1 free
	Bovine serum albumin	Bos d 6	1	66.3	583	Minor	S and T phosphorylations/K succinylations/K methylation	Albumin family	17 + 1 free
	Immunoglobulins	Bos d 7	3	160		Minor	Immunoglobulin family	Immunoglobulin	Variable
	Lactoferrin		<1	80	703	Minor		Transferrin	16

References: Wal (2002), Chapman (2008), Hochwallner et al. (2014) and Villa et al. (2018)

19.1.1 Bovine Milk Proteins

Bovine milk contains approximately 30–35 g of protein per litre, distributed among more than 25 different proteins, many of which are allergens (Wal 2002). Bovine milk allergens can be divided into two main fractions, based on their solubility at pH 4.6 at 20 °C, namely casein and whey proteins (Wal 2002; Villa et al. 2018), as shown in Table 19.1. The casein fraction (*Bos d 8*), which comprises 80% of the bovine milk proteins, consists of four proteins designated α_{S1}-casein (*Bos d 9*), α_{S2}-casein (*Bos d 10*), β-casein (*Bos d 11*) and K-casein (*Bos d 12*). All four proteins are major allergens, defined as an allergen that causes IgE responses in more than 50% of the cow's milk allergic patients (Chapman 2008). The whey protein fraction comprises 20% of the bovine milk proteins and consists of the allergens α-lactalbumin (ALA, *Bos d 4*), β-lactoglobulin (BLG, *Bos d 5*), bovine serum albumin (BSA, *Bos d 6*), immunoglobulins (Igs, *Bos d 7*) and lactoferrin (LF, *Bos d lactoferrin*). Whereas ALA and BLG are regarded as major allergens, BSA, Igs and LF are regarded as minor allergens (Hochwallner et al. 2014; IUIS 2019).

Caseins in milk form well-organised complexes called micelles that bind essential minerals for easy uptake (Marchesseau et al. 2002). The function of BLG is unclear but is thought to bind hydrophobic ligands such as cholesterol and vitamin D2 (Wal 2002; Kontopidis et al. 2004; Adams et al. 2006). ALA is a Ca^{2+}-binding protein that acts as a regulatory component in the synthesis of lactose (Permyakov and Berliner 2000; Wal 2002) and is capable of interacting with lipid membranes (Cawthern et al. 1996). The three-dimensional (3D) conformation of the individual caseins is generally unknown due to their flexible and unstructured from (Wal 2002). Whereas caseins are unstructured proteins consisting of random coils stabilised by hydrophobic interactions, the whey protein BLG is a globular protein consisting of nine β-barrels and three α-helixes, stabilised by two disulphide bonds (Wal 2002; Kontopidis et al. 2004) and ALA is likewise a globular protein consisting of a large α-helical domain at the N-terminus and a short β-sheet domain at the C-terminus flanked by a calcium-binding loop, stabilised by four disulphide bonds (Stuart et al. 1986; Farkas et al. 2005).

19.1.2 Bovine Milk Epitopes

In spite of bovine milk allergens being divided into either minor or major allergens, individuals with CMA react very heterogeneously, and no one allergen can be accounted for as the most allergenic (Wal et al. 1995). In fact, the majority of cow's milk allergic patients react to multiple allergens (Bleumink and Young 1968; Wal et al. 1995), and no specific structure or function seem to be associated with bovine milk protein allergenicity, as both the unstructured caseins and globular whey proteins are highly allergenic (Wal 2002). Also, at the level of the antibody recognition sites—the epitopes—cow's milk allergic patients react very heterogeneously, greatly varying in both specificity and diversity, though some immunodominant epitopes seems to be present on bovine milk proteins (Spuergin et al. 1996; Sélo et al. 1999). Epitopes can be divided into linear and conformational epitopes, based on the vicinity of the involved amino acids in the primary structure comprising the epitope (Arnon and Van Regenmortel 1992; Van Regenmortel 1998, 2009; Broekman et al. 2015). Linear epitopes are comprised of amino acids juxtaposed in the protein primary structure, whereas conformational epitopes are comprised of amino acids from two or more stretches distant from each other in the primary sequence but brought together by the structural folding of the protein (Barlow et al. 1986; Van Regenmortel 1996; Roggen 2006; Broekman et al. 2015). Studies have suggested a particular important role of linear epitopes in food allergy, as IgE epitopes of the linear version may still be present in processed food, such as hydrolysed or heat-treated foods, while conformation epitopes may be lost by such processing, due to loss of the structural folding of the protein. Furthermore, linear epitope diversity has been suggested as a biomarker for persistent and severe CMA (Chatchatee et al. 2001; Järvinen et al. 2001, 2002; Wang et al. 2010).

19.1.3 *Bovine Milk Proteolysis*

Proteolysis of dairy products, and thereby bovine milk proteins, is found in many processes and products, often in order to reduce the product allergenicity. Proteolytic procedures involve the irreversible breakdown of proteins into peptide fragments and amino acids, although the degree of breakdown depends on the specificity and amount of enzymes used. The breakdown often results in the direct loss of linear epitopes due to the location of a cleavage site, and thereby destruction of the epitope, but also in the loss of most conformational epitopes, due to changes in the structural folding of the protein. Consequently, proteolytic processing most often leads to a reduction of the product allergenicity, although in rare cases proteolysis may also result in increased allergenicity due to exposure of new epitopes (Bøgh and Madsen 2016; Villa et al. 2018).

19.2 Digestion

Resistance to gastrointestinal (GI) proteolysis has received much attention in the last decades, as it is generally believed that for a protein to sensitise or elicit allergic reactions in the GI tract, it must survive the GI digestion process as intact protein or as large peptide fragments (Astwood et al. 1996; Taylor and Hefle 2001; Bannon 2004; Mills et al. 2004; Bøgh and Madsen 2016). Thus, the longer a protein survives the digestion process in a form with sufficient size and structure, the more easily it will be recognised and taken up by cells of the immune system in the GI tract, allowing for an allergic response.

Astwood et al. (1996) was the first to compare the proteolytic stability of food allergens with that of non-allergenic food proteins in a simulated human gastric digestion process, and showed that the food allergens were in general more resistant to digestion than were the non-allergenic dietary proteins. Thus, this led to the conclusion that there exists a correlation between food protein allergenicity and stability to digestion. Furthermore, this prompted the inclusion of a pepsin resistance test as a decision parameter in the weight-of-evidence approach used in the allergenicity safety assessment of novel proteins in genetically modified foods (FAO/WHO 2001; CAC 2003; EFSA 2010). Since the study by Astwood et al., many studies have been performed in order to investigate the relationship between resistance to digestion and allergenic potency (Fu et al. 2002; Takagi et al. 2003; Thomas et al. 2004; Herman et al. 2007; Ofori-Anti et al. 2008; Bøgh and Madsen 2016); however, most of those could not confirm such a strict correlation between stability to digestion and allergenicity.

19.2.1 Digestion Process

Ingestion of dairy products induces a complex sequence of digestive processes with the aim of extracting the nutrients essential for maintenance of the health. Bovine milk proteins are digested by proteolytic enzymes located in the stomach, pancreas, and small intestine (Erickson and Kim 1990). At first, the proteins are exposed to digestion by pepsins in the stomach, which have a wide specificity, preferentially cleaving peptide bonds between hydrophobic and aromatic amino acid residues with an activity optimum around pH 3.5 (Tang 1963, 2013; Inouye and Fruton 1967; Etherington and Taylor 1970; Erickson and Kim 1990). After an average transit time of around 1–2 h (Marciani et al. 2001), proteolysis continues in the small intestine with proteases and peptidases produced by the pancreas, such as trypsin and chymotrypsin, or produced by the brush border of the intestinal mucosa (Etherington and Taylor 1970; Olsen et al. 2004). While trypsin cleaves peptide bonds at the carboxyl side of the basic amino acids arginine and lysine, chymotrypsin cleaves peptide bonds where the carbonyl group is aromatic, as in the amino acids phenylalanine, tyrosine and tryptophan (Etherington and Taylor 1970; Olsen et al. 2004). Trypsin and chymotrypsin have optimal activity around pH 8 (McLaren and Estermann 1957; Sipos and Merkel 1970; Asgeirsson and Bjarnason 1991).

Several in vivo studies of human digestion have been performed (Sandberg et al. 1981; Troost et al. 2001; Polovic et al. 2007). However, due to technical and ethical challenges associated with human in vivo studies (Wickham et al. 2009), many attempts have been made to develop in vitro models simulating the proteolytic process occurring in the average human GI tract (Vieths et al. 1999; Mouécoucou et al. 2004; Moreno et al. 2005b; Dupont et al. 2010; Dupont and Mackie 2015). These studies have shown that in vitro methods simulating the gastro-duodenal digestion may indeed present a good approximation to the real human digestion of bovine milk proteins (Sanchón et al. 2018). However, various models have been presented, greatly varying in the digestion conditions, with the use of different pH values, enzyme to protein ratios and digestion time, with inclusion of either solely the gastric phase or a gastric phase in combination with a duodenal phase, and with surfactant present or not. However, it is important to keep in mind that no single in vitro digestion model can be made that perfectly represents all humans, as the digestion process differs significantly from person to person and is influenced by factors such as age and health status (Untersmayr et al. 2005; Untersmayr and Jensen-Jarolim 2008).

19.2.2 Digestibility of Bovine Milk Proteins

Several studies have been performed to investigate the susceptibility of bovine milk proteins to digestion, with simulation of either the gastric digestion phase alone or the gastric in combination with the duodenal digestion phase (Bøgh and Madsen 2016). While studies investigating solely the resistance to pepsinolysis—taking into account only the digestion taking place in the stomach and often uses non-physiological conditions—are frequently used as a measure of the general biochemical stability, more and more studies are being conducted in order to mimic the physiological digestion process, taking into account both the gastric as well as the duodenal digestion process in physiologically relevant conditions. Such a simulated in vitro human GI digestion process combines the use of pepsin with the use of chymotrypsin and trypsin and aims to generate digestion products that, to a great extent, should reflect the protein/peptide profile to which the immune system in the GI tract is exposed (Bøgh and Madsen 2016).

Table 19.2 provides an overview of the stability of individual bovine milk proteins to simulated in vitro gastric or gastro-duodenal digestion, as well as the digestion condition used. No absolute definition seems to exist on when to define a protein as stable and when to define a protein as labile to simulated in vitro digestion. Thus, in line with Bøgh and Madsen (2016), a protein is defined as being stable when residual intact protein survived the digestion process with the given digestion conditions. As can be seen from the Table 19.2, though some inconsistencies exist, ALA, BSA, lactoferrin, α-casein and β-casein are susceptible to pepsinolysis, while BLG and IgG are resistant to pepsinolysis. The same pattern exists for the simulated gastro-duodenal digestion, where ALA and β-casein are easily digested whereas BLG is in general regarded as being resistant.

On the other hand, Takagi et al. (2003) showed that, while BLG was resistant to pepsinolysis in simulated gastric fluid (SGF), it was labile to digestion by pancreatin in simulated intestinal fluid (SIF). In contrast, while BSA was labile to digestion in SGF, it was resistant to digestion in SIF, indicating that resistance to digestion depends on the enzymes used. However, when Mandalari et al. performed a pepsin test identical to the one performed by Fu et al., they found BLG to be a labile protein in contrast to the latter authors, who found it to be resistant, indicating the significance of differences in the exact digestion procedure or in the techniques used for

Table 19.2 Overview of resistance of bovine milk allergens to gastric or gastro-duodenal digestion

Allergen	Allergen name	Gastric pH	Pepsin: allergen MW ratio	Gastric digestion time	Duodenal pH	Trypsin/chymotrypsin: allergen MW ratio	Duodenal digestion time	Resistance to digestion	References
α_{S1}-Casein	Bos d 9	1.2	12.8	120	–	–	–	No	Fu et al. (2002)
β-Casein	Bos d 11	1.2	12.8	120	–	–	–	No	Mandalari et al. (2009a)
		1.2	19	60	–	–	–	No	Astwood et al. (1996)
		2	0.05	60	–	-	–		Benedé et al. (2017)
		2.5	0.05	60	–	–	–	No	Mandalari et al. (2009a)
		2.5	0.05	60	–	–	–	No	Macierzanka et al. (2009)
		2.5	0.05	60	–	–	–	No	Dupont et al. 2010)
		3	0.007	60	–	–	–	No	Dupont et al. (2010)
		2.5	0.05	60	6.5	0.0025/0.01	30	No	Dupont et al. (2010)
		2.5	0.05	60	6.5	0.0025/0.01	30	No	Mandalari et al. (2009a)
		2.5	0.05	60	6.5	0.0025/0.01	30	No	Macierzanka et al. (2009)
		3	0.007	60	6.5	0.00025/0.001	30	No	Dupont et al. (2010)
α-Lactalbumin	Bos d 4	1.2	12.8	120	–	–	–	No	Fu et al. (2002)
		2	0.02	90	–	–	–	No	Schmidt et al. (1995)
		2.5	0.05	120	–	–	–	No	Moreno et al. (2005a)
		3	0.02	90	–	–	–	No	Schmidt et al. (1995)
		4	0.02	90	–	–	–	Yes	Schmidt et al. (1995)
		2.5	0.05	120	6.5	0.0025/0.01	120	No	Moreno et al. (2005a)

(continued)

Allergen	Allergen name	Gastric pH	Pepsin: allergen MW ratio	Gastric digestion time	Duodenal pH	Trypsin/chymotrypsin: allergen MW ratio	Duodenal digestion time	Resistance to digestion	References
β-Lactoglobulin	Bos d 5	1.2	0.94	60	–	–	–	Yes	Yagami et al. (2000)
		1.2	3	60	–	–	–	Yes	Thomas et al. (2004)
		1.2	3	30	–	–	–	Yes	Zeece et al. (2008)
		1.2	3.05	60	–	–	–	Yes	Ofori-Anti et al. (2008)
		1.2	11	30	–	–	–	Yes	Zeece et al. (2008)
		1.2	12.8	120	–	–	–	Yes	Fu et al. (2002)
		1.2	12.8	120	–		–	No	Mandalari et al. (2009a)
		1.2	16	60	–	–	–	Yes	Misra et al. (2009)
		1.2	19	60	–	–	–	Yes	Astwood et al. (1996)
		1.5	1.25	120				Yes	Schulten et al. (2011)
		1.5	3	60	–	–	–	Yes	Lucas et al. (2008)
		2	0.02	90	–	–	–	Yes	Schmidt et al. (1995)
		2	0.05	60	–	–	–	Yes	Benedé et al. (2014)
		2	0.05	60	–	–	–	Yes	Martinez et al. (2016)
		2	3	60	–	–	–	Yes	Thomas et al. (2004)
		2	3.04	60	–	–	–	Yes	Takagi et al. (2003)
		2.5	0.05	60	–	–	–	Yes	Dupont et al. (2010)
		2.5	0.05	60	–	–	–	Yes	Macierzanka et al. (2009)
		2.5	0.05	60	–	–	–	Yes	Mandalari et al. (2009b)
		2.5	0.05	60	–	–	–	Yes	Mandalari et al. (2009a)
		2.5	0.05	120	–	–	–	Yes	Bøgh et al. (2013)
		3	0.007	60	–	–	–	Yes	Dupont et al. (2010)
		3	0.02	90	–	–	–	Yes	Schmidt et al. (1995)
		4	0.02	90	–	–	–	Yes	Schmidt et al. (1995)
		2	0.05	60	6.5	0.0025/0.01	30	No	Benedé et al. (2014)
		2.5	0.05	60	6.5	0.0025/0.01	30	Yes	Mandalari et al. (2009a)
		2.5	0.05	60	6.5	0.0025/0.01	30	Yes	Mandalari et al. (2009b)
		2.5	0.05	60	6.5	0.0025/0.01	30	No	Macierzanka et al. (2009)
		2.5	0.05	60	6.5	0.0025/0.01	30	Yes	Dupont et al. (2010)
		2.5	0.05	120	6.5	0.0025/0.01	15	Yes	Bøgh et al. (2013)
		3	0.007	60	6.5	0.00025/0.001	30	Yes	Dupont et al. (2010)

Allergen	Allergen name	Gastric pH	Pepsin: allergen MW ratio	Gastric digestion time	Duodenal pH	Trypsin/chymotrypsin: allergen MW ratio	Duodenal digestion time	Resistance to digestion	References
Bovine serum albumin	Bos d 6	1.2	0.025	120	–	–	–	No	Yamada et al. (2006)
		1.2	3	60	–	–	–	No	Thomas et al. (2004)
		1.2	3.05	60	–	–	–	No	Ofori-Anti et al. (2008)
		1.2	12.8	120	–	–	–	No	Fu et al. (2002)
		1.2	13	60	–	–	–	No	Polovic et al. (2007)
		1.2	16	60	–	–	–	No	Pantoja-Uceda et al. (2004)
		1.2	16.7	60	–	–	–	No	Murtagh et al. (2002)
		1.2	16.7	60	–	–	–	No	Murtagh et al. (2003)
		1.2	19	60	–	–	–	No	Astwood et al. (1996)
		2	0.02	90	–	–	–	No	Schmidt et al. (1995)
		2	3	60	–	–	–	No	Thomas et al. (2004)
		2	3.04	60	–	–	–	No	Takagi et al. (2003)
		3	0.02	90	–	–	–	No	Schmidt et al. (1995)
		4	0.02	90	–	–	–	No	Schmidt et al. (1995)
Immunoglobulins	Bos d 7	2	0.02	90	–	–	–	No	Schmidt et al. (1995)
		3	0.02	90	–	–	–	Yes	Schmidt et al. (1995)
		4	0.02	90	–	–	–	Yes	Schmidt et al. (1995)
Lactoferrin		1.2	3.05	60	–	–	–	No	Fu et al. (2002)

Included in the table are digestibility results of only purified cow's milk allergens digested at the physiologically relevant 37 °C, for which it was possible to identify molecular weight (MW) enzyme to allergen ratios. –; information not available

analysis of any residual amount of protein left after termination of the digestion process (Fu et al. 2002; Mandalari et al. 2009a).

While all studies evaluated the stability of intact bovine milk proteins, only few have evaluated the stability of peptide fragments produced during the simulated digestion processes, and even fewer the peptide size distribution profile of the peptide fragments in the final digestion product. However, it is well recognised that, even though digestion in most situation reduces the allergenicity of the bovine milk proteins, even small peptide fragments may retain allergenicity (Bøgh and Madsen 2016).

19.2.3 Digestion Parameters Affecting the Digestibility of Bovine Milk Proteins

As shown in Table 19.2, variations in the digestibility conditions may influence the outcome of a digestibility assay, designating the bovine milk protein as either a stable or a labile protein. Parameters related to the digestion conditions such as digestion time, pH, enzyme to protein ratio and use of surfactant may impact the outcome of the digestion assay. Also, parameters related to the bovine milk proteins such as processing may influence whether the protein is degraded or not.

Protease to Protein Ratio

A wide range of protease to protein ratios have been used (see Table 19.2), with, for example, a ratio corresponding to 12.8 being far from physiologically relevant, and mostly used as a measure of the general biochemical stability, while a ratio of 0.05 is designed to simulate physiologically relevant conditions. The protease to protein ratio will greatly influence the stability of bovine milk proteins to digestion; Mandalari et al. (2009a), for example, showed that, while BLG was resistant to simulated gastric digestion at physiologically relevant pepsin to protein ratios, it was easily digested by increasing the pepsin to protein ratio 256-fold.

pH

Substantial variation also exists in the pH used, especially in the gastric digestion phase (see Table 19.2), where pH ranges from 1.2 to 4.0. No single pH will ever mimic the gastric digestion process taking place in the stomach, as the pH varies greatly between individuals (Dupont et al. 2010) and during the day, dependent on the time since last meal and the content of that meal (James and Baron 1962; Sandberg et al. 1981). In a study by Schmidt et al. bovine milk whey proteins were digested in a simulated gastric digestion process at pH 2, 3 and 4, which clearly showed that, while both the ALA and IgG were easily digested at a pH of 2, both were stable at pH 4 (Schmidt et al. 1995). This obviously shows that the susceptibility of bovine milk proteins to gastric digestion is greatly influenced by the pH, probably due to the pH-dependent activity optimum of pepsin and the denaturation of target proteins (Bøgh and Madsen 2016).

Surfactant

Addition of surfactants such as bile salts or phosphatidylcholine (PC) may also impact on the digestibility of bovine milk proteins. The addition of PC was shown to have a protective effect on the bovine milk proteins ALA and BLG, decreasing the digestibility rate, whereas it had no effect on the digestibility of the bovine β-casein (Moreno et al. 2005a; Mandalari et al. 2009a; Dupont et al. 2010). Addition of bile salts also affects the digestibility of bovine milk proteins, where it was shown to greatly increase the digestion of BLG and BSA (Gass et al. 2007). Collectively, this shows that addition of surfactant can increase, decrease or have no effect on the resistance to simulated digestion, probably dependent on the specific surfactant and protein of interest.

Matrix

Investigations of the digestibility of purified proteins, without considering the potential effect of the food matrices, may be misleading and provide results which may differ from those achieved with digestion of the whole food. The presence of a food matrix prevented simulated gastric digestion of BLG, indicating that other food components such as carbohydrates, fat or other proteins affect the rate of digestion (Schulten et al. 2011). A clear matrix effect on the digestibility of bovine milk proteins was shown by Macierzanka et al., who showed that β-casein and BLG in an emulsified form had increased susceptibility to digestion compared to digestion of β-casein and BLG without emulsion (Macierzanka et al. 2009). The authors suggested that emulsification led to an adsorption-induced change in the folding and/or flexibility of the bovine milk proteins, rendering them more easily digested in simulated gastro-duodenal digestion (Macierzanka et al. 2009). Studies of the matrix effect by addition of polysaccharides, which are used in the food industry to improve texture and stability, showed that they might induce a significantly reduced digestibility of BLG in simulated gastro-duodenal digestion, although the degree of reduction was dependent on the exact type of polysaccharides (Mouécoucou et al. 2003, 2004; Nacer et al. 2004). Sletten et al. showed that varying the fat content of food matrices did not influence the digestibility of either BLG or caseins (Sletten et al. 2008).

Processing

Processing of bovine milk proteins may affect their digestibility. It was, for example, shown that the susceptibility of BLG was significantly increased by heat treatment (Sletten et al. 2008; Morisawa et al. 2009; Peram et al. 2013). The degree of increased digestibility was, however, dependent on the exact heating temperature, where heat treatment at 100 °C induced a higher susceptibility to digestion compared to 80 °C (Morisawa et al. 2009). The effect of heat treatment is probably due to unfolding of the proteins, making them more accessible for proteolysis. Similarly, ultra-high temperature (UHT) processing of bovine milk (a food processing technology used for sterilisation of bovine milk, with heating for a few seconds at temperatures above 135 °C in order to obtain a greatly extended shelf life), significantly increased the digestibility of BLG (Sletten et al. 2008). Similar results could also be seen for K-casein, but not for α-caseins, where heat treatment was shown to have no effect or increase the stability to digestion (Sletten et al. 2008; Morisawa et al. 2009). This clearly shows that different bovine milk proteins react very different to heat treatment and that heat treatment may impact on the digestibility in various ways even on proteins from the same food.

High pressure (HP) treatment is a cold pasteurisation technique by which packaged milk products are subjected to high level of isostatic pressure, in the range 300–600 MPa, transmitted by water. While HP treatment of BLG was shown to increase the digestibility by pepsin in comparison to no treatment (Peñas et al. 2006; Zeece et al. 2008), the digestibility of ALA was unaffected (Peñas et al. 2006). The exact pressure used was found to impact the degree of susceptibility, e.g., treatment at 400 MPa did not influence the digestibility of BLG, whereas HP treatment at 600 and 800 MPa resulted in rapid digestion of BLG (Peñas et al. 2006; Zeece et al. 2008), showing that the proteolytic optimum differed between the proteases, with an optimum at 100 MPa for chymotrypsin, 200 MPa for trypsin and 300 MPa for pepsin (Peñas et al. 2006).

Microwave (waves ranging from 1 m to 1 mm) treatment reportedly also increases the digestibility of BLG in a wattage-dependent manner, where treatment at 50 and 100 W did not increase proteolysis of BLG by pepsin, whereas that at 200 W did, probably due to structural changes induced by the microwave irradiation at this power level (El Mecherfi et al. 2015).

Glycation

Glycation is the non-enzymatic process of covalent linking of a carbohydrate to a protein, which may both increase and decrease the stability of proteins. In a study by Corzo-Martínez et al. (2010), it was shown that high levels of glycation with galactose, tagatose or dextran via Millard reactions at 50 °C, but not at 40 °C, impaired the digestion of BLG. On the other hand, both BLG and ALA became more susceptible to digestion after glycation with oligoisomaltose (Li et al. 2013), indicating that the type of glycation may be decisive for the effect on digestion.

19.2.4 Evaluation of Digestibility

Assessing if residual intact protein and/or large peptide fragments formed during the digestion process are present after termination of the digestion process are most often performed with the use of Sodium Dodecyl Sulfate-Polyacrylamide Gel Electrophoresis (SDS-PAGE) and/or western blotting, but other methods may also be used (EFSA 2010). The exact technique used for evaluation of residual intact allergen and emerging peptide fragments may greatly influence the results and thereby the conclusion on whether the specific bovine milk protein is regarded as a stable or labile allergen. In a study by Mandalari et al., residual intact BLG left after simulated gastro-duodenal digestion was evaluated by SDS-PAGE as well as with Reversed-Phase High-Performance Liquid Chromatography (RP-HPLC) (Mandalari et al. 2009b). While SDS-PAGE indicated that no intact BLG was left after the digestion process, RP-HPLC clearly demonstrated that residual intact BLG had survived the digestion process. These studies infer that assessment of residual intact protein with other protein-chemical methods than SDS-PAGE should be performed. Such assay may include immunoassays, and frequently involve the use of Enzyme Linked Immunosorbent Assays (ELISAs), chromatographic or mass spectrometry (MS) techniques (Villa et al. 2018).

19.2.5 Evaluation of Residual Allergenicity of Digestion Products

Generally, it is considered that digestion-prone proteins are less allergenic than digestion-resistant proteins, and thus that digestion of proteins into peptide fragments and amino acids reduces their allergenicity. However, the relationship between susceptibility to digestion and allergenicity is not straight forward (Bøgh and Madsen 2016). Peptide fragments resulting from a digestion process have been shown to have reduced, similar or even increased allergenicity compared to the parent protein (Bøgh and Madsen 2016). When investigating the allergenicity of proteolytic products of bovine milk, three different aspects should be assessed: the IgE-binding, and the eliciting and sensitising capacities (Aalberse 2000).

Many studies have evaluated the residual IgE-binding capacity of digested bovine milk proteins. For example, assessment of the IgE-binding capacity of digestion products from proteolysis of BLG revealed an IgE-binding capacity similar to the parent intact BLG in one study (Sélo et al. 1999), a reduced IgE-binding capacity in another study (Benedé et al. 2014), and an even higher IgE-binding capacity in another study (Haddad et al. 1979). This clearly demonstrates that proteolysis does not always reduce the allergenicity of proteins, but may also increase the allergenicity, which is suggested to be a result of exposure of new epitopes after digestion that are not accessible in the folded intact protein. Epitope-mapping studies have shown that peptides from bovine milk proteins may indeed maintain the ability of bind IgE (Järvinen et al. 2002).

Evaluation of the sensitising capacity of bovine milk products have all been performed in animal studies. Several studies have evaluated the sensitising capacity of BSA in mice, showing that digestion abolished the sensitising capacity of BSA, indicating no sensitising capacity in comparison to the intact counterpart, which sensitised the mice (Dosa et al. 1979; Ferguson et al. 1983; Michael 1989). In a study by Bøgh et al. it was shown that the sensitising capacity of BLG was heavily reduced after gastro-duodenal digestion, though it retained some sensitising capacity (Bøgh et al. 2013). However, when removing the intact BLG from the digestion product, the sensitising capacity was abolished.

19.3 Infant Formulas

Breast milk is the first choice of nutrition for infants; however, when breastfeeding in not possible or is insufficient, an alternative nutritional source is needed. Bovine milk-based infant formulas (IF) are the obvious alternative and recommended as the substitute for breastmilk as IFs are designed to meet the nutritional requirement of infants and small children. However, for cow's milk allergic children or children at risk of developing CMA, alternatives to these conventional bovine milk-based IFs may be needed and consequently numerous commercially dietary substitutes are available, based on either hydrolysed bovine milk proteins or proteins from other sources than milk (Parekh and Bahna 2016). Besides being nutritionally adequate, a main focus is to manage or prevent CMA, avoiding allergic reactions to bovine milk proteins. While hypoallergenic (HA) IFs based on hydrolysed bovine milk proteins are normally the preferred choice, these are not well-tolerated by all children, and substitutes for bovine milk products such as soybean, rice or potato formulas have been suggested as plant-based alternatives. Also, milk from other mammalian species such as buffalo, sheep, goat, camel, mare, reindeer, donkey and yak have been suggested as alternatives to bovine milk-based IFs (Villa et al. 2018). However, while plant-based IF may introduce new allergies, milk form other mammals may cause allergic reaction due to cross-reactivity to bovine milk proteins (Parekh and Bahna 2016; Villa et al. 2018).

19.3.1 Hypoallergenic Infant Formulas

HA IFs are used for the management and prevention of CMA and are typically based on bovine milk proteins which have been hydrolysed to destroy the allergenic epitopes that cause allergic reactions. Examples of commercially available IFs are shown in Table 19.3 (Golkar et al. 2019). HA IF products are predominantly based on either the casein or whey milk protein fractions, and only few are based on whole milk proteins. HA IFs are divided into extensively (eHF) and partially (pHF) hydrolysed IF based on the degree of protein hydrolysis and the molecular weight (MW) of the peptides present (Bindels and Boerma 1994; Høst et al. 1999). Although no formal standardisation exists, eHFs generally consist of peptides more than 95% of which have a MW of less than 3 kDa, while pHFs generally contain peptides where a higher percentage have a MW between 3 and 10 kDa and a much lower percentage have a MW below the 3 kDa (Von Berg 2009, 2013).

Table 19.3 Examples of commercially available infant formulas (Golkar et al. 2019)

Brand	Type	Producer
Novalac Premium	CMF	Novalac
Baby & me 1 and 2	CMF	Arla
Allomin 1 and 2	CMF	Semper
Nutrilon Premium 1	CMF-C	Nutricia
Infacare® Gold 1	CMF-W	Aspen Nutritionals
NAN 1 Protect Start	CMF-W	Nestle
Karicare Sensikare HA	pHF-C	Nutricia
Gentlease	pHF-C	Mead Johnson
Good Start Supreme	pHF-C	Nestle
Humana HA 1, 2 and 3	pHF-W	Humana
Beba HA	pHF-W	Nestle
Good Start	pHF-W	Nestle
Similac® Advance HA	pHF-W	Abbott
Novalac® HA	pHF-W	Novalac
Infacare® Nuture HA Comfort	pHF-W	Aspen Nutritionals
Nutramigen	eHF-C	Mead Johnson
Similac® Alimentum	eHF-C	Abbott
Aalfa-Re®	eHF-W	Nestle
Nutrilon Pepti Junior HE	eHF-W	Nutricia
Hypolac Alimento	eHF-W	ALK
Hipp HA	eHF-W	AGRANA
Pepticate	eHF-W	Nutricia
Allernova	eHF-W	Novalac
Peptidi-Tutteli	eHF-W	Valio

Reproduced with permission from Taylor & Francis Ltd (www.tandfonline.com)
CMF cow's milk formula, *CMF-C* casein-based formula, *CMF-W* whey-based formula, *eHF-C* extensively hydrolysed casein-based formula, *eHF-W* extensively hydrolysed whey-based formula, *pHF-C* partially hydrolysed casein-based formula, *pHF-W* partially hydrolysed whey-based formula

19.3.2 Production Process

IF production is usually performed in several steps, with the protein ingredient (whey or caseinate) as the base material (Guo and Ahmad 2014), see Fig. 19.1. HA IF manufacturing commonly starts by hydrolysis of proteins using an enzymatic process. This is followed by the addition of other nutrients (vitamins, minerals, sugars, and fats) required by regulators for optimal infant nutrition. The final product is then heat-treated, followed by spray-drying before packaging. Some products may be UHT-treated if packaged in a liquid form (Guo and Ahmad 2014).

Though most production follows the same overall process, precise information on enzymes and exact processing parameters used in the production of commercial HA IFs are not widely available, likely due to company trade secrets. However, some information can be drawn from public patents. An early 1991 patent describes

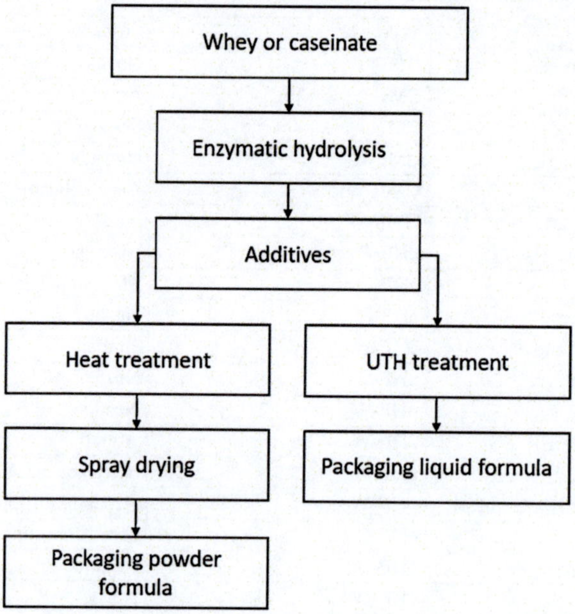

Fig. 19.1 Main steps in the production of hydrolysed infant formula

the hydrolysis of whey using pancreatic enzymes (trypsin, chymotrypsin or pancreatin) (Jost et al. 1991). The process includes up to two rounds of hydrolysis separated by a heat treatment, partly to denature the remaining intact proteins. The resulting product is then UHT-treated before being spray- or freeze-dried. Another patent from the year 2000 describes the hydrolysis of whey protein using Alcalase, Neutrase and pancreatin at 55 °C (Zdenek and Jean-Claude 2000). After the addition of other nutrients (e.g., minerals, vitamins, and sugars) the hydrolysate is again heat-treated followed by spray- or freeze-drying.

Laboratory-scale studies using other enzymes and enzyme combinations report reduction of bovine milk protein allergenicity (Villa et al. 2018; Golkar et al. 2019). Alcalase, papain or trypsin treatment of whey proteins has been shown to reduce the allergenicity of BLG and ALA (Wróblewska et al. 2004; Zheng et al. 2008). Furthermore, treatment of whey protein with trypsin was found to reduce allergenicity when assessed by the development of whey-specific IgE and T cell responses in mice (Duan et al. 2014). Trypsin was similarly able to hydrolyse and reduce the allergenicity of isolated BLG and α-caseins when assessed by histamine release by passively sensitised basophils (Morisawa et al. 2009). Treatment of caseinate by papain has been shown to decrease allergenicity of α-casein and β-casein (Liu et al. 2012). A more recent approach to hydrolysis used *Enterococcus faecalis* WB63F isolated from raw bovine milk (Biscola et al. 2016). This bacterial strain expresses proteolytic properties and was found to reduce the allergenicity of α-casein and β-casein, but had only minor effects on BLG and ALA.

19.3.3 Clinical Use of eHF and pHF

The use of eHF in the management of infant CMA is well-established and recommended in guidelines published by several recognised medical societies (Fiocchi et al. 2018). Specifically, the guidelines recommend eHF to infants not in risk of or no previous anaphylactic reactions, whereas infants with previous anaphylactic reactions are recommended to receive amino acid-based formulae. eHF has been shown to be tolerated in 90–95% of allergic infants, which can be attributed to the highly hydrolysed nature of the product (Ragno et al. 1993; Bahna 2008). More than 95% of the eHF peptides are below 3 kDa, making them too small to cross-link IgE molecules (requires peptides >3 kDa, of approx. 30 amino acids) and thereby are too small to elicit an allergic reaction (Huby et al. 2000). pHFs are unsuitable for disease management since they contain peptides able to elicit an allergic response (3–10 kDa), which is reflected by clinical studies reporting allergic reactions to pHF in 33–55% allergic infants (Ragno et al. 1993; Bahna 2008).

However, the reduced allergenicity of pHF, compared to regular infant formula, make pHF useful in the management of infants in risk of developing CMA (Crittenden and Bennett 2005; Bahna 2008). Studies in animal models support the clinical findings and recommendations by demonstrating reduced allergenicity of eHF compared to pHF, and both in comparison to regular formula (Fritsché and Bonzon 1990; Niggemann et al. 2001). Other studies in animals found that pHF, but not eHF, was able to induce to oral tolerance to bovine milk, probably due to the higher immunogenicity of pHF (Fritsché et al. 1997; Fritsché 2009; Van Esch et al. 2011). However, these findings conflict with clinical studies indicating that both eHF and pHF can be used in the prevention of CMA (Crittenden and Bennett 2005).

19.3.4 Allergenicity Evaluation

Evaluation of HA IF allergenicity is important when developing a new product or assuring product quality when in production. Parameters include degree of hydrolysis and peptide distribution, as well as residual intact protein and allergenicity. A recent study of several commercially available HA IF products and production batches indicated the need for more stringent quality control (Eigenmann et al. 2018). Several whey- and casein-based eHFs were found to contain residual BLG (0.01–0.3 mg/kg) and casein (0.2–1.1 mg/kg). Furthermore, some product brands had a significant batch-to-batch variation, ranging from undetectable intact protein to several times the detection limit. These variations may have clinical implications and call for better routine assessment of product allergenicity in the industry. The sections below provide an overview of protein-chemical assays, in vitro and in vivo methods to evaluate the eliciting, sensitising, preventive, and desensitising properties of HA IFs. Such methods could form routine or initial evaluation of IFs before the completion of carefully conducted clinical testing.

Eliciting Capacity

The eliciting capacity of a hydrolysate is linked to the degree of hydrolysis and the presence of residual intact protein. Standard analytical methods in protein chemistry can be applied to evaluate these parameters. SDS-PAGE, in combination with immunoblotting, and HPLC can be used to estimate the degree of hydrolysis and the presence of specific intact proteins (Rosendal and Barkholt 2000). Inhibitory ELISAs using sera from animals immunised with intact BLG have been used to determine the presence of intact epitopes in BLG and whey protein hydrolysates, since the absence of intact epitopes are unable to block the binding of IgG1 to intact BLG (Bøgh et al. 2015). However, these methods do not directly describe the capacity of the hydrolysate to elicit an allergic response, since this requires the binding and cross-linking of allergen-specific IgE molecules. IgE-binding can be evaluated using sera from patients or sensitised animals using immunoblotting (Meulenbroek et al. 2014). Furthermore, residual IgE-binding can be quantified using ELISA (Bøgh et al. 2015). The basophil activation test (BAT) has recently emerged as an important in vitro measure of IgE cross-linking (Hemmings et al. 2018), which correlated well with allergy severity in patients (Song et al. 2015). The assay uses flow cytometry to measure activation marker expression on basophils, and has been adapted for the evaluation of eliciting capacity of HA IFs in allergic patients (Meulenbroek et al. 2014) and sensitised mice (Iwamoto et al. 2016). Finally, oral challenge of sensitised mice and subsequent assessment of clinical symptoms by scoring and measurement of body temperature represents the most biological representative methods of evaluating the allergic eliciting capacity of HA IFs (Iwamoto et al. 2016).

Sensitising Capacity

The methods mentioned above used to evaluate the eliciting capacity of HA IFs can also be applied in estimating the sensitising capacity, since these methods indicate if intact allergenic proteins and epitopes may be present. However, studies have shown that different animal models may be useful in determining the sensitising capacity of HA IF. A study in BALB/c mice of seven different formulas found a correlation between the degree of hydrolysis and the sensitising capacity following intraperitoneal immunisations using adjuvant (Niggemann et al. 2001). Later studies found that Brown Norway rats can be used to evaluate the sensitising capacity of hydrolysates by measuring IgE and performing skin prick tests following intraperitoneal immunisations (Bøgh et al. 2015).

Primary Preventive Capacity

Animal models can also be used to evaluate the primary preventive capacity (i.e. the development of oral tolerance). Here, the hydrolysates are given before sensitisation to the intact protein, and assessment of the degree of sensitisation. Models have been developed in both mice (Van Esch et al. 2011) and rats (Fritsché et al. 1997; Jensen et al. 2019), which consistently reported a preventive capacity of pHF, but not eHF.

Desensitising Capacity

Assessment of the desensitising capacity of HA IFs can be used to evaluate the potential application in treatment of sensitised individuals. A Brown Norway rat model has been used to compare the desensitising capacity of eHF, pHF and BLG following sensitisation to BLG (Jensen et al. 2019). Only BLG was found to have a desensitising effect by reducing BLG-specific IgE levels. These finding suggest that hydrolysed products have limited application in treatment of CMA.

19.4 Conclusions

CMA is a health problem of growing concern. Enzymatic hydrolysis of bovine milk proteins in general reduces their allergenicity. Thus, hydrolysis represents an important industrial process technology in the production of HA IF for the management, treatment and prevention of CMA. The degree of hydrolysis is though influenced by parameters such as the specific enzymes used, hydrolysis duration, enzyme to protein ratio, matrix and processing. Several molecular methods can be used to determine the degree of hydrolysis of bovine milk proteins; however, animal models represent an important tool for evaluation of residual allergenicity, particularly in the analysis of sensitising and primary prevention capacities.

References

Aalberse RC (2000) Structural biology of allergens. J Allergy Clin Immunol 106:228–238

Adams JJ, Anderson BF, Norris GE et al (2006) Structure of bovine beta-lactoglobulin (variant A) at very low ionic strength. J Struct Biol 154:246–254

Arnon R, Van Regenmortel MH (1992) Structural basis of antigenic specificity and design of new vaccines. FASEB J 6:3265–3274

Asgeirsson B, Bjarnason JB (1991) Structural and kinetic properties of chymotrypsin from Atlantic cod (Gadus morhua). Comparison with bovine chymotrypsin. Comp Biochem Physiol B 99:327–335

Astwood JD, Leach JN, Fuchs RL (1996) Stability of food allergens to digestion in vitro. Nat Biotechnol 14:1269–1273

Bahna SL (2008) Hypoallergenic formulas: optimal choices for treatment versus prevention. Ann Allergy Asthma Immunol 101:453–459

Bannon GA (2004) What makes a food protein an allergen? Curr Allergy Asthma Rep 4:43–46

Barlow DJ, Edwards MS, Thornton JM (1986) Continuous and discontinuous protein antigenic determinants. Nature 322:747–748

Benedé S, López-Expósito I, Giménez G et al (2014) Mapping of IgE epitopes in in vitro gastroduodenal digests of β-lactoglobulin produced with human and simulated fluids. Food Res Int 62:1127–1133

Benedé S, López-Fandiño R, Molina E (2017) Assessment of IgE reactivity of β-Casein by Western blotting after digestion with simulated gastric fluid. In: Methods in molecular biology. Humana Press, Clifton, pp 165–175

Bindels JG, Boerma JA (1994) Hydrolysed cow's milk formulae. Pediatr Allergy Immunol 5:189–190
Biscola V, Tulini FL, Choiset Y et al (2016) Proteolytic activity of Enterococcus faecalis VB63F for reduction of allergenicity of bovine milk proteins. J Dairy Sci 99:5144–5154
Bleumink E, Young E (1968) Identification of the atopic allergen in cow's milk. Int Arch Allergy Appl Immunol 34:521–543
Bøgh KL, Madsen CB (2016) Food allergens: is there a correlation between stability to digestion and allergenicity? Crit Rev Food Sci Nutr 56:1545–1567
Bøgh KL, Barkholt V, Madsen CB (2013) The sensitising capacity of intact β-lactoglobulin is reduced by co-administration with digested β-lactoglobulin. Int Arch Allergy Immunol 161:21–36
Bøgh KL, Barkholt V, Madsen CB (2015) Characterization of the immunogenicity and allergenicity of two cow's milk hydrolysates - a study in brown Norway rats. Scand J Immunol 81:274–283
Broekman HCH, Eiwegger T, Upton J, Bøgh KL (2015) IgE – the main player of food allergy. Drug Discov Today Dis Model 17–18:37–44
CAC (2003) Codex Alimentarius Commission - Alinorm 03/34: Joint FAO/WHO Food Standard Programme, Codex Alimentarius Commission, Twenty-Fifth Session, Rome, Italy 30 June - 5 July, 2003
Cawthern KM, Permyakov E, Berliner LJ (1996) Membrane-bound states of α-lactalbumin: implications for the protein stability and conformation. Protein Sci 5:1394–1405
Chapman MD (2008) Allergen nomenclature. Clin Allergy Immunol 21:47–58
Chatchatee P, Järvinen KM, Bardina L et al (2001) Identification of IgE and IgG binding epitopes on beta- and kappa-casein in cow's milk allergic patients. Clin Exp Allergy 31:1256–1262
Corzo-Martínez M, Soria AC, Belloque J et al (2010) Effect of glycation on the gastrointestinal digestibility and immunoreactivity of bovine β-lactoglobulin. Int Dairy J 20:742–752
Crittenden RG, Bennett LE (2005) Cow's milk allergy: a complex disorder. J Am Coll Nutr 24:582S–591S
Dosa S, Pesce AJ, Ford DJ et al (1979) Immunological properties of peptic fragments of bovine serum albumin. Immunology 38:509–517
Duan C, Yang L, Li A et al (2014) Effects of enzymatic hydrolysis on the allergenicity of whey protein concentrates. Iran J Allergy Asthma Immunol 13:231–239
Dupont D, Mackie AR (2015) Static and dynamic in vitro digestion models to study protein stability in the gastrointestinal tract. Drug Discov Today Dis Model 17–18:23–27
Dupont D, Mandalari G, Molle D et al (2010) Comparative resistance of food proteins to adult and infant in vitro digestion models. Mol Nutr Food Res 54:767–780
EFSA (2010) Scientific Opinion on the assessment of allergenicity of GM plants and microorganisms and derived food and feed. EFSA J 8:1700
Eigenmann P, Madrazo-de la Garza JA, Nutten S, O'Mahony L (2018) Towards optimised management of cow's milk protein allergy - European medical journal. EMJ Allergy Immunol 3:50–59
El Mecherfi K-E, Rouaud O, Curet S et al (2015) Peptic hydrolysis of bovine beta-lactoglobulin under microwave treatment reduces its allergenicity in an *ex vivo* murine allergy model. Int J Food Sci Technol 50:356–364
Erickson RH, Kim YS (1990) Digestion and absorption of dietary protein. Annu Rev Med 41:133–139
Etherington DJ, Taylor WH (1970) The pepsins from human gastric mucosal extracts. Biochem J 118:587–594
FAO/WHO (2001) Evaluation of Allergenicity of Genetically Modified Foods. Report of a Joint FAO/WHO Expert Consulation on Allergenicity of Food Derived from Biotechnology, 22-25 january 2001. Food and Agriculture organisation of the United Nations (FAO), Rome
Farkas V, Vass E, Hanssens I et al (2005) Cyclic peptide models of the Ca2+-binding loop of alpha-lactalbumin. Bioorg Med Chem 13:5310–5320
Ferguson TA, Peters T, Reed R et al (1983) Immunoregulatory properties of antigenic fragments from bovine serum albumin. Cell Immunol 78:1–12

Fiocchi A, Schünemann HJ, Brozek J et al (2010) Diagnosis and Rationale for Action Against Cow's Milk Allergy (DRACMA): a summary report. J Allergy Clin Immunol 126:1119–1128

Fiocchi A, Schunemann H, Ansotegui I et al (2018) The global impact of the DRACMA guidelines cow's milk allergy clinical practice. World Allergy Organ J 11:2

Fritsché R (2009) Utility of animal models for evaluating hypoallergenicity. Mol Nutr Food Res 53:979–983

Fritsché R, Bonzon M (1990) Determination of cow milk formula allergenicity in the rat model by in vitro mast cell triggering and in vivo IgE induction. Int Arch Allergy Appl Immunol 93:289–293

Fritsché R, Pahud JJ, Pecquet S, Pfeifer A (1997) Induction of systemic immunologic tolerance to beta-lactoglobulin by oral administration of a whey protein hydrolysate. J Allergy Clin Immunol 100:266–273

Fu T-J, Abbott UR, Hatzos C (2002) Digestibility of food allergens and nonallergenic proteins in simulated gastric fluid and simulated intestinal fluid-a comparative study. J Agric Food Chem 50:7154–7160

Gass J, Vora H, Hofmann AF et al (2007) Enhancement of dietary protein digestion by conjugated bile acids. Gastroenterology 133:16–23

Golkar A, Milani JM, Vasiljevic T (2019) Altering allergenicity of cow's milk by food processing for applications in infant formula. Crit Rev Food Sci Nutr 59:159–172

Guo M, Ahmad S (2014) Formulation guidelines for infant formula. In: Human milk biochemistry and infant formula manufacturing technology. Woodhead Publishing, Cambridge, pp 141–171

Haddad ZH, Kalra V, Verma S (1979) IgE antibodies to peptic and peptic-tryptic digests of betalactoglobulin: significance in food hypersensitivity. Ann Allergy 42:368–371

Hemmings O, Kwok M, McKendry R, Santos AF (2018) Basophil activation test: old and new applications in allergy. Curr Allergy Asthma Rep 18:77

Herman RA, Woolhiser MM, Ladics GS et al (2007) Stability of a set of allergens and non-allergens in simulated gastric fluid. Int J Food Sci Nutr 58:125–141

Hochwallner H, Schulmeister U, Swoboda I et al (2014) Cow's milk allergy: from allergens to new forms of diagnosis, therapy and prevention. Methods 66:22–33

Høst A, Koletzko B, Dreborg S et al (1999) Dietary products used in infants for treatment and prevention of food allergy. Joint Statement of the European Society for Paediatric Allergology and Clinical Immunology (ESPACI) Committee on Hypoallergenic Formulas and the European Society for Paediatric Gastroenterology, Hepatology and Nutrition (ESPGHAN) Committee on Nutrition. Arch Dis Child 81:80–84

Huby RD, Dearman RJ, Kimber I (2000) Why are some proteins allergens? Toxicol Sci 55:235–246

Inouye K, Fruton JS (1967) Studies on the specificity of pepsin. Biochemistry 6:1765–1777

IUIS (2019) IUIS allergen nomenclature (http://www.allergen.org)

Iwamoto H, Matsubara T, Nakazato Y et al (2016) Evaluation of the antigenicity of hydrolyzed cow's milk protein formulas using the mouse basophil activation test. Toxicol Lett 242:53–59

James AH, Baron JH (1962) Discussion on the physiological and clinical assessment of gastric function. Proc R Soc Med 55:71–75

Järvinen KM, Chatchatee P, Bardina L et al (2001) IgE and IgG binding epitopes on alpha-lactalbumin and beta-lactoglobulin in cow's milk allergy. Int Arch Allergy Immunol 126:111–118

Järvinen K-M, Beyer K, Vila L et al (2002) B-cell epitopes as a screening instrument for persistent cow's milk allergy. J Allergy Clin Immunol 110:293–297

Jensen LH, Larsen JM, Madsen CB et al (2019) Preclinical brown Norway rat models for the assessment of infant formulas in the prevention and treatment of cow's milk allergy. Int Arch Allergy Immunol 178:307–314

Jost R, Meister N, Monti JC (1991) Preparation of a hypoallergenic whey protein hydrolyzate and food. US Patent US5039532A

Kattan JD, Cocco RR, Järvinen KM (2011) Milk and soy allergy. Pediatr Clin North Am 58:407–426

Kontopidis G, Holt C, Sawyer L (2004) Invited review: β-Lactoglobulin: binding properties, structure, and function. J Dairy Sci 87:785–796

Li Z, Luo Y, Feng L, Liao P (2013) Effect of Maillard reaction conditions on antigenicity of β-lactoglobulin and the properties of glycated whey protein during simulated gastric digestion. Food Agric Immunol 24:433–443

Liu X, Luo Y, Li Z (2012) Effects of pH, temperature, enzyme-to-substrate ratio and reaction time on the antigenicity of casein hydrolysates prepared by papain. Food Agric Immunol 23:69–82

Lucas JSA, Cochrane SA, Warner JO, Hourihane JOB (2008) The effect of digestion and pH on the allergenicity of kiwifruit proteins. Pediatr Allergy Immunol 19:392–398

Macierzanka A, Sancho AI, Mills ENC et al (2009) Emulsification alters simulated gastrointestinal proteolysis of β-casein and β-lactoglobulin. Soft Matter 5:538–550

Mandalari G, Adel-Patient K, Barkholt V et al (2009a) In vitro digestibility of beta-casein and beta-lactoglobulin under simulated human gastric and duodenal conditions: a multi-laboratory evaluation. Regul Toxicol Pharmacol 55:372–381

Mandalari G, Mackie AM, Rigby NM et al (2009b) Physiological phosphatidylcholine protects bovine beta-lactoglobulin from simulated gastrointestinal proteolysis. Mol Nutr Food Res 53(Suppl 1):S131–S139

Marchesseau S, Mani J-C, Martineau P et al (2002) Casein interactions studied by the surface plasmon resonance technique. J Dairy Sci 85:2711–2721

Marciani L, Gowland PA, Spiller RC et al (2001) Effect of meal viscosity and nutrients on satiety, intragastric dilution, and emptying assessed by MRI. Am J Physiol Gastrointest Liver Physiol 280:G1227–G1233

Martinez MJ, Martos G, Molina E, Pilosof AMR (2016) Reduced β-lactoglobulin IgE binding upon in vitro digestion as a result of the interaction of the protein with casein glycomacropeptide. Food Chem 192:943–949

McLaren AD, Estermann EF (1957) Influence of pH on the activity of chymotrypsin at a solid-liquid interface. Arch Biochem Biophys 68:157–160

Meulenbroek LAPM, Oliveira S, den Hartog Jager CF et al (2014) The degree of whey hydrolysis does not uniformly affect *in vitro* basophil and T cell responses of cow's milk-allergic patients. Clin Exp Allergy 44:529–539

Michael JG (1989) The role of digestive enzymes in orally induced immune tolerance. Immunol Invest 18:1049–1054

Mills ENC, Jenkins JA, Alcocer MJC, Shewry PR (2004) Structural, biological, and evolutionary relationships of plant food allergens sensitizing via the gastrointestinal tract. Crit Rev Food Sci Nutr 44:379–407

Misra A, Prasad R, Das M, Dwivedi PD (2009) Probing novel allergenic proteins of commonly consumed legumes. Immunopharmacol Immunotoxicol 31:186–194

Moreno FJ, Mackie AR, Mills ENC (2005a) Phospholipid interactions protect the milk allergen alpha-lactalbumin from proteolysis during in vitro digestion. J Agric Food Chem 53:9810–9816

Moreno FJ, Mellon FA, Wickham MSJ et al (2005b) Stability of the major allergen Brazil nut 2S albumin (Ber e 1) to physiologically relevant in vitro gastrointestinal digestion. FEBS J 272:341–352

Morisawa Y, Kitamura A, Ujihara T et al (2009) Effect of heat treatment and enzymatic digestion on the B cell epitopes of cow's milk proteins. Clin Exp Allergy 39:918–925

Mouécoucou J, Sanchez C, Villaume C et al (2003) Effects of different levels of gum arabic, low methylated pectin and xylan on in vitro digestibility of beta-lactoglobulin. J Dairy Sci 86:3857–3865

Mouécoucou J, Villaume C, Sanchez C, Méjean L (2004) Beta-lactoglobulin/polysaccharide interactions during in vitro gastric and pancreatic hydrolysis assessed in dialysis bags of different molecular weight cut-offs. Biochim Biophys Acta 1670:105–112

Murtagh GJ, Dumoulin M, Archer DB, Alcocer MJ (2002) Stability of recombinant 2 S albumin allergens in vitro. Biochem Soc Trans 30:913–915

Murtagh GJ, Archer DB, Dumoulin M et al (2003) In vitro stability and immunoreactivity of the native and recombinant plant food 2S albumins Ber e 1 and SFA-8. Clin Exp Allergy 33:1147–1152

Nacer A, Sanchez C, Villaume C et al (2004) Interactions between beta-lactoglobulin and pectins during in vitro gastric hydrolysis. J Agric Food Chem 52:355–360

Niggemann B, Nies H, Renz H et al (2001) Sensitizing capacity and residual allergenicity of hydrolyzed cow's milk formulae: results from a murine model. Int Arch Allergy Immunol 125:316–321

Ofori-Anti AO, Ariyarathna H, Chen L et al (2008) Establishing objective detection limits for the pepsin digestion assay used in the assessment of genetically modified foods. Regul Toxicol Pharmacol 52:94–103

Olsen JV, Ong S-E, Mann M (2004) Trypsin cleaves exclusively C-terminal to arginine and lysine residues. Mol Cell Proteomics 3:608–614

Pantoja-Uceda D, Palomares O, Bruix M et al (2004) Solution structure and stability against digestion of rproBnIb, a recombinant 2S albumin from rapeseed: relationship to its allergenic properties. Biochemistry 43:16036–16045

Parekh H, Bahna SL (2016) Infant formulas for food allergy treatment and prevention. Pediatr Ann 45:e150–e156

Peñas E, Préstamo G, Luisa Baeza M et al (2006) Effects of combined high pressure and enzymatic treatments on the hydrolysis and immunoreactivity of dairy whey proteins. Int Dairy J 16:831–839

Peram MR, Loveday SM, Ye A, Singh H (2013) In vitro gastric digestion of heat-induced aggregates of β-lactoglobulin. J Dairy Sci 96:63–74

Permyakov EA, Berliner LJ (2000) Alpha-Lactalbumin: structure and function. FEBS Lett 473:269–274

Polovic N, Blanusa M, Gavrovic-Jankulovic M et al (2007) A matrix effect in pectin-rich fruits hampers digestion of allergen by pepsin in vivo and in vitro. Clin Exp Allergy 37:764–771

Prescott SL, Pawankar R, Allen KJ et al (2013) A global survey of changing patterns of food allergy burden in children. World Allergy Organ J 6:21

Ragno V, Giampietro PG, Bruno G, Businco L (1993) Allergenicity of milk protein hydrolysate formulae in children with cow's milk allergy. Eur J Pediatr 152:760–762

Roggen EL (2006) Recent developments with B-cell epitope identification for predictive studies. J Immunotoxicol 3:137–149

Rosendal A, Barkholt V (2000) Detection of potentially allergenic material in 12 hydrolyzed milk formulas. J Dairy Sci 83:2200–2210

Sampson HA (2004) Update on food allergy. J Allergy Clin Immunol 113:805–819

Sanchón J, Fernández-Tomé S, Miralles B et al (2018) Protein degradation and peptide release from milk proteins in human jejunum. Comparison with in vitro gastrointestinal simulation. Food Chem 239:486–494

Sandberg AS, Andersson H, Hallgren B et al (1981) Experimental model for in vivo determination of dietary fibre and its effect on the absorption of nutrients in the small intestine. Br J Nutr 45:283–294

Schmidt DG, Meijer RJ, Slangen CJ, Van Beresteijn EC (1995) Raising the pH of the pepsin-catalysed hydrolysis of bovine whey proteins increases the antigenicity of the hydrolysates. Clin Exp Allergy 25:1007–1017

Schulten V, Lauer I, Scheurer S et al (2011) A food matrix reduces digestion and absorption of food allergens in vivo. Mol Nutr Food Res 55:1484–1491

Sélo I, Clément G, Bernard H et al (1999) Allergy to bovine beta-lactoglobulin: specificity of human IgE to tryptic peptides. Clin Exp Allergy 29:1055–1063

Sicherer SH, Sampson HA (2014) Food allergy: epidemiology, pathogenesis, diagnosis, and treatment. J Allergy Clin Immunol 133:291–307

Sipos T, Merkel JR (1970) An effect of calcium ions on the activity, heat stability, and structure of trypsin. Biochemistry 9:2766–2775

Skripak JM, Matsui EC, Mudd K, Wood RA (2007) The natural history of IgE-mediated cow's milk allergy. J Allergy Clin Immunol 120:1172–1177
Sletten GBG, Holden L, Egaas E, Faeste CK (2008) Differential influence of the degree of processing on immunogenicity following proteolysis of casein and β-lactoglobulin. Food Agric Immunol 19:213–228
Song Y, Wang J, Leung N et al (2015) Correlations between basophil activation, allergen-specific IgE with outcome and severity of oral food challenges. Ann Allergy Asthma Immunol 114:319–326
Spuergin P, Mueller H, Walter M et al (1996) Allergenic epitopes of bovine alpha S1-casein recognized by human IgE and IgG. Allergy 51:306–312
Stuart DI, Acharya KR, Walker NP et al (1986) Alpha-lactalbumin possesses a novel calcium binding loop. Nature 324:84–87
Takagi K, Teshima R, Okunuki H, Sawada J (2003) Comparative study of in vitro digestibility of food proteins and effect of preheating on the digestion. Biol Pharm Bull 26:969–973
Tang J (1963) Specificity of pepsin and its dependence on a possible "hydrophobicbinding site". Nature 199:1094–1095
Tang J (2013) Pepsin A. In: Handbook of proteolytic enzymes. ScienceDirect, pp 27–35
Taylor SL, Hefle SL (2001) Will genetically modified foods be allergenic? J Allergy Clin Immunol 107:765–771
Thomas K, Aalbers M, Bannon GA et al (2004) A multi-laboratory evaluation of a common in vitro pepsin digestion assay protocol used in assessing the safety of novel proteins. Regul Toxicol Pharmacol 39:87–98
Troost FJ, Steijns J, Saris WH, Brummer RJ (2001) Gastric digestion of bovine lactoferrin in vivo in adults. J Nutr 131:2101–2104
Untersmayr E, Jensen-Jarolim E (2008) The role of protein digestibility and antacids on food allergy outcomes. J Allergy Clin Immunol 121:1301–1308
Untersmayr E, Poulsen LK, Platzer MH et al (2005) The effects of gastric digestion on codfish allergenicity. J Allergy Clin Immunol 115:377–382
Van Esch BCAM, Schouten B, de Kivit S et al (2011) Oral tolerance induction by partially hydrolyzed whey protein in mice is associated with enhanced numbers of Foxp3+ regulatory T-cells in the mesenteric lymph nodes. Pediatr Allergy Immunol 22:820–826
Van Regenmortel MHV (1996) Mapping epitope structure and activity: from one-dimensional prediction to four-dimensional description of antigenic specificity. Methods 9:465–472
Van Regenmortel MHV (1998) Mimotopes, continuous paratopes and hydropathic complementarity: novel approximations in the description of immunochemical specificity. J Dispers Sci Technol 19:1199–1219
Van Regenmortel MHV (2009) What is a B-cell epitope? Methods Mol Biol 524:3–20
Vieths S, Reindl J, Müller U et al (1999) Digestibility of peanut and hazelnut allergens investigated by a simple in vitro procedure. Eur Food Res Technol 209:379–388
Villa C, Costa J, Oliveira MBPP, Mafra I (2018) Bovine milk allergens: a comprehensive review. Compr Rev Food Sci Food Saf 17:137–164
Von Berg A (2009) Modified proteins in allergy prevention. In: Microbial host-interaction: tolerance versus allergy. Karger, Basel, pp 239–250
Von Berg A (2013) Dietary interventions for primary allergy prevention - what is the evidence? World Rev Nutr Diet 108:71–78
Wal J-M (2002) Cow's milk proteins/allergens. Ann Allergy Asthma Immunol 89:3–10
Wal JM, Bernard H, Créminon C et al (1995) Cow's milk allergy: the humoral immune response to eight purified allergens. Adv Exp Med Biol 371B:879–881
Wang J, Lin J, Bardina L et al (2010) Correlation of IgE/IgG4 milk epitopes and affinity of milk-specific IgE antibodies with different phenotypes of clinical milk allergy. J Allergy Clin Immunol 125:695–702
Wickham M, Faulks R, Mills C (2009) In vitro digestion methods for assessing the effect of food structure on allergen breakdown. Mol Nutr Food Res 53:952–958

Wróblewska B, Karamać M, Amarowicz R et al (2004) Immunoreactive properties of peptide fractions of cow whey milk proteins after enzymatic hydrolysis. Int J Food Sci Technol 39:839–850
Yagami T, Haishima Y, Nakamura A et al (2000) Digestibility of allergens extracted from natural rubber latex and vegetable foods. J Allergy Clin Immunol 106:752–762
Yamada C, Yamashita Y, Seki R et al (2006) Digestion and gastrointestinal absorption of the 14-16-kDa rice allergens. Biosci Biotechnol Biochem 70:1890–1897
Yu W, Freeland DMH, Nadeau KC (2016) Food allergy: immune mechanisms, diagnosis and immunotherapy. Nat Rev Immunol 16:751–765
Zdenek K, Jean-Claude M (2000) Infant formula containing sweet whey protein. Patent EP1048226A1
Zeece M, Huppertz T, Kelly A (2008) Effect of high-pressure treatment on in-vitro digestibility of β-lactoglobulin. Innov Food Sci Emerg Technol 9:62–69
Zheng H, Shen X, Bu G, Luo Y (2008) Effects of pH, temperature and enzyme-to-substrate ratio on the antigenicity of whey protein hydrolysates prepared by Alcalase. Int Dairy J 18:1028–1033

Chapter 20
Future Opportunities and Challenges in Dairy Enzymology

Alan L. Kelly, Shane V. Crowley, Aisling Crotty, Fanyu Meng, and Lotte Bach Larsen

20.1 Introduction

As the preceding chapters in this book have explored in detail, the field of dairy enzymology is wide and complex, and enzymes from multiple sources can influence the quality of dairy products, in some cases positively and in some negatively; others, such as coagulants, are intrinsic to the production of whole families of dairy products, like cheese. In addition, while enzymes have been a part of dairy processing for millennia, directly or indirectly, today they offer potential for future innovation and product development, whether by creation of new textures through cross-linking of proteins or by addressing health-related issues through proteolysis-aided reduction of allergenicity or lactose intolerance.

The purpose of this final short chapter is to consider future possibilities related to research and innovation in dairy enzymology, and to also discuss some enzymes which have not been discussed in earlier chapters.

A. L. Kelly (✉) · S. V. Crowley · A. Crotty · F. Meng
School of Food and Nutritional Sciences, University College Cork, Cork, Ireland
e-mail: a.kelly@ucc.ie; shane.crowley@ucc.ie; 114475822@umail.ucc.ie; f.meng@umail.ucc.ie

L. B. Larsen
Department of Food Science, Faculty of Science and Technology, Aarhus University, Aarhus, Denmark
e-mail: lbl@food.au.dk

A. L. Kelly, L. B. Larsen (eds.), *Agents of Change*, Food Engineering Series,
https://doi.org/10.1007/978-3-030-55482-8_20

20.2 Future Challenges in Indigenous and Endogenous Milk Enzymology

With every passing year, new research findings bring the complexity of milk enzymology into sharper relief. For example, advances in proteomics and peptidomics have greatly enhanced our understanding of proteolysis pathways and protein fragments in bovine and human milk, indicating a much more complex system than the long-standing paradigm that emphasised the activity of plasmin. Recent years have seen a vast multiplication of the number of known proteolytic enzymes derived from psychrotrophic bacteria, among which some remarkably heat-stable enzymes have been identified, many of which exhibit residual activity following some of the most severe thermal processes applied to milk, such as UHT treatment. Furthermore, the interactions between indigenous and endogenous enzymes drive new areas of research and understanding of this complex area, as exemplified by the aforementioned relationships between survival of endogenous enzymes and product quality, and new emerging knowledge on the importance and significance of these heat-stable enzymes and their interactions with indigenous enzymes, like the plasmin system and somatic cell proteases.

One thing that is evident from the Chapters within this book is the never-ending need for harmonization of methods for determining enzyme activity. Furthermore, there is a clear need for specific and precise definitions of units for enzyme activities, and it is important to address whether *enzyme activity* or the *result of enzyme activity* is addressed in any specific context. One such example is the proteolytic enzymes in milk and dairy products, where enzyme activities are determined by the use of chromogenic substrates, which assess the net level of the specific active enzyme present, while the determination of the level of peptides generated is an assessment of the net result of the present proteolytic activity, and thereby also the accessibility of substrate (here, the milk proteins).

Kelly and Fox (2006) discussed future research requirements in the area of indigenous enzymes in milk, and their key points are listed in Table 20.1, along with a summary of the current state of knowledge as has advanced in the intervening 15 years, and summarised in the chapters of this book.

20.3 Processing and Milk Enzymes

Today, milk is fractionated into a wide range of ingredients, and innovative varieties, particularly those based on milk proteins, continue to emerge as new dairy product categories. These include milk protein concentrates (MPCs), with varying protein:lactose ratios, micellar casein concentrate (MCCs), the coproduct of which is native whey, and more specialised protein fractions, such as the β-casein concentrate generated by cold dissociation of this protein from native micelles in milk during MCC production, followed by micellization of β-casein to enable its fractionation from native whey (O'Mahony et al. 2014).

Table 20.1 Progress in 15 years of research into indigenous milk enzymes? An update of the research requirements outlined by Kelly and Fox (2006)

Gaps observed by Kelly and Fox (2006)	Current status
Limited and sporadic research in enzymes of non-bovine dairying species	More done on ovine, caprine and camel milk, e.g., in relation to somatic cell count, but still many gaps
Need for standardisation of methods for assaying indigenous enzymes	Still needed. Strict definition of units in studies still needed
Impact of lactation and farm-related factors on a broader range of enzymes	Greater understanding has progressively developed
Impact of processing on milk enzyme activity and distribution (e.g., enzymes of MFGM, and impact of membrane separation technologies) and implications for fractionated dairy products	Updates in knowledge, particularly in relation to MFGM, especially using mass spectrometry methods have occurred, also in relation to partition relative to processing and fractionation. The area of extracellular vesicles has emerged as a new area from the former definition (skim milk membranes) and their origin and significance is being intensively studied
Roles of lysosomal enzymes and link to total or differential somatic cell count, and levels of SCC at which impact becomes significant	Differential somatic cell count equipment more accessible due to development of fast methods. Still gaps in knowledge in relation to secretions of lysosomal enzymes from somatic cells. New knowledge gained on expression side from qPCR
Thermal inactivation of milk enzymes	Great knowledge, in particular to bacterial protease stability
Physiological roles of indigenous enzymes Indicators of efficacy of emerging technologies, both thermal (e.g., very rapid heating by steam infusion) and non-thermal (e.g., high-pressure processing)	More knowledge, e.g., in relation to the role of proteases in human milk and their potential role in pre-digestion of proteins. Glycosidases has become an emerging area, of significance for bovine and human milk oligosaccharides Some studies published in these areas
Biological and neutraceutical effects of milk enzymes and their products	Public milk bioactive peptide database established. Still a lack of knowledge in relation to *in vivo* dose: response relationships of bioactive peptides
Role of plasmin in generation of cheese flavour	Not many studies specifically focused on this topic
Impact of indigenous enzymes on possible future secretion of transgenic products in milk	Not much development of knowledge
Lack of commercial availability of potentially industrially valuable enzymes (e.g., lactoperoxidase, glycosidases)	Potential in relation to expression and GMO
Enzymes in colostrum	Still limited knowledge

(continued)

Table 20.1 (continued)

Gaps observed by Kelly and Fox (2006)	Current status
Effect of genetics, husbandry practices and animal breeding on milk enzymes	Some new knowledge in relation to lactation stages, dry periods and extended lactation, but otherwise knowledge still scarce, e.g., in relation to cow/calf physiology, involution etc.
Veterinary significance of the activity and measurement of milk enzymes	Most done on NAGase, potential uses for other enzymes, both proteolytic and other glycosidases and other enzyme classes

However, it must be considered how such apparently mild (in terms of changes in temperature and pH) processes may affect enzyme distribution across different fractions. For example, membrane filtration of milk through microfiltration membranes, which retain casein micelles, but not serum (whey) proteins, could result in concentration of plasmin associated with micelles, while removing inhibitors in the permeate, such that the resultant ingredients have essentially up-regulated proteolytic potential. This was shown for a range of dairy protein ingredients by Gazi et al. (2014), who reported that milk protein concentrates and milk protein isolates had higher plasmin activity than calcium or sodium caseinate; the ingredients showed significant activation of plasminogen and proteolysis during storage. Such retained activity which could lead to stability and flavour issues when later incorporated into products like beverages; as milk protein ingredients are increasingly used in high protein beverages treated using UHT conditions to give a long ambient shelf-life, such activity may also lead to difficulties such as age gelation on storage (as discussed in Chap. 13).

The same has been observed for β-casein concentrates (BCC), which have promise as potential ingredients in products like infant formulae, as shown in the following unpublished data. A number of different methods for enrichment or isolation of β-casein have been published in recent years (Crowley et al. 2018a, b; McCarthy et al. 2017; Thienel et al. 2018; Schäfer et al. 2019), which typically involve manipulating the association state of β-casein to facilitate its size-based separation from other dairy proteins. For example, cooling of milk leads to partial dissociation of β-casein from casein micelles, which permits the enrichment of native whey with β-casein during microfiltration (McCarthy et al. 2017), while warming of this native whey forms β-casein micelles that can be concentrated relative to whey proteins by further microfiltration (O'Mahony et al. 2014). In both cases, temperature is adjusted to control the strength of hydrophobic interactions, which influences the extent to which the protein permeates a membrane.

In a study of such a process, BCC was produced using a 'warm then cold' (WTC) microfiltration process (Christensen and Holst 2014; Schäfer et al. 2019). A micellar casein concentrate was first fractionated from the whey proteins using warm microfiltration (~25 °C), followed by cold microfiltration (<5 °C) to generate a relatively pure β-casein stream (permeate) that was concentrated by ultrafiltration. An additional 'clean-up' step was then applied, in which the concentrated β-casein solution was passed through a very loose membrane with a negative charge. The solution was acidified before this step to reduce the zeta-potential of the whey proteins and (theoretically) promote their permeation relative to β-casein [more detail on this process can be found at Crowley (2016, pp. 225)].

The BCC powder was reconstituted with DI water at room temperature to a final casein concentration at 1% (w/w) with 0.02% (w/w) sodium azide addition to prevent microbial spoilage. The pH of the BCC solution was then adjusted to pH 6.7 before incubation at 37 °C. To a second portion of similarly-reconstituted powder, the plasmin inhibitor aprotinin (from bovine lung, Sigma-Aldrich Ireland Ltd., Wicklow, Ireland) was reconstituted with DI water to 10 mg mL^{-1} aprotinin solution and this was added to protein solutions to a final aprotinin concentration at 0.1 mg mL^{-1}. Samples were then collected during incubation and were frozen at −20 °C until electrophoresis. Urea-polyacrylamide gel electrophoresis (Urea-PAGE) was performed according to the method of Andrews (1983) with the modifications of Shalabi and Fox (1987). The gels were stained directly with Coomassie Brilliant Blue G-250 by the method of Blakesley and Boezi (1977) and destained using distilled water.

From Fig. 20.1, it can be seen that the sample incubated without aprotinin readily hydrolysed during incubation, with the distinctive pattern of γ-caseins being produced from β-casein; such hydrolysis was completely inhibited on addition of

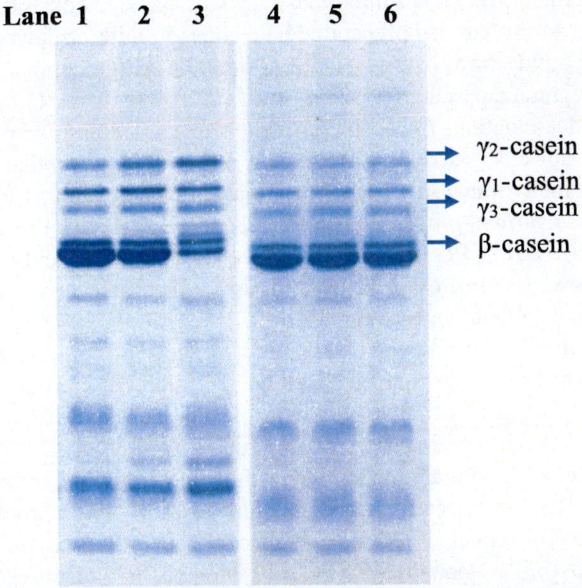

Fig. 20.1 Urea–polyacrylamide gel electrophoretograms of β-casein concentrate produced from 'warm than cold' microfiltration without aprotinin addition (Lanes 1–3) or with aprotinin addition (Lanes 4–6) during incubation at 37 °C for 0 (Lane 1 and 4), 1 (Lane 2 and 5), and 3 (Lane 3 and 6) days

aprotinin, showing that the concentration process used to isolate β-casein co-enriched active plasmin, which is presumably a certain fraction of that present in the original milk, which dissociated from the casein micelles during filtration. The key question is then whether the BCC could lead to physical instability (or cause flavor issues such as bitterness) when later incorporated into a formulated food product.

Thus, it seems that there is a research gap in bridging the gap between laboratory studies and industrial practice, to support the development of such ingredients and their applications in terms of study of the fractionation of enzymes, substrates and inhibitors, leading to their separation or combination in ways which profoundly affect their relative activity and influence on product quality. In the example given above, it may become important to understand how transfer of plasmin into BCC could be mitigated, ideally by enhancing its retention in the casein-rich retentate.

20.4 Emerging Enzymes

20.4.1 Protein Glutaminase

Deamidation of proteins, polypeptides and free amino acids can occur, both as a result of process-induced deamidation (non-enzyme induced) (Le et al. 2017) or as the result of enzyme-catalyzed reactions.

Protein glutaminase (PG) is a microbial enzyme capable of converting glutamine residues, either in protein or polypeptide backbones in milk or dairy products, by converting the amide form of the residue or amino acid (glutamine, Glu, Q) into glutamic acid (glutamate, Glu, E) (Miwa et al. 2010). Compared to other methods of enzymatic deamidation, for example using transglutaminase or proteases, PG does not cause any side reactions to occur (e.g., cross-linking, peptide hydrolysis).

PG was originally found in a culture supernatant from the Gram-negative bacterium *Chryseobacterium proteolyticum* (Yamaguchi et al. 2001). According to Hashizume et al. (2011), PG is a monomeric protein 185 amino acids in length; PG and TG both have the same catalytic triad located at their active site. It has become commercially available in recent years.

As a result of PG-induced deamidation, and the subsequent increase in negative carboxyl groups, there is greater electrostatic repulsion between proteins which, coupled with the exposure of regions of hydrophobicity due to deamidation, results in substantially altered protein properties and functionality.

In dairy systems, caseins are reported to be the preferred substrates for PG, as the enzyme's reactivity is affected by the globular state of whey proteins such as β-lactoglobulin. The effect of PG on proteins in skim milk was investigated by Miwa et al. (2010). PG deamidated skim milk proteins very effectively, causing a decrease in turbidity, which suggests that the casein micelles are dissociated in the process, as micelles are the main light-scattering particles in skim milk (Miwa et al. 2010). Particle size measurements indicated an increase in the number of smaller particles (13.5 nm), with a simultaneous decrease in the larger particles (100 nm). Miwa et al. (2010) proposed that the κ-casein on the surface of the casein micelle is deamidated by PG first, which introduces a negative charge that reduces the integrity of the micelle, and exposes the caseins inside the micelle, which are then deamidated as the PG reaction proceeds.

The impact of PG on the properties was investigated in studies at University College Cork (unpublished to date). The PG used was supplied by Amano (Amano Enzyme Inc., Japan), kept refrigerated until use, and added at a dosage of 0.1% PG for treatment of pasteurized skim milk, followed by incubation for up to 168 h at 50 °C (the optimum temperature for PG; Miwa et al. 2010, Miwa et al. 2014).

Hunter L*a*b* values were measured at various time intervals during incubation using a Konica Minolta Chroma Meter CR-400. PG treatment led to a rapid decrease in whiteness (L-value), with a value of close to 40, similar to that obtained for whey, being reached in 24 h (Fig. 20.2, top). This transformation into a whey-like clarity was also visibly evident (Fig. 20.2, bottom), with the PG-treated milk being mainly coloured by the riboflavin present in the milk. As the milk treated in this case was skim milk, in which the principal light-scattering colloidal particles are casein micelles, this confirms a very high degree of dissociation of these micelles into particles too small to scatter visible light, as reported by Miwa et al. (2010).

Throughout incubation, PG-treated milk samples and samples incubated without PG (controls) were also analysed by Urea-polyacrylamide gel electrophoresis (PAGE) using the method of Summer et al. (2009) with modifications and gels were stained directly by the method of Blakesley and Boezi (1977). It can be seen (by comparing patterns in the first box superimposed on the gel) from the resulting electrophoretograms (Fig. 20.3) that PG treatment resulted in very significant changes in the electrophoretic mobility of caseins, with both β-casein and α_{s1}-casein apparently moving to positions significantly lower than their initial location on the gel (conclusion drawn from their being the two bands of greatest intensity on the gel). Such a change in electrophoretic mobility is consistent, as urea-PAGE separates proteins on the basis of size-charge ratio, with increased negative charge on the protein molecules, due to deamidation, and this having occurred to a greater extent in β-casein than α_{s1}-casein, due to a greater number of substrate residues in the former. This change in protein mobility is essentially unique to PG treatment to the best of the authors' knowledge.

In addition, it can be seen from Fig. 20.3 that, after prolonged incubation (168 h) the control samples showed extensive hydrolysis of β-casein, and production of γ-caseins, the hallmark of the action of native plasmin in the milk (Chap. 2). In the PG-treated milk, however, there appeared to be a significant level of remaining β-casein at the end of incubation, while the electrophoretic mobility of the γ-caseins had clearly profoundly shifted, and it is as yet unclear where on the gel these bands may have migrated. Such a possible change in proteolysis in milk due to PG treatment could reflect an effect of deamidation on protein substrate or an effect on the enzyme itself, or both. The reason for this impact on proteolysis in milk, and its significance for dairy product stability, are currently being investigated.

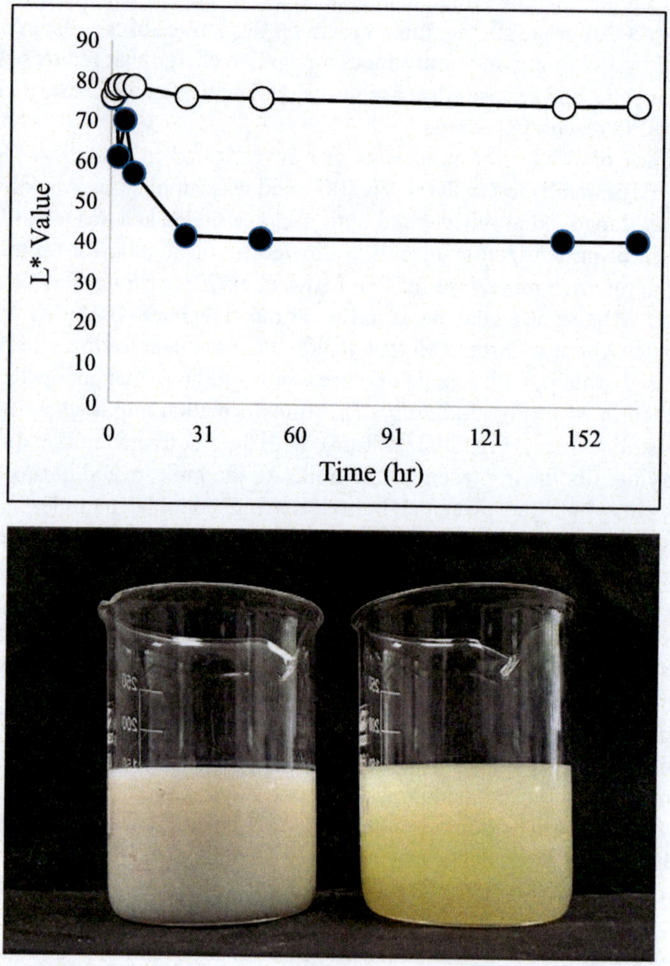

Fig. 20.2 (Top) Changes in lightness (L-value) of skim milk (open symbols) or skim milk with 0.1% added protein glutaminase (PG, closed symbols) during 168 h at 50 °C. (Bottom) Appearance of PG-treated (right) and control (left) skim milk at the end of incubation

Due to the changes in protein charge, the isoelectric point of the caseins decreases, allowing the proteins to be more soluble in acid solutions (Kikuchi et al. 2008). This is an important characteristic as food proteins commonly have an acidic isoelectric point, and the pH of many food systems is acidic. Protein glutaminase also enhances the emulsifying function of skim milk. Although caseins are naturally good emulsifiers due to their amphiphilic and flexible structure, PG enhances emulsifying properties through deamidation, which opens the structure of casein up further.

Miwa et al. (2010) also studied the SDS-PAGE pattern for skim milk following treatment with the enzyme and noted that the bands representing the whey proteins

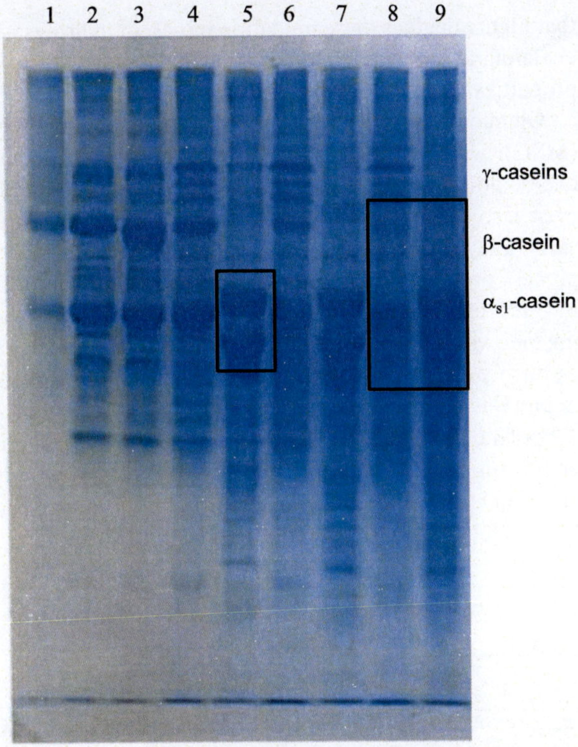

Fig. 20.3 Urea–polyacrylamide gel electrophoretograms of Lane 1: Sodium caseinate; lanes 2, 4, 6, 8: Skim milk with 0.05% sodium azide after 0, 24, 72, 168 h incubation at 50 °C, respectively; lanes 3, 5, 7, 9: Skim milk with 0.05% sodium azide plus 0.1% protein glutaminase after 0, 24, 72, or 168 h incubation at 50 °C, respectively

had shifted following deamidation, indicating that the main whey proteins (α-lactalbumin and β-lactoglobulin) may be affected as well as the caseins.

The application of PG in the dairy industry is an area that needs more detailed research. Miwa et al. (2014) examined the effect of PG on set yoghurt, in particular in the context of prevention of syneresis of low-fat yoghurt. The deamidation of milk proteins prior to formation of the yoghurt gel by microbial acidification influenced the "creaminess" of the yoghurt, and decreased syneresis.

20.4.2 Glutaminase

Whereas protein glutaminase deamidates glutamine residues as part of protein chains, the related enzyme glutaminase is able to act on free glutamine residues, to convert these to glutamate, with ammonia being produced in parallel. Compared to transglutaminase, or even protein glutaminase, there have been few reports of the impact of glutaminase on milk or dairy products. Li and Zhao (2011) reported on the generation of deamidated casein hydrolysates with use of glutaminase and noted that higher degrees of deamidation of casein hydrolysates resulted in greater metal-chelating ability and also *in vitro* ACE (angiotensin converting enzyme) inhibition properties.

Glutamate, in its form of a salt, e.g., with sodium, forms monosodium glutamate (MSG), which is responsible for the desirable umami taste, e.g., in cheese and cheese powder, but also other dairy products and ingredients. Glutamate can, however, also form other salts, e.g., with calcium, potassium, ammonium or magnesium, and thereby is able to widely contribute to taste in our foods. Steering this process provides a huge potential of diversification of dairy proteins and products into new ingredients, providing desirable aroma components, and potentially elevating their value; however, the contribution of glutamate to umami taste is expected to be mainly due to glutamate as its free amino acid and not when tied up as a residue in proteins or polypeptides. Therefore, there is increasing interest in understanding the potential significance of glutaminase and protein glutaminase for development of umami taste in dairy products in relation to protein composition, processing parameters, as well as influences of proteolysis, and the susceptibility of free and bound glutamine to the reactions and their significance.

20.4.3 Lactose Oxidase

Lactose oxidase (LO) is an oxidoreductase that can oxidize the disaccharide lactose to lactobinic acid (LBA). It has high specificity towards oligo- and polymeric long-chain carbohydrates, with malto- or cello-tetraoses as the preferred substrates (Nordkvist et al. 2007).

The most widely applied and well described LO is the carbohydrate: acceptor oxidoreductase (COX) containing one flavin adenine dinucleotide (FAD) (Kulys et al. 2000; Xu et al. 2001; Ahmad et al. 2004; Nordkvist et al. 2007). It has a molecular mass at 55 kDa and an isoelectric point of 9, and was first characterised and cloned from *Microdochium nivale* by Xu et al. (2001). To improve the yield, LO can be produced by a genetically modified strain of *Fusarium venenatum* followed by a submerged fermentation and a concentration process (Ahmad et al. 2004). The safety of this enzyme was examined by Ahmad et al. (2004), who reported no significant toxicological effects in a rat experiment and no induction of gene mutations or chromosomal aberrations *in vitro*. The kinetics and operational stability of oxidation of lactose to LBA by COX has been measured by Nordkvist et al. (2007), who also found that LO was protected against inactivation by bases at high lactose content.

A novel LO has recently been identified from *Myrmecridium flexuosum* NUK-21, which contains two zinc ions rather than FAD, and it was shown to have a higher substrate specificity towards lactose than the COX from *M. nivale* (Lin et al. 2019). Other oxidoreductases oxidizing lactose, which are all flavoproteins, include the COX purified from *Paraconiothyrium* sp. KD-3 (PCOX, Kiryu et al. 2008), cellooligosaccharide oxidase (COOX) obtained from *Sarocladium oryzae* (Lee et al. 2006), and glucooligosaccharide oxidase (GOOX) obtained from *Acremonium strictum* T1 (Lin et al. 1991).

In industry, LO (COX) is applied as a liquid enzyme concentrate to convert lactose from whey protein concentrate (WPC) to LBA as an environmental friendly process compared with the chemical process (Ahmad et al. 2004). The product, LBA, is a highly valuable lactose derivative that can be applied in medical, cosmetic, chemical, and food applications, due to its antioxidant, chelating, amphiphilic, and non-toxic properties (Alonso et al. 2013).

LO itself has potential to be applied in an enzyme-based preservation technology in the dairy industry. As LO can increase the level of H_2O_2 in bovine milk to activate the lactoperoxidase system (LS), an antimicrobial system naturally present in milk (Chap. 3), it could therefore improve the shelf life of milk (Lara-Aguilar and Alcaine 2019a). The addition of LO at a concentration of 0.12 g L^{-1} decreased the growth of *Pseudomonas fragi* in pasteurized milk significantly at 6 °C (Lara-Aguilar and Alcaine 2019a). Combined with sodium thiocyanate, LO also enhanced the inhibition of bacterial growth in raw milk at 21 °C, and it performed better than the recommended method for preservation of raw milk by LS (FAO/WHO 1991) (Lara-Aguilar and Alcaine 2019a). The effect of LO on other foodborne pathogens and spoilage microorganisms, including *Escherichia coli* O157:H7, *Listeria monocytogenes*, *Salmonella enterica* ser. Typhimurium, *Staphylococcus aureus*, and *Penicillium chrysogenum* has also been studied and a higher concentration of LO (12 g L^{-1}) was required to inhibit the growth of all of these (Lara-Aguilar and Alcaine 2019b).

20.4.4 Phospholipase

There has been significant interest in recent years in using phospholipase to modify the functionality of dairy products. Phospholipases (PL) act by hydrolysing phospholipids associated with the milk fat globule membrane (e.g., phosphatidylethanolamine, phosphatidylcholine, phosphatidylinositol and sphingomyelin), with concomitant production of lysophospholipids, free fatty acids, phosphatidates and other products. There are four known classes of PL, called A (divided into A1 and A2), B, C and D, that differ in their precise substrate specificity. PLA1 and A2 cleave the SN1 and SN2 acyl chains of phospolipids, respectively, PL B cleaves both SN1 and SN2 acyl chains, and PL C and D cleave before and after the phosphate group respectively; PL C and D are phosphodiesterases.

For example, phospholipase A1 hydrolyses phosphatidylcholine at the SN1 ester bond, leading to release of lysophospholipid and a fatty acid; lysophospholipid reduces the surface tension of fluids and can act as an emulsifying or foaming agent.

There have been reports of indigenous phospholipase D activity in milk, but this remains somewhat controversial (Fox and Kelly 2006); PL action in milk can disrupt the MFGM and facilitate hydrolysis by lipases (Chrisope and Marshall 1976). Psychotrophic bacteria in raw milk can produce heat-stable PL (Koka and Weimer

2001; Vithanage et al. 2016), the impact of which can be controlled by nitrogen gas flushing (Munsch-Aalatossava et al. 2018); contaminating yeasts in dairy products can also produce PL (Melville et al. 2011). The action of phospholipase C has been shown to inhibit the development of oxidised flavour in milk (O'Mahony and Shipe 1972).

Treatment with PLA1 (such as that from *Fusarium venetatum*) has been shown to result in increased yield of Mozzarella cheese, due to better retention of water and fat in the cheese curd, and reduced loss of phospholipids and fat globules in the whey; the reason for this effect appears to be due to the lysosphospholipids acting as surface-active emulsifying agents in the curd, perhaps re-emulsifying fat which has been released from mechanically damaged fat globules, or perhaps lyso-phospholipids interacting with casein through hydrogen bonding (Hoeir et al. 2006; Lilbaek et al. 2006). Increases of up to 1% in yield, after moisture adjustment, have been reported (Hoier et al. 2006). Such enzymatic treatment does not affect functionality of the cheese (e.g., flowability, stretchability and browning), nor does it impact on sensory properties (Hoeir et al. 2006; Lilbaek et al. 2006). Synergistic effects of PL treatment and use of exopolysaccharide (EPS)-producing bacteria on yield and fat and moisture retention of Chihuahua cheese have been reported (Transeco-Reyes et al. 2014). Fungal PL A1 has been shown to modify the functional properties of milk and whey, specifically in terms of reducing surface tension and improve foaming properties, as well as improving the heat stability of whey protein concentrates (Havn et al. 2006; Lilbaek et al. 2007). PL can also be used to modify the functionality of phospholipids, such as emulsifying properties (Borlieu et al. 2009). The heat stability of recombined concentrated milk emulsions has also been enhanced by addition of lecithin hydrolysed by PL (Kasinos et al. 2014).

Commercial preparations of phospholipase are available due to this interest. One such preparation is YIELDMAX® PL (Chr. Hansens), a preparation of Phospholipase A1 (obtained from a genetically modified *Aspergillus oryzae* species, producing a *Fusarium venenatum* phospholipase) which is added to cheesemilk prior to coagulation to improve cheesemaking efficiency (particularly for Paste Filata-type cheese, with a 1.8% average yield increase reported for Low Moisture Part Skim Mozzarella). This enzyme has been reported to hydrolyse >90% of the PL in milk to produce lyso-phospholipids (John Lyne, Chr. Hansen, Personal Communication). The enzyme is inactivated by conventional pasteurisation treatments, and so is readily inactivated in whey during later processing.

20.5 Application of Molecular Methods: Omics

Molecular methods have also increasingly found their applicability in the field of dairy enzymology. These methods include the application of peptidomic methods for the understanding of action and contribution of proteolytic enzymes (indigenous, endogenous, exogenous) to milk quality (Larsen et al. 2010; Guerrero et al.

2015), dairy product quality (Ebner et al. 2016; Nielsen et al. 2018) and nutritional aspects of human milk (Dallas et al. 2015).

Furthermore, molecular methods like Single or Multiple Reaction Monitoring, SRM and MRM, respectively, represent new Triple Q-based mass spectrometry based methods for absolute quantification of milk enzymes and their molecular forms. Overall, SRM and MRM involve the selection of one or more representative peptides from the primary structure of the enzyme in question, and then using this or these peptides as reporter peptides for the quantification of the protein after enzymatic degradation, often by trypsin (Le et al. 2017). The selected reporter peptides are then chemically synthesized (stable isotope-labelled or not) and measured as well. Their signal relative to known amounts measured then provides the basis of the absolute quantification. As, in many cases, the absolute levels and the molecular forms present in milk or dairy products relative to, e.g., physiological states or processing steps, are not known or clear, this seems like a promising method to obtain more basis knowledge of the milk enzymatic battery.

20.6 Conclusion: Remaining Gaps in Our Knowledge and Future Outlook

Even though much has been achieved, the knowledge across areas is heterogeneous. In some areas the knowledge is comprehensive and mature, while in others the knowledge is deep in some aspects though scarce in others.

Kelly and Fox (2006) reviewed what were then considered to be outstanding challenges in the field of indigenous milk enzymology, and in Table 20.1, some of the progress, or even lack of progress, made in 15 years of research since that year into indigenous enzymes is summarized.

One prominent example for the very evident need for harmonizing the methods for assessment of enzyme activities is the example of assessment of rennet coagulation properties of milk (Chap. 14, this book), where parameters characterizing the second stage of rennet-induced milk coagulation are monitored simultaneously on several samples, and therefore can be used in screenings and larger studies, e.g., in relation to bovine genetics. Many different methods have been employed, since the use of the former widely used Formagraph method (Bittante et al. 2012), where the lack of basic materials for performing the method (special thermo-sensitive paper for the output) has led to many different suggestions for alternatives. These include, for example, measurement of turbidity in microtiter plates, and the development of different more or less advanced, thermostated equipment, like ReoRox, Lactodynamograph, ElastoSense and Rotem equipments, based on different combinations of viscoelastic and oscillation principles.

Another area with considerable recent attention is the area of UHT milk quality and stability (Anema 2019). Even though non-enzymatic physico-chemical processes are still agreed to be significant, the involvement of microbial enzymes in the gelation, sedimentation, creaming and generation of off-flavours is emerging, and being more and more documented. It is important to acknowledge that the level of heat-stable protease required, e.g., for gelation of UHT milk and the degree of proteolysis necessary for gelation to occur is not yet known; thus, the prediction of gelation of UHT milk is still a great challenge to UHT processors, partly due to the diverse nature and wide variation in biochemical characteristics of the heat-stable bacterial proteases. Therefore, differences in proteases from different strains and prediction of secretion relative to stimuli, as well as methodologies for prediction of suitability of silo milk for use for UHT treatment and products, are needed.

In outlook, generally the increasing focus on sustainability and climate effects points to potential new directions in many areas of scientific research, including in relation to milk enzymology. Future scenarios may be *in vitro* production of milk components by isolated mammary epithelial cells (from udder or from milk somatic cell pool), or, in the light of The ReThinkX report (https://www.rethinkx.com/food-and-agriculture), even production of animal-free milk proteins in microorganisms (bacteria, yeast). This would provide new matrices for applying enzymes on milk constituents, including milk proteins.

It is a question for the future if the climate agenda may mitigate some of our reservations towards e.g., GMO, and thereby create new potential for milk substances as matrices for enzymes. This will give considerations on which milieus the enzymes are working in, as well as use of specifically developed cultures secreting desired exogenous enzymes, like lactases lipases and proteases. Furthermore, eventually moving towards production of milk proteins in single cells like bacteria or yeast and providing new matrices for the enzymes to work gives both new opportunities and challenges for changes in food products based on "dairy" proteins. Such developments may give rise to other considerations, such as the presence of allergens in fermentation media, or the Halal or Kosher status of materials used, which need to be borne in mind when considering new methods for developing enzyme sources and technologies, as do regulatory considerations (e.g., whether enzymes used in dairy applications are classified as processing aids or ingredients). In many cases, science can proceed at a faster pace than regulation, and technological application of newly identified promising enzymes or applications may be constrained by the need to ensure compliance with relevant legislation. Nonetheless, it appears certain that ongoing programmes of enzyme discovery by enzyme and biotechnology companies will lead in the coming years to an ever expanding portfolio of enzymes available to processors to tailor the functionality, stability and sensory properties of dairy, or dairy-containing, food products.

Finally, the sustainability agenda points to a need for continued focus on importance of enzymes in relation to optimization of product functionality, sensory properties and raw material utilization, to reduce waste and increase gain in the whole dairy value chain.

Acknowledgements The authors wish to acknowledge helpful discussions and contributions from Mr. John Lyne (Chr. Hansen Inc., Milwaukee, U.S.A.) and Ms. Ger Behan (Kerry Group, Charleville, Ireland) and technical assistance from Ciara O'Connell in preparing figures.

References

Ahmad SK, Brinch DS, Friis EP, Pedersen PB (2004) Toxicological studies on lactose oxidase from *Microdochium nivale* expressed in *Fusarium venenatum*. Regul Toxicol Pharmacol 39:256–270

Alonso S, Rendueles M, Díaz M (2013) Bio-production of lactobionic acid: current status, applications and future prospects. Biotechnol Adv 31:1275–1291

Andrews AT (1983) Proteinases in normal bovine milk and their action on caseins. J Dairy Res 50: 45-55

Anema SG (2019) Age gelation, sedimentation, and creaming in UHT Milk: a review. Compr Rev Food Sci Food Saf 18(1):140–166

Bittante G, Penasa M, Cecchinato A (2012) Invited review: genetics and modelling of milk coagulation properties. J. Dairy Sci 95:6843–6870

Blakesley RW, Boezi JA (1977) New staining technique for proteins in polyacrylamide gels using Coomassie Brilliant Blue G250. Analytical Biochem 82: 580-582

Borlieu C, Bouhallab S, Lopez C (2009) Biocatalyzed modification of milk lipids: applications and potentialities. Trends Food Sci Technol 20:458–469

Chrisope GL, Marshall RT (1976) Combined action of lipase and microbial phospholipase C on a model fat globule emulsion and raw milk. J Dairy Sci 59:2024–2030

Christensen J, Holst HH (2014) WO 2014/114709 A2: method of producing beta-casein compositions and related products

Crowley SV (2016) Physicochemical characterisation of protein ingredients prepared from milk by ultrafiltration or microfiltration for application in formulated nutritional products. PhD thesis University College Cork, Ireland

Crowley SV, Burlot E, Silva JVC, McCarthy NA, Wijayanti HB, Fenelon MA, Kelly AL, O'Mahony JA (2018a) Rehydration behaviour of spray-dried micellar casein concentrate produced using microfiltration of skim milk at cold or warm temperature. Int Dairy J 81:72–79

Crowley SV, Burlot E, Silva JVC, McCarthy NA, Wijayanti HB, Fenelon MA, Kelly AL, O'Mahony JA (2018b) Rehydration behaviour of spray-dried micellar casein concentrates produced using microfiltration of skim milk at cold or warm temperatures. Int Dairy J 81:72–79

Ebner J, Baum F, Pischetsrieder M (2016) Identification of sixteen peptides reflecting heat and/or storage induced processes by profiling of commercial milk samples. J Proteomics 147: 66-75

FAO/WHO (Food and Agriculture Organization of the United Nations/World Health Organization) (1991) Guidelines for preservation of raw Milk by use of the Lactoperoxidase system. FAO/WHO, Rome

Fox PF, Kelly AL (2006) Indigenous enzymes in milk: overview and historical aspects – part 1. Int Dairy J 16:500–516

Gazi I, Vilalva IC, Huppertz T (2014) Plasmin activity and proteolysis in milk protein ingredients. Int Dairy J 38:208–212

Guerrero A, Dallas DC, Contreras S, Bhandari A, Canovas A, Islas-Trejo A, Medrano JF, Parker EA, Wang M, Hettinga K, Chee S, German JB, Barile D, Lebrilla CB (2015) Peptidomic analysis of healthy and subclinically mastitic bovine milk. Int Dairy J 46:46–52

Hashizume R, Maki Y, Mizutani K, Takahashi N, Matsubara H, Sugita A, Sato K, Yamaguchi S, Mikami B (2011) Crystal structures of protein Glutaminase and its pro forms converted into enzyme-substrate complex. J Biol Chem 286:38691–38702

Havn SS, Ipsen R, Nielsen PM, Lilbaek HM (2006) Improving the foaming properties and heat stability of whey protein concentrates by phospholipase treatment. Milchwissenshaft 61:188–191

Hoier E, Lilbaek HM, Broe ML, Sorensen NK (2006) Enhancing cheese yield by phospholipase treatment of cheese milk. Aust J Dairy Technol 61:179–182

Kasinos M, Goñi ML, Nguyen MT, Sabatino P, Martins JC, Dewettinck K, Van Der Meeren P (2014) Effect of hydrolysed sunflower lecithin on the heat-induced coagulation of recombined concentrated milk emulsions. Int Dairy J 38:187–194

Kelly AL, Fox PF (2006) Indigenous enzymes in milk: a synopsis of future research requirements. Int Dairy J 16:707–715

Kikuchi Y, Itaya H, Date M, Matsui K, Wu LF (2008) Production of Chryseobacterium proteolyticum protein-glutaminase using the twin-arginine translocation pathway in Chryseobacterium glutamicum. Applied Microbiol Biotechnol 78: 67-74.

Kiryu T, Nakano H, Kiso T, Murakami H (2008) Purification and characterization of a carbohydrate: acceptor oxidoreductase from *Paraconiothyrium* sp. that produces lactobionic acid efficiently. Biosci Biotechnol Biochem 72:833–841

Koka R, Weimer BC (2001) Influence of growth conditions on heat-stable phospholipase activity in Pseudomonas. J Dairy Res 68:109–116

Kulys J, Tetianec L, Schneider P (2000) The development of an improved glucose biosensor using recombinant carbohydrate oxidase from *Microdochium nivale*. Analyst 125:1587–1590

Lara-Aguilar S, Alcaine SD (2019a) Lactose oxidase: a novel activator of the lactoperoxidase system in milk for improved shelf life. J Dairy Sci 102:1933–1942

Lara-Aguilar S, Alcaine SD (2019b) Screening inhibition of dairy-relevant pathogens and spoilage microorganisms by lactose oxidase. J Dairy Sci 102:7807–7812

Larsen LB, Hinz K, Jørgensen LW, Møller HS, Wellnitz O, Bruckmaier RM, Kelly AL (2010) Proteomic and peptidomic study of proteolysis in quarter milk after infusion with lipoteichoic acid from *Staphylococcus aureus*. J Dairy Sci 93:5613–5626

Le TT, Deeth HC, Larsen LB (2017) Proteomics of major milk proteins: novel insights. Int Dairy J 67:2–15

Lee MH, Lai WL, Lin SF, Liu Y, Hsu YH, Tsai YC (2006) Purification and characterization of a novel cellooligosaccharide oxidase from rice pathogen *Sarocladium oryzae*. Enzyme and Microb Technol 39:85–91

Li D, Zhao XH (2011) Glutaminase-induced deamidation and hydrolysis of casein and metal-chelating or ACE-inhibitory activity of the hydrolysates *in vitro*. Int J Food Sci Technol 46:324–332

Lilbaek HM, Fatum TM, Ipsen R, Sorensen NK (2006) Improving the yield of mozzarella cheese by phospholipase treatment of milk. J Dairy Sci 89:4114–4125

Lilbaek HM, Broe ML, Hoier E, Fatum TM, Ipsen R, Sorensen NK (2007) Modification of milk and whey surface properties by enzymatic hydrolysis of milk phospholipids. J Agric Food Chem 55:2970–2978

Lin SF, Yang TY, Inukai T, Yamasaki M, Tsai YC (1991) Purification and characterization of a novel glucooligosaccharide oxidase from Acremonium strictum T1. Biochim Biophys Acta Protein Struct Mol Enzymol 1118:41–47

Lin SF, Li CK, Chung YP (2019) Identification of a novel lactose oxidase in *Myrmecridium flexuosum* NUK-21. FEBS Open Bio 9:364–373

McCarthy NA, Wijayanti BH, Crowley SV, O'Mahony JA, Fenelon MA (2017) Pilot-scale ceramic membrane filtration of skim milk for the production of a protein base ingredient for use in infant milk formula. Int Dairy J 73:57–62

Melville PA, Benites NR, Ruz-Peres M, Yokoya E (2011) Proteinase and phospholipase activities and development at different temperatures of yeasts isolated from bovine milk. J Dairy Res 78:385–390

Miwa N, Yoyoyama K, Wakabayashi H, Nio N (2010) Effect of deamidation by protein-glutaminase on physicochemical and functional properties of skim milk. Int. Dairy J 20:393–399

Miwa N, Nio N, Somomoto K (2014) Effect of enzymatic deamidation by protein-glutaminase on the textural and microstructural properties of set yoghurt. Int. Dairy J 36:1–5

Munsch-Aalatossava P, Kakela R, Ibarra D, Youbi-Idrissi M, Alatossava T (2018) Phospholipolysis caused by different types of bacterial phospholipases during cold storage of bovine raw milk is prevented by N_2 gas flushing. Front Microbiol 9:1307

Nielsen SD, Zhao D, Le TT, Rauh V, Sørensen J, Andersen HJ, Larsen LB (2018) Proteolytic side-activity of lactase preparations. Int. Dairy J 78:159–168

Nordkvist M, Nielsen PM, Villadsen J (2007) Oxidation of lactose to lactobionic acid by a *Microdochium nivale* carbohydrate oxidase: kinetics and operational stability. Biotechnol Bioeng 97:694–707

O'Mahony JA, Smith KE, Lucey JA (2014). US 2014/8889208 B2: Purification of beta casein from milk. US Pat. No. US 2014/8889208 A1

O'Mahony JO, Shipe WF (1972) Hydrolysis of lipoprotein fractions of milk by phospholipase-C. J Dairy Sci 55: 408

Schäfer J, Schubert T, Atamer Z (2019) Pilot-scale β-casein depletion from micellar casein via cold microfiltration in the diafiltration mode. Int Dairy J 97:222–229

Shalabi SI, Fox PF (1987) Electrophoretic analysis of cheese – comparison of methods. Irish J Food Sci Technol 11: 135-151

Summer H, Grämer R, Dröge P (2009) Denaturing urea polyacrylamide gel electrophoresis (urea-PAGE). J Vis Exp 32:1485

Thienel KJF, Holder A, Schubert T, Boom RM, Hinrichs J, Atamer Z (2018) Fractionation of milk proteins on pilot scale with particular focus on β-casein. Int Dairy J 79:73–77

Transeco-Reyes N, Gutierrez-Mendez N, Sepulveda DR, Hernandez-Ochoa LR (2014) Assessing the yield, microstructure, and texture properties of miniature Chihuahua-type cheese manufactured with a phospholipase A(1) and expolysaccharide-producing bacteria. J Dairy Sci 97:598–608

Vithanage NR, Dissanayake M, Bolge G, Palombo EA, Yaeger TR, Datta N (2016) Biodiversity of culturable microbiota in raw milk attributable to refrigeration conditions, seasonality and their spoilage potential. Int. Dairy J 57:80–90

Xu F, Golightly EJ, Fuglsang CC, Schneider P, Duke KR et al (2001) A novel carbohydrate: acceptor oxidoreductase from *Microdochium nivale*. Eur J Biochem 268:1136–1142

Yamaguchi S, Jeenes DJ, Archer DB (2001) Protein-glutaminase from *Chryseobacterium proteolyticum*, an enzyme that deamidates glutaminyl residues in proteins. Eur J Biochem 268:1410–1421

Index

A. L. Kelly, L. B. Larsen (eds.), *Agents of Change*, Food Engineering Series,
https://doi.org/10.1007/978-3-030-55482-8

P